T0251250

# Electromagnetic Compatibility

# ELECTRICAL AND COMPUTER ENGINEERING

*A Series of Reference Books and Textbooks*

FOUNDING EDITOR

*Marlin O. Thurston*
Department of Electrical Engineering
The Ohio State University
Columbus, Ohio

ı

1. Rational Fault Analysis, *edited by Richard Saeks and S. R. Liberty*
2. Nonparametric Methods in Communications, *edited by P. Papantoni-Kazakos and Dimitri Kazakos*
3. Interactive Pattern Recognition, *Yi-tzuu Chien*
4. Solid-State Electronics, *Lawrence E. Murr*
5. Electronic, Magnetic, and Thermal Properties of Solid Materials, *Klaus Schröder*
6. Magnetic-Bubble Memory Technology, *Hsu Chang*
7. Transformer and Inductor Design Handbook, *Colonel Wm. T. McLyman*
8. Electromagnetics: Classical and Modern Theory and Applications, *Samuel Seely and Alexander D. Poularikas*
9. One-Dimensional Digital Signal Processing, *Chi-Tsong Chen*
10. Interconnected Dynamical Systems, *Raymond A. DeCarlo and Richard Saeks*
11. Modern Digital Control Systems, *Raymond G. Jacquot*
12. Hybrid Circuit Design and Manufacture, *Roydn D. Jones*
13. Magnetic Core Selection for Transformers and Inductors: A User's Guide to Practice and Specification, *Colonel Wm. T. McLyman*
14. Static and Rotating Electromagnetic Devices, *Richard H. Engelmann*
15. Energy-Efficient Electric Motors: Selection and Application, *John C. Andreas*
16. Electromagnetic Compossibility, *Heinz M. Schlicke*
17. Electronics: Models, Analysis, and Systems, *James G. Gottling*
18. Digital Filter Design Handbook, *Fred J. Taylor*
19. Multivariable Control: An Introduction, *P. K. Sinha*
20. Flexible Circuits: Design and Applications, *Steve Gurley, with contributions by Carl A. Edstrom, Jr., Ray D. Greenway, and William P. Kelly*
21. Circuit Interruption: Theory and Techniques, *Thomas E. Browne, Jr.*
22. Switch Mode Power Conversion: Basic Theory and Design, *K. Kit Sum*
23. Pattern Recognition: Applications to Large Data-Set Problems, *Sing-Tze Bow*

*Additional Volumes in Preparation*

# Electromagnetic Compatibility

## Principles and Applications

### Second Edition, Revised and Expanded

**David A. Weston**
*EMC Consulting Inc.*
*Merrickville, Ontario, Canada*

MARCEL DEKKER, INC.          NEW YORK · BASEL

Library of Congress Cataloging-in-Publication Data

Weston, David A.
    Electromagnetic compatibility : principles and applications / David A. Weston.—2nd
ed., rev. and expanded.
        p. cm.—(Electrical and computer engineering ; 112)
    Includes bibliographical references and index.
    ISBN 0-8247-8889-3 (alk. paper)
      1. Electromagnetic compatibility. 2. Electronic apparatus and appliances—Design and
construction.   I. Title.   II. Electrical engineering and electronics ; 112.

TK7867.2.W46 2000
621.382′24—dc21

00-047591

This book is printed on acid-free paper.

**Headquarters**
Marcel Dekker, Inc.
270 Madison Avenue, New York, NY 10016
tel: 212-696-9000; fax: 212-685-4540

**Eastern Hemisphere Distribution**
Marcel Dekker AG
Hutgasse 4, Postfach 812, CH-4001 Basel, Switzerland
tel: 41-61-261-8482; fax: 41-61-261-8896

**World Wide Web**
http://www.dekker.com

The publisher offers discounts on this book when ordered in bulk quantities. For more information, write
to Special Sales/Professional Marketing at the headquarters address above.

Current printing (last digit):
10 9 8 7 6 5 4 3

# Preface

This second, revised and updated, edition of the book contains approximately 65% more information than the first edition. This includes a review of computer modeling programs, a new chapter on PCB layout, and additional commercial and military EMI test methods. New data on cable radiation and coupling to cables is included, extending out to 12 GHz, and on EMI enclosure shielding.

All electronic and electrical equipment is a potential source of electromagnetic interference (EMI). Similarly, such equipment will not function as designed at some level of electromagnetic ambient. The problems associated with EMI can range from simple annoyance (e.g., static on telecommunications equipment or increased bit error rates on digital equipment) to catastrophe (e.g., inadvertent detonation of explosive devices).

Electromagnetic compatibility (EMC) can be achieved by evaluating the electromagnetic environment (often characterized by standards or requirements) to which equipment/systems is exposed and then designing and building equipment/systems to function correctly in the operational environment without itself creating EMI.

This book is written for the design/systems engineer, technologist, technician, or engineering manager who designs, maintains, or specifies equipment either to meet an electromagnetic compatibility requirement specification or to function safely in a given electromagnetic environment.

Many engineers do not have, or need, radio frequency (RF) experience. However, in operation, digital control or switching power equipment functions as an RF system. Therefore, an understanding of the high-frequency characteristics of components, simple radiators, and wave theory is imperative in achieving an understanding of EMC.

One aim of the book is to teach EMI prediction and enable the reader to build EMC into equipment and systems without overdesign. By achieving EMC, the designer averts the program delay and additional cost of fixing EMI after the equipment is built. With the recognition that EMI problems exist, we present EMI diagnostic techniques and cost effective solutions with practical implementation and options.

The book discusses typical sources of EMI and characteristics of the radiated and conducted emissions that might be expected in a given electromagnetic environment and reviews ways of decreasing electromagnetic emissions as well as the susceptibility of equipment to EMI. Some books on EMI/EMC contain equations that are theoretically sound but may not be useful in practical EMI/EMC problems. All equations in this book have been found to be invaluable in EMI prediction and EMC design. In most instances, theory is substantiated by measured data, and where anomalies exist most probable reasons are offered. Where the reader may wish to pursue a given subject area further, numerous references are provided. Worked examples of the equations are given in predictions and case studies throughout the book. Physical geometry and frequency limitations exist in the application of all wave or circuit theory, including the effect of parasitic components, and these limits are discussed.

The apparent anomalies that have given EMC a reputation for "black magic" are explained. For example, the case where the addition of shielding, a filter, or grounding increases either the level of EMI emissions or the susceptibility during EMC tests is examined. The major reason these results are apparently inexplicable is that the underlying theory is not well understood. The approach used in the book is to provide an understanding of the theory with an emphasis on its applicability in the practical realization of EMC design and EMI solutions, including implementation and maintenance.

The intent is that information contained herein have a practical application or be required for an understanding of the principles of EMC. For example, calculated or published data on attenuation or shielding effectiveness is of little use unless its application is explained. Therefore, it must be used in conjunction with the worst-case levels of radiated or conducted noise that may be expected in a given environment. Any practical limitation on the achievable attenuation or shielding must then be accounted for, after which the noise levels applied to the system or circuit and its immunity may be predicted. The aim has been to avoid the overly simplistic cookbook approach with its inherent errors, and yet to limit the mathematics to that used by the practicing engineer or technician.

Simple measurement techniques that are possible with standard electronic measurement equipment are described. These are useful for EMI diagnostic measurements as well as a "quick look" at equipment that must meet EMC requirements such as the commercial FCC, DO-160, VCE, and EN, and the military/aerospace MIL-STD-461. Also, the correct measurement techniques and possible errors encountered using more sophisticated equipment required for certification and qualification EMC testing are introduced.

The book is based on experience gained in EMC consulting and on the course notes of one- to four-day EMC seminars presented over a 12-year period. Many questions posed by attendees of the seminars and clients have been answered in this book.

I am very grateful to David Viljoen, who made a significant contribution to the preparation of the contents of the first edition of the book (i.e., the layout of the book, drafting the majority of the figures, editing Chapters 1 to 5, and writing the computer programs). Without the attention to detail, hard work, and high-quality effort of Mr. Viljoen, this book would not have been possible in its present form.

I am also indebted to the late Mr. Jabez Whelpton, of Canadian Astronautics Ltd., who was of great assistance in reading and correcting the content of those chapters that contain information on wave theory and antennas.

For the second edition of the book I wish to thank Mr. Chris Ceelen, who made many of the additional EMI measurements, and Ms. Lianne Boulet, who helped prepare the text and figures.

Finally, I wish to thank the organizations specifically acknowledged beneath figures and in the text, especially the Canadian Space Agency.

*David A. Weston*

# Contents

# Electromagnetic Compatibility

# 1

# Introduction to EMI and the Electromagnetic Environment

## 1.1 INTRODUCTION TO ELECTROMAGNETIC INTERFERENCE (EMI)

This chapter introduces a few of the important coupling modes involved in electromagnetic interference (EMI) as well as some EMI regulations. Both of these topics will be addressed in more detail in subsequent chapters. In addition, this chapter presents data on average and worst-case electromagnetic emissions found in the environment. This data is useful in evaluating the severity of a given electromagnetic environment and in predicting electromagnetic compatibility (EMC) for equipment operated in these environments.

### 1.1.1 Effects of Electromagnetic Interference

The effects of EMI are extremely variable in character and magnitude, ranging from simple annoyance to catastrophe. Some examples of the potential effects of EMI are

- Interference to television and radio reception
- Loss of data in digital systems or in transmission of data
- Delays in production of equipment exhibiting intraunit, subsystem, or system-level EMI
- Malfunction of medical electronic equipment (e.g., neonatal monitor, heart pacemaker)
- Malfunction of automotive microprocessor control systems (e.g., braking or truck anti-jackknife systems)
- Malfunction of navigation equipment
- Inadvertent detonation of explosive devices
- Malfunction of critical process-control functions (e.g., oil or chemical industry)

To correct EMI problems that occur after equipment is designed and in production is usually expensive and results in program delays, which may adversely affect the acceptance of a new product. It is preferable to follow good EMC engineering practice during the equipment design and development phases. Our goal should be to produce equipment capable of functioning in the predicted or specified electromagnetic environment and that does not interfere with other equipment or unduly pollute the environment—that is, to achieve EMC.

The techniques of EMC prediction described in subsequent chapters will aid in meeting the goal of EMC when applied at the design stage. These same techniques of analysis and modeling are applicable to EMI control and problem solving or in the location of out-of-specification emissions. It is in the area of emission reduction where analysis is most likely to be supplemented by measurement and diagnostic intervention. However, the value of simple EMI

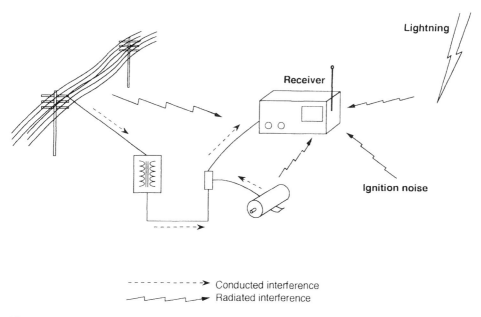

**Figure 1.1**   Possible sources of ambient noise and how they may be coupled into a receiver.

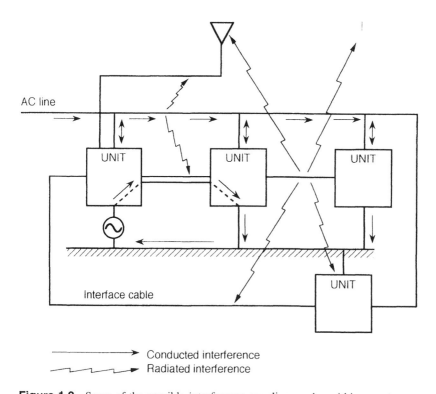

**Figure 1.2**   Some of the possible interference coupling modes within a system.

measurements made as early as feasible in the design, breadboard, and prototype phases cannot be emphasized enough.

### 1.1.2 Electromagnetic Interference Coupling Modes

For EMI to exist there must be a source of emission, a path for the coupling of the emission, and a circuit, unit, or system sensitive to the received noise. Figures 1.1–1.3 illustrate that two modes of coupling, radiated and conducted, can exist. In the near field, the radiated coupling may be either a predominantly magnetic (H) field coupling or an electric (E) field coupling, whereas the coupling in the far field will be via electromagnetic waves exhibiting a fixed ratio of the E to H field strengths. A more rigorous definition of near and far field is presented in Section 2.2.1. Suffice it to say here that the near field is in close proximity to a source and the far field is beyond some determined distance from the source.

For circuits and conductors in close proximity we consider the coupling, or crosstalk, to be via mutual inductance and intercircuit capacitance, although one of these modes usually predominates. The source of noise may be power lines, signal lines, logic (especially clocks and data lines), or current-carrying ground connections. The conducted path may be resistive or contain inductance or capacitance, intentional or otherwise, and it is often a combination of these. The reactive components often result in resonances, with their concomitant increase or decrease in current at the resonant frequencies.

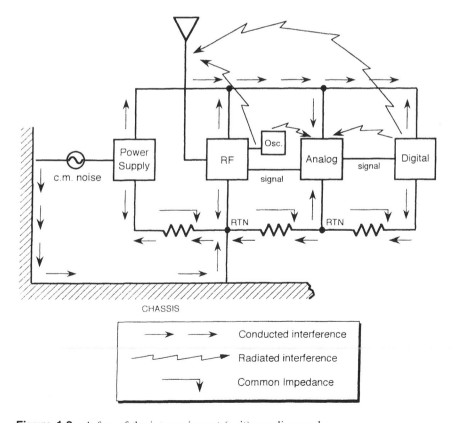

**Figure 1.3** A few of the intraequipment (unit) coupling modes.

## 1.2  INTRODUCTION TO ELECTROMAGNETIC
## INTERFERENCE REGULATIONS

The level of immunity built into equipment depends on how critical the correct functioning of the equipment is and on the electromagnetic environment in which it is designed to operate. Many EMI requirements take the criticality and environment into account by classifying equipment and by imposing different susceptibility test levels on the different classes.

EMI can be considered a form of environmental pollution; in order to reduce the impact of this pollution, some control on the environmental level of conducted and radiated emissions of noise is necessary.

Many countries impose commercial regulations on the emissions from data-processing equipment; industrial, scientific, and medical (ISM) equipment; vehicles; appliances; etc. In some instances standards are developed by a nongovernmental agency, such as the Society of Automotive Engineers (SAE), and are not necessarily mandatory. The majority of military regulations and standards, and some commercial specifications, also require that equipment be demonstrated immune to susceptibility test levels.

Chapter 9 describes the typical EMI regulations and requirements and EMI measurement techniques.

### 1.2.1  Military Regulations

The options are limited for manufacturers of equipment that must meet specified requirements. The military requirements are intended to be tailored to the specific electromagnetic environment by the procuring agency; however, this is seldom implemented. Should equipment fail specified military requirements and, after analysis or measurement, the environment be found more benign than the specified levels indicate, then the possibility exists for the procuring agency to grant a waiver on the specification limit. A more satisfactory approach is to specify realistic limits in the first place. The difficulty here is that the requirements are location dependent. That is, the proximity of equipment to transmitting antennas or other equipment or the number of units connected to the same power supply varies from case to case. Where equipment is intended for operation in a known location, the limits may be readily tailored to the environment.

### 1.2.2  Commercial Regulations

The manufacturers of equipment that must meet commercial requirements are seldom if ever awarded a waiver, and the limits are inflexible. To date only the countries of the European Union (EU) require immunity testing. Some manufacturers who want to market in non-EU countries may consider this an advantage until the equipment is found to be susceptible in a typical environment.

### 1.2.3  Unregulated Equipment

For the manufacturers of equipment to which no regulations apply but who want to achieve EMC either for the sake of customer satisfaction or safety or to minimize the risk of a lawsuit, the choice is either to design for a realistic worst-case environment or to define the environment with an existing EMI standard. We define a realistic worst case as either a measured maximum environment in a large sample of similar environments or a predicted maximum where all the mitigating factors have been considered.

In an ideal world, specified limits would be close to the realistic worst-case environment, whereas, as we shall see in Section 1.3, this is not always true.

## 1.3 ELECTROMAGNETIC ENVIRONMENT

The information in this section is intended to provide a comparison of the various worst-case environments and to provide guidelines to equipment designers and those writing procurement specifications.

Sources of EMI can be divided into natural and manmade, with, in most cases, the natural sources of radiation present at a much lower level than the manmade. The majority of unintentional emissions occupy a wide range of frequencies, which we may call *broadband* in a nonstrict sense of the term. Intentional emissions, such as radio and television transmissions, are termed *narrowband* and in the strictest sense of the term are emissions that occur at a single frequency or are accompanied by a few frequencies at the sidebands. The strict definitions of narrowband and broadband, as used in EMI measurements, and addressed in Chapter 9, are dependent on both receiver bandwidth and the pulse repetition rate of the source.

Electric field strength is measured in volts/m, as described in Section 2.1. Another unit of measurement is the dBμV/m. The unit of broadband field strength as used in military standards is dBμV/m/MHz. Here the reference bandwidth of 1 MHz is included in the unit. Another unit is the dBμV/m/kHz, where the reference bandwidth is 1 kHz.

For the sake of comparison to other sources of broadband noise, we will arbitrarily use the military MIL-STD 461 RE02 limit for broadband emission from equipment measured at a distance of 1 meter from the source. Figure 1.4 is a reproduction of the RE02 limit for spacecraft. The RE02 limit is more stringent than commercial limits on broadband noise. For example, the West German Commercial Regulation contained in VDE 0875 for broadband limits, when scaled to a 1-meter measuring distance and converted to the 1-MHz reference bandwidth, imposes a limit of 78.5 dBμV/m/MHz from 30 to 300 MHz. This limit is 13.5 dB above RE02 at 200 MHz and 4 dB and 8.5 dB above at 30 MHz and 300 MHz, respectively. These limits are presented in greater detail in Section 9.4.

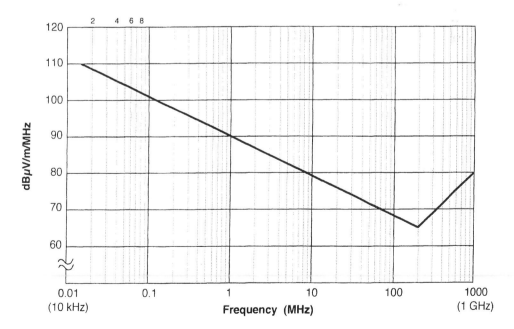

**Figure 1.4** RE02 broadband emission limits for spacecraft.

The remainder of this chapter deals with radiated and conducted components of the electromagnetic environment. The radiated electromagnetic environment is treated in Sections 1.3.1–1.3.3 and the conducted electromagnetic environment in Section 1.3.4.

### 1.3.1   Natural Sources of Electromagnetic Noise

Natural sources of electromagnetic noise are

- Atmospheric noise produced by electrical discharges occurring during thunderstorms
- Cosmic noise from the Sun, Moon, stars, planets, and galaxy

Atmospheric noise is produced predominantly by local thunderstorms in the summer and by tropical-region thunderstorms in the winter. The electromagnetic emissions from thunderstorms are propagated over distances of several thousand kilometers by an ionospheric skywave, and thus potential EMI effects are not localized. In the time domain, atmospheric noise is complex, but it may be characterized by large spikes on a background of short random pulses or smaller pulses on a higher continuous background.

Upper and lower limits for atmospheric radio noise displayed in the frequency domain are shown in Figure 1.5(D), ranging in level from a maximum of 108 dBμV (0.25 V)/m/MHz

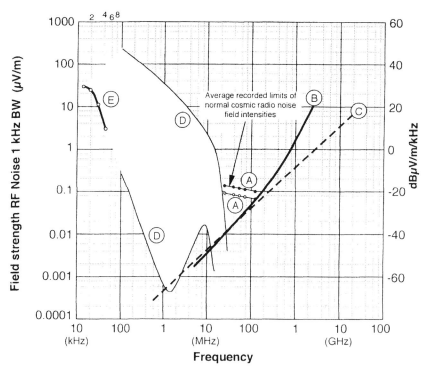

A   Average daily upper and lower limits of normal cosmic radio noise field intensities
B   Noise field intensities corresponding to internal noise of well-designed receiver
C   Noise field intensities (in one plane of polarization) produced by "black-body" radiation at 300 K
D   Upper and lower limits of atmosphere noise intensities (Nat. Bureau of Standards Circ. No. 462)
     Also radio propagation unit report RPU-5.
E   Atmospheric radio noise intensities measured in the Arctic

**Figure 1.5**   Atmospheric, cosmic, and thermal noise levels. (From Ref. 2.)

to a minimum of −6 dBµV (0.5 µV)/m/MHz at 100 kHz. Figure 1.5(E) also plots the atmospheric noise measured in the Arctic, which describes a yearly variation of approximately 7 dB as well as a systematic daily and seasonal variation. From measurements in Canada, the daily and seasonal variation is from 91 dBµV/m/MHz to 106 dBµV/m/MHz. Compared to our arbitrarily chosen RE02 broadband reference limit, the upper-limit atmospheric noise may be close to the RE02 limit at 100 kHz and 20 dB below the limit at 10 MHz.

Cosmic noise is a composite of noise sources comprised of sky background radio noise, which is caused by ionization and synchrotron radiation (which undergoes daily variation), and solar radio noise, which increases dramatically with an increase in solar activity and the generation of solar flares. Secondary cosmic noise sources are the Moon, Jupiter, and Cassiopeia-A. At 30 MHz, the average cosmic noise is 34 dB below the RE02 limit.

A comparison of the relative intensities of atmospheric and cosmic sources and the frequency range of emissions are shown in Figure 1.6. The average daily upper and lower limits of the normal cosmic noise are shown in Figure 1.5(A). Additional sources of emissions exist at lower levels, including the thermal background. The theoretical thermal background from the Earth's surface is shown in Figure 1.5(C) with the internal thermal noise of a well-designed receiver for comparison purposes.

The EMI effect on radio communications is often described as ''static'' due to the impulsive nature of atmospheric noise. A second source of transient EMI, which may be incorrectly attributed to atmospheric noise, is precipitation static discharge in the proximity of the receiving antenna. Static discharge on the ground is caused by a buildup of charge on the surface, resulting in a corona discharge. Correct grounding and bonding of conductive elements, use of high-breakdown-voltage dielectrics, static discharge coating, and transient suppression devices are some of the methods used to avoid static charge buildup and protect the receiver input circuit.

Because of the wide frequency span over which natural emissions occur, they may cause EMI in HF/VHF/UHF/SHF transmissions.

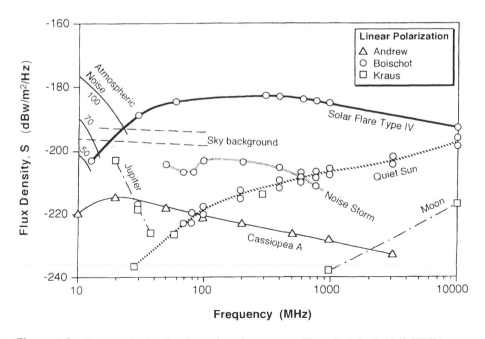

**Figure 1.6**  Comparative levels of cosmic noise sources. (From Ref. 1, © 1969 IEEE.)

### 1.3.2 Manmade Electromagnetic Noise

Some of the major sources of manmade electromagnetic noise are

- Arc welders
- RF heaters
- Industrial, scientific, and medical (ISM) equipment
- AC high-voltage transmission line
- Automotive ignition
- Fluorescent lamps
- Microwave ovens
- Hospital equipment
- Diathermy equipment
- Communication transmitter intentional and spurious radiation
- Electric motors

Each of these sources will be discussed with reference to the RE02 specification limit.

Heliarc welders use an RF arc at a typical fundamental frequency of 2.6 MHz. The spectrum occupancy of the heliarc welder emission covers the frequency range from 3 kHz to 120 MHz and thus contains frequencies lower than 2.6 MHz. The typical EMI effect on radio is a "frying" noise. Representative levels of radiation from a substantial population of RF-stabilized arc welders measured at a distance of 305 m are (1)

| Frequency | Radiation level |
|-----------|-----------------|
| 0.7 MHz | 75 dBμV/m/MHz |
| 25 MHz | 82 dBμV/m/MHz and |
| | 80 dBμV/m/MHz (10 mV/m/MHz) |
| 30 MHz | 70 dBμV/m/MHz |

The level is 4 dB above the RE02 limit at 30 MHz and is naturally much higher at distances closer than 305 m. For example, at a distance of 2 m from an arc welder, the E field level at 30 MHz is approximately 124 dBμV/m/MHz (1.5 V/m), which is 50 dB (316 times) above the RE02 limit. In extrapolating for distance at these frequencies, either a $1/d$ or $1/d^{1.5}$ law is used ($d$ = distance), depending on frequency and other criteria. From measurements at 14 manufacturing plants, the peak level of emission from arc welders was found to be 0.1 V/m at a measuring distance of 1–3 m (5). Unfortunately, the bandwidth of the measuring instrument was not given, so it is not possible to convert the measured level to a broadband unit of measurement.

In measurements described in Ref. 3, a heliarc welder exhibited significant emissions from 14 kHz to 240 MHz. However, when measured at a distance of a few miles, only the fundamental frequency of 2.6–3 MHz remained at a significant level. In measurements at different locations on 152 welders, comprising 54 different models, the highest emission measured at 305 m was 64 dBμV/m. Of the 152 welders, 31 produced emissions above 40 dBμV/m and only 6 above 54 dBμV/m. Some welders generated levels as low as 0 dBμV (1 μV)/m. The low-level emitters were characterized by one or more of the following: short welding leads; low-impedance ground; shielded wiring, including supply; and enclosure in a shielded building. High-level emitters, in contrast, were characterized by one or more of the following: poor grounding; unshielded wires; and proximity to power lines, which picked up the emissions and reradiated them.

The Federal Communications Commission (FCC), in Document 47 CFR Part 18, place a limit of 10 $\mu$V/m at 1600 m on industrial heaters and RF-stabilized arc welders below 5.725 MHz. Converting this to the 305-m distance used in the preceding survey and applying a $1/d$ law results in a limit of 34.4 dB$\mu$V/m. From the survey it is clear that many existing arc welder installations do not comply with the FCC limit.

The fields generated by induction and dielectric-type RF heaters are principally narrowband, with peaks extending up to about the ninth harmonic. Induction heaters are used for forging, case hardening, soldering, annealing, float zone refining, etc., while dielectric-type heaters are typically used to seal plastic packages. The fundamental operating frequency for induction heaters is from 1 kHz to 1 MHz, and for dielectric heaters is from 13 MHz to 5.8 GHz. In measurements on 36 induction heaters, the minimum and maximum emissions at a distance of 30 m varied from 30 dB$\mu$V (31.6 $\mu$V)/m to 114 dB$\mu$V (0.75 V)/m (4). Since these emissions are narrowband (NB), the 1-MHz reference bandwidth is omitted in the measurement unit used.

We have so far used the relatively low-level military broadband radiated emission limit (RE02) as a reference when comparing natural and manmade emission levels. To continue to use this comparison for narrowband noise, the narrowband RE02 limit may be used.

The emissions from four different manufacturers and 10 different models of dielectric heater when measured at a distance of 30 m ranged from a maximum of 98.8 dB$\mu$V (87 mV)/m and a minimum of 75 dB$\mu$V/m (5.6 mV) at the fundamental (27 MHz) and reduced to a maximum of 84 dB$\mu$V (15.8 mV) and a minimum of 38 dB$\mu$V (79 $\mu$V) at the sixth harmonic (162 MHz).

No measurements of the conducted noise placed on the power line by these devices are available. However, the radiated emissions at 30 m were found to be above the typical ambient levels of 0.15–0.9 mV/m measured in offices, electronic laboratories, and computer facilities, but they do not pose a severe EMI threat. This is illustrated by a cursory survey of the 35,434 complaints of interference with radio communication lodged with the UK regulatory authority over a 12-month period. This survey reveals that 143 complaints were attributed to ISM sources, 11 to medical apparatus, and 66 to RF devices not tuned to designated frequencies, but not one complaint was attributed to induction or dielectric heating equipment.

The levels of noise measured at 14 manufacturing sites from a variety of ISM equipment is shown in Figure 1.7. The 14 manufacturing plants included discrete and continuous production plants, an automotive tool and die shop, a chemical plant, a heavy equipment manufacturer, an aerospace manufacturing plant, newspaper printers, paper and pulp plants, and metal smelting plants. The levels are given in volts per meter, with no information as to the bandwidth of the measuring instrument. Assuming the bandwidth used was narrow enough to capture only one of the spectral lines of emission, then the electromagnetic field levels contained in Figure 1.7 will have the same magnitude as a narrowband emitter of the same field strength, at the specified frequency of maximum emission.

When a broadband field is expressed in broadband units, the magnitude is invariably higher than the same field expressed in a narrowband unit. Here we use the term *broadband field* loosely to indicate a field comprised of a number of frequencies. A broadband unit of measurement uses a reference bandwidth, typically 1 MHz (e.g., dB$\mu$V/m/MHz), whereas a narrowband unit does not specify a bandwidth (e.g., dB$\mu$V/m). However, the bandwidth used in a narrowband measurement is always less than 1 MHz. In a broadband measurement of broadband noise, many of the random or harmonically related spectral lines are captured in the receiver bandwidth. In a narrowband measurement, only one of the spectral lines is captured in the receiver bandwidth.

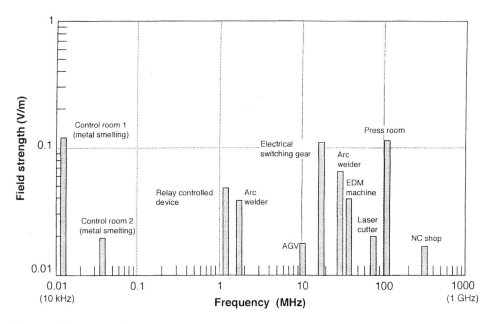

**Figure 1.7**  Peak field measurements from ISM sources. (From Ref. 4.)

For example, assume a broadband signal source generating harmonically related spectral lines at 1 kHz intervals apart with a constant amplitude over a 1-MHz span. Assume the field strength at some distance from the source, for the sake of our example, is 50 dBµV/m/MHz. If we were to make a narrowband measurement with a 1-kHz measuring bandwidth, then only one spectral line is measured and the field amplitude is reduced to −10 dBµV/m, expressed in narrowband terms. In this way, coherent broadband noise decreases as a function of 20 dB/decade of bandwidth. It is therefore incorrect to directly compare the magnitude of narrowband sources, for example, those ''intentional emitters'' contained in Section 1.3.3, to broadband sources. The subject of broadband and narrowband measurements are dealt with in Chapter 9 and sources of broadband noise in Chapter 3.

The susceptibility of equipment and cables to an impinging broadband or narrowband field is dependent, among other factors, on cable and enclosure resonance effects, the bandwidth of the equipment (including signal interfaces), and the transient response of the cable and structure. These factors are considered in subsequent chapters.

It is common to experience EMI to AM reception in cars that are driven in close proximity to, or under, high-voltage transmission lines. Figure 1.8 illustrates the spectrum occupancy of transmission line noise with, as expected, a maximum at the power line frequency of either 50 or 60 Hz. The curves in Figure 1.8 are from several sources, and the numbers in brackets indicate the distance (meters) from the line at which measurements were made.

Ignition noise level is dependent on traffic density and proximity. In the time domain, ignition noise is characterized by bursts of short- (ns) duration pulses with a burst duration of from a microsecond to milliseconds. The repetition rate of the bursts is dependent on the RPM, the number of cylinders of the motor, and the number of cars. When measured in close proximity to the road, these bursts vary in amplitude and direction with traffic flow. In cosmopolitan areas, ignition noise is a major contributor to the electromagnetic environment. The average of measurements taken in three cities at a roadside location for a traffic flow of 30 autos per minute are 60 dBµV/m/MHz at 100 MHz decreasing to 50 dBµV/m/MHz at 1 GHz. Ignition noise is

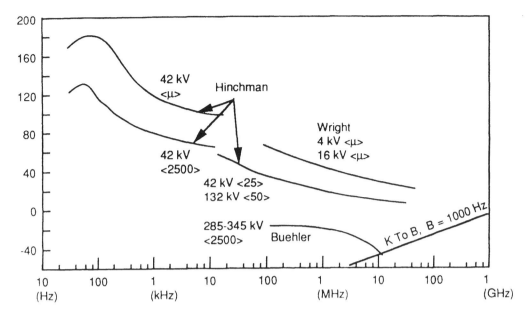

• Observation point location relative to the sources in feet <25>, <50>, etc. ; <μ> beneath conductor

**Figure 1.8** Transmission line noise (numbers in brackets are measuring distances in meters). (From Ref. 1, © 1969 IEEE.)

also present from 10 kHz upwards, and levels as high as 80 dBμV/m/MHz have been measured at 10 kHz.

Fluorescent and gaseous discharge tubes produce impulsive radio noise similar to power line noise in its waveform characteristics. The level of noise from 10 kHz to 1 MHz was measured in a shielded room with the lights on and the ambient noise with the lights off. The highest fluorescent light emission was at 300 kHz, measured at 89 dBμV/m/MHz, and the RE02 BB limit at 300 kHz is 98 dBμV/m/MHz. Thus, although the ambient is just below the specification limit, all subsequent measurements were made with the fluorescent lights off. For MIL STD 461 measurements, the ambient should be at least 6 dB below the specification limit. The remaining incandescent lamps in the shielded room did not add to the electromagnetic ambient. The average level of radiation from fluorescent lamp fittings measured at a distance of 1 meter is shown in Figure 1.9.

Field-strength measurements of microwave ovens operating in the ISM band at 915 MHz were measured in the laboratory at distances of 3.05 m and 305 m. Measurements were also made outside a large condominium containing 385 ovens. The measurements were made on one make of oven and similar models. The maximum field strength measured in the laboratory was 1.5 V/m at 3.05 m and 11 mV/m at 305 m. The field strength measured outside the condominium building would have been altered by some shielding due to the building structure and reflections from the ground and nearby buildings. The maximum field strength measured was 8.9 mV/m at 920 MHz at a measuring distance of 152 m from the two buildings comprising the condominium (6).

The electromagnetic environment in hospitals has been of growing concern as the number of EMI problems experienced in hospitals has increased. Measurements have been made in ten American hospitals in a number of locations, including operating rooms, intensive care

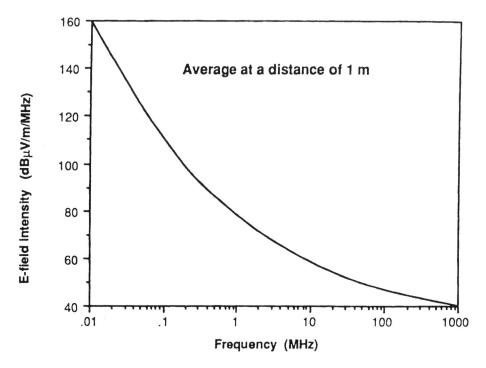

**Figure 1.9** Average level of radiation from fluorescent lamps.

units, chemistry labs, special procedure rooms, and physical therapy facilities. In the majority of locations, E fields as high as 3 V/m/MHz were measured. It should be emphasized that the measurement results are displayed in broadband units. If the measurements had been made with a narrowband bandwidth and displayed in narrowband units, the majority of emissions would have reduced in amplitude but with peaks at 70 kHz and 1–5 MHz remaining at approximately 3 V/m. These peak emissions are from diathermy or electrosurgical units, which use high-frequency currents to achieve bloodless surgery and are generators of narrowband noise. Figure 1.10 is a composite of the worst-case field levels measured in all locations at the 10 hospitals surveyed (7).

When these unintentional manmade high-level sources of radiation are compared to our standard RE02 level, the sources may have intensities of 60 dB above (i.e., 1,000 times) the RE02 limit.

In addition to RF electromagnetic fields, magnetic fields at power line frequencies are ever-present. Due to the concern for the potential health hazard of magnetic fields, a number of measurements were made in a home and in an office building. The magnetic field strength in one room on the 1st floor in the office building was sufficient to result in distortion of computer monitor displays. The displays were susceptible to magnetic fields with magnitudes of 1.3–3 A/m. The magnetic fields in this 1st floor varied from a low 0.016 A/m to the maximum of 3 A/m at 60 Hz. On the 5th floor of the same building, the magnetic field varied from 0.00485 A/m to 0.0428 A/m. The highest magnetic fields measured in the basement of the same building was 1.63 A/m, close to a power transformer. A survey was also conducted at several locations in a home; the result of these measurements are shown in Table 1.1.

One technique for reducing 60-Hz magnetic fields from wiring is to keep line and neutral in any two-phase circuit or the A, B, C, and neutral conductors in close proximity or preferably

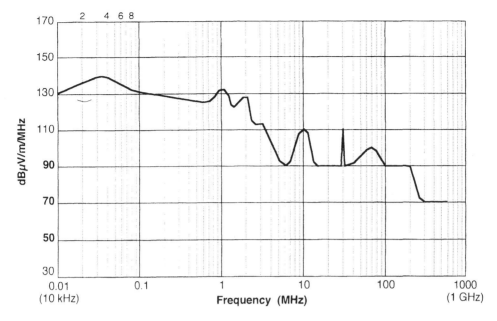

**Figure 1.10**   Composite worst-case field levels in hospitals. (From Ref. 2. © 1975 IEEE.)

twisted together. Section 2.1.3 discusses the reduction effect seen with twisted-pair cable in detail. An alternative is to run the cables in a high-permeability conduit. Measurements were made on a galvanized cold-rolled-steel seamless conduit with a wall thickness of approximately 1.2 mm (15/32″), and an attenuation of approximately 16 dB at 60 Hz was seen. Section 6.2 describes low-frequency magnetic field shielding in detail.

### 1.3.3   Intentional Emitters

The electromagnetic environment is also crowded with the intentional emissions from radio, television, and radar transmitters, all of which can interfere with equipment that is not intended for any form of reception as well as receivers tuned to a different frequency. One of the most frequent causes of EMI is electromagnetic fields produced by radio transmitters. The EMI effect may be confined to annoyance, for example, due to the spurious operation of a garage door

**Table 1.1**   Magnitude of Magnetic Fields Measured in a Home

| Location of magnetic field measurements in a home | Magnetic field (A/m) |
| --- | --- |
| 20 cm from top of stove | 4.86 |
| 20 cm from oven door | 0.4 |
| 20 cm from electric food mixer | 4 |
| 20 cm from electrical panel | 0.82 |
| 20 cm from 120–250-V autotransformer | 46.2 |
| 2 m from hi-fi setup | 0.1 |
| Center of kitchen; all appliances off except for refrigerator | 0.1 |
| 20 cm from toaster | 0.81 |

caused by the use of a CB radio. There also exists documentation of far more serious effects, such as the destruction of a warship attributed indirectly to EMI (due to interference with communication) or the crash of an aircraft as the direct result of EMI (flying in close proximity to a high-power transmitter).

Radio transmissions are narrowband, with limited emissions at the harmonics of the fundamental continuous wave (CW). If these waves are modulated, then the resulting sidebands are also transmitted. In addition to frequencies related to the fundamental, a transmitter may radiate the local oscillator frequency and broadband noise generated within the stages of the transmitter. The composite of spurious, broadband, and harmonically related noise from a transmitter is typically at least 70 dB down from the fundamental. The radiated fields from transmitting antennas are dependent on proximity, transmitter output power, directivity of the antenna, relative height between antenna and measuring point, and proximity of reflecting or intervening absorbing material or structures, etc. Measurements have been made in two major urban centers in Canada (Montreal and Toronto, Ref. 7). From a large number of measurements made close to ground level, a typical field-strength value developed at usual transmitter–receiver distances was computed. The results are shown in Table 1.2. The maximum typical E field at a fixed distance of 100 m is shown in Figure 1.11.

An electromagnetic RF ambient survey is often made before erecting a receiving antenna or before the installation of potentially susceptible equipment. Examples of peak emitters encountered in such measurements are a radar signal measured at a site in Goosebay, Labrador, that exhibited a frequency of 1280 MHz and a level of $-21$ dB-W/m$^2$ (i.e., an E field of 1.73 V/m); an AM radio transmitter in Alice Springs, Australia, generated E fields of 4 V/m at 4.83 MHz incident on nearby receiving equipment; and, at a proposed site in Hong Kong, E fields of 3.75 V/m at 12 MHz were measured at the proposed location of receiving equipment. Many apartment and office buildings have antennas mounted on the roof. In one apartment building, Bell cellular antennas were mounted on the roof, and the E fields, at 880 MHz, measured in two apartments were 7.86 V/m and 6.48 V/m, respectively. On the balconies of these same apartments, the maximum E field was at 12 V/m and 12.9 V/m, respectively. The ambient inside a household that contains an amateur radio transmitter with the antenna mounted on the roof was measured at 30 V/m inside the home. Transceivers (walkie-talkies) can generate fields as high as 55 V/m at 823 MHz at a distance of 12 cm from a 5-W transceiver (8). The power generated by a cellular phone at a distance of 2.5 cm from the human head is absorbed by the head at the same rate as an incident field of 41 V/m (10). In Ref. 10, ambient measurements were made on apartment buildings in 15 different cities in the United States. The median exposure for

**Table 1.2**  Typical Ambient Field
Strengths (Ref. 8)

| Frequency (MHz) | Field strength[a] (V/m) |
|---|---|
| 0.5–1.6 | 0.6 |
| 26.9–27.4 | <0.1 |
| 54–88 | 0.07 |
| 88–108 | 0.15 |
| 108–174 | 0.05 |
| 174–216 | 0.07 |

[a] Typical field strength in major urban centers
  from broadcast transmitters.
*Source*: Ref. 8.

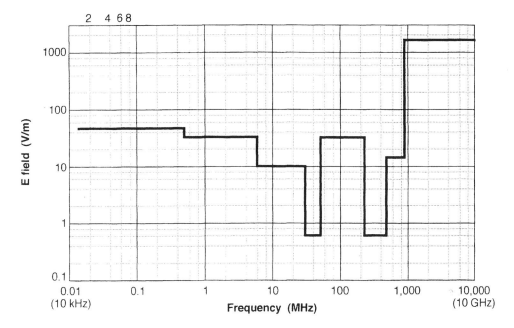

**Figure 1.11**   Average and peak fields from radio transmitters at a distance of 100 m.

the inhabitants of all cities was 0.137 V/m, and 99.9% of the inhabitants were exposed to fields of less than 1.94 V/m. Thus we can conclude that ambients with magnitudes of volts per meter are uncommon, but are increasingly the cause of EMI in apartment and office buildings. Electric fields having volts-per-meter magnitudes are potential sources of interference to equipment even when the passband of the equipment is far from the interfering frequency. Case studies for this type of EMI are contained in Chapter 12.

The preceding summary of measured spectrum occupancy and radiated emission levels from unintentional and intentional sources is intended to provide guidelines to manufacturers whose equipment must achieve EMC in these harsh environments. One important piece of information presented with the data is the measuring distance from the source. In measurements it has been observed that the low-frequency component falls more rapidly with increasing distance than those at VHF and UHF. The mechanism for E field reduction with distance and frequency is discussed in Chapter 2. When predicting the electromagnetic ambient level at a given location, not only is the distance from potential sources of EMI an important consideration but also the presence of conductive structures close to the emission source, such as walls, buildings, and ground topology. When structures are behind or to the side of the source, they act as reflectors; they act as shields when the structures are intervening. Chapter 6 discusses the shielding effectiveness of structures and soil.

### 1.3.4   Conducted Noise on Power Lines

The electromagnetic environment is confined not only to electromagnetic fields but also to signals/noise existing in a transmission medium. Thus the noise conducted on equipment power lines must be accounted for when characterizing an electromagnetic ambient. For example, a computer on a machine shop floor, located at some distance from numerical control machines, may be immune to the radiated field levels but susceptible to conducted noise. Therefore, in addition to immunity to electromagnetic fields, achieving EMC entails immunity to conducted noise, present predominantly on power lines but also on signal interfaces and grounds. To predict susceptibility for equipment on which no EMC requirements are placed, data on expected maximum amplitude, frequency content, and waveform type (e.g., CW or spike) that may exist on the power supply in a given electromagnetic environment is required.

The conducted noise currents flowing out of medical equipment into a known load connected to the power line was measured in the survey made in 10 hospitals. The impedance of the load is not given in Ref. 11; however, a typical value is 50 Ω. A 400-Hz highpass filter was included in the measurement setup to reduce the magnitude of load current measured at 60 Hz and its harmonics by the current probe. Figure 1.12 is a composite of the maximum conducted noise currents measured in all locations at the 10 hospitals (7). In other measurements in a hospital environment, voltage spikes as high as 3,000 V have been measured on the AC power line.

Unfortunately, the amount of data available from measurements of conducted noise current or voltage is limited. An alternative approach is to use the specified susceptibility test levels described in Chapter 9 for the relevant type of equipment (i.e., military, ISM, or vehicle-mounted equipment). Where susceptibility test levels have not been developed for the equipment type, then the regulations concerning conducted emission limits for equipment such as digital, household appliance, portable tools, etc. may be used to predict the conducted power line noise level. When using emission limits in a prediction, some correction must be made for the specific or worst-case power line impedance and the number of devices sharing the same power line. For example, in a laboratory environment, several computers, hand tools, RF sources, and a refrigerator may share a common power line. A worst case AC, 120-V, 60-Hz power line source impedance plotted against frequency is shown in Figure 5.28a. Taking a composite of the noise currents from all sources, a composite noise voltage developed across the worst-case source impedance may be calculated, to arrive at susceptibility test/design levels.

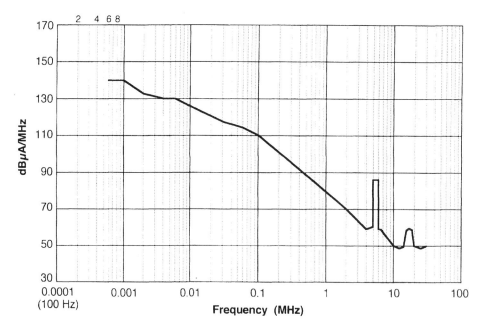

**Figure 1.12** Composite of conducted AC power line noise currents measured at 10 hospitals. (From Ref. 2. © 1975 IEEE.)

## REFERENCES

1. E.N. Skomal. The dimension of radio noise. 1969 IEEE Symposium on EMC, Vol. 69, C3-EMC, June 17–19.
2. H.V. Cottony. Radio noise of ionospheric origin. Science, Vol. 111, No. 2872, p. 41; January, 1950.
3. H.V. Cottony, J. R. Johler. Cosmic radio noise intensities in the VHF band. Proceeding of the IRE, September 1952, pp. 1050–1060.
4. R. Nielson. Ethernet performance in harsh industrial environments. Control Engineering 1988, Vol. 35, No. 10.
5. A.S. McLachlan. Radio frequency heating apparatus as a valuable tool of industry and a potential source of radio interference. 7th International Zurich Symposium and Technical Exhibition on Electromagnetic Compatibility, Zurich, 1987.
6. A. Tell. Field strength measurements of microwave oven leakage at 915 MHz. IEEE Transactions on Electromagnetic Compatibility, May 1978, Vol. EMC-20, No. 2.
7. Government of Canada Department of Communications. EMCAB 1, Issue 2.
8. John Adams. Electric field strength measured near personal transceivers. National Institute of Standards and Technology. IEEE EMC Symposium record 1993.
9. Paolo Bernardi, Marta Cavagnaro, Stefano Pisa. Evaluation of the SAR distribution in the human head for cellular phones used in a partially closed environment. IEEE transactions on Electromagnetic Compatibility, August 1996, Volume 38, Number 3.
10. Daniel D. Holihan. A technical analysis of the United States environmental protection agency's proposed alternatives for controlling public exposure to radiofrequency radiation. IEEE EMC Symposium record 1987.
11. R.J. Hoff. EMC measurements in hospitals. 1975 IEEE International Symposium on Electromagnetic Compatibility.

# 2
# Introduction to E and H, Near and Far Fields, Radiators, Receptors, and Antennas

Judging by the response from attendees of seminars, this chapter may be the least popular with design and system engineers.

The concept of radiation from, and coupling to, interface cables, PCB tracks, wiring, etc. is generally foreign to engineers, despite involvement with equipment containing digital, analog, RF, and control circuits. The reason may be that it is difficult to envisage interconnections as antennas, or circuit elements, or to see the potential for crosstalk between conductors. This is particularly true when equipment exhibits EMI or fails an EMC requirement and the engineer is under the pressure of schedule to find the quick fix usually demanded by management. In order to make simple EMC predictions or solve an EMI problem efficiently, an understanding of the principles of radiation and coupling, including frequency dependency and resonance effects, is essential.

Even the choice of an effective diagnostic test or the evaluation of a problem for either a radiated or a conducted source are difficult without this understanding. The title of this chapter may appear to imply that radiators and receptors are somehow different from antennas. Of course this is not true. One antenna engineer, when asked to define an *antenna*, replied, ''What is not an antenna?'' It used to be said by TV repair technicians, not totally in jest, of a high-field-strength location that a piece of wet string would serve as an antenna.

One investigator has examined the resonant frequency and receiving properties of leaves and fir cones, modeled as log periodic antennas, and questioned the effects of RF currents flowing, as a result of electromagnetic fields, on their surfaces. The differentiation made in this chapter between radiators, receptors, and antennas is merely to distinguish between structures intentionally designed to radiate and receive and those not. It will be seen that much of the antenna theory is applicable to nonintentional antennas, for example, in investigating the magnitude of current flow caused by an incident field on a structure and the radiation from a current-carrying conductor. Many good books exist on the subject of wave theory and antennas. Then why the need for this chapter? The magnitude of an electromagnetic field at some distance from an AC source varies with time. In the measurement and prediction of field magnitude for EMI analysis or in meeting a limit, it is the peak magnitude that is required. Therefore it has been possible to simplify the field equations presented in this book by eliminating the time-dependent terms. In EMC we are interested both in highly efficient resonant antennas and in the nonintentional radiation and coupling, to nonresonant structures, with impedances not matched to the termination impedance. Most antenna books are confined to the analysis of antennas that are highly efficient and relatively narrowband. However, many of the equations developed for antennas are quite applicable to any conductive structures.

The formulas contained in this chapter are used in the predictions contained in subsequent chapters and are presented as *magnitudes* (i.e., they may be used directly). Derivations for the equations are omitted and may be found in the reference material. In order to make life as simple as possible, the equations for coupling to and radiation from wires and cables as well as crosstalk between cables have been incorporated into computer programs written in BASIC and provided at the end of this and other chapters. However, an understanding of the correct model to use and the limits on its applicability is still required.

In addition, this chapter will enable the reader to make, calibrate, and correctly use simple magnetic field probes and E field antennas, and it introduces the meaning of terms, such as *gain* and *antenna factor*, used with EMI measuring antennas.

## 2.1  STATIC AND QUASI-STATIC FIELDS

In order to understand electromagnetic waves in proximity to current-carrying conductors as well as in free space, it is helpful to examine static electric and magnetic fields. In addition, at low frequency and at close distances to the source, it is often the reactive near field or quasi-static field that couples strongly, and these fields can be computed using the following DC analyses.

### 2.1.1  DC Electric Field

An electric field exists between the plates of a capacitor to which a potential $V_0$ is applied (Figure 2.1). The electric field intensity is

$$E = \frac{V_0}{h} \quad \text{[V/m]} \tag{2.1}$$

where $h$ is the distance between the plates, measured in meters. Thus $E$ is a magnitude per unit length. For example, if 10 V were applied to plates 10 cm apart, the field strength between the plates would be 100 V/m. The electric field has electric lines of force associated with it that are tangent to the electric field and are proportional in strength to the electric field strength. The electric field intensity can be measured by inserting a small dipole into the field (Figure 2.1). The component of the electric force tangential to the two conductive arms of the probe moves electrons in the conductors, and a voltage appears across the arms.

The voltage induced into a thin probe is substantially independent of the radius and proportional to the length of the probe. In practice, the conductive wires connected to the probe would disturb the electric field and Eq. (2.1) would not be strictly correct. The field between two infinite parallel plates is uniform, and in the field between the plates we find

$$V_1 = Eh_d \tag{2.2}$$

where

$\qquad E$ = electric field strength [V/m]

$\qquad h_d$ = effective dipole length

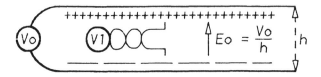

**Figure 2.1**   A cross section between parallel plates.

In a nonuniform field, the electric probe measures the average intensity of the field occupied by the probe; hence $h_d$ should be as small as possible. The electric field probe described is the simplest example of a receiving antenna. Electric field lines may start on a positive charge and end on a negative charge, as shown in Figure 2.1, or they may start on a positive charge and end at infinity or start at infinity and end on a negative charge or with time-varying fields form closed loops that neither start nor end on a charge.

### 2.1.2 DC Magnetic Field

An electric current is surrounded by a magnetic field of force. The magnetic field around a very long wire carrying a constant current is given by

$$H = \frac{I}{2\pi r} \quad [\text{A/m}] \tag{2.3}$$

and illustrated in Figure 2.2.

Magnetic field lines always form closed loops around current-carrying conductors because magnetic charges do not exist. The magnetic field is, like an electric field, a magnitude per unit length. For example, the magnetic field 10 cm from a wire carrying 10 A is 16 A/m and reduces to 1.6 A/m at a distance of 1 m. Expressed in another way, if the circumference of the field line at 10-cm distance is 0.628 m, the magnetic field strength is then 10 A/0.628 m = 16 A/m. Equation (2.3) may be found by dividing up the conductor into infinitesimal lengths (current elements) through which the current flows. The total field is then the sum of the contributions from all of the current elements stretching out to infinity on both sides of the measuring point. When the length of the wire is not infinite but much longer than the measuring distance $r$, Eq. (2.3) may still be used with a normally acceptable small error in the computed field strength. If the wire changes direction, then the equation for computing the field from a current loop may be applicable.

In practice it is uncommon to find short lengths of isolated current-carrying conductors. One exception may be where a conductor enters a space through a shield and then exits the space after a short distance via a second shield. The approximate magnetic field from this specific example is then

$$H = \frac{I}{4\pi r^2} \quad [\text{A/m}] \tag{2.4}$$

It is very common to find a second return conductor in close proximity to the supply conductor, with the return conductor carrying exactly equal and opposite current to the supply.

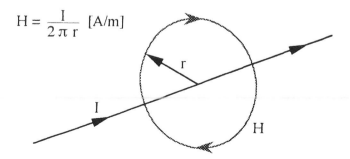

$$H = \frac{I}{2\pi r} \quad [\text{A/m}]$$

**Figure 2.2** Magnetic field around a long conductor.

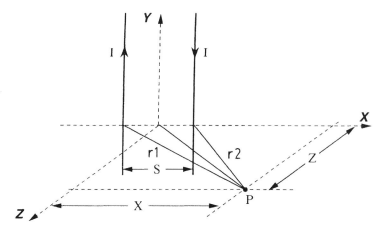

**Figure 2.3a** Configuration of a two-wire line.

This configuration is shown in Figure 2.3a. Because the field magnitude for the two conductors is not the same in the plane of the loop, the x-axis, as it is tangential to the plane of the loop, the z-axis, two field magnitudes $H_z$ and $H_x$ must be computed.

The derivation of the following equations for $H_z$ and $H_x$ may be found in Ref. 1 as well as in other textbooks on electromagnetic theory:

$$H_z = \frac{I}{2\pi}\left(\frac{x + S/2}{r_1^2} - \frac{x - S/2}{r_2^2}\right) \tag{2.5}$$

$$H_x = \frac{I}{2\pi}\left(\frac{z}{r_1^2} - \frac{z}{r_2^2}\right) \tag{2.6}$$

All distances, such as $x$, $z$, $r_1$, and $r_2$, are in meters.

From Eqs. (2.5) and (2.6) we see that the field on the x-axis at the center of the two conductors is zero and on the z-axis it is twice the magnitude computed for the z-component either side of the center. This may be seen graphically in Figure 2.3b, where the fields from the two conductors are in antiphase on the x-axis and in phase on the z-axis at the center of the conductors.

When probing around the two conductors with a magnetic field measuring probe, the change in measured field may be seen with change in probe location and orientation in the x- and z-axes around the conductors.

Twin conductor cables carrying 50–20-kHz AC power are potential sources of high magnetic fields at close proximity to the conductor. The closer together the conductors are, the lower the resultant H field. Where the conductors are separated and, for example, routed around the inside of an enclosure, the magnetic field is at a maximum, and the correct prediction method for magnetic field from a current loop is Eq. (2.7).

### 2.1.3  Twisted-Pair Wires

In order to minimize the generation of magnetic fields from conductors, twisted-pair wire is used. The magnetic fields from each loop of the twisted wires are in antiphase and therefore tend to cancel. At the exact center of the loop and equidistant from all loops, assuming all loops

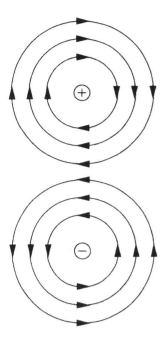

**Figure 2.3b**   Field distribution around a two-wire line.

are of equal size, the total magnetic field is zero. The realistic situation is that not all loops are equidistant from the measuring point and that the sizes of the loops are not exactly equal. Figure 2.4 illustrates the resultant fields and their directions, which may be computed for any location by computing the individual fields from each loop. Then the total field may be graphically estimated from the vector product or vector analysis may be used.

The magnetic field from a twisted pair contains a radial component, $x$, a component down the axis of the pair, $z$, and a circumferential component, $\phi$, due to the helical nature of the twist. The field at some distance from the wire, $p$, is dependent not only on $p$ but on the pitch distance, $h$ (distance for one cycle of twist), as well as the distance between the center of the twisted pair and the center of one conductor of the pair.

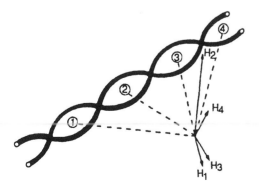

**Figure 2.4**   Resultant fields and their directions for an individual loop of a twisted pair.

At distances very close to the twisted pair, the field may be higher than for the two-wire line. However, the field very quickly reduces in magnitude with increasing distance $p$. From measurements in Ref. 2 it was found that the magnetic field strength from a twisted pair reduces more rapidly, as a function of $1/r^3$, than the field from an untwisted pair or a coaxial cable in which the center conductor is not concentric or on which an unbalanced current flows, which reduce as a function of $1/r^2$.

We use the term $1/r^n$ throughout this chapter. In order to illustrate the meaning, imagine that the magnetic field has been measured at a distance of 1 cm from a twisted wire pair and that the field at 10 cm is required. The change in distance between measuring point and predicted point is 10 cm/1 cm = 10, the reduction in field strength is therefore $1/10^3 = 1/1000$.

From measurements and predictions on twisted-pair cable with a large pitch distance of 3 inches (i.e., only 0.33 twists per inch) from Ref. 3, the reduction in field for the twisted-pair cable is approximately $1/r^{10}$, or 60 dB, for a doubling in distance $p$. From the same reference, the reduction in field for an untwisted cable is 12 dB (i.e., $1/r^2$) for a doubling of distance. A comparison between the reduction in field versus distance for both a twisted pair and an untwisted pair is shown in Figure 2.5. In Ref. 4, generalized curves using the ratios of $a/h$ and $p/h$ have been constructed for $B_r$, $B_\phi$, and $B_z$. These generalized curves are reproduced in Figure 2.6. The $y$-axis of Figure 2.6 is magnetic field strength, in decibels/gauss for 1 ampere of current flow. To calculate the field components from the figure:

1. Find the number of decibels for the ratio $a/h$.
2. Add the decibels for the ratio of $p/h$.
3. For the $B_r$ and $B_z$ fields, add $20 \log(1/2.54xh)$ [cm] or $1/h$ [inches]) to the number of decibels obtained in step 2. For the $B_\phi$ field, add $20 \log(1/2.54p)$ [cm] or $l/p$ [inches] to the number of decibels obtained in step 2.
4. To correct for the actual current flow $I$, add $20 \log I/l$ to the number obtained in step 3.

The field magnitude at some point or points in close proximity to a twisted pair is of interest when the voltage induced into either a loop or a wire, which crosses the pair at some angle, is required. Where the concern is not the electromagnetic field incident on a point but

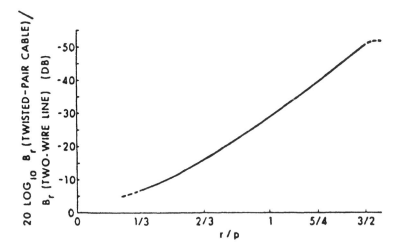

**Figure 2.5** Comparison of the reduction in field vs. distance curves for both a twisted pair and an untwisted pair. (© 1968 IEEE.)

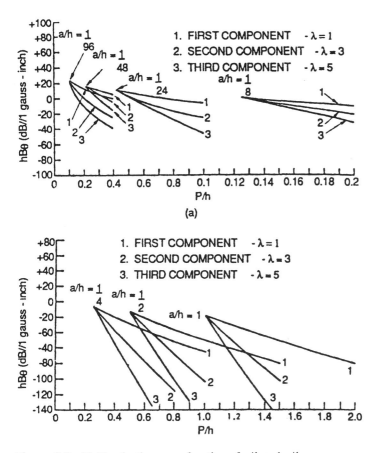

**Figure 2.6** Field reductions as a function of $a/h$ and $p/h$.

on a cable or wire running the length of the twisted wire pair, then a current is induced into the receptor wire. Again these currents tend to cancel, but not exactly due to varying distance and orientation between the twisted wire source and the receptor wire, unequal loop areas, and an odd number of loops. End effects where large loops may exist due to the termination of the twisted pair must also be considered. For the evaluation of crosstalk, the worst-case magnetic field generated down the length of the twisted pair may be estimated for a given distance from the twisted pair and used with the cable crosstalk equations and computer program found in Chapter 4. This program allows the prediction of coupling to either a single wire above a ground plane, a two-wire line, or a shielded cable.

The susceptibility and radiation from twisted pairs from 20 kHz to GHz (i.e., above the frequency where the quasi-static equations are valid for long distances from the source) is considered in Section 2.2.6.

### 2.1.4  DC and Quasi-Static Fields from a Loop

The radial magnetic field from a loop (i.e., in the $x$-axis) at a point coaxial to the loop at DC and low frequency in close proximity is given by

$$H_x = \frac{Ir^2}{2(r^2 + d^2)^{1.5}} \quad [\text{A/m}] \tag{2.7}$$

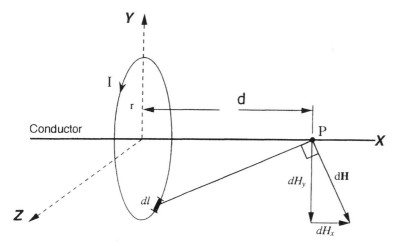

**Figure 2.7** Configuration for the magnetic field from a loop.

where

$d$ = distance from the loop [m]

$r$ = radius of the loop [m]

$I$ = current carried by the loop [A]

The loop and measuring point are shown in Figure 2.7. The components of the magnetic field in the y-axis at point $p$ cancel, and thus $H_y$ is zero.

## 2.2 ELECTRIC WAVES ON WIRES AND IN FREE SPACE

Connecting a sinewave generator to parallel conductors of length greater than the wavelength, $\lambda$, results in the generation of a wave along the length of the conductors. Parallel conductors can be considered chains of small inductors in series and small capacitors in parallel. We thus expect a delay in the transmission of voltage down the length of the conductors. The wave profile is shown in Figure 2.8.

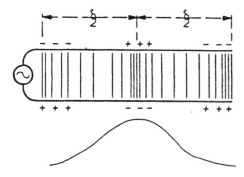

**Figure 2.8** Distribution of charge and waves on parallel plates.

The distance between the crests of the sinewave is called the *wavelength*, $\lambda$. If the wave moves down a conductor so that the time between successive crests is $t$, then the velocity of the wave is

$$v = \frac{\lambda}{t}$$

The frequency is given by

$$f = \frac{1}{t}$$

Hence,

$$v = \lambda f$$

The velocity of a wave in free space or air is a constant $3.0 \times 10^8$ m/s, and the relationship between frequency and wavelength is

$$f\ [\text{MHz}] = \frac{300}{\lambda\ [\text{m}]}$$

$$\lambda\ [\text{m}] = \frac{300}{f\ [\text{MHz}]}$$

The velocity for a medium is given by

$$\frac{1}{\sqrt{\mu\mu_r \epsilon \epsilon_r}}$$

where

$\mu$ = permeability of free space = $4\pi \times 10^{-7}$ [H/m] = 1.25 [$\mu$H/m]

$\mu_r$ = relative permeability

$\epsilon$ = dielectric constant of free space = $1/36\pi \times 10^9$ = 8.84 [pF/m]

$\epsilon_r$ = relative permittivity

The wave velocity of a transmission line is given by

$$\frac{1}{\sqrt{LC}}$$

where

$L$ = inductance of the line per unit length

$C$ = capacitance per unit length

Figure 2.8 shows the distribution of charge in the form of electric lines of force. Lines of force are considered to be tangential to the electric field $E$. Clusters of dense lines of force exist where the charge density is high, and in the next cluster the lines of force are oppositely directed. Loops of current are also present, formed partly on the conductors and partly in the space between the plates.

Thus a magnetic field is present around the parallel plates. The current flow in the space between the conductors is named the *displacement current*, and, as we shall see in Section 7.6.3.1, it is important when considering the common mode current on a line. The open circuit

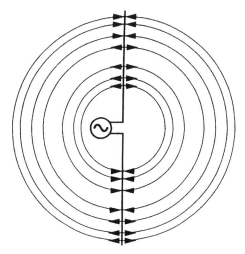

**Figure 2.9**  Lines of force around divergent conductors.

transmission line of Figure 2.8 results in standing waves along the length of the line, due to reflection of the wave arriving at the end of the line. If the conductors are not parallel but rather divergent, then the lines of force are as shown in Figure 2.9.

### 2.2.1  Radiation

The dipole antenna depicted in Figure 2.10 is said to be short, because charge reaches the ends of the antenna in much less than a period; expressed differently, the length of the antenna is much less than a wavelength. The current flow in conductors of the two-wire transmission line driving the antenna are of the same magnitude but 180° out of phase, and standing waves are created on the transmission line. When the length of the antenna is less than the wavelength, the standing wave current along the length of each arm of the antenna is in phase and the fields radiated from both arms will reinforce. As the discharge begins, the lines of force diminish and after a half period are at zero. During this period, the lines of force reaching out to $P$ are

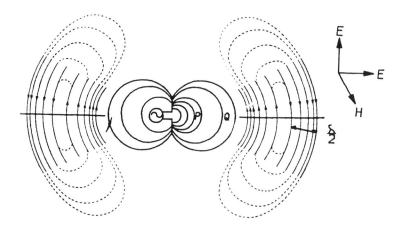

**Figure 2.10**  Detached lines of force around a short dipole antenna.

cancelled; however, a fraction of the lines of force spread out to $Q$, and this cluster of lines becomes detached at the end of the first half-period. The cluster of lines of force move on, and new lines of force take their place. The width of the configuration of lines remains $\lambda/2$, with the area over which the lines of force spread increasing with increasing distance $r$ from the antenna. The area can be proven to equal $\pi\lambda r$.

In accordance with our conception of lines of force, the density of the lines of force is proportional to the electric intensity. The number of lines of force issuing from the antenna is proportional to the charge and, therefore, to the current in the generator. The fraction of the lines of force detached from the antenna is proportional to the length $2l$ of the antenna. Magnetic and electric fields exist in the equatorial plane of the antenna (Figure 2.10). In the equatorial plane, the magnetic field, in amps per meter, is proportional to

$$\frac{2Il}{\pi\lambda r}$$

The radial electric field is the field that has the same direction as the static or quasi-static field, and it is given, in volts per meter, by

$$\frac{2Il}{4\pi r^2}$$

and therefore diminishes as a function of $1/r^2$ with distance $r$. The electric field in the meridian (i.e., in the plane of the antenna arms) is given by

$$\frac{2IlZ_w}{\pi\lambda r}\cos\theta$$

where $\theta$ is the angle between the direction of the antenna arm and the measuring point.

The ratio of $E/H$ has the dimension of an impedance and is often referred to as the wave impedance, $Z_W$, used in the preceding expression. The electric intensity $E$ and the magnetic intensity $H$ in free space take the place of the voltage and current at the terminals of a circuit. Therefore, $E = Z_W H$ and $H = E/Z_W$.

The physical dimensions of $E$ and $H$ are those of voltage and current per unit length. The ratio of $E/H$ is dependent on the proximity of the wave to the source of emission. Very close to the source, where the field contains radial components, it is called the *induction field*. Further away, where some field components decrease as $1/r^3$, it is called the *fresnel region*. Still further from the source, where the fields fall off as $1/r$, it is called the *far-field* or *fraunhoffer region*. The fields in the fresnel and fraunhoffer regions both radiate. At low frequency and in close proximity to the dipole antenna, the reactive or quasi-static field exists and does not radiate.

The distances at which far-field conditions occur are dependent, among other factors, on the size of the antenna. Where the size $D$ of an antenna is less than $\lambda/2$, then the near-field/far-field interface distance is defined as

$$r = \frac{\lambda}{2\pi} \tag{2.8}$$

where $D > \lambda/2\pi$, the interface distance is $r = D/2\pi$.

This definition is useful in EMC because it describes the distance at which the magnetic and electric fields from an electric current element or current loop begin to reduce as a function of $1/r$. When $D$ is much less than $\lambda/2\pi$, the antenna approximates a point source, and the phase is a function of the distance $r$ from the source. When the dimension $D$ approaches $\lambda/2\pi$, a correction for phase error may be made, as described in Section 2.2.2. The phase error expressed, in degrees, can be defined as the difference in the following path lengths: between the closest point of the antenna to the measuring point and the furthest point of the antenna to the measuring point.

For a high-gain antenna, the interface distance is often considered to be

$$r > \frac{4D^2}{\lambda}$$

at which distance the phase error is less than 22.5°. For the equation to be useful, the aperture dimension of the antenna, $D$, should be large compared to the wavelength. In the far-field region, $Z_W$ is equal to the intrinsic impedance of free space, which is commonly expressed as either ($R_cZ_c$, $n$ or $p$).

The intrinsic impedance of free space is equal to

$$\sqrt{\frac{\mu_o}{\epsilon_o}}$$

which is 376.7 $\Omega$, or very nearly $120\pi$ or 377 $\Omega$. As we have seen, $\mu_o$ is the permeability of free space ($4\pi \times 10^{-7}$ [H/m]) and $\epsilon_o$ is the dielectric constant of free space ($1/36\pi \times 10^9$ [F/m]).

At large distances from a short antenna, phase differences can be ignored and the resultant field can be termed a *plane wave* with an electric field intensity proportional to

$$\frac{2(377)Il}{\pi\lambda r}$$

For a plane wave, the electric and magnetic field intensities in free space are always in phase and perpendicular to each other and to the direction of propagation. Often the electric wave from an antenna is more complex with regard to phase and doubly so in the near or induction field of an antenna.

In problems of EMC, unintentional sources of electromagnetic emissions are primarily from current loops or current elements, whether from PCB tracks, intraunit, interunit, or system wiring and cables. Thus we shall concentrate on fields from current loops and current elements and in the current induced into loops and short lengths of conductor by electromagnetic fields.

### 2.2.2 Current Elements as Radiators

A *current element* is defined, for the purpose of EMC, as an electrically short length (i.e., $l < \lambda$) of current-carrying conductor connected to the current source at one end and disconnected, or at least connected via a high impedance, from ground at the other end. Strictly, the current source should be disconnected from ground at both ends. The practical use of an isolated wire to which no generator or return current path is connected as a model in EMC predictions may appear extremely limited. However, a practical example is in the modeling of a shielded cable on the center conductor of which a current flows and returns on the inside of the shield. In our example, the shielded cable is connected to a well-shielded enclosure at each end. If the shield of the cable is not perfect, then some small fraction of the current flowing on the inside of the shield will diffuse through the shield to the outside. If the length of the cable is less than the wavelength, $\lambda$, then the current flow on the outside of the shield will reduce to zero at the termination with the enclosures at both ends and remain relatively constant over the length of the cable.

For certain cable and measuring distance configurations, discussed later, the electric current element model may be used with either one or both ends connected to a return path. In addition, the electric current element model is strictly valid when the distance from a plane conductor or ground plane is much greater than the wavelength. When the proximity is closer, the current element equations may be used and corrected for the effect of reflections in the plane conductor, as discussed in Section 9.3.2. When the proximity of the ground plane is much less than the wavelength, $\lambda$, and the length of the current element is equal to or greater than $\lambda$, then the equation for radiation from a transmission line contained in Section 7.6 is applicable.

The current on a current element is assumed to be constant over its length. When applying the formulae for the current element to an electrically short length of wire or cable (i.e., $l <$ 0.1$\lambda$) disconnected from ground at both ends, then the current distribution is as shown in Figure 2.11a, and the electromagnetic radiation from the short wire is approximately half of the value of the current element. The current distribution for a length of wire which is resonant (i.e., $l =$ 0.5$\lambda$) is sinusoidal, as shown in Figure 2.11b, and the radiation is approximately 0.64 that of the current element. A more accurate approach that is particularly useful when the length of the wire is greater than the wavelength, as shown in Figure 2.11c, is to break up the wire into a number of current elements and compute the composite field from all sources. If the distance from the wire, at which the magnitude of the field is required, is much closer than the wavelength, then the current element equations may be used with no correction, because the contribution to the total field from the current elements at some distance from the point of calculation is negligibly small. Even where the wire is terminated at one or both ends, the current element equation

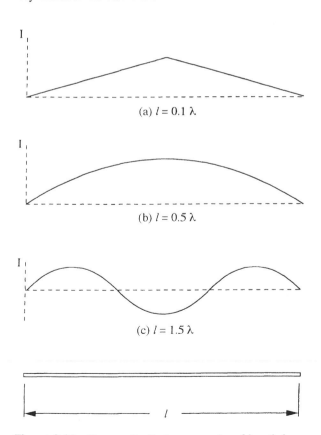

(a) $l = 0.1\ \lambda$

(b) $l = 0.5\ \lambda$

(c) $l = 1.5\ \lambda$

**Figure 2.11**   Current distribution on a wire of length $l$.

may still be used at close proximity, as long as the return current path is at a large electrical distance from the wire. If, as often happens, the return path is in close proximity, then the formulae for fields from either a transmission line or a current loop, whichever is applicable, should be used. In calculating the field strength, the magnitude of the current flow on the wire must be known. The current may be calculated; however, a simpler approach is to measure by means of a current probe.

Thus we see from the foregoing discussion that the current element, as a model, has more practical applications than may at first be realized.

### 2.2.3 Current Loops

The current loop is a valuable model for use in predicting radiation and coupling. The cable connecting two units in a rack or two units mounted one upon the other often forms a loop. Likewise, where a cable is routed along the ground or in a cable tray and then is connected to equipment, a loop is often formed. The wiring inside an enclosure (or the tracks on a printed circuit board, PCB) often takes the shape of a loop. Another use of the current loop model is in the design of a simple antenna, which may be constructed and used as either a source of, or to measure, magnetic fields.

### 2.2.4 Spherical Waves

To understand spherical waves (i.e., those present close to an antenna or an electric current element or loop), we shall use the spherical coordinate system. Figure 2.12 shows the field vector around an electric current element, sometimes referred to as an *infinitesimal dipole*. The simplified field equations for a nondissipative medium such as air are

$$E_\theta = \left( jRc \, \frac{I_s}{2\lambda r} \right) \left( 1 - \frac{\lambda^2}{4\pi^2 r^2} - j \, \frac{\lambda}{2\pi r} \right) \sin \theta \qquad (2.9)$$

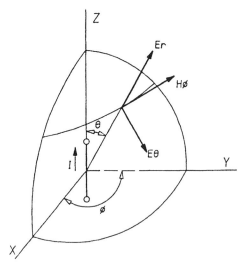

**Figure 2.12**  Field vector around electric current element.

$$H_\phi = \left( j \frac{I_S}{2\lambda r} \right) \left( 1 - j \frac{\lambda}{2\pi r} \right) \sin \theta \tag{2.10}$$

$$E_r = \left( \frac{Z_W I_S}{2\pi r^2} \right) \left( 1 - j \frac{\lambda}{2\pi r} \right) \cos \theta \tag{2.11}$$

Expressed as magnitudes, these become

$$|E_\theta| = 377 \frac{I_S}{2\lambda r} \sqrt{1 - \left( \frac{\lambda^2}{4\pi^2 r^2} \right)^2 \left( \frac{\lambda}{2\pi r} \right)^2} \sin \theta \tag{2.12}$$

$$|H_\phi| = \frac{I_S}{2\lambda r} \sqrt{1 - \left( \frac{\lambda}{2\pi r} \right)^2} \sin \theta \tag{2.13}$$

$$|E_r| = 377 \frac{I_S}{(2\pi r)^2} \sqrt{1 - \left( \frac{\lambda}{2\pi r} \right)^2} \cos \theta \tag{2.14}$$

In the far field of the electric current element, where we assume $r/\lambda \gg \pi/2$, and using the intrinsic impedance of free space (377) for $Z_W$, the field equations simplify to

$$E_\theta = j \left( \frac{60\pi I_S}{\lambda r} \right) \sin \theta \tag{2.15}$$

$$H_\phi = j \left( \frac{I_S}{2\lambda r} \right) \sin \theta \tag{2.16}$$

$$E_r = 60 \frac{I_S}{r^2} \cos \theta \tag{2.17}$$

In the far field it can be seen that $E_\theta$ and $H_\phi$ are the predominant radiation components, with $E_r$ as the induction (or reactive) component that soon decays at large distances from the current element.

The open-circuit voltage induced in the loop is

$$V_{ab} = -j2\pi f \, \mu_o H_\phi S \tag{2.18}$$

where

$\mu_o$ = permeability or inductivity of air or vacuum, $4\pi \times 10^{-7}$ [H/m]

$S$ = area of the loop

$\mu_o H_\phi = B$ = flux density

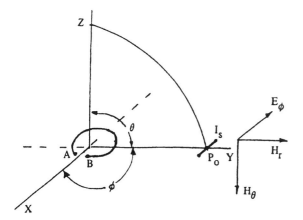

**Figure 2.13**  Field vector around a current loop depicts a small loop with a current element at position $P_o$.

The value of $H_\phi$ can be found from the current $I$ flowing in the current element at position $P_o$ in Figure 2.13 and from Eqs. (2.10) and (2.18) as follows:

$$V_{ab} = -j\omega\mu\left(\frac{j\beta I_s}{4\pi r}\right)\left(1 + \frac{1}{j\beta r}\right)S \tag{2.19}$$

where $\beta = 2\pi/\lambda$. As a magnitude, this is

$$|V_{ab}| = 2\pi f\mu\left(\frac{\beta I_s}{4\pi r}\right)\sqrt{1 + \left(\frac{1}{\beta r}\right)^2}\, S \tag{2.20}$$

### 2.2.5  Receiving Properties of a Loop

From Eq. (2.18), the receiving properties of the loop can be expressed as

$$V = 2\pi f\mu_o H_\theta S \cos\theta = \frac{2\pi Z_W H_\theta S \cos\theta}{\lambda} = \frac{2\pi E_\phi S \cos\theta}{\lambda} \tag{2.21}$$

where $\theta$ is the angle between the normal, or perpendicular, to the plane of the loop and the magnetic intensity $H_\theta$, as shown in Figure 2.13.

Equation (2.21) gives the open-circuit receiving properties of a loop. The impedance of the loop and the impedance across which the voltage is developed must be accounted for when predicting either the voltage developed across the load or the current flowing in the loop. In Chapter 7, on cable coupling, the EMI voltage induced in a loop formed by a wire or cable by a field is examined further.

The inductance of a loop of wire or a wire above a ground plane, which forms a loop with its electromagnetic image in the ground plane, is given by

$$L = \frac{\mu_0}{\pi}\left[l\ln\frac{2hl}{a(l+d)} + h\ln\frac{2hl}{a(h+d)} + 2d - \frac{7}{4}(l+h)\right] \tag{2.22}$$

where $a$ is the radius of the wire, in meters, $l$ is the length of the wire, in meters, $h$ is either the distance between the wires or twice the distance between the wire above a ground plane to account for the electromagnetic image of the wire in the ground plane, and

$$d = \sqrt{(h^2 + l^2)}$$

The impedance of the loop and load as a magnitude is

$$Z = \sqrt{(Z_L^2 + 2\pi f L^2)}$$

With current applied to the loop and assuming even current distribution around the loop (i.e., $S \ll \lambda$), the field components at any point around the loop can be found from

$$E_\phi = \frac{R_c \beta^2 IS}{4\pi r} \left(1 + \frac{1}{j\beta r}\right) e^{-j\beta r} \sin\theta \qquad (2.23)$$

$$H_\theta = \frac{\beta^2 IS}{4\pi r} \left(1 + \frac{1}{j\beta r} - \frac{1}{\beta^2 r^2}\right) e^{-j\beta r} \sin\theta \qquad (2.24)$$

$$H_r = \frac{j\beta IS}{2\pi r^2} \left(1 + \frac{1}{j\beta r}\right) e^{-j\beta r} \cos\theta \qquad (2.25)$$

Expressed as magnitudes, these become

$$|E_\phi| = 377\frac{\beta^2 IS}{4\pi r} \sqrt{1 + \frac{1}{\beta^2 r^2}} \sin\theta \qquad (2.26)$$

$$|H_\theta| = \frac{\beta^2 IS}{4\pi r} \sqrt{\left(1 + \frac{1}{\beta^2 r^2}\right)^2 + \frac{1}{\beta^2 r^2}} \sin\theta \qquad (2.27)$$

$$|H_r| = \frac{\beta IS}{2\pi r^2} \sqrt{\left(1 + \frac{1}{\beta^2 r^2}\right)} \cos\theta \qquad (2.28)$$

These equations are useful in computing the fields generated by a current loop in the induction field, in the near field, and in the far field. For example, Eq. (2.28) may be used to obtain the approximate value of the radial induction/near field when the radius of the loop is less than the measuring distance. Equation (2.28) should always be used when the measuring distance is greater than six times the radius of the loop. At low frequencies and for distances closer than the radius of the loop, Eq. (2.7) or (11.1), which provides the DC or quasi-static radial field, should be used.

In order to examine the ratio between $E$ and $H$ fields generated by a loop, as shown in Figure 2.13, the values of $E$ and $H$ fields from a 0.01-m circular loop with a generator current

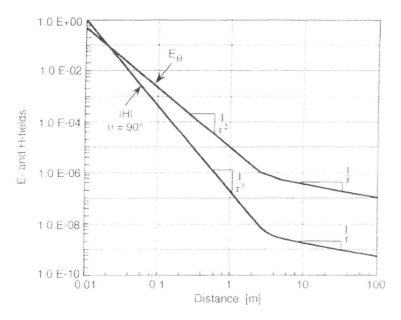

**Figure 2.14** $E$ and $H$ fields vs. distance.

of 1 mA at 10 MHz is plotted against distance from the loop in Figure 2.14. In Figure 2.16 the ratio $E/H$ (i.e., the wave impedance) has been plotted. It can be seen that the near-field/far-field interface for the small current loop is at approximately 4 m, which corresponds well with the definition of $\lambda/2\pi = 30$ m$/6.28 = 4.7$ m. The wave impedance for the small current loop is given by

$$Z_W = \frac{Z_c 2\pi r}{\lambda} \leq 377 \ \Omega \tag{2.29}$$

(e.g., at 0.1 m, $Z_W = 7.9$, and this is called a *magnetic* or *low-impedance field*).

Also, from Figures 2.14 and 2.15 it can be seen that in the near field, $H_\theta$ reduces as a function of $1/r^3$, as does $H_r$, whereas $E_\phi$ reduces as a function of $1/r^2$ (where $r$ = distance). Had we been examining a small dipole or current element (i.e., $1 \ll \lambda$), then $E_\theta$ would have followed a $1/r^3$ law with distance and $H_\phi$ a $1/r^2$. The wave impedance would then equal

$$Z_W = \frac{Z_c \lambda}{2\pi r} \geq 377 \ \Omega \tag{2.30}$$

and the field is then called an *electric* or *high-impedance field*. It appears from Figure 2.16 that the wave impedance increases above 377 close to the $\lambda/2\pi$ transition zone, which is not true. Instead a transition zone exists, and the $(1/r^2)$-to-$(1/r)$- and the $(1/r^3)$-to-$(1/r)$ transitions are not abrupt. Numerous measurements on a single current-carrying cable and on multiple cables located on a nonconductive table 1 m above the floor in a shielded room have confirmed that the current element is the correct model for the cable configuration. The measurements have also shown that the transition zone exists. It has been seen from a number of the measurements that the circumferential magnetic field $H_\theta$ from a cable reduces approximately as a function of $1/r^2$ up to a certain distance, thereafter reducing by $1/r^{1.5}$ up to a certain distance, and thereafter reducing by $1/r$.

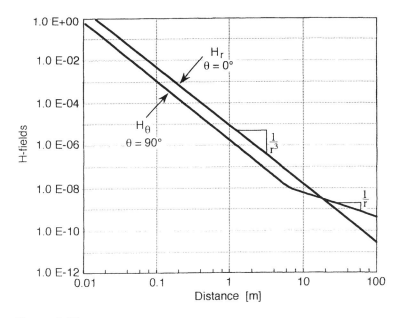

**Figure 2.15** $H_\theta$ and $H_r$ vs. distance.

Chapter 9 and Section 9.3.2 discuss measurement techniques and compare predictions, based on cable current magnitude, to measured levels and reduction with distance. Equations 2.26, 2.27, and 2.28 are valid when the loop is electrically small (i.e., the perimeter is much less than the wavelength).

For a large loop, where the perimeter approaches one-half the wavelength, standing waves

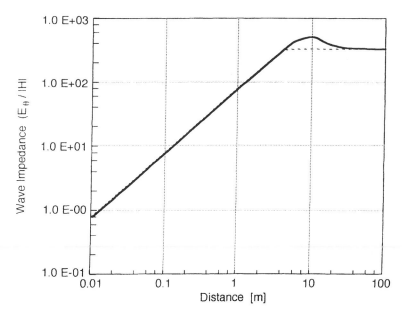

**Figure 2.16** Wave impedance vs. distance.

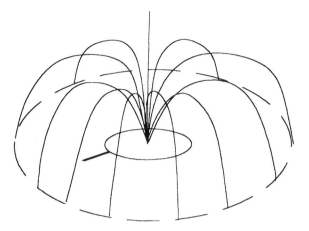

**Figure 2.17a**  Radiation pattern for a small loop, $E_\phi$, $P < 0.1\lambda$.

are generated on the loop and the current is no longer constant around the loop. At frequencies where the ratio $P/\lambda$, where $P$ is the perimeter of the loop, in meters is 0.5, 1.5, 2.5, . . . , the loop exhibits the characteristics of a parallel resonant circuit (termed an *antiresonance*), and the input current and radiation from the loop are greatly reduced.

At frequencies where the ratio $P/\lambda$ is 1, 2, 3, . . . , the loop behaves as a series resonant circuit, and the current is limited only by the DC and radiation resistance.

The input impedance of the loop is a minimum at resonance and a maximum at an antiresonance. For example, when the circumference of the loop $P$ is equal to $\lambda$, the input impedance is approximately 100 $\Omega$, and when $P = 2\lambda$, the input impedance is approximately 180 $\Omega$. At the first antiresonance, when $P = 0.5\lambda$, the input impedance is approximately 10 k$\Omega$; and when $P = 1.5\lambda$, the input impedance is from 500 $\Omega$ to 4000 $\Omega$, depending on the ratio of the radius of the loop to the radius of the conductor.

In addition, the radiation pattern changes, with reduced radiation from the plane of the large loop and maximum radiation from the plane of the small loop; these are the $E_\phi$ field of Figure 2.13. Figures 2.17a and 2.17b illustrate the $E_\phi$ radiation pattern for the small loop and

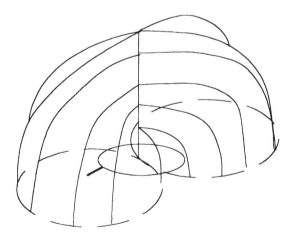

**Figure 2.17b**  Radiation pattern for a large loop, $E_\phi$, $P = \lambda$.

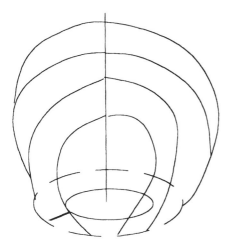

**Figure 2.17c** Radiation pattern for a large loop, $E_\theta$, $P = \lambda$.

the large loop, where $P = 0.1\lambda$ and $\lambda$, respectively. With the large loop in addition to the $E_\phi$ field, which is small in the vertical plane, a second component $E_\phi$, which has a maximum value in the vertical plane and is zero in the horizontal plane, is generated. Figure 2.17c shows the radiation pattern for the $E_\theta$ field. The directivity in the vertical plane (at 90° to the plane of the loop, where $\theta = 0$) with sinusoidal current distribution around it is provided in Table 2.1. The power gain of the loop, when the ratio of the radius of the loop to the radius of the conductor is 30, is 1.14 when $P = 0.6\lambda$ and 3.53 when $P = 1.2\lambda$.

The radiation resistance of the large loop, compared to $31{,}200\ (\pi a^2/\lambda^2)^2$ for a small circular or rectangular loop, is shown in Figure 2.18 from Ref. 5.

### 2.2.6 Far-Field Radiation from a Twisted Wire Pair

The EMC literature contains a model for the prediction of radiation from a twisted pair based on the model of a helix described in Ref. 5. For a helix whose dimensions are much less than the wavelength, the radiation is primarily from the sides of the helix, and the axial-mode radiation is at a minimum.

**Table 2.1** Directivity of the $E_\theta$
Field for a Large Loop

| $P/\lambda$ | Directivity |
| --- | --- |
| 0.2 | 0.7 |
| 0.4 | 0.9 |
| 0.6 | 1.12 |
| 0.8 | 1.3 |
| 1.0 | 1.5 |
| 1.2 | 1.58 |
| 1.4 | 1.67 |
| 1.6 | 1.58 |
| 1.8 | 1.25 |
| 2.0 | 0.75 |

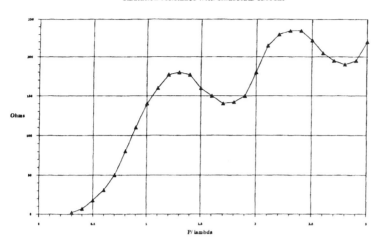

Radiation resistance with sinusoidal current

Ohms

P/ lambda

**Figure 2.18**   Radiation resistance of large loop.

The helix may be modeled as a series of small current loops and current elements, as shown in Figure 2.19a, and this model may be of use in an EMC prediction where a single wire takes the form of a helix. In this case, the far-field radiation is found from the sum of the loop and element sources. The physical layout of a twisted pair is shown in Figure 2.19b, from which it is seen that the twisted pair forms a bifilar helix in which both helices are identical but displaced in position axially down the length of the pair. The axial displacement is determined by the distance between the two wires. An important parameter in controlling the radiation from the pair is the distance between the wires and the pitch of the twist. Decreasing both the distance and the pitch will decrease the radiation. For commercially available twisted-pair wire made up as a cable, the distance between the two wires is approximately 0.3 cm, and the pitch is from 3.8 to 4.4 cm. The current elements in the equivalent circuit are therefore longer than the diameter of the loops in this practical example.

Because the two helices are axially in the same plane and the current flow is in opposite directions, the equivalent circuit is as shown in Figure 2.19c. Assuming an even number of twists with the total length of the pair less than $0.5\lambda$, we may see from the equivalent circuit

**Figure 2.19a**   Equivalent circuit of a helix comprised of current elements and current loops.

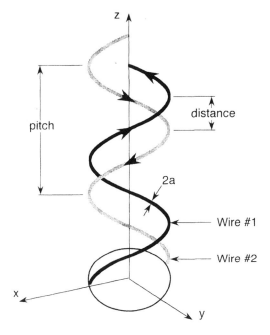

**Figure 2.19b**   Physical layout of a twisted wire pair illustrating the bifilar helix form of the cable.

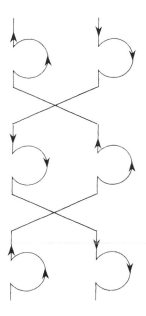

**Figure 2.19c**   Equivalent circuit of a twisted pair.

that the fields from the current elements and current loops tend to cancel at a considerable distance from the source. This is not surprising, based on the physical layout of the pair. Measurements from a twisted pair have confirmed that the far-field radiation is lower than for an untwisted pair.

Where the total length of the twisted-pair line is greater than $0.5\lambda$, the current flow in the two wires changes direction at some point down the length of the pair, and the cancellation of fields from the loops and elements is not as complete. In the limiting case where the pitch is equal to the wavelength, the axial radiation from two adjacent loops and current elements tends to add, and the level of radiation is higher than for an untwisted-pair cable. This occurs for the majority of twisted pairs in the gigahertz frequency range, which is typically, but not exclusively, above the frequency of emissions from unintentional sources, such as logic and converters.

The major source of radiation from the twisted pair is not from loops and elements but from the common-mode current, which inevitably flows on both conductors of the cable. One source of common-mode current is displacement current, which flows in the space between the two conductors; this source, along with others, is discussed in Section 7.6.

Likewise, when a field cuts a twisted-pair cable, the induced differential-mode currents are at a relatively low magnitude, whereas the common-mode current is identical to that of an untwisted pair. Calculations for differential-mode voltage and common-mode current flow resulting from an incident field and the effect of unbalanced input impedance on the noise voltage developed are discussed in subsequent chapters.

## 2.3  RADIATED POWER

In addition to $E$ and $H$ fields, we may consider antenna transmitted and received power. *Radiated power* is defined as the average power flow per unit area and is expressed in watts per square meter. In a perfect, lossless, isotropic antenna, the power is radiated equally in all directions and must pass through every sphere of space. The relationship between the power and the $E$ and $H$ fields is given by the Poynting vector cross-product

$$\vec{P} = \vec{E} \times \vec{H}$$

The Poynting vector is merely a convenient method of expressing the movement of electromagnetic energy from one location in space to another, and it is not rigorous in every instance when applied to antennas. It has been shown that in the far field, the ratio $E/H$ is a constant and is termed the *wave impedance*. In the far field, the $E$ and $H$ fields are mutually perpendicular and perpendicular to the direction of propagation. Therefore,

$$P = EH = \frac{E^2}{Z_W}$$

Because the E and H fields each decay by the factor $1/r$ in the far field, the power decays by the factor $1/r^2$ and the radiated power is then

$$W = \frac{P_{in}}{4\pi r^2} \quad [\text{W/m}^2] \tag{2.33}$$

The $1/4\pi r^2$ term may be visualized by considering the power flow as spreading out from the source in the form of a sphere. If the radiation pattern were truly in the shape of a sphere, then the source would be an isotropic radiator. No practical antenna is an isotropic radiator; the least directive radiator is a current element and a current loop, both of which have the same radiation pattern.

The antenna gain is a measure of the "directivity" of the antenna. In antenna terminology, the term *directivity* is reserved to describe the ratio of power at beam peak, assuming no losses, to the power level realized with the same input power radiated isotropically. Where ohmic losses in the antenna are low and cable/antenna impedances are matched, the directivity $D$ and the gain $G$ are virtually the same. The gain is defined as the ratio

$$\frac{4\pi W_{max}}{P_{in}}$$

where $P_{in}$ is the antenna input power. The gain of the perfect isotropic antenna is thus 1. Neither the equation for radiated power nor that for antenna gain is valid in the near field of the antenna. The received power of a receiving antenna is likewise not constant in the near field nor equal to the gain in the far field, because near-field contributions radiating from the antenna, which would be in phase in the far field, can cancel in the near field. The radiated power at a distance $r$ from an antenna with gain $G$ is given by

$$W = \frac{GP_{in}}{4\pi r^2} \tag{2.34}$$

The input voltage to the antenna for a given input power is

$$V_{in} = \sqrt{P_{in}R_{in}}$$

where $R_{in}$ is the antenna input or radiation resistance.

The $E$ field generated by the antenna, using $V_{in}$ and assuming the antenna input impedance is 50 $\Omega$ and equals the radiation resistance, is given by

$$E = \frac{V_{in}}{r\sqrt{628/Z_W G}} \tag{2.35}$$

where $r$ is the distance from the antenna, in meters.

The *radiation resistance* is defined as that part of the antenna impedance that contributes to the power radiated by an antenna. In the case of a resonant antenna, such as a dipole, the antenna input impedance equals the radiation resistance. For the tuned dipole, the input impedance equals 70–73 $\Omega$, and for a tuned monopole it equals 30–36 $\Omega$, depending on the radius of the conductors used and assuming the conductor radius is small compared to the antenna length. We use the term *tuned* to denote that the antenna length has been adjusted so that the antenna is resonant at the frequency of interest.

For an electrically small dipole or monopole antenna, where $l$, which is the length of one arm, is much less than $\lambda/4$, the input impedance is predominantly capacitive and is higher than the resonant value. The radiation resistance is then only a small fraction of the input impedance.

The *effective area* of an antenna ($A_e$) is defined as the ratio of the power received at the antenna load resistance ($P_r$) to the power per unit area of incident wave ($W$). If the antenna impedance is matched to the load impedance, then $P_r = WA_e$.

The effective capture area, or *aperture*, of a receiving antenna ($A_e$) is dependent on the actual area and efficiency of the antenna. The physical area of a rectangular-aperture antenna, such as the horn antenna, is the length times the width. For a circular antenna, such as a dish, the area is $\pi r^2$. The effective aperture is the physical aperture times the aperture efficiency, $\eta_a$. For example, $\eta_a$ is approximately 0.5 for a dish antenna. The effective area can be expressed in terms of gain, thus

$$A_e = \frac{G\lambda^2}{4\pi} \tag{2.36a}$$

So

$$G = \frac{4\pi A_e}{\lambda^2} \quad \text{and} \quad P_r = \frac{WG\lambda^2}{4\pi}$$

The $\lambda^2/4\pi$ term does not imply that with increasing frequency, waves decrease in magnitude. It means that at higher frequency, the area over which a given power flow occurs is smaller.

The equation for gain must also include a correction for losses. In the specified aperture efficiency for an antenna, losses due to reflections caused by a mismatch between the transmission line driving the input terminals of the antenna and the antenna input impedance should be included. Where aperture efficiency or realized gain is not specified and Eq. (2.36) is used to determine the gain of an antenna, any loss due to transmission-line-to-antenna impedance mismatch must be calculated to arrive at the realized gain ($G_{re}$). $G_{re}$ equals $G \times L$, where $L$ is the loss. The loss may be found from the transmission line/antenna reflection coefficient, $K$, using $L = 1 - K^2$:

$$K = \frac{1 - Z_{ant}/Z_L}{1 + Z_{ant}/Z_L} \tag{2.36b}$$

where

$Z_{ant}$ = antenna input impedance

$Z_L$ = impedance of the transmission line connected to the antenna

For two antennas in free space in the far-field region, both polarization matched and ignoring losses in cables baluns and due to impedance mismatch, the received power is given by combining Eq. (2.34) for transmitted power, using the antenna input power, with the equation for received power (Eq. 2.36a), which gives

$$P_r = \left(\frac{G_t P_{in}}{4\pi r^2}\right)\left(\frac{G_r \lambda^2}{4\pi}\right) \tag{2.37}$$

which may be simplified to Eq. (2.38):

$$P_r = P_{in} G_t G_r \left(\frac{\lambda}{4\pi r}\right)^2 \tag{2.38}$$

where

$P_r$ = received power at the input of the receiving device

$P_{in}$ = input power to transmitting antenna

$G_t$ = gain of transmitting antenna

$G_r$ = gain of receiving antenna

If we express the received power in dbW and the antenna gains in decibels, then Eq. (2.38) is

$$P_r = P_t + G_t + G_r - 20 \log \left( \frac{4\pi r}{\lambda} \right) \tag{2.39}$$

The received power is readily converted to voltage at the input of the receiver when the receiver input impedance is known (assuming the cable impedance matches the terminating impedance) by $V = \sqrt{P_r Z_{in}}$. The power equation is useful in calculating free-space propagation loss in the basic antenna-to-antenna EMC prediction. For this prediction, additional factors due to the unintentional nature of the coupling have to be accounted for. Some of these additional factors, which are discussed in Chapter 10, are:

Non line-of-sight coupling
Antenna polarization and alignment losses
Intervening atmospheric effects
Frequency misalignment losses

## 2.4 UNITS OF MEASUREMENT

The unit dBW was used to describe power; the following are some definitions of commonly used units.

**dB**: The decibel (dB) is a dimensionless number that expresses the ratio of two power levels. It is defined as

$$dB = 10 \log \frac{P_2}{P_1}$$

The two power levels are relative to each other. If power level $P_2$ is higher than $P_1$, then dB is positive; vice versa, dB is negative. Since

$$P = \frac{V^2}{R}$$

when voltages are measured across the same or equal resistors, the number of decibels is given by

$$dB = 20 \log \frac{V_2}{V_1}$$

A rigid voltage definition of dB has no meaning unless the two voltages under consideration appear across equal impedances. Thus above some frequency where the impedance of wave-guides varies with frequency, the decibel calibration is limited to power levels only.

**dBW**: The decibel above 1 W (dBW) is a measure for expressing power level with respect to a reference power level $P_1$ of 1 W. Similarly, if the power level $P_2$ is lower than 1 W, the dBW is negative.

**dBm**: The decibel above 1 mW in 50 Ω. dBm = 1 mW = 225 mV (i.e., 225 mV²/50 = 1 mW). Since the power level in receivers is usually low, dBm is a useful measure of low power.

**dB μV**: The decibel above 1 μV is a dimensionless voltage ratio in decibels referred to a reference voltage of 1 μV and is a commonly used measure of EMI voltage.

**μV/m**: Microvolts per meter are units used in expressing the electric field intensity.

**dB μV/m**: The decibel above 1 μV/m (dB μV/m) is also used for field intensity measurement.

**μV/m/MHz**: The microvolt per meter per megahertz is a broadband field intensity measurement.

**dB μV/m/MHz**: The decibel above 1 μV/m/MHz.

**μV/MHz**: Microvolts per megahertz are units of broadband voltage distribution in the frequency domain. The use of this unit is based on the assumption that the voltage is evenly distributed over the bandwidth of interest.

The following log relationships, which have been used in this chapter to convert magnitude to decibels, are useful to remember:

$$\log(AB) = \log A + \log B$$

$$\log\left(\frac{A}{B}\right) = \log A - \log B$$

$$\log(A^n) = n \log A$$

Appendices 2 and 3 show the conversion between electric fields, magnetic fields, and power densities.

## 2.5   RECEIVING PROPERTIES OF AN ANTENNA

In accordance with the reciprocity theorem, certain characteristics of a receiving antenna, such as pattern shape and return loss, remain the same when the antenna is used for transmitting. Likewise, the power transfer between two antennas, not necessarily identical, is the same regardless of which is transmitting and which receiving. One important consideration when applying the reciprocity theorem is that it applies to the terminals to which the voltage is applied in the case of the source, and to the terminals to which the voltage is measured in the case of the receptor. The pattern of power reradiated by a receiving antenna is different from its radiation when used as a transmitting antenna. In problems of EMI it is often the current flow on a structure, cable, or wire that is required, and it should be remembered that the current distribution on a radiating structure is often different from that on the receptor structure, even when both structures are identical, and the power level most certainly is. An explanation for the coupling of an E field to a wire receiving antenna follows.

A field incident on the receiving antenna is assumed to have an angle of incidence that results in a voltage developed across the terminals of the antenna. This open circuit voltage $V$ is proportional to the effective height, sometimes referred to as the effective length, of the antenna, which is seldom equal to the physical height of the antenna; thus, $V = h_{eff}E$. For aperture antennas, the ratio between the power developed in the antenna load and the incident power is known as the *effective aperture* of the antenna.

When the antenna input impedance equals the load impedance, a division of 2 occurs in the antenna received voltage. When the antenna impedance is greater than the load impedance, the voltage division is greater than 2. The loss due to antenna/load impedance mismatch is accounted for in the equation for the gain of the antenna, which is a term in the equation for $h_{eff}$.

A second potential source of loss is due to a mismatch between the wave impedance and the radiation resistance of the antenna, and this is accounted for in determining the effective height of the antenna. The physical length of the wire antenna and the wavelength of the incident field both determine the radiation resistance and, therefore, the voltage developed.

### 2.5.1 Conversion of Power Density to Electric Field Intensity

For EMI measurements, the value of E and H field intensity is normally required, and it is possible to find these intensities from the power density at some distance from the source. To convert from power density $P_d$ in the far field to electric field intensity at the measuring point:

$$P_d = \frac{E^2}{Z_W} \quad [\text{W/m}^2] \tag{2.40}$$

and

$$E = \sqrt{Z_W P_d} = \sqrt{377 P_d} \tag{2.41}$$

### 2.5.2 Conversion of Power Density to Electric Field Intensity in Terms of Antenna Gain

A receiving antenna is often used in EMI as a measuring device to find either the power density or the incident E field. The power density may be found from the received power developed in the antenna load resistance using Eq. (2.23):

$$P_d = \frac{4\pi P_r}{\lambda^2 G_r} \tag{2.42}$$

where

$P_d$ = power density at the receiving antenna

$P_r$ = power into the receiver

$G_r$ = gain of the receiving antenna

The value of the E field may be found from the power density using $P_d = E^2/Z_W$; the E-field-to-power relationship is then $E = \sqrt{P_d Z_W}$. Using Eq. (2.42) for $P_d$ and assuming $Z_W = 377$ $\Omega$, the electric field intensity $E$ in the far field is given by Eq. (2.24):

$$E = \frac{68.77}{\lambda} \sqrt{\frac{P_r}{G_r}} \quad [\text{V/m}] \tag{2.43}$$

Assuming the field intensity measuring instrument has an input impedance of 50 $\Omega$ and $V$ is the voltage measured by the instrument, $P_r$ can be found from

$$P_r = \frac{V^2}{50}$$

Then $E$ can be expressed as

$$E = \frac{9.7V}{\lambda} \sqrt{\frac{1}{G_a}} \quad [\text{V/m}] \tag{2.44}$$

### 2.5.3 Antenna Factor

Antenna factor AF or $K$ is an important calibration term; it is defined as the ratio of the electric field to the voltage developed across the load impedance of the measuring antenna, as follows:

$$AF = \frac{E}{V} \qquad (2.45)$$

where

AF = antenna factor numeric

$E$ = field strength, in volts/meter

$V$ = voltage developed across the load

The voltage developed is given by

$$V = \frac{Eh_{\text{eff}}}{(Z_{\text{ant}} + Z_L)/Z_L} \qquad (2.46)$$

Thus for an antenna where $Z_{\text{ant}} = 50\ \Omega$, $V = Eh_{\text{eff}}/2$. For a resonant dipole where $Z_{\text{ant}} = 72\ \Omega$, $V = Eh_{\text{eff}}/2.46$; and for a resonant monopole, $Z_{\text{ant}} = 42\ \Omega$, $V = Eh_{\text{eff}}/1.84$. The equation for $h_{\text{eff}}$ is

$$h_{\text{eff}} = \frac{Z_{\text{ant}} + Z_L}{Z_L} \sqrt{\frac{A_{\max} R_r}{Z_w}} \qquad (2.47)$$

where

$A_{\max}$ = maximum effective aperture = $\dfrac{G\lambda^2}{4\pi}$

= $0.135\lambda^2$ for a resonant dipole

= $0.119\lambda^2$ for a small dipole                                          (2.48)

$Z_w$ = wave impedance

$R_r$ = radiation resistance

At resonance, the radiation resistance equals the antenna impedance. For a short dipole, from Ref. 5,

$$R_r = 197 \left(\frac{2H}{\lambda}\right)^2$$

And for a monopole,

$$R_r = \frac{197\ (2H/\lambda)^2}{2}$$

$H$ is the physical height of one arm of the dipole or of the rod of the monopole. The gain in Eq. (2.48) must include the loss correction for any $Z_{\text{ant}}$-to-$Z_L$ impedance mismatch. This is true when $Z_{\text{ant}}$ is higher than $Z_L$. However, when the load impedance is higher than the antenna impedance, the voltage developed across the load is either the same magnitude or higher than when $Z_{\text{ant}}$ and $Z_L$ are matched, even though reflections occur. In determining the AF, it is the ratio $E/V$ that is of interest and not reflected power. Therefore when the load impedance is higher than the antenna impedance, $K$ (given by Eq. 2.36b) is set to zero, the loss is zero, and the gain is unchanged.

The $Z_{\text{ant}} + Z_L$ term is included in both Eq. (2.46) for $V$ and Eq. (2.47) for $h_{\text{eff}}$, and thus Eq. (2.47) may be expressed as

$$V = E \sqrt{\frac{A_{max}R_r}{Z_W}} \qquad (2.49)$$

Assuming the antenna is matched to the terminating impedance,

$$V = \frac{h_{eff}E}{2} \qquad (2.50)$$

In some textbooks, either the antenna-to-load mismatch is ignored or an unterminated antenna is examined, in which case $V = h_{eff}E$. Assuming $R_r = Z_{ant} = Z_{term}$ (i.e., impedance of the antenna termination), for a dipole,

$$h_{eff} = 2 \sqrt{\frac{A_{max}Z_{term}}{Z_W}} \qquad (2.51)$$

Thus for $Z_{term} = 50\ \Omega$ and $Z_W = 377\ \Omega$,

$$h_{eff} = \frac{\lambda\sqrt{G}}{4.87} \qquad (2.52)$$

For a monopole, the effective height is half the value given by Eqs. (2.51) and (2.52). Using Eqs. (2.44), (2.46), and (2.52) for antenna factor and antenna height gives us

$$AF = \frac{9.73}{\lambda\sqrt{G}} \qquad (2.53)$$

Expressed in decibels this is

$$AF_{dB} = 20 \log\left(\frac{9.73}{\lambda}\right) - G \qquad [dB]$$

and, from Eq. (2.53),

$$G = \left(\frac{9.73}{AF\lambda}\right)^2$$

or

$$G = 20 \log\left(\frac{9.73}{\lambda}\right) - AF_{dB} \qquad [dB]$$

The relationship between AF and frequency is plotted in Figure 2.20a for a number of constant-gain antennas. One example of a constant-gain antenna is the half-wave tuned resonant dipole in which the physical length of the arms are adjusted to ensure that the total length of both arms is nearly $\lambda/2$.

For example, the effective height of a half-wave dipole is given by

$$h_{eff} = \frac{\lambda}{\pi}$$

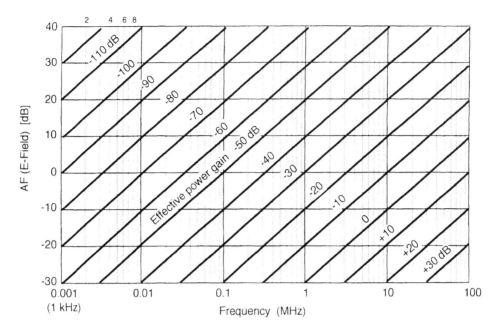

**Figure 2.20a**　AF versus $f$ for constant-gain antennas.

For a monopole, it is

$$h_{\text{eff}} = \frac{\lambda}{2\pi}$$

where $h$ is the physical height of both arms of the antenna $= \lambda/2$ in free space (assuming sinusoidal current distribution and an unterminated antenna).

When the physical height of an antenna is much smaller than $\lambda/2$, the current distribution is not sinusoidal but rather of a triangular shape, with maximum current at the center and zero current at either end, and the effective height is one-half of the physical length of both arms of the dipole. For a monopole, $h_{\text{eff}}$ is one-half of the physical length of the monopole rod.

For a true electric current element or infinitesimal dipole, $h_{\text{eff}} = h$. An equation that may be used when the physical length of the antenna is not $\lambda/2$ but is approaching this length is

$$h_{\text{eff}} = \frac{\lambda}{\pi} \tan\left(\frac{H}{2}\right) \tag{2.54}$$

where

$H = \beta h = 2\pi h/\lambda$

$h$ = physical length of the dipole antenna or the length of the rod of the monopole antenna

$H_{\text{eff}}$ for the monopole is one-half the value given by Eq. (2.54). The foregoing values of $h_{\text{eff}}$ for nonterminated antennas do not include losses due to impedance mismatch, so Eq. (2.47), using gain and maximum effective area, has a more general application, especially in EMC prediction.

Consider the following example: At 100 MHz, the effective height of a half-wave dipole is 3 m/3.142 = 0.955. From Eqs. (2.45) and (2.46), and assuming far-field conditions and $Z_{term} = 50\ \Omega$ and $R_r = Z_{ant} = 70\ \Omega$,

$$AF = \frac{2.4}{h_{eff}} = 2.5$$

which, expressed in decibels, is 8 dB. Figure 2.20b shows AF versus frequency for a $\lambda/2$ dipole; here the loss from the mismatch between the antenna impedance (70 $\Omega$) and the terminating impedance (50 $\Omega$) has been included. Manufacturers of antennas designed for EMI measurements usually publish gain and AF figures, whereas antennas designed for communications are typically supplied with gain calibration figures only. The antenna factor graphs published by manufacturers of broadband antennas are measured either on an open field site or in a semi-anechoic chamber. In either case, the results have been corrected for the effects caused by the reflection from the ground or the floor of the chamber. However, should the antenna then be used in a shielded room, as recommended in MIL STD 462 EMC test methods, where the antenna is positioned 1 m away from the equipment under test (EUT) and the ground plane, then the actual antenna factor can be greatly different from the published figures. This error is caused primarily by reflections from ceiling, walls, and floor, standing waves in the shielded room, and capacitive loading on the antenna due to ground plane proximity. Section 9.5.1 includes a description of techniques available to reduce these errors.

Where large standing waves exist, the ratio of E to H fields may vary from nearly zero to very high values (theoretically from zero to infinity), depending on the region (enclosure or cavity). In practice, the value of magnetic field measured in a shielded room is more constant than the E field.

One method of measuring the antenna factor is to use two identical antennas in the test setup shown in Figure 2.21. Two paths for the received voltage are obtained, one directly via the matching network into 50 $\Omega$ ($V_{50dir}$) and the second the radiated path ($V_{50rad}$). For two identical antennas, where $G_t = G_r$, and using Eq. (2.37), the product of the two gains is given by

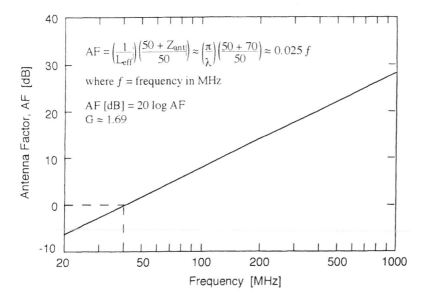

$$AF = \left(\frac{1}{L_{eff}}\right)\left(\frac{50 + Z_{ant}}{50}\right) \approx \left(\frac{\pi}{\lambda}\right)\left(\frac{50 + 70}{50}\right) \approx 0.025\,f$$

where $f$ = frequency in MHz

AF [dB] = 20 log AF
$G \approx 1.69$

**Figure 2.20b**   AF versus $f$ for a $\lambda/2$ dipole and a 50-$\Omega$ receiver.

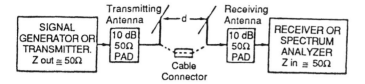

**Figure 2.21**   Test setup for the two-antenna method of determining AF. (From Ref. 20.)

$$G_t G_r = \frac{P_{\text{rad}}}{P_{\text{dir}}} \left( \frac{4\pi r}{\lambda} \right)^2$$

where

$P_{\text{rad}}$ = radiated power measured by the receiver

$P_{\text{dir}}$ = power measured by direct connection

Because $G_t = G_r$, the antenna realized gain, $G_{\text{re}}$, is

$$G_{\text{re}} = \sqrt{\frac{P_{\text{rad}}(50)}{P_{\text{dir}}(50)} \left( \frac{4\pi r}{\lambda} \right)^2}$$

Now

$$\frac{P_{\text{rad}}}{P_{\text{dir}}} = \frac{V_{\text{rad}}(50)^2}{V_{\text{dir}}(50)^2}$$

where $P(50)$ is the power into 50 Ω and $V(50)$ is the voltage developed across 50 Ω. Therefore,

$$G_{\text{re}} = \frac{V_{\text{rad}}(50)}{V_{\text{dir}}(50)} \frac{4\pi r}{\lambda}$$

and from Eq. (2.29),

$$\text{AF} = \frac{9.73}{\lambda\sqrt{G_{\text{re}}}}$$

If this calibration is conducted in a shielded room where subsequent EMI measurements are to be made and the receiving antenna is located at a fixed and marked location, which is then used for subsequent measurements, then the measured AF will have compensated, to a large extent, for the errors caused by reflections inherent in the test location.

Because the realized gain has been determined using this method, no further compensation due to wave-impedance-to-load-impedance mismatch is required and the near-field AF for the antenna may be obtained. It is therefore important that the distance between the two antennas, used in obtaining the AF, be duplicated when using the calibrated antenna for EMI measurements, for at different distances the AF calibration is not valid. Any loss due to the mismatch between the antenna impedance and load is also included in the measured $G_{\text{re}}$. It is important to ensure that the generator is able to produce the same voltage across the antenna impedance as across the 50-Ω load. When this is not the case, and the reason is a high-input-impedance antenna, then a 50-Ω load at the generator end of the cable feeding the transmitting antenna may be included during the radiated path measurement and removed during the direct measurement. This ensures that the generator load is approximately the same for the radiated and direct path measurements. When the antenna impedance is high, the receiving and transmitting antenna connecting cables should be kept short and of equal length. For maximum accuracy, the same cable should be used in measurements with the antenna. Any attenuation due to the cables is not included in the two-antenna measurement technique, and the cables should be calibrated separately.

The antenna factor for a tuned dipole using the two-antenna calibration method in a shielded room, with the manufacturer's AF curve, is plotted in Figure 2.22. By plotting the AF curve over a narrower frequency range, Figure 2.23, the effect of resonances, antiresonances, and reflections may be clearly seen, and the AF curve may be used to calculate the field magnitude with higher accuracy than when the manufacturer's curve is used. The accuracy of the AF curve is sensitive to exact antenna location and to the presence of equipment or personnel in the shielded room. Techniques that are helpful in achieving good repeatability include the installation of a plumb bob directly above the receiving antenna location used in the two-antenna test setup. This location should be chosen for its usefulness. For example, MIL-STD-462 test requirements specify a distance for the measuring antenna of 1 m from the EUT (equipment under test), which is placed 0.05 m from the edge of a table covered with a ground plane that is bonded to the shielded room wall. Thus where MIL-STD measurements are made, one or more calibration points down the length of the table at a distance of 0.95 m from the edge of the ground plane would be useful.

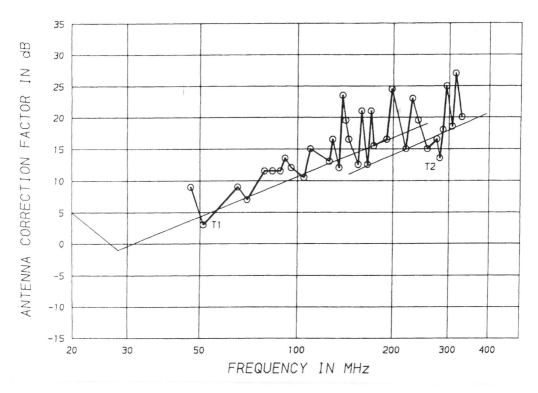

**Figure 2.22** AF, in decibels, of a tuned dipole measured in a shielded room.

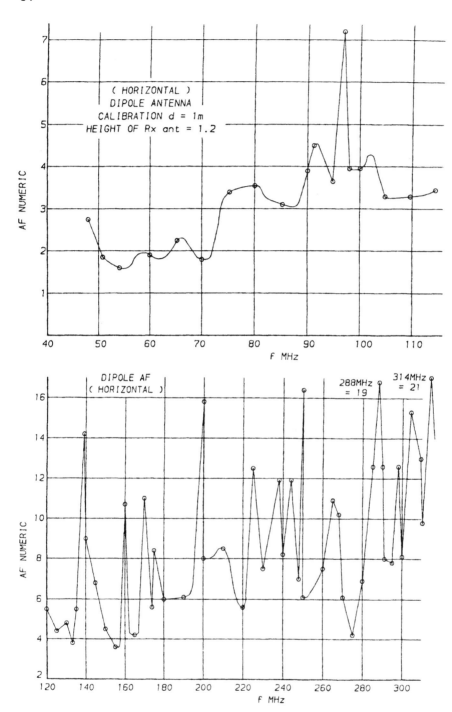

**Figure 2.23** AF, numeric, for the same tuned dipole.

The antenna calibration and subsequent measurements using the antenna should be made in a room clear of equipment and personnel, or at least the same room configuration should be maintained for both.

If the two-identical-antenna test method is used for calibrating antennas to be used on an open area test site or in an anechoic chamber, then the calibration should be conducted on an open area test site. The site should preferably be without a ground plane and with the area between the two antennas covered in a combination ferrite tile and foam absorber material to absorb the reflected ground wave. Horizontal polarization of the antenna should be used to reduce the impact of the antenna feed cables, which are oriented vertically. The 6–12-dB 50-$\Omega$ attenuators should be included, as shown in Figure 2.21. Both antennas should be placed at least 4 m above the ground to reduce antenna-to-ground coupling. The distance between the antennas should be 10m to eliminate antenna-to-antenna coupling.

Alternative antenna calibration techniques are the reference Open Area Test Site (OATS) and the reference dipole calibration contained in ANSI C63.5 and described in Section 9.4.1. If correctly constructed, the four reference diopoles, designed to cover the 25–1000-MHz frequency range, will have antenna factors within 0.3 dB of theoretical. An alternative test method uses the GTEM cell, as described in Section 9.5.2.2. The antenna factor can also be calculated using either the ''induced-emf method,'' numerical electromagnetics codes (NEC), MININEC, GEMACS, or the site attenuation. In the induced-emf method, mutual coupling between antennas is calculated by formulas given by S. A. Schelkunoff and H. E. King. These formulas usually replace the presence of the ground plane by images of the transmit and receive antenna. NEC, MININEC, and GEMACS use the moment method (MOM) to calculate the coupling between antennas, and with GEMACS using the MOM and GTD hybrid, the presence of a ground plane can be included. Reference 16 compares the accuracy of these analysis techniques to OATS, anechoic chamber, and GTEM measurements for symmetrical dipoles and horn and waveguide antennas. With symmetrical dipoles in horizontal polarization, the maximum deviation between the site-attenuation model and measurements was 3 dB. The maximum deviation between measurements and the induced emf, MOM, MININEC, and other calculations was 2.7 dB.

### 2.5.4  Receiving Properties of an Isolated Conductor/Cable

An *isolated cable* is defined as one disconnected from ground at both ends and at some distance from a ground plane. Although this configuration used as a model has limited practical use, one practical, albeit unusual, example of this cable configuration was, in a case of EMI, caused by coupling to a shielded cable. The cable was connected to a helicopter at one end and to equipment contained in an enclosure towed beneath the helicopter at the other end. With the helicopter airborne, the cable was isolated from ground at both ends. The more common cable configurations are discussed in detail in Chapter 7, on cable coupling; however, for the sake of completeness, here we examine coupling to an isolated cable. When a cable is disconnected from ground at both ends, the E field component of a wave that cuts the cable at an angle of 90° to its axis does not induce a current flow. In the case where a wave is incident on the cable such that the magnetic field component cuts it at an angle of 90° and the E field is in the plane of the cable, a current is induced. The magnitude of the current is determined by the length and impedance of the cable and the wavelength of the field. If the length of the cable is less than $0.1\lambda$ (i.e., nonresonant), then the average flow per second in the conductor of the cable is given approximately by

$$\frac{4\pi fBl}{Z_c}\left(\frac{\pi}{\lambda}\right)^2 \frac{(l/2)^2 - \frac{1}{2}l^2}{2}$$

where

$f$ = frequency, in hertz

$B$ = magnetic flux density

$l$ = physical length of the conductor

$Z_c = \sqrt{R^2 + 2\pi fL^2}$

and where

$R$ = total resistance of the conductor

$L$ = total inductance of the conductor

Values of resistance and inductance for single conductors and for the shields of shielded cables are provided in Tables 5.1 and 4.2, respectively.

When the length of the cable equals $\lambda/2$, it is resonant and the characteristics of a resonant short-circuit dipole, sometimes referred to as *a parasitic element*, may be used. In a resonant dipole terminated in its radiation resistance and ignoring losses, half of the received power is delivered to the load and half the power is reradiated from the antenna. In a shorted resonant dipole, four times the power is reradiated, compared to the antenna matched dipole. The radiation resistance of the resonant dipole is approximately 70 $\Omega$, and this is true regardless of the load. The reradiated power from the dipole is therefore $I^2 70$. Because the reradiated power for the short-circuit dipole is four times that for the matched dipole, the current flow is twice the value of that for a matched dipole.

The receiving characteristics of the dipole, described in the preceding section, may be used to find the current $I$ in the load of a dipole. The average current in the same length of isolated cable (i.e., the short-circuit dipole) at the same frequency then has the value $2I \times 0.64$.

### 2.5.5   Monopole Antenna as a Measuring Device and in Prediction of Electromagnetic Compatibility

Few practical cable, wire, or structure configurations look like a dipole antenna, and thus it is not particularly useful as a model in EMC predictions. The monopole antenna is in principle a vertical wire above a ground plane, with the ground plane sometimes referred to as a *counterpoise*. The monopole receiving properties and to a lesser extent its transmitting properties are useful in EMC prediction. A cable connected to a metal enclosure, with the cable routed over a nonconductive table, may be represented by a monopole antenna as long as the far end of the cable does not connect to ground. That the cable is horizontal to the floor of the room does not change the validity of the antenna model. The far end of the cable may terminate in a small metal enclosure, in which case the model is similar to a top-loaded monopole antenna. If the far end of the cable loops down to a conductive floor, such as is found in shielded rooms, or terminates in equipment connected to ground, then the monopole model is not appropriate. Instead the model described in Section 7.4 for coupling to loops should be used. The cable, which we model as the rod of the monopole antenna, may be a shielded cable, with the shield terminating on the enclosure or a twin or multiconductor cable that terminates inside the enclosure. Whichever it is, we require the termination impedance in order to apply the monopole model. One additional good reason to examine the receiving properties of a monopole antenna is its usefulness as an EMC measurement antenna and its ease of construction. A 41-inch (1-m) monopole or rod antenna over a square-meter counterpoise is a common measurement antenna. Its useful frequency range is 14–30 MHz when it is not used as a resonant antenna. When the rod is adjustable from 1 m to approximately 0.2 m, the antenna may be used as a resonant monopole over the range 75–375 MHz. To cover the 300–1500 MHz, it is more convenient to manufacture a monopole with a smaller counterpoise, typically 30 cm by 30 cm, and a rod adjustable from 30 cm to 5 cm. For use in EMC predictions and in EMC measurements, the resonant and nonresonant characteristics of the monopole are required. When the length of the cable or rod is equal to $\lambda/4$, where $\lambda$ is the wavelength of the incident field, then the antenna is resonant, the input impedance is 35–42 $\Omega$, and the gain is 1.68 numeric. When the length of the rod is much less than $\lambda$, the gain is the same as the current element, which is 1.5, and the antenna input impedance is high.

For an electrically short monopole, the input impedance is approximately

$$Z_o = 10H^2 - j\frac{30}{H}\left(\frac{\omega - 2}{1 + \dfrac{2\ln 2}{\omega - 2}}\right)$$

or, as a magnitude,

$$|Z_o| = \sqrt{(10H^2)^2 + \left[\frac{30}{H}\left(\frac{\omega - 2}{1 + \dfrac{2\ln 2}{\omega - 2}}\right)\right]^2}$$

where

$$\omega = 2\ln\left(\frac{2h}{a}\right)$$

and $h$ is the physical height of the rod and $a$ is its radius. $\beta h$ is defined as $2\pi h/\lambda$. As $\beta h$ approaches 1, the antenna impedance tends to become resonant and at values of $\beta h = 1.5, 2.5, 3, 3.5, 4, 4.5, \ldots$ the monopole is resonant. Curves for the reactance and resistance of the dipole, for $\beta h = 0.5$–7, are provided in Fig. 2.24 and 2.25 from Ref. 6. The reactance and resistance of a monopole are one-half of the values for a dipole. The magnitude of the impedance of the monopole is

$$Z_o = \sqrt{Z_{oim}^2 + Z_{ore}^2}$$

where

$Z_{oim}$ = reactance of the antenna

$Z_{ore}$ = resistance of the antenna

The maximum open-circuit gain of the resonant monopole is 1.68; for the short monopole it is 1.5. By using Eq. (2.36b) to obtain the reflection coefficient due to the antenna-to-load mismatch, the realized gain $G_{re}$ may be obtained. From Eq. (2.48) and using $G_{re}$, the value of $A_{max}$ may be determined. Using $A_{max}$ with the wave and radiation resistance in Eq. (2.49) gives us the voltage developed across the load, from which the load current may be calculated. The average current flow in the rod of the monopole is then equal to one-half the load current for the short monopole and 0.64 times the load current for the resonant monopole.

Consider an example of the calculations of antenna factor for a small tunable rod antenna at a frequency of 550 MHz. The AF of the antenna is

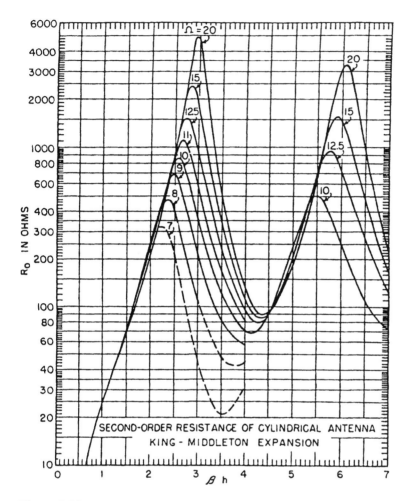

**Figure 2.24**  Input reactance of a dipole (monopole is half the dipole value). (From Ref. 6.)

$$\frac{E}{V} = \frac{1}{h_{\text{eff}}} = \frac{1}{\lambda/2\pi}$$

Therefore the AF at 550 MHz is $1/(0.545/12.56) = 23 = 27$ dB.

The antenna factor of a tunable monopole over the 300–800-MHz frequency range is plotted in Figure 2.26. The calibration was made in a shielded room using the two-antenna test method. At 550 MHz, the measured AF = 26.2 dB, whereas the predicted AF is 27 dB. At other frequencies, a difference of $\pm 3$ dB between predicted and measured AF is apparent. The measured antenna factor and gain for a 1-m rod antenna, calibrated at 1-m distance from a second identical antenna, over a 10 kHz to 30 MHz frequency range are shown in Figures 2.27 and 2.28. Consider the antenna factor at 20 MHz, where the antenna impedance is approximately 500 $\Omega$ and the wave impedance at a distance of 1 m is approximately 900 $\Omega$. The reflection coefficient, from Eq. (2.36b) is

$$K = \frac{1 - 500/50}{1 + 500/50} = \frac{-9}{11} = -0.818$$

The loss is $1 - (-0.818)^2 = 0.33$, so the gain is $1.5 \times 0.33 = 0.496$.

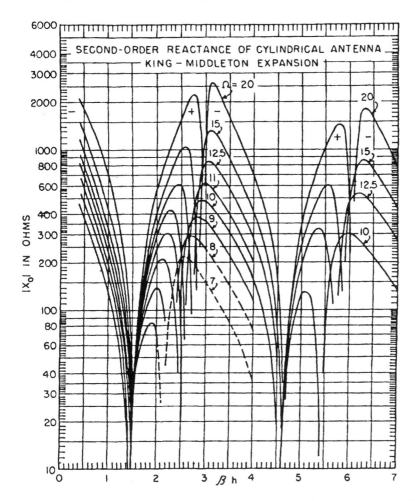

**Figure 2.25**  Input resistance of a dipole (monopole is half the dipole value). (From Ref. 6.)

The radiation resistance of the 1-m monopole at 20 MHz is

$$\text{radiation resistance} = \frac{197(2/15)^2}{2} = 1.75\ \Omega$$

The effective height, from Eq. (2.47), is

$$\text{effective height} = \frac{\sqrt{8.88 \times 1.75/99}}{2} = 0.065$$

The first term in Eq. (2.47) appears in the equation for the AF and is, in our example, omitted from both equations. Therefore, from Eqs. (2.45) and (2.49) the AF (E/V) is 1/0.065 = 15.4. From the plot of the measured antenna factor for a monopole, Figure 2.27, the AF at 20 MHz is 6.2 and the difference between measured and predicted AF is 8 dB. The measurements were made in a shielded room, and thus some difference between measured and predicted AF is to be expected due to the proximity of the rod of the antenna to the conductive ceiling and due to reflections. An additional potential source of error is that the measurements

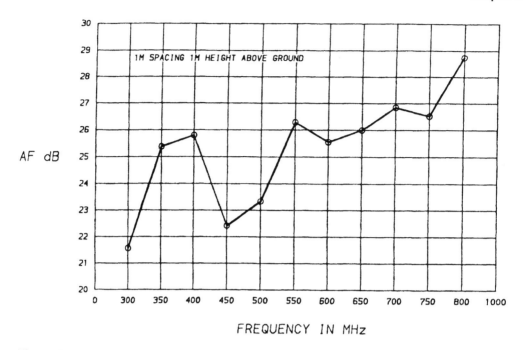

**Figure 2.26**   Measured AF, in decibels, of a resonant monopole.

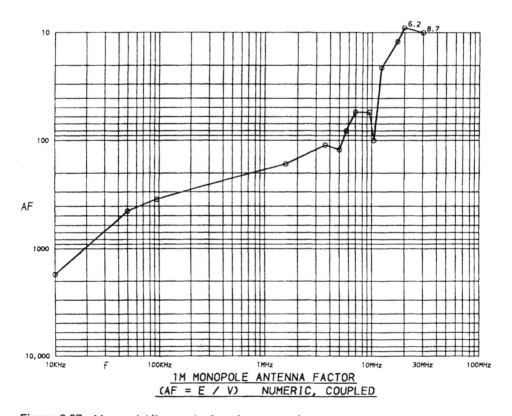

**Figure 2.27**   Measured AF, numeric, for a 1-m monopole.

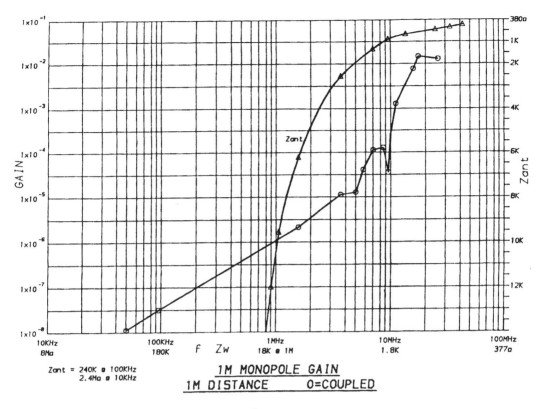

**Figure 2.28**   Measured gain of a 1-m monopole.

were made in the near field, where curvature of the radiated E field will reduce the coupling to the receiving antenna.

The monopole in our example is neither an effective E field measurement nor a communication antenna, due to the large mismatch between the antenna input impedance and the 50-$\Omega$ load impedance. The inclusion of a high-input impedance, e.g., FET, buffer, or amplifier positioned under the counterpoise and connected directly at the base of the antenna, will greatly increase the sensitivity of the antenna at low frequency. The two monopoles, located 1 m apart during measurements, are in the near field, and thus the quasi-static radial field terminates on the counterpoises of both antennas as well as on the rod of the receiving antenna.

The quasi-static coupling is via the very low ($\cong 0.5$ pF) mutual capacitance of the antennas. The mutual reactance of the antennas is therefore very high at the frequencies of interest, and minimal voltage is developed across the 50-$\Omega$ receiving antenna load impedance, due to the radial E field. When the antenna termination impedance is high, the quasi-static field will develop an appreciable voltage across the load and change the AF calibration of the antenna.

When the monopole model is used to predict the current flow on a shielded cable connected to a shielded enclosure, the load impedance is equal to the termination impedance of the shielded cable, typically the sum of the shield-to-backshell, male-to-female connector, and connector-to-bulkhead interface impedances. This termination impedance may then be used in the equations for the short nonresonant antenna to find the load voltage, from which the load current may be found. The current flowing on the shield of the cable is then one-half of the load current.

Where the shielded cable length is resonant, the characteristics of a resonant short-circuit monopole may be used to find the shield current. First the receiving properties of the resonant antenna into a matched load is used to find the load current. The average matched antenna current is then 0.64 times the load current, and the current flow on the short-circuit resonant antenna (cable shield) is twice that for the matched antenna.

A simple circuit comprising a MOSFET input stage followed by a single transistor gain stage, with an input impedance of 100 kΩ, was built and powered by a 9-V battery. The increase in gain at 10 kHz and 100 kHz was measured at 83 dB and 78 dB, respectively. The noise floor increased by 55 dB at 10 kHz and 42 dB at 100 kHz, with a resultant increase in signal-to-noise ratio of 28–36 dB.

## 2.6 SIMPLE, EASILY CONSTRUCTED E AND H FIELD ANTENNAS

This section presents the design of a number of easily constructed antennas for the measurement of either E or H fields. When considering time-varying waves, E and H fields are always present and either may be measured. One important aspect of the design of a measurement antenna is that the antenna differentiates between E and H fields. Thus an antenna designed to measure H fields should reject the influence of the E field component of the electromagnetic wave.

The advantage of physically small antennas is that the localized E and H fields in close proximity to sources such as enclosures, apertures, and cables may be readily measured. The larger dipole and commercial broadband antennas also have a place in measuring the composite field from all sources. Measurement techniques using both types of antenna are discussed in Chapter 9.

### 2.6.1 Shielded Loop Antenna

The simple loop antenna connected to a shielded cable is unbalanced with respect to the shield of the cable and therefore responds to both E and H fields. One technique used to reduce the influence of the E field is to shield the loop.

The schematic of a shielded loop antenna is shown in Figure 2.29a, with a photograph

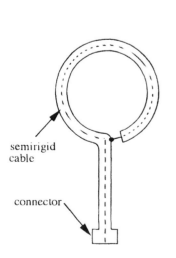

(a) Single shielded loop antenna

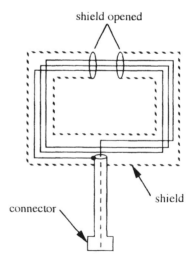

(a) Multiturn shielded loop antenna

**Figure 2.29** Shielded loop antennas.

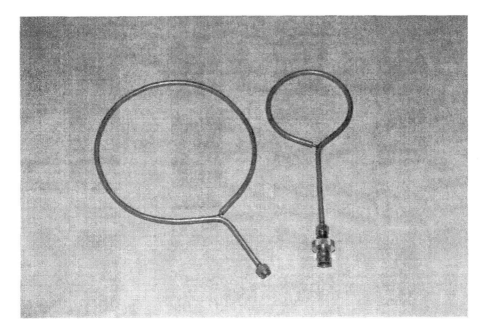

**Figure 2.30**  Photo of shielded loop antennas.

of an 11-cm and a 6-cm antenna given in Figure 2.30. The antenna is constructed from semirigid cable that is available in a number of different diameters. The shielded loop antenna becomes resonant above a frequency determined by the inductance of the loop and the capacitance between the center conductor and the outer sheath. The lower the capacitance of the semirigid cable used in the manufacture of the loop, the higher the useful frequency response. Calibration curves for the 6-cm loop, which has a useful upper frequency of greater than 15 MHz, and the 11-cm loop, which has a useful upper frequency of 10 MHz, are shown in Figures 2.31 and 2.32. The receiving properties of the 6-cm and 11-cm loops are close to the predicted properties when terminated in a 50-Ω load, and so calibration is necessary only to determine the upper frequency limit or where maximum accuracy is required. A word of warning: The measurement technique, unless it is carefully controlled, may result in less accuracy than the use of the predicted characteristics alone. Where the sensitivity of the single loop at low frequency is not sufficient, a multiturn shielded loop may be constructed. The multiturn antenna may be shielded by wrapping the turns with a conductive tape. The tape is connected to the outer surface of the coaxial connector as shown in Figure 2.29b. The shield must contain a gap around the circumference of the loop in order to avoid the shielding of the loop against magnetic fields.

The multiturn loop exhibits a lower resonance frequency than a single turn, due to its higher inductance and intrawinding capacitance.

### 2.6.2  Balanced Loop Antenna

A second extremely useful H field antenna is a 6-cm balanced loop antenna, the schematic of which is shown in Figure 2.33. The balanced loop antenna contains a balun constructed from a two-hole ferrite bead. Baluns are used in a number of balanced antennas, both E and H field types, to match the antenna to the unbalanced coaxial cable. The cable and associated equipment are unbalanced because the shield of the cable and the outer shell of the coaxial cable are

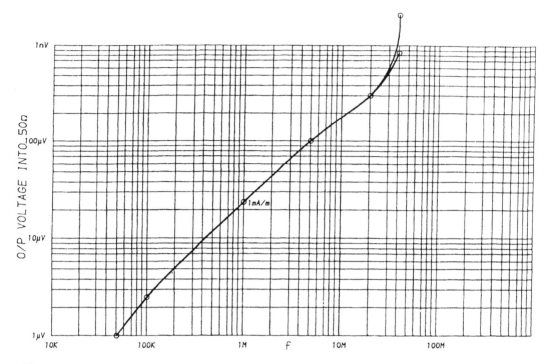

**Figure 2.31**  Calibration curves for a 6-cm shielded loop antenna, upper limit 15 MHz.

connected to ground, via the enclosure that is connected to safety ground on the majority of equipment. In the balanced loop antenna, the E field induces equal and opposite voltages into the primary (loop side) of the balun, and thus, ideally, zero E-field-induced voltage appears across the secondary of the balun.

In practice, some capacitive imbalance exists both in the intrawinding capacitance of the balun and between the loop and the metal enclosure housing the balun. Any imbalance results in an incomplete cancellation of the E-field-induced voltage. In addition, the capacitance, with the inductance of the loop, determines the resonant frequencies of the loop. In a carefully de-signed antenna, the connections from the loop to the primary of the balun should have a charac-teristic impedance of 200 Ω. The secondary connection to the coaxial connector mounted on the enclosure should have an impedance of 50 Ω. The center tap of the primary of the balun is connected to ground via the enclosure. Ideally this connection should be made via a low impedance, such as a short length of wide PCB material. The calibration curve of a balanced loop antenna is shown in Figure 2.34. This antenna was not constructed with maximum care in the layout, and some of the kinks in the calibration curve may be the result. Nevertheless, the measured characteristics are within 4 dB of the predicted. The sensitivity of this simple antenna is approximately 46 dB above that of the Hewlett-Packard HP 11940A near-field mag-netic field probe, which costs almost 100 times as much. It should be added that the HP probe, due to its very narrow tip, is invaluable for locating emissions from printed circuit board tracks and integrated circuits. In addition, the HP probe covers the wider, 30–1000-MHz frequency range, compared to the 20–200 MHz for the simple loop. One potential source of measurement error in the small loop antennas is produced by E-field-induced current flow on the shield of the coaxial interconnection cable. The shield current induces a voltage in the center conductor of the cable, which adds to the signal from the antenna.

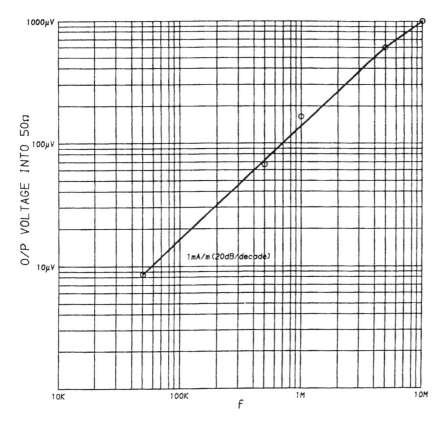

**Figure 2.32** Calibration curve for an 11-cm shielded loop antenna, upper limit 10 MHz.

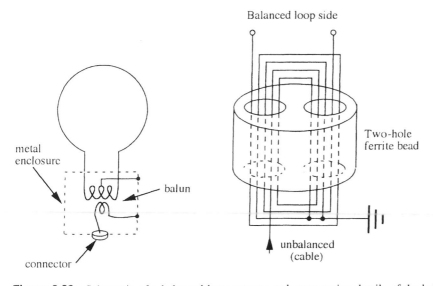

**Figure 2.33** Schematic of a balanced loop antenna and construction details of the balun.

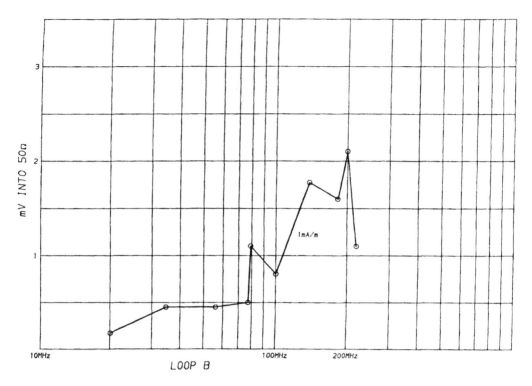

**Figure 2.34** Calibration curve of a balanced loop antenna.

In other chapters we will see that H field measurements are typically more reliable in locations where reflections cause measurement errors. We shall see that measurement errors may be reduced by computing the magnitude of the E field, at some distance from the measuring point, from the magnitude of the H field at the point. The calibration of the example H field antennas are in mV/mA/m. However, an alternative is to use the H field antenna factor, defined as H(A/m)/V.

### 2.6.3 E Field Bow Tie Antenna

The balun incorporated into the balanced loop antenna may also be used to match a broadband E field "bow tie" antenna to the coaxial cable. The schematic diagrams of a bow tie antenna for the 20–250-MHz and 200–600-MHz frequency ranges are shown in Figure 2.35. The impedances of the 20–250-MHz and 200–600-MHz antennas with 2/1 voltage- and 4/1 impedance-ratio baluns are shown in Figure 2.36 and are derived from Ref. 7.

The bow tie is useful in measuring both the radiation from an EUT in close proximity as well as the level of the E field in a radiated susceptibility (immunity) test. The small bow tie, with 10-cm-long elements, although designed for 200–600 MHz, has been calibrated from 25 MHz to 1000 MHz. At low frequency, the antenna factor is high, in common with all physically small antennas, and great care has to be exercised, in both calibration and use, to ensure that the coupling to the feed cable is not higher than that to the antenna. To reduce the coupling, use a double-braid shielded cable with many ferrite baluns mounted on the cable; and when monitoring susceptibility test levels, orient the cable vertically for horizontally polarized fields and antenna, and orient it horizontally for vertically polarized fields and antenna. Figure 2.37

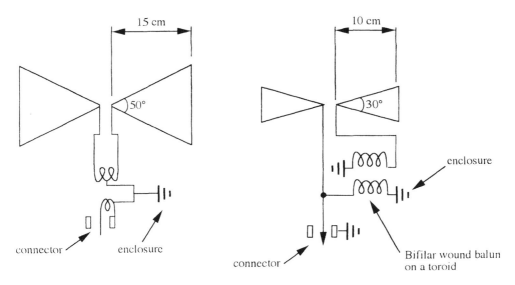

(a) 20–250-MHz bow tie antenna          (b) 200–600-MHz bow tie antenna

**Figure 2.35**   Bow tie antennas.

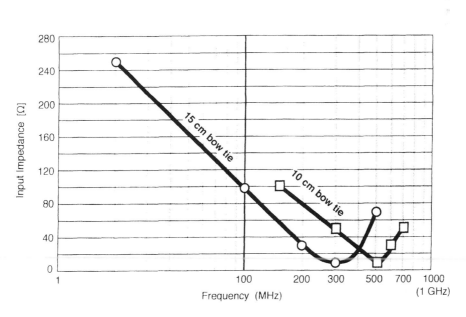

**Figure 2.36**   Input impedance calibration curves for the 15-cm and 10-cm bow tie antennas. (From Ref. 7.)

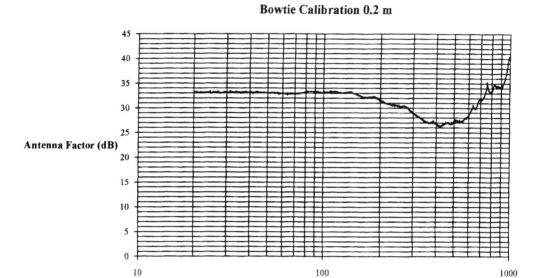

**Figure 2.37**  Bow tie antenna factor at 0.2 m.

shows the 10-cm bow tie calibration at a distance of 0.2 m; Figure 2.38 shows it at a distance of 1 m, with the antenna horizontally polarized, and 1 m above a ground plane. The antenna was constructed on a rectangular-shaped PCB with the two bow tie elements etched out of the PCB material, with a BNC connector mounted in the center of the antenna and with the balun located close to the connector as shown in Figure 2.38.

### 2.6.4   Monopole Antennas

The 1-m and the tunable resonant monopole antennas described in Sections 2.5.5 and 2.5.6 cover the 14-kHz–30-MHz and 300–800-MHz frequency ranges, respectively, and the simple construction of these antennas is illustrated in Figure 2.39. The monopoles with ground plane do not require a balun to match the antenna to the cable, which is one reason for their widespread use.

### 2.6.5   Tuned Resonant Dipole Antennas

The schematic of the dipole with a typical balun used over the 40–300-MHz frequency range is shown in Figure 2.40. The calibration curve for this antenna in a shielded room and the theoretical open field test site calibration is shown in Figure 2.22. The construction details of a similar balun, showing how the coaxial cable is wound on formers and attached to the connector, are presented in Figure 2.41. The construction details are reproduced by courtesy of the Electrometrics Corporation.

### 2.6.6   Helical Spiral Antennas

Helical spiral antennas, which cover the 800-MHz–18-GHz frequency range, as shown in Figure 2.42, are relatively simple to construct. The larger antenna is 26.5 cm long and 12 cm in diameter

**Bowtie Calibration 1m**

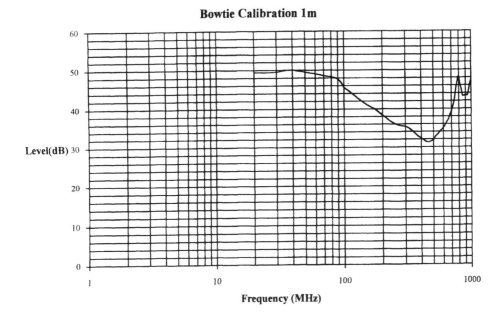

**Figure 2.38**   Bow tie antenna factor at 1 m and 1 m above a ground plane.

at the base, tapering to 10 cm at the top. The larger antenna is wound as a helical log periodic, with the angle between the base of the antenna and the spiral at approximately 20°. The distances between the windings of the spiral are not critical, and in the larger antenna, shown in Figure 2.42, the ratio of the distance between the windings of the lower spiral and the distance between the windings of the next higher spiral is approximately 1.25. The important criterion is that the

**Figure 2.39** 1-m and resonant monopole antennas.

two antennas be made as similar as possible when the two-antenna test method is used for calibration.

The calibration curve for the 26.5-cm log periodic helix is shown in Figure 2.43. The design frequency range for the antenna is 1–4.77 GHz, based on the circumferences of the lowest and highest spiral. From the calibration curve it is seen that the antenna is usable down to 800 MHz; a recent recalibration was made up to 4 GHz.

In order to simplify the construction of the two smaller antennas, covering the 3–18-GHz frequency range, only the diameter of the helix is changed from top to bottom. The intermediate-size spiral has a maximum circumference of 8.96 cm and a minimum of 3.49 cm; thus the design frequency range is 3.34–8.6 GHz. The distance between the spirals on the intermediate-sized antenna is a constant 0.476 cm (3/16 in.). The small antenna has a design frequency range of 8–18 GHz. The material on which the helix is wound should exhibit a low permittivity and low loss at high frequency; here, expanded polystyrene is a good choice. However, the antennas illustrated were wound on cardboard formers impregnated with epoxy, and acceptable results were achieved in the calibration of the antenna.

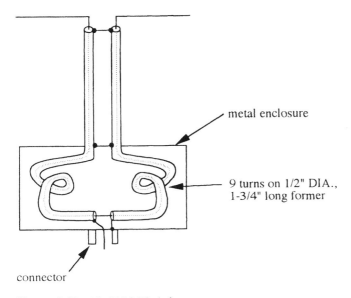

metal enclosure

9 turns on 1/2" DIA.,
1-3/4" long former

connector

**Figure 2.40** 30–500-MHz balun.

**Figure 2.41** Construction details of a similar balun. (Reproduced courtesy of Electrometrics Corporation.)

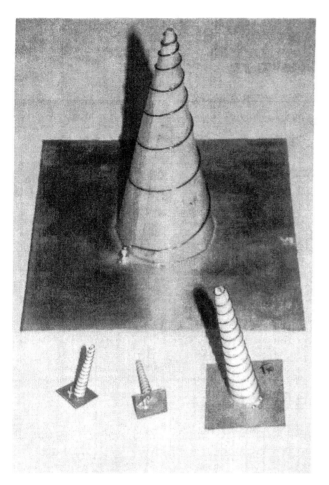

**Figure 2.42**   Helical antennas covering the 800-MHz–18-GHz frequency range.

### 2.6.7   Small Toroid Antenna

An inductor wound on a toroid core does not contain the field inside the core particularly well and, as described in Section 5, any inductor wound on a core will have a current induced in it when an electromagnetic field impinges on the inductor. Thus in certain filter applications it is imperative to shield the inductor to minimize this radiated coupling effect. This coupling effect is used in the toroid antenna, which is constructed from a small iron-dust toroid type T12-3 from Micrometals. The T12-3 core has a diameter of only 3.18 mm and an $A_L$ value of 6 nH/N$^2$; to construct the antenna, ten turns of magnet wire are wound on the core. The toroid is mounted at the end of a brass tube, with one side of the inductor connected to the tube and the second side connected to a center conductor in the tube. At the other end of the tube, an SMA connector is soldered, as shown in Figure 2.44. The toroid antenna shows no resonances over the 20–1000-MHz frequency range. It is sensitive to very localized fields and is invaluable for measuring PCB emissions down to a single trace or to differentiate between emissions from the PCB and the integrated circuits (ICs). It is used at very close proximity to sources, where the field reduces rapidly with distance, for comparative measurements, and so it makes no sense

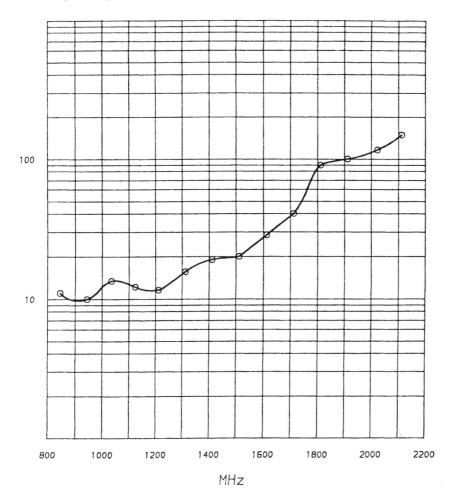

**Figure 2.43** Calibration curve for the 26.5-cm helical log spiral antenna.

to calibrate the antenna. If quantative measurements are required, then the balanced loop or bow tie antenna at distances of approximately 0.2 m are the correct antennas to use.

### 2.6.8 Calibration

When care is taken in the construction of two antennas to make them as similar as possible, then the two-antenna test method described in Section 2.5.3 is recommended. Alternative antenna calibration techniques are the reference open area test site (OATS) and the reference dipole calibration, both of which are contained in ANSI C63.5 and described in Section 9.4.1. If correctly constructed, the four reference dipoles, described in ANSI C63.5 and designed to cover the 25–1000-MHz frequency range, will have antenna factors within 0.3 dB of theoretical. An alternative test method uses the GTEM cell, as described in Section 9.5.2.2. The antenna factor can also be calculated using either the "induced-emf method," numerical electromagnetics codes (NEC), MININEC, or GEMACS or from site attenuation. In the induced-emf method, mutual coupling between antennas is calculated via formulas given by S. A. Schelkunoff and

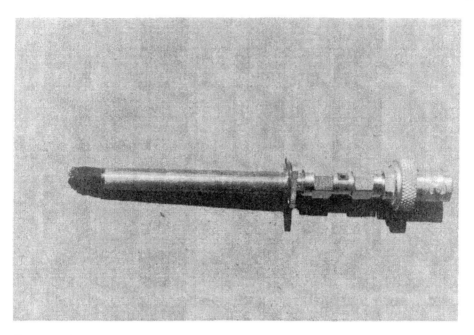

**Figure 2.44**   Small toroid antenna.

H. E. King. These formulas usually replace the presence of the ground plane by images of the transmit and receive antennae. NEC, MININEC, and GEMACS use the moment method (MOM) to calculate the coupling between antennas; with GEMACS, using the MOM and GTD hybrid, the presence of a ground plane can be included. Reference 21 compares the accuracy of these analysis techniques to OATS, anechoic chamber and GTEM measurements for symmetrical dipoles and horn and waveguide antennas. With symmetrical dipoles in horizontal polarization, the maximum deviation between the site attenuation model and measurements was 3 dB. The maximum deviation between measurements and the induced-emf, MOM, MININEC, and other calculations was 2.7 dB.

One alternative test method is to use a calibrated-gain transmitting antenna. A measured and controlled input power is applied, and therefore a known E field is generated at the distance from the transmitting antenna at which the receiving antenna is placed. The measured output voltage from the antenna under calibration can then be used, with the value of the E field, to obtain the antenna factor, from which the gain of the antenna under calibration may be calculated. The second test method is to use the antenna under calibration as the transmitting antenna and the antenna factor of the calibrated receiving antenna to calculate the gain of the antenna under test, from which the antenna factor may be calculated. The mutual coupling between antennas can be minimized by using a 10-m distance between the antennas and, as described in section 2.5.3, the effect of the ground reflected wave has to be either minimized or calculated.

One technique for the calibration of H field antennas is shown in Section 9.6.7.8. Although the calibration is for an RS01 transmitting antenna, it is identical to the calibration of a receiving loop antenna, if the reference transmitting loop is constructed.

## 2.7 NONIONIZING ELECTROMAGNETIC FIELD EXPOSURE SAFETY LIMITS

The term *nonionizing radiation* is used to denote radiation from such sources as microwave ovens, antennas, and wires carrying high-RF currents. Infrared and visible light are also nonionizing radiation. Ionizing radiation, which includes x-rays, gamma rays, and electromagnetic particles, is so named because it can cause ionization of gas molecules, whereas nonionizing radiation cannot.

The maximum allowable or recommended levels of exposure to nonionizing radiation are not the same for every country or for different organizations within the same country. One reason for the inconsistency may be that, unlike ionizing radiation, the amount of data on the biological effects on human beings and other warm-blooded creatures is limited; also, different criteria are used by the various regulatory organizations to arrive at safe levels. Generally speaking, the effects of nonionizing radiation on living tissue depends upon the wavelength (frequency), which, with the conductivity of the tissue, determines the depth of tissue penetration. As a general guideline for those working with high-power microwave radiation, special precautions must be taken at frequencies where an ungrounded person is half a wavelength or less from a radiating source.

**CAUTION: Although every effort has been made to verify the accuracy and the currency of the safety levels published in this section, they are intended for information only. If any of these limits are to be applied, the responsible organization must be contacted to obtain the current frequency band and levels. Also, separate limits apply to controlled environments and to RF workers.**

### 2.7.1 Clinical Studies on Human Beings

Since the first edition of this book, many articles and books have been published on human exposure to electromagnetic fields, especially the fields generated by 50–60-Hz power lines. Many of the data are contradictory, and so a review of the data is well beyond the scope of this edition. Information on this topic may be obtained from the Internet and from the database used in the IEEE C95.1—1991 document.

### 2.7.2 Canadian Limits

In the early limits, both Welfare Canada and the IEEE American National Standard confined themselves to the thermal effect in the interaction of RF and microwave fields with biological systems.

The basal metabolic rate (BMR), or the dissipation of an average man resting, is 100 W, and this figure can double or even triple during activity. It was assumed that the human body could easily cope with a 100% increase in the BMR (i.e., an additional 100 W). The earliest widely adopted limit in Western Europe, the United States, and Canada was 10 mW/cm$^2$ based on exposure of one-half the cross-sectional area of an average man to an incident plane wave with a field intensity of 10 mW/cm$^2$. This was chosen because it is the level sufficient to increase the dissipation by an additional 100 W.

By calculation and experimentation, using saline-filled human phantoms, it was subsequently found that the average-sized man, grounded, resonates at 31–34 MHz (the resonant frequency for a 6-foot-high shorted monopole). The resonance frequency is, of course, higher for children. Taking into account the resonance and other frequency effects as well as the possibility of reflection (8) the Canadian exposure limit for the general public was revised down to

**Table 2.2** Canada Safety Code 6, 1999: Exposure Limits for Persons Other than RF Workers (the General Public)

| 1<br>Frequency<br>(MHz) | 2<br>Electric field<br>strength rms<br>(V/m) | 3<br>Magnetic field<br>strength rms<br>(A/m) | 4<br>Power<br>density<br>(W/m$^2$) | 5<br>Averaging<br>time<br>(min) |
|---|---|---|---|---|
| 0.003–1 | 280 | 2.19 | | 6 |
| 1–10 | 280/f | 2.19/f | | 6 |
| 10–30 | 28 | 2.19/f | | 6 |
| 30–300 | 28 | 0.073 | 2* | 6 |
| 300–1500 | 1.585$f^{0.5}$ | 0.0042$f^{0.5}$ | f/150 | 6 |
| 1500–15000 | 61.4 | 0.163 | 10 | 6 |
| 15000–150000 | 61.4 | 0.163 | 10 | 616.000/$f^{1.2}$ |
| 150000–300000 | 0.158$f^{0.5}$ | 4.21 ×<br>10$^{-4}f^{0.5}$ | 6.67 × 10$^{-5}f$ | 616.000/$f^{1.2}$ |

\* Power density limit is applicable at frequencies greater than 100 MHz.

*Notes*:

1. Frequency, $f$, is in MHz.
2. A power density of 10 W/m$^2$ is equivalent to 1 mW/cm$^2$.
3. A magnetic field strength of 1 A/m corresponds to 1.257 microtesla (μT) or 12.57 milligauss (mG).

1 mW/cm$^2$ continuous in the frequency range 3 MHz–100 GHz. In the latest version of this document (Ref. 10) the power limit is further reduced to 2 mW/m$^2$ (0.2 mW/cm$^2$) from 30–300 MHz and is $f$(MHz)/150 (mW/m$^2$) from 300 to 1500 MHz and 10 mW/m$^2$ from 1500 to 300,000 MHz.

The detailed derivation of the limits in Ref. 10 that are reproduced in Table 2.2 for the general public has not been provided. Reference 10 contains different exposure limits for RF workers and persons other than RF workers. The power limit shall not be exceeded when averaged over any 0.1-hr period and when spatially averaged.

Where the electromagnetic radiation consists of frequencies from more than one frequency band in the first column of Table 2.2, the fraction of the actual radiation (power density or square of the field strength) in a frequency band relative to the value given in column 2, 3, or 4 shall be determined and the sum of all fractions of all frequency bands shall not exceed unity when time and spatially averaged.

Time-averaged values need only be calculated from multiple measurements if the field changes significantly (more than 20%) within a period of 0.1 hr; otherwise a single measurement suffices. Reference 10 provides formulas for time averaging. Spatial averaging refers to the uniformity of the exposure field. For portable transmitters and other devices which produce highly spatially nonuniform fields, the levels in Table 2.2 may be exceeded, but the following values shall not be exceeded:

| Conditions (w/KG) | SAR limit |
|---|---|
| The SAR* averaged over whole body mass | 0.08 |
| The local SAR* for head, neck, and trunk, averaged over any<br>  one gram (g) of tissue | 1.6 |
| The SAR* in the limbs, as averaged over 10g of tissue | 4 |

\* SAR = specific absorption rate = power absorbed per weight of tissue, in watts per kilogram.

**Table 2.3** Canada Safety Code 6 Induced and Contact-Current Limits for Persons Other Than RF Workers (the General Public)

| 1<br>Frequency<br>(MHz) | 2<br>Rms induced current<br>(mA)<br>through | | 3<br>Rms contact current<br>(mA)<br>hand grip and<br>through each foot | 4<br>Averaging<br>time |
|---|---|---|---|---|
| | Both feet | Each foot | | |
| 0.003–0.1 | 900$f$ | 450$f$ | 450$f$ | 1 s |
| 0.1–110 | 90 | 45 | 45 | 0.1 h (6 min) |

*Notes*:
1. Frequency, $f$, is in MHz.
2. The above limits may not adequately protect against startle reactions and burns caused by transient spark discharges for intermittent contact with energized objects.

Although not a requirement of the Code, it is suggested that whenever possible, the organ-averaged SAR for the eye shall not exceed 0.2 W/kg.

In addition, a limit is placed on contact current. For free standing individuals (no contact with metallic objects) the current induced into the human body by electromagnetic radiation shall not exceed the value in column 2 of Table 2.3. No object with which a person may come in contact shall be energized by electromagnetic radiation in the frequency band listed in column 1 of Table 2.3 to such an extent that the current flowing through a circuit having the impedance of the human body exceeds the value given in column 3 of Table 2.3 as measured with a contact current meter, where the electromagnetic radiation consists of frequencies from more than one frequency band of column 2 of Table 2.3, the ratio of the square of the measured current at each frequency to the square of the limit at that given frequency shown in column 2 or 3 (whichever is applicable).

### 2.7.3 American Standards

One of the present standards is contained in Ref. 9, IEEE C95.1—1991. The present standards include revisions in the limits from the 1982 to the 1998 version based on additional data. From 1988 to 1991 the revisions resulted in an expanded frequency range, limits on induced body current to prevent radio frequency (RF) shock or burn, a relaxation of limits on exposure to magnetic fields at low frequencies, and exposure limits and averaging time at high frequencies that are compatible at 300 GHz with existing infrared maximum permissible exposure (MPE) limits.

Different MPEs apply to controlled environments, where exposure is incurred by persons who are aware of the potential for exposure, such as RF workers, and to uncontrolled environments, where exposure to individuals who have no knowledge or control of their exposure occurs, such as the general public. The following limits are for the uncontrolled environment.

The MPEs refer to exposure values obtained by spatially averaging over an area equivalent to the vertical cross section of the human body (projected area). In the case of partial-body exposure, Ref. 9 describes some relaxation. In nonuniform fields, spatial peak values of field strengths may exceed the MPEs if the spatially averaged value remains within the specified limits. The MPEs may also be relaxed by reference to SAR limits in Ref. 9. For near-field exposure at frequencies less than 300 MHz, the applicable MPE is in terms of rms electric and magnetic field strength. The MPE refers to values averaged over any 6–30-min period for

**Table 2.4**  IEEE C95.1 MPE Limits for Uncontrolled Environments

| 1 Frequency range (MHz) | 2 Electric field strength $(E)$ (V/m) | 3 Magnetic field strength $(H)$ (A/m) | 4 Power density $(S)$ E-Field, H-Field $(mW/cm^2)$ | 5 Averaging time (min) $E^2$, $S$ or $H^2$ | |
|---|---|---|---|---|---|
| 0.003–0.1 | 614 | 163 | 100, 100,000[b] | 6 | 6 |
| 0.1–1.34 | 614 | $16.3/f$ | 100, $1000/f^{2b}$ | 6 | 6 |
| 1.34–3.0 | $823.8/f$ | $16.3/f$ | $180/f^2$, $1000/f^{2b}$ | $f^2/0.3$ | 6 |
| 3.0–30 | $823.8/f$ | $16.3/f$ | $80/f^2$, $1000/f^{2b}$ | 30 | 6 |
| 30–100 | 27.5 | $158.3/f^{1.668}$ | 0.2, $940,000/f^{3.336}$ | 30 | $0.0636f^{1.337}$ |
| 100–300 | 27.5 | 0.0729 | 0.2 | 30 | 30 |
| 300–3000 | 27.5 | | $f/1500$ | 30 | |
| 3000–15,000 | | | $f/1500$ | $90,000/f$ | |
| 15,000–300,000 | | | 10 | $616,000/f^{1.2}$ | |

Electromagnetic fields[a]

[a] The exposure values in terms of electric and magnetic field strengths are the values obtained by spatially averaging values over an area equivalent to the vertical cross section of the human body (projected area).

[b] These plane-wave equivalent power density values, although not appropriate for near-field conditions, are commonly used as a convenient comparison with MPEs at higher frequency and are displayed on some instruments in use.

*Source*: Ref. 9.

frequencies up to 3000 MHz and over shorter periods for higher frequencies down to 10 s at 300 GHz, as indicated in Table 2.4.

For mixed or broadband fields at a number of frequencies for which there are different values of the MPE, the fraction of the MPE [in terms of $E^2$, $H^2$, or power density $(S)$] incurred within each frequency interval should be determined and the sum of all such fractions should not exceed unity. In a similar manner for mixed or broadband induced currents, the fraction of the induced current limits (in terms of $I^2$) incurred within each frequency interval should be determined and the sum of all such fractions should not exceed unity. For exposures to pulsed radio frequency fields in the range 0.1–300,000 MHz, the peak (temporal) value of the MPE in terms of the E field is 100 kV/m.

For exposures to pulsed radio frequency fields of pulse durations less than 100 ms and for a single pulse, the MPE is given by the MPE from Table 2.4 (E field equivalent power density) multiplied by the averaging time, in seconds, and divided by 5 times the pulse width, in seconds. That is:

$$\text{Peak MPE} = \frac{\text{MPE} \times \text{Avg time (seconds)}}{5 \times \text{Pulsewidth (seconds)}}$$

A maximum of five such pulses, with a pulse repetition rate of at least 100 ms, is permitted during any period equal to the averaging time. If there are more than five pulses during any period equal to the averaging time, or if the pulse durations are greater than 100 ms, normal averaging time calculations apply, except that during any 100-ms period, the energy density is limited by the preceding formula, viz

$$\text{Peak MPE} \times \text{Pulsewidth (seconds)} = \frac{\text{MPE} \times \text{Avg time (seconds)}}{5}$$

**Table 2.5** Induced and Contact RF Currents for Uncontrolled Environments[a]

| Frequency range (MHz) | Maximum current (mA) | | |
|---|---|---|---|
| | Through both feet | Through each foot | Contact |
| 0.003–0.1 | 900$f$ | 450$f$ | 450$f$ |
| 0.1–100 | 90 | 45 | 45 |

$f$ = frequency, in MHz.
[a] The current limits given may not adequately protect against startle reactions caused by transient discharges when contacting an energized object; see Ref. 9 for additional comment.

At frequencies between 100 kHz and 6 GHz, the MPE in uncontrolled environments for electromagnetic field strengths may be exceeded if: the exposure conditions can be shown by appropriate techniques to produce SARs below 0.08 W/kg, as averaged over the whole body, and spatial peak SAR values not exceeding 1.6 W/kg, as averaged over any 1 g of tissue (defined as a tissue volume in the shape of a cube), except for the hands, wrists, feet, and ankles, where the spatial peak SAR shall not exceed 4 W/kg, as averaged over any 10 g of tissue, and the induced currents in the body conform to Table 2.5.

A second source of limits is the FCC RF exposure guidelines contained in 47 C.F.R. 1.1310, Ref. 17. The criteria listed in Table 2.6 shall be used to evaluate the environmental impact of human exposure to radio frequency (RF) radiation as specified in 1.1307 (b), except in the case of portable devices, which shall be evaluated according to the provisions of 2.1093 of the 47 C.F.R. document. Further information on evaluating compliance with these limits can be found in the FCC's OET/OET Bulletin Number 65, "Evaluating Compliance with FCC-Specified Guidelines for Human Exposure to Radiofrequency Radiation. The FCC limits are generally based on recommended exposure guidelines published by the National Council on Radiation Protection and Measurements (NRCP) in "Biological Effects and Exposure Criteria for Radiofrequency Electromagnetic Fields," NCRP Report No. 86, Sections 17.4.1, 17.4.1.1, 17.4.2, and 17.3. Copyright NRCP, 1986, Bethesda, MD 20814. In the frequency range from 100 MHz to 1500 MHz, exposure limits are also generally based on guidelines recommended by the American National Standards Institute (ANSI) in section 4.1 of IEEE Standard for Safety Levels with Respect to Human Exposure to Radio Frequency Electromagnetic Fields, 3 kHz to 300 GHz, ANSI/IEEE C95.1—1992.

### 2.7.4 East European, European, and other Standards

The philosophy of standard setting in East European countries, at least in 1977, appears to be that any effect due to microwave radiation, whether subjective, objective, or reversible, should be avoided, and thus the Russian and Czechoslovakian standards were as low as 1 μW/cm² (i.e., 1/2000th of the Canadian level).

In May 1997 the Hungarian Standards Institution released the European Pre-standard (ENV 50166-1/2) as Hungarian Pre-standard MSZ ENV 50166-1/2. It is the general policy of the Hungarian Parliament that confirmed EU standards shall be considered as Hungarian National Standards.

The Russian limits have remained the same over 40 years and are based on intensity and time of exposure: 10 μW/cm² for 8 hours, 100 μW/cm² up to 2 hrs, and 1000 μW/cm² for up to 20 minutes per working day.

New general-public limits were issued in Poland in August 1998 and, although increased, are still lower than the proposals of CENELEC, IRPA or ANSI for the same frequency range (0.1 Hz–300 GHz). It would almost certainly be difficult to increase the public values too drastically, at least initially.

The European Commission has proposed basic restrictions on SAR designed to prevent whole-body heat stress and excessive localized heating of tissues. The basic restrictions are set to account for uncertainties related to individual sensitivities, environmental conditions, and for the fact that the age and health status of members of the public varies. The SAR limits over the 10 MHz to 10 GHz frequency range are: 0.08 W/kg for the whole body; the localized SAR for the head and trunk of 0.08 W/kg and for the limbs 4 W/kg. CENELEC (The European Committee on Electrotechnical Standardization) accepted the ENV 50166 at the end of November 1994 for provisional application for three years, which has been extended to the year 2000.

In 1997 the 26th EMF Ordinances came into effect in Germany and are based on the International Radiation Protection Association (IRPA), WHO, and ICNIRP guidelines. The EMF Ordinances is limited to the frequency range 10 MHz to 300,000 MHz at the levels shown in Table 2.6.

In the UK the National Radiological Protection Board (NRPB) revised its advice in 1993 and published the new advice 8.

**Table 2.6** ICNIRP Reference Levels for General Public Exposure to Time-Varying Electric and Magnetic Fields

| Frequency range | E-field strength $(Vm^{-1})$ | H-field strength $(Am^{-1})$ | B-field $(\mu T)$ | Equivalent plane wave power density $S_{eq}(Wm^{-2})$ |
|---|---|---|---|---|
| up to 1 Hz | — | $3.2\times 10^4$ | $4 \times 10^4$ | — |
| 1–8 Hz | 10,000 | $3.2 \times 10^4/f^2$ | $4 \times 10^4/f^2$ | — |
| 8–25 Hz | 10,000 | $4,000/f$ | $5,000/f$ | — |
| 0.025–0.8 kHz | $250/f$ | $4/f$ | $5/f$ | — |
| 0.8–3 kHz | $250/f$ | 5 | 6.25 | — |
| 3–150 kHz | 87 | 5 | 6.25 | — |
| 0.15–1 MHz | 87 | $0.73/f$ | $0.92/f$ | — |
| 1–10 MHz | $87/f^{1/2}$ | $0.73/f$ | $0.92/f$ | — |
| 10–400 MHz | 28 | 0.073 | 0.092 | 2 |
| 400–2,000 MHz | $1.375f^{1/2}$ | $0.0037f^{1/2}$ | $0.0046f^{1/2}$ | $f/200$ |
| 2–300 GHz | 61 | 0.16 | 0.20 | 10 |

[a] *Note*:

1. $f$ as indicated in the frequency range column.
2. Provided that basic restrictions are met and adverse indirect effects can be excluded, field strength values can be exceeded.
3. For frequencies between 100 kHz and 10 GHz, $S_{eq}$, $E^2$, $H^2$, and $B^2$ are to be averaged over any 6-min period.
4. For peak values at frequencies up to 100 kHz see Table 4, note 3.
5. For peak values at frequencies exceeding 100 kHz see Figs. 1 and 2. Between 100 kHz and 10 MHz, peak values for the field strengths are obtained by interpolation from the 1.5-fold peak at 100 kHz to the 32-fold peak at 10 MHz. For frequencies exceeding 10 MHz it is suggested that the peak equivalent plane wave power density, as averaged over the pulse width does not exceed 1,000 times the $S_{eq}$ restrictions, or that the field strength does not exceed 32 times the field strength exposure levels given in the table.
6. For frequencies exceeding 10 GHz, $S_{eq}$, $E^2$, $H^2$, and $B^2$ are to be averaged over any $68/f^{1.05}$-min period ($f$ in GHz).
7. No E-field value is provided for frequencies <1 Hz, which are effectively static electric fields, perception of surface electric charges will not occur at field strengths less than 25 kVm$^{-1}$. Spark discharges causing stress or annoyance should be avoided.

In Japan the Ministry of Posts and Telecommunications (MPT) issued revised rules to establish Radio Radiation Protection for Human Exposure to Electromagnetic Fields from 10 kHz to 300 GHz. These guidelines contain levels which are a mix of the Canada safety code 6 and the IEEE C95.1 but are not the same as either. In 1997 the MPT slightly revised the guidelines in changing the local SAR limit from 1.6 W/kg for 1 g of tissue to 2 W/kg for 10 g of tissue for the general environment.

The Australasian Standards Committee created a joint Australian and New Zealand standard and published a draft of NZS/AS2772 Part 1 in 1998. At present the UHF reference level remain capped at 200 $\mu$W/cm$^2$, which is more conservative than the ICNIRP which allows 450 $\mu$W/cm$^2$ at 900 MHz.

The Italian limits for the general public are 60 V/m and 0.2 A/m from 100 kHz to 3 MHz, 20 V/m and 0.05 A/m from 3 MHz to 3 GHz and 40 V/m and 0.1 A/m in the frequency range 3–300 MHz. These limits are to be averaged over 6 min. "Precautionary levels" are 6 V/m and 0.016 A/m for any frequency.

### 2.7.5  ICNIRP, CENELEC, IRPA, and CEU Limits

In 1998 the International Commission on Non-Ionizing Radiation Protection (ICNIRP) published guidelines for limiting exposure to time-varying electric, magnetic, and electromagnetic fields. They concluded that any established biological and health effects in the frequency range from 10 MHz to a few GHz are consistent with responses to a body temperature rise of more than 1°C. This level of temperature increase results from exposure of individuals under moderate environmental conditions to a whole-body SAR of approximately 4 W/kg for about 30 minutes. A whole-body average of 0.4 W/kg has been chosen as the restriction that provides adequate protection for occupational exposure. An additional safety factor of 5 has been introduced for exposure of the public, giving an average whole-body SAR limit of 0.08 W/kg. The lower basic restriction for exposure of the general public take into account the fact that their age and health status may differ from those of workers. The reference levels (E, H, and power density) are obtained from the basic SAR restrictions by mathematical modeling and by extrapolation from the results of laboratory investigations at specific frequencies. These reference levels for the general public are provided in Table 2.6. The ICNIRP developed basic restriction guidelines with the International Non-Ionizing Radiation Association (IRPA) and these are shown in Table 2.7 for the exposure of workers.

The guideline limit values for the European Committee for Electrotechnical Standardization (CENELEC) contained in the European pre-standard ENV 50166, the International Radiation Protection Agency (IRPA) and the Commission of the European Communities (CEU) are provided in Table 2.7. These values of SAR are applicable to the exposure of workers only, over a full working day, and include limits on the exposure to fields at 50/60 Hz.

However, in common with all other organizations, the committee did not take into account the effects of:

Drugs
Chemical or physical agents in the environment
Modulated microwave effects
Possible nonthermal very long-term effects
High-level low-repetition-rate sources.

### 2.7.6  Measurement of Electromagnetic Field Levels

A broadband isotropic probe is an instrument designed for the measurement of EM fields. However, such instruments are calibrated in the far field (i.e., illuminated with a near plane wave).

**Table 2.7** CENELEC ENV 50166, IRPA, and CEU Guidelines for Workers (Not the General Public) over a Full Working Day

| (a) power frequency 50/60 Hz | | CENELEC | IRPA | CEU |
|---|---|---|---|---|
| Induced current density, head and trunk [mA/m$^2$] | | 10 | 10 | 10 |
| Electric field strength | [kV/m] | 10 | 10 | 19.6;12.3;6.1 |
| Magnetic flux density | [mT] | 1.6 | 0.5 | 0.64;0.4;0.2 |
| Contact current | [mA] | 3.5 | | 1.5 |

| (b) high frequencies | | CENELEC | IRPA | CEU |
|---|---|---|---|---|
| Frequency range | [Hz] | $1.0 \times 10^4$–$3.0 \times 10^{11}$ | $1.0 \times 10^7$–$3.0 \times 10^{11}$ | $1.0 \times 10^5$–$3.0 \times 10^{11}$ |
| Specific absorption rate SAR [W/kg] | | | | |
| Whole body | | 0.4 | 0.4 | 0.4 |
| Extremities (averaged over 10 g tissue) | | 20 | 20 | 20 |
| Head and trunk (av. over 10 g) | | 10 | 10 | 10 |
| Peak specific absorption rate sa [mJ/kg] | | 10 | | 10 |
| Contact current (0.1–3 (100) MHz) [mA] | | 35 | (50) | 50 |

When measuring in the near field it is important to use the correct probe, and a magnetic (H) field or electric (E) field probe is available for most instruments. The near-field/far-field transition for radiation from a point source is defined as $\lambda/2\pi$; however, for an antenna the transition is dependent on antenna aperture. For example, the transition is at 25 ft for a specific helical antenna (11) that has an operating frequency of 250–400 MHz.

As discussed in preceding sections and of use as a guideline only, the type of field generated by a dipole is predominantly an E field, as is the field from a current-carrying cable, and the field from a loop is predominantly an H field. In the case of one type of helical antenna, neither the electric nor the magnetic fields predominate in the proximity of the antenna; therefore, either an E field or H field probe may be used for measurements (Ref. 11). To illustrate the importance of using the correct probe: If a magnetic field probe is used to measure the near field of a dipole and it registers the maximum recommended level of magnetic field (i.e., 0.073 A/m) and we assume at the measurement location that the wave impedance $Z_W$ is 6000 Ω, then the electric field would be

$$Z_W \times \text{A/m} = \text{V/m}$$
$$6000 \ \Omega \times 0.073 \ \text{A/m} = 438 \ \text{V/m} \qquad \text{(i.e., 16 times the recommended level)}$$

Potentially hazardous fields can be generated from unlikely, or at least unexpected, sources. In the course of EMI investigations, a broadband E field of 200 V/m from a shielded box (without apertures) and its power cable have been measured. Similarly, narrowband magnetic fields as high as 7 A/m at 30 MHz have been measured in the vicinity of an instrument containing a plasma source.

### 2.7.7  DC and Power Frequency Fields

Those living in an industrial society are constantly exposed to electric and magnetic fields from power wiring, appliances, and high-voltage transmission lines at frequencies in the 50–60-Hz range. For the sake of efficiency, overhead power lines have increased in voltage over the years, with a recent increase from 345 kV to 765 kV. As a result of public concern, a number of studies on possible hazards have been undertaken.

It is clear from earlier studies that 60-Hz electric and magnetic fields can affect biological systems; however, due to the conflicting results of the recent studies on the effects on humans, the author considers that no conclusions can be drawn at the time of publication of this book. Health and Welfare Canada does not specify a maximum exposure to magnetic fields below 10 kHz. However, a contact there has provided a reference to an IRPA report entitled "Interim Guidelines for Exposure to 60-Hz magnetic fields," published in the *Health Physics Journal* in 1990. The recommended maximum exposure over a period of 24 hr is 1000 mG = 79.36 A/m. (Flux density limits in gauss or tesla can be converted to magnetic field strength when the relative permeability is 1. The relationship is 1 tesla = $7.936 \times 10^5$ A/m and 1 tesla = $10^4$ gauss. Therefore, 1 G = 79.36 A/m.)

The International Commission on Nonionizing Radiation Protection has provided "Guidelines on Limits of Exposure to Static Magnetic Fields." These guidelines have resulted in the limits for exposure to static magnetic fields shown in Table 2.8. The guidelines state that those with cardiac pacemakers should be discouraged from inadvertently entering areas with fields large enough in dimension to include most of a person's trunk at magnetic flux densities greater than 0.5 mT (396 A/m).

Reports (Refs. 12–14) have raised the issue of the extremely high LF or DC electric fields (up to 10,000 V/M) in the proximity of video display terminals (VDTs), whereas one report (Ref. 13) indicates that ionizing radiation measured from a number of VDTs is at an insignificantly low level. Several manufacturers have produced conductive antistatic plastic material that may be placed under the bezel surrounding a cathode ray tube screen. One disadvantage of this material is that when the screen is curved, wrinkles form in the material. Another disadvantage with the use of this material are reflections, which can be alleviated if the material is antiglare. The antistatic material is ineffective unless it is grounded to some metal structure within the monitor that is in turn connected to AC safety ground. As typically only one side of the material is conductive, it is important to choose the correct side when making the ground connection.

**Table 2.8**  Guidelines for Exposure to Static Magnetic Fields[a]

| Exposure characteristics | Magnetic flux density |
|---|---|
| **Occupational** | |
| Whole working day (time-weighted average) | 200 mT ($1.58 \times 10^5$ A/m) |
| Ceiling value | 2 T |
| Limbs | 5 T |
| **General public** | |
| Continuous exposure | 40 mT ($3.17 \times 10^4$ A/m) |

[a] **Caution**: People with cardiac pacemakers and other implanted electrically activated devices, or with ferromagnetic implants, may not be adequately protected by the limits given here. The majority of cardiac pacemakers are unlikely to be affected from exposure to fields below 0.5 mT. People with some ferromagnetic implants or electrically activated devices (other than cardiac pacemakers) may be affected by fields above a few milliteslas.

When magnetic flux densities exceed 3 mT, precautions should be taken to prevent hazards from flying objects.

Analog watches, credit cards, magnetic tapes, computer disks, etc. may be adversely affected by exposures to 1 mT, but this is not a safety concern for humans.

General public. Occasional access of members to special facilities where magnetic flux densities exceed 40 mT can be allowed under appropriately controlled conditions provided that the appropriate occupational exposure limit is not exceeded.

Far better are antiglare/antistatic screens that are specifically designed for different makes of monitor. The screens are available from computer accessory suppliers and, according to the advertising material from one manufacturer, reduce the field strength by 80–90%.

## 2.8  COMPUTER PROGRAMS

The following computer programs were written using Microsoft's QuickBASIC Version 3.0. Unlike other common versions of compiled BASIC, the QuickBASIC compiler does not require line numbers and allows the use of alphanumeric labels.

The programs incorporate some of the equations contained in this chapter and Chapter 7 and are useful in EMC prediction and problem solving as well as antenna calibration using a PC. The programs are available as freeware from EMC Consulting Inc.

### 2.8.1  Computer Programs for Radiation from Wires

These programs may be used for calculation of the electric field or magnetic field or power radiated from a wire or cable, depending on the equations used in the program. The equations cover the radiation from a current element or electric dipole, a current loop, and a wire above a ground plane. These equations may be used to model the following cable configurations:

1.  An electrically short length of wire/cable either that is some distance from a ground plane or for which the fields in close proximity are required. The wire/cable must be disconnected at one or both ends.
2.  A wire or cable forming a loop with either a return conductor or its electromagnetic image in a ground plane. The length of the wire and the distance between the wire and the return path must be much less than a wavelength. When modeling the wire or cable above a ground plane, enter twice the height of the wire or cable above a ground plane for the width of the frame antenna.
3.  A transmission line comprised of a wire/cable and a return path. The return path may be either a conductor or formed by the wire/cable electromagnetic image in a ground plane. The cable/wire must be of a resonant length, i.e., 1/4, 3/4, 1-1/4, . . . wavelength when only one end is connected to ground and 1/2, 1, 1-1/2, . . . wavelength when both are connected to ground. The distance between the wires/cables must be much less than a wavelength. When the return path is the electromagnetic image in a ground plane, use twice the height of the wire above the ground plane when entering the distance between conductors.

2.8.1.1  Computer Program for the Electric Field, Magnetic Field, and Wave
          Impedance for a Current Element (Electric Dipole)

```
'Initialize constants
C=3E+08
PI=3.14159
'Initialize input variables
F=2000: CUR=1: R=1: THETA=90
GOTO ComputeCurrentElement:
CurrentElementMenu:
CLS
PRINT
PRINT "CURRENT ELEMENT
```

```
PRINT "Schelkunoff & Friis, Antennas, Theory and Practice
PRINT "Wiley, 1952, p120
PRINT
PRINT "INPUTS:"
PRINT "[F] FREQUENCY (F):",F;" Hz"
PRINT "[C] LOOP CURRENT (I): ",CUR;" amps"
PRINT "[D] DISTANCE FROM ELEMENT:",R;" meters"
PRINT "[A] ANGLE:",,THETA;" degrees"
PRINT "[X] End program"
PRINT "Which variable do you wish to change (F,C,etc)?"
PRINT
PRINT "OUTPUTS:"
PRINT "MAGNETIC FIELD     :",HH,"A/m"
PRINT "ELECTRIC FIELD     :",EE,"V/m"
PRINT "WAVE IMPEDANCE      :",ZW,"Ohms"
CurrentElementLoop:
SEL$=INKEY$
IF LEN(SEL$)>0 THEN LOCATE 12,1: PRINT SPACE$(70): LOCATE 12,4
IF (SEL$="F") OR (SEL$="f") THEN INPUT " Enter frequency ",F
IF (SEL$="C") OR (SEL$="c") THEN INPUT " Enter loop current ",CUR
IF (SEL$="D") OR (SEL$="d") THEN INPUT " Enter distance from element ",R
IF (SEL$="A") OR (SEL$="a") THEN INPUT " Enter angle ",THETA
IF (SEL$="X") OR (SEL$="x") THEN END
IF LEN(SEL$)>0 THEN GOTO ComputeCurrentElement:
GOTO CurrentElementLoop:
ComputeCurrentElement:
  LAMBDA=C/F : BETA=2*PI/LAMBDA
  ERE=1-LAMBDA^2/(4*PI^2*R^2) : EIM=LAMBDA/(2*PI*R)
  EMAG=SQR(ERE^2+EIM^2)
  ET=377*CUR/(2*LAMBDA*R)*EMAG*SIN(THETA*PI/180)
  XMAG=SQR(1+(LAMBDA/(2*PI*R))^2)
  HH=CUR/(2*LAMBDA*R)*XMAG*SIN(THETA*PI/180)
  ER=377*CUR/(2*PI*R^2)*XMAG*COS(THETA*PI/180)
  IF THETA=90 THEN ZW=ET/HH ELSE ZW=0
  EE=SQR(ET^2+ER^2)
GOTO CurrentElementMenu:
```

2.8.1.2   Computer Program for the Electric Field, Magnetic Field, and Wave
          Impedance for a Current Loop (Frame Antenna)

```
'Initialize constants
C=3E+08: RC=377: PI=3.14159
'Initialize variables
F=20000: I=1: L=1: W=1: R=1: THETAD=90
GOTO ComputeCurrentLoop:
CurrentLoopMenu:
CLS
PRINT
PRINT "ELECTRIC FIELD FROM FRAME ANTENNA"
```

```
PRINT "Schelkunoff, Antennas, Theory and Practice, pg. 320"
PRINT
PRINT "INPUTS:"
PRINT "[F] FREQUENCY:",,F;" Hz"
PRINT "[C] LOOP CURRENT:",,I;" amps"
PRINT "[L] LENGTH OF FRAME ANTENNA:",,';" m"
PRINT "[W] WIDTH OF FRAME ANTENNA:",W;" m"
PRINT "[D] DISTANCE FROM FRAME TO MEAS. PT.:",R;" m"
PRINT "[A] ENTER FRAME ANGLE:",,THETAD;" degrees"
PRINT "[X] End program
PRINT "Which variable do you wish to change (F,C,etc)?"
PRINT
PRINT "OUTPUTS:"
PRINT "ELECTRIC FIELD=",EPHI;" V/m"
PRINT "MAGNETIC FIELD=",HMAG;" A/m"
PRINT "WAVE IMPEDANCE=",ZW;" Ohms"
'
CurrentLoopLoop:
SEL$=INKEY$
IF LEN(SEL$)>0 THEN LOCATE 13,1
IF LEN(SEL$)>0 THEN PRINT SPACE$(70)
LOCATE 13,4
IF (SEL$="C") OR (SEL$="c") THEN INPUT " Enter loop current ", I
IF (SEL$="F") OR (SEL$="f") THEN INPUT " Enter frequency ",F
IF (SEL$="L") OR (SEL$="l") THEN INPUT " Enter length of frame antenna ",L
IF (SEL$="W") OR (SEL$="w") THEN INPUT " Enter width of frame antenna ",W
IF (SEL$="D") OR (SEL$="d") THEN INPUT " Enter distance ",R
IF (SEL$="A") OR (SEL$="a") THEN INPUT " Enter frame angle ",THETAD
IF (SEL$="X") OR (SEL$="x") THEN END
IF LEN(SEL$)>0 THEN GOTO ComputeCurrentLoop:
GOTO CurrentLoopLoop:
ComputeCurrentLoop:
S=L*W
BETA=2*PI*F/C
THETAR=THETAD*PI/180
EPHI=RC*BETA^2*I*S*SQR(1+(1/(BETA*R)^2))*SIN(THETAR)/(4*PI*R)
HTHETA                    =                    BETA^2*I*S*SQR((1-
(1/(BETA*R)^2))^2+(1/(BETA*R)^2))*SIN(THETAR)/(4*PI*R)
HR=BETA*I*S*SQR(1+1/(BETA*R)^2)*COS(THETAR)/(2*PI*R^2)
HMAG=SQR(HTHETA^2+HR^2)
IF THETAD=90 THEN ZW=EPHI/HMAG ELSE ZW=0
GOTO CurrentLoopMenu:
```

2.8.1.3   Computer Program for the Radiation from a Resonant
          Transmission Line

```
WireAboveGroundPlane:
'CLS:LOCATE 4,1
'PRINT "RADIATION FROM A TRANSMISSION LINE"
'PRINT
```

```
'PRINT " P=30*BETA^2*BB^2*I^2"
'PRINT " H=SQR(P*K/(4*PI*R^2*ZW)"
'PRINT " BETA=2*PI*F/C
'PRINT " PI=3.14159
'PRINT " K = DAVID'S CONSTANT = 1.5
'PRINT " SPEED OF LIGHT IN VACUUM, (C = 3E08 m/s)"
'PRINT
'GOSUB Spacebar:
GOTO ComputeWireAboveGroundPlane:
WireAboveGroundPlaneMenu::
CLS: PRINT
PRINT "WIRE ABOVE A GROUND PLANE
PRINT
PRINT "INPUTS:"
PRINT
PRINT "[F] FREQUENCY (F):",,F;" Hz"
PRINT "[R] DIAMETER OF WIRE (D):",,D;" metres"
PRINT "[C] LINE CURRENT (I):",,I;" amps"
PRINT "[D] DISTANCE FROM WIRE TO MEAS. PT.:",R;" metres"
PRINT "[H] DISTANCE BETWEEN CONDUCTORS"
PRINT "   OR HEIGHT ABOVE GROUND PLANE:",BB;" metres"
PRINT "[X] Exit to Main Menu"
PRINT "Which parameter would you like to change (A,B,..)?"
PRINT
PRINT "OUTPUTS:"
PRINT
PRINT "POWER= ",,P;"W/m^2"
PRINT "MAGNETIC FIELD= ",H;" A/m"
PRINT "ZW= ",,ZW;"Ohms"
PRINT
PRINT


WireAboveGroundPlaneLoop::
SEL$=INKEY$
IF LEN(SEL$)>0 THEN LOCATE 12,1
IF LEN(SEL$)>0 THEN PRINT SPACE$(70)
LOCATE 12,4
IF (SEL$="F") OR (SEL$="f") THEN INPUT " Enter frequency ",F
IF (SEL$="C") OR (SEL$="c") THEN INPUT " Enter loop current ",I
IF (SEL$="R") OR (SEL$="r") THEN INPUT " Enter diameter of wire ",D
IF (SEL$="D") OR (SEL$="d") THEN INPUT " Enter distance ",R
IF (SEL$="H") OR (SEL$="h") THEN INPUT " Enter distance or height ",BB
IF (SEL$="X") OR (SEL$="x") GOTO MainMenu:
IF LEN(SEL$)>0 GOTO ComputeWireAboveGroundPlane:
GOTO WireAboveGroundPlaneLoop::

ComputeWireAboveGroundPlane:
GOSUB CheckVariables:
K=1.5
```

```
LAMBDA=C/F
IF R>LAMBDA/(2*PI) THEN
  ZW=377
ELSE
  ZO=138*(LOG((4*BB)/D)/LOG(10))
  ZW=(R/(LAMBDA/(2*PI)))*(ZO-377)+377
END IF
P=30*BETA^2*BB^2*I^2
H=SQR(P*K/(4*PI*R^2*ZW))
GOTO WireAboveGroundPlaneMenu:
```

## 2.8.2   Computer Programs for Coupling to Wires/Cables

These programs calculate the current flow in the victim wire/cable based on an incident E or H field. The equations used in the programs cover the following cable configurations:

1.  A wire or cable forming a loop with either a return conductor or the electromagnetic image of the cable or wire in the ground plane. Where a wire above a ground plane is modeled, enter the height of the cable above the ground plane. For a wire or cable, enter one-half the distance between the two conductors for the height. Where the height ($H$) or length ($L$) is much less than the wavelength, use the inductance of a rectangular loop and enter (I) in the program. (I) may also be used to calculate the receiving properties of a loop antenna when the load impedance is set to zero. Where the length is much greater than the height and the height is less than half the wavelength, use transmission line theory and enter (T) in the program.
2.  Where a cable is less than one-fourth of a wavelength in length, is disconnected from ground at one end, and terminates on a metal enclosure at the other end, enter (M) in the program. (M) may be used to calculate the receiving properties of a nonresonant monopole antenna.

A computer program for the coupling between wires or cables is provided at the end of Chapter 4, on cable coupling and crosstalk.

### 2.8.2.1   Computer Program for the Receiving Properties of a Loop (I)

```
'Initialize constants
C=3E+08: RC=377:PI=3.14159
'Initialize variables
Z0=50: Z1=50: H=1: LL=1: A=.006: F=20000: IO=1
GOTO INDcompute:
InductanceInputs:
CLS
PRINT "LINE CURRENTS USING INDUCTANCE"
PRINT "Equation 8 from Taylor & Castillo, IEEE, Vol. EMC-20, No. 4, Nov. 1978"
PRINT "INPUTS:
PRINT "[F] FREQUENCY",,F;"Hz
PRINT "[S] SOURCE IMPEDANCE",,Z0;"Ohms
PRINT "[T] LOAD IMPEDANCE",,Z1;"Ohms
PRINT "[Z] HEIGHT",,,H;"metres
PRINT "[L] LENGTH",,,LL;"metres
PRINT "[D] DIAMETER OF WIRE",,A;"metres
```

```
IF XX$="?" THEN
    PRINT "[C] CURRENT IN TERMINATION (Amps or ?):"," ? = Refer to OUTPUT"
    PRINT "[H] MAGNETIC FIELD",,MF;"Amps/m"
ELSE
    PRINT "[C] CURRENT IN TERMINATION (Amps or ?):", IO;" Amps
END IF
PRINT "[X] End program"
PRINT "Which parameter would you like to change (A,B,..)?"
PRINT
PRINT
PRINT "OUTPUT:"
PRINT "INDUCTANCE OF RECTANGULAR LOOP = ",L;"Henrys"
PRINT "FLUX DENSITY        = ",B;"Webers/m^2"
MF=B/UO
IF XX$="?" THEN PRINT "CURRENT IN TERMINATION (Amps) = ",IO;"A"
IF XX$<>"?" THEN PRINT "MAGNETIC FIELD       = ",MF;"A/m"
INDLoop:
SEL$=INKEY$
IF LEN(SEL$)>0 THEN LOCATE 13,1: PRINT SPACE$(70):LOCATE 13,1
IF (SEL$="S") OR (SEL$="s") THEN INPUT "Enter source impedance ",Z0
IF (SEL$="T") OR (SEL$="t") THEN INPUT "Enter source impedance ",Z1
IF (SEL$="B") OR (SEL$="b") THEN GOTO CNGEXX::
IF (SEL$="Z") OR (SEL$="z") THEN INPUT "Enter height ",H
IF (SEL$="L") OR (SEL$="l") THEN INPUT "Enter length ",LL
IF (SEL$="D") OR (SEL$="d") THEN INPUT "Enter diameter of wire ",A
IF (SEL$="F") OR (SEL$="f") THEN INPUT "Enter Frequency ",F
IF (SEL$="C") OR (SEL$="c") THEN GOTO CNGEXX:
IF (SEL$="H") OR (SEL$="h") THEN INPUT "Enter magnetic field ",MF
IF (SEL$="X") OR (SEL$="x") THEN END
IF LEN(SEL$)>0 THEN GOTO INDcompute:
GOTO INDLoop:
CNGEXX:
INPUT "Enter CURRENT IN TERMINATION (Amps or ?): ",XX$
IF XX$="?" THEN
    LOCATE 13,1
    PRINT "Current value for MAGNETIC FIELD is:";MF;"Amps/m"
    INPUT "Enter new value:";MF
    GOTO INDcompute:
END IF
IO=VAL(XX$)
GOTO INDcompute:
CNGEMF:
    LOCATE 13,1
    PRINT "Current value for MAGNETIC FIELD is:";MF;"Amps/m"
    INPUT "Enter new value:";MF
GOTO INDcompute:
INDcompute:
    UO=4*PI*.0000001
    D=SQR((2*H)^2+LL^2)
```

```
AA=LL*LOG(4*H*LL/(A*(LL+D)))
BB=2*H*LOG(4*H*LL/(A*(2*H+D)))
CC=2*D-7/4*(LL+2*H)
L=UO/PI*(AA+BB+CC)
ZZ=SQR((2*(Z0+Z1))^2+(2*PI*F*L)^2)
IF XX$="?" THEN
   B=MF*UO
   IO=B*2*PI*F*4*H*LL/ZZ
ELSE
   B=IO*ZZ/(2*PI*F*4*H*LL)
   END IF
GOTO InductanceInputs:
```

2.8.2.2   Computer Program for Calculating the Receiving Properties
          of a Transmission Line (T)

```
'Initialize constants
C=3E+08: RC=377:PI=3.14159
'Initialize variables
IO=1: EO=1: ZI=50: H=1: F=20000: A=.003: IsItCurrent=1
GOTO ComputeTransmissionLine:
TransmissionLineInputs:
CLS
PRINT "LINE CURRENTS USING TRANSMISSION LINE THEORY"
PRINT "Taylor & Castillo: Eqn. 6,10 from IEEE, Vol. EMC-20, No. 4, Nov, 1978"
PRINT
IF IsItCurrent=1 THEN
   PRINT "[A] WIRE CURRENT (Amps or ?) :";IO
ELSE
   PRINT "[A] ELECTRIC FIELD (V/m or ?) :";EO
END IF
PRINT "[R] RESISTANCE OF LINE :"; ZI;"Ohms"
PRINT "[H] HEIGHT :";H;"metres"
PRINT "[F] FREQUENCY :";F;"Hz"
PRINT "[D] DIAMETER OF WIRE :";A;"metres"
PRINT "[X] End program
PRINT "Which parameter would you like to change (A,B,..)?"
PRINT
PRINT
PRINT "OUTPUT:"
IF IsItCurrent=1 THEN PRINT "ELECTRIC FIELD (V/m) = ";EO;" V/m"
IF IsItCurrent=0 THEN PRINT "WIRE CURRENT = ";IO;" Amps"
PRINT "CHARACTERISTIC IMPEDANCE = ";ZC;"Ohms"
TLoop:
SEL$=INKEY$
IF LEN(SEL$)>0 THEN LOCATE 10,1: PRINT SPACE$(70): LOCATE 10,1
IF (SEL$="A") OR (SEL$="a") THEN GOTO ALTERXXX:
IF (SEL$="R") OR (SEL$="r") THEN INPUT "Enter impedance of line ",ZI
IF (SEL$="H") OR (SEL$="h") THEN INPUT "Enter height ",H
IF (SEL$="F") OR (SEL$="f") THEN INPUT "Enter frequency ",F
```

```
IF (SEL$="D") OR (SEL$="d") THEN INPUT "Enter diameter ",A
IF (SEL$="X") OR (SEL$="x") THEN END
IF LEN(SEL$)>0 THEN GOTO ComputeTransmissionLine:
GOTO TLoop:
ALTERXXX:
LOCATE 10,1
IF IsItCurrent=1 THEN GOTO ALTERXXX2:
INPUT "ELECTRIC FIELD (V/m or ?): ",XXX$
IF XXX$="?" THEN
  PRINT "Current value for WIRE CURRENT is:";IO;"Amps"
  INPUT "Enter new value:",IO
  IsItCurrent=1
ELSE
  EO=VAL(XXX$)
END IF
GOTO ComputeTransmissionLine:
ALTERXXX2:
LOCATE 10,1
INPUT "WIRE CURRENT is (Amps or ?): ",XXX$
IF XXX$="?" THEN
  PRINT "Current value for ELECTRIC FIELD is:";EO;"V/m"
  INPUT "Enter new value:",EO
  IsItCurrent=0
ELSE
  IO=VAL(XXX$)
END IF
GOTO ComputeTransmissionLine:
ComputeTransmissionLine:
UO=4*PI*.0000001
LE=(UO/PI)*LOG(H/A)
K=2*PI*F/C
ZC1=(ZI/1000+2*PI*F*LE)*2*PI*F*LE/(K^2)
ZC=SQR(ZC1)
IF IsItCurrent=1 THEN EO=IO*ZC/4/H
IF IsItCurrent=0 THEN IO=4*EO*H/ZC
GOTO TransmissionLineInputs:
```

2.8.2.3   Computer Program for Calculating the Receiving Properties
          of a Nonresonant Monopole Antenna

```
CLS
PRINT "Open circuit voltage or current into a load caused by an E field
PRINT "incident on a short length of wire/cable (length ≪ lambda) located
PRINT "at least 1 m from a ground plane and terminated to ground at one end
PRINT "only.
GOSUB Spacebar:
'Initialize constants
C=3E+08: PI=3.14159
'Initialize variables
EO=1: F=2000: H=1: A=.001: RL=50: ZW=377
```

```
GOTO Compute:
Inputs:
CLS
PRINT " INPUTS:
PRINT " [A] Electric field",EO;"V/m"
PRINT " [B] Frequency",,F;"Hz"
PRINT " [C] Length of wire",H;"metres"
PRINT " [D] Diameter of wire",A;"metres"
PRINT " [E] Load Resistance",RL;"Ohms"
PRINT " [F] Wave Impedance",ZW;"Ohms"
PRINT " [X] End program
PRINT " Please select variable you wish to change (A,B,etc)
PRINT "
PRINT " OUTPUTS:"
IF 2*PI*H/LAMBDA<1 THEN GOSUB Outputs1:
IF (2*PI*H/LAMBDA>1) AND (2*PI/LAMBDA<2.75) THEN GOSUB Outputs2:
IF 2*PI*H/LAMBDA>2.75 THEN GOSUB Outputs3:
LOCATE 9,3
SelectLoop:
SEL$=INKEY$
IF LEN(SEL$)>0 THEN PRINT SPACE$(80)
LOCATE 9,3
IF (SEL$="A") OR (SEL$="a") THEN INPUT "Enter electric field",EO
IF (SEL$="B") OR (SEL$="b") THEN INPUT "Enter Frequency",F
IF (SEL$="C") OR (SEL$="c") THEN INPUT "Enter Length of wire",H
IF (SEL$="D") OR (SEL$="d") THEN INPUT "Enter Diameter of wire",A
IF (SEL$="E") OR (SEL$="e") THEN INPUT "Enter Load Resistance",RL
IF (SEL$="F") OR (SEL$="f") THEN INPUT "Enter Wave Impedance",ZW
IF (SEL$="X") OR (SEL$="x") THEN END
IF LEN(SEL$)>0 THEN GOTO Compute:
GOTO SelectLoop:
Compute:
SEL$=""
LAMBDA=C/F
BETA=2*PI/LAMBDA
HH=BETA*H
OMEGA=2*(LOG(2*H/A))/LOG(10)
RR=197.5*((2*H)/LAMBDA)^2
ZORE=10*(HH^2)
TEMP1=(OMEGA-2)^2
TEMP2=2*(LOG(2)/LOG(10))
ZOIM=(30/HH)*(TEMP1/(1+TEMP2))
ZO=SQR(ZORE^2+ZOIM^2)
K=(1-(ZO/RL))/(1+(ZO/RL))
ALPHA=1-K^2
G=1.5*ALPHA
Amax=G*(LAMBDA^2/(4*PI))
HE1=(Amax*RR)/(2*ZW)
HE2=(EO*LAMBDA)/(2*PI)*TAN(PI*H/LAMBDA)
```

```
IF RL>1000000! THEN VO=EO*HE2 ELSE VO=EO*HE1
IF RL>1000000! THEN HE=HE2 ELSE HE=HE1
VL=EO*HE1
IL=VL/RL
IS=VO*TAN(2*PI*H/LAMBDA)/60
GOTO Inputs:
Outputs1:
PRINT " Open circuit voltage",ABS(VO);"Volts"
PRINT " Short circuit current",ABS(IS);"Amps"
PRINT " Effective height (He)",ABS(HE);"metres"
PRINT " Self impedance (Zo)",ABS(ZO);"Ohms"
PRINT " Current in load (IL)",ABS(IL);"Amps"
PRINT " Voltage across load (VL)",ABS(VL);"Volts"
RETURN
Outputs2:
PRINT "        **RESONANT**
PRINT "   See Transmission Lines, Antennas and Waveguides
PRINT "       (pp. 98,166)
PRINT "
PRINT " Omega=";OMEGA
PRINT " Beta=";HH
PRINT " Open Circuit Voltage=";VO
RETURN
Outputs3:
PRINT "
PRINT "   The frequency and length of wire are such that:"
PRINT "       2 *PI*H/LAMBDA>2.75
PRINT
PRINT "   This program is not applicable for such conditions.
RETURN
Spacebar:
LOCATE 15,1
PRINT "Press the spacebar to continue or X to end
SpacebarLoop:
Spcbr$=INKEY$
IF Spcbr$=CHR$(32) THEN GOTO Returner:
IF (Spcbr$="x") OR (Spcbr$="X") THEN END
GOTO SpacebarLoop:
Returner:
RETURN
```

## REFERENCES

1. Mathew Zaret. Outline of Electromagnetic Theory. Regents, New York, 1965.
2. J.E. Bridges. Study of low-frequency fields for coaxial and twisted pair cables. Proceedings of the 10th Tri-Service Conference on Electromagnetic Compatibility, Chicago, IL, Nov. 1964.
3. J. Moser. Predicting the magnetic fields from a twisted pair cable. IEEE trans. on EMC, Vol. EMC 10, No. 3, Sept. 1968.
4. S. Shenfield. Magnetic fields of twisted wire pairs. IEEE Trans. on EMC Vol. EMC 11, No. 4, Nov. 1969.

5.  A. Richtscheid. Calculation of the radiation resistance of loop antennas with sinusoidal current distribution. IEEE Transactions on Antennas and Propagation, November 1976.

6.  R.W.P. King, H.R. Mimno, and A.H. Wing. Transmission lines, antennas, and wave guides. Dover, New York, 1965.

7.  G.H. Brown, and O.M. Woodward Jr. Experimentally determined radiation characteristics of conical and triangular antennas. RCA Review 13, No. 4, Dec. 1952.

8.  Health Aspects of Radio Frequency and Microwave Exposure. Part 2. Health and Welfare Canada, 78-EHD-22, Ottawa, 1978.

9.  IEEE Standard for Safety Levels with Respect to Human Exposure to Radio Frequency Electromagnetic Fields, 3 kHz to 300 GHz, IEEE C95.1—1991.

10. Safety Code 6. Canada Health and Welfare, 99-EHD-237 (1999).

11. Private Communication with N. Sultan, formerly with Canadian Astronautics.

12. B. Spinner, J. Purdham, and K. Marha. The Case for Concern about Very Low Frequency Fields from Visual Display Terminals: The Need for Further Research and Shielding of VDT's. From the Canadian Centre for Occupational Health and Safety, Hamilton, Ontario, Canada.

13. K. Marha. Emissions from VDTs: Possible Biological Effects and Guidelines. Canadian Centre for Occupational Health and Safety, Hamilton, Ontario, Canada.

14. K. Marha. VLF—Very Low Frequency Fields near VDTs and an Example of Their Removal. Canadian Centre for Occupational Health and Safety, Hamilton, Ontario, Canada.

15. Additional information on antenna factor and calibration is available from: Ezra B. Larsen. NBS, Calibration and Meaning of Antenna Actors and Gain for EMI Antennas. ITEM 1986.

16. A comparison of electric field-strength standards for the frequency range of 30–1000 MHz. Heinrch Garn. IEEE Transactions on Electromagnetic Compatibility, November 1997, Vol. 39, No. 4.

17. Code of federal regulations 47. Chapter 1, Part 1, 1.1310. October 1, 1997.

• Additional current element and current loop equations derived from: Schelkunoff and Friis. Antennas Theory and Practice. Wiley, New York, 1952.

# 3
# Typical Sources and Characteristics of Radiated and Conducted Emissions

## 3.1 INTRODUCTION TO NOISE SOURCES

Some typical sources of electromagnetic emissions are transmitters, pulse generators, oscillators, digital logic circuits, switching power supplies and converters, relays, motors, and line drivers. The majority of unintentional emissions are composed of a great number of frequencies typically caused by switching and transient noise. Design engineers usually measure digital and impulsive signals in the time domain with an oscilloscope or logic analyzer, whereas EMI specifications and measurements are made in the frequency domain. Either Fourier analysis or Laplace transform may be used to compute the amplitude of the frequencies contained within the impulsive noise (the spectrum occupancy), or a spectrum analyzer or receiver may be used to measure the frequency components. Making a measurement is easier and more accurate, especially when the waveform is complex in shape.

During breadboarding and when the first prototype of equipment is available, an assessment of the level and frequencies generated by the circuit is important, both in achieving intra-equipment EMC and in meeting EMC specifications. The same information is also useful when locating the source in an EMI investigation. Assuming a spectrum analyzer is not available, some simple technique for assessing the frequency occupancy based on the time-domain signal is desirable. This chapter illustrates the frequency occupancy of common waveforms with simple formulae for calculating the amplitude versus frequency characteristics, introduces a ''quick transform technique,'' and presents a case study of measurements made on a voltage converter (which is a common source of noise).

### 3.1.1 Harmonically Related Noise from Single and Periodic Pulses

For a periodic pulse, the low-frequency spectrum is comprised of the pulse repetition rate (PRR) and its harmonics. Figure 3.1 shows a symmetrical trapezoidal periodic pulse train where $t_o + t_r = T/2$. Figure 3.2 shows the envelope approximation of the spectral lines shown at $1/T$, $3/T$, and $5/T$ (the spectral lines at $2/T$ and $4/T$ are at zero amplitude).

Figure 3.3 shows the spectral lines for a pulse where $t_o + t_r = T/5$. The negative amplitudes of the $\sin x/x$ envelope shown in Figures 3.2, 3.3, and 3.4 may be inverted, because we are only interested in the amplitude and frequency and an envelope approximation made of the decrease in the amplitude of the spectral line with increasing frequency, as shown in all the subsequent figures.

The spectrum analyzer inverts the negative amplitude and, depending on the resolution bandwidth and sweeptime settings, may display the spectral lines. When the pulse train is unmodulated, the scan starts at 0 Hz.

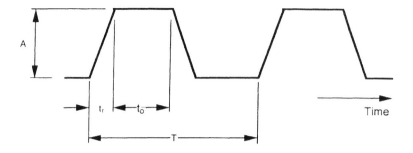

By Fourier analysis the amplitude at the frequency of the
nth harmonic is given by:

$$C_n = 2\,A \cdot \frac{(t_o + t_r)}{T} \cdot \frac{\sin\left[\pi n\,(t_o + t_r)/T\right]}{\pi n\,(t_o + t_r)/T} \cdot \frac{\sin\left(\pi n t_r/T\right)}{\pi n t_r/T} \quad \text{the argument of sine is in radians}$$

**Figure 3.1**   Symmetrical trapezoidal periodic pulse. (From Ref. 1.)

The pulse width can be found from the frequency at the sidelobe corresponding to $1/(t_r + t_o)$ and the pulse repetition rate from the spacing between the spectral lines corresponding to $1/T$, as shown in Figure 3.3. Assuming $T = 1$ μs for the pulse illustrated in Figure 3.1, $t_r + t_o = 0.5$ μs (i.e., $T/2$) and the sidelobe frequency will be 2 MHz (i.e., 1/0.5 μs).

One common error is often made in finding the PRR from a spectrum analyzer measurement for a waveform with a mark-to-space ratio of unity, that is, the symmetrical pulse of Figure 3.1. The sine $x/x$ function envelope is shown for convenience only in Figure 3.1 and is not displayed in a spectrum analyzer measurement; therefore the location of the sidelobe is not clear from a spectrum analyzer display. The spectral lines at the sidelobes are at zero amplitude, and the negative line at $3/T$ is inverted by the spectrum analyzer. Thus the PRR is commonly taken as the spacing between the two positive spectral lines, which is $2/T$, i.e., a PRR twice as high as the signal under measurement. Section 9.2.2 deals with the use of a spectrum analyzer in some detail. The best method of familiarization with the spectrum analyzer and its representation of a periodic pulse train, and one I know the customer support people at Hewlett-Packard have used, is to apply a known signal from a pulse or square-wave generator and examine the resultant spectral lines.

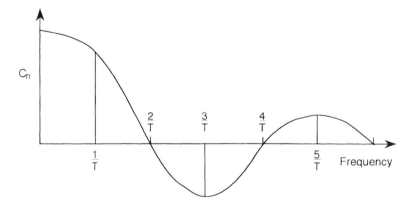

**Figure 3.2**   Low-frequency harmonic spectrum for $t_r + t_o = T/2$ trapezoidal pulse. (From Ref. 1.)

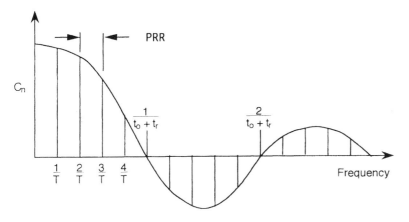

where pulse repetition rate (PRR) $= \frac{1}{T}$ and $t_o + t_r = \frac{T}{5}$

**Figure 3.3** Low-frequency harmonic spectrum for trapezoidal periodic pulse (narrowband). (From Ref. 1.)

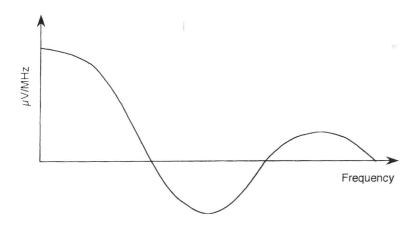

To convert to coherent broadband:

Let $t_o + t_r = d$

$f = \frac{n}{T}$

$\frac{\text{narrowband } C_n}{\text{PRR}} = 2 \, A \, d \, \frac{\sin \pi f d}{\pi f d}$ broadband

or in dB and MHz:

dB$\mu$V (narrowband) - 20 log (PRR in MHz) = dB$\mu$V/MHz (broadband)

**Figure 3.4** Low-frequency harmonic envelope for single-pulse broadband. (From Ref. 1.)

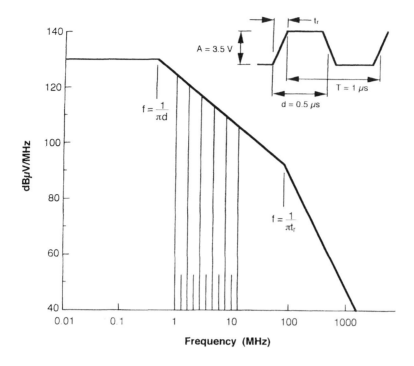

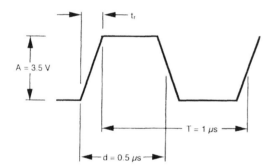

**Figure 3.5** High-frequency spectrum of a TTL-level 1-MHz clock pulse.

The periodic pulse train illustrated in Figure 3.1 generates narrowband frequencies harmonically related to the PRR. The equation that may be used to find the amplitude at the frequency of the $n$th harmonic is shown under the waveform in Figure 3.1. An example of the use of this equation is given in calculating the harmonic components of the 1-MHz transistor–transistor logic (TTL) clock pulse illustrated in Figure 3.5.

A single pulse and pulses with low repetition rates, relative to the pulse width, generate coherent broadband noise. This noise is coherent because the frequencies generated are related to the rise and fall times and width of the pulse. A definition of *coherent* is a signal or emission in which the neighboring frequency elements are related or well defined in both amplitude and phase.

The definition of *broadband* as a single pulse or low-repetition-rate pulse is sometimes used in commercial EMC specifications with a specified maximum repetition rate, typically 10 kHz, below which a source is considered broadband.

A commonly accepted definition of *broadband* relates to a measurement bandwidth: noise that has a spectrum broad in width compared to the nominal bandwidth of a receiver or spectrum analyzer. Chapter 9 discusses the definition and measurement of narrowband and broadband noise in some detail.

The equation shown in Figure 3.4 allows the conversion of the narrowband harmonic amplitude from the periodic pulse into the broadband coherent amplitude, measured in µV/MHZ, generated by a single pulse.

Figure 3.5 shows the high-frequency spectrum of a trapezoidal periodic pulse train. The first breakpoint comes at $1/({}^1t_r + t_o)$ and this can also be seen in Figure 3.4; the second breakpoint comes at $1/{}^1t_r$. The amplitude at frequencies below $1/({}^1t_r + t_o)$ is equal to $(2A) \times (t_r + t_o)/T$ whereas for the single pulse it is $(2 \text{ A}) \times (t_r + t_o)$.

The high-frequency spectrum of a 3.5-V 1-MHz clock with 3.5-ns rise and fall times, which is a typical TTL level, is illustrated in Figure 3.5. We shall use this 1-MHz clock as an example in calculating the amplitude at the harmonics and converting them into a coherent broadband-equivalent level. It should be noted that due to the relatively high PRR, the clock is almost certainly classified as a narrowband source, under the definitions given in Chapter 9. From the following equation, the amplitude of the narrowband harmonics may be calculated:

$$C_n = (2 \text{ A}) \frac{(t_o + t_r)}{T} \cdot \frac{\sin[\pi n(t_o + t_r)/T]}{\pi n(t_o + t_r)/T} \cdot \frac{\sin(\pi n t_r/T)}{\pi n t_r/T} \qquad \text{(arguments of sine in radians)}$$

The narrowband amplitude ($C_n$) at the first harmonic (the fundamental) using this equation is 2.226 V. The equations in Figure 3.4 are used to find the broadband-equivalent amplitude as follows: from narrowband $C_n$/PRR we obtain 2.226 V/1 MHz = 2.226 µV/Hz. To express the broadband level in the common broadband unit of dBµV/MHz, we convert 2.226 µV to dBµV (i.e., 6.9 dBµV) and add a 120-dB correction for the conversion from the 1-Hz to the 1-MHz reference bandwidth, which results in 127 dBµV/MHz.

A second equation using dBµ V (narrowband) and PRR in megahertz is as follows: dBµV (narrowband) − 20 log (PRR in MHz). The amplitude at 1 MHz (the first harmonic) equals 2.226 V, which, expressed in dBµV, is 127 dBµV. The broadband level is therefore 127 dBµV − 20 log 1 = 127 dBµV/MHz. The broadband amplitudes at other harmonics are 53 dBµV/MHz at the second harmonic, 117 dBµV/MHz at the third harmonic, 110 dBµV/MHz at the seventh harmonic, and 53 dBµV/MHz at the eighth harmonic, as plotted along with the envelope in Figure 3.6.

The equations for the calculation of the spectrum occupancy for square, rectangular, and many other single- or low-repetition-rate event pulse shapes are shown in Figures 3.7–3.13. The curves in these figures result from connecting the maxima of the spectral lines envelope, as shown in Figure 3.6. These figures also contain information useful in achieving a design that limits the amplitude of the low- and high-frequency emissions from pulses. Figure 3.8, for example, shows that the low-frequency component of a trapezoidal pulse may be reduced by decreasing the duration of the pulse. However, when the pulse width is decreased, the maximum value of the rise and fall times may be limited by the pulse width. Increasing the rise and fall times decreases the frequency of the breakpoint and reduces the amplitude of the high-frequency components.

To summarize: for any pulse shape, the interference level at low frequencies depends only on the area under the pulse, while at higher frequencies the level depends on the number (PRR) and steepness (rise and fall times) of the slopes.

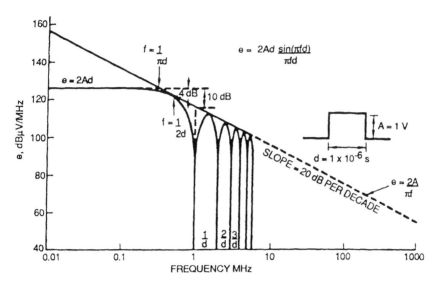

**Figure 3.6**  Interference level for a 1-V 1-μs rectangular pulse. (From Ref. 1.)

Figure 3.9 illustrates the rate of reduction in amplitude for frequencies above the breakpoint for eight common pulse shapes. The Gaussian pulse has the most rounded contours and exhibits the steepest slope in amplitude reduction with increasing frequency beyond the breakpoint. Thus it is not the rise time alone that controls the amplitude of the high-frequency components.

A logic or similar pulse may often exhibit rounding at the start and end of the rising/falling edge, with a subsequent reduction in high-frequency emissions, compared to a pure trape-

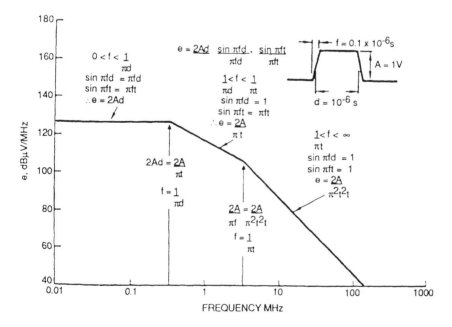

**Figure 3.7**  Interference level for a 1-V 1-μs trapezoidal pulse (From Ref. 1.)

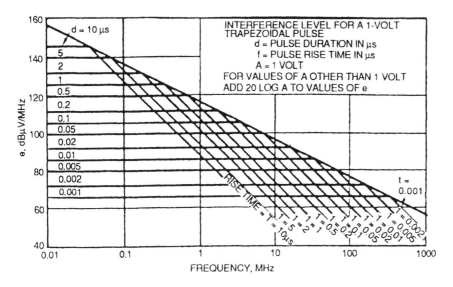

**Figure 3.8** Trapezoidal pulse interference. (From Ref. 1.)

zoidal pulse. Conversely, a pulse with overshoot and/or undershoot or with a section of the transition that exhibits a faster rate of $dV/dt$ than the overall edge will result in high-frequency emissions at greater amplitude. Where possible, the edges of pulses should be rounded and the rise and fall times increased. In a printed circuit board (PCB) loaded with logic, this is a near-impossible task, and other reduction techniques described in subsequent chapters may be applicable. However, for a clock line or data bus with a high fan out (i.e., driving a large number of

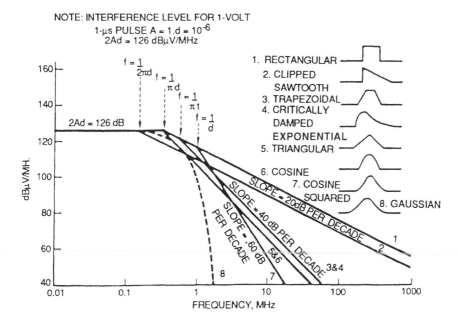

**Figure 3.9** Interference levels for eight common pulse shapes (dBμ V/MHz versus $f$). (From Ref. 1.)

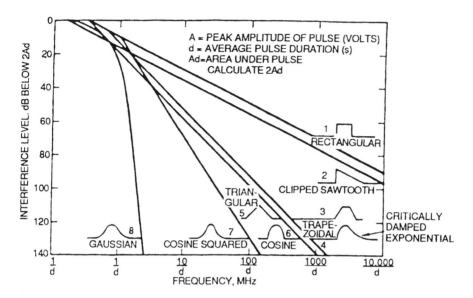

**Figure 3.10** Interference levels for various pulse shapes (dB below $2Ad$ vs. $f$). (From Ref. 1.)

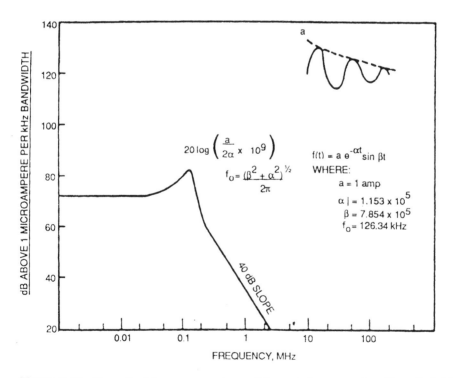

**Figure 3.11** Normalized frequency spectrum of damped sinewave pulse. (From Ref. 1.)

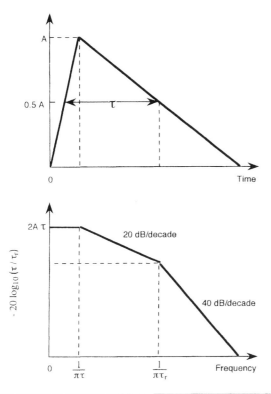

| Source | $\tau_r$ | $\tau$ | $1/\pi\tau$ | $1/\pi\tau_r$ | A | $2A\tau$ |
|---|---|---|---|---|---|---|
| EMP | 10 ns | 250 ns | 1.3 MHz | 30 MHz | 50 kV/m | 25 kV/m/MHz (25 V/m/kHz) |
| Space plasma arc discharge (MIL-STD-1541) | 15 ns | 40 ns | 8 MHz | 20 MHz | 1200 V (at 30 cm) | 96 V/m/MHz |
| Human body ESD Model (MIL-STD-883c) | 15 ns | 100 ns | 3.2 MHz | 20 MHz | 6 A | 1.2 A/MHz |
| Lightning (MIL-STD-5087B) | 2 $\mu$s | 25 $\mu$s | 12.7 kHz | 159 kHz | 200 kA | 10 kA/kHz |

**Figure 3.12**   Time- and frequency-domain relations due to lightning, EMP, ESD, and space plasma arc discharge.

logic inputs), it may be practical to slow down and round edges by the inclusion of a series resistor or a series resistor-parallel capacitor combination. Adding the resistor in TTL-type logic will decrease the ''high''-level voltage and increase the ''low''-level and must be limited in value to typically 100Ω or less to ensure the correct operation of the logic and maintain some level of DC noise immunity. An alternative to the resistor is a ferrite bead, which introduces some inductance. Assuming the edges are not slowed down to the extent that the maximum ''high'' and minimum ''low'' levels are not reached at the repetition rate of the signal, the inclusion of the ferrite bead will not degrade the ''high'' and ''low'' logic magnitudes. One disadvantage with the bead is the possibility of resonance and peaking in the logic levels. The

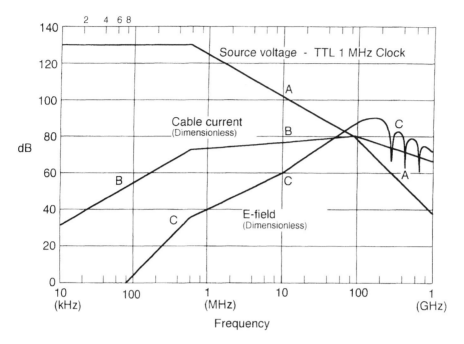

**Figure 3.13** Relative spectral content of a 1-MHz TTL clock source, the resultant interface cable common-mode current flow, and the E field generated by the cable.

addition of a 1 to 10$\Omega$ resistor in series with the inductor or a resistor across the inductor will reduce the inductor Q and minimize peaking. The characteristic of a ferrite bead, resonance, and peaking and the choice of resistor value to minimize these effects are discussed in Chapter 5.

Figure 3.10 is the same curve as in Figure 3.9 except that the $y$-axis is in ''dB below 2/$Ad$'' and the $x$-axis is plotted in frequency as a function of $d$. Thus Figure 3.9 may be conveniently used for any pulse amplitude and pulse duration. The curves of Figure 3.10 may be used for other waveshapes by considering the waveshape as made up of a number of different waveshapes, such as rectangular and triangular, and finding the total area to obtain the low-frequency content and the number and steepness of the slopes to find the high-frequency content. From the envelopes of typical waveshapes it is clear that the high-frequency components are at a lower magnitude than the fundamental but can exist at amplitudes that may be problematic to other sections of a system or equipment. It is therefore not surprising that designers are often amazed when equipment fails radiated or conducted emission requirements at frequencies orders of magnitude above the fundamental frequencies used in the equipment. When we look at the level of emissions from a 1-MHz clock with a typical TTL waveshape at frequencies of 100–300 MHz, they are indeed 40–46 dB below the fundamental.

Then why are measured emissions often at a maximum in this 100–300-MHz frequency range? Typically a combination of factors is involved. One is that common-mode noise developed across a PCB tends to increase with increasing frequency as PCB track/ground plane impedance increases. In addition, the coupling of common-mode noise may be via the capacitance between the board and the enclosure, and this coupling path decreases in impedance with increasing frequency. Where radiated emissions are caused by common-mode current flow on a cable, the radiation will typically increase with increasing frequency until cable resonances occur, and at the same time the common-mode voltage driving the current increases, due to increasing circuit impedance. Where a shielded interface cable is used, the shielding effectiveness typically decreases with increasing frequency, and the radiation, caused by current flow on the outside of the shield, increases. These coupling mechanisms, including the effect of PCB and cable resonance, are discussed in subsequent chapters.

Consider a simple model of an unshielded cable exiting a piece of equipment and terminating in a second piece of equipment, as illustrated in Figure 7.25. Let us assume that the common-mode voltages in each piece of equipment, which cause the current flow, and its coupling to the enclosure result in a 26-dB increase in current per decade increase in frequency. The envelope of the frequency components of the cable current flow, plotted for the sake of comparison against the source voltage of Figure 3.5, is as illustrated in Figure 3.13. The radiated emissions from the cable increase by approximately 20–30 dB per decade increase in frequency, dependent on the frequency, until the resonant frequency of the length of cable is reached, after which emissions either level out or reduce, as discussed in Section 7.6. This may be confirmed by use of the loop equations given in Chapter 2. The envelope of the frequency components of the resultant E field at a fixed distance from the cable is plotted for the sake of comparison in Figure 3.13. It should be emphasized that the current and E field plots are dimensionless.

Figure 3.13 illustrates that the maximum E field in the 50–500-MHz frequency range may be at an amplitude 36 dB above the fundamental, something commonly seen in practice. Often, circuit and cable resonances occur and we see one or more frequencies at which the measured field is higher than surrounding frequencies.

Sometimes the repetition rate of the impulsive noise coincides with a resonance frequency, and a sinewave or damped sinewave is generated. When a continuous sinewave is generated in this manner it gives rise to an emission that is considered narrowband in nature. It is not uncommon to find narrowband and broadband emissions together resulting from a single source of noise.

We have assumed thus far that the source of noise is the logic voltage transition. Although this voltage plays an important role in determining the radiated emission and in the crosstalk and conducted noise voltage generated by the logic, it is often the current pulse that is the predominant source of emissions.

High-current spikes flow in the logic signal and return interconnections, as the output device changes state and charges/discharges the input capacitance of the load device and removes the charge from either a base/emitter junction or a diode. During the transition, the load on the output device is therefore predominantly capacitive and, depending on the geometry of the interconnections and the logic type, the source of radiated emissions is typically a low impedance (i.e., a predominantly magnetic field source). Even though the current pulse is the predominant source of emissions, those created by the voltage transition should not be ignored in an EMC prediction, especially when the load is a high impedance with low capacitance.

Thus, in predicting the radiated emission from a PCB track or the voltage drop in either signal power or return connections, the typical magnitude of the current spikes must be known. The switching currents from various logic types and the rise and fall times with single input load are shown in Table 3.1.

Using the current pulse for a fanout of 1, as shown in Table 3.1, the current spike due to a maximum fan-out of 15 on a 74-series TTL output device is 97 mA. Large-scale integrated circuits typically incorporate buffered clock and data inputs; however, input current spikes as high as 500 mA at the logic transition have been measured. When the noise source is the current spikes at the transition and the current flow during the steady-state logic level is relatively low, the peak current pulse duration is short (typically 7 ns for TTL), and the first breakpoint in

**Table 3.1**  Input Current During Switching Transient, Output Rise and Fall Times, and Typical Input Capacitance of Different Logic Types

| Characteristic | TTL | CMOS 5 V[a] | CMOS 10 V[a] | CMOS 15 V[a] | TTL 74 L | TTL 74 S | TTL 74 H | ECL 10 K | ECL 100 K |
|---|---|---|---|---|---|---|---|---|---|
| Propagation delay (ns) | 10 | 150 | 65 | 50 | 33 | 3 | 6 | 2 | 0.75 |
| Approximate rise and fall times (ns) | 3.5 | 100 | 40 | 30 | 7 | 1.7 | 2 | 1.8 | 0.75 |
| Input current pulse (mA) (fan-out = 1) | 6.5 | 0.16 | 0.9 | 1.6 | 1 | — | 9.2 | 16 | 14 |
| Max. input capacitance (pF) | 6 | 4 | 4 | 4 | 6 | 4 | 6 | 0.1[b] | 0.1[b] |

[a] B-type CMOS.
[b] Dependent on PCB layout.

Figure 3.5, at $f = 1/\pi d$, moves from 636 kHz to 45 MHz. At the same time, the low-frequency amplitude reduces by 37 dB. This represents, therefore, another reason for the relatively high magnitude of emissions at high frequencies (above the fundamental).

### 3.1.2  Frequency Spectrum Occupancy of a Step Function

From the Laplace transform, the amplitude generated by a step function (i.e., a single event such as caused by a control line or a switch changing state) is given, for the case where the frequency of interest, $f$, is higher than $1/t_r$, approximately by

$$V = \frac{0.05 V_1 - V_2}{t_r f^2} \qquad \text{for } f > \frac{1}{t_r} \tag{3.1}$$

where

$\quad V_1$ = initial voltage

$\quad V_2$ = voltage after step

$\quad t_r$ = rise time [s]

$\quad f$ = frequency [Hz]

The envelope of the resultant curve ($V$ vs. $f$) decreases at a rate proportional to the frequency squared (40 dB/decade), the magnitude is in $V/Hz$, in order to express this in the common broadband unit of dBμ V/MHz, 120 dB must be added to the magnitude expressed in dBμV. For example, if Eq. (3.1) yields a value of 0.1 μV/Hz, then this can be converted to dBμ V/Hz:

$$0.1 \ \mu V/Hz = 20 \log\left(\frac{0.1}{1}\right) = -20 \ dB\mu V/Hz$$

and then into dBμ V/MHz:

$$-20 \ dB\mu V/Hz + 120 \ dB = 100 \ dB\mu V/MHz$$

### 3.2  FOURIER TRANSFORM METHODS AND COMPUTER PROGRAMS

''Practicing engineers tend to shun the application of complex mathematics. In EMC work, this tendency surfaces in a reluctance to perform Fourier Transforms when investigating complex waveforms'' (2). This book has been written for ''practicing engineers,'' and by one for whom this reluctance is well understood. The figures and equations contained in the previous sections of this chapter are useful in obtaining the frequency occupancy of fairly simple waveforms. What is needed, however, is a simple method of obtaining the same information for the complex waveforms so often encountered in practice. One such method is the ''quick transform procedure'' by Toia (2). This is a graphical method of obtaining the harmonics and their amplitudes from a complex time-domain waveform. The following is merely a brief review of this method. In order to use the method, the reader must, almost certainly, refer to the paper by Toia (2).

In this method, the time-domain waveform is sketched over time in cycles, the $x$-axis, with the amplitude of the waveform plotted on the $y$-axis. The time-derivative spikes are then sketched and, from a number of simple properties of delta chains, a harmonic table is constructed. In summary, the steps involved in the ''quick transform'' procedure are:

1. Sketch the time-variant waveform.
2. If it contains no spikes, sketch its time derivative.
3. Continue sketching derivatives until spikes appear. Record the highest order of derivatives sketched as ''$m$.''
4. In tabular form, record the relative amplitude and phase of each harmonic using the known spectral property of delta-function chains (i.e., for the positive chain that all frequencies are present and of the same magnitude and for the negative chain the same but at a different position in time).
5. Add spectral components of the table vectorially.
6. Divide the $n$th harmonic's amplitude by $n^m(2\pi/T)$ for each harmonic.
7. Retard the $n$th harmonics phase by $(n \times 90°)$ for each harmonic.
8. Remove the spikes from the sketch. If a residue remains, continue this procedure.

An example for a waveform more complex than a square wave is given in Ref. 2 along with an example for a smooth waveform that contains no spikes. Although the quick transform possesses shortcomings, it is a useful procedure that reduces the math to a matter of addition.

A number of programs exist for use on both mainframe and personal computers that perform fast Fourier transforms (FFTs). A typical program enables the fast Fourier transform of a vector of real data representing measurements at regular intervals in the time domain. The result is a vector of complex coefficients in the frequency domain. The phase information is retained, and thus the inverse Fourier transform can be used to convert from the frequency domain back into the time domain. The real data contains magnitude and phase or time information. The program can also accept complex data containing real and imaginary magnitudes, and the phase information is then implicit in the complex data. An inverse Fourier transform of the resultant vector of data representing values in the frequency domain returns a vector representing values in the time domain.

Common sources of noise at the unit level are AC-to-DC converters, DC-to-DC converters, and switching power supplies.

## 3.3 CASE STUDY 3.1: NOISE LEVELS GENERATED BY DC-TO-DC CONVERTERS

This section supplies data on the magnitude and type of waveform, including frequency-domain components, of the noise generated by DC-to-DC converters. The emissions from a typical converter are then compared to the military standard MIL-STD-461 limits on conducted and radiated emissions. In the investigation, the input-noise current and output-noise voltage were measured for small 15–25-W converters from four different manufacturers. The characteristics and magnitude of noise from the different manufacturers' converters were relatively similar. Differences were confined to the fundamental converter switching frequency and the maximum frequency at which emissions were still significant. As expected, all the converters produced coherent harmonically related emissions covering frequencies from the fundamental up to 490 MHz. One converter produced a single narrowband emission at 14.65 kHz and no further emissions until 2.5 MHz. However, in measurements made in the investigation and in subsequent measurements, this converter is the exception to the rule. The measured results from two typical converters from the same manufacturer are presented in the remainder of this section.

### 3.3.1   General Test Setup and Method

Two small converters, a $+ 5$-V 15-W and a $\pm 15$-V 15-W, were tested for the following:

> Conducted emission from the input terminals
> Conducted emission from the output terminals
> Radiated emission from a 2-m-long cable connected to the output terminals

Conducted emissions were tested in accordance with the methods and to the limits specified by MIL-STD-462/1 CE01 and CE03. For radiated emissions, MIL-STD-462/1 RE02 was used. The CE01 and CE03 specifications represent limits on the noise current produced by a device under test (DUT) at its input connections.

A number of different filters were tried at the input and output of the converters and to compare filter performance; although these are referenced in the following text, the attenuation performance of the filters is discussed in the continuation of this case study in Chapter 5 in the section on filters.

The converters were mounted upside down on a conducting copper ground plane. A low-impedance connection between the metal case of the converter and the ground plane was made via an EMI gasket of the "silver-coated aluminium particles in elastomer" type. The pins of the converter were then pointing up and were brought through holes in a piece of double-sided PCB material. The PCB material was used as a second ground plane that was connected to the main ground plane via 1-cm-long and 15-cm-wide tinned-copper braid. The EMI power-line filters, which were tested at both the input and output of the converters, were mounted through holes in the PCB ground plane. The PCB ground plane was soldered to the case pin of the converter. Pressure was applied between the PCB ground plane/converter/copper ground plane by clamps, which ensured a low-impedance connection at the two interfaces. The mounting method was intended to be used for initial evaluation purposes and only until an improved method was devised.

It was found during testing that the quality of the electrical connection of the converter case to the PCB ground plane was an important factor in the level of attenuation achieved in the high-frequency components of the output noise. With the case clamped to the PCB ground plane with a moderate clamping pressure, the output-noise voltage with the least effective output filter in the circuit was 40 mV. At a higher clamping pressure, the noise voltage reduced to 5 mV. For the most effective filter, the GK2AA-SO8, the high-frequency component reduced to 0.25 mV, and this residual level may be attributed to radiated coupling between the input and output wiring.

The MIL-STD-462 CE01-1 test method was used to measure the conducted input noise with and without input filters. The current flowing out of each of the input connections is measured into a specified impedance between the input connection and the ground plane on which the DUT is mounted. In the majority of military or space EMC requirements, the impedance is that of a 10-μF RF feedthrough capacitor, as described in MIL-STD-462.

The CE01 limit is applicable to narrowband emissions only in the frequency range 30 Hz to 15 kHz. CE03 requirements are for narrowband and broadband emissions, and separate narrowband and broadband limits are imposed. Different measurement bandwidths are used in the narrowband and broadband measurements, and any out-of-specification emission is then characterized as either narrowband or broadband. The narrowband limit applies only to those emissions that have been determined as narrowband, and the broadband limit applies to those emissions that have been determined to be broadband. Chapter 9 describes four measurement methods commonly used in determining whether an emission is narrowband or broadband. Typical broadband and narrowband measurement bandwidths are given in Table 9.20.

The measurements were made in the frequency domain by use of a spectrum analyzer and in the time domain by use of an oscilloscope. The CE01 and CE03 test setup and test equipment is shown in Figure 3.14. Two current probes were used, a Genisco GCP 5120, to cover the frequency range from 30 Hz to 15 MHz, and the Genisco GCP 5160, used from 15 MHz to 50 MHz.

A preamplifier that provides a gain of 26 dB ±3 dB (14 kHz to 1 GHz) was connected between the current probe and the spectrum analyzer. By first omitting the preamplifier and connecting the probe directly to the spectrum analyzer, it was ascertained that the predominant low-frequency emissions are above 100 kHz; thus the lower-frequency limit of the preamplifier (14 kHz) does not affect the test results.

As a test of the type of noise emitted by the converter, the resolution bandwidth of the spectrum analyzer was increased. As expected, the measured level also increased, which indicates that the noise may be classified as broadband. Two +5-V converters from the same manufacturer were tested, with similar levels of measured noise.

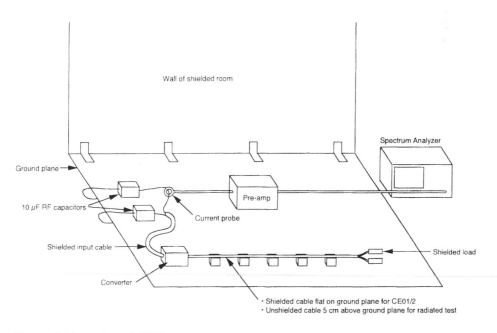

**Figure 3.14** CE01 and CE03 test set up.

### 3.3.2   Summary of +5-V-Converter Conducted Emissions (Input) CE01 and CE03 Test Results

Plot 3.1 shows a peak value of conducted emission of $-26$ dBm at 200 kHz measured using the current probe. Converting to millivolts and dividing by the gain of the preamp gives

$$\frac{11 \text{ mV}}{20} = 0.55 \text{ m}$$

The current probes are calibrated in transfer impedance versus frequency, where the transfer impedance is defined as $Z_t = V/I$, $V$ is the current-probe output voltage, and $I$ is the current under measurement. The transfer impedance of the low-frequency probe is almost 0 dB above $1\Omega$; thus the current measured is

$0.55 \text{ mA} = 55 \text{ dB}\mu\text{A}$

The CE01 narrowband conducted limit is 40 dB$\mu$A at 100 kHz, so the converter is outside the CE01 limit. Plot 3.2 shows the high-frequency broadband noise at 37.6 MHz to be $-27.3$ dBm. By the same methodology as just applied, and using a $Z_t$ of $-10$ dB ($0.316\Omega$) for the high-frequency probe, the current is found to be 64 dB$\mu$A.

The resolution bandwidth of the spectrum analyzer during tests for broadband noise was set at 100 kHz, so converting to the broadband standard bandwidth of 1 MHz entails adding 20 dB to the measured current. Therefore, the current is 84 dB$\mu$A/MHz. Since the CEO3 broadband limit is 50 dB$\mu$A/MHz at 37 MHz, the measured current is well above the specification (i.e., by 34 dB$\mu$A/MHz).

### 3.3.3   ± 15-V Converter: Summary of Conducted Emission CE01 and CE03 Test Results

Plot 3.3 shows the conducted CE01 emissions for the $\pm$15-V converter, which are 20 dBm higher than for the $+5$-V converter. The CE03 emissions, shown in Plot 3.4, are comparable to the $+5$-V converter levels.

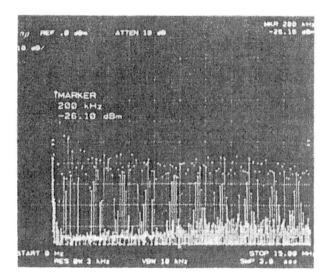

**Plot 3.1**   CE01 conducted emissions, $+5$-V supply, NB.

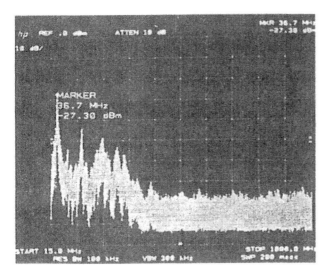

**Plot 3.2**   CE03 conducted emissions, +5-V supply, BB.

### 3.3.4   Differential-Mode and Common-Mode Noise Conducted Noise at the Output of the +5-V Converter

We shall use the terms *common-mode* and *differential-mode voltage* or *current* throughout the book. The concepts of *differential mode* and *common mode* are often difficult to comprehend, but an understanding of the difference between the two modes is of great importance when attempting to analyze a source or in applying reduction techniques, such as those described in Chapters 5 and 11.

We shall use an isolating DC-to-DC converter to illustrate the different modes of noise voltages, since such converters are notorious generators of both modes of noise. The converter shown in Figure 3.15 has two output connections, both of which are isolated from the enclosure

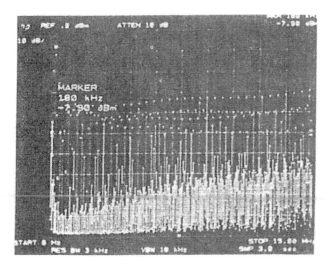

**Plot 3.3**   CE01 conducted emissions, +15-V supply, NB.

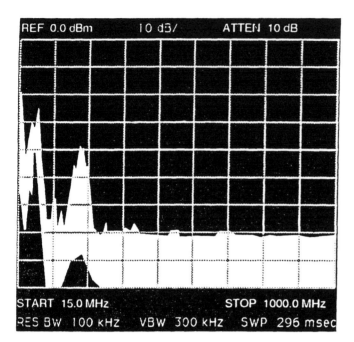

**Plot 3.4**  CE03 conducted emissions, +15-V supply, BB.

of the converter and from the chassis ground. Had one or the other of the output connections been made to the chassis ground, and assuming an infinitesimally low-impedance ground connection, then no common-mode voltage would exist. A common-mode voltage exists between each output connection and the chassis. In Figure 3.14, this common-mode voltage is shown between the lower connection and the chassis only. In addition, a differential-mode noise voltage exists between the two output connections. Thus the total noise voltage appearing on the output connections of Figure 3.14 is the sum of the common-mode voltage appearing between the connection and the chassis and the differential noise appearing between the two connections.

The test setup used to measure common-mode and differential-mode noise voltage is the same as shown in Figure 3.14, with the exception that the shielded wire connected to the output connections of the converter was replaced with a two-conductor cable, 2 m long, located at a height of 5 cm above the ground plane (this setup was also used for radiated tests). Two 50 Ω

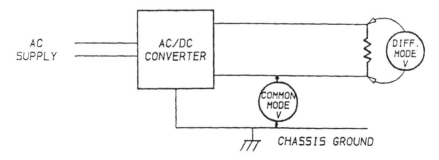

**Figure 3.15**  Illustration of common-mode and differential-mode voltage at the output of a converter.

loads were connected in parallel and located at the far end of the cable with reference to the converter. The total load current on the converter was 0.2 A (1 W).

During measurements, one of the 50 Ω loads at the end of the 2-m-long cable was replaced by the 50 Ω input impedance of the oscilloscope/spectrum analyzer for differential-mode noise measurements. The connection from the load end of the 2-m-long two-conductor cable to the measuring equipment was via 50 Ω coaxial cable. Common-mode noise was measured between one side of the load and the ground plane using the high-impedance (1 MΩ) input of either the oscilloscope or the spectrum analyzer. The characteristic impedance of the two-conductor cable is approximately 88 Ω. Changing the load resistance on this 88 Ω cable from 25 to 50 Ω produced no discernable effect on the frequency characteristics of the conducted noise, but a slight change in the peak level of emission was seen. All subsequent tests were made with a 25 Ω load. In practice, the mismatch between the load and cable impedance is likely to be higher than that used in the test setup, because the load on the converter output contains capacitance.

Plot 3.5 shows the differential-mode conducted noise measured in the time domain by use of an oscilloscope. The noise is characterized by a complex wave (Plot 3.6) that appears at the positive and negative switching transitions of the converter, i.e., = 266 kHz, which is, as expected, twice the converter fundamental switching frequency. The time base of the oscilloscope used to make the measurement shown in Plot 3.7 is 20 ns/division.

From Plot 3.6, it may be seen that the transitions occurring in the complex waveform are as fast as 5 ns, thus the high-frequency content of the noise is expected to be significant. Plot 3.7 shows the common-mode conducted noise.

The noise in the frequency domain is shown in Plot 3.8, from 0 Hz to 15 MHz, using a narrowband setting of the resolution bandwidth of the spectrum analyzer.

The noise from 15 MHz to 1 GHz is shown in Plot 3.9, using a broadband setting of the resolution bandwidth. The converter noise may almost certainly be classified as broadband in nature and, as expected, contains frequencies up to 490 MHz.

### 3.3.5  Radiated Emissions

The level of radiated emissions from the secondary of the converters is dependent on the geometry of the wiring between the output of the converter and the load.

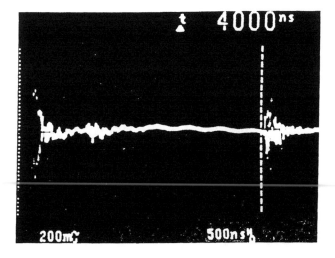

**Plot 3.5**  Differential-mode conducted noise, unfiltered, time domain, +5-V supply (output).

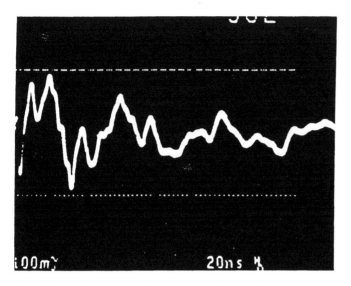

**Plot 3.6**  Differential-mode conducted noise, unfiltered, time domain +5-V supply (output) (complex wave).

The MIL-STD-462 (RE02) radiated-emissions test setup (14 kHz to 10 GHz) requires that interface cabling be included in the test setup and that any cables be located 5 cm from the front of the ground plane at a height of 5 cm above the ground plane and run for a length of at least 2 m.

The radiated-emissions test setup complied with the basic RE02 configuration, with the exception that the radiated emissions from the input wiring were minimized by the use of shielded cable between the 10-μF RF input capacitors and the input of the converter. The input wiring was further shielded by taping it to the ground plane with conductive-adhesive aluminum

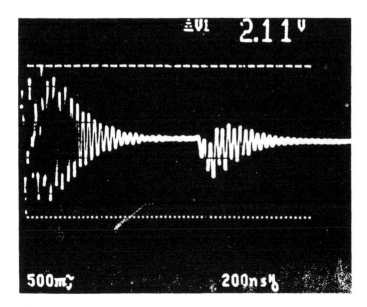

**Plot 3.7**  Common-mode conducted noise, unfiltered, time domain, +5-V supply (output).

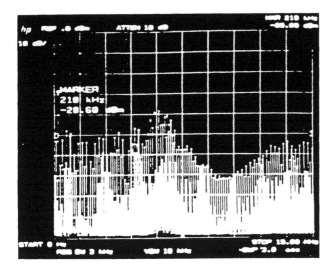

**Plot 3.8** Differential-mode conducted noise, unfiltered, frequency domain, 0 Hz to 15 MHz, NB, +5-V supply (output).

tape. The two-/three-conductor wiring used between the output of the converter/filters and the load forms a transmission line with its electromagnetic image in the ground plane.

The major contributor to radiated emissions is the common-mode currents flowing on the two-/three-wire line, with negligible contribution from the differential-mode current flow. Thus by measuring the common-mode current and using the equation for radiation from a resonant transmission line, the E field at a distance of 1 m from the line may be predicted. A second measurement was made using an 11-cm balanced loop antenna, which measures H field. Once the H field at 1-m distance from the line is known, the E field may be calculated. The MIL-STD-462 (RE02) test method requires the use of an E field antenna located 1 m from the source.

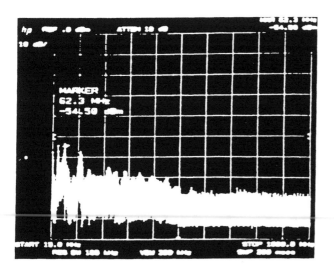

**Plot 3.9** Differential-mode conducted noise, unfiltered, frequency domain, 15 MHz to 1 GHz, BB, +5-V supply (output).

However, as described in Chapter 9, on MIL-STD-462 tests, the measurement technique is prone to large measurement errors unless the antenna AF has been calibrated at the exact spot in the room used for the measurement or other techniques are used to compensate for room resonances and antiresonances. The methods using prediction and H field measurement, in contrast, result in a relatively good correlation, as described in Chapter 9.

In summary, the level of radiated emissions from the line was found by

1. Measurement using an 11-cm balanced loop antenna
2. Measurement of common-mode current flow with a current probe around the line and use of a program for calculation of radiation from a resonant transmission line.

Plot 3.10 shows the ambient EM level in the shielded room measured with the 11-cm balanced loop antenna with the source of emissions and the spectrum analyzer in the shielded room. Plot 3.11 illustrates the radiated emissions from the +5-V converter from 10 MHz to 1 GHz. Peaks were detected at 34 MHz, 228 MHz, and 79 MHz.

Shielding the 2-m-long line reduced the radiated emission level to below either the ambient level of the shielded room or the measuring instrument's sensitivity. An obvious question is, why not connect the return of either the +5-V or ±15-V supply to the ground plane and thereby reduce the common-mode noise and radiation from the line? Indeed this may be practical and result in a significant reduction in radiation. However, where many converters are used and the returns are connected together, multiple connections of the returns to the chassis are made. This may violate the grounding scheme or result in unacceptably high levels of RF current induced via the loops formed between the commoned returns and chassis.

In the case study, a requirement existed for a single-point ground that was made at some considerable distance from the equipment. An additional requirement was for a reference ground linking each piece of equipment and isolated from chassis at the equipment. The reference ground was used to connect the signal ground to the single-point ground. The +5-V and ±15-V return was connected to signal ground and thus must be isolated from chassis to comply with the ground isolation requirement. This type of grounding scheme is illustrated in Chapter 8, on grounding.

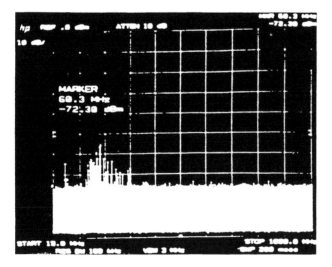

**Plot 3.10**   11-cm balanced loop antenna 1 m from cable, power off (ambient).

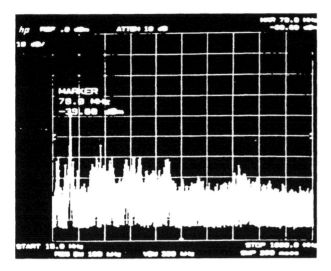

**Plot 3.11**   Balanced loop antenna 1 m from cable, +5-V supply.

A total of 17 similar converters were used in an instrument, with five located in close proximity in a compartment in one of the pieces of equipment making up the instrument. Based on the investigation it was decided, for the sake of intraequipment EMC and to limit the levels of conducted noise placed on the common +28V supply, that the converter inputs and outputs should be filtered. The choice of filter and attenuation or gain achieved when six different filters were tested are described in a continuation of this case study in Section 5.1.10.7, on power-line filters.

Digital logic and converters are common and major sources of noise, and so these have been the focus in preceding sections. However, the equations and figures presented are applicable to other sources of transient intraequipment noise, such as stepper motors, CRT deflection drivers, relays, and switches. For an EMC prediction external to any equipment (e.g., at the subsystem or system level), the electromagnetic environment may be found from the typical or maximum levels from ISM sources, examined in Chapter 1, as well as the typical levels of E field at the carrier frequency of transmitters.

Additional types of noise are generated by transmitters, as described in the following section.

**Table 3.2**   Comparison Between MIL-STD RE02 Broadband Emission Levels and Measured Levels

| Frequency (MHz) | RE02 limit (dBµV/m/MHz) | Measured (dBµV/m/MHz) | Difference between measured and limit values (ΔdB) |
|---|---|---|---|
| 34 | 62 | 85 | 23 |
| 79 | 60 | 87 | 27 |
| 228 | 55 | 73 | 18 |

## 3.4  TRANSMITTER-GENERATED NOISE

A transmitter is designed typically to generate a specific frequency or a number of adjacent frequencies. When the transmitter is used to convey information, one or more modulation techniques are used that result in sidebands (i.e., frequency bands on either side of the carrier). For example, when amplitude modulation is used, the upper sideband is the carrier frequency plus the modulating frequencies and the lower sideband is the carrier frequency minus the modulating frequencies. In addition to the frequencies used in the desired transmission, a transmitter generates a number of unwanted, spurious emissions. These include harmonics of the carrier and sidebands and the master oscillator. In addition, the transmitter generates nonharmonically related frequencies and broadband noncoherent noise.

Broadband noncoherent noise includes *thermal noise*, sometimes called *Gaussian noise* or *white noise*. The assumption made is that the thermal noise exhibits uniform power spectral density. An additional source of noise is known as *flicker* or $1/f$ noise. Thermal noise, $1/f$ noise, and popcorn noise are of particular concern in analog and video circuits; Section 5.3.3 discusses typical levels and evaluation techniques. The major potential source of EMI to equipment and cabling in proximity to an antenna is at the carrier frequency, especially when the E field at the equipment location is in the volts-per-meter range. For receiving equipment tuned to a different frequency and incorporating a filter after the antenna, immunity at the carrier and sideband frequencies generated by a transmitter may be achieved. Case Study 10.2 describes an EMI problem due to equipment, including a receiver, that is located in close proximity to a transmitter. When the transmitter is licensed, the frequencies required by the transmitter are clearly denied to a receiver in the area. In addition, channels may be denied due to the spurious emissions from a transmitter. Typical spurious transmitter output spectrums are contained in Ref. 3.

Transmitters in the past almost exclusively used tuned amplifiers in the driver and final output stage, and thus an effective output filter existed. With the advent of broadband solid-

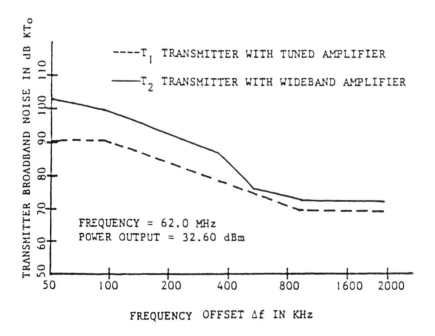

**Figure 3.16**  Transmitter broadband noise measurements on transmitters T1 and T2. (© 1984, IEEE.)

state amplifiers, the useful frequency range of the amplifier was extended to decades. For example, amplifiers are available with frequency ranges of 1–100 MHz, 10 kHz–1 GHz, and 1–10 GHz.

The use of wideband amplifiers has typically resulted in increased levels of noncoherent broadband noise. The only filtering effect may be due to the reduction in out-of-band transmitting antenna gain, which is discussed in Chapter 10, on system-level EMC. Modern resonant antennas, which exhibit a high gain at the resonant frequency and a rapid reduction in gain on either side of resonance, may mitigate the EMI potential due to broadband noise. The effect of broadband noise and other spurious emissions is to increase the level of ambient noise and possibly deny a number of additional channels in the area. When the noise level at the output of the amplifier, the gain of the transmitting antenna, the distance and relative location and gain of the receiving antenna, reflections, or shielding caused by buildings or topography are all known, the potential for EMI may be predicted. An example of an antenna-to-antenna EMI prediction and subsequent measurement at the site is included in Case Study 10.2. The level of thermal noise power, $P_n$, generated by a transmitter, ignoring input-phase noise and internally generated $1/f$ noise, may be found from the following equation:

$$P_{n(\text{out})} = F(\text{KT})BG \text{ W}$$

where

$F$ = transmitter noise figure (numeric)
KT = thermal noise power density
      (at 290 K, KT = $4 \times 10^{-21}$ W/Hz = $-144$ dBW/Hz)
$B$ = bandwidth under consideration, in hertz
$G$ = gain of the amplifier = $P_{\text{out}}/P_{\text{in}}$ ($P_{\text{out}}$ and $P_{\text{in}}$ in watts)

Figure 3.16 illustrates measured transmitter broadband noise from a tuned amplifier transmitter and a transmitter with a wideband amplifier for which noise levels almost 10 dB (power) above the levels for the tuned amplifier transmitter are apparent.

Some of the measures that may alleviate EMI caused by transmitter broadband noise is the inclusion of an antenna output filter, the choice of a low-noise power amplifier or a low-gain amplifier with a corresponding higher-level, low-noise, input.

## REFERENCES

1. E. Kann. Design Guide for Electromagnetic Interference (EMI) Reduction in Power Supplies, MIL-HDBK-241B. Power Electronics Branch, Naval Electronic Systems Command, Department of Defense, Washington, DC, 1983.
2. M. J. Toia. Sketching the Fourier transform: a graphical procedure. 1980 IEEE International Symposium on Electromagnetic Compatibility. New York, 1980.
3. P. N. A. P. Rao. The impact of power amplifiers on spectrum occupancy. Theoretical models and denied channel calculations. International Conference on Electromagnetic Compatibility, London, IEEE, 1984.

# 4

# Crosstalk and Electromagnetic Coupling Between PCB Tracks, Wires, and Cables

## 4.1 INTRODUCTION TO CROSSTALK AND ELECTROMAGNETIC COUPLING

*Crosstalk* is defined as the interference to a signal path from other localized signal paths. The term *crosstalk* is often limited in its use to circuits/conductors in close proximity, where the coupling path may be characterized by the mutual capacitance and mutual inductance of the circuits. This restricted definition will be used throughout this chapter. Crosstalk prediction applies irrespective of whether the source is an intentional signal or noise; thus *crosstalk* is used in this chapter in a broader sense, to describe the coupling from either a signal or noise source into a signal path. The terms *electromagnetic coupling* and *crosstalk* are used interchangeably in the chapter. However, in general, *electromagnetic coupling* includes the coupling of an electric, magnetic, or electromagnetic field from one conductor into another, which is not necessarily in close proximity.

As with all EMC predictions or EMI investigations, the first step is to identify the source of emissions and to determine if the path is conducted, radiated, or crosstalk. Crosstalk should be a prime suspect in an EMI investigation, or a candidate for examination in a prediction, if high transient current or fast rise-time voltages are present on conductors in close proximity to signal-carrying conductors. An example is a cable that contains conductors that carry control or drive levels to a relay motor, lamps, etc. or digital transmission and also contains conductors used for analog, RF, or video signals. The sharing of a cable with digital signals or supply as well as analog signals is also likely to result in crosstalk. Likewise, when a PCB track carrying control or logic levels is in close proximity to a second track, which is used for low-level signals, over a distance of 10 cm or more, crosstalk may be expected. Case Study 11.1 in Chapter 11 (Section 11.9 on PCB layout case studies), provides an example of crosstalk between a logic supply track on a PCB and an analog signal track that are in close proximity over a distance of only 6 centimeters.

Long cables that carry several serial channels or parallel bits of high-speed data or telemetry are also prime candidates for a crosstalk prediction or candidates in an EMI investigation.

Crosstalk between conductors or cables in close proximity can be caused by an electric field coupling via mutual capacitance or a magnetic field coupling via mutual inductance, as illustrated in Figure 4.1. A third form of coupling is via common impedance, which is covered in Section 8.8.3. It is important to discriminate between EMI caused by common impedance coupling and that resulting from electromagnetic coupling or crosstalk. An example in which either common impedance or crosstalk may be the culprit is as follows: Assume high current flows on a conductor driving filament lamps and that an adjacent signal conductor shares the same return connection. When a common connection of the two returns is made at the lamp

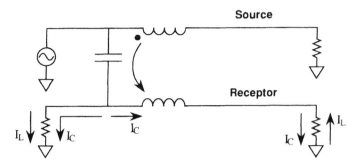

**Figure 4.1**  Electric field (capacitive) and magnetic field (inductive) crosstalk.

end of the cable, EMI may result either from crosstalk or due to the high voltage drop caused by the high lamp current flowing in the DC resistance of the returns.

Before attributing EMI to crosstalk, it is important to ensure that the data on the receptor line is noise free at the source end and that the input circuit, at the far end of the line from the source, does not generate noise.

Another potential cause of incorrect reception of data on an interface are reflections due to a line-to-load impedance mismatch, which may be mistaken for crosstalk.

One practical method of determining the presence of crosstalk-induced EMI is to separate the source and receptor conductors. Where this is not feasible, the temporary slowing down of the edges of the source waveform or a reduction in the frequency may be feasible. Where intermittent EMI exists and the suspected cause is crosstalk, the source signal may be increased in amplitude, frequency, or speed, or an additional noise source may be injected via, for example, a transformer or a capacitor. When the incidence of EMI is thereby increased, the likelihood exists that the source of crosstalk has been found.

The crosstalk values in the following sections may be computed, or they may be found in tables or plots, where they are given for specified frequencies or rise times. Generally when the frequency is given, the rise time for which the same crosstalk occurs is given by $1/\mu f$, likewise, when the rise time is specified, the frequency may be found from $1/\mu t$.

The crosstalk predictions in this chapter assume lossless transmission lines. One source of losses in transmission lines is *skin effect* (i.e., the current flows predominantly in the surface of the conductor). The skin effect in a current-carrying conductor is characterized by the AC resistance of the conductor. The confinement of the current flow to a small fraction of the cross-sectional area of a conductor is increased in the case where two conductors carrying currents in opposite directions are in close proximity. The current flow then tends to crowd away from the area close to the adjacent conductor, which increases the AC resistance by a factor of 0.6–0.7. A second loss mechanism is in the dielectric material separating the conductors. In practice, the loss due to skin effect tends to predominate in cable transmission lines, such as twisted shielded pairs.

The error due to the exclusion of transmission line loss in the prediction of crosstalk is significant only with long cables and fast rise times and is generally much less than the prediction error. For example, the error due to ignoring loss in the crosstalk prediction for a 60-m-long cable with a 100-ns rise-time source is only 7%, whereas the prediction techniques presented in this chapter do not result in errors of much less than 15%.

When source and receptor circuits are not in close proximity, the equations presented in Sections 2.2.4 and 2.2.5 for the generation and reception of electromagnetic fields may conveniently be used to predict the coupling, as described in Section 4.4.5.

When considering crosstalk between PCB tracks, conductors in a cable, or wires and cables in close proximity, it is important to determine if it is predominantly electric field (capacitive) or magnetic field (inductive) coupling. If neither mode predominates, then either both must be examined separately or, more conveniently, the characteristic impedance between the source and receptor circuit and the receptor circuit and ground may be used in a crosstalk prediction, as described in Section 4.4.1. Which of the two modes of coupling predominates depends on circuit impedances, frequency, and other factors. A rough guideline for circuit impedance (in ohms) is as follows:

> When source and receptor circuit impedance products are less than $300^2\Omega$, the coupling is primarily magnetic field.
> When the products are above $1000^2\Omega$, the coupling is predominantly electric field.
> When the products lie between $300^2\Omega$ and $1000^2\Omega$, the magnetic or electric field coupling may predominate, depending on geometry and frequency.

However, these guidelines are not applicable in all situations (e.g., crosstalk between PCB tracks located over a ground plane). When the characteristic impedance of the tracks above a ground plane is relatively low (e.g., $100\Omega$) and the load and source impedances on the receptor track are higher than the characteristic impedance, the crosstalk is predominantly capacitive.

## 4.2  CAPACITIVE CROSSTALK AND ELECTRIC FIELD COUPLING BETWEEN WIRES AND CABLES

Figure 4.2 shows a four-wire arrangement of parallel signal lines. If the circuits are unbalanced (i.e., when the return conductor in each circuit is at the same ground potential), then the four-wire arrangement may be replaced by two wires above a ground plane for prediction purposes. The capacitance between the wire connected to one circuit and the ground plane we shall call $C_1$, and the wire connected to the second circuit and the ground plane we shall call $C_2$, as shown in Figure 4.2. When the distances between each wire and its ground plane are equal, $C_1 = C_2$

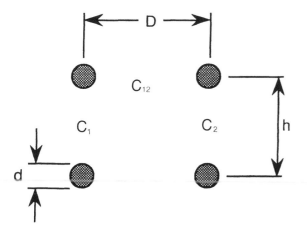

**Figure 4.2**  Four-wire capacitive arrangement.

and $L2 = L1$. The capacitance can be obtained from the inductance of a wire above a ground plane, upcoming Eq. (4.8), as shown in Ref. 1:

$$C = \frac{3.38}{L1} \tag{4.1}$$

$$\therefore\ C = \frac{3.38}{0.14\ \log\left(\dfrac{4\cdot h}{d}\right)}\quad \text{pF/m}$$

where $h$ is the height above the ground plane and $d$ is the diameter of the wire. Any unit may be used for the height and the diameter as long as they are both expressed in the same units.

The mutual capacitance between the two-wire circuits, $C_{12}$, is computed from the mutual inductance $(M)$, upcoming Eq. (4.9), and the equation for the inductance of a wire above a ground plane, Eq. (4.8), in accordance with Ref. 1:

$$Cm = \left[\frac{0.07\cdot\log\left[1 + \left(\dfrac{2\cdot h}{D}\right)^2\right]}{0.14\cdot\log\left(\dfrac{4\cdot h}{d}\right)\,0.14\cdot\log\left(\dfrac{4\cdot h}{d}\right)}\right]\,3.28\quad \text{pF/m} \tag{4.2}$$

where

$D$ = distance between the two circuits
$h$ = height above the ground plane
$d$ = diameter of the wires

Although any unit of measurement may be used, $D$, $h$, and $d$ must be in the same units for all of these computations. Figure 4.3 shows how the geometry of the circuits affect the

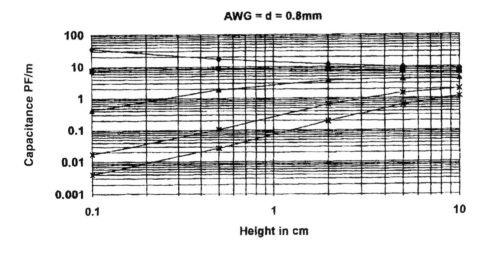

**Figure 4.3** Variation of $C_1$, $C_2$, and $C_{12}$ with $h$ and $D$. (From Ref. 1.)

self-capacitance and mutual capacitance for 22 AWG wires, which have a diameter of 0.8 mm (0.032″). For small $h/D$ ratios, $C_{12}$ grows slowly with an increase in the height of the wire above the ground plane, $h$. For large values of $h/D$, $C_{12}$ remains more or less unchanged and then begins to decrease with the wire height. It can be seen that the self-capacitance of the receptor circuit, $C_2$, plays a role in the coupling. This can be shown by considering the coupling coefficient, $K$:

$$K = \frac{C_{12}}{C_2} + C_{12} \tag{4.3}$$

From Figure 4.3 it can be seen that $C_2$ increases with a decrease in height, and from Eq. (4.3) it can be seen that $K$ decreases with an increase in $C_2$. Thus, the level of crosstalk can be decreased by:

> Reducing the height of the wires above the ground plane or decreasing the distance between wires in the same circuit
> Increasing the distance between the two circuits

Figure 4.4a shows two unbalanced circuits exhibiting capacitive crosstalk, with the equivalent circuit in Figure 4.4b. The self-capacitance of the source circuit does not play a role in the crosstalk and is therefore ignored. The self-capacitance of the receptor circuit, $C_2$, is in parallel with the source impedance, $R_S$, and load impedance, $R_L$, of the receptor circuit.

When the line length is longer than the wavelength (i.e., $l \gg \lambda$), the voltage down the length of the line is not constant and a more complex analysis is required. However, for the simple case, where the wavelength is such that the voltage is assumed to be constant for the entire length of line, the crosstalk voltage $V_c$ is given by

$$V_c = V_1 \frac{Z_2}{Z_1 + Z_2} \tag{4.4}$$

where

$$Z_2 = XC_2 \text{ in parallel with } R_L \text{ and } R_S$$

$$Z_1 = XC_{12} = \frac{1}{2\pi f C_{12}}$$

The total resistance $R_t$ equals $R_L$ in parallel with $R_S$ and is given by

$$R_t = \frac{1}{1/R_L + 1/R_S} \tag{4.5}$$

$$Z_2 = \frac{1}{\sqrt{(1/R_t)^2 + 2\pi f C_2^2}} \tag{4.6}$$

If the receptor wire of Figure 4.4a is replaced by a twisted pair in which one conductor is connected to the ground plane at each end, as shown in Figure 4.4c, then the capacitive crosstalk is the same as for an untwisted pair of receptor wires.* Assuming an untwisted pair of receptor wires in close proximity, the untwisted pair will exhibit reduced crosstalk, compared to a single wire above a ground plane, simply because of the increase in capacitance $C_2$ due to

---

* This is because one wire is connected to ground and the induced voltage is close to zero whereas the second wire is connected to a source and load impedance and a crosstalk voltage is induced.

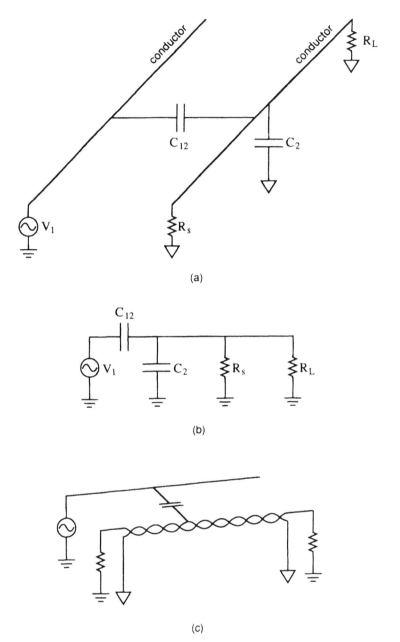

**Figure 4.4** Unbalanced circuit exhibiting crosstalk.

the close proximity of the wires. If the twisted pair is connected to a balanced circuit, illustrated in upcoming Figure 4.8, then the capacitive crosstalk is theoretically zero. In practice, the level of crosstalk is very sensitive to irregularity in the twist and variation in the distance between the source and receptor wires down the length of the wires. A twisted pair connected to a balanced circuit typically exhibits a lower level of crosstalk than an untwisted pair. The level of crosstalk voltage induced in an untwisted pair connected to a balanced circuit is dependent on the relative distances between the source wire and the two receptor wires, as discussed later in the evaluation of crosstalk in a four-conductor cable. If the distances between source and receptor wires are known or if a worst case can be assumed, then the mutual and self-capacitances of the wires may be calculated, an equivalent circuit constructed, and the capacitive crosstalk predicted. When crosstalk occurs between conductors inside a shielded cable, the capacitance between the center conductors and the shielding may become a significant factor in assessing the level of crosstalk. For example, consider a four-conductor shielded cable connected to two unbalanced circuits in which the return conductors are connected at some point to the shield of the cable (i.e., the two return conductors and the shield are at the same potential). The cable configuration and capacitances are shown in Figure 4.5a, and the equivalent circuit of the unbalanced arrangement is shown in Figure 4.5b. In this configuration, conductor 1 is the source-signal wire, conductor 2 is the receptor-signal wire, and conductors 3 and 4 are the signal return conductors. The crosstalk coefficient $K$ of this arrangement is given by

$$K = \frac{C_1}{C_1 + C_2 + C_3 + C_4} \tag{4.7}$$

where

$C_1$ = mutual capacitance between signal conductors of the coupled pair
$C_2, C_3$ = capacitances between the receptor signal wire and the signal returns
$C_4$ = shield capacitance

Equations (4.4)–(4.6) may be used for the shielded cable by replacing $C_2$ by the sum of capacitances $C_2$, $C_3$, and $C_4$. A practical example of crosstalk in a four-conductor shielded cable is given below.

The cable used in this example contains 20-gauge stranded (27 × 34 AWG) conductors with a diameter of 0.035″ (0.89 mm), an insulation thickness of 0.0156″ (0.4 mm), and an interaxial spacing diagonally between the insulated conductors of approximately 0.008″. The relative permittivity of the polyethylene insulation is 2.3. The mutual capacitance between conductors 1 and 2, 2 and 4, 4 and 3, and 3 and 1 is 11 pF, diagonally between 1 and 4, and 2 and 3 it is 5 pF. The conductor to shield capacitance is 20 pF. Assuming that the cable is used for two unbalanced circuits, as shown in Figure 4.4a, that the receptor source and load impedances are each 20 kΩ, and that 1 V AC is applied to the source circuit, the crosstalk plotted against frequency is as shown in Figure 4.6. In our example circuit, we see that the crosstalk voltage increases linearly with frequency until approximately 160 kHz, after which the slope decreases until the crosstalk voltage levels out at 1 MHz. With the load and source resistors of the receptor circuit reduced to 2 kΩ each, the high-frequency breakpoint occurs at 10 MHz. The amplitude above the breakpoint is given by the crosstalk coefficient and is independent of frequency. The crosstalk coefficient in our example circuit is, from Eq. (4.7),

$$K = \frac{11 \text{ pF}}{11 \text{ pF} + 11 \text{ pF} + 5 \text{ pF} + 20 \text{ pF}} = 0.234$$

The maximum crosstalk voltage is therefore 1 V × 0.234 = 0.234 V, or 23% of the voltage applied to the source circuit. Measurements on the four-conductor cable are plotted against the theoretical values in Figure 4.6 for the $R_L$ = 10 kΩ case.

When the source voltage is a step function and the rise or fall time is equal to or less than 0.2 times the time constant of the receptor circuit, the crosstalk peak voltage may be found from the crosstalk coefficient. The time constant is the sum of the cable capacitances times the

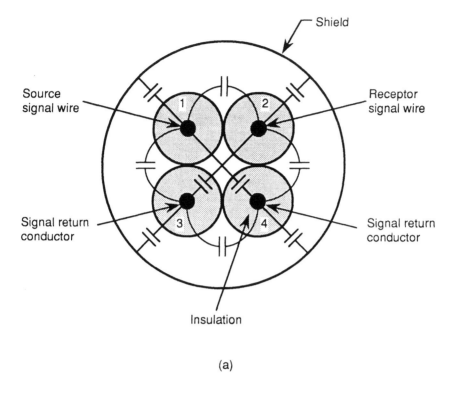

(a)

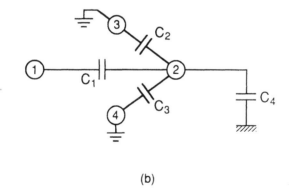

(b)

**Figure 4.5**   Four-conductor shielded-cable configuration.

receptor circuit load and source resistors in parallel. Thus, in our example cable, the time constant is

$$\tau = \frac{20\ \text{k}\Omega \times 20\ \text{k}\Omega}{20\ \text{k}\Omega + 20\ \text{k}\Omega}\ (11\ \text{pF} + 11\ \text{pF} + 5\ \text{pF} + 20\ \text{pF}) = 0.47\ \mu\text{s}$$

When the rise or fall time of the source step function is greater than the time constant, the peak crosstalk voltage is given approximately by $K\tau/t$. The rise or fall time of the crosstalk voltage is defined in this case as the time between the 10% and 90% voltage levels and is given

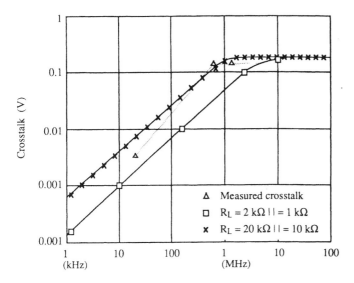

**Figure 4.6** Measured and predicted crosstalk on the four-conductor shielded cable in an unbalanced circuit.

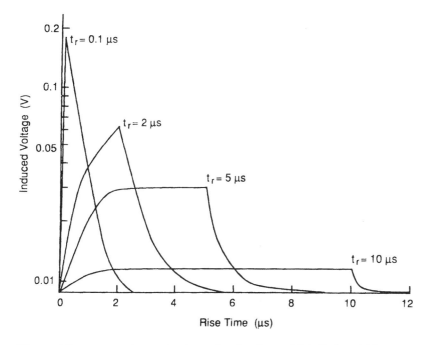

**Figure 4.7** Crosstalk-induced voltages for rise times of 0.1, 2, 5, and 10 μs.

approximately by $2\tau$. The duration of the crosstalk voltage is equal to the source rise time, $t_r$. Figure 4.7 illustrates the measured crosstalk voltage for $\tau = 0.47$ μs, $R_L = R_S = 20$ kΩ, and $t = 0.1$ μs, 2 μs, 5 μs, and 10 μs.

Using the equivalent-circuit approach for multiconductor cables becomes unwieldy. A simpler method is to use published data, which includes both the inductive and capacitive cross-

**Figure 4.8**  Balanced circuit configuration.

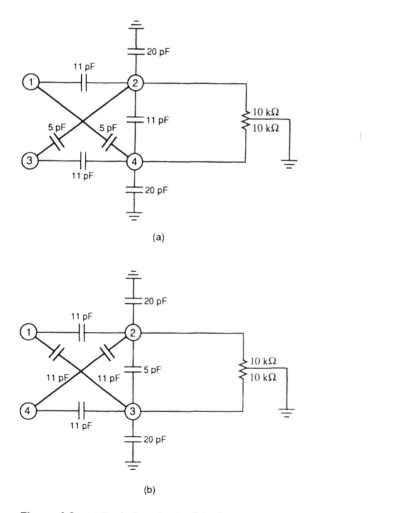

**Figure 4.9**  (a) Equivalent circuit of the four-conductor cable in a balanced circuit. (b) Equivalent circuit of the cable when the diagonal conductors are used for the source and receptor circuits.

talk contributions, as described in Section 4.4.2. A worst-case example of capacitive crosstalk in a multiconductor shielded cable is when two conductors are located in the center of the cable (i.e., furthest from the shield) and are surrounded by other signal conductors. Assuming two centrally located 24AWG signal conductors connected to unbalanced circuits and surrounded by other 24AWG signal conductors in a 50-conductor cable with a mutual capacitance between conductors of 14 pF, the crosstalk coefficient is approximately 25%. These very high levels of crosstalk may be considerably reduced by the correct choice of cable, as described in Section 4.4.2. The choices include twisted pair, cross-stranded twisted pair, and shielded twisted pair. The use of balanced circuits, as shown in Figure 4.8, will theoretically reduce the crosstalk to zero. However, when using our example four-conductor cable in balanced circuits we see from the equivalent circuit of Figure 4.9a that this will not be the case. The measured coupling coefficient for this configuration is 0.1, or 10%, which is only 13% better than for the unbalanced circuits. By using conductors 1 and 4 in the source circuit and 2 and 3 in the receptor circuit, the theoretical crosstalk is zero and the measured crosstalk coefficient is 6.8 mV, or 0.68%. Figure 4.9b illustrates the arrangement and why the voltages at conductors 2 and 3 should be at zero. In practice, crosstalk voltages do occur on the receptor conductors due to capacitive imbalances between the four conductors. The crosstalk may be considerably reduced by use of the balanced circuit with a cable that contains twisted pairs, as discussed in Section 4.4.2.

## 4.3 INDUCTIVE CROSSTALK AND MAGNETIC FIELD COUPLING BETWEEN WIRES AND CABLES

At high frequencies it is normally the capacitive coupling that predominates. However, if either the source or receptor lines, or both, are shielded, with the shield connected to ground at both ends, as is often the case, then the coupling will be magnetic. Also at low frequency, and where low circuit impedances are encountered, the crosstalk is likely to be inductive. The inductance of a wire over a ground plane is given by

$$L = 0.14 \log \left( \frac{4h}{d} \right) \qquad [\mu H/ft] \qquad (4.8)$$

where

$h$ = height above a ground plane
$d$ = diameter of the wire.

The mutual inductance between two wires over a ground plane is given by

$$M = 0.07 \log \left[ 1 + \left( \frac{2h}{D} \right)^2 \right] \qquad [\mu H/ft] \qquad (4.9)$$

where D is the distance between the two wires.

Figures 4.10 and 4.11 show mutual and self-inductance as a function of $h/D$ and $h/d$. When the height is much greater than the diameter but the diameter of the cable is large, the inductance of the wire may be limited by the value for a wire far removed from ground, as given in Section 5.1.2.

A practical circuit, which we will use to demonstrate magnetic coupling between unshielded wires, is shown in Figure 4.12. Here, cable capacitances are ignored and the lengths of the wires are less than a wavelength. The generator voltage $e_1$ causes a current flow $i_1$ in the emitter loop. The current, $i_1$, is determined by the voltage $e_1$, the source impedance $R_s$, the reactance of $L_1$, and the load impedance $R_b$.

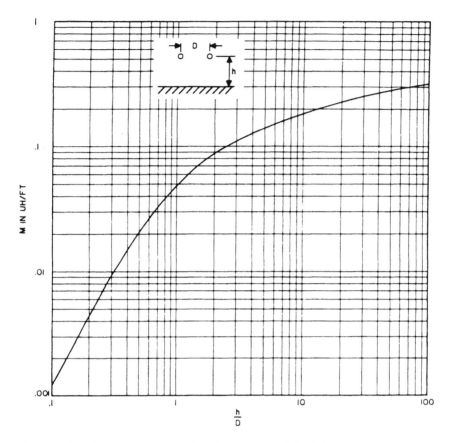

**Figure 4.10**  Inductance between the wires. (From Ref. 2, © 1967 IEEE.)

The voltage induced in the victim loop is given by

$$e_2 = j\omega M i_1 \tag{4.10}$$

and the voltage across the receiver load resistance $R_d$ is given by

$$V_d = 2\pi f M i_1 \frac{R_d}{R_c + R_d} \frac{1}{\left(1 + \dfrac{j\omega L_2}{R_c + R_d}\right)} \tag{4.11}$$

The last term is replaced in the following discussion by the factor $aL_2$, giving

$$V_d = 2\pi f M i_1 \frac{R_d}{R_c + R_d} aL_2$$

Equation (4.11) expressed as a magnitude is

$$|V_d| = 2\pi M i_1 \frac{R_d}{R_c + R_d} \frac{1}{\sqrt{1 + \dfrac{\omega^2 L_2^2}{(R_c + R_d)^2}}} \tag{4.12}$$

From Eq. (4.12) we see that the load voltage is divided not merely by the source and load

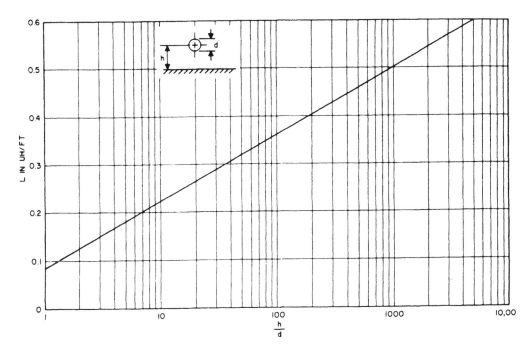

**Figure 4.11** Inductance of a wire above a ground plane. (From Ref. 2, © 1967, IEEE.)

resistance but by the inductive reactance of the victim wire $L_2$. The interfering source voltage $e_1$ may be a sinewave, in which case the open-circuit voltage $e_2$ induced in the receptor loop is

$$e_2 = 2\pi f M i_1 \tag{4.13}$$

or a transient interference when

$$e_2(t) = M \frac{I_1}{\tau} A e^{-t/\tau} \tag{4.14}$$

where

$M$ = mutual inductance between circuits
$i_1$ = source AC current
$I_1$ = peak value of source transient voltage
$\tau$ = time constant of source current
$a, A$ = factors that account for the division of induced voltage

Where shielded cable is used, factors $a$ and $A$ must be included in Eq. (4.15) and (4.16) to account for the additional attenuation of the shields. The peak voltage across the receiver load resistance $V_d$ for the transient case is

$$V_{d(max)} = M \frac{I_1}{\tau} \frac{R_d}{R_c + R_d} \frac{1}{1 - \dfrac{L_2}{(R_c + R_d)\tau}} [e^{-t/\tau} - e^{-(R_c + R_d)t/L_2}] \tag{4.15}$$

The equation can be simplified by the use of $aL_2$ and $AL_2$ factors so that

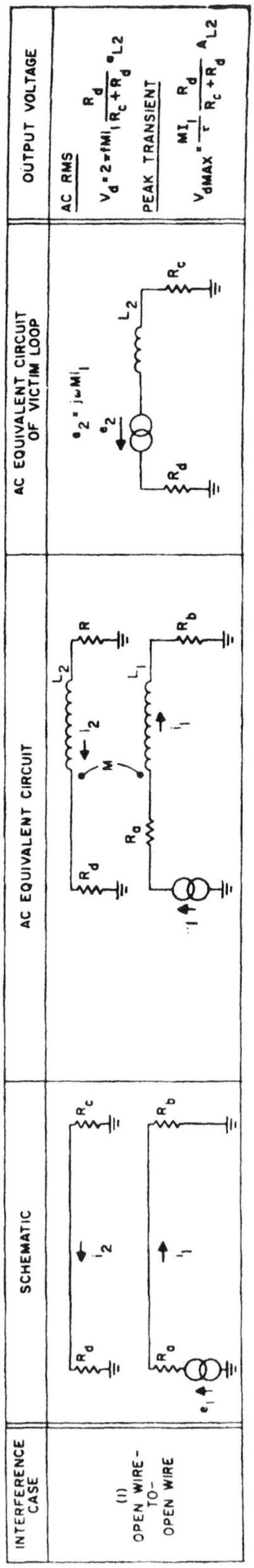

**Figure 4.12** Magnetic coupling, schematic, and equivalent circuit. (From Ref. 2, © 1967, IEEE.)

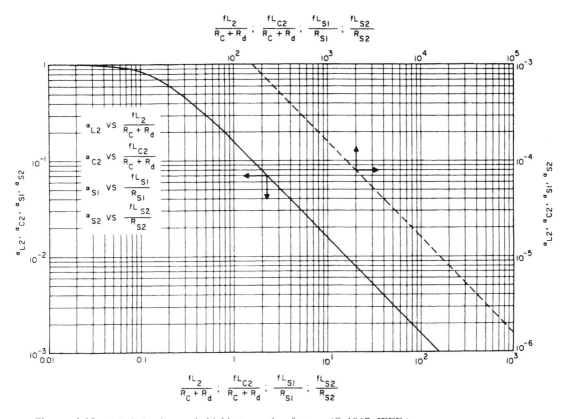

**Figure 4.13** AC, inductive, and shield attenuation factors. (© 1967, IEEE.)

$$V_{d(\text{max})} = M \frac{I_1}{\tau} \frac{R_d}{R_c + R_d} AL_2$$

The factors $aL_2$ and $AL_2(t)$ are plotted in Figures 4.13 and 4.14, from Ref. 2, along with the factors for the shielded case. In the transient case, the induced voltage has a peak value equivalent to

$$M \frac{I_1}{\tau} \qquad \text{occurring at } t = 0$$

If the time constant of the receptor circuit is very short compared to the time constant of the source circuit, that is, if

$$\frac{L_2}{R_c + R_d} \ll \tau$$

then the induced voltage will effectively divide between only $R_c$ and $R_d$. When it is not much smaller than $\tau$, the voltage drop across $L_2$ must be considered.

Where the coupling is from an unshielded current-carrying wire into a shielded wire, the return current in the victim circuit via the ground plane is attenuated by the factor

$$\frac{1}{1 + j\left(\dfrac{\omega L_s}{R_s}\right)} \tag{4.16}$$

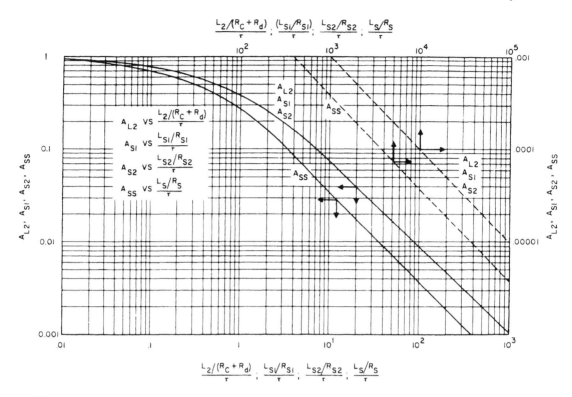

**Figure 4.14** Transient, inductive, and shield attenuation factors (© 1967, IEEE.)

where

$L_S$ = shield inductance
$R_S$ = resistance

Thus, at high frequency, the return path for a large proportion of the current induced in the shielded victim cable is via the shield. In Table 4.1, the values of the shield inductance $L_S$ for several types of shielded cable located 2 inches above a ground plane have been computed based on cable diameter, from Ref. 2. Table 4.1 also shows the values of core inductance $L_i$ and the difference, which is the cable inductance $L_c(L_c = L_i - L_s)$, as well as the resistance of the shield for lengths of 1 foot. The equivalent circuit of the unshielded-to-shielded wire coupling is shown in Table 4.2. For convenience, the current flow $i_1$ is shown attenuated by the attenuation factor of the shielded cable:

$$\frac{i_1}{1 + j\left(\dfrac{\omega L_{S2}}{R_{S2}}\right)} \tag{4.17}$$

whereas in reality the attenuation takes place in the receptor loop, not at the source, and the mechanism is as discussed in Section 7.2.1, on shielded cables.

Due to the mutual inductance between the shield and the center conductor, equal noise voltages are induced in both, and therefore the noise current through $R_d$ is reduced as compared to the unshielded case. The voltage induced in the load resistor $R_d$ is now divided not by the

**Table 4.1** Coaxial Cable and Shielded Wire Parameters

| Type RG( )/U | $Z_O$ ($\Omega$) | $d_s$ (inch) | $d_i$ (inch) | $L_s$* ($\mu$H feet) | $L_i$* ($\mu$H feet) | $L_c$ ($\mu$H feet) | $R_s$ (m$\Omega$ feet) | $L_s/R_s$* ($\mu$s) |
|---|---|---|---|---|---|---|---|---|
| **Single Braid** | | | | | | | | |
| 8A | 50.0 | 0.285 | 0.086 | 0.20 | 0.27 | 0.07 | 1.35 | 150 |
| 58C | 50.0 | 0.116 | 0.035 | 0.26 | 0.33 | 0.07 | 4.70 | 55 |
| 141 | 50.0 | 0.116 | 0.036 | 0.26 | 0.33 | 0.07 | 4.70 | 55 |
| 122 | 50.0 | 0.096 | 0.029 | 0.27 | 0.34 | 0.07 | 5.70 | 47 |
| 188 | 50.0 | 0.060 | 0.019 | 0.30 | 0.36 | 0.07 | 7.60 | 40 |
| 29 | 53.5 | 0.116 | 0.032 | 0.26 | 0.33 | 0.07 | 4.75 | 55 |
| 59A | 75.0 | 0.146 | 0.022 | 0.24 | 0.36 | 0.12 | 3.70 | 65 |
| 180 | 95.0 | 0.103 | 0.011 | 0.26 | 0.40 | 0.14 | 6.00 | 43 |
| A" | - | 0.057 | 0.023 | 0.30 | 0.36 | 0.06 | 16.00 | 19 |
| B" | - | 0.056 | 0.030 | 0.30 | 0.34 | 0.04 | 6.20 | 48 |
| **Double Braid** | | | | | | | | |
| 9B | 50.0 | 0.280 | 0.085 | 0.20 | 0.28 | 0.08 | 0.80 | 250 |
| 5B | 50.0 | 0.181 | 0.053 | 0.23 | 0.31 | 0.08 | 1.20 | 190 |
| 55A | 50.0 | 0.116 | 0.035 | 0.26 | 0.33 | 0.07 | 2.55 | 100 |
| 142 | 50.0 | 0.116 | 0.039 | 0.26 | 0.33 | 0.07 | 2.20 | 118 |
| 223 | 50.0 | 0.116 | 0.035 | 0.26 | 0.33 | 0.07 | 2.55 | 100 |
| 13A | 75.0 | 0.280 | 0.043 | 0.20 | 0.32 | 0.10 | 0.95 | 210 |
| 6A | 75.0 | 0.185 | 0.028 | 0.23 | 0.34 | 0.11 | 1.15 | 200 |

\* Valid for h = 2 inches. For h $\neq$ 2 inches use Figure 4.17
A" and B" are shielded hook-up wires.

core inductance but by the difference between the shield inductance $L_s$ and the core inductance $L_i$ (i.e., by $L_c$). The voltage $V_d$ is given by

$$V_d = 2\tau f M i_1 \frac{R_d}{R_c + R_d} a_{S2} a_{C2} \tag{4.18}$$

where

$a_{S2}$ = attenuation due to the shielded cable
$a_{C2}$ = factor due to the division of voltage in the victim loop

Both of these factors are plotted in Figure 4.13.
For the transient case, $V_d$ is given by

$$V_{d(max)} = M \frac{I_1}{r} \frac{R_d}{R_c + R_d} A_{S2} \tag{4.19}$$

$A_{S2}$ is plotted in Figure 4.14.
Table 4.2 includes the additional equivalent circuits for shielded-to-unshielded and shielded-to-shielded coupling modes and the equations for both the $ac_{RMS}$ and peak transient voltages. The voltage induced in the victim loop versus frequency is shown in Figure 4.15 for both calculated and experimental values. The voltage for an open-wire to open-wire coupling mode increases linearly with the frequency due to the $2\pi f M$ expression in the equation. The induced voltages for open wire to two types of shielded cable are shown. Here, above a certain

**Table 4.2**   Summary of Open-Wire and Shielded-Wire Induced Interference. (From Ref. 2, ©1967 IEEE)

| INTERFERENCE CASE | SCHEMATIC | AC EQUIVALENT CIRCUIT | AC EQUIVALENT CIRCUIT OF VICTIM LOOP | OUTPUT VOLTAGE |
|---|---|---|---|---|
| (1) OPEN WIRE-TO-OPEN WIRE | | | $e_2 = j\omega M i_1$ | AC RMS<br>$V_d = 2\pi f M i_1 \left|\dfrac{R_d}{R_c + R_d}\right| a_{L2}$<br><br>PEAK TRANSIENT<br>$V_{dMAX} = \dfrac{M I_1}{\tau}\dfrac{R_d}{R_c + R_d} A_{L2}$ |
| (2) SHIELDED WIRE-TO-OPEN WIRE | | | $e_2 = j\omega M i_1 \dfrac{1}{1 + j\frac{\omega L_{S1}}{R_{S1}}}$ | AC RMS<br>$V_d = 2\pi f M i_1 \left|\dfrac{R_d}{R_c + R_d}\right| a_{S1} a_{L2}$<br><br>PEAK TRANSIENT<br>$V_{dMAX} = \dfrac{M I_1}{\tau}\dfrac{R_d}{R_c + R_d} A_{S1}$ |
| (3) OPEN WIRE TO-SHIELDED WIRE | | | $e_2 = j\omega M i_1 \dfrac{1}{1 + j\frac{\omega L_{S2}}{R_{S2}}}$ | AC RMS<br>$V_d = 2\pi f M i_1 \dfrac{R_d}{R_c + R_d} a_{S2} a_{C2}$<br><br>PEAK TRANSIENT<br>$V_{dMAX} = \dfrac{M I_1}{\tau}\dfrac{R_d}{R_c + R_d} A_{S2}$ |
| (4) SHIELDED WIRE-TO-SHIELDED WIRE | | | $e_2 = j\omega M i_1 \dfrac{1}{1 + j\frac{\omega L_{S1}}{R_{S1}}}\dfrac{1}{1 + j\frac{\omega L_{S2}}{R_{S2}}}$ | AC RMS<br>$V_d = 2\pi f M i_1 \left|\dfrac{R_d}{R_c + R_d}\right| a_{S1} a_{S2} a_{C2}$<br><br>PEAK TRANSIENT<br>$V_{dMAX} = \dfrac{M I_1}{\tau}\dfrac{R_d}{R_d + R_c} A_{SS}$ |

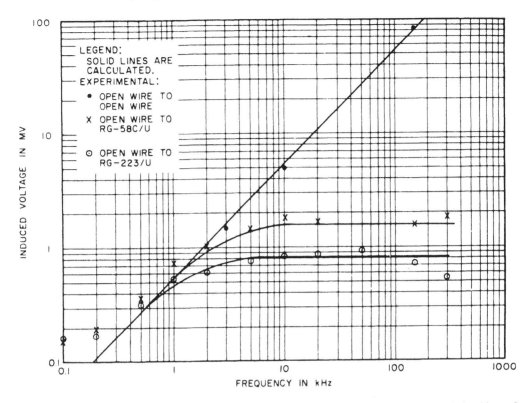

**Figure 4.15** Coupling from open-wire to open-wire, or RG58C/U to RG223/U, shielded cables. (© 1967, IEEE.)

frequency a large proportion of the induced current flows on the shield; the inductive reactance of the shield is high and so the $2\pi fM$ term is balanced by the term

$$\frac{1}{1 + j\left(\dfrac{\omega L_S}{R_S}\right)}$$

and the induced voltage levels out.

The difference between the two types of shielded cable is that the RG58C/U is a single-braid cable exhibiting a resistance of 4.70 mΩ/ft and the RG223/U is a double braid with a resistance of 2.55 mΩ/ft. The experimental and calculated results for the shielded cables begin to diverge above 200 kHz, because the transfer impedance of the cables no longer equals the DC resistance, due to skin effect and braid leakage factors. The preceding equations may be used for shielded cables above 200 kHz by modeling them as unshielded and finding the current flow on the shield. The load and source resistances are then the shield-termination impedances. The current is found from the voltages developed across the low impedances. Once the current flow is known, the transferred voltage, $V_t$, may be found from the transfer impedance of the cable as described in Section 7.2.2.

In the inductive crosstalk calculations, no account has been made for the effect of current flow in the receptor circuit on the source circuit. Little error will result in ignoring the interaction between the source magnetic field and the receptor magnetic fields when the circuits are not closely coupled. However, when close coupling exists, the additional load represented by the

**Table 4.3** Induced Interference Due to Transient Current Open Wire to Coaxial Cable, and Comparison of Measurements with Calculations

| Cable type | $\tau$ ($\mu$s) | Calculated[a] | | Measured[a] | |
|---|---|---|---|---|---|
| | | $e_m$ (mV) | $t_m$ ($\mu$s) | $e_m$ (mV) | $t_m$ ($\mu$s) |
| **Open wire to coaxial cable** | | | | | |
| RG58C/U | 1 | 2.7 | 4 | 2.7 | 3 |
| | 18 | 1.8 | 31 | 1.9 | 28 |
| | 34 | 1.5 | 44 | 1.6 | 40 |
| | 50 | 1.3 | 55 | 1.4 | 50 |
| RG223/U | 1 | 1.6 | 4.6 | 1.6 | 5.5 |
| | 18 | 1.2 | 39 | 1.2 | 35 |
| | 34 | 1 | 58 | 1 | 50 |
| | 50 | 0.9 | 72 | 0.9 | 65 |
| **Coaxial to coaxial cable** | | | | | |
| RG58C/U | 0.2 | 1.16 | 50 | 1.2 | 50 |
| | 8 | 1.08 | 60 | 1.2 | 50 |
| | 17 | 1.06 | 77 | 1.1 | 60 |
| | 26 | 1 | 89 | 1 | 70 |
| | 54 | 0.86 | 108 | 0.8 | 80 |
| RG223/U | 0.2 | 0.65 | 80 | 0.7 | 70 |
| | 8 | 0.64 | 96 | 0.7 | 80 |
| | 17 | 0.61 | 114 | 0.6 | 80 |
| | 26 | 0.6 | 133 | 0.6 | 90 |
| | 54 | 0.54 | 167 | 0.56 | 130 |

[a] $l = 24$ inches, $h = 2$ inches, $D = 1$ inch, $I = 1$ A; $e_m$ is the induced voltage and $t_m$ is the pulse width.

receptor circuit may be reflected back into the source. Using the equations presented here for close coupling to a receptor circuit, with a low load and source impedance or a shielded cable, may result in a predicted current that is above the source current!

A very approximate limit to the magnitude of the receptor circuit current is given by measurements using the MIL-STD-462 RS02 test setup, in which the source cable is wrapped around the receptor cable for a distance of 2 m. In performing the RS02 test, currents of approximately 50% of the source current have been measured in the receptor circuit.

Table 4.3 compares the measured and calculated transient induced voltage for the two types of shielded cable previously considered, and Figure 4.16 shows the induced waveforms for transient currents of various rise times. It should be noted that the inductances of the two cables are identical; only the DC resistances are different (refer to Table 4.1).

Due to the number of calculations, the numerous attenuation factors, and the amount of cable data involved in calculating the induced voltage in the victim load resistance, this type of prediction lends itself to the use of a computer program such as the one at the end of this chapter (Section 4.4.6). The limitations on the use of these equations are as follows:

When the cable length is long compared to 0.4 times a wavelength

When the frequency is above 200 kHz for shielded cables, after which the cable may be treated as unshielded and the shield current may be computed

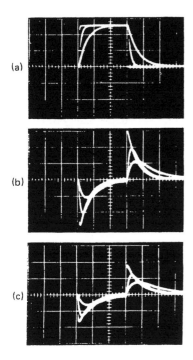

**Figure 4.16** Scope traces of induced voltage, open wire to coaxial cable. (© 1967, IEEE.) 100 μs/cm on *x*-axis. Waveforms for 0.8-, 13-, and 50-μs rise times.
(a) Open-wire input voltage across 50 Ω, 20 V/cm on *y*-axis.
(b) Open-circuit induced voltage in RG58C/U, 1 mV/cm on *y*-axis.
(c) Open-circuit induced voltage in RG233/U, 1 mV/cm on *y*-axis.

## 4.4 COMBINED INDUCTIVE AND CAPACITIVE CROSSTALK

Referring to Figure 4.1, reproduced below for convenience, we see that the current flow due to capacitive coupling in the resistor of the receptor circuit close to the source-voltage end of the lines (near end) adds to the inductively induced current, whereas, in the resistor at the far end of the receptor circuit, the capacitive and inductively induced currents tend to cancel. The reason for this difference can be found by considering the capacitive and inductive coupling modes separately. The voltage induced in the receptor circuit by capacitive coupling results in a current

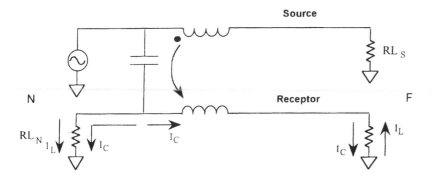

that flows in both directions on the line, due to the line-to-ground capacitance distributed down the length of the line, as described in Section 4.4.3. When examining inductive coupling, we may consider the source conductor as the primary of a transformer and the receptor conductor as the secondary. The transformer coupled voltage appears down the length of the receptor conductor and develops a current flow in the load, composed of the near-end and far-end resistances. The inductively induced current thus flows in one direction only in the receptor circuit.

When the coupling impedance, denoted as $Z_{12}$, which equals $\sqrt{L_{12}/C_{12}}$, equals $\sqrt{RL_s\,RL_N}$, the inductively and capacitively induced currents through the far-end receptor load resistor exactly cancel.

Combining the equations for capacitive and inductive coupling, described in Sections 4.2 and 4.3, gives the following expressions for the total interference in the circuit at either end of the receptor circuit and for AC and transient coupling:

For the AC coupling:

$$V_c = 2\pi f M 1 I_1 \left(\frac{R_c}{R_c + R_d}\right) \times \left[(3.38 \times 10^{-6})\left(\frac{R_b R_d}{L_1 L_2}\right) - 1\right]$$

$$V_d = 2\pi f M 1 I_1 \left(\frac{R_d}{R_c + R_d}\right) \times \left[(3.38 \times 10^{-6})\left(\frac{R_b R_c}{L_1 L_2}\right) + 1\right]$$

For the transient case:

$$V_{c(pk)} = \left(\frac{M 1 I_2}{\gamma}\right)\left(\frac{R_c}{R_c + R_d}\right) \times \left[(3.38 \times 10^{-6})\left(\frac{R_b R_d}{L_1 L_2}\right) - 1\right]$$

$$V_{d(pk)} = \left(\frac{M 1 I_2}{\gamma}\right)\left(\frac{R_d}{R_c + R_d}\right) \times \left[(3.38 \times 10^{-6})\left(\frac{R_b R_c}{L_1 L_2}\right) + 1\right]$$

Where

$V_c, V_d$ = AC voltages across $R_c$ and $R_d$, respectively

$V_{c(pk)}, V_{d(pk)}$ = peak transient voltages across $R_c$ and $R_d$ respectively

$f$ = source frequency, in MHz

M = mutual inductance, in µH/ft

$L_1, L_2$ = self-inductance of source and receptor lines, in µH/ft

$I_1$ = AC current in source circuit, in amps

$I_2$ = peak value of current in source circuit, in amps

$\gamma$ = time constant of source current, in microseconds

### 4.4.1  Use of the Characteristic Impedance of PCB Tracks and Wires Over a Ground Plane for Predicting Crosstalk

One conductor of a pair of conductors located over a ground plane exhibits a characteristic impedance to ground when both conductors are driven at the same potential, termed the even-mode impedance, $Z_{oe}$. When driven with different potentials, the impedance of one conductor to ground is termed the odd-mode impedance, $Z_{oo}$, which has a different value than $Z_{oe}$. The characteristic impedance between an isolated conductor and ground $Z_o$, may be found from Appendix 1 for a number of different conductor configurations. When the conductors are terminated with an impedance that approximates their characteristic impedance, $Z_o$, the magnitudes

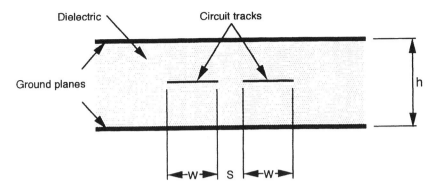

**Figure 4.17** Coupled-stripline configuration.

$Z_{oo}$ and $Z_{oe}$ may be used to calculate crosstalk for AC voltages using the following relationship (Ref. 3):

$$C = \frac{\dfrac{Z_{oe}}{Z_{oo}} - 1}{\dfrac{Z_{oe}}{Z_{oo}} + 1} \tag{4.20}$$

Consider the configuration shown in Figure 4.17, which is termed an open *coupled stripline* by the microwave community. The practical application of the coupled stripline for the nonmicrowave engineer is in crosstalk predictions in ribbon cables or in a multiconductor printed circuit boards, where the signal conductors are immersed in a homogeneous dielectric and are sandwiched between ground planes. Admittedly this configuration is not too practical in a PCB, due to the difficulty of tracing signals on the board. However, sandwiching circuit tracks in a closed stripline is a very effective method of reducing radiation from a board. We may extend the usefulness of the model by including the case where signal tracks are sandwiched between a plane connected to a supply rail and a ground plane, when a low impedance exists between the supply and ground, typically via decoupling capacitors. The crosstalk found from Eq. (4.20) applies only to near-end crosstalk in the coupled stripline, since the capacitively and inductively coupled voltages exactly cancel at the far end.

The characteristic impedance of the coupled stripline may be found from the relationship

$$Z_0 = \sqrt{Z_{oo}Z_{oe}}$$

The relative permittivity of epoxy glass printed circuit boards is approximately 4, and the relative permittivity of the insulation of a ribbon cable ranges from 2 to 5. Typical PCB dimensions are 0.02″ (0.5 mm) for the width of the tracks, $W$, 0.03″ (0.75 mm) for the distance between the tracks, $S$, 0.034″ (0.86 mm) for the height of a single-layer board, and 0.005″ (0.12 mm) for the height of the layers in a multilayer board. The near-end crosstalk, $C$, for a number of ratios of $S/h$ and $W/S$ and $\varepsilon_r = 4$ and 2 may be obtained from Table 4.4.

A more common PCB configuration, in which the tracks are on the surface of the PCB material, may be modeled by the coupled microstrip shown in upcoming Figure 4.27. The impedances $Z_{oo}$ and $Z_{oe}$ for permittivities of 2 and 4 and the crosstalk as a coefficient, $C$, are shown in Table 4.5. From Eq. (4.20), as demonstrated in Tables 4.4 and 4.5, it is seen that the crosstalk coefficient is independent of the values of $\varepsilon_r$ and may be obtained simply from the ratios $W/h$ and $S/h$.

**Table 4.4** $Z_{oo}$, $Z_{oe}$, and Crosstalk, $C$, for the Stripline Configuration

| | | $\varepsilon_r = 4$ | | | $\varepsilon_r = 2$ | | |
|------|------|----------|----------|------|----------|----------|------|
| $W/h$ | $S/h$ | $Z_{oe}$ | $Z_{oo}$ | $C$ | $Z_{oe}$ | $Z_{oo}$ | $C$ |
| 0.3 | 0.01 | 360 | 104 | 0.55 | 253 | 73 | 0.55 |
| 0.3 | 0.10 | 340 | 172 | 0.38 | 239 | 121 | 0.38 |
| 0.3 | 0.50 | 284 | 236 | 0.09 | 200 | 166 | 0.09 |
| 0.5 | 0.01 | 266 | 92 | 0.48 | 187 | 65 | 0.48 |
| 0.5 | 0.10 | 244 | 140 | 0.27 | 172 | 99 | 0.27 |
| 0.5 | 0.50 | 216 | 186 | 0.08 | 152 | 130 | 0.08 |
| 1.0 | 0.01 | 154 | 74 | 0.35 | 108 | 52 | 0.35 |
| 1.0 | 0.10 | 150 | 100 | 0.20 | 106 | 70 | 0.20 |
| 1.0 | 0.50 | 138 | 124 | 0.05 | 97 | 87 | 0.05 |

Table 4.4 and 4.5 provide the near-end crosstalk. Unlike the stripline, the far-end crosstalk in the microstrip is not zero, due to the inhomogeneous dielectric composed of both the air and the PCB material. Far-end crosstalk is typically, but not always, lower than near-end crosstalk and may be zero, depending on the track layout, which influences the magnitudes of $Z_{oo}$ and $Z_{oe}$, and the receptor and source load resistor values. Typical values of far-end crosstalk are a factor of 1.5–0.37 of the near-end values for microstrip lines (Ref. 4).

A comparison of Tables 4.4 and 4.5 illustrates that, for the same dimensions of $h$, $S$, and $W$, the crosstalk is lower for the stripline configuration, due to the increased capacitance between

**Table 4.5** $Z_{oo}$, $Z_{oe}$, and Crosstalk, $C$, for the Coupled Microstrip

| | | $\varepsilon_r = 4$ | | | $\varepsilon_r = 2$ | | |
|------|------|----------|----------|------|----------|----------|------|
| $W/h$ | $S/h$ | $Z_{oe}$ | $Z_{oo}$ | $C$ | $Z_{oe}$ | $Z_{oo}$ | $C$ |
| 0.2 | 0.01 | 360 | 104 | 0.55 | 253 | 73 | 0.55 |
| 0.2 | 0.05 | 202 | 56 | 0.56 | 310 | 86 | 0.56 |
| 0.2 | 0.20 | 190 | 77 | 0.42 | 292 | 119 | 0.42 |
| 0.2 | 0.50 | 171 | 98 | 0.27 | 263 | 150 | 0.27 |
| 0.2 | 1.00 | 157 | 116 | 0.15 | 240 | 177 | 0.15 |
| 0.5 | 0.05 | 141 | 45 | 0.52 | 216 | 69 | 0.52 |
| 0.5 | 0.20 | 136 | 60 | 0.38 | 208 | 93 | 0.38 |
| 0.5 | 0.50 | 125 | 73 | 0.26 | 191 | 113 | 0.26 |
| 0.5 | 1.00 | 116 | 85 | 0.15 | 177 | 130 | 0.15 |
| 1.0 | 0.05 | 97 | 37 | 0.45 | 149 | 56 | 0.45 |
| 1.0 | 0.20 | 95 | 47 | 0.33 | 145 | 73 | 0.33 |
| 1.0 | 0.50 | 89 | 56 | 0.23 | 137 | 86 | 0.23 |
| 1.0 | 1.00 | 84 | 63 | 0.14 | 129 | 96 | 0.14 |
| 2.0 | 0.20 | 61 | 28 | 0.37 | 94 | 43 | 0.37 |
| 2.0 | 0.50 | 61 | 35 | 0.26 | 92 | 53 | 0.26 |
| 2.0 | 0.50 | 58 | 40 | 0.18 | 88 | 61 | 0.18 |
| 2.0 | 1.00 | 55 | 43 | 0.12 | 85 | 66 | 0.12 |

the center conductors and the ground planes. For either configuration, increasing the distance between the tracks (the width of the tracks) or decreasing the height between the tracks and ground will decrease the crosstalk. Increasing the spacing between tracks and inserting a shield track, which is connected to ground, will decrease the level of crosstalk. However, measurements have shown that the presence of the shield conductor reduces the crosstalk by only 6–9 dB. The wider the shield conductor, the greater the reduction. Use of Eq. (4.20) is not limited to PCBs but may be used in a wire-to-wire crosstalk prediction. The odd- and even-mode impedances for pairs of wires above a ground plane may be obtained from the following equations:

$$Z_{oo} = 276 \log\left(\frac{2D}{d}\right)\sqrt{\frac{1}{1 + \left(\frac{D}{2h}\right)^2}}$$

$$Z_{oe} = 69 \log\left(\frac{4h}{d}\right)\sqrt{\frac{1}{1 + \left(\frac{2h}{D}\right)^2}}$$

where

$h$ = height above the ground plane
$D$ = distance between the wires
$d$ = diameter of the wires

The same units must be used for $h$, $D$, and $d$. The crosstalk may be found by use of Eq. (4.20).

### 4.4.2 Crosstalk in Twisted-Pair, Cross-Stranded Twisted-Pair, Twisted-Shielded-Pair, and Ribbon Cables

The very high levels of crosstalk measured in cables manufactured from untwisted conductors and used in either balanced or unbalanced circuits may be greatly reduced by replacement with twisted-pair or shielded twisted-pair cables. To reduce problems due to reflections, such as pulse degradation, long cables used for communications or cables used for transmission of data with short rise and fall times are usually terminated in the characteristic impedance of the lines. These terminated cables are the focus of attention here. In a properly designed, appropriately terminated communication signal interface, the characteristic impedance of the cable will be extended to the back plane inside the equipment and even to the tracks of the PCB on which the interface circuits are mounted. The characteristic impedance of a number of different cables and PCB track configurations may be computed from the equations contained in Appendix 1. A statistical model has been developed for predicting the near-end and far-end crosstalk in cross-stranded multiple twisted-pair cable versus measured crosstalk (Ref. 5). A cross-stranded multiple twisted-pair cable comprises a number of twisted pairs randomly positioned within the cable during manufacture. The advantage of the cross-stranded over the more typical construction, in which the pairs rotate around a fixed point, is that the same pairs are not always in close proximity down the length of the cross-stranded cable. Assuming that the group twist frequency is the same (that is, the twist period is the same for all twisted pairs in the cable), the near-end crosstalk isolation varies between 60 and 70 dB and the far-end crosstalk isolation between 45 and 57 dB for a 10-pair cross-stranded cable.

The crosstalk, in decibels, is defined as either the ratio of the crosstalk-induced voltage to the source voltage, in which case the crosstalk is negative, or the ratio of the source voltage to the induced voltage, in which case the crosstalk is positive and is more accurately termed the *crosstalk isolation*. The latter of these two definitions will be used here.

By twisting the pairs within the cable at a different twist frequency, the crosstalk may be further reduced; however, this type of construction is rarely available in commercial cables. Measured data is available on crosstalk in unshielded twisted-pair cable and shielded twisted-pair cable (Ref. 6). The CR-CS test method, contained in U.S. government document CR-CS-0099-000, was used in the test. The CR-CS method uses the cable in a balanced circuit configuration, so the test results are valid only for balanced circuits.

The mean values of near-end and far-end crosstalk for 19-pair shielded and unshielded cables and 27-pair shielded and unshielded cables are reproduced in Figure 4.18. The pairs in the shielded type B cable have a characteristic impedance of 50 $\Omega$ and a polypropylene insulation with a thickness of 0.01″. In type C cable, the pairs have a characteristic impedance of 100 $\Omega$ and a foam polyethylene insulation with a thickness of 0.023″.

Ribbon or flat cables exhibit a stable characteristic impedance, a controlled distance between conductors, and, depending on type, a low crosstalk. The following crosstalk figures for various types of flat or ribbon cable are published by, and reproduced here with the kind permission of, the Belden Cable Company.

The characteristic impedance of the cables are specified either with one conductor connected to ground and the adjacent conductor used for the signal (GS configuration) or with the signal conductor sandwiched between two ground conductors (GSG configuration). The typical unbalanced crosstalk is expressed in percentages, with a rise time of between 3 and 7 ns. The frequency at which the unbalanced frequency-domain crosstalk equals the pulse crosstalk may be very approximately obtained by the relationship $f = 0.5\,\pi t$.

The Belden 9L280XX Series is a flat 0.05-pitch unshielded cable with a GS characteristic impedance of 150 $\Omega$ and a GSG impedance of 105 $\Omega$. The cable is available with 10–64 conductors. The cable dimensions and adjacent conductor unbalanced pulse crosstalk results are reproduced in Figure 4.19. The balanced crosstalk in the frequency domain is provided in Figure 4.20.

The 9L283XX Series is a flat 0.05-pitch shielded cable with a GS impedance of 45 $\Omega$ and a GSG impedance of 50 $\Omega$. The cable is available with 9–64 conductors. The cable dimensions and adjacent conductor pulse crosstalk are reproduced in Figure 4.21.

The 9GP10XX Series is a 0.05-pitch flat cable that incorporates a copper ground plane and is available with 20–60 conductors, of which one conductor is ground. The cable has a

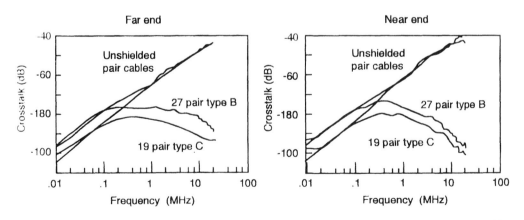

**Figure 4.18**  Mean values of near-end and far-end crosstalk for shielded and unshielded multipair cables.

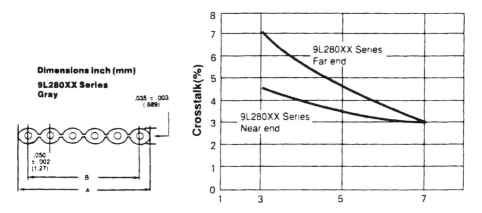

**Figure 4.19**   Belden 9L280XX flat cable dimensions and unbalanced pulse crosstalk.

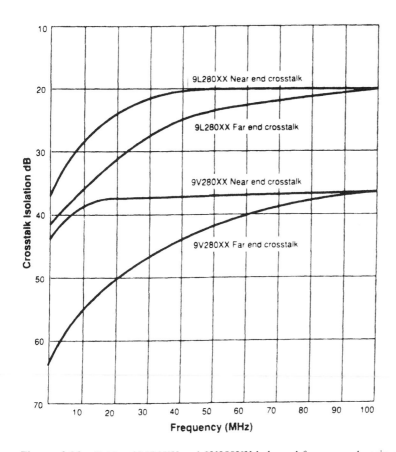

**Figure 4.20**   Belden 9L280XX and 9V280XX balanced frequency-domain crosstalk.

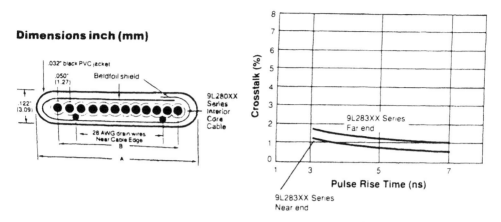

**Figure 4.21**  Belden 9L283XX cable dimensions and unbalanced pulse crosstalk.

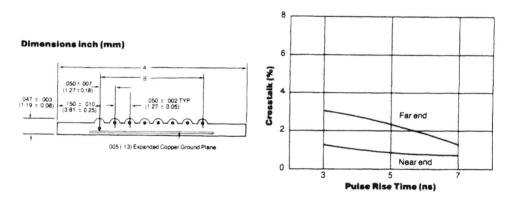

**Figure 4.22**  Belden 9GP10XX dimensions and unbalanced pulse crosstalk.

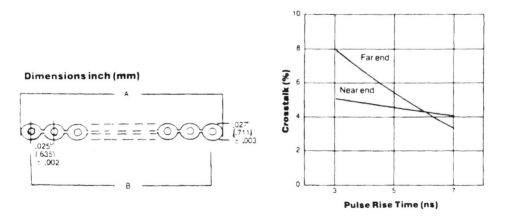

**Figure 4.23**  Belden 9L320XX unbalanced pulse crosstalk.

**Dimensions inch (mm)**

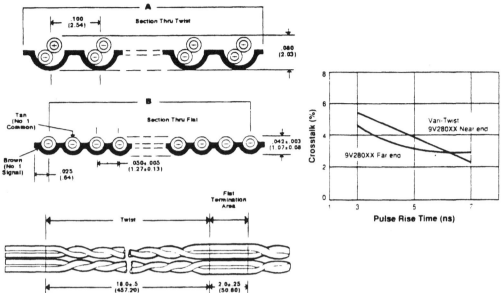

**Figure 4.24**    Belden 9V280XX dimensions and unbalanced pulse crosstalk.

GSG impedance of 60 Ω. The dimensions and unbalanced pulse crosstalk are given in Figure 4.22.

The 9L320XX Series is a 0.025-pitch cable available with 41 or 53 conductors. The GS impedance is 135 Ω, and the GSG impedance is 93 Ω. The dimensions and pulse crosstalk are reproduced in Figure 4.23.

The 9V280XX is a varitwist flat cable in which conductors are twisted as pairs for 18 inches followed by a 2-inch flat section provided for termination and then followed by a further 18 inches of twist. The cable has an 0.05 pitch and is available with 5–32 conductors. The dimensions and pulse crosstalk are illustrated in Figure 4.24, with the balanced frequency-domain crosstalk provided in Figure 4.20.

Ribbon coaxial cables containing from 4 to 25 coaxial cores are available. The Belden 9K50, 9K75, and 9K93 Series have 50-, 75-, and 93-Ω characteristic impedances, respectively. The near-end and far-end crosstalk for these types of cable is less than 0.1% for 3-, 5-, and 7-ns rise times.

### 4.4.3  Crosstalk on Lines with a Long Propagation Delay Relative to the Rise Time of the Source

With the increased use of high-speed logic involving emitter-coupled logic (ECL) and Schottky and gallium arsenide technology, designers have realized the importance of designing the interconnections as transmission lines terminated in their characteristic impedance. When the rise time of the logic is equal to or shorter than the propagation delay of the line, the crosstalk prediction is somewhat different than described thus far in this chapter. These high-speed terminated source transmission lines are the focus here.

The propagation delay of two PCB tracks side by side is given by

$$t_{pd} = 1.017 \sqrt{0.475\varepsilon_r + 0.67} \qquad [\text{ns/ft}]$$

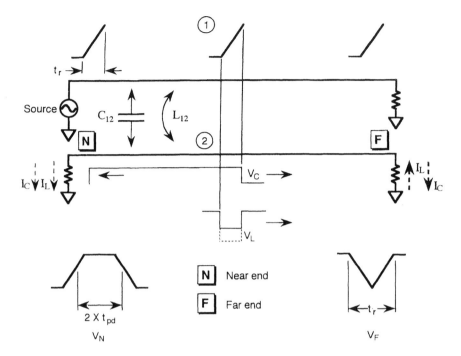

**Figure 4.25** Near-end and far-end crosstalk in a line where the propagation delay is longer than $t_r$.

where $\varepsilon_r$ is the relative dielectric constant of the PCB. Thus, for a typical glass epoxy PCB, for which the dielectric constant ($\varepsilon_r$) ranges from 4 to 5, the propagation delay ($t_d$) is 1.62–1.77 ns/ft, which is approximately half of the rise time of TTL, equals the rise time of Schottky TTL, and is approximately twice the rise time of ECL100K.

Figure 4.25 illustrates the inductive and capacitive crosstalk between two coupled lines. Consider a positive pulse, with rise time $t_r$ and duration much greater than the propagation delay of the line, that has just begun its propagation down the source line. A capacitively induced positive voltage causes a current flow in the receptor line back toward the near end, $N$, and also begins to propagate down towards the far end of the line, $F$. The inductively induced voltage is negative and results in a current that flows in the $N$ direction and thus adds to the capacitively induced current. The same inductively induced current, $I_L$, subtracts from the capacitively induced current, $I_C$, flowing toward the far end. When the source and receptor lines are terminated in their characteristic impedances, $I_L$ is greater than $I_C$ and the voltage propagating toward $F$, the forward voltage, has a negative polarity. As the pulse propagates further down the source line, the voltages induced in the receptor line add cumulatively to the forward crosstalk voltage, which is also propagating toward the far end of the line, assuming the propagation delays of the two lines are identical. Thus the negative-polarity forward crosstalk voltage, $V_l$, increases in direct proportion to the length of the line and has a pulsewidth approximately equal to the rise time of the source voltage. The backward voltage propagated toward the $N$ end of the line results in a continuous, constant current flow, so the voltage developed across the near-end termination impedance is constant, regardless of the length of the line, and of a duration longer than the rise time of the source voltage. As the source voltage reaches the end of the source line, the forward crosstalk voltage reaches the $F$ end of the receptor line and appears across the $F$ termination impedance. However, the backward crosstalk voltage must still propagate toward the $N$ end of the line in a time equal to the propagation delay of the line. Therefore, the voltage appearing across the terminating resistor at the *N* end has a duration equal to twice the propagation delay of the line, since it was developed from the moment the source voltage began its propagation down the line until the last backward voltage returned from the far end of the line.

Logic does not typically respond to an input voltage pulse with a duration equal to the rise time of the logic, even when the voltage level is above the logic threshold. The input-pulse duration must be several times the rise time of the logic before an output response occurs. It is for this reason that near-end crosstalk is more likely to result in EMI than is far-end crosstalk. A method of computing crosstalk in high-speed digital interfaces when the mutual inductance, mutual capacitance, and propagation delay of the line are known is provided in Ref. 7. An alternative to this is the use of one of the computer programs described in the next section.

### 4.4.4   Crosstalk Computer Programs

Many of the latest computer-aided design (CAD) PCB layout programs do allow crosstalk, and these include the ''Incases'' and ''Applied Simulation Technology'' programs described in Section 11.8

In addition to full PCB layout programs, other programs allow the characteristics of high-speeds transmission on PCB tracks, such as impedance, crosstalk, ringing, and time delay. One such program is GREENFIELD2 TM, from Quantic Laboratories. Also μWave SPICE, from EESOF, allows the computation of crosstalk, delay, and ringing. This program was used to compile the tables, presented in this section, of near-end and far-end crosstalk voltage for a number of PCB layouts and pulse rise times. Clayton R. Paul has developed a simple model that may be used with SPICE or PSPICE to predict crosstalk in coupled transmission lines. SPICE allows the transient analysis of transmission lines with a specified characteristic impedance and propagation delay but not the analysis of coupled transmission lines. The model developed by C. R. Paul extends the SPICE capability to coupled lines and allows the modeling of a line immersed in an inhomogeneous medium, such as tracks on a PCB (Ref. 11). A comparison of the crosstalk prediction with measured crosstalk data from PCBs shows excellent prediction accuracy.

In order to examine the effect on crosstalk voltage, a number of PCB configurations as well as source and receptor circuit impedances were modeled, using μWave SPICE. The initial configuration was that described in Ref. 9 and shown in Figure 4.26, in which the PCB track length was 7860 mils (20 cm), the track width, *W*, and distance between tracks, *S*, were 100 mils (2.5 mm), the PCB material had a relative permittivity of 5, and the height of the tracks above the ground plane was 62 mils (1.6 mm). The predicted near-end and far-end crosstalk voltage and circuit coupling parameter (Ref. 9), the measured voltage (Ref. 10), and the predicted crosstalk using μWave SPICE are shown in Figure 4.27. The rise time of the source signal is 50 ns, with a peak-to-peak voltage of 2.5 V. The source-circuit termination impedance is 50 Ω, and the receptor-circuit termination impedances are 50 Ω in case 3 and 1000 Ω in case 4. A very good correlation exists between the predicted crosstalk (Ref. 9) and the measured crosstalk, whereas an acceptable correlation exists between the μWave SPICE predicted values and the measured. In the following μWave SPICE plots V[1] is the near-end source waveform, V[2] is the far-end source waveform, V[3] is the near-end receptor waveform, and V[4] is the far-end receptor waveform.

The waveforms for correctly terminated (50-Ω) source and receptor circuits with a source rise time of 2 ns and a peak voltage of 0.8 V are shown in Figure 4.28. The positive near-end crosstalk exhibits an amplitude approximately twice the value of the negative far-end crosstalk, with a slightly greater pulsewidth, as expected.

Decreasing the distance *S* between the two tracks to 13 mil (0.33 mm) not only increases

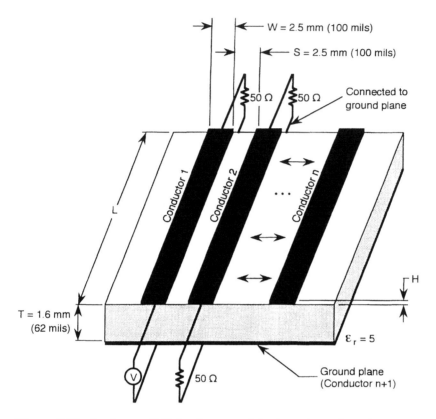

**Figure 4.26** Tracks on a PCB forming a coupled microstrip configuration.

the crosstalk voltage but increases the near-end and far-end crosstalk voltage pulsewidths, as shown in Figure 4.29. The pulsewidth increase is due to a modification in the microstrip characteristic impedance caused by the proximity of the two tracks.

Retaining the 50-Ω terminations but removing both source and receptor tracks from close proximity to the ground plane also changes the characteristic impedance of the line, resulting in an increase in the amplitude and pulsewidths of the near-end and far-end crosstalk, as shown in Table 4.6. Figure 4.30 illustrates the waveforms. The rise time of the voltage at the far end of the source line is degraded by reflections due to the mismatch between the line and the load impedance.

The case where the characteristic impedance of the source line matches the source load impedance (50 Ω) but the receptor terminations are 1000 Ω and are mismatched to the line is shown in Figure 4.31. Here we see that although the far-end crosstalk is initially negative, it changes sign and equals the near end crosstalk in amplitude. Due to the multiple reflections, which are clearly visible, the pulsewidths at 30 ns are much greater than in the matched case.

The potential for EMI to analog-signal lines may not be high with a 30-ns crosstalk pulsewidth. However, when the pulse repetition rate is in megahertz, a very real potential for EMI exists.

In examining the original matched configuration and extending the length of the line from 20 cm to 40 cm we see that the far-end crosstalk amplitude and the near-end crosstalk pulsewidth are both twice the value of the 20-cm-long line. Table 4.6 contains the predicted crosstalk for a number of risetimes, receptor-circuit termination impedances, distances between tracks, and

|  | Predicted crosstalk (mV) | Measured crosstalk* (mV) | Predicted crosstalk (μWave SPICE) (mV) |
|---|---|---|---|
| **50Ω Termination** |  |  |  |
| VNEXT(max) | 4.98 | 4.5 | 5.2 |
| VFEXT(max) | -2.35 | -2 | -2.5 |
| **1000Ω Termination** |  |  |  |
| VNEXT(max) | 24.3 | 25 | 29 |
| VFEXT(max) | 24.1 | 25 | 22 |
| **Per-Unit length Parameters** |  |  |  |

$C_{11} = 137.7pF/m \quad C_{21} = 5.15pF/m$
$C_{12} = 5.15pF/m \quad C_{22} = 137.7pF/m$

$L_{11} = 0.314\mu H/m \quad L_{21} = 0.036\mu H/m$
$L_{12} = 0.036\mu H/m \quad L_{22} = 0.314\mu H/m$

**Figure 4.27** Coupled-microstrip parameters with predicted and measured crosstalk for 50-ns rise time and a 2.5-V transition.

heights above the ground plane. As expected, the crosstalk is greater where the tracks are closer together and/or higher above the ground plane.

The microstrip model terminated in passive loads, used thus far, is useful in evaluating factors that mitigate crosstalk, such as the use of ground planes under coupled tracks, increasing the distance between tracks, and matching the termination impedance to the line impedance. In addition, the crosstalk with 1000-Ω receptor-circuit termination impedance may apply to the prediction of crosstalk from a logic-to-analog circuit, because analog-circuit input and source impedances are often close to 1000 Ω. The crosstalk voltage may be scaled when the source voltage is other than 0.8 V. For example, with a logic voltage of 3.6 V at a 2-ns risetime, a 1000-Ω termination impedance at both ends of the receptor track, $l = 20$ cm, $W = 2.5$ mm, $S = 2.5$ mm, the far-end crosstalk voltage is $(3.6/0.8) \times 0.6$ V = 2.7 V.

Further information on modeling crosstalk on printed circuit boards is provided in Ref. 10. Reference 13 describes a finite element method that may be used for modeling crosstalk on multiple PCB tracks in either a microstrip or stripline configuration. The author of Ref. 13 has described a computer program that can model crosstalk between 5 tracks when run on a PC and up to 50 tracks when run on a mainframe computer.

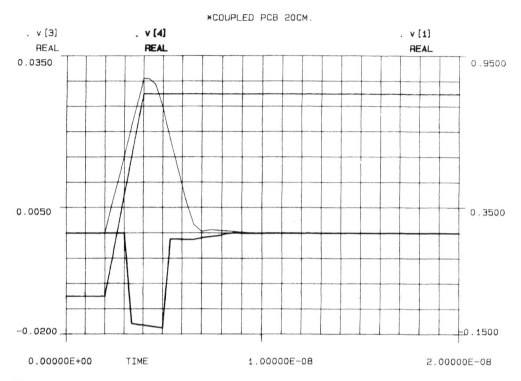

**Figure 4.28**  Crosstalk with $V_s = 0.8$ V, $t_r = 2$ ns, $W = 100$ mil, $S = 100$ mil, $l = 7860$ mil, $h = 62$ mil, $t = 0.001$ mil, $R_L = R_S = 50$ $\Omega$.

In predicting the crosstalk to a line connected to IC gates, the line source and load impedances are not the same and are complex (i.e., with real and imaginary components). Also, they change with input or output voltage level. Consider the receptor circuit shown in Figure 4.32, in which the input of a TTL gate is connected at the near end of the line and the output of a TTL gate is connected at the far end. For the positive excursion of the source voltage step, the Figure 4.32 configuration may be considered the worst case for crosstalk. The input of a TTL gate sources current that flows out of the gate and is sinked by the driver gate at the far end of the line. Thus, a TTL input tends to rise to a "1" level when disconnected from a low impedance. The positive near-end crosstalk is therefore developed across an impedance that is effectively connected to the logic supply and not to ground. The far-end impedance is low referenced to ground, for it is formed by the "on" impedance of a transistor.

μWave SPICE has the advantage that discrete, logic, and analog ICs and reactive loads may be modeled directly, whereas GREENFIELD offers only a SPICE interface. The worst-case TTL crosstalk configuration shown in Figure 4.32 was modeled using μWave SPICE. To ensure that this is indeed a worst-case configuration, the crosstalk was compared to that predicted for a negative source-voltage step and when the driver and input gate locations were interchanged. The Figure 4.32 configuration was found to be indeed the worst case. This configuration was used for cases where the track widths were fixed at 12 mil (0.3 mm) and the distance between the tracks was varied, as were the lengths for two values of board thickness (i.e., 34 mil [0.86 mm] and 71 mil [1.8 mm]).

The near-end and far-end waveforms for two 96-cm tracks located 0.86 mm above the ground plane with a spacing of 0.6 mm are shown in Figure 4.33. The amplitude of the near-

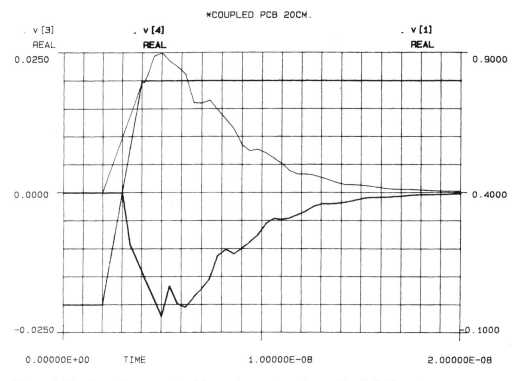

**Figure 4.29** Crosstalk as a result of decreasing track spacing to 13 mil (0.33 mm).

**Table 4.6** Crosstalk Values for the Coupled Microstrip

| W [mils] | S [mils] | L [mils] | L [cm] | H [mils] | $R_{L1}$ [$\Omega$] | $R_{L2}$ [$\Omega$] | $t_r$ [ns] | $N_E$ [V] | $F_E$ [V] |
|---|---|---|---|---|---|---|---|---|---|
| Source voltage = 0.8 V into Ro = 50 $\Omega$, $\varepsilon_r$ = 5 | | | | | | | | | |
| 100 | 100 | 7,860 | 20 | 62 | 50 | 50 | 2 | +0.03 | −0.019 |
| 100 | 13 | 7,860 | 20 | 62 | 50 | 50 | 2 | +0.13 | −0.11 |
| 100 | 13 | 7,860 | 20 | 62 | 1000 | 1000 | 2 | +0.23 | +0.2 |
| 100 | 100 | 15,720 | 40 | 62 | 50 | 50 | 2 | +0.03 | −0.027 |
| 100 | 100 | 7,860 | 20 | 62 | 1000 | 1000 | 2 | +0.6 | −0.3 |
| | | | | | | | | | +0.5 |
| 100 | 500 | 7,860 | 20 | 62 | 50 | 50 | 2 | +0.0024 | −0.0025 |
| 100 | 100 | 7,860 | 20 | 62 | 50 | 50 | 4 | +0.02 | −0.01 |
| 100 | 100 | 7,860 | 20 | 62 | 50 | 50 | 8 | +0.01 | +0.005 |
| 100 | 100 | 7,860 | 20 | 500 | 50 | 50 | 2 | +0.12 | −0.08 |
| 100 | 100 | 7,860 | 20 | 5,000 | 1000 | 1000 | 2 | +0.4 | +0.06 |
| Source voltage = 3.6 V into Ro = 50 $\Omega$, $\varepsilon_r$ = 5 | | | | | | | | | |
| 100 | 100 | 7,860 | 20 | 62 | 50 | 50 | 2 | +0.14 | −0.08 |

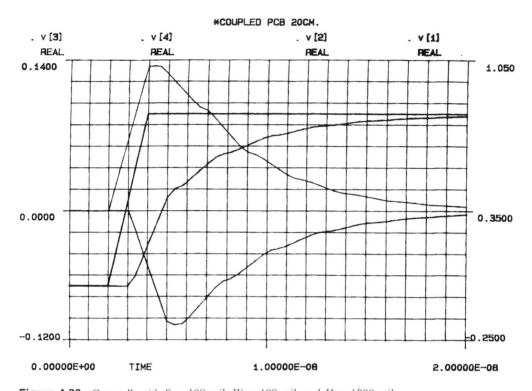

**Figure 4.30**   Crosstalk with $S = 100$ mil, $W = 100$ mil, and $H = 1000$ mil.

end crosstalk is higher than that of the far end, and the far end exhibits both a negative and positive voltage excursion, which is common with an unmatched load impedance. The far-end positive excursion is clamped to between 0.4 V and 0.6 V by the "on" transistor in the driver output circuit. In the μWave SPICE plot, V[7] is the output of the near-end TTL gate. A summary of the near-end and far-end crosstalk for line lengths up to 192 cm and tracks as close as 0.3 mm are shown in Table 4.7. Very often the receptor track is sandwiched between two or more source tracks, especially true for a data bus, and the impact of this configuration is shown in Table 4.7, where the far-end crosstalk voltage is seen to be approximately 1.62 times and the near-end 1.76 times that for a single source track. When the receptor track is on the outside of the two source tracks, the far-end crosstalk is 1.62 times and the near end 1.17 times that for a single source track.

Electromagnetic interference does not occur until the length of the track is 384 cm for a distance between tracks of 0.3 mm, as shown in Figure 4.34. The input voltage is 1.7 V, with a pulsewidth of 45 ns. The measured input current spike is 11 mA, so the input noise energy at which EMI occurs, given by $tIV$ (where $t$ is the pulsewidth, $I$ is the current, and $V$ is the voltage), is 0.87 nJ. At a line length of 192 cm, the input level is well above the TTL DC "1" level voltage of 1 V; but with a pulse width of only 26 ns, the TTL gate does not have time to respond. Had we modeled a faster type of logic, the permissible line length before EMI occurs would almost certainly be less than 384 cm. The near-end crosstalk voltage is independent of line length, and the 1.7-V level at a pulse width of 13 ns, developed with a 96-cm line length, may upset some fast logic types, such as 74H and FACT.

When the ground plane is removed from under the source and receptor tracks, source

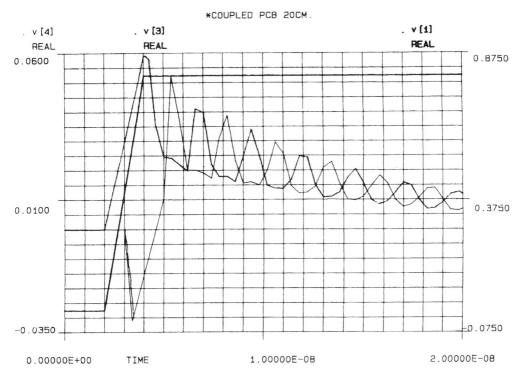

**Figure 4.31** Crosstalk with $W = 100$ mil, $S = 100$ mil, $H = 62$ mil, and $Z_\circ = 50\ \Omega$ but with the receptor load resistors changed to 1000 $\Omega$.

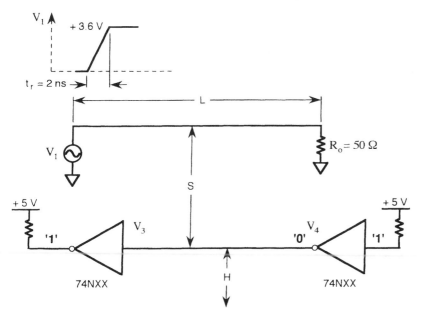

**Figure 4.32** Worst-case configuration for crosstalk to TTL gates.

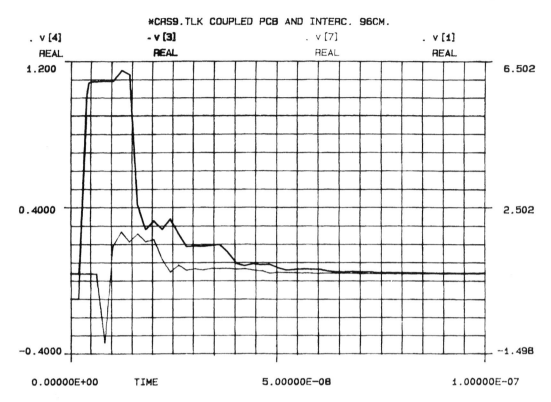

**Figure 4.33** Crosstalk to TTL gates with $S = 0.6$ mm, $l = 96$ cm, $h = 0.86$ mm, $V_s = 5$ V, and $t_r = 2$ ns.

line–to-load impedance mismatch is created. This results in an increase in the rise time of the source voltage developed across the far-end termination impedance and in an increase in the crosstalk pulsewidth, as illustrated in Figure 4.35 for 96-cm tracks with 2.5-mm spacing and a 5-V source voltage with a 2-ns risetime.

### 4.4.5  Electromagnetic Coupling

When the distance between conductors, or the distance between the signal conductors and a return conductor or ground plane, is large, the coupling prediction may be more accurately accomplished using the transmitting and receiving properties of electric current elements and loops, as described in Sections 2.2.2–2.2.6. Signal tracks on a PCB that are terminated in a high impedance and are far removed from a ground plane may be conveniently modeled by the electric current element using either the electric field receiving or the electric and magnetic field radiation characteristics of the element. When a track forms a loop, either circular or rectangular, then the receiving or transmitting properties of the loop may be used in a coupling prediction. The coupling may be loop to loop, loop to element, or element to element. When two loops are formed on the same PCB surface, the loop coupling is coplanar; when the coupling is from a loop on one printed circuit board to a loop on a second, adjacent board, the coupling is coaxial. The impedance of the loop and the load impedance must be included when predicting the loop current flow and the voltages developed around the loop. Chapter 7 addresses electromagnetic field-to-cable coupling and provides an equation for the inductance of a loop that may be used

**Table 4.7** Crosstalk on a TTL Signal Track in a Microstrip Configuration

| H [mm] | L [cm] | S [mm] | $N_E$ [V] | $N_F$ [$t_{pw}$ ns] | $F_E$ [V] | $F_F$ [$t_{pw}$ ns] |
|---|---|---|---|---|---|---|
| 0.86 | 96 | 0.3 | +1.7 | 13 | −0.37 | 4 |
|  |  |  |  |  | +0.38 | 13 |
| 0.86 | 96 | 0.6 | +1.0 | 13 | −0.46 | 4 |
|  |  |  |  |  | +0.2 | 13 |
| 0.86 | 96 | 7.2 | +0.65 | 13 | −0.35 | 4 |
|  |  |  |  |  | — | — |
| 0.86 | 96 | 2.5 | +0.42 to +0.70 | 13 | −0.25 | 4 |
|  |  |  |  |  | +0.05 | 13 |
| 0.86 | 96 | 12.5 | +0.008 | 13 | −0.0035 | 4 |
|  |  |  |  |  | +0.0015 | 13 |
| 0.86 | 192 | 0.3 | +1.7 | 26 | −0.8 | 4 |
|  |  |  | (2.1 pk) |  | +0.38 | 26 |
| 0.86 | 384 | 0.3 | +1.7 | 47 | −0.82 | 4 |
|  |  |  |  |  | +0.4 | 47 |
| 1.8 | 96 | 0.3 | +2.3 | 13 | −0.3 | 4 |
|  |  |  |  |  | +0.35 | 13 |
| 1.8 | 96 | 1.2 | +1.1 | 13 | −0.3 | 4 |
|  |  |  |  |  | +0.15 | 13 |
| 1.8 | 96 | 2.5 | +0.4 to +0.6 | — | −0.25 | 4 |
|  |  |  |  |  | +0.05 | 13 |
| Two source tracks, receptor on outside |  |  |  |  |  |  |
| 0.86 | 96 | 0.3 | +2 |  | −0.6 |  |
|  |  |  |  |  | +0.4 |  |
| Two source tracks, receptor on inside |  |  |  |  |  |  |
| 0.86 | 96 | 0.3 | +3 |  | −0.6 |  |
|  |  |  |  |  | +0.6 |  |

to calculate the loop impedance. Often, parallel impedances formed by IC loads or decoupling capacitors exist around the loop, in which case the loop may be divided into a number of loops sharing a load impedance. One rather idealized example of loop-to-loop coupling, with the loop dimensions, is illustrated in Figure 4.36. The receiving loop layout of Figure 4.36 is optimized for minimum error in the predicted received voltage by making the width of the loop small and the radius of the loop follow the radius of equipotential electric and magnetic fields from the source loop.

Equations (2.26)–(2.28) are used to compute the radiation from a current loop, and Eq. (2.21), which gives the receiving properties of a loop, is used to predict the induced voltage. Figure 4.37 shows the measured and calculated values for receptor loops short-circuited at one end and open at the other (Ref. 11). The resonance at approximately 70 MHz would typically be caused by cable-to-source loop impedance mismatch. The more realistic situation, where a driver output impedance and a receiver input impedance form the loads on either end of the loop, is found in Eq. (2.22).

The electric and magnetic quasi-static near fields, as we have seen in Chapter 2, are extremely nonuniform, reducing as a function of $1/r^2$ or $1/r^3$ with distance from the source. When the receptor loop is in the form of a circle or square with an area larger than the source loop, the nonuniformity of the field across the receptor loop and the change in the angle of the incident field as it cuts the receptor loop results in an error in the predicted voltage received by the loop.

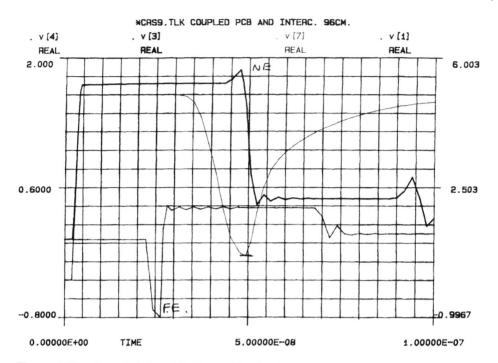

**Figure 4.34**  Crosstalk-induced EMI to a TTL-signal track. $S = 0.3$ mm, $h = 0.86$ mm, $l = 384$ cm, $V = 5$ V, $t_r = 2$ ns.

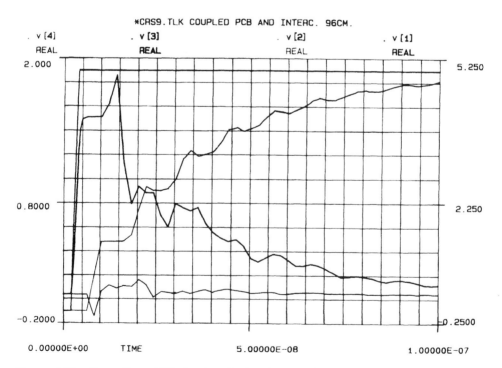

**Figure 4.35**  Crosstalk to a TTL-signal track when the ground plane is removed from under both source and receptor tracks.

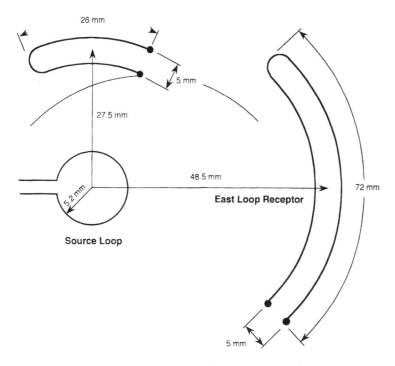

**Figure 4.36** Dimensions of source and receptor current loops.

The error for both coaxially and coplanar-oriented loops has been calculated (Ref. 12). Figure 4.38 presents the error in the open-circuit voltage developed at the terminals of the receiving loop as a function of the ratio $R/R_o$, where $R_o$ is the distance between the center of the loops and $R$ is the radius of the receiving loop, assuming the source loop is much smaller. The error is defined as

$$e = \frac{V_{near} - V_{uniform}}{V_{uniform}}$$

where

$V_{near}$ = voltage induced in a nonuniform near field
$V_{uniform}$ = voltage induced into the loop in a uniform field

The measured voltage must be corrected as follows:

$$V = \frac{V_{near}}{1 + e}$$

When the error is negative, which is true of the coaxially oriented loop, $V$ is greater than $V_{near}$; when the error is positive, which is true for the coplanar-oriented loops, $V$ is less than $V_{near}$.

### 4.4.6 Computer Program for Evaluating Shielded and Unshielded Cable/ Wire–to–Cable/Wire Coupling

The following computer program was written using Microsoft's QuickBASIC Version 3.0. As with the programs in Chapter 2, there are no line numbers, and alphanumeric labels are used

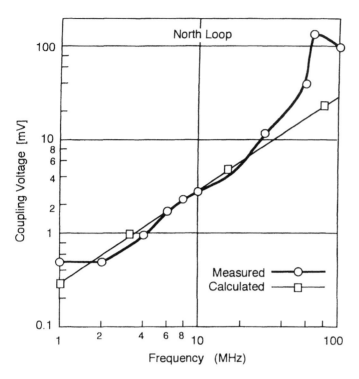

**Figure 4.37**  Measured and calculated coupling voltages. (© 1985, IEEE.)

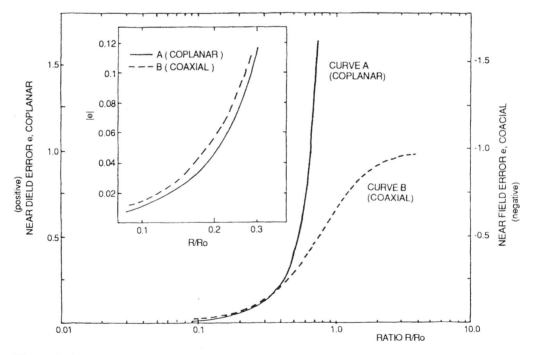

**Figure 4.38**  Near-field coupling error for coaxial and coplanar loops as a function of $R/R_o$. (© 1989, IEEE.)

to identify subroutines, etc.
This program is available from EMC Consulting Inc.


'Initialize constants
C=3E+08: RC=377:PI=3.14159
'Initialize variables used in EMC program
I=1: F=100000!: L=1: W=1: R=1: THETAD=90: CUR=1: THETA=90
BB=1: CCABLE$="RG5B/U": VCABLE$="10AWG": ACOTEXT$="AC": ICAC=1
VARIABLE1$="Frequency": VARIABLE2$="Current": VAR1=F: VAR2=ICAC
UNITS1$="Hz": UNITS2$="Amps": RCC=200: D=1: LW=1: ZO=200: ZI=50:
LL=1: A=.006: IO=1: EO=1: RL=50: RD=50: H=1: ZW=377
CZ1=50: DS1=.181: DI1=.053: RS1=1.2: CZ2=1: DS2=.1019: DI2=.1019: RS2=1.018
ICP=1: TR=.001
H=H*39.37        'Convert from meters to inches
D=D*39.37
LW=LW*3.28      'Convert from meters to feet


GOTO ComputeCoupling:
WireToWireMenu:
CLS
PRINT "WIRE TO WIRE COUPLING–R.J. Mohr, IEEE, Vol.EMC-9 No.2, Sept, 1967:
PRINT " 'Coupling Between Open and Shielded Wire Lines Over a Ground Plane' "
PRINT
PRINT" [A] CULPRIT wire: ",,CCABLE$
PRINT "   [B] VICTIM wire: ",,VCABLE$
PRINT "   [T] Type of analysis: ",,ACOTEXT$
PRINT "     [1] ";VARIABLE1$;"=":LOCATE 7,43:PRINT VAR1;UNITS1$
PRINT "     [2] ";VARIABLE2$;"=":LOCATE 8,43:PRINT VAR2;UNITS2$
PRINT " [S] Source Impedance:",,RCC;" ohms"
PRINT " [L] Load Impedance:",,RD;" ohms"
PRINT " [H] Height above ground plane:",H;" inches"
PRINT " [D] Distance between wires:",D;" inches"
PRINT " [C] Coupled Length:",,LW;" feet"
PRINT " [X] End program"
PRINT " Which parameter would you like to change (A,B,C...)?"
PRINT "----------------------------------------------------------------"
PRINT "   ";OUTtext$
PRINT "   ";IDTEXT$;"=";OUT1;"V"
PRINT "----------------------------------------------------------------"


IF PEAKI>ICAC THEN
PRINT "TIGHTLY COUPLED, VICTIM SHIELD CURRENT APPROACHES CULPRIT CUR-
RENT"
'IF UTU$="Y" THEN LOCATE 21,1: PRINT "VICTIM SHIELD CURRENT =",PEAKI
END IF


'Test for f>200kHz
FTEST=F/1000/100
IF FTEST>2 THEN

```
PRINT "* WARNING, METHOD NOT VALID ABOVE 200kHz FOR SHIELDED
CABLES *"
'PRINT "Victim shield current calculated using unshielded-unshielded coupling
PRINT "Load and source impedance are the victim shield termination impedance"
END IF


IF (CCABLE$=VCABLE$) AND (CCNUM<16) THEN
PRINT "   *PROGRAM ASSUMES SAME CABLE FOR SHIELDED TO SHIELDED
CASE*
END IF


CouplingInputLoop:
SEL$=INKEY$
IF LEN(SEL$)>0 THEN LOCATE 15,1
IF LEN(SEL$)>0 THEN PRINT SPACE$(70)
LOCATE 15,4
IF (SEL$="A") OR (SEL$="a") THEN GOSUB CulpritCable:
IF (SEL$="B") OR (SEL$="b") THEN GOSUB VictimCable:
IF (SEL$="T") OR (SEL$="t") THEN
INPUT "Enter A for AC or T for Transient ",ACOT$
IF (ACOT$="a") OR (ACOT$="A") THEN ACOT$="A" ELSE ACOT$="T"
END IF
IF (SEL$="S") OR (SEL$="s") THEN INPUT "Enter source impedance ",RCC
IF (SEL$="L") OR (SEL$="l") THEN INPUT "Enter load impedance ",RD
IF (SEL$="H") OR (SEL$="h") THEN INPUT "Enter height above ground plane ",H
IF (SEL$="D") OR (SEL$="d") THEN INPUT "Enter distance between wires ",D
IF (SEL$="C") OR (SEL$="c") THEN INPUT "Enter coupled length ",LW
IF (ACOT$="A") AND (SEL$="1") THEN INPUT "Enter frequency ",F
IF (ACOT$="A") AND (SEL$="2") THEN INPUT "Enter current ",ICAC
IF (ACOT$="T") AND (SEL$="1") THEN INPUT "Enter peak current ",ICP
IF (ACOT$="T") AND (SEL$="2") THEN INPUT "Enter risetime ",TR
IF (SEL$="X") OR (SEL$="x") THEN GOTO WireToWireExit:
IF LEN(SEL$)>0 THEN GOTO ComputeCoupling:
GOTO CouplingInputLoop:


CulpritCable:
GOSUB Database:
PRINT "Choose the CULPRIT wire from the above list [Number]:";CCNUM
INPUT X$:IF LEN(X$)>0 THEN CCNUM=VAL(X$)
RESTORE
FOR N=1 TO 27
READ CCABLE$,CZ1,DS1,DI1,RS1
IF CCNUM=N THEN GOTO ComputeCoupling:
NEXT N
PRINT "Must be a number from 1 to 27"
GOTO CulpritCable:


VictimCable:
GOSUB Database:
PRINT "Choose the VICTIM wire from the above list [NUMBER]:";VCNUM
```

```
INPUT X$:IF LEN(X$)>0 THEN VCNUM=VAL(X$)
RESTORE
FOR N=1 TO 27
READ VCABLE$,CZ2,DS2,DI2,RS2
IF VCNUM=N THEN GOTO ComputeCoupling:
NEXT N
PRINT "Must be a number from 1 to 27"
GOTO VictimCable:
RETURN

ComputeCoupling:
IF ACOT$="A" THEN
ACOTEXT$="AC"
VARIABLE1$="Frequency"
VAR1=F
UNITS1$="Hz"
VARIABLE2$="Current"
VAR2=ICAC
UNITS2$="Amps"
OUTtext$="Peak AC Voltage"
ELSE
ACOTEXT$="Transient"
VARIABLE1$="Peak Current"
VAR1=ICP
UNITS1$="Amps"
VARIABLE2$="Interference signal rise time"
VAR2=TR
UNITS2$="sec"
OUTtext$="Peak Transient Voltage"
END IF

THAU=TR/2.2
RS1C=RS1*LW*.001
RS2C=RS2*LW*.001
M=7E-08*LOG(1+(2*H/D)^2)/LOG(10)*LW

IF CCNUM<16 THEN
IF VCNUM<16 THEN
VCABLE$=CCABLE$
CZ2=CZ1: DS2=DS1: DI2=DI1: RS2=RS1
GOTO ShieldedToShielded:
ELSEIF VCNUM>15 THEN
GOTO ShieldedToOpen:
END IF

ELSEIF CCNUM>15 THEN
IF VCNUM<16 THEN
GOTO OpenToShielded:
ELSEIF VCNUM>15 THEN
GOTO OpenToOpen:
```

```
END IF
END IF

OpenToOpen:
IF ACOT$="T" THEN GOTO OpenToOpenTransient:
IDTEXT$="AC UNSHIELDED-UNSHIELDED Coupling"
LS2=1.4E-07*LOG(4*H/DI2)/LOG(10)
AL2=1/(SQR(1+(2*PI*F*LS2/(RCC+RD))^2))
VD=2*PI*F*M*ICAC*(RD/(RCC+RD))*AL2
PEAKI=VD/RD
OUT1=VD
GOTO WireToWireMenu:

OpenToOpenTransient:
IDTEXT$="Transient UNSHIELDED-UNSHIELDED Coupling"
LS2=1.4E-07*LOG(4*H/DI2)/LOG(10)
TM=THAU*LOG((RCC+RD)*THAU/LS2)/((RCC+RD)*THAU/LS2-1)
AAL2=1/(1-LS2/((RCC+RD)*THAU))*(EXP(-TM/THAU)-EXP(-(RCC+RD)*TM/LS2))
VDMAX=M*ICP/THAU*RD/(RCC+RD)*AAL2
OUT1=VDMAX
GOTO WireToWireMenu:

OpenToShielded:
IF ACOT$="T" THEN GOTO OpenToShieldedTransient:
IDTEXT$="AC UNSHIELDED-SHIELDED Coupling"
LC2=1.4E-07*LOG(4*H/DI2)/LOG(10)*LW
LS2=1.4E-07*LOG(4*H/DI2)/LOG(10)
AS2=1/(SQR(1+(2*PI*F*LS2/(RS2C))^2))
AC2=1/(SQR(1+(2*PI*F*LC2/(RCC+RD))^2))
VD=2*PI*F*M*ICAC*(RD/(RCC+RD))*AS2*AC2
OUT1=VD
GOTO WireToWireMenu:

OpenToShieldedTransient:
IDTEXT$="Transient UNSHIELDED-SHIELDED Coupling"
LS2=1.4E-07*LOG(4*H/DS2)/LOG(10)*LW
TM=THAU*LOG(THAU/(LS2/RS2C))/(THAU/(LS2/RS2C)-1)
AAS2=1/(LS2/RS2C/THAU-1)*(EXP(-RS2C*TM/LS2)-EXP(-TM/THAU))
VDMAX=M*ICP/THAU*RD/(RCC+RD)*AAS2
OUT1=VDMAX
GOTO WireToWireMenu:

ShieldedToOpen:
IF ACOT$="T" THEN GOTO ShieldedToOpenTransient:
IDTEXT$="AC SHIELDED-UNSHIELDED Coupling"
LS1=1.4E-07*LOG(4*H/DI1)/LOG(10)
AL2=1/(SQR(1+(2*PI*F*LS2/(RCC+RD))^2))
AS1=1/(SQR(1+(2*PI*F*FS1/RS1C)^2))
VD=2*PI*F*M*ICAC*(RD/(RCC+RD))*AL2*AS1
```

```
OUT1=VD
GOTO WireToWireMenu:

ShieldedToOpenTransient:
IDTEXT$="Transient SHIELDED-UNSHIELDED Coupling"
LS1=1.4E-07*LOG(4*H/DI1)/LOG(10)
TM=THAU*LOG(THAU/(LS1/RS1C))/(THAU/(LS1/RS1C)-1)
AAS1=1/(LS1/RS1C/THAU-1)*(EXP(-RS1C*TM/LS1)-EXP(-TM/THAU))
VDMAX=M*ICP/THAU*RD/(RCC+RD)*AAS1
OUT1=VDMAX
GOTO WireToWireMenu:

ShieldedToShielded:
IF ACOT$="T" THEN GOTO ShieldedToShieldedTransient:
IDTEXT$="AC SHIELDED-SHIELDED Coupling"
LS2=1.4E-07*(LOG(4*H/DS2)/LOG(10))*LW
LS1=1.4E-07*(LOG(4*H/DS1)/LOG(10))*LW
AS1=1/(SQR(1+(2*PI*F*LS1/(RS1C*RS1C))))
AS2=1/((1+(2*PI*F*LS2/RS2C^2))^.5)
AC2=1/((1+(2*PI*F*LS2/((RCC+RD))^2))^.5)
VD=2*PI*F*M*ICAC*(RD/(RCC+RD))*AS1*AS2*AC2
OUT1=VD
GOTO WireToWireMenu:

ShieldedToShieldedTransient:
IDTEXT$="Transient SHIELDED-SHIELDED Coupling"
LS2=1.4E-07*LOG(4*H/DS2)/LOG(10)*LW
E2#=0
TM=LS2/RS2C
TransientConvergence:
DD#=1-THAU*RS2C/LS2
E#=EXP(-RS2C/LS2*TM)
E2NEW#=M*ICP*RS2C/LS2*((THAU*RS2C/LS2/DD#^2)*(EXP(-TM/THAU)-
E#)+TM*(RS2C/LS2)/DD#*E#)
IF     E2NEW#-E2#>0    THEN     TM=TM+.05*LS2/RS2C    ELSE    GOTO
OutOfConvergenceLoop:
E2#=E2NEW#
GOTO TransientConvergence:
OutOfConvergenceLoop:
VDMAX=RD/(RCC+RD)*E2#
OUT1=VDMAX
GOTO WireToWireMenu:
Database:
CLS:PRINT" DATA BASE": PRINT
PRINT "Number RG#/U Type     Number ##AWG Type
PRINT "[1]    RG5B/U    [16]  10AWG"
PRINT "[2]    RG6A/U    [17]  12AWG"
PRINT "[3]    RG8A/U    [18]  14AWG"
PRINT "[4]    RG9B/U    [19]  16AWG"
```

```
PRINT "[5]    RG13A/U    [20]  18AWG"
PRINT "[6]    RG29/U     [21]  22AWG"
PRINT "[7]    RG55A/U    [22]  24AWG"
PRINT "[8]    RG58C/U    [23]  26AWG"
PRINT "[9]    RG59A/U    [24]  28AWG"
PRINT "[10]   RG122/U    [25]  30AWG"
PRINT "[11]   RG141/U    [26]  32AWG"
PRINT "[12]   RG142/U    [27]  34AWG"
PRINT "[13]   RG180/U"
PRINT "[14]   RG188/U"
PRINT "[15]   RG223/U":PRINT ""
RETURN


'CABLE DATA FOR PROGRAM
'RG(#)/U,Char. Imp.,Shield Diameter,Inner Conducter Diameter,Resistance
'
   DATA RG5B/U,50,.181,.053,1.2
   DATA RG6A/U,75,.185,.028,1.15
   DATA RG8A/U,50,.285,.086,1.35
   DATA RG9B/U,50,.280,.085,.8
   DATA RG13A/U,75,.280,.043,.95
   DATA RG29/U,53.5,.116,.032,4.75
   DATA RG55A/U,50,.116,.035,2.55
   DATA RG58C/U,50,.116,.035,4.7
   DATA RG59A/U,75,.146,.022,3.7
   DATA RG122/U,50,.096,.029,5.7
   DATA RG141/U,50,.116,.036,4.7
   DATA RG142/U,50,.116,.039,2.2
   DATA RG180/U,95,.103,.011,6.0
   DATA RG188/U,50,.06,.019,7.6
   DATA RG223/U,50,.116,.035,2.55
   DATA 10AWG,1,.1019,.1019,1.018
   DATA 12AWG,1,.0808,.0808,1.619
   DATA 14AWG,1,.0641,.0641,2.575
   DATA 16AWG,1,.0508,.0508,4.099
   DATA 18AWG,1,.0403,.0403,6.510
   DATA 20AWG,1,.0320,.0320,10.35
   DATA 22AWG,1,.0253,.0253,16.46
   DATA 24AWG,1,.0201,.0201,26.17
   DATA 26AWG,1,.0159,.0159,41.62
   DATA 28AWG,1,.0126,.0126,66.17
   DATA 30AWG,1,.0100,.0100,105.2
   DATA 32AWG,1,1,.203,
   DATA 34AWG,1,1,.160,
   DATA ***
WireToWireExit:
H=H/39.37 'Convert from meters to inches
D=D/39.37
```

LW=LW/3.28 'Convert from meters to feet
GOTO ENDER:

Spacebar:
PRINT "Press the spacebar to continue"
SBLOOP:
SPCBR$=INKEY$
IF SPCBR$=" "THEN GOTO Returner:
GOTO SBLOOP:
Returner:
CLS
RETURN

ENDER:
END

## REFERENCES

1. R. J. Mohr. Coupling between open wires over a ground plane. IEEE EMC Symposium Record. Seattle, WA, July 23–25, 1968.
2. R. J. Mohr. Coupling between open and shielded wire lines over a ground plane. IEEE Trans. on Electromag. Compat. Vol. EMC-9, September 1967.
3. G. L. Matthaei, L. Young, and E. M. T. Jones. Microwave Filters, Impedance-Matching Networks and Coupling Structures. Artech House, Dedham, MA, 1980.
4. J. A. Defalco. Predicting crosstalk in digital systems. Computer Design, June 1973.
5. N. Holte. A crosstalk model for crosstranded cables. International Wire and Cable Symposium Proceedings, 1982.
6. J. A. Krabec. Crosstalk and shield performance specifications for aluminum foil shielded twisted pair cable (MIL-STD-49285). International Wire and Cable Symposium Proceedings, 1987.
7. A. Feller, H. R. Kaupp, and J. J. Digiacomo. Crosstalk and reflections in high speed digital systems. Proceedings Fall Joint Computer Conference, 1965.
8. C. R. Paul. A simple SPICE model for coupled transmission lines. Proceedings of the IEEE EMC Symposium. Atlanta, GA, 1988.
9. R. L. Khan and G. I. Costache. Considerations on modeling crosstalk on printed circuit boards. Proceedings of the IEEE EMC Symposium, 1987.
10. C. R. Paul and W. W. Everett. Modeling crosstalk on printed circuit boards. RAdc-TR-85-107. Rome Air Development Center, Griffiss AFB, New York, Phase Report, July 1985.
11. W. J. Adams, J. G. Burbano, and H. B. O'Donnell. SGEMP induced magnetic field coupling to buried circuits. IEEE EMC Symposium Record. Wakefield, MA, 1985.
12. S. Iskra. H field sensor measurement error in the near field of a magnetic dipole source. IEEE Trans. on Electromag. Compat. Vol. 31, No. 3, August 1989.
13. R. L. Khan and G. I. Costache. Finite element method applied to modeling crosstalk problems on printed circuit boards. IEEE Trans. on Electromag. Compat. Vol. 31, No. 1, February 1989.

# 5

# Components, Emission Reduction Techniques, and Noise Immunity

## 5.1 COMPONENTS

### 5.1.1 Introduction to the Use of Components in Electromagnetic Compatibility

Components are used either to reduce noise current by increasing the impedance of the current path or to reduce noise voltage by decreasing the impedance of a shunt path. A combination of these functions is often used in the design of a filter.

All components and interconnections contain unintentional (parasitic) circuit elements, such as inductance, capacitance, and resistance, and often a combination of these. An understanding of the magnitude of these parasitic elements and the characteristics of components over a range of frequencies will ensure the correct choice and application of a component. Some components are lossy and have the property of converting RF noise energy into heat. Loss may be preferable to shunting the noise energy into an alternative path. All components and interconnections resonate, and an understanding of the effect of resonance is often vital in achieving EMC or solving an EMI problem.

Other uses of components in EMC are to slow down the edges of a waveform (thereby to reduce emissions) and to clamp voltages to a specific level and thus suppress transients.

### 5.1.2 Impedances of Wires, Printed Circuit Board Tracks, and Ground Planes

A common mistake in an EMC/EMI evaluation is not including the impedance of a conductor, especially when the conductor is used to make a ground connection or a cable shield termination. Conductors exhibit an intrinsic or internal impedance composed of an internal inductance, due to the internal magnetic flux, an AC resistance, due to the concentration of current near the outer periphery of the conductor, known as the *skin effect*, and a DC resistance. In addition, conductors exhibit an external inductance, giving rise to an external magnetic flux. This external inductance is often referred to as *self-* or *partial inductance* and, when it is unintentional, as *parasitic inductance*. Above some frequency, the AC resistance is higher than the DC resistance of the conductor and increases as the square of the frequency. At the same frequency, the internal inductance begins to reduce by the square of the frequency. The external inductance is frequency independent.

Wires with a circular cross section are commonly used as interconnections. In circular wires the frequency above which the AC resistance begins to increase and the internal inductance decrease is when the skin depth approaches the radius of the wire. The AC resistance of a copper wire is given by the following equation:

$$R_{AC} = (0.244d\sqrt{f} + 0.26)R_{DC}$$

where

$d$ = diameter of the wire [cm]

$f$ = frequency [Hz]

$R_{DC}$ = DC resistance [$\Omega$]

Wires invariably exhibit higher inductive reactance, due to self-inductance, than resistance, even at low frequencies. The self-(external) inductance of a wire is usually higher than the internal inductance. The impedance of a wire with a circular cross section at least 15 cm from a ground structure is given by

$$L = 0.002l\left(2.303 \log\left(\frac{4l}{d}\right) - 0.75\right) \quad [\mu H] \tag{5.1}$$

where

$l$ = length of wire [cm]

$d$ = diameter [cm]

Equation (5.1) assumes that the relative permeability of the conductor is 1. When $\mu_r$ is not 1, the self-inductance is given by

$$L = 0.002l\left(2.303 \log\left(\frac{4l}{d}\right) - 1 + \frac{\mu_r}{4}\right) \quad [\mu H] \tag{5.2}$$

In comparing the inductances of a square conductor and a circular conductor we find that when the diameter of a circular wire equals the dimension of the sides of a square conductor, the square conductor exhibits a lower self-inductance and AC resistance. When the cross-sectional area of the square and circular conductors are the same, the inductance is the same. The inductance of a square conductor is approximately

$$L = 0.002l\left(2.303 \log\left(\frac{2l}{W + t}\right) + 0.498\right) \quad [\mu H] \tag{5.3}$$

where

$W$ = width of the conductor [cm]

$t$ = thickness of the conductor [cm]

Equation (5.3) applies approximately to a rectangular conductor. For a PCB or similar ground plane where $t < 0.025W$, the inductance approximately equals

$$L = 0.002l\left(2.303 \log\left(\frac{2l}{W}\right) + 0.5\right) \quad [\mu H] \tag{5.4}$$

We see that replacing a circular wire with a strap that has a thickness approximating the diameter of the wire but that has a much greater width will result in reduced inductance.

Table 5.1 illustrates the impedance of some straight copper wires with circular cross section. At 10 Hz, the impedance equals the DC resistance of the wire; but even at frequencies as low as 10 kHz, the impedance is significantly higher than the DC resistance due to the self-inductance. The impedance values in Table 5.1 are for wires at some great distance from ground and below the frequency at which the length of the wire equals a quarter wavelength ($\lambda/4$). In practice, conductors such as wires, PCB tracks, and ground straps are typically close to grounded conductive structures or other wires. One effect on the inductance of the conductor due to proximity effects is to reduce the inductance, as shown in Figure 4.11.

Published data, even in the most recent publications, on the impedance of a square ground plane based exclusively on the intrinsic or metal impedance of the 1-oz copper foil conductor gives the impedance as follows (the unit $\Omega/sq$ is explained in Section 6.1.3):

| | |
|---|---|
| 10 Hz | 0.812 m$\Omega$/sq |
| 10 MHz | 1.53 m$\Omega$/sq |
| 100 MHz | 3.72 m$\Omega$/sq |

To disregard the self-inductance of the plane, which presents an impedance to an AC current flowing from one end of the plane to the other, is to minimize the impedance. Table 5.2

**Table 5.1** Impedance of Straight Circular Copper Wires

| | Impedance of wire ($\Omega$) | | | | | | | |
|---|---|---|---|---|---|---|---|---|
| | AWG #2, $d = 0.65$ cm | | | | AWG #10, $d = 0.27$ cm | | | |
| Freq. | 1 cm | 10 cm | 1 m | 10 m | 1 cm | 10 cm | 1 m | 10 m |
| 10 Hz | 5.13 μ | 51.4 μ | 517 μ | 5.22 m | 32.7 μ | 327 μ | 3.28 m | 32.8 m |
| 1 kHz | 18.1 μ | 429 μ | 7.14 m | 100 m | 42.2 μ | 632 μ | 8.91 m | 116 m |
| 100 kHz | 1.74 m | 42.6 m | 712 m | 10 | 2.66 m | 54 m | 828 m | 11.1 |
| 1 MHz | 17.4 m | 426 m | 7.12 | 100 | 26.6 m | 540 m | 8.28 | 111 |
| 5 MHz | 87.1 m | 2.13 | 35.5 | 500 | 133 m | 2.7 | 41.3 | 555 |
| 10 MHz | 174 m | 4.26 | 71.2 | — | 266 m | 5.4 | 82.8 | — |
| 50 MHz | 870 m | 21.3 | 356 | — | 1.33 | 27 | 414 | — |
| 100 MHz | 1.74 | 42.6 | — | — | 2.66 | 54 | — | — |
| 150 MHz | 2.61 | 63.9 | — | — | 4.0 | 81 | — | — |
| | Impedance of wire ($\Omega$) | | | | | | | |
| | AWG #22, $d = 0.065$ cm | | | | AWG #26, $d = 0.04$ cm | | | |
| Freq. | 1 cm | 10 cm | 1 m | 10 m | 1 cm | 10 cm | 1 m | 10 m |
| 10 Hz | 529 μ | 5.29 m | 52.9 m | 529 m | 1.33 m | 13.3 m | 133 m | 1.33 |
| 1 kHz | 531 μ | 5.34 m | 53.9 m | 545 m | 1.38 m | 14 m | 144 m | 1.46 |
| 10 kHz | 681 μ | 8.89 m | 113 m | 1.39 | 1.81 m | 21 m | 239 m | 2.68 |
| 100 kHz | 4.31 m | 71.6 m | 1.0 | 12.9 | 6.1 m | 90.3 m | 1.07 | 14.8 |
| 1 MHz | 42.8 m | 714 m | 10 | 129 | 49 m | 783 m | 10.6 | 136 |
| 5 MHz | 214 m | 3.57 | 50 | 645 | 241 m | 3.86 | 53 | 676 |
| 10 MHz | 428 m | 7.14 | 100 | — | 481 m | 7.7 | 106 | — |
| 50 MHz | 2.14 | 35.7 | 500 | — | 2.4 | 38.5 | 530 | — |
| 100 MHz | 4.28 | 71.4 | — | — | 4.8 | 77 | — | — |
| 150 MHz | 6.42 | 107 | — | — | 7.2 | 115 | — | — |

shows the impedance for a number of ground straps and a square 1-oz (0.03-mm-thick) copper ground plane based on Eq. (5.4) and the measured values. The voltage across these structures, due to a controlled current flow at frequencies of 2 MHz, 10 MHz, and 20 MHz, when measured differentially, results in good correlation between predicted and measured values. The measurement setup must ensure that the conductors carrying current to and from the conductor under test are shielded and are far removed from the conductor under test. In a practical PCB layout, current may flow in a conductor that is in close proximity to the ground plane and return in the ground plane. In this configuration the return current tends to be confined to the ground plane immediately below the supply conductor. This current concentration may be expected to increase the impedance of the ground plane, whereas above approximately 2 MHz the impedance is lower than when the supply conductor is far removed from the ground plane.

Although inductance is defined only for complete loops, the assigning of partial inductance values to a section of a current loop, such as the ground plane, and signal trace inductances of PCB layouts is described in Ref. 8. In Ref. 9 the inductance of the ground plane with a signal trace located above it is determined using Maxwell's equations. The value of ground plane inductance when it is used as the signal return is given by Eq. (5.4b).

$$L_{\text{ground}} = (0.2 \times l) \times ln\left(\frac{h \times \pi}{w} + 1\right) \tag{5.4b}$$

Where $l$ is the length of the ground plane under the trace (i.e., the current return path), in meters $h$ is the height, in metres, and $w$ is the width of the ground plane in meters.

The inductance of the trace above the ground plane can be found by replacing the width of the ground plane ($w$) with the width of the trace, assuming the thickness of the trace is much smaller than the width. Equation (5.4b) is used to calculate the impedance of the ground plane with the trace located at 1 cm and 1 mm above the ground plane. The calculated impedance at 1 cm is close to the measured impedance. A measurement made at 2 mm does not agree well

**Table 5.2**  Impedance of Ground Straps and a 1-oz Copper Ground Plane

| Frequency (MHz) | Measured impedance [$\Omega$][b] | Calculated impedance [$\Omega$] | Measured 1 cm above GP[a] | Calculated 1 cm above GP | Calculated trace 1 mm above GP |
|---|---|---|---|---|---|
| | 1-oz copper ground plane 12.6 cm long by 10 cm wide | | | | |
| 2 | 0.312 | 0.45 | | | |
| 10 | 1.67 | 2.25 | 0.64 | 0.43 | 0.045 |
| 20 | 2.46 | 4.5 | 1.14 | 0.86 | 0.09 |
| 50 | | | 1.76 | 2.15 | 0.225 |
| | 10-oz copper ground plane 3.2 cm long by 3.2 cm wide | | | | |
| 10 | 0.36 | 0.47 | | | |
| 20 | 0.56 | 0.94 | | | |
| | Impedance of a 12.6-cm-long by 2-cm-wide 3-mm-thick braid | | | | |
| 10 | 2.84 | 4.77 | | | |
| 20 | 7.3 | 9.5 | | | |
| | Impedance of a 12.6-cm-long by 0.8-cm-wide, 1.5-mm-thick tinned copper wire | | | | |
| 10 | 5.5 | 6.2 | | | |
| 20 | 14.3 | 12.4 | | | |

[a] Measured inductance with the supply conductor 1 cm above the ground plane.
[b] Measured impedance with the supply conductor far removed from the ground plane.

with the calculation, the reason is almost certainly due to radiated coupling to the measurement probe in the test setup, resulting in a higher measured voltage across the ground plane.

The reason for the reduction in the ground plane impedance is the mutual inductance between the supply conductor and the ground plane, which facilitates the return current flow in the plane. Table 5.2 illustrates the reduction in the impedance of a 12.6-cm-long by 10-cm-wide ground plane with the supply conductor 1 cm and 1 mm above the plane compared to the return conductor far removed from the ground plane. The reduction in impedance due to the proximity of a wire, used for either a supply or signal, to a return wire is discussed in further detail in Chapter 7, on cable shielding.

The importance of including conductor impedance in any EMC prediction or EMI investigative work is illustrated in Case Study 12.1 in Chapter 12. One common practice is the temporary grounding of circuits or structures by use of an approximately 3-ft-long wire in order to determine if a ground connection is required. If no detectable change occurs as a result of grounding, the reason may be that the impedance of the conductor, which may be up to 300 $\Omega$, is just too high to function as an effective ground connection.

The influence of conductor impedance on the performance of a component can also be beneficial. For example, high-frequency attenuation in a filter may be higher than predicted, due to the self-inductance of a wire. It can also be detrimental, as in the case of increased radiation from an enclosure when a small-diameter high-impedance ground wire is replaced with a lower-impedance ground connection. Thus, for the purpose of EMC, all conductors should be considered as components exhibiting impedance.

In equipment designed for space applications, the need may arise for a high thermal resistance between structures and, at the same time, a low-impedance bond. One technique that may be applicable when the structures are close together is the use of a shorted transmission line and increased parasitic capacitance between the two structures. A practical example is shown in Figure 5.1. Decreasing the distance between the tracks of the line and increasing the length of

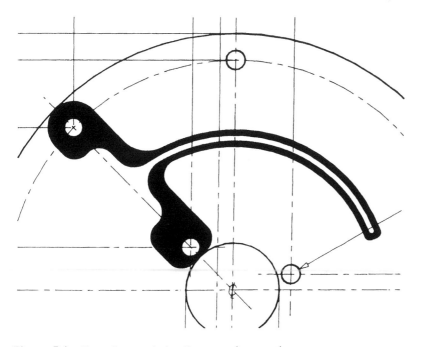

**Figure 5.1**   Shorted transmission-line ground connection.

the adjacent filled-out sections of the line reduces the line impedance and increases the parasitic capacitance, respectively. Unfortunately, the thermal resistance decreases as well. An equation for the impedance of a short-circuited transmission line is (Ref. 1)

$$Z_{sc} = Z_o \tan(\pi l f \sqrt{\varepsilon_r} / 150) \qquad (5.5)$$

where

$Z_o$ = characteristic impedance of the line

$l$ = length of the line

$f$ = frequency [MHz]

$\varepsilon_r$ = relative permittivity of the material between the conductors of the line

$\sqrt{\varepsilon_r}$ may be replaced by $(3 \times 10^8) \sqrt{LC}$.

The measured impedance of the capacitor and shorted transmission line with a characteristic impedance of 215 Ω and the geometry as shown in Figure 5.1 is compared to the same length of straight track in the following table:

| Frequency [MHz] | Impedance (Ω) | |
| --- | --- | --- |
| | Shorted transmission line | Straight track |
| 15 | — | 19 |
| 54 | 20 | — |
| 63 | — | 50 |
| 108 | 40 | — |
| 162 | — | 447 |
| 287 | — | 350 |
| 364 | 200 | — |
| 1000 | — | 125 |
| 1100 | 51 | — |
| 1200 | 10 | — |

A conductor used as an interconnection is usually only part of a larger circuit that contains a return path. Intentional or parasitic capacitance, typically across a load or between the conductor and a ground structure, may form a series resonant circuit with the inductance of the conductor. The input impedance of a series resonant circuit is low, and the output voltage may be higher than the input voltage, depending on the Q of the circuit. The Q will be high when the load resistance is high. An equation for the calculation of the Q of a series resonant circuit is provided in upcoming Figure 5.29. Due to the high currents and high voltages present, a series resonance is very often the cause of high levels of radiated or conducted emissions.

Conductors in close proximity to ground may also form a quasi-parallel resonant circuit that exhibits a higher impedance of the wire, and these effects may be measured on a network analyzer. When the distance between a conductor and a return path is not much less than the length of the conductor, the circuit may be modeled as a loop. In practice it is the parasitic capacitance to a grounded structure or across a load included in series with the loop that dictates the resonant frequency. However, assuming a loop far removed from a grounded structure and terminated in a low value of resistance, the resonance may coincide with the distance around the perimeter of the loop. When the perimeter of a loop reaches a resonant length, the input impedance varies from a very high to a very low value, as described in Section 2.2.5. The resonant frequency and impedance of $LC$ circuits with and without resistance is provided in Appendix 4.

When the return path, or a grounded structure, is closer to a conductor than the length of the conductor, a transmission line is formed. For most practical applications the transmission line can be considered lossless, above approximately 100 kHz, the characteristic impedance is

$$Z_o - \sqrt{\frac{L}{C}}$$

The characteristic impedance of a number of conductor and PCB track configurations is provided in Appendix 1.

The velocity of a wave traveling down a transmission line is

$$v = \frac{1}{\sqrt{LC}} \quad \text{or} \quad \frac{3 \times 10^8}{\sqrt{\varepsilon_r}} \quad [\text{m/s}]$$

where $\varepsilon_r$ is the relative permittivity of the material between the conductors of the line.

The propagation delay of a line is

$$\tau = l\sqrt{LC}$$

where $l$ is the line length, in meters. And the propagation constant per unit length of the line is

$$\gamma = \sqrt{LC}$$

When a transmission line is terminated in its characteristic impedance, $Z_o$, the input impedance equals $Z_o$.

When a transmission line is formed unintentionally, the terminating impedance is unlikely to equal the characteristic impedance of the line. The input impedance of a short-circuited line is given by Eq. (5.5). When the short-circuited line length equals $\lambda/4$, $3\lambda/4$, $1.25\lambda$, . . . , the input impedance is theoretically infinite, and when the line length is $\lambda/2$, $\lambda$, $1.5\lambda$, . . . , it is zero. The measured impedance of the shorted transmission line is lower at frequencies that correspond to a line length of $\lambda/4$ or multiples thereof because of the intentionally increased capacitance at the input of the line.

The input impedance of an open-circuited line is

$$Z_{oc} = \frac{Z_o}{\tan\left(\dfrac{\pi l f\sqrt{\varepsilon_r}}{150}\right)}$$

The input impedance is theoretically zero when the line length equals $\lambda/4$ and infinite when the line length is $\lambda/2$. When the termination impedance of a transmission line is higher than the characteristic impedance, the output voltage will be higher than the input voltage when the line length equals $\lambda/4$ or $3\lambda/4$, etc.

For a transmission line terminated in an impedance $Z_t$ that does not equal $Z_o$, the input impedance of the line, $Z_i$, equals

$$Z_i = Z_o \frac{Z_t + Z_o \tan(\pi l f\sqrt{\varepsilon_r}/150)}{Z_o + Z_t \tan(\pi l f\sqrt{\varepsilon_r}/150)}$$

Low-impedance power distribution and returns on PCBs are important for both digital and analog circuits. For example, a lower connection impedance results in a smaller voltage developed by noise currents flowing on the line. A useful method of achieving a low-impedance power distribution for digital logic, plus some built-in decoupling capacitance, is by the use of a "Q Pac" bus bar containing $+V_{CC}$ and $0$-$V$ tracks soldered either perpendicular to the PCB or horizontally under the ICs. The use of such bus bars does not fully eliminate the need for decoupling capacitors located close to ICs and designed to reduce the area of the "current loop." However, the number of decoupling capacitors required may be reduced.

### 5.1.3  General Wiring Guidelines

These wiring guidelines are useful for design engineers and for incorporation into EMC control plans but are no substitute for a detailed prediction, which should be carried out for potentially critical designs and layouts. Many exceptions to the following guidelines exist!

### 5.1.4  Circuit Classification

Circuits that interface between units or subsystems should be allocated a classification. For example, the following classification scheme might be used:

| Class | Type of signal |
|-------|----------------|
| I     | Digital, low current control, and filtered/regulated power |
| II    | Analog or video |
| III   | High current control, e.g., relay switching |
| IV    | AC and DC unfiltered line voltage |

### 5.1.5  Wire Separation

Wire bundles of differing EMC classification should be physically separated from each other. Where no metallic barrier exists between them, the following minimum separation distance should be maintained:

A 10-cm minimum between Class III wire bundles and Class II wire bundles
A 5-cm minimum between all other categories of wire bundles.

Maximum separation must be maintained between "clean" wires and "noisy" wires. This is especially important in wiring of filters, where the input and output wires of the filter must be isolated as much as possible.

### 5.1.6  Internal Unit/Equipment Wiring

The area of current loops on PCBs, board interconnections, and unit interconnections should be the practical minimum. The design aim should be that the maximum loop area be less than 4 cm$^2$. Larger loops should be subdivided into smaller loops arranged so that their generated fields cancel. Loops and wires should cross at right angles to reduce coupling. Analog wiring, digital wiring, and RF wiring should not be bundled together and should be separated as far as possible. Power lines should be twisted two-wire circuits or in some cases twisted shielded pair. Care must be taken to ensure that the distance between the twisted pairs is at least 1.5 times the twist length for proper isolation. *Twist length* is defined as the reciprocal of the number of full twists per inch. Compliance with this guideline usually requires that separate harnesses be used for each of the twisted pairs. If the distance cannot be increased, this criterion can be met by increasing the number of twists per inch, thereby reducing the twist length.

### 5.1.7 External Unit/Equipment Wiring

Circuits that interconnect equipment should provide minimum loop coupling and maximum field cancellation by twisting the return with the high side wires. Both signal and power circuits should be twisted with their respective returns and assigned adjacent pins in connectors when safety permits. Multiple circuits using a common return should be twisted as a group. Twists, where possible, should be as high as 16 per foot. Wires between units should follow the most direct route, since interference coupling is a direct function of length. Twisted-pair wiring should be used for balanced or quasi-balanced circuits to reduce magnetic field coupling, as well as in energized circuits requiring a high DC current, in order to reduce the DC magnetic field. Coupling caused by common-mode current flow is not reduced by twisting; instead, source reduction or shielding must be used.

### 5.1.8 Wire Shielding

When unit/subsystem designers use wire shielding to meet emission requirements or to protect susceptible circuits, these shields should cover the twisted pair or twisted group rather than individual wires. The shield should be connected concentrically around the backshell to minimize radio frequency (RF) potentials at the cable termination. If one outer shield is used (e.g., twisted shielded pair), the shield should be connected to chassis at both ends. High E field susceptibility circuits may include an additional inner electrostatic shield connected to signal ground at the analog source end only, but isolated from the chassis. No shield should intentionally carry current except for coax cables used for RF.

High-voltage and high-frequency high-power circuits are likely to be E field generators. Circuits that carry large currents at low frequencies are likely to be H field generators. These circuits should be shielded to control emissions. Sensitive circuits having low-current, high-impedance inputs are likely to be susceptible to E fields and should be shielded. Shielding effectiveness is a positive function of a shield's thickness, conductivity, and percent coverage, as described in Chapter 7.

When circuits are transformer coupled, common-mode noise can be reduced by using transformers with shielded primary and secondary windings. The reduction in common-mode noise due to the use of a single shield connected to case ground is approximately 20 dB from 100 Hz to 1 MHz. With three shields connected in a single transformer—one connected to the input ground, the second to case ground, and the third to output ground—the reduction in common-mode noise may be as high as 80 dB at 20 kHz, reducing to 35 dB at 1 MHz. The shield must be connected to a quiet ground, such as a chassis, for if a single shield is used that is connected to a noisy signal ground, the common-mode coupling may increase with the shield in place.

Cable shield continuity should be maintained over the entire length of the cable. When a shielded wire goes through a junction box, the wire should be shielded inside the junction box. Unshielded segments of shielded wires should be kept to an absolute minimum. Peripheral shield termination hardware at the connector or conductive epoxy potting of connector backshells can be used to obtain this objective.

### 5.1.9 Radio Frequency Shielding

Where high-RF currents have to be carried and it has been determined that an EMI emission problem may arise, the following shielding recommendation should be followed. A second shield over the shield carrying RF current return should be used, as is the case in a triax cable, or the RF return may be via a second center conductor, as in a twinax cable. Multiple connections of triax shield introduce ground current loops or common-mode currents and therefore should not be used to shield sensitive input circuits. However, reduced emissions may result for high-power RF when the connections to the ground plane are made at distances less than λ/4. If multiple grounding of triax is impractical, apparent continuous RF grounding may be possible. Apparent continuous RF grounding of a triax cable is achieved by routing the triax cable, with its insulating outer sheath, close to case and structure, thus causing significant capacitance between the triax shield and structure.

Cables connecting sensitive RF input circuits should be shielded by use of semirigid coax where possible. If this is not possible, then a triax cable should be used with the outer shield connected to chassis at both ends. The inner shield should be used for the signal current return. When such cables are routed external to a unit and are located unavoidably close to a high-power RF emitter (e.g., an antenna), the cable should be routed as close to a metal ground plane as feasible.

### 5.1.10 Components Used in Emission Control and to Increase Noise Immunity

#### 5.1.10.1 Capacitors

Capacitors are used for filtering, for control of rise and fall times, and to provide a low-impedance path at high frequencies. The impedance of an ideal capacitor decreases linearly with increasing frequency. The AC equivalent circuit of a practical capacitor, shown in Figure 5.2, contains an equivalent series resistance (ESR) and an equivalent series inductance (ESL). At some frequency, a capacitor is a series resonance circuit with an impedance equal to the ESR. Above this resonance, the impedance of the capacitor increases with frequency due to the ESL. The length of the leads is an important factor in assessing the self-inductance, and thus the resonant frequency, of a capacitor. The following are examples of the typical self-inductance for a number of low-inductance-type capacitors.

Metallized polyester lead length = 6 mm each, $L = 12$ nH
MKT Radial polyester (Mylar) capacitor stacked film

| Lead spacing (mm) | 7.5 | 10 | 15 |
|---|---|---|---|
| Self-inductance (nH) | 5 | 6 | 7 |

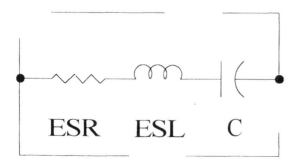

**Figure 5.2** AC equivalent circuit of a capacitator.

Polypropylene capacitors—radial

| | | | |
|---|---|---|---|
| Capacitor's body height (h) | 9 | 12 | 15 |
| Self-inductance (nH) | 15 | 17 | 20 |

High rel. polyester (Mylar)

Lead length =3 mm each, $L = 20$ nH

The self-resonant frequencies (SRF) of a number of values of ceramic CCR series capacitors, with lead length of 1.6 mm each, were measured, with the following results:

| Capacitance | | SRF [MHz] | $L$ [nH] |
|---|---|---|---|
| 1 | μF | 1.7 | 9 |
| 0.1 | μF | 4 | 16 |
| 0.01 | μF | 12.6 | 16 |
| 3300 | pF | 19.3 | 16 |
| 1800 | pF | 25.5 | 16 |
| 1100 | pF | 33 | 21 |
| 820 | pF | 38.5 | 21 |
| 680 | pF | 42.5 | 21 |
| 560 | pF | 45 | 22 |
| 470 | pF | 49 | 22 |
| 390 | pF | 54 | 22 |
| 330 | pF | 60 | 21 |

The equivalent series resistance of a capacitor is given by the Q of the capacitor, which is typically quoted for a frequency of 1 MHz. The Q of a capacitor is given by $X_c$/ESR, where $X_c$ equals $1/2\pi fC$, with $f$ in hertz and $C$ in farads or $f$ in MHz and $C$ in μF. Thus for a typical 30-pF capacitor with a Q of 931 at 1 MHz, the ESR is 5.7 Ω, for a typical 0.01 = μF capacitor with a Q of 168 at 1 MHz, the ESR is 95 mΩ.

When the noise voltage to be attenuated covers a wide frequency range, the impedance of a single capacitor may not be sufficiently low, especially above and below the resonant frequency of the capacitor. The use of two, three, or even four different capacitances in parallel will ensure a low impedance over a wide range of frequencies. For example, a 10-μF tantalum, a 1-μF ceramic, a 1000-pF ceramic, and a 200-pF ceramic all in parallel will ensure an impedance lower than 50 Ω from 350 Hz to 100 MHz. The resonant frequencies of capacitors are valid for the extremely short lead lengths quoted. As seen in Section 5.1.2, the impedance of a conductor due to self-inductance is appreciable even when the length is short. Thus, the inductance of conductors used to connect capacitors will reduce the resonant frequency of the capacitor in circuit and therefore increase the high-frequency impedance. When an EMI/EMC problem occurs at a specific frequency, the value of the capacitor should be chosen so that the resonant frequency of the capacitor and its connections matches the problem frequency, thereby achieving the lowest possible impedance.

The traces of a PCB should go to the capacitor connection and then continue from there as shown in the correct connection of Figure 5.3. In the incorrect connection of Figure 5.3, the PCB trace inductance is in series with the capacitor self-inductance, which lowers the resonant frequency. Better still is the connection in which the capacitor is directly between the traces, as shown for surface-mount capacitors in Figure 5.4, without the use of vias. If vias must be

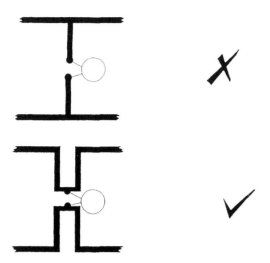

**Figure 5.3**   Correct and incorrect PCB trace connections for a capacitor.

used, they should have a large diameter. Surface-mount chip capacitors may have extremely low inductance when the traces connecting them are properly laid out, as shown in Figure 5.4.

The total inductance of a chip capacitor is determined by its length and width ratio and by the mutual inductance coupling between its electrodes. Thus a 1210 chip size has a lower inductance than a 1206 chip. An 0805 chip size, NPO-type ceramic chip capacitor with values from 0.5 pF to 0.033 μF manufactured by AVX has a resonant frequency of 150 MHz for a 1000-pF value; thus the self-inductance is 1.2 nH. A 100-pF 0805 capacitor has a resonant frequency of 380 MHz; thus the self-inductance is 1.76 nH. Both the resonant frequency and the impedance above resonance varies with chip size. Thus the 1210 has a lower impedance

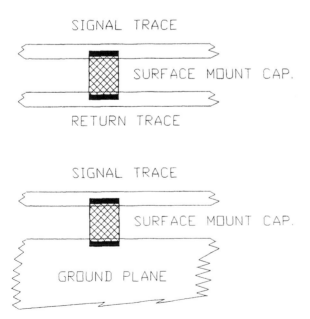

**Figure 5.4**   Ideal PCB connection, shown for a surface-mount capacitor.

above resonance than does the 1206 size. The ceramic formulation also affects the impedance at resonance, so the NSPO 1000-pF 0805-size chip has an impedance of 0.1 $\Omega$ at the resonant frequency of 150 MHz, whereas an 1000-pF with X7R ceramic formuation has the same resonant frequency, but the impedance is 0.4 $\Omega$.

AVX has manufactured a low-inductance capacitor, the low-inductance chip array (LICA); the self-resonant frequency and the approximate inductance of the LICA versus other capacitor designs are shown in Figure 5.5.

Multilayer chip (MLC) capacitors are constructed of a monolithic block of ceramic containing two sets of offset, interleaved planar electrodes that extend to two opposite surfaces of the ceramic dielectric. The great advantage of the MLC capacitor is its low ESR and low ESL, making it ideal for use in the input and output filters of switching-power supplies. High-value MLC capacitors have an inductance of approximately 3 nH. A 24-$\mu$F MLC capacitor manufactured by AVX has an ESR of 0.001 $\Omega$ at the resonant frequency of 50 kHz compared to an ESR of 0.1 $\Omega$ for a low-ESR tantalum. Low-inductance MLC capacitors have a resonant frequency of 25 MHz for a 0.1-$\mu$F with an ESR of 0.06 $\Omega$ at resonance. The typical inductance of a 0.1-$\mu$F with a 0.508 chip size is 0.6 nH, and for the 0612 chip size it is 0.5 nH. Thus these 0.1-$\mu$F low-inductance MLC capacitors are an ideal choice as decoupling capacitors.

Microwave capacitors are designed with extremely high resonant frequencies and therefore extremely low self-inductance. Although the series resonant frequency is invariably lower, a

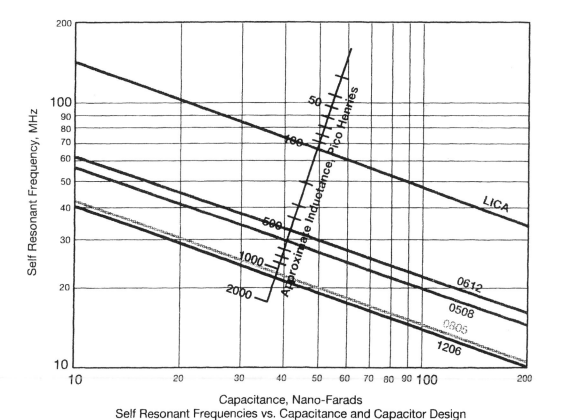

Capacitance, Nano-Farads
Self Resonant Frequencies vs. Capacitance and Capacitor Design

**Figure 5.5**  Self-resonant frequency and inductance for the LICA versus other chip capacitors. (Reproduced by kind permission of AVX.)

parallel resonant frequency also exists. Figure 5.6 shows the variation of impedance wth capacitor value, chip size, and ceramic formulation.

For an AVX AQ11-14 capacitor with a 100-pF capacitance, the series inductance is 0.4 nH, and the ESR is 0.034 $\Omega$ at 100 MHz and 0.062 $\Omega$ at 500 MHz. Figure 5.7 shows the series resonant frequency of a single-layer (SLC) capacitor, which has a series inductance of an incredibly low 0.035 nH. ATC manufactures a low-ESR microwave capacitor with a series inductance of 0.129 nH and, for a 20-pF capacitor, an ESR of 0.04 $\Omega$ at 500 MHz and 0.055 $\Omega$ at 1000 MHz.

It has been found empirically that when two similar surface-mount capacitors, for which the capacitance values and resonant frequencies are close together, are mounted either one on top of the other or side by side, the upper resonant frequency disappears. However, if the two capacitors are moved further apart, the two distinct resonant frequencies can be seen. Figure 5.8 shows the series resonant frequency of SLC µWave capacitors from 1 pF to over 5000 pF.

The feedthrough type of capacitor exhibits characteristics closest to those of an ideal capacitor. The construction and insertion loss for tubular- and discoidal-type feedthroughs are shown in Figures 5.9–5.12. The insertion loss is valid only when used with a 50-$\Omega$ series resistance. When the noise voltage source impedance is low, the attenuation will be much reduced and is limited at resonance by the ESR of the capacitor. In these situations, a series resistance and/or inductance may be required to achieve sufficient attenuation. When both source and load impedances are low, the capacitor may not be the correct component to use to achieve a significant attenuation. When using a capacitor to attenuate low-frequency noise or noise from a low-impedance source, values of one to hundreds of microfarads may be required.

The feedthrough capacitor, due to its construction, has a very low self-inductance. The correct method of mounting the feedthrough is to bolt or solder the case directly to the chassis, bulkhead, etc. When a length of wire is used to connect the case to ground, the performance is severely compromised, as shown in Figure 5.13.

The discoidal capacitor exhibits a lower impedance than the feedthrough type and has found a use in filtered connectors.

For modern LSI or VLSI IC packages, the use of surface-mount capacitors between VCC planes and ground planes, as discussed in Chapter 11, is the correct decoupling method.

Figure 5.6 shows that the use of a 0.1-µF capacitor with an X7R formulation has a low impedance and very low ESR over the 20–100-MHz frequency range and would be ideal if used as decoupling for digital circuits with clock and data frequencies operating over this range. For other materials and capacitors with higher inductance, a lower value of capacitance should be used over the same frequency range, but this will result in an increase in ESR.

Sufficient decoupling capacitors, at least one per three ICs when they are in close proximity and one per IC when they are spaced apart, should be the goal.

For surface-mount capacitors connected directly, or through vias, to the VCC and ground plane, an added trace between the capacitor and the VCC pin of the IC can reduce radiation from the IC and reduce ground bounce if the trace is long enough. The additional trace does not reduce the resonant frequency of the capacitor. Although this practice is controversial, in many practical cases it has been slightly beneficial. The addition of a resistor between the capacitor and the VCC pin of the device will increase the bounce in the VCC at the IC but will reduce ground bounce and reduce emissions from the IC. Figure 5.14a illustrates the recommended decoupling capacitor connection and Figure 5.14b the resistor.

Choice of the correct value of decoupling capacitor is important both in reducing supply and return-line noise voltage and in reducing radiation from the supply and return connections. Consider a case where the combined inductance of the supply and return tracks is 100 nH and the combined series resistance is 10 m$\Omega$. Two TTL gates are connected at the end of the supply

### IMPEDANCE VS. FREQUENCY
0612 0.1μF - X7R

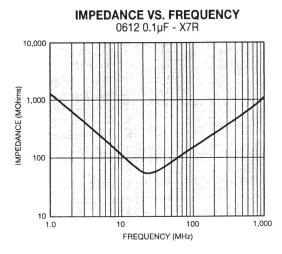

### IMPEDANCE VS. FREQUENCY
0805 VS. 0508 0.1μF - Y5V

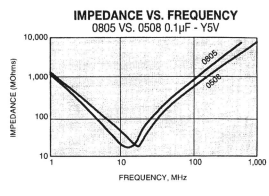

### ESR VS. FREQUENCY
0612 0.1μF - X7R

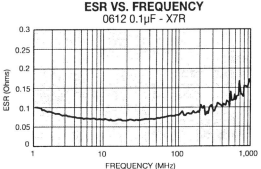

### TYPICAL CAPACITANCE VS. TEMPERATURE

### TYPICAL INDUCTANCE FOR 100nF CAPACITANCE

| Chip Size | pH |
|-----------|-----|
| 0508 | 600 |
| 0612 | 500 |

### RECOMMENDED SOLDER LANDS

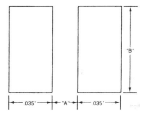

| Chip Size | "A" | "B" |
|-----------|--------|--------|
| 0508 | .020" | .080" |
| 0612 | .030" | .120" |

**Figure 5.6** Variation of impedance with capacitor value, chip size, and ceramic formulation from AVX. (Reproduced by kind permission of AVX.)

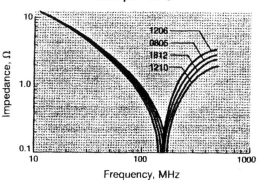

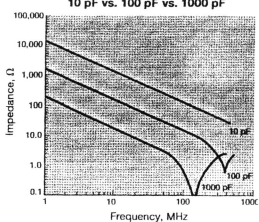

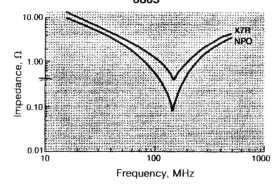

**Figure 5.7**   Series resonant frequency versus capacitance for an AVX SLC. (Reproduced by kind permission of AVX.)

and return connections. The following table shows the influence of the decoupling capacitor on the magnitude of a 5-ns current pulse, $i_{S/R}$, and of a 5-ns negative voltage spike, $v_{S/R}$, on the supply and return connections that is generated by the TTL gates changing states at the same time:

| $C$ | ESL [nH] | $i_{S/R}$ [mA] | $v_{S/R}$ [V] |
|---|---|---|---|
| — | — | 16 | 0.15 |
| 300 pF | 22 | 16 | 0.13 |
| 0.1 μF | 16 | 2 | 0.03 |

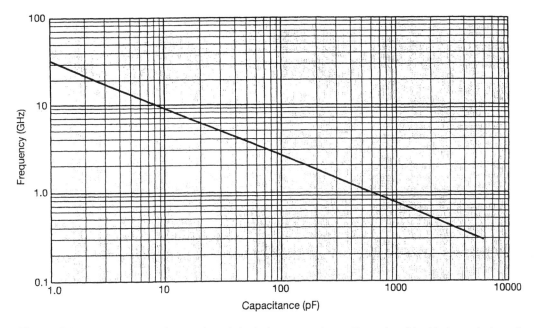

**Figure 5.8** Series resonant frequencies of single-layer capacitors. (Reproduced by kind permission of AVX.)

The 0.1-μF decoupling capacitor supplies the remaining 14 mA during the gate transition, and the reduced current flow in the supply connections results in less radiation from the circuit.

The importance of locating a decoupling or reservoir capacitor close to the IC, which requires or supplies the transient current, is illustrated in the following example. A MOSFET analog switch connects a 3-V supply to a device that has a capacitance of 5800 pF. A large transient on the 3-V supply is observed as the analog switch connects the 3 V to the load device. The circuit and transient voltages are shown in Figure 5.15a. The reason for the transient was that the reservoir capacitor was located approximately 10 cm from the analog switch. By moving the capacitor to the pins of the analog switch, the transient voltage was reduced, as shown in Figure 5.15b, and the load device functioned correctly.

The feedthrough capacitor is designed for use between compartments of equipment or as the power line connection through the case of a filter, as shown in Figure 5.16a. However, the

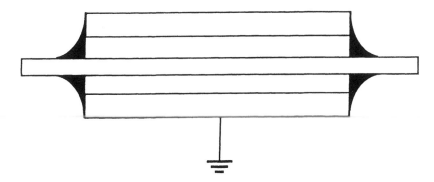

**Figure 5.9** Tubular ceramic feedthrough capacitor construction.

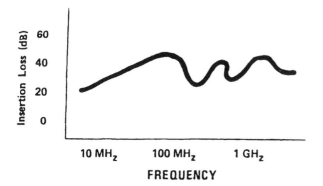

**Figure 5.10**  Insertion loss for typical ceramic feedthrough capacitor. (Reproduced courtesy of Erie Technological Products.)

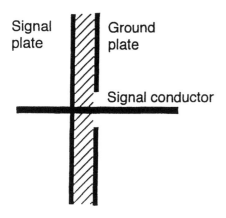

**Figure 5.11**  Typical construction of a discoidal capacitor.

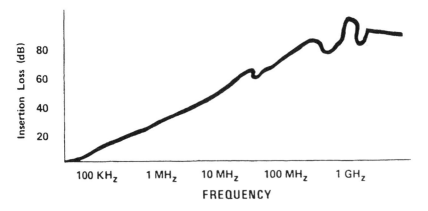

**Figure 5.12**  Insertion loss of a discoidal capacitor in a 50-Ω system. (Reproduced courtesy of Erie Technological Products.)

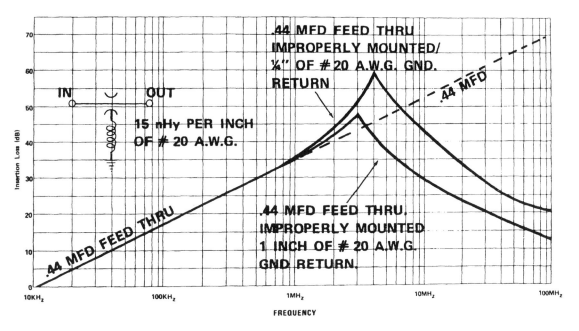

**Figure 5.13** Effect of adding a lead to a feedthrough capacitor. (© MuRata Erie. Reproduced courtesy of Erie Technological Products.)

low impedance of the feedthrough capacitor may be applied to an RF ground connection, as shown in Figures 5.16 b, c, and d. The ground plane may be a copper PCB or the metal wall of the enclosure, and the electrical connection is made to one terminal of the feedthrough only. When use of a feedthrough is impractical, an axial or radial lead capacitor, as shown in Figure 5.17a, may be used. Figures 5.17b and 5.17c illustrate incorrect connection of the capacitor, since the impedance of the wire between the power and signal ground reduces the effectiveness of the capacitor. In Figure 5.17a the impedance of the wire is in series with the noise currents and increases the effectiveness of the capacitor. In reality the inductances of the wires form a T-type filter with the capacitor. Figure 5.3 and 5.4 show the correct and incorrect PCB trace layout for a capacitor. Erie has manufactured a three-wire capacitor in which the inductance of the wires is further increased by the inclusion of ferrite beads. The extremely low self-inductance of MLC, SLC, and microwave chip capacitors can be lost if the traces to the capacitor are not as shown in Figure 5.4.

Capacitors are available in miniature feedthrough types with capacitances from 2 pF to 2µF. Feedthrough capacitors designed for use on AC or DC power lines may have capacitance values as high as 10 µF rated at up to 600 Vdc/440 Vac and a current of 100 A with a size of 7 cm by 8.2 cm. The current rating of the feedthrough refers to the current-carrying capability of the bolt, which goes through the filter, and the terminations. A 100-V, 1.5-µF, 50-amp capacitor feedthrough, or a filter feedthrough, may conveniently be used as the terminations of a power supply. When a feedthrough cannot be accommodated in an existing design, a PCB may be used to connect capacitors between the terminals and the metal case of the supply. Figure 5.18 illustrates this method, which achieves the lowest possible impedance.

A capacitor does not dissipate noise energy; rather, it shunts it to another path, typically ground. When the ground impedance is high, an additional noise voltage appears across the ground. When the connection is made to an imperfectly shielded enclosure, the radiation from

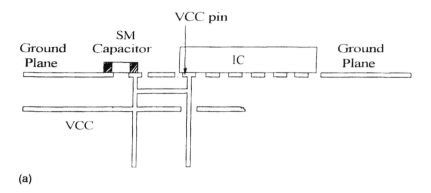

**(a)**

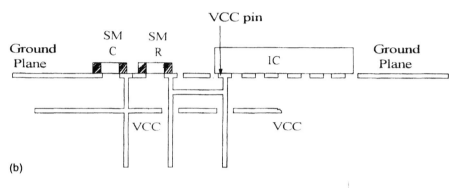

**(b)**

**Figure 5.14** (a) Adding a trace between the VCC pin and the decoupling capacitor. (b) Adding a resistor between the VCC pin and the decoupling capacitor.

the enclosure may increase. Thus, the inclusion of a capacitor between power or signal ground and chassis, to reduce common-mode voltages, may not be without a secondary effect on EMC.

### 5.1.10.2   Inductors

The impedance of an ideal inductor increases linearly with increasing frequency. The AC equivalent circuit of a practical inductor, shown in Figure 5.19, contains series resistance and parallel capacitance. At some frequency, an inductor becomes a parallel resonant circuit with a very high impedance. Above this resonance, the impedance decreases due to the parallel capacitance.

The inductance ($L$), resistance ($R$), DC current-carrying capacity ($I_{DC}$), and the resonant frequency ($f_0$) of a number of heavy-duty hash chokes wound on molded powdered iron forms are as follows:

| $L$ [$\mu$H] | $R[\Omega]$ | $f_0$ [MHz] | $I_{DC}$ [A] |
|---|---|---|---|
| 3.35 | 0.010 | 45 | 20 |
| 8.8 | 0.021 | 28 | 10 |
| 68 | 0.054 | 5.7 | 5.0 |
| 125 | 0.080 | 2.6 | 3.5 |
| 500 | 0.260 | 1.17 | 2.0 |

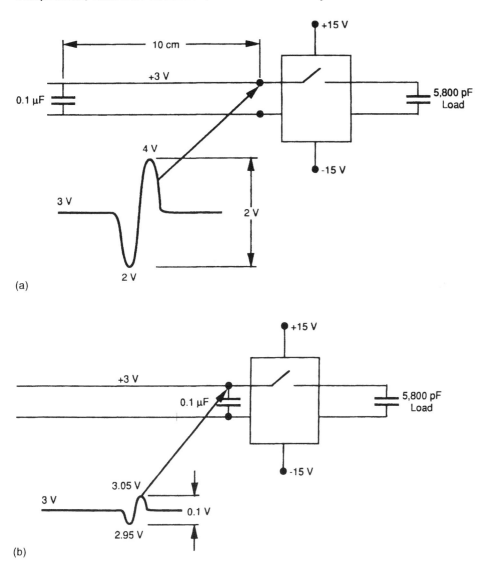

**Figure 5.15** (a) Reservoir capacitor far removed from an analog switch. (b) Reservoir capacitor located close to an analog switch.

Iron powder is invariably used in preference to ferrite, because its permeability, although initially lower than ferrite's, remains high at high levels of DC magnetizing force and high levels of AC flux density, as shown in Section 5.1.10.6 on power-line filters.

However, when the inductor is wound as a common-mode choke, as shown in Figure 5.20, the permeability of the material, the number of turns, and therefore the inductance of the choke may be very high without the danger of saturating the core material due to very high DC currents. Both supply and return conductors are threaded through the inductor. Therefore, the DC flux generated by the supply and return tend to cancel in the core material. The disadvantage with the common-mode inductor is that no attenuation of differential-mode noise currents is achieved when bifilar wound; i.e., both conductors wound together through the core. If the

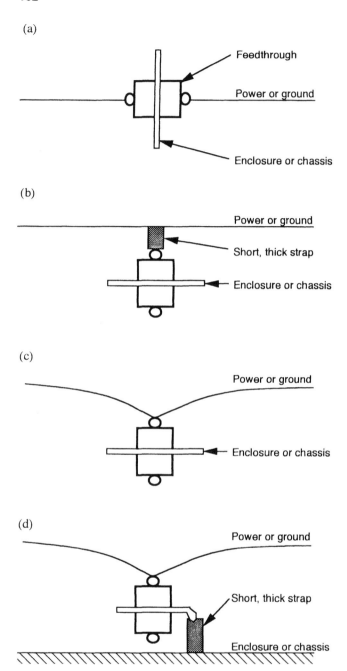

**Figure 5.16** Proper installation configurations for a feedthrough capacitor.

conductor is enamel-coated magnet wire, then the insulation of the enamel may not be considered adequate when the inductor is used on AC or DC power lines. In reality this is true only if the enamel has become abraded during the manufacturing process. To increase the insulation, thread the magnet wire through a Teflon tube or use Teflon-insulated wire. If the common-mode windings are separated around the circumference of the core, then the increased insulation has been

(a) Standard capacitor correct connection

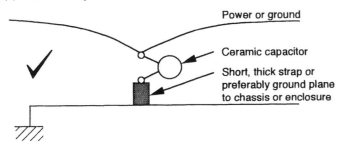

Power or ground

Ceramic capacitor

Short, thick strap or
preferably ground plane
to chassis or enclosure

(b) Incorrect connection

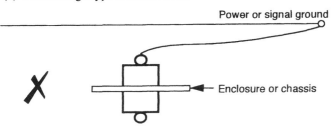

Power or signal ground

(c) Feedthrough type incorrect connection

Power or signal ground

Enclosure or chassis

**Figure 5.17** Correct and incorrect connection of axial, radial lead, and feedthrough capacitors.

achieved and some differential-mode (leakage) inductance. Be aware that with very high inductance C/M coils wound on high-permeability cores and used on high-DC-current supplies that the leakage inductance may be enough to reduce the permeability of the core. This is because the DC flux is not totally cancelled unless the supply and return conductors are kept very close together using a bifilar winding. Some leakage inductance can be introduced when the windings are spaced apart around the toroid. This will add some differential-mode inductance but make the core inductance more sensitive to DC current and can degrade high-frequency signals.

The use of a coil wound in sections, each exhibiting a different inductance and, therefore, a different capacitance, has the effect of increasing the useful frequency range of the inductor. The alternative is to use a number of individual inductors in series, each with a different inductance value and different resonance frequency. When the problem frequency or frequencies are known, the resonant frequency of the inductor/s may be chosen or tailored by additional capacitance in order to match the problem frequency and achieve maximum impedance.

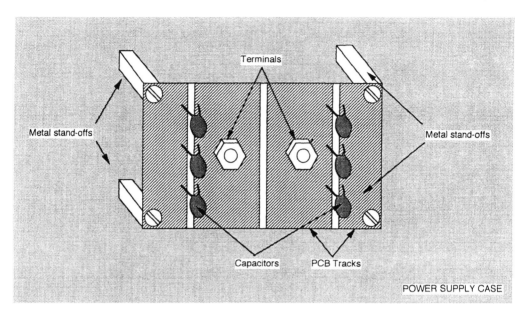

**Figure 5.18**  Low-impedance connection of capacitors between terminals and chassis.

Molded RF coils are available in a form similar to a resistor. These use either phenolic (air), iron, or ferrite as the core material and are available from 0.022 µH (resonance frequency 50 MHz, DC resistance 0.01 Ω, and current capacity of 3.8 A) to 10 mH (0.25 MHz, 72 Ω, and 48 mA).

The permeability of core material is often quoted in inductance per 100 turns or inductance per turn. To obtain the number of turns for a given inductance when the inductance is specified per 100 turns, use the following formula:

$$T = 100 \sqrt{\frac{\text{Required inductance [µH]}}{\text{Inductance per 100 turns [µH]}}}$$

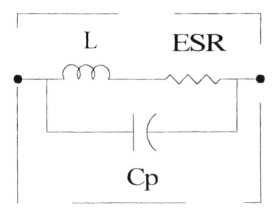

**Figure 5.19**  AC equivalent circuit of an inductor.

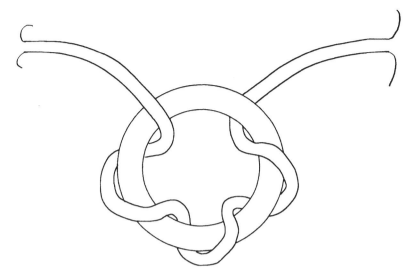

**Figure 5.20**   Common-mode choke.

When the inductance is specified in mH/1000 turns, use

$$T = 1000 \sqrt{\frac{\text{Required inductance [mH]}}{\text{Inductance per 1000 turns [mH]}}}$$

When the inductance is specified per turn or inductance per $N^2$, use

$$T = \sqrt{\frac{\text{Required inductance [µH]}}{\text{Inductance per turn [µH]}}}$$

When the source and load impedances are high, the attenuation of the inductor may not be adequate and a high-value resistor or a combination of series resistor and shunt capacitor or series inductor and shunt capacitor may be more effective. This type of filter is described in detail in Section 5.1.10.6.

The inductor will reduce noise-current flow and may be used in reducing ground or supply currents. However, a low-loss high-Q inductor does not dissipate noise energy but rather develops a noise voltage across it. Thus, the inclusion of an inductor may increase the noise voltage at the source of the noise. When the source is digital logic or load switching and the inductor is in the supply line, the potential increase in noise voltage across the logic or load may result in EMI. When the inductor forms a filter with load or on-board decoupling capacitance, the LC combination may exhibit insertion gain, as described in Section 5.1.10.6 on filters.

### 5.1.10.3   Ferrite Beads

Ferrite material exhibits a frequency-sensitive permeability and quality factor ($Q = X_L/R_S$). In this section, beads or baluns are assumed to have either a wire passing through the material or a limited number of turns of wire through the material. As frequency is increased, the reactance of the bead increases, as does its residual core loss represented by a resistance. Above a given frequency, the inductive reactance (permeability) $X$ decreases and the resistance $R$ continues to increase, and so does the impedance Z. Figure 5.21 illustrates this change with frequency. This figure is valid only for a single wire through the center of the bead, which we will refer to as a *half-turn*. The ferrite bead is one of the few components in which the noise energy is converted into heat within the component. Figures 5.22a–f show the impedance-versus-frequency characteristics for a number of ferrite materials made by the Steward Manufacturing Company and Fair-Rite Products.

As the number of turns in the core is increased, there is an increase in the interwinding capacitance and the impedance at higher frequencies is reduced. Figure 5.22c shows impedance versus frequency for 2-1/2 turns of wire as compared to a half-turn and 1-1/2 turns. The permeability of ferrite, in addition to its frequency dependence, is very sensitive to line DC current, this is shown in Figure 5.23.

Due to the relatively low impedance (30–800 Ω) at most frequencies above 1 MHz, ferrite beads are most effective in low-impedance circuits or in conjunction with shunt capacitance. Also at low frequency, the losses are small and the Q of the bead may result in the resonance effects described in Section 5.1.10.6:

The inductance of a single wire through a bead is given by

$$L = \frac{\mu_r \, \mu_o \, 0.01 \, s}{2\pi} \ln\left(\frac{r_2}{r_1}\right)$$

where $s$ = length of the bead [cm]

   $r_1$ = inner diameter [cm]

   $r_2$ = outer diameter [cm]

   $\mu_o = 4\pi \times 10^{-7}$ [H/m]

   $\mu_r$ = relative permeability

The physical parameters included in this equation are shown in Figure 5.24. The magnetic force within the bead is given by

$$H = \frac{I}{P}$$

where

   $I$ = magnetizing current [mA or A]

   $P = r_2 - r_1$ [cm or m]

To obtain the magnetizing force in oersteds, divide the force in amps per meter by 79.6.

The number of ferrite beads on a line also effects the high-frequency performance, as can be seen from Figure 5.25. Although the low-frequency impedance of the line with multiple beads is higher than that for the single bead, the impedance at higher frequency is much reduced.

One material that should have a very useful application at microwave frequencies is ''lossy sleeving,'' manufactured by Capcon International Inc. This material is flexible and can be used over a temperature range from −55° to +250°C. It comes with an internal diameter of from 0.04 inches to 1.25 inches and in a shielded or unshielded version. The specification shows an attenuation of 45 dB at 1 GHz and 100 dB at 9 GHz for a 1-inch length of the unshielded type and 100 dB at 1 GHz for a 1-foot length and remaining at 100 dB up to 100 GHz. Although this material should have many applications at gigahertz frequencies, it has not proven so in practice. In one case, a 1/4-inch length was placed over a pair of conductors, used for a differential signal, to function as a balun to reduce common-mode currents at 2.5–10 GHz. This was not as effective as a 1/4-inch bifilar wound and twisted magnet wire balun, with the magnet wire wound with one turn through a two-hole core made of a material rated to be effective up to

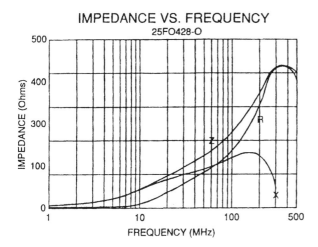

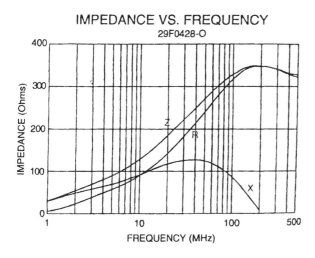

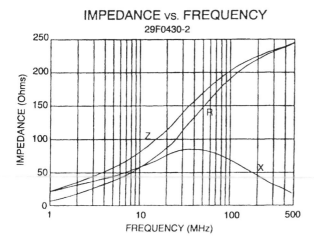

**Figure 5.21a** Frequency dependence of the impedance, inductance, and resistance of a ferrite material. (Reproduced by kind permission of Steward.)

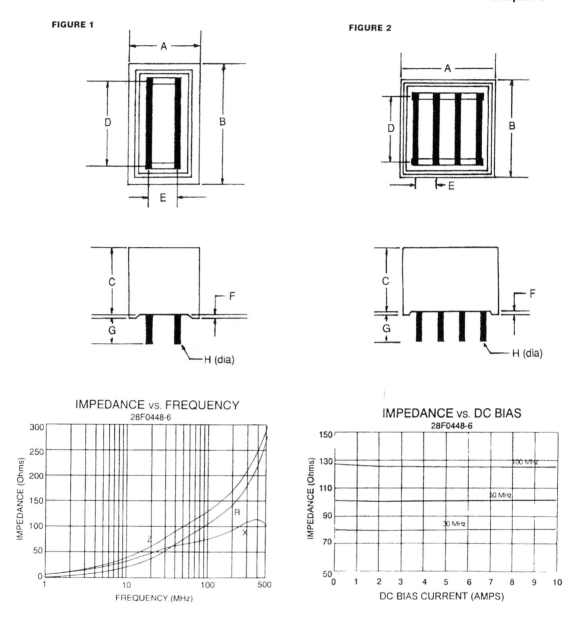

**Figure 5.21b** Steward common-mode power/data line EMI suppression ferrite. (Reproduced by kind permission of Steward.)

only 3 GHz. One problem with the use of baluns at gigahertz frequencies is coupling around the balun. Ideally the traces up to PCB-mounted baluns used at high frequency should be in stripline and the connections should be routed through a PCB-mounted local shield, typically used to shield one section of a circuit from a different section (either higher or lower power or a different frequency).

### 5.1.10.4   Baluns

The *balun* is a wideband transformer used to transfer energy from a balanced to an unbalanced line. A common application for a balun is in matching a balanced antenna to an unbalanced

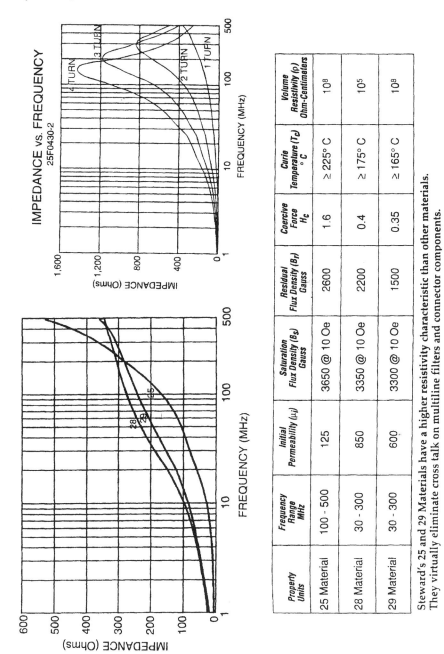

| Property Units | Frequency Range MHz | Initial Permeability ($\mu_i$) | Saturation Flux Density ($B_s$) Gauss | Residual Flux Density ($B_r$) Gauss | Coercive Force $H_c$ | Curie Temperature ($T_c$) ° C | Volume Resistivity ($\rho$) Ohm-Centimeters |
|---|---|---|---|---|---|---|---|
| 25 Material | 100 - 500 | 125 | 3650 @ 10 Oe | 2600 | 1.6 | ≥ 225° C | $10^8$ |
| 28 Material | 30 - 300 | 850 | 3350 @ 10 Oe | 2200 | 0.4 | ≥ 175° C | $10^5$ |
| 29 Material | 30 - 300 | 600 | 3300 @ 10 Oe | 1500 | 0.35 | ≥ 165° C | $10^8$ |

Steward's 25 and 29 Materials have a higher resistivity characteristic than other materials. They virtually eliminate cross talk on multiline filters and connector components.

**Figure 5.21c** Comparison of Steward ferrite materials and the effect of increasing the number of turns on a 25F0430-2. (Reproduced by kind permission of Steward.)

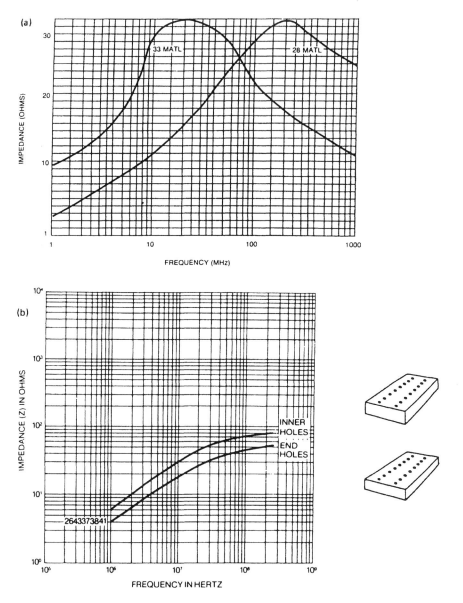

**Figure 5.22** Impedance-versus-frequency characteristics for different ferrite materials and configurations. (b–f) Effect of different types of material used in a bead with similar physical dimensions. (Reproduced with kind permission of Fair-Rite Products Corp.)

cable. In this section we use the term *balun* to describe a special use of a ferrite or metal oxide bead or toroid. The simple case we will discuss is a balun core with two wires passing through its center. When the two wires carry differential currents, the impedance to these currents at all frequencies is effectively zero. Thus, where the differential currents are caused by a signal, neither the signal amplitude nor rise time is affected by the presence of the core. However, common-mode current due to either current induced by an incident electromagnetic field or noise voltage present between the grounds of two systems is reduced by the series impedance of the balun. A differential-input amplifier with common-mode noise rejection is often compro-

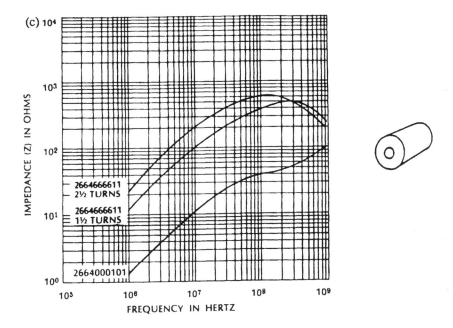

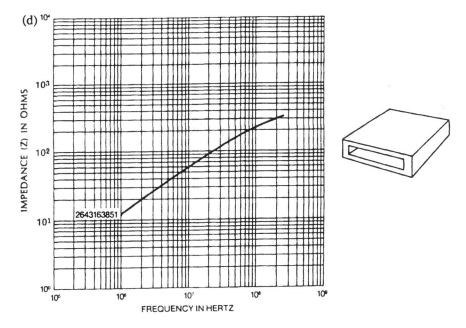

**Figure 5.22**  Continued

mised in its rejection by the conversion of common-mode (C/M) currents to differential-mode (D/M) voltage. For example, when twisted shielded pair cable is used to carry the differential signal, an imbalance in the pair to shield impedance may result in the conversion of C/M noise to D/M noise. When the circuit impedances are sufficiently low, use of a balun will reduce the C/M current and thereby any D/M induced noise. It is imperative that the two wires be threaded

(e)

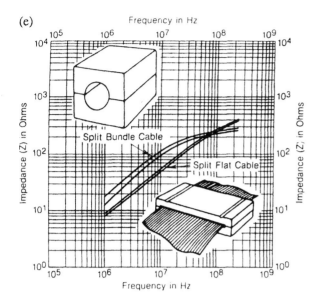

(f)

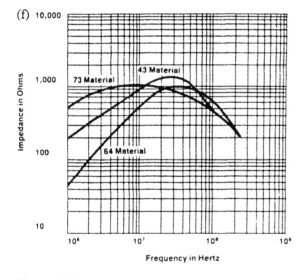

**Figure 5.22** Continued

through the same hole in the bead when using a multihole bead. When a single hole is used, the differential signal remains unaffected by the presence of the bead up to 1 GHz and above. However, when the two wires are threaded through different holes, signals with repetition rates as low as 1 MHz are degraded by inclusion of the balun. The high-frequency common-mode noise rejection of an amplifier is invariably compromised by capacitive imbalance to ground on input wiring and PCB tracks and by the poor common-mode noise rejection of operational amplifiers at high frequency. Here the addition of a balun over input wiring may cure the problem by a decrease in the C/M current flow. When the input impedance of the differential circuit is too high, the inclusion of two capacitors, matched as closely as possible in value, located after the balun and between each of the inputs and ground may be the solution.

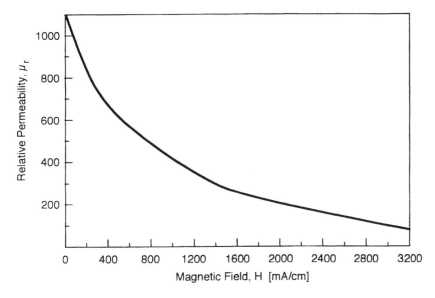

**Figure 5.23** Permeability versus DC current. (Reproduced courtesy of Electronic Design, June 7, 1969.)

Figure 5.26 illustrates the use of the balun where the input is unbalanced and the common-mode current flowing in the ground results in a low voltage whereas the current flow in the high input impedance of the amplifier results in a much higher voltage (i.e., a differential-mode voltage is developed). Decreasing the magnitude of common-mode current at the source is a viable option in the configuration shown in Figure 5.26a. Also, increasing the number of turns to between 1 and 5 may improve the performance of the balun. The effectiveness of the balun on the shielded cable in Figure 5.26b will be higher than that of the Figure 5.26a configuration due to the low impedance of the shield to the equipment case connection below resonant frequencies.

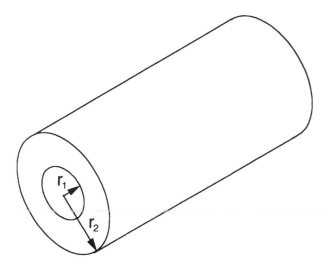

**Figure 5.24** Relevant dimensions of a cylindrical bead. (Reproduced with kind permission from Electronic Designs, June 7, 1969.)

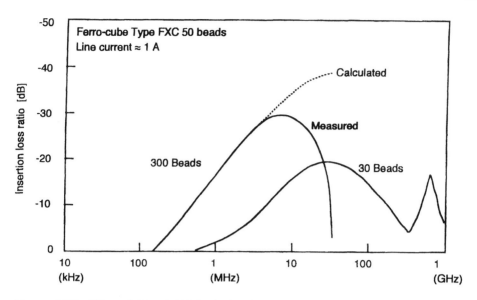

**Figure 5.25**   Effect of 30 and 300 ferrite beads on a wire. (Reproduced with kind permission from Electronic Designs, June 7, 1969.)

### 5.1.10.5   Resistors

Resistors are also frequency dependent, exhibiting capacitance and resistance as shown in Figure 5.27.

The inductance and capacitance of several types of resistors, reproduced from Ref. 10, is shown in Table 5.3. The inductance of a normal metallic film (MR25) resistor is reported to be 20 nH, of the 1206 chip resistor 2 nH, and of the 0603 chip resistor 0.4 nH. For low resistance values, the series inductance has the predominant effect of modifying the impedance at high frequency; and for high resistance values, the parallel capacitance has the largest effect. For example, a 50-$\Omega$ MR25 resistor will have an impedance of approximately 170 $\Omega$ at 1 GHz, and a 50-$\Omega$ 1206 chip resistor has an impedance of approximately 62 $\Omega$ at 1 GHz due to inductance. Due to the parasitic capacitance, a 2.2-k$\Omega$ metal oxide resistor will have an impedance of approximately 320 $\Omega$ at 1 GHz, whereas a typical chip resistor will maintain an impedance just below 2.2 k$\Omega$.

### 5.1.10.6   Power-Line Filters

Power-line filters are available in a number of different configurations, including feedthrough types used for a single AC or DC line and chassis-mounted types for either single- or double-DC-line or two- or three-phase AC applications. The supply connection may be via an integral AC line socket or connector or via terminals. Figure 5.34 illustrates the packaging of typical power entry filters. The current ratings of standard filters range from 0.2 A to 200 A, with voltage ratings from 50 Vdc to 440 Vac.

The attenuation of a line filter is usually measured in accordance with MIL-STD-220A, which specifies 50-$\Omega$ source and load impedance. MIL-STD-220A allows either a common-mode or differential-mode attenuation measurement and does not require that the attenuation measurement be made with the supply current flowing through the filter. It is typically easier to achieve a high level of C/M attenuation by use of a high-permeability common-mode inductor of small dimension, and the manufacturers data sheet may not indicate if the specified attenuation is for differential- or common-mode noise.

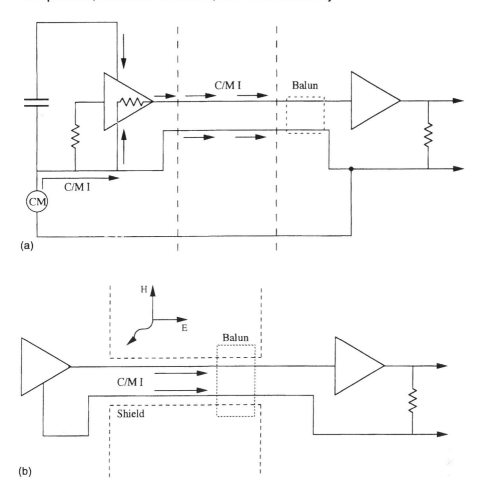

**Figure 5.26**  Uses of a balun.

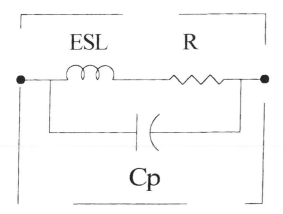

**Figure 5.27**  AC equivalent circuit of a resistor.

**Table 5.3**  Parasitic Capacitance and Inductance of
Different Types of Resistor

| Type of resistor | Inductance (nH) | Capacitance (pF) |
|---|---|---|
| Wire wound, axial leads | 100–1000 | 0.5–1 |
| Composite, axial lead | 2–10 | 0.05–0.3 |
| Carbon film, axial lead | 5–200 | 0.3–1 |
| Metal film, axial lead | 5–200 | 0.3–1 |
| Metal oxide, chip | 0.5–2 | 0.05–0.1 |

*Source*: Ref. 10.

When a manufacturer of a filter of small dimensions and a current capacity of 5 A or greater claims a greater than 20-dB level of attenuation over the low-frequency range (10–50 kHz), the attenuation may apply to common-mode noise only. When a manufacturer does not specify attenuation above 30 MHz, one of two assumptions may be made. The first is that the filter is specifically designed to meet a commercial conducted-emission requirement, such as contained in FCC Part 15, in which the maximum specified test frequency is 30 MHz. The second possibility is that the filter exhibits a parallel resonance close to or just below 30 MHz, and so attenuation above 30 MHz decreases linearly.

H. M. Schlicke quite rightly exhorts us to throw away our conventional filter books, because they illustrate filter performance in highly idealized and simplified conditions that are not valid for many EMC situations (Ref. 3). Figures 5.28a, b, and c illustrate how far load and source impedances may deviate from 50 Ω. The graphs in Figure 5.28 are based on statistical data obtained from many hundreds of outlets and devices used in households, factories, and laboratories and aboard naval vessels. We see that impedances vary from 0.1 Ω to above 1000 Ω; with the exception of common-mode loads, low Q's are dominant.

Other measurements on the source impedance of AC power lines above 10 MHz have resulted in a maximum of 1000 Ω, average of 100 Ω, and minimum of 25 Ω at 15 MHz decreasing to a maximum of 100 Ω, average of 40 Ω, and minimum of 10 Ω at 400 MHz. A more realistic filter source and load impedance for testing and analysis purposes is 1 Ω and 100 Ω, respectively.

During MIL-STD-461 CE01 and CE03 conducted emission measurements, the equipment power-supply source impedance below 50 kHz is typically the output impedance of the supply connected to the input of the equipment under test. Above 50 kHz the impedance is limited by the impedance of the 10-μF capacitors specified between the line and the ground plane on which the equipment under test is mounted and between the return and the ground plane. As the frequency of the emissions increases, the source impedance decreases. The filters procured or designed for equipment that must meet MIL-STD-461 requirements must therefore function adequately with the normal variation of supply impedances and the very low impedances of the 10-μF capacitors at RF. Adding lower than 10-μF common-mode or differential-mode capacitors on filters that will be tested using the 10-μF capacitor has no value, for the RF high-quality 10-μF capacitor in the test setup will carry the bulk of the current, and it is this noise current that is the subject of the test. Adding low-value C/M capacitors (approximately 0.1 μF) does have a value, because the power lines between the filter and the 10-μF test capacitor exhibit some impedance at high frequencies, and the filter capacitors will shunt the RF current at the filter. This is a case of designing a power-line filter around a test setup and incorporating the 10 μF into the design of the filter. This is surely valid, because the specified test setup should be representative of the real world. An alternative to the 10-μF capacitor, which is proposed for the

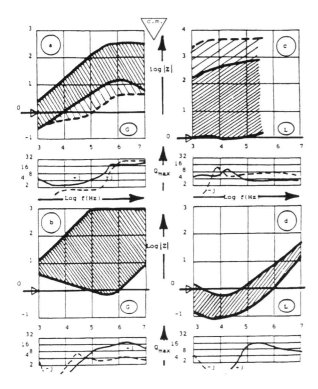

**Figure 5.28a**  Differential-mode interface impedances (U.S.) 120 V, 60 Hz. (a) Unregulated outlets (generators). (b) Regulated outlets (generators). (c) 60-Hz loads (with the top less densely shaded area representing high-voltage, low-current power supplies; the middle range pertaining to motors, TV sets, and so on; and the low limit dominated by high-current devices). (d) Regulated power supplies.

latest version of MIL-STD-461 CE01 and CE03 measurements, is a line impedance stabilization network (LISN) with an impedance of 50 Ω above some specified frequency.

This section is intended to aid in either the design or the selection of a filter that is adequate for the job but not overdesigned. The required level of attenuation for impulsive noise generated by digital logic and converters, for example, may only be achieved by careful selection of a filter design tailored to the specific type and level of emissions and susceptibility test levels. When designing or procuring a filter for equipment it is advisable to wait for the breadboard or engineering model and then measure the level and type of emissions as well as the response to the susceptibility levels. The filter may then be an optimum EMC design and of minimum size and weight. In general, series inductors are used for low source or load impedance. Thus, the following types of filter are most suitable for the various combinations of source and load impedance:

| Impedance | | Filter |
|---|---|---|
| Source | Load | type |
| Low | Low | T |
| Low | High | L |
| High | Low | π |

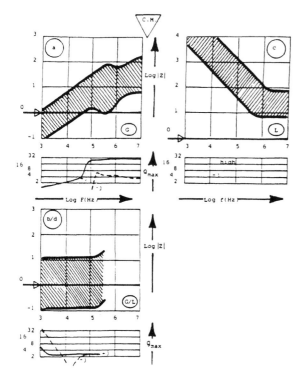

**Figure 5.28b**   Common-mode interface impedances (U.S.) 120 V, 60 Hz. (a) 60-Hz outlets (generators). (b/d) Outlets and loads in heavily filtered installations. (c) Loads.

In designing filters it must be understood that components do not maintain their identities over wide frequency ranges. Components have natural resonant frequencies beyond which they no longer follow their low-frequency impedance characteristics. Capacitors become inductive and inductors become capacitive beyond their natural resonant frequencies.

It is usually impossible to achieve a power-line filter design that achieves high attenuation over a frequency range of 10 kHz to 1 GHz with a single filter section. A combination of sections or filters is usually required. Feedthrough capacitors and low-value bypass capacitors and inductors must be used at high frequencies in order to reduce the self-capacitance of inductors and the self-inductance of capacitors. The Q of the filter must be low under all load and source impedances to minimize insertion gain. When a transient noise source is applied to a filter with a high Q and the transient is at the resonance frequency or a harmonic thereof, the filter output waveform may be an undamped sinewave!

The equivalent circuit and equations for Q for a single-section filter are shown in Figure 5.29 (Ref. 2). In a multisection filter, where the series inductances have different values and the shunt capacitances have different values, the equations will provide an approximate value for Q when the highest value of inductance ($L$) and capacitance ($C$) are used.

The use of an AC and transient circuit analysis program like SPICE is invaluable in the design or selection of filters. The program allows the modeling of the filter and noise sources and is especially useful in the analysis of complex load and source impedances. As an example, consider a single-section filter with a 500-μH inductance, a 2-μF capacitance, and a 72-Ω load. The filter response to an AC waveform, supplied from a low-impedance source, peaks (resonates) at 5 kHz. At this frequency it exhibits 13 dB of insertion gain. Changing the filter to a

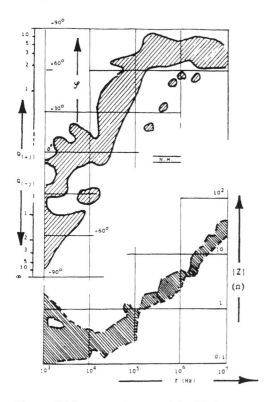

**Figure 5.28c** Impedances and Q of DC power supplies loaded and unloaded.

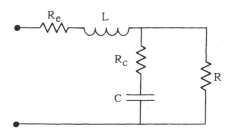

$$Q = \frac{R_o}{R_{te}} \quad \text{degree of damping at } \omega_o$$

where $R_o = \sqrt{\dfrac{L}{C}}$

$\omega_o = \dfrac{1}{2\pi\sqrt{LC}}$

$R_{te} = R_e + R_c + \dfrac{R_o^2}{R}$

= total effective damping resistance

**Figure 5.29** Q of an LC filter. (From Ref. 2.)

three-section type with equal-value inductances and capacitances does not change the insertion gain but changes the frequencies at which insertion gain is a maximum (i.e., 2–6.3 kHz). When a second section is added, with a 100-μH inductance and a 1-μF capacitance, the original resonance occurs at 4 kHz, with a second resonance at 20 kHz. In filter selection the magnitude of the load capacitance must be considered. For example, a low-Q filter may exhibit a high Q when used at the output of a supply that is loaded by a number of on-board decoupling capacitors.

The Q of a filter such as the one in Figure 5.29 may be reduced by increasing the value of $R_e$ or $R_c$ via the addition of a resistor or via placing a resistor across the inductor, $L$. Increasing $R_c$ is the best option, for it does not degrade the attenuation performance of the filter. In high-current, low-voltage supply applications, the voltage drop across the additional resistor, $R_e$ is probably unacceptable. However, in applications with low current requirements and voltages above 12 V, a loss of 0.5–1 V may be acceptable. Another practical option is shown in Figure 5.30. Here, a second, larger-value capacitor with an appreciable value of added series resistance is placed in parallel with the existing capacitor. Very often the combined load capacitance formed by decoupling capacitors on PCBs is higher than the highest value of filter capacitor that will fit into the filter available footprint. In this case an additional series inductor with the damping resistor in parallel with the inductor is very effective, although the filtering effectiveness of the inductor is lost. The sole purpose of the inductor is to reduce the DC voltage drop and to present a higher impedance than the resistor at the resonant frequency of the filter. If damping cannot be incorporated into the filter design, make sure that the resonant frequency is no higher than 1/10 of the frequency of switching-power supplies (typically 100 kHz) and at least 10 times the AC power-line frequency. Also, if the frequency of high-level noise sources sharing the same power line are known, try to shift the filter resonance away from these frequencies.

The reason all filters perform well in a MIL-STD-220 test is that the input attenuator effectively increases the value of $R_e$ and decreases the Q of the filter. Although a ferrite bead has a low Q at high frequency, when used to filter a supply on a board that contains decoupling capacitors, the Q of the resultant $LC$ circuit at low frequency may result in a sinewave impressed on the board supply. Here, the addition of a series resistor or replacement of the bead by a resistor may be the solution.

Manufacturers of commercial filters are invariably willing to provide the component values and schematics of their filters. From this information and a knowledge of the minimum and maximum load and source impedances, a filter with a low Q and, therefore, either minimum or no insertion gain may be chosen using either the equations of Figure 5.29 or a circuit analysis program. The design of a filter exhibiting common-mode attenuation and a low Q when used with a supply or load isolated from enclosure ground is difficult. In this application, the common-mode $LC$ combination is not damped by a load resistance (due to the isolation requirement).

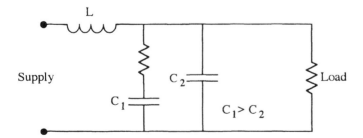

**Figure 5.30**  Reducing the Q of a filter with an additional capacitor and damping resistor.

Very high levels of undamped sinewave may then be generated due to the high Q of the filter. One solution is to move the resonance to a frequency at which the load is least susceptible or at which the conducted emission limit is at a maximum.

Lossy line or dissipative filters perform well at high frequency (500 MHz to 10 GHz) and have the great advantage of removing the noise energy in the form of heat. In lumped $\pi$ or L-type filters, the noise energy flows in the capacitors and into the structure or power or signal ground, and this in turn can cause EMI/EMC problems. Where T-type filters are used, the magnitude of capacitor ground current is reduced. A potential disadvantage is that the noise energy generated—by logic, for example—typically results in a higher, and sometimes unacceptably high, level of noise voltage across the load (noise source) due to the input or output impedance of the T-type filter. Some examples of lossy components and filters are

> Ferrite beads
> Ferrite rods
> EMI suppressant tubing (permeable)
> "Lossy line" EMI absorptive filters

The disadvantage of lossy filters are their typically poor performance at low frequency.

The remainder of this section provides a set of general filter guidelines. Supply-line filters should have their output lines shielded from the supply line, preferably by mounting the filter/s outside the unit enclosure or within a shielded compartment inside the unit enclosure.

The first choice in filter selection is the type that has built-in connectors for input power and terminals for connection to the load. With this type, the filter case may be mounted on the exterior of the enclosure or to the rear wall, with the connectors accessible through an aperture in the wall. When the filter must be located inside the enclosure, shielded cable should be used, with the shield connected to the filter case at one end and to the wall of the enclosure, as close to the rear of the connector as feasible, at the other end. In designing a filter, the output terminals should be feedthrough capacitors or a feedthrough filter with a low value of inductance.

Inverters are a major cause of noise, as described in Chapter 2. Filters for use with power regulators employing inverters should be designed or chosen specifically to match the characteristics of the inverter they are used with.

Filters should be designed to damp self-resonances, which cause problems with impulsive-type noise. Resonant peaking in the filter frequency response should, where possible, be no greater than 3 dB.

When incorporating AC power-line filters, consideration must be given to a potential change in the power factor as a result of the filter inductors and capacitors. When the load is reactive, the power factor of the load and the filter combination should be calculated. The values of the inductances and capacitances in the filter may then be chosen to achieve a power factor as low as feasible. A second consideration is the level of power-line frequency current flowing in the ground due to the parallel filter capacitance. Capacitors used on AC power lines must be of the correct X or Y type and rated for the AC power line voltage as well as the frequency. Some X- or Y-type capacitors are rated at 250 V AC, 60 Hz and are not suitable for use at 250 V and 400 Hz. Some EMC requirements limit the value of filter capacitance used on AC power lines. When the supply to a DC power-line filter is switched via a relay or contactor and the filter input component is an inductor, it is good practice to incorporate a diode or rectifier between the supply line and the case ground to clamp the voltage spike as the input power is interrupted.

Where the filter contains capacitors connected to the metal enclosure of the filters, designed to reduce common-mode noise voltages, the enclosure of the filter must be connected back to the ground at the noise source via a low-impedance connection.

Filters used with switching-power supplies or converters may compromise the stability of the supply unless certain precautions are taken. The input power of a linear regulator power supply increases with increasing input voltage, unlike a converter, where the input power remains relatively constant. Figure 5.31 compares the input characteristics and illustrates that the converter input is represented by a negative resistance.

When proposing the addition of a filter to the input of a converter, or where the addition of a filter results in a change in the converter line or load regulation, the stability criteria may be checked by referring to Appendix B in Ref. 2.

The usefulness of power-line filters is not confined to reducing conducted noise levels. When RE02, DO-160, or similar radiated emission levels are exceeded due to radiation from shielded cables, either a filter is required in order to reduce the radiation from the cable or source reduction must be used. We shall see in Chapter 7 that the use of double-braid, braid-and-foil, or triple-braid shielded cables brings only a limited reduction in radiation compared to a single-braid cable. The commercially available and relatively inexpensive power-entry filters are designed to reduce conducted emissions over the range 150 kHz to 30 MHz, and the attenuation is typically specified over this range. However, these power-line filters do provide significant

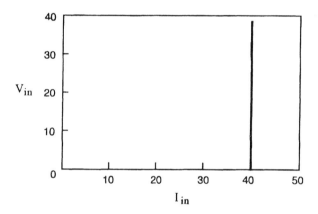

(a) Linear mode regulator.

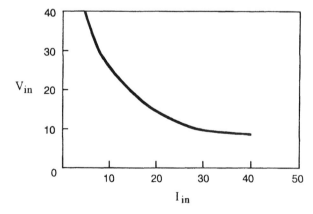

(b) Switch mode regulator.

**Figure 5.31**   Input characteristics of linear-mode and switch-mode regulators.

attenuation at higher frequencies, and if this is not adequate and the noise has been determined as C/M, then an additional ferrite balun on the wires inside the equipment to the filter may be all that is required. Also, the transfer impedance of the connector and its interface to the enclosure contributes to the radiation from the cable. One case is where noise is generated on a shared power line by equipment contained in one enclosure or rack, with the power line connected to equipment in a second rack. The resultant noise currents flowing on the inside of the braid of the shielded cable diffuse through to the outside or couple through the apertures, resulting in an above-specification level of radiated emissions. The solution is one, or more power-line filters. Only one filter may be required when the noise source is confined to equipment in one enclosure. When both enclosures contain equipment that places noise on the shared power line, the use of one filter may still suffice. The filter should be of the L- or T-type and contained, where possible, in the enclosure/rack that houses the source of the highest noise. The inductors in the filter should be the load presented to the equipment in enclosure 2, without filter.

The common-mode noise currents generated in enclosure 2, which flow out on the power-line conductors and return on the inside of the shielded cable, are reduced in magnitude due to the high impedance presented by the filter inductors.

Filters are commercially available that are extremely effective at reducing switching-power supply output noise. This enables a switching-power supply to be used in sensitive electronic equipment, where in the past only a linear supply provided sufficiently low noise. Typical applications for this high level of filtering has been found in satellite communication, lasers, sonar buoys, and microwave simulators.

Also by use of these filters a single switching-power supply can be used for both digital boards and low-signal-level analog and RF circuits. In this case the supply output is used directly for the digital boards and the filter is placed between the supply and the sensitive. These filters will typically attenuate noise spikes (as high as 40 V) and ripple (as high as 1.5 V) to an output ripple of 30 μV at the switching frequency of the converter. No measurable high-frequency component is seen when measured with a number of different manufacturers' switching-power supplies. The full attenuation effectiveness of the chassis-mounted filter can be achieved only if it is mounted directly to a dividing bulkhead or shielded cables are used for both input and output connections. Figure 5.32 shows the use of a bulkhead-mounted filter to maintain isolation between a ''noisy compartment'' and a ''quiet compartment'' in a piece of equipment. The PCB-mount version must have the input and output connections via striplines at different levels

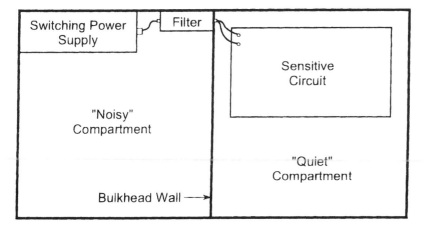

**Figure 5.32** Bulkhead mounting of a filter to maintain compartmentalization of equipment.

in the PCB to achieve maximum effectiveness. The effectiveness of these filters is measured by mounting them to the wall of a shielded room, with the signal source outside the room and the measuring device inside the room. The room is used to isolate the input test signal from the filter output level.

If common-mode noise appears on the power return, then this may require it be filtered. Figure 5.33 shows the topography of typical filters. If no impedance is possible between the unfiltered supply-power return and the filtered-power return, then only differential-mode filtering is required, as shown in the single-line and two-line D/M filters of Figure 5.33. Also, if the power return is connected to chassis at the power supply as well as on the load PCB, or more likely on a signal interface connection, then adding return-line filtering would be redundant. If, however, the supply return is isolated from chassis at the switching power, then power return filtering not only is possible but should be included, as shown in the two-line D/M and C/M filter and the three-line D/M and C/M filter. Very often the value of the C/M capacitor is limited, due to the required voltage rating and the limit on leakage current and because nonpolarized capacitors may not be allowed, even on DC power lines, since the polarity of the C/M DC voltage is typically unknown. This excludes the use of high-value tantalum or aluminum electrolytic capacitors unless they are unpolarized or placed back to back. Luckily, the permeability of the C/M inductor core is not affected by the power supply current, especially when the inductor is bifilar wound. Typically, a high-permeability ferrite core is used for the C/M inductor/s, resulting in a high (500 µH to 10 mH) inductance, and the C/M components' resonant frequency can be optimized.

A surface-mount pi filter that may be used as a power-line or signal-line filter is manufactured by Tusonix. This 4700 series filter is available with capacitors from 100 pF to 8200 pF and with a typical inductance of 100 nH. The performance of all filters is compromised by coupling between the input and output connections. The Tusonix filter is designed to be shielded by, typically, a PCB-mount shielded compartment, as illustrated in Figure 5.35. The insertion

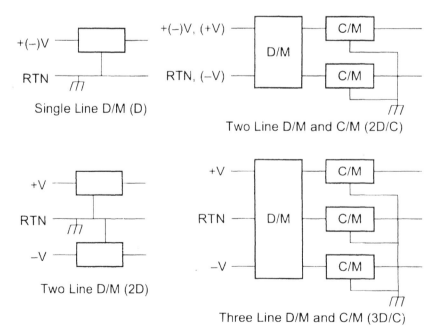

**Figure 5.33**  Number of secondary power lines filtered and attenuation mode.

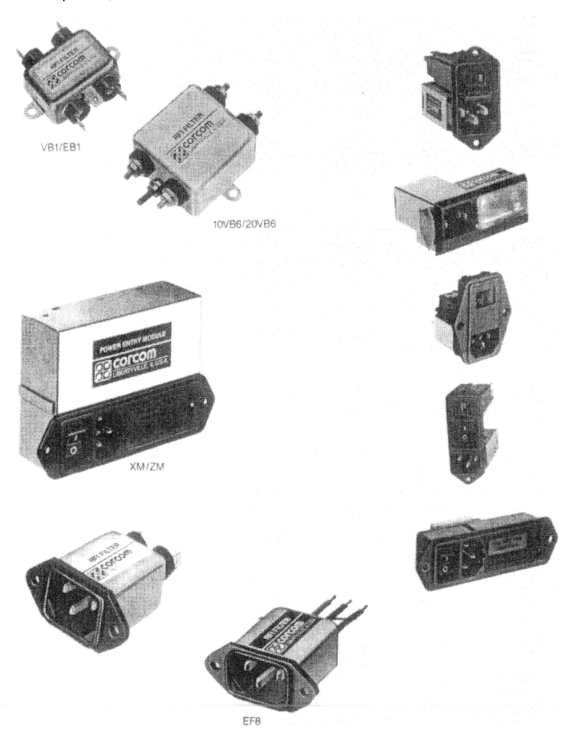

**Figure 5.34**  Typical power entry modes. (Reproduced by kind permission of Corcom.)

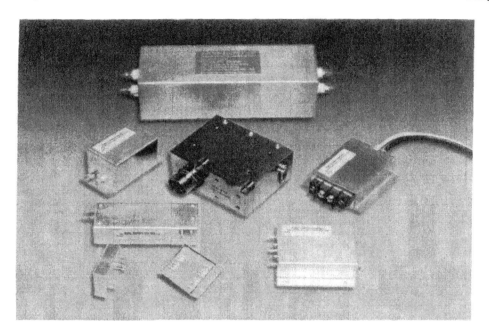

**Figure 5.35**   Connectorized, terminal, and PCB-mount filter configurations.

loss of the shielded filter is shown in Figure 5.36 and for the unshielded filter in Figure 5.37. A potential insertion loss of 70 dB above 1 GHz for the 8200-pF, 5000-pF, and 2000-pF versions when shielded is degraded to 30 dB when the filter is used unshielded.

### 5.1.10.7   Case Study 5.1: Filter Design

We shall examine the design of a +28-V power-line filter designed to attenuate the MIL-STD-461 conducted susceptibility test levels CS01, CS02, and CS06 and to provide adequate reduction of transient noise from 17 converters and 12 boards of logic in order to meet CE01, CE03 conducted emission levels. The filter schematic is shown in Figure 5.38. The design was made with the aid of the SPICE computer program using the equivalent circuit of the filter shown in

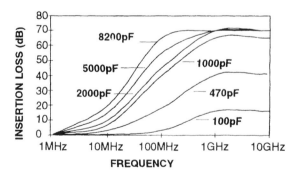

**Figure 5.36**   Typical shielded insertion loss versus frequency for the Tusonix 4700 series surface-mount filter.

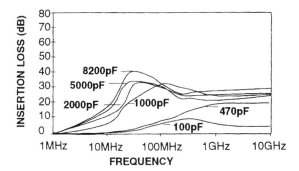

**Figure 5.37** Typical unshielded insertion loss versus frequency for the Tusonix 4700 series surface-mount filter.

Figure 5.39. A large ferrite bead with 1-inch length and ¹/₂-inch diameter was used on the +28-V input line and another on the return line. The first iron dust toroidal-core inductor was wound with 70 turns of No. 18 AWG magnet wire with a method designed to reduce interwinding capacitance. The capacitance of this core is 30 pF, and the permeability at a DC current of 6 A (magnetizing force = 81 oersteds) results in a minimum inductance of 230 μH. The self-resonant frequency of this inductor is 1.2 MHz.

Figure 5.40a illustrates how the permeability of the core decreases with an increase in DC current flow (the permeability of a core may increase with high levels of AC current flow). If the permeability of the core reduces to below 50% at the maximum current, the use of a core with a lower initial permeability that is less sensitive to DC current may be advisable. When used with AC power, the permeability increases at high AC flux density, and this increase is applicable at all frequencies, not just the power-line frequency. This means the inductance may be higher than predicted and any filter resonance lower. Figure 5.40b shows the variation in permeability with AC flux density for the Micrometals cores. Although iron dust cores exhibit a high permeability at high frequency with some materials, the permeability does reduce. Figure 5.40c shows the permeability at high frequency for the micrometals cores.

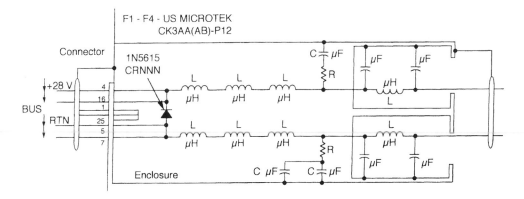

**Figure 5.38** +28-V DC line filter schematic and construction.

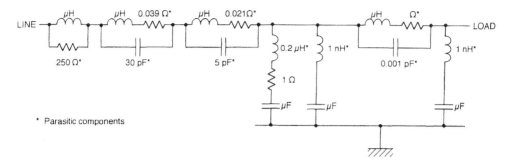

**Figure 5.39**  Equivalent circuit of one line of the +28-V power-line filter.

The second inductor exhibited 30–42 µH of inductance and a self-resonant frequency of 17 MHz and was wound in a manner to achieve the low self-capacitance of 3 pF. The feed-through filter is either a Capcon LMP-50 500 or a US µTeK GK3AA-P11, both of which exhibit a high level of attenuation at high frequencies (up to 1 GHz).

The filter was analyzed with the configurations corresponding to the MIL-STD CS01, 2, 6, CE01, 3, and MIL-STD-220 test setups. These configurations, as well as the attenuation characteristics or transient response as computed using SPICE, are shown in Figures 5.41–5.48 and in the nearby tables. In the conducted emission CE01 configuration, the transient response to a typical load noise waveform was analyzed. The test setup is shown in Figure 5.43. A summary of the filter performance is given in the nearby tables.

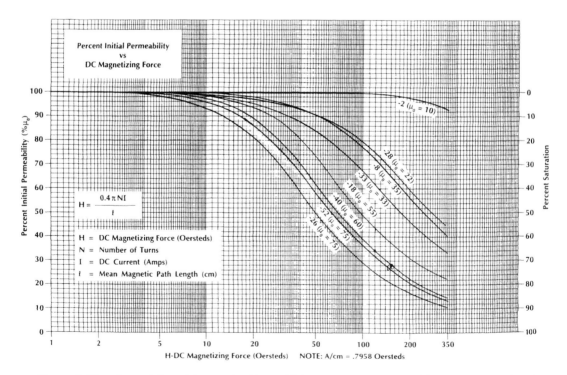

**Figure 5.40a**  Permeability of the metal oxide toroidal-core versus DC magnetizing force. (Reproduced by kind permission of Micrometals.)

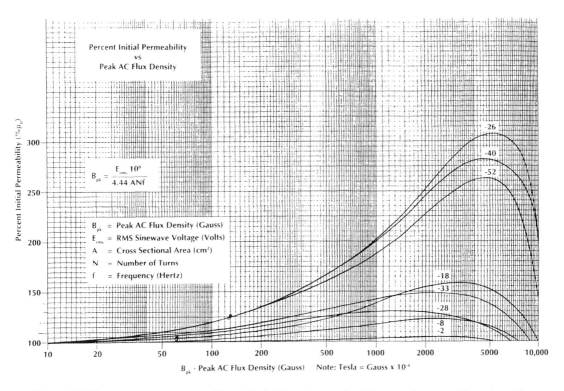

$$B_{pk} = \frac{E_{rms} \; 10^8}{4.44 \; ANf}$$

Percent Initial Permeability
vs
Peak AC Flux Density

$B_{pk}$ = Peak AC Flux Density (Gauss)
$E_{rms}$ = RMS Sinewave Voltage (Volts)
A = Cross Sectional Area (cm²)
N = Number of Turns
f = Frequency (Hertz)

$B_{pk}$ - Peak AC Flux Density (Gauss)    Note: Tesla = Gauss x 10⁻⁴

**Figure 5.40b** Variation in permeability with AC flux density for Micrometals cores. (Reproduced by kind permission of Micrometals.)

MIL-STD-462-CS01 Conducted
Susceptibility Test Levels (5-Vrms
input, 30 Hz–1.5 kHz)

| Frequency | Output referenced to input [dB] | Output [V] |
|---|---|---|
| 30 Hz | 0.0 | 5.0 |
| 960 Hz | 2.4 | 6.6 |
| 1.3 kHz | 0.0 | 5.0 |
| 1.5 kHz | −2.0 | 4.0 |
| 50 kHz | −40 | 0.05 |

MIL-STD-462-CS02 Conducted
Susceptibility Test Levels (1-Vrms
input)

| Frequency | Output referenced to input [dB] | Output |
|---|---|---|
| 50 kHz | −50 | 10 mV |
| 400 MHz | −99 | 11 μV |

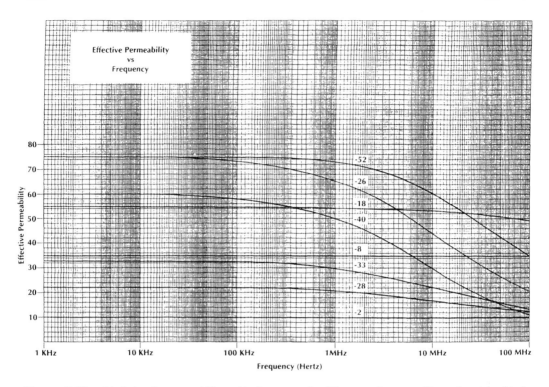

**Figure 5.40c** Variation in permeability with frequency for Micrometals cores. (Reproduced by kind permission of Micrometals.)

MIL-STD-220A Insertion Loss
(Attenuation) Test

| Frequency | Insertion loss [dB] |
|---|---|
| 100 Hz | 4 |
| 500 Hz | 15 |
| 1 kHz | 21 |
| 10 kHz | 28 |
| 100 kHz | 40 |
| 1 MHz | 80 |
| 100 MHz | 90 |
| 1 GHz | 94 |

The attenuated conducted emission level, measured in accordance with CE01 and CE02 test methods, from the simulated input current from one 15-W converter power supply of 0.2 mA into 10 μF gives 46 dBμA. The input current from 17 converters must be attenuated. Assuming a worst-case addition of input currents gives 59 dBμA. The specification limit at the frequency of maximum emission is 90 dBμA. Thus, the filter provides adequate attenuation for this source of emission.

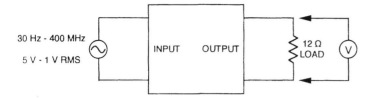

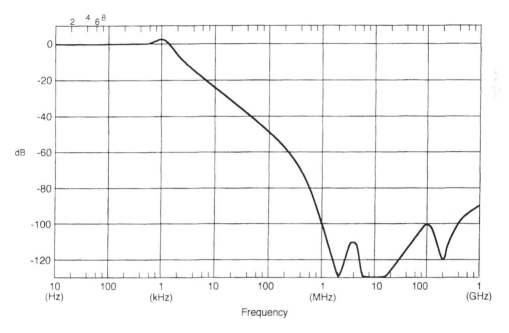

**Figure 5.41** Configuration for analysis of the +28-V power-line filter response to CS01 and CS02 input levels using ''SPICE.''

**Figure 5.42** CS01 and CS02 frequency response of the power-line filter.

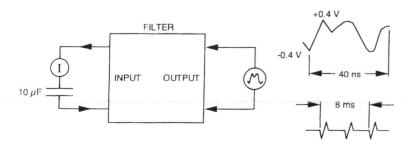

**Figure 5.43** Configuration for analysis of the +28-V power-line filter response to converter-generated O/P noise in accordance with the CE01, 3 test setup using ''SPICE.''

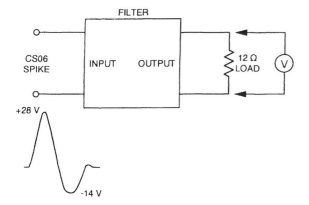

**Figure 5.44**  Configuration for analysis of the +28-V power-line filter response to the CS06 +28-V spike using "SPICE."

We see that the attenuation of the filter used in a realistic test setup exhibits an insertion gain of 2.4 dB, whereas in the MIL-STD-220 test an attenuation of 21 dB is predicted at the same frequency.

### 5.1.10.8   Continuation of Case Study 3.1

The first part of Case Study 3.1, in Chapter 3, presented the measured waveforms and noise currents generated by +5-V and ±15-V DC-to-DC converters. In order to choose a suitable commercially available filter, six different miniature feedthrough filters were tested. During the test, the filters were connected through the sides of a small enclosure on which the converter was mounted, as shown in Figure 5.49.

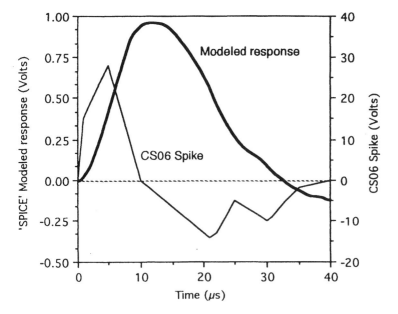

**Figure 5.45**  Result of the "SPICE" transient analysis of the +28-V power-line filter to the CS06 spike.

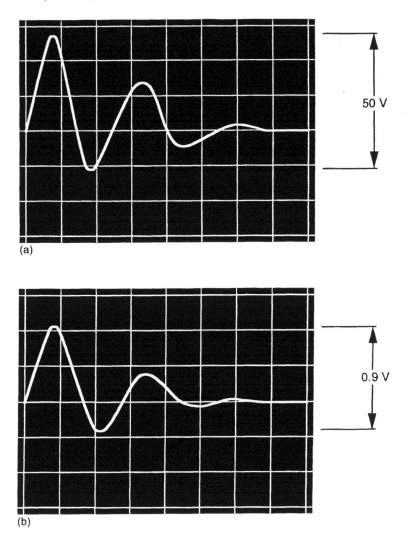

**Figure 5.46** (a) Measured CS06 input to the filter. (b) Measured output response from the power-line filter.

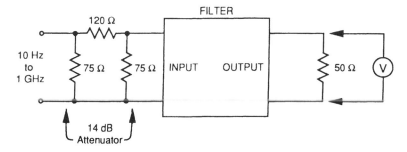

**Figure 5.47** Configuration for analysis of +28-V power-line filter response to the MIL-STD-220A test using ''SPICE.''

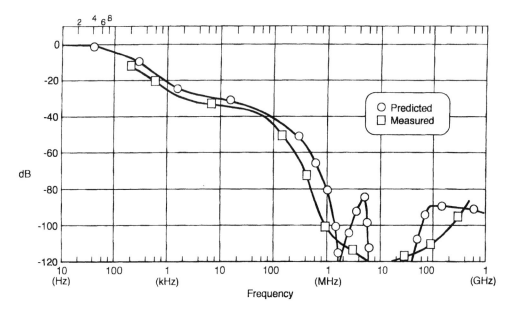

**Figure 5.48**  Result of the "SPICE" AC analysis of the +28-V power-line filter response to the MIL-STD-220-A test.

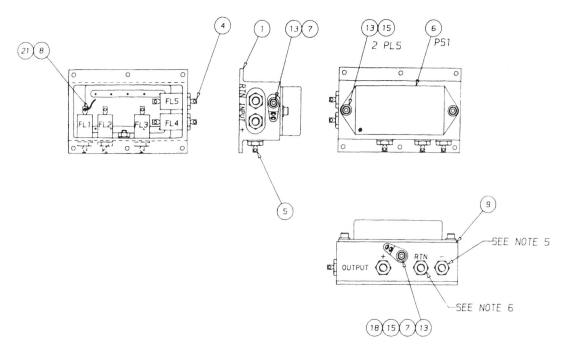

**Figure 5.49**  Mounting method for the filters described in Case Study 3.1. (Reproduced courtesy of Canadian Astronautics Ltd.)

**Table 5.4**  Published Minimum Insertion Loss of Filters Measured in Accordance with MIL-STD-220A

| Filter | Current A | Frequency | | | | | | |
|---|---|---|---|---|---|---|---|---|
| | | 30 kHz | 150 kHz | 300 kHz | 1 MHz | 10 MHz | 1 GHz | 10 GHz |
| 1 | 3 | 9 | 25 | 45 | 70 | 70 | 70 | — |
| 2 | 15 | 15 | 25 | 34 | 44 | 60 | 70 | — |
| 3 | 15 | 20 | 32 | 36 | 46 | 64 | 80 | 80 |
| 4 | 1 | 18 | 50 | 66 | 80 | 80 | 80 | 80 |
| 5 | 1 | 15 | 31 | 42 | 63 | 70 | 70 | — |

The input noise current with and without the filters in the circuit was measured in accordance with MIL-STD-462 CE01 and CE03. In addition, the differential- and common-mode noise at the output of the supply with filters was measured in both the frequency domain and the time domain. The test setup is described in Section 3.3.1. The published minimum insertion loss of the filters measured in accordance with MIL-STD-220A is shown in Table 5.4.

In the CE01 and CE03 input test on filters 1, 2, 3, and 6, only certain frequencies of input noise current were reduced at or close to the major frequency of emission, which was 250 kHz (twice the switching frequency of the converter). Only filters 4 and 5 reduced the emissions close to the switching frequency and at or close to 36 MHz. The magnitude and frequency components of the noise currents without filters are shown in Plots 1–4 of Section 3.3.3.

Table 5.5 provides a summary of the performance for the six filters. Only filters 4 and 5 achieved any common-mode insertion loss at 250 kHz; the remaining filters exhibited insertion gains as high as 12 dB. The output of the majority of the filters measured in the time domain was either an undamped or a damped 250-kHz sinewave. These measurements on available filters highlights the difficulty of choosing a suitable filter based on the manufacturer's published data, whereas using the equations of Figure 5.29 with a 12.5- or 25-$\Omega$ load resistor, as used in the measurement setup, results in an accurate prediction of those filters that exhibited insertion gain.

### 5.1.10.9  Signal-Line Filters

Commercial filters are available covering frequencies from audio up to tens of gigahertz. These filters are designed and specified for a specific load impedance, typically 50 $\Omega$, 75 $\Omega$, 125 $\Omega$, or 600 $\Omega$. The filters may be active or passive and of the following types: *LC* high pass, low

**Table 5.5**  Measured Insertion Loss When Used at Input and Output Terminals of a 15-W + 5-V Converter

| Filter | Input | | | Output | |
|---|---|---|---|---|---|
| | CE01 210 kHz | CE03 36 MHz | D/M 206–263 MHz | C/M @ 250 kHz | Wideband |
| 1 | | | | −11 dB | 30 dB |
| 2 | | | | −12 dB | |
| 3 | | | | −5 dB | |
| 4 | 38 dB | 30 dB | 6.5 dB to −3.8 dB | 8.7 dB | 39 dB |
| 5 | 18 dB | 34 dB | 0 dB | 27 dB | |
| 6 | | | | −2.6 dB | 42 dB |

pass, bandpass, bandsplitter, diplexer, band reject or notch, tunable notch, tunable preselectors, narrowband helical resonator filters, comb-line, interdigital cavity resonator, and digital electronic. The amplitude and delay response of filters may be chosen and include: delay equalized low pass, Bessel, Butterworth, elliptic, Chebyshev, transitional Gaussian low pass, transitional Gaussian linear-phase bandpass, Gaussian linear-phase bandpass, and delay equalized bandpass. The relative merits and weaknesses of the different filter types are described in books on filter design. Simple filter facts are contained in a catalog of precision $LC$ filters published by Allen Avionics, Inc. In addition to books, computer programs and nomographs are readily available and ensure a suitable design for all but complex and variable source and load impedances. Very often a single-section or double-section $LC$ circuit is unnecessary and a simple $RC$ filter will suffice.

At frequencies higher than the maximum operating frequency of the circuit, the noise immunity of control, analog, video, or digital interface circuits may be increased by inclusion of an $RC$ filter without the resonance problems encountered with the $LC$ filter. For differential or balanced input circuits, the values of resistance and capacitance in each of the input lines should be matched, to minimize the conversion of common-mode noise to differential at the input.

As with all filters, the radiated coupling between the input wiring/PCB tracks and the output wiring/PCB tracks should be minimized to the extent feasible. With surface-mount filters, the coupling between input and output traces can be reduced by imbedding the traces in a stripline PCB configuration. Regardless of filter type, in applications where the filter is designed to exhibit high insertion loss, the components should be contained in a metal enclosure with either a central compartment and feedthrough capacitor or the use of a feedthrough capacitor mounted in the enclosure wall to bring the ''clean'' signal out of the enclosure. The ''noisy'' input signals should be connected via a shielded cable, or, where the environment is noisy, the ''clean'' signal should be shielded.

### 5.1.10.10   Case Study 5.2

Case Study 5.2 illustrates the effectiveness of a simple $RC$ filter. In this case we have a 50-kW transmitter, at a frequency of either 4.8 MHz or 2.2 MHz, located approximately 40 m from a 10–14-kHz Omega timecode receiver antenna. A 4-V/m E field incident on the timecode rod antenna was measured. The antenna is connected at the base to a high-input-impedance preamplifier. The preamplifier gain reduces sharply above 50 kHz, and the preamplifier normally functions as an effective filter. However, the voltage generated at the input of the preamplifier by the interfering transmitter is approximately 26 V peak to peak, which is above the preamplifier supply voltage. The interfering voltage resulted in compression in the first stage of the preamplifier and amplitude modulation of the timecode signal, thereby destroying its integrity.

The antenna impedance is 1.18 MΩ at 10 kHz, and the input impedance of the preamplifier is 3 MΩ; thus a standard 50-Ω low-pass filter cannot be used in this application. The $RC$ filter shown in Figure 5.50 achieved an attenuation of 22 dB at 2.2 MHz and 28 dB at 4.83 MHz. The filter is contained in a small metal enclosure connected between the base of the antenna and the preamplifier. The inside of the filter was compartmentalized to avoid radiated coupling between the input and output, which was almost inevitable due to the very high level of the interfering signal as well as the high input impedance and low capacitance of the preamplifier. The feedthrough is of a standard type, with no intentional capacitance. However, the measured parasitic capacitance was 1.5 pF, and this was accounted for in the filter design along with the 1.5 pF of preamplifier input capacitance. An additional 4.7-pF capacitor was included in the filter, resulting in the approximately 7.7-pF capacitance shown in the schematic. The insertion

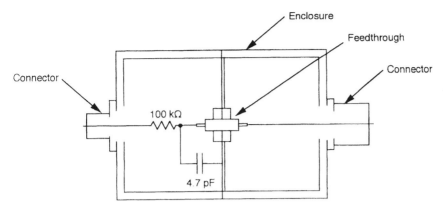

**Figure 5.50** Signal-line filter and construction.

loss at the timecode signal frequency of 10 kHz was an acceptable 0.3 dB, and the EMI problem was solved by inclusion of the filter.

### 5.1.10.11 Filter Connectors

Filter connectors are connectors with intrinsic filter components such as capacitors or capacitors and inductors in $T$, $L$ or $\pi$ configurations or simply bulk ferrite forming a balun. Filter connectors may be used to filter either signal or power up to 15 A and 200 Vdc or 120 Vac. The connector types available include D-type and the majority of military types, such as MIL-C-389999, 5015, 83723, 26482, TKJL, TKJ, TKJA, manufactured by Amphenol and ITT Cannon, as well as BNC and TNC types manufactured by Erie. Filter connectors are available with moderate levels of attenuation at low frequency, suitable for power or control lines, or with minimum attenuation at 1 MHz, 10 MHz, or 100 MHz, resulting in negligible pulse degradation in signals.

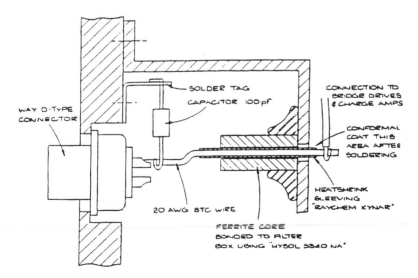

**Figure 5.51** Filter box enclosing the rear of a D-type connector.

Spectrum Control, Inc., manufacturers D-type filtercons or filter adapters that, depending on type, result in good signal pulse performance at 0.13, 0.16, 0.26, 0.64, 1.3, 2.8, 3.4, 6.4, and 130 MHz. One disadvantage of the filter connector when used in balanced, differential circuits is the 20% or so tolerance on the capacitor value, which effectively unbalances the input and converts common-mode noise to differential mode.

There are filter adapters available that are inserted between the existing unfiltered plug and socket and are useful in a diagnostic test or to cure an EMI problem. In the MIL-STD-462 test setup for the radiated emission (RE02) test, the equipment under test in the shielded room may be controlled by a test set, computer, or ground-based equipment located outside of the room. The RE02 limits are very often exceeded when the interface cables entering the shielded room are connected to the test set outside the room but disconnected from the equipment under test (EUT). The reason is typically common-mode noise on signal or supplies sourced by the test set, resulting in radiation from the interface cables, which are usually shielded. The solution is to use D-type connectors on the test set and include filter adapters where required.

When filter connectors are too expensive or unavailable "space qualified," or the capacitance must be matched closely for differential circuits, the filter box arrangement illustrated in Figure 5.51 may be used. Flexible wafers containing filter surface mount (SM) components, available from TRW and Metatech, fit over the pins of existing connectors.

## 5.2  EMISSION REDUCTION TECHNIQUES

### 5.2.1  Signal and Power Generation Characteristics

Pulsewidths and rise times of signals as well as power control and conversion should be selected so as to use no more of the frequency spectrum than required. This means that pulsewidths should be as narrow as possible and that the rise and decay times should be as long as possible within the limits of the required signal characteristics. High current pulses due to switching charge in diodes and other semiconductor devices as well as the rapid charge and discharge of capacitors should be minimized, typically by inclusion of series impedance.

The rise and fall times of switching transistor and rectifier diodes, usually tailored for the fastest possible switching times to improve efficiency, will produce an incredibly rich frequency spectrum of emissions. Selection of "soft" fast-recovery rectifiers and slowing down the transistor rise and fall times (at the expense of efficiency) will significantly suppress high-frequency (greater than 1 MHz) components. Where possible, the use of thyristors or triacs with their associated high $dV/dt$ or $dI/dt$ characteristics should be replaced with transistors or power MOSFETS, where the transition from "off" to "on" may be slowed down.

The heater switching circuit shown in Figure 5.52, without the additional 0.01-µF miller capacitor between Q16 base and collector, switched "on" in 1.5 µs and "off" in 1.4 ms. The capacitor slowed the switching speed to 4 ms "on" and 4 ms "off". The reduction in the level of broadband conducted noise on the line is shown in Figure 5.53.

Broadband noise is often generated by linear power supplies at the AC power-line frequency and appears on both the input and output terminals. The reason for the noise voltage/current is the commutation of the rectifiers from a conducting to a nonconducting state. This commutation may be speeded up by the use of fast-recovery rectifiers; however, some time is still required to extract the charge from the rectifier junction during turnoff. During the charge switching time, a short circuit appears across the AC input terminals. The use of series inductance, as shown in Figure 5.54a, will very effectively reduce the input-current spike causing broadband noise.

The inclusion of a series inductor between the bridge and the smoothing capacitor may

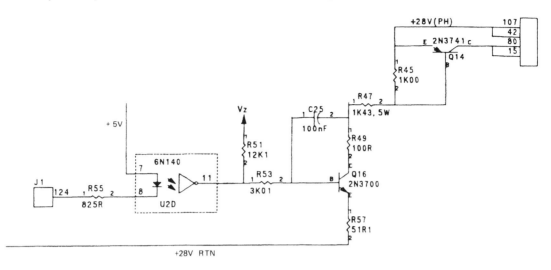

**Figure 5.52** Heater switching circuit.

be used to isolate the supply from current spikes generated by the load switching or to reduce the input ripple current, as shown in Figure 5.54b. The trade-off is that the inclusion of the inductor reduces the maximum voltage developed across the smoothing capacitor. The ripple current at the input power frequency may be further reduced by adding a capacitor across the inductor and thereby creating a parallel tuned circuit at the power-line frequency.

It is understood that at very high switching frequency it may not be possible to increase rise and fall times. For example, the trend toward converter designs with switching frequencies in the megahertz region requires short rise and fall times to ensure high efficiency.

However, there are techniques available to the designer to reduce the generation, radiation, and conduction of unwanted noise. Four of these techniques are discussed next.

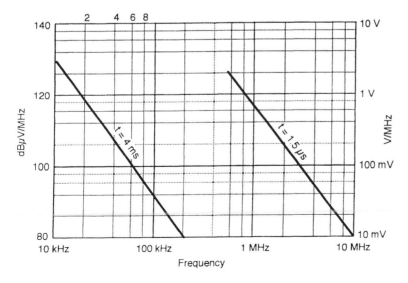

**Figure 5.53** Reduction in the level of broadband emissions due to the inclusion of C2 (0.01 μF).

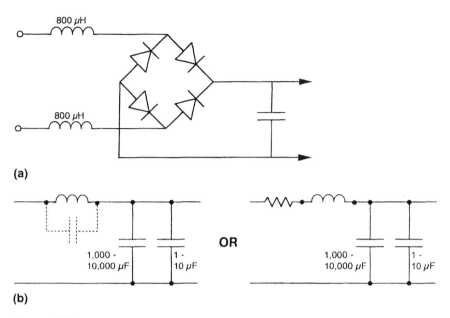

**(a)**

**(b)**

**Figure 5.54** Inductor between bridge and capacitor used to reduce ripple and transient current spikes.

### 5.2.1.1 Circuit Topology

All switching-power supply topologies will produce conducted and radiated emissions, but some are better choices than others. For example: (a) Forward converters are better than flybacks, because peak currents are lower and secondary AC currents are much lower. (b) Topologies that operate at a duty cycle greater than 50% (e.g., pulsewidth modulation or push–pull) also result in lower peak currents. The use of input power factor correction will reduce input spikes on AC power.

### 5.2.1.2 Reservoir and Decoupling Capacitors

It is better to provide transient current from a capacitor, with a low ESL and low ESR, mounted as close to the switching transistors or MOSFET as feasible, than from the supply. Series resistance or inductance may further isolate the supply to the converter from transient current requirements. The conductive paths for transient current flow should cover as small an area as possible.

### 5.2.1.3 Heatsinks

The major cause of high common-mode noise in converters is the switching waveform on the collector/source of the switching transistor/MOSFET, which is usually connected to the case of the device. The case of the switching device is then typically heatsunk to the enclosure. Thus the noise source is referenced to the enclosure via the high-quality capacitance formed between the case of the switching transistor or MOSFET and the enclosure, with the heatsink insulator acting as the dielectric. The capacitance between the device and the enclosure may be reduced by using a beryllium oxide heatsink insulator (capacitance 18 pF, 41 pF assembled) instead of a mica insulator (capacitance 150 pF, 156 pF assembled). A better solution is the use of an insulator that contains an imbedded shield. With the shield connected to the return of the power supply of the switching devices, the capacitively coupled transient currents flow back into the supply and not into the enclosure, and the common-mode noise voltages are considerably reduced.

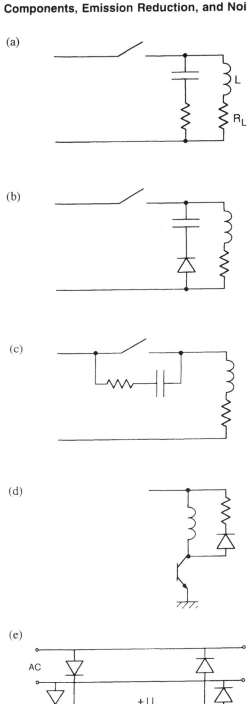

**Figure 5.55** Snubber circuits.

### 5.2.1.4  Circuit Layout

Confining high transient or RF current-carrying conductors and devices into a small area with the use of short low-impedance connections will reduce both conducted and radiated emissions. The circuit should be as far removed from the enclosure as feasible in order to reduce capacitive coupling. The use of a small secondary enclosure, shielding the source of noise, or compartmentalization with the use of feedthrough capacitors or filters may be required for sources of high radiation. When switching inductive loads by either a relay or semiconductor device, high voltages may be generated during switch-off. The magnitude of the voltage is determined by the rate of change of current as follows: $V = L(dI/dt)$. The high voltage developed when $dI/dt$ is high may couple capacitively to circuit conductors, to the inside of an enclosure, or across the capacitance of the semiconductor device. When the inductive load is switched by a relay, damage to the relay contacts may result due to arcing. The snubber circuits shown in Figure 5.55 may be used to reduce the voltage generated. Omission of the resistor that is in series with the diode in Figures 5.55b and 5.55d decreases the voltage further but will result in a high transient current flowing in the diode, which in turn may result in an EMI problem. When switching a load with high in-rush current, such as a converter or filament lamps, a series inductor may be used to reduce the magnitude of the transient.

It is not uncommon in an EMI/EMC investigation to find that in switching a load, the resultant voltage or current transients are the cause of EMI or emissions above a specification limit. However, the following case concerns a load that would not be expected to cause a problem. A piece of commercial equipment made up of a mix of TTL and CMOS logic on a single PCB contained in a nonshielded enclosure failed FCC radiated-emission requirements, but only when a beeper sounded. The cause of the problem was the piezoelectric beeper. When the transistor switching the beeper at an audio frequency switched off, the beeper resonated at a frequency above 30 MHz and a 100-Vpk-pk voltage was developed across the beeper. Adding a resistor in parallel with the beeper dampened the resonance but reduced the audio output to a lower, albeit acceptable, level. Chapter 11 describes PCB layout for reduced emissions and further circuit-level reduction techniques.

## 5.3  NOISE IMMUNITY

### 5.3.1  Interface-Circuit Noise Immunity

It is important to assess the EM environment before deciding on the level of differential-mode and common-mode immunity required of an interface. It is not unusual for common-mode noise between the grounds of systems separated by appreciable distances to reach levels of several volts. This type of noise can be considerably reduced by the correct grounding scheme, carefully realized.

Digital equipment communicate with low error using standard interfaces in an office environment. However, when a computer is located in such an electromagnetically benign environment and interfaces to equipment located in a hostile environment, such as the shop floor in a process and control facility, the noise immunity of standard interfaces may be insufficient. When the computer is located in the hostile environment, EMI is likely caused by power-line transients.

As described in Chapter 7, the level of achievable shielding of an incident field with flexible shielded cables is limited. It is therefore not uncommon to experience failures during a radiated susceptibility test due to interface circuits with insufficient noise immunity. When EMI occurs in the passband of the interface, filtering as a solution is precluded.

After the EM environment has been estimated, using the examples of Chapter 2, the specified susceptibility test limits, or the techniques described in Chapter 11, an interface can be chosen that, in addition to meeting speed and power consumption constraints, will meet the EMI immunity requirements.

Manufacturers of commercial equipment exported to the European Community must make the EMC Declaration of Conformity, and this will typically be made after emission and immunity testing. One of the signal cable tests is the C/M capacitive injection of an electrical fast transient (EFT) pulse of 500 V for domestic equipment and 1 kV for nondomestic equipment. The EFT pulse when measured into a 50-Ω load has a rise time of 5 nS and a pulse width of 50 nS at the 50% amplitude value. The pulsewidth induced in a cable may be significantly longer than 50 nS, dependent not only on the C/M load seen by the EFT generator but on the EFT generator design.

Another test is bulk C/M injection into a cable of a 1-kHz 80% am signal from 150 kHz to 100 MHz.

Standard EIA interfaces using unshielded cables, or shielded cables with a poor shield termination technique, commonly fail the EFT test, and less commonly the C/M signal injection. It is important to ensure that sufficient noise immunity is built into the interfaces. Low-frequency unbalanced inputs may require high-frequency filtering and high frequency. Balanced input lines should have equal path lengths and present equal impedances to ground, to very fast pulses.

The noise immunity of standard TTL, CMOS, and ECL logic is shown in Table 5.6. The noise immunity of standard logic can be increased by the use of low-impedance drivers and terminating networks as well as Schmitt trigger receivers, as shown in Figure 5.56.

**Table 5.6**  Noise Immunity of Standard TTL, CMOS, and ECL Logic

|  | TTL 64 | HTL | CMOS 5 V | CMOS 10 V | CMOS 15 V | TTL 74L | TTL 74H | ECL[a] 10 k | ECL[a] 100 k |
|---|---|---|---|---|---|---|---|---|---|
| Typical propagation delay (ns) | | | | | | | | | |
| tPHL | 12 | 85 | 150 | 65 | 50 | 30 | 8 | 2 | 0.5 |
| tPLL | 17 | 130 | 150 | 65 | 50 | 35 | 9 | 2 | 0.5 |
| DC-noise margin (minimum) | | | | | | | | | |
| VNL (V) | 0.4 | 5 | 1 | 2 | 2.5 | 0.4 | 0.4 | 0.155 | 0.155 |
| VNH (V) | 0.4 | 4 | 1 | 2 | 2.5 | 0.4 | 0.4 | 0.125 | 0.125 |
| Typical output impedance (Ω) | | | | | | | | | |
| Low | 30 | 140 | 400 | 400 | 400 | 30 | 30 | 8 | 8 |
| High | 140 | 1.6 k | 400 | 400 | 400 | 510 | 55 | 6 | 6 |
| Typical noise energy immunity | | | | | | | | | |
| ENL (nJ) | 1.7 | 60 | 1 | 3.7 | 7.2 | — | — | — | — |
| Pulsewidth (ns) | 20 | 125 | 155 | 70 | 50 | — | — | — | — |
| ENH (nJ) | 1 | 5 | 0.9 | 3.1 | 8.5 | — | — | — | — |
| Pulsewidth (ns) | 25 | 145 | 280 | 90 | 75 | — | — | — | — |

[a] $R_L = 50\ \Omega$.

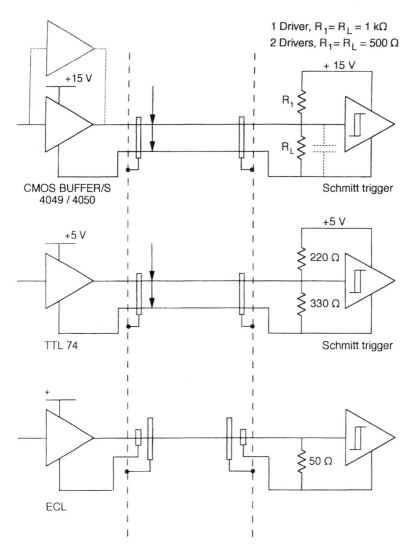

**Figure 5.56**   Standard TTL, CMOS, and ECL line drivers/receivers.

**5.3.1.1   Standard Line Drivers and Receivers, RS232, RS423A, RS422A, RS485, RS449, RS530, IEEE (488-1978) (GPIB), 10BASE TX, 100BASE TX, 155Mb/s ATM, and MIL-STD-1553 Interfaces**

Line driver and receiver ICs are available that have some or all of the following features:

   Built-in termination resistors
   Clamp diodes at input
   Clamp diodes at output
   High current outputs
   Individual frequency response control
   Input hysteresis
   High DC noise margin

High common-mode noise rejection ratio
EIA standard
GPIB standard
Single-ended transmission
Differential data transmission
Input threshold reference input

### 5.3.1.2 Receivers and Drivers

Typical differential line driver and receiver configurations are shown in Figure 5.57. When more than one transceiver shares the line shown in Figure 5.57, this configuration is termed a *party line* or *bus* system. The transmission line is typically a twisted pair individually shielded or with an overall shield over a number of twisted pairs. The differential configuration has the advantage of high common-mode noise immunity below 5 MHz, increased immunity to crosstalk from pair to pair, and the possibility of added "in band" common-mode noise filtering. The input impedance to ground in the lower circuit in Figure 5.57 is typically in the kiliohm range

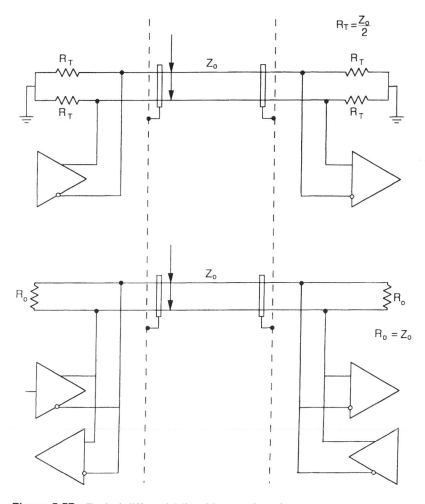

**Figure 5.57** Typical differential line drivers and receivers.

due to the relatively low input impedance of most commercially available line receiver ICs. Thus these devices are not suitable for the isolation of equipment–equipment or system–system grounds. Conversely, the input impedance to ground of the device in the lower circuit of Figure 5.57 is too high to achieve effective C/M filtering by the use of a balun. The impedance of a ferrite balun is approximately 50—1000 $\Omega$, and with a C/M input impedance of 1–10 k$\Omega$, the C/M attenuation is likely to be less than 6 dB. Baluns are often placed on line driver and receiver connections to the interface cable to reduce radiated emissions, and these will have very little effect on increasing C/M noise immunity when used in the Figure 5.57 circuit. However, when the terminating resistor is split as shown in the upper circuit of figure 5.57, significant C/M attenuation is achieved. For example, assume $Z_o$ is 100 $\Omega$ and the impedance of the balun shown in Figure 5.58a is 500 $\Omega$ over a wide frequency range. The impedance line to ground $R_T$ is 50 $\Omega$, and the C/M noise voltage attenuation is 21 dB. The C/M noise current divides between the two lines and flows in the two 50-$\Omega$ impedances, so the total impedance seen by the C/M noise source is 25 $\Omega$. The C/M power noise immunity is therefore high.

Noise voltages are typically generated C/M due to a ground noise potential between the driver and receiver ground and due to common-mode injection onto the signal cable.

Differential receiver inputs that are well balanced to ground and exhibit high C/M noise rejection over the in-band frequencies should be chosen. Where out-of-band signals are a problem, the addition of a high-pass filter for low-frequency noise or a low-pass filter for high-frequency noise can be added to the circuit, as shown in Figure 5.58b.

The noise immunity to in-band noise of a single-ended system is much lower than the differential input, because C/M-induced noise is converted to D/M noise at the receiver input

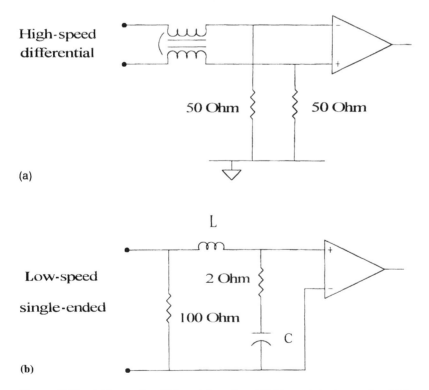

**Figure 5.58** (a) Increasing differential receiver C/M input noise immunity. (b) Low-speed single-ended receiver input filter.

by the unbalanced impedance of the input. Additional filtering can be added to a single-ended receiver input for out-of-band noise rejection.

A receiver's differential input impedance is typically dependent on the input frequency and the input signal. Adding a termination resistor to the receiver input will reduce the reflections and improve the signal quality and often the noise immunity of the circuit. This increased noise immunity is achieved because noise spikes with voltages lower than the signal level and impressed on an undistorted signal are less likely to exceed the receiver input switching threshold than spikes impressed on a distorted signal. The noise input power required to generate voltages above the threshold is also much higher for a 45–460-Ω input impedance than a 1–10-kΩ input impedance. The increased noise immunity achieved with a termination resistor is offset to some extent by the reduction in signal level due to the loading. To maintain noise immunity, choose a driver with a low output impedance. In a party line system, the driver output is often open collector with a low-value pull-up resistor. The pull-up resistor should be no higher than 20 Ω when the receiver input resistance is 50 Ω. If the interface is not party line, then an active pull-up should be used, because as a ''high'' output is pulled down, more current flows out of the active driver.

The DC noise margin and the power noise margin shown in Table 5.6 assumes no AC or DC loss between the driver and receiver, which may be true for very short line lengths. Where longer line lengths are used at high frequency, the cable attenuation plays an important role in reducing signal level and decreasing the signal/noise ratio. The 50-Ω impedance RG188A/U and the unshielded twisted pair have a higher attenuation than the 50-Ω RG58/U, and even lower is the 75-Ω RG59/U.

The DS26LS32 receiver, manufactured by National Semiconductor, has a differential- or common-mode input range of ±7 V and an input sensitivity of ±0.2 V and is therefore ideal for the EIA RS-422, RS-423, RS-499, RS-530, and federal standards 1020 and 1030 for balanced and unbalanced data transmission. The DS26LS32 has a ±15-V differential- and common-mode voltage range and a ±0.5-V differential sensitivity, and this device, although it does not conform to the EIA standards, will exhibit a much higher noise immunity. Table 5.7 shows the characteristics of drivers and Table 5.8 those of receivers, reproduced from the AN781A application note from Motorola Inc.

Figure 5.58a illustrates the common-mode filtering circuit. For low-frequency single-ended signals, such as the RS232C, the high-frequency noise immunity can be increased by adding a filter or filter connector at the receiver input. Filter connectors containing a π-circuit can be selected with a 3-dB frequency appropriate for the data rate of the signal. The only problem with filter connectors is the capacitive imbalance between the pins. Very high levels of C/M noise applied to the connector are converted to differential noise by this imbalance.

Figure 5.58b illustrates a filter suitable for a low-speed single-ended receiver. The input impedance is maintained at a low 100 Ω to increase the amount of input noise power required

**Table 5.7**  EIA-Compatible Line Drivers from the Motorola Application Note AN781A

| RS Std. | Device # | Drivers per package | Power supplies | Prop. delay | Rise/fall time | Hi-Z output comments |
|---------|----------|---------------------|----------------|-------------|----------------|----------------------|
| 232-C | MC1488 | 4 | ±9 to ±13.2 V | 175/350 nS | Adjustable | No inverting |
| 232-C | MC3488 | 2 | ±10.8 to ±13.2 V | 20 nS | Adjustable | No inverting |
| 422 | MC3487 | 4 | +5 V | 20 nS | 20 nS | Yes |
| 423 | MC3488 | 2 | +10.8 to 13.2 V | — | Adjustable | No inverting |
| 485 | SN75172 | 4 | +5 V | 25/65 nS | 75 nS | Yes |
| 485 | SN75174 | 4 | +5 V | 25/65 nS | 75 nS | Yes |

**Table 5.8** EIA-Compatible Line Receivers from Motorola AN781A

| RS Std. | Device # | Drivers per package | Input hysteresis | Input threshold | Prop. delay | Comments |
|---------|----------|---------------------|------------------|-----------------|-------------|----------|
| 232-C   | MC1489   | 4                   | 0.25–1.15 V      | 3 V             | 85 nS       |          |
| 422/423 | MC3486   | 4                   | 30 mV typical    | 200 mV          | 35 nS       | ±7 V C/M input range |
| 485     | SN7513   | 4                   | 50 mV typical    | 200 mV          | 35 nS       |          |

at in-band frequencies before an upset occurs. The values of $L$ and $C$ are chosen to produce a low-pass filter with a frequency response that does not degrade signal quality but may increase the signal rise and fall times within acceptable limits. The input impedance of the receiver is high, and so the 2-$\Omega$ resistor is required to damp the filter resonance. The resonance frequency of the filter should be far removed from the maximum clock rate of the signal.

Where in-band differential-mode noise at a differential receiver input, or common-mode noise, which is converted to D/M at a single-ended receiver input, is causing a problem, the solution is to use a shielded cable for the interface. If the cable is already shielded, then it is important to ensure the shield termination is made via a low impedance to the equipment enclosure or to use a double-braid shielded cable or at least a foil/braid shielded cable.

Line drivers are often capable of driving a large number of receivers and have sink and source capabilities of typically 40 mA and up to 200 mA. Load impedances of 78–250 $\Omega$ are common, and the interface may be used with a twisted pair that exhibits an impedance of 80–130 $\Omega$ or a twisted shielded pair of 70–120-$\Omega$ impedance. The DS26S31 line driver meets all the requirements of EIA RS-422, RS-530, and RS-449 and federal standard 1020. The DS26S31 can supply a minimum high output voltage of 2.5 V at an output current of $-20$ mA and a maximum low voltage of 0.5 V at 20 mA. The output short-circuit current is limited to between $-30$mA and a maximum $-150$mA. Table 5.6 shows the characteristics of the EIA standard interfaces. The RS449 standard was developed to provide a new definition for RS232-C interconnection lines. RS449 uses 30 lines in a 37-pin connector, whereas RS232 uses 20 lines in a 25-pin connector. RS449 does not define electrical characteristics, because those defined for RS422, and RS423 are to be used. RS530 is just like RS422 and uses differential signaling on a DB25–RS232 format.

To decrease crosstalk in a cable containing a number of interface circuits, a maximum $dV/dt$ is specified, as is the transition time.

The maximum cable length for any interface is dependent on the attenuation characteristics of the cable: higher resistance shortens the maximum permissible length and higher capacitance shortens the maximum length at the higher baud rates. The required noise immunity and maximum bit error rate also limits either the cable length or the baud rate. The maximum cable lengths recommended in RS232, RS422, and RS423 are recommendations only; longer lengths are possible with high-quality cable and/or using signal waveshaping, signal distortion compensation, or interface conditioning techniques, which may also be used to improve noise immunity. The min/max cable lengths for the RS422 in Table 5.9 is based on the use of twisted-pair, 24AWG cable and a balanced interface.

Table 5.10 shows the characteristics of a typical bidirectional instrumentation bus (IEEE-488, GPIB) transceiver. The input termination resistors of 2.4 k$\Omega$ to VCC and 5 k$\Omega$ to ground do not represent a low impedance input, and thus the coupling, either capacitively line to line or from an incident electric field, is higher than expected for a low- (50–130-$\Omega$)-input-impedance receiver.

**Table 5.9**  EIA Standard Interfaces

| | Specification | RS232-C 1969 | RS423-A 1978 | RS422-A 1978 | RS485 1983 |
|---|---|---|---|---|---|
| **G E N E R A L** | Type | Single-ended | Single-ended | Differential | Differential |
| | Line length | 50 ft (15 m) | 1200 m (4000 ft) | 1200 m (4000 ft) | Application dependent (4000 ft) |
| | Max. frequency | 20 kbaud | 10 Mb at 10 m (33 ft); 80 kbaud at 1200 m (3937 ft) | 100 kbaud | 10 baud |
| | Transition time | <4% of γ or <1 mS (in undefined area between "0" and "1") | Lesser of 0.3γ and 300 µS (time for 10–90% of final value) | Greater than 20 nS or <0.1γ (time for 10–90% of final value) | <0.3γ (54-Ω, 50-pF load (time for 10–90% of final value) |
| **D R I V E R S** | No. of Drivers | 1 | 1 | 1 | 32 |
| | Open-circuit output | $3.0\,V < |V_o| < 15\,V$ | $4.0\,V < |V_o| < 6\,V$ | $|V_o| < 6\,V\ |V_{oa}|, |V_{ob}| < 6\,V$ | $1.5\,V < |V_t| < 6\,V$ |
| | $V_t$ (loaded output) | $5\,V < |V_o| < 15\,V$ | $|V_t| > 0.9|V_o|450\text{-}\Omega\ \text{load}$ | $> 2\,V$ or $\tfrac{1}{2}V_o < |V_t| < 6\,V$ 100 Ω balanced | $1.5\,V < V_t < 5\,V$ |
| | Short-circuit $I$ | < 500 mA | < 150 mA | < 150 mA | < 250 mA |
| | Output Z | — | <50 Ω | <100 Ω balanced | — |
| | Driver load | 3–7 kΩ | 460 Ω | 100 Ω | 54 Ω |
| **R E C E I V E R S** | No. of receivers | 1 | 10 | 10 | 32 |
| | Input range | +15 V | +12 V | −7 V to +7 V | −7 V to +12 V |
| | Min. receiver response | >±3.0 V | 200 mV differential | 200 mV differential | 200 mV differential |
| | Input resistance | 3–7 kΩ, 2500 pF | 4.0 kΩ min. | 4.0 kΩ min. | 15 kΩ dynamic 12 kΩ (+12 V) 8.75 kΩ at −7 V |
| | C/M voltage for balanced receiver | — | — | $7.0\,V, V_{cm} < 7\,V$ | $7.0\,V, V_{cm} < 7\,V$ |
| | DC noise margin | 2 V | 3.4 V | 1.8 V | 1.3 V |
| | Power noise margin | 1.3 (mW) | 25 (mW) | 32 (mW) | 31 (mW) |

**Table 5.10**  Characteristics of Typical Bidirectional Instrumentation Bus
(IEEE-488, GPIB) Transceiver

| GPIB input (receiver) | GPIB output (driver) | Logic state | Voltage or current | |
|---|---|---|---|---|
| | | | Minimum | Maximum |
| V low | | "1" | −0.6 V | +0.8 V |
| V high | | "0" | +2.0 V | +5.5 V |
| | V low | "1" | 0.0 V | +0.4 V |
| | V high | "0" | +2.4 V | +5.0 V |
| I low | | "1" | | −1.6 mA |
| I high | | "0" | | +50 μA |
| | I low | "1" | +48 mA | |
| | I high | "0" | | −5.2 mA |

Input impedance terminated = 1.6 kΩ. Input hysteresis = 600 mV.
Typical noise margin: high to low = 1.5 V; low to high = 1.2V.


Although the GPIB is a single-ended transmission line, the immunity of the cable and receiver to magnetic fields is surprisingly high. Using a 6.7-m single shielded GPIB cable with plastic backshells, the differential-mode noise detected at the receiver input with the cable exposed over a 1-m length to 0.7 A/m at 30 MHz was approximately 600 mV (i.e., 200 mV above the receiver noise threshold). When the GPIB cable was changed to a double-braid, single-foil type with metal backshell (so called EMI hardened), the differential-mode induced voltage was reduced to 150 mV (i.e., −12 dB) and the receiver was immune to the reduced level of combined common and differential noise. The local area network "Ethernet" has been tested and shown immunity to high levels of E field (20 V/m from 10 kHz to 1 GHz) (Chapter 4, Ref. 4).

The MIL STD 1553B data bus exhibits a common-mode noise immunity of 10 Vpk-pk, 1 Hz to 2 MHz, and a differential-mode noise immunity of 140 mVrms (Gaussian 1 kHz to 4 MHz) for a transformer-coupled stub and 200 mVrms for a direct-coupled stub.

The 10BASE-TX interface transmits ≈2.5 Vpk Manchester encoded data at 10 Mb/s. For 100BASE-TX, a 1 Vpk MLT-3 encoded data stream at 100 Mb/s is transmitted. A suitable receiver circuit for either 10BASE-TX or 100BASE-TX is the National Semiconductor DP83223 TWISTER High Speed Networking Transceiver Device. The DP83223 allows links of up to 100 meters over shielded twisted pair (Type-1A STP) and category-5 datagrade unshielded twisted pair (Cat-5 UTP) or equivalent. The receiver is specified with a maximum 700-mV signal detect assert threshold. The signal detect threshold is a measure of the pk–pk differential signal amplitude at the RXI ± inputs at the RJ45 connector of Figure 5.59a required to cause the signal detect differential output to assert. The signal detect de-assert threshold is minimum 200 mV and the signal de-assert is defined as a measure of the pk–pk differential signal amplitude at the RX1 ± inputs at the RJ45 connector of Figure 5.59a required to cause the signal detect differential output to de-assert. The typical hysteresis is therefore 500 mV. The RXI differential input resistance is 7–9 kΩ, and the common-mode impedance is not defined. A typical receiver circuit recommended by National is the DP83223 for a 100BASE-TX physical medium dependent (PMD) circuit, shown in Figure 5.59a. The data stream is AC coupled from the RJ45-8 media connector to the DP3223 twisted-pair receiver by an isolation transformer. The DP83223 then equalizes the receiver signal to compensate for signal degradation caused by the nonideal transmission line properties of the twisted-pair cable. The resistor values shown in the schematic are chosen to match an insertion loss of approximately 0.4 dB, which with the 12.1-Ω and 37.9-Ω dividers will result in a final amplitude of approximately 1.45 V pk–pk differential as seen across the RXI inputs for zero meters of cable and with a standard 2-V input. The noise immunity at the RXI input is 0.75 V and at the RJ-45 input is 1.03 V for zero meters of cable. An 0.01-μF capacitor is included as a common-mode noise filter. This, along with the 37.9-Ω resistors and the isolating transformer primary to secondary capacitance will determine the level of common-mode voltage developed at the receiver input. If available, an isolating transformer with an electrostatic shield between primary and secondary and connected to ground will reduce the capacitive coupling significantly. The alternative receiver circuit described by National Semiconductor and required to allow full support of autonegotiation is shown in Figure 5.59b. In this circuit the common-mode termination is immediately after the RJ45 connector. The impedance to common-mode currents presented by the transformer winding on the RJ45 connector side is low, for the C/M currents flow out of phase through the transformer. The total receive C/M impedance is therefore a 49.9-Ω resistor and a 0.01-μF capacitor. The C/M noise power required to generate a specific voltage is almost certainly much higher than Figure 5.59a, where the C/M input impedance through the primary-secondary capacitance should be high. The problem with the Figure 5.59b circuit is that the receiver-to-ground impedance is unknown, and probably high, which means that neither the isolating transformer nor the common inductor is as effective at reducing C/M noise as in the Figure 5.59a circuit. At first glance, the transformer and C/M inductor in Figure 5.59b should effectively attenuate C/M noise currents flowing out on the interface, and this is true when the 0.01-μF 2-kV capacitor is connected to the enclosure close to the RJ45 connector. On no account should the 0.01-μF capacitor be connected to digital ground, which will have C/M noise impressed upon it and through the 0.01-μF and 49.9-Ω resistor out on the twisted-pair cable.

The DP83223 driver transmit current is from 38.2 mA to 41.8 mA, and this is a measure of the total differential current present at the TXO $\pm$ outputs into a standard 50-Ω differential load. In Figure 5.59a, the $R_{ref}$ resistor is adjusted to achieve the standard specific 2.0-V differential TXO $\pm$ output amplitude as measured across the RJ45 transmit pins.

DP83223A Twister can also be used for binary or three-level signaling at data rates up to 155 Mb/s. The ATM Forum Physical Layer Subworking Group has drafted a working document that specifies the protocol for 155-Mb/s STS-3c signaling over 100-Ω Cat-5 UTP (unshielded twisted pair) or 150-Ω STP (shielded twisted pair) cabling and the DP83223A TWISTER transceiver is compliant with this document. Figure 5.59c provides a typical circuit used for either 100-Ω UTP or 150-Ω STP cable. The step-up transformer in the 150-Ω version of the receiver matches the 100-Ω receiver input resistance and the 100-Ω transmitter output impedance to the 150-Ω STP cable. The receiver must exhibit a C/M noise immunity of 1 V pk–pk from 0 to 155 MHz when injected via a 75-Ω resistor into the center tap of the receiver side of the isolation transformer. With the 1-V C/M test level, the bit error rate shall be less than $1 \times 10^{-10}$. The transmitter differential output with the 100-Ω UTP load shall be from 960 mV to 1060 mV and for the 150-Ω STP load shall be from 1150 mV to 1300 mV. A channel reference model is used to charactize the link between the transmitter and the receiver, and the receiver shall function and meet link attenuation and a maximum 20-mV near-end crosstalk requirements with this mode. For the category-5 100-Ω UTP system, the channel reference model is defined as 90 meters of category-5 UTP cable, 10 meters of category-5 flexible cords, and four category-5 connectors. The channel reference model for the 150-Ω STP is defined as 90 meters of STPA cable, 10 meters of STP-A patch cord, and four STP-A connectors. The unshielded UTP system can meet FCC and CISPR Class B limits when the transmitted signal imbalance (C/M component) is minimized, when the category-5 cable is used, and when each line is referenced to a low-noise ground. As with all interfaces, C/M noise current flowing out on the cable can be reduced by adding the balun at the transmitter, as shown in Figure 5.59a. The C/M immunity of the receiver can be improved by adding the balun at the receiver, as shown in Figure 5.59c. Additional receiver C/M noise immunity and reduced C/M current on the interface can be achieved by connecting all unused pairs in the cable to a low-noise chassis

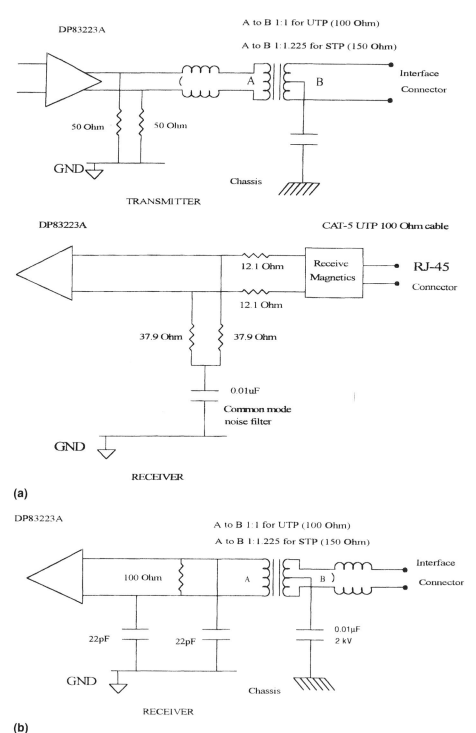

**Figure 5.59** (a) 100Base-TX driver and receiver connections to a Cat-5 UTP 100-$\Omega$ cable. (b) 100Base-TX driver and receiver circuit for full support of autonegotiation. (c) 155-Mb/s ATM recommended transmitter and receiver using the National Semiconductor DP83223. (Reproduced with the kind permission of National Semiconductor.)

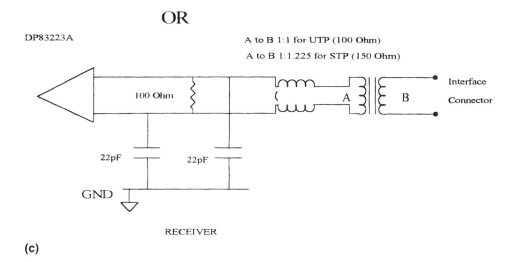

**(c)**

ground via a capacitor. This configuration is also effective at reducing radiation due to C/M currents on the cable.

Where a very high-EM ambient is predicted or where there is a requirement for ground-to-ground isolation in the megaohms in circuits with multiple interfaces, either interface circuit with very high input impedance and high noise immunity or fiber-optic links must be used.

There are commercially available fiber-optic links that provide a direct conversion of a single serial TTL data channel into a fiber-optic link and conversion back to a TTL level at the receiver. Other fiber-optic links are available with additional associated electronics that are transparent to the GP-IB or RS232 data buses. No current will flow on a fiber-optic link due to radiated or conducted noise. However, if the EM environment is very severe, it may be important to reduce the radiated coupling through the aperture in the enclosure provided for the fiber-optic cable. This reduction may be achieved by making the aperture into a waveguide below cutoff, as described in Chapter 6 on shielding.

The following examples are of nonfiber-optic analog and digital interface circuits. The general requirements for these interfaces are:

Low power consumption
Very high common-mode noise immunity (for use in a high-noise environment or over long distances)
Good input isolation and in some instances well-balanced differential input and output circuits

All of the circuits are short-circuit proof to 0 V, with some circuits, designated *protected*, short-circuit protected to +10 V at 200 mA. The circuits operate at different speeds, ranging from DC-50 kHz, DC-200 kHz, or DC-2 MHz to 150 kHz–17 MHz.

### 5.3.1.3  VMOSFET Driver and Optocoupler Receiver PLD2 and PLR2 DC-50 kHz Interface

The circuit in Figure 5.60 provides the maximum degree of isolation between grounds with 1.5 G$\Omega$ at 1500 V, input-output capacitance of 1 pF, and a common-mode transient noise immunity of 1000 V/$\mu$s. Although designed to use 125-$\Omega$ characteristic impedance twisted-shielded-pair cable, the interface circuit receiver load impedance is not an exact match to 125 $\Omega$. However,

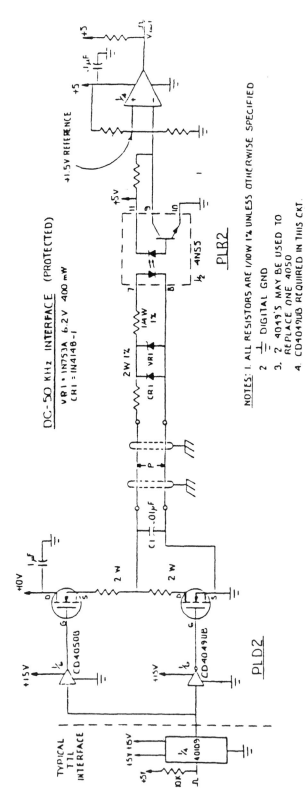

**Figure 5.60** PLR2 and PLD2 circuit. (Reproduced with kind permission of Canadian Astronautics Ltd. and the Canadian Space Agency.)

because of the slow and controlled driver rise and fall times, no reflections were seen when using the interface circuit with 100 m of deliberately mismatched 75-$\Omega$ impedance cable. The relatively slow driver rise and fall times of 2 $\mu$s have the secondary advantage of reducing radiated emission from the interface cable. The PLR2/PLD2 interfaces are ideally suited for critical control lines or slow data buses used in a very high-noise environment or over long distances. The interface has been tested with 100 m of interconnection cable and remains within the following specifications over the temperature range 0°C to + 70°C.

## PLD2 and PLR2 Typical Performance

Note: All parameters tested from 0°C to +70°C.

*Common-mode noise immunity*: A Gaussian noise generator was connected between the driver and receiver grounds in an attempt to induce errors in the output signal from the LM 139 comparator. The output waveform in particular was examined to ensure that "double edging" did not occur at the output waveform transition points. With the following peak-to-peak noise voltages covering the corresponding bandwidths, neither double edging nor jitter was observed.

*Noise bandwidth pk–pk voltage measured between Tx and Rx Grounds*:
| | |
|---|---|
| 20 kHz | 25 Vpk-pk |
| 500 kHz | 14 Vpk-pk |
| 5 MHz | 9 Vpk-pk |

*Maximum speed*: 100 kHz with $C_1 = 0.01$ $\mu$F and 100 m of cable

*Driver output*: Single ended into 300-$\Omega$ impedance
PLD2 S/C protected to +10 V and ground

*Receiver Input*: Opto-coupler PLR2 S/C protected to +10 V.

### 5.3.1.4 Transformer Coupled 10-MHz Interface HAD-1 and HAR-1

The HAD-1 and HAR-1 interface, shown in Figure 5.62 is transformer coupled and thus the data/clock must achieve a specific mark–space ratio to avoid saturation of the input transformer. The CLC103AM device is no longer available and can be replaced by a video amplifier IC followed by one of the many high speed video buffer ICs which are specifically designed to drive coaxial cables. This interface is suitable for transmission of pulse width modulated or Manchester coded data over long distances.

### 5.3.1.5 VMOSFET Driver, Differential Input Receiver, 2 MHz Interface. HD-1, HR-1, and PHR-1

This circuit, shown in Figure 5.61, provides for high-speed transmissions over long distances with high noise immunity. A well-balanced input providing 1 M$\Omega$ isolation in parallel with approximately 10 pF and 1 k$\Omega$ in series to ground, and common mode voltage capability of +10 V characterizes this circuit. It is importaant to use 120–130 $\Omega$ twisted shielded pair cable to ensure minimal line reflections as both the source and load impedances of this circuit are 125 $\Omega$. Maximum cable length tested was 100 m. Because of the inherent high current capacity of the 2N6661 VMOSFET, it is necessary only to increase the wattage rating of the two 62 $\Omega$ resistors to 2 W to achieve protection against a S/C to +10 V as well as to ground. The maximumspeed of operation with 100 m of cable is 2 MHz with an adequate safety margin. The measured rise and fall times of the driver output were both 20 ns when loaded with the

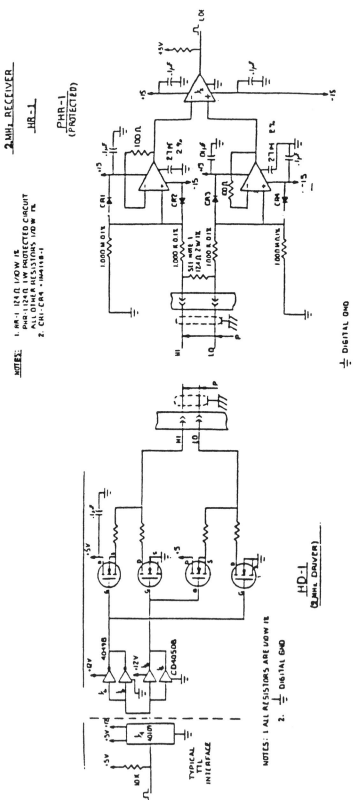

**Figure 5.61** HD-1 and HR-1 circuit. (Reproduced with kind permission of Canadian Astronautics Ltd.)

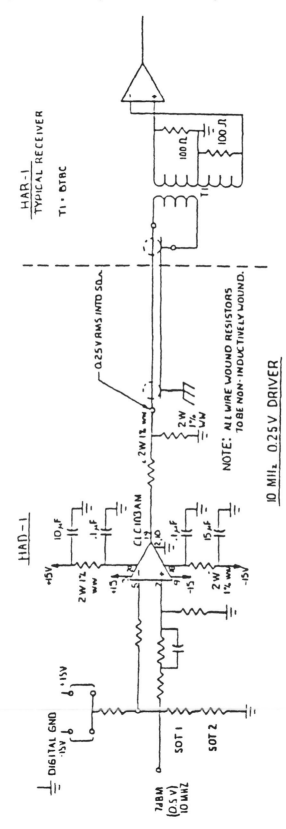

**Figure 5.62** HAD-1 and HAR-1 circuit. (Reproduced with kind permission of Canadian Astronautics Ltd. and the Canadian Space Agency.)

receiver input via 3 m of 120 Ω twisted shielded pair cable. The receiver contributes 150 ns to the overall transmitter input-receiver output delay, which was measured at 180 ns.

*Differential mode voltage noise immunity*:
   High = 2.5 V
   Low = 2.5 V

Input power noise immunity is 50 mW

*Common mode voltage noise immunity*: The common mode noise level at which double edging occurred is recorded in the following table (the cause of double edging was a combination of the common mode and approximately 100 mV of different mode noise present).

*Broadband noise voltage level*:

| Noise bandwidth | Pk-Pk Voltage |
|---|---|
| 20 kHz | 20 V |
| 500 kHz | 10 V |
| 5 MHz | 8 V |

Driver output, differential above ground into 125 Ω load impedance protected against +10 V and short circuited to ground with 62 Ω resistors rated at 2 W. The 2 MHz receiver may be used as an analog receiver with the LM119 replaced with a differential amplifier. A differential video amplifier version of the interface has been built, with a measured CM noise rejection of 40 dB at 30 MHz.

## 5.3.2 Typical Integrated Circuit Response to Noise and Immunity Test Levels

The available data on the immunity level of ICs describes the input voltage or power at which upset occurs, i.e., the device malfunctions or some parameter such as DC offset, gain, frequency changes, or a spurious response occurs.

With the conducted C/M RF and radiated RF immunity tests specified in EN 50082-1 and EN 50082-2 the equipment is exposed to an incident RF field modulated at 1 kHz and 80% and to a C/M RF (bulk current) cable injected test level, also modulated at 1 kHz and 80%. Either of these injection methods will typically result in a common-mode (C/M) RF current flow across a PCB. Despite the C/M nature of the current, IC junctions can be affected by the radiation set up by the current flow, which in turn results in current flow in the semiconductor junction. Also C/M current is easily converted to D/M voltage when the circuit is unbalanced, and the C/M current can induce a noise voltage into the device.

The spurious response is typically demodulation of the audio frequency used to modulate the incident field or the bulk RF current injected into cables. In RF circuits, the spurious response appears as either FM of the received or transmitted carrier, or ''close in'' or ''far out'' (from the carrier frequency) responses or an increase in broadband noise, or receiver desensitization. Often a combination of these problems is seen.

Any semiconductor device can rectify an RF signal, and devices can be characterized by their rectification efficiency. For example, bipolar devices rectification efficiency is proportional to $1/\sqrt{2\pi f}$ whereas MOS devices exhibit an efficiency of $(1/2\pi f^2)$. Therefore, due to the higher rectification efficiency of the bipolar device for the same emitter perimeter size, it tends to be more susceptible than MOS devices.

**Table 5.11** Power Induced in Microstrip and Transmission Line PCB Layouts with an Incident 3-V/m Field

| f (MHz) | Power (W) | |
|---|---|---|
| | T.X. Line | Microstrip |
| 30 | $9.6 \times 10^{-8}$ | $9.6 \times 10^{-10}$ |
| 160 | $84 \times 10^{-6}$ | $8.4 \times 10^{-8}$ |
| 1000 | $89 \times 10^{-6}$ | $1.1 \times 10^{-6}$ |

The rectification effect is also a function of the $f_T$ of the device ($f_T$ is the frequency at which the small-signal gain of the device equals unity). Above $f_T$, EMI through the rectification effect of the device manifests itself as audio rectification, in which the output of the device follows the envelope of the input RF.

Also, frequencies well above the $f_T$ of the device may merely be transmitted through the device and can cause EMI to a device further down the chain.

If a PCB is not contained inside a shielded enclosure, the EN 50082 radiated immunity test field is incident on the PCB. We assume an 18-cm-long PCB layout in the form of a transmission line (without ground plane) and an 18-cm-long microstrip PCB (signal trace above a ground plane). If the incident E field is 3 V/m amplitude modulated at 1 kHz and 80%, what is the effect on typical digital and analog devices? The first step is to predict the power level and voltage level induced in the transmission line and microstrip PCB layout. Table 5.11 shows the induced power calculated from the measured gain of the 18-cm microstrip and the 18-cm transmission line. Table 5.12 shows the induced voltage based on the induced power, the structure, and the load impedance. The load impedance affects the voltage induced in the circuit. The load impedance for the microstrip is calculated at 50 Ω at 30 MHz, 47 Ω at 160 MHz, and 40 Ω at 1 GHz, which is approximately correct for RF and video circuits but is almost certainly too low for analog circuits. Table 5.12 shows the induced voltage for both the ≈50-Ω load and for a 10-kΩ load in parallel with a 0.4-pF capacitor. The 10 kΩ in parallel with the 0.4-pF load has an impedance of 10 kΩ at 30 MHz, 2 kΩ at 160 MHz, and 390 Ω at 1 GHz.

**Table 5.12** Voltage Induced in Microstrip and Transmission Line PCB Layouts

| f (MHz) | T.X. Line | Microstrip |
|---|---|---|
| **Voltage induced in 50-Ω load** | | |
| 30 | 2.3 mV | 0.26 mV |
| 160 | 20 mV | 1.4 mV |
| 1000 | 6.3 mV | 9 mV |
| **Voltage induced in 10-kΩ ‖ 0.4 pF** | | |
| 30 | 31 mV | 0.26 mV |
| 160 | 320 mV (520 mV)[a] | 13 mv |
| 1000 | 80 mV | 19 mV |

[a] This is a corroborating value obtained by the use of a high-level electromagnetic analysis program.

### 5.3.3  Digital Logic Noise Immunity

Logic noise immunity is usually expressed as a DC voltage, which is a valuable measure of noise immunity for a nonimpulsive low-frequency induced noise. The voltage noise immunity of a device is often specified as the voltage at which the device becomes susceptible, and therefore immunity is achieved at levels below this voltage.

Where the induced noise consists of voltage spikes with a duration equal to or just a little longer than the rise and fall times of the logic (which is typical of the crosstalk between logic tracks on a PCB), the input noise energy is a useful immunity criterion. In addition, the immunity of logic to an RF signal is often required. The input noise energy immunity (defined as the voltage multiplied by the current for a specific pulse duration, i.e., that required to change the logic output state) is an important factor in comparing the noise immunity of different logic types. Typically, voltages much higher than the DC noise immunity level are required for short-duration pulses before an output response is evoked. Table 5.6 shows DC noise margin, the noise energy immunity, and signal-line impedance or the input/output impedance in parallel.

The energy noise immunity expressed in Table 5.6 is valid only for the specified pulsewidth. For example, from the predictions of crosstalk to a TTL gate in Chapter 4, the gate responded to a 2.5-V, 80-nS pulse, which produced a 1.6-mA current at the TTL input. Thus, the input noise energy immunity of the gate with an 80-ns input pulsewidth is 0.24 nJ, whereas, from Table 5.6, the input noise energy immunity at a 20-ns pulsewidth is 1.7 nJ.

From measurements, the immunity of a CMOS gate, type CD4013B, and an advanced low-power Schottky, type 54ALS74A, to an RF sinewave signal from 1.2 MHz to 200 MHz has been determined (Ref. 6). The RF was injected into the Vcc pin, the data input pin, and the clock pin of the IC. In the case of the data and clock pins, the RF was combined with data and clock waveforms, so the RF was riding on these waveforms. The ICs were included in a test setup that ran a functional test on the IC. At the same time, the level of RF voltage injected into the IC was increased until the functional test program failed. Table 5.13 shows the RF voltage at which an upset occurred (i.e., a change in the normal operating mode of the IC). Two units of each device type were tested for RF upset susceptibility.

Figure 5.63 shows both the calculated and the measured input power required to upset digital devices with different values of $f_T$.

The induced power calculated for the transmission line and microstrip PCB layouts shown in Table 5.11 is also plotted in Figure 5.63. The predicted induced power is far below that required to cause an upset to TTL or ECL 10K; however, for VHSIC TTL or ECL, the power induced into the transmission line or into the microstrip PCB at 3 V/m could cause an upset. The major difference between the standard TTL and the phase I VHSIC is that the TTL has an emitter perimeter of 12.5 microns and the VHSIC I devices have emitter perimeters of 1.25 microns. The phase II VHSICs have linewidths of 0.5 microns. To reduce the probability of an upset to VHSICs at 3 V/m, reduce the length of the microstrip or use a stripline PCB layout.

Open pins on logic ICs degrade noise immunity. For example, with a 2-V 2-ns-wide pulse applied to the input of a 7420 device, the output pulse is 2.2 V with a 10-ns pulsewidth. Clamping the unused inputs reduces the output pulse to 1 V with an 8-ns pulsewidth. For standard saturated logic types, the power induced into the 18-cm-long PCBs when illuminated at 3 V/m will not result in an upset.

Some of the latest logic types include: ABT, LVT, LCX, LVX, LVO, AC, ACT, ACQ, ACTQ, VHC, HC, HCT, GTLP, and FAST. No RF or noise energy immunity data is readily available for these types of logic; however, we have seen the importance of the emitter perimeter size in determining noise immunity. If a number of devices are suitable for an application and noise immunity is a potential problem, a comparative test of the noise energy immunity can be

**Table 5.13** RF Upset Voltage of a CMOS and ALS Devices

| | ALS 74 | | CD4013 | |
|---|---|---|---|---|
| $f$ [MHz] | S/N1 [V] | S/N4 [V] | S/N1 [V] | S/N2 [V] |
| *Vcc pin upset voltage* | | | | |
| 1.2 | 0.44 | 0.34 | 0.006 | 0.006 |
| 5 | 0.4 | 0.4 | 0.4 | 0.28 |
| 10 | 0.44 | 0.4 | 0.44 | 0.4 |
| 50 | 0.44 | 0.4 | 1.1 | 1.1 |
| 100 | 0.6 | 0.6 | 2.8 | 3.4 |
| 200 | 0.44 | * | * | * |
| *Data pin upset voltage* | [mV] | [mV] | [mV] | [mV] |
| 1.2 MHz | 0.0011 | 0.0010 | 0.0004 | 0.0005 |
| 5 MHz | 0.0022 | 0.0022 | 0.0009 | 0.0012 |
| 10 MHz | 0.0017 | 0.0031 | 0.0024 | 0.0028 |
| 50 MHz | 0.0044 | 0.0040 | 0.0120 | 0.0140 |
| 100 MHz | 0.0070 | 0.0100 | 0.0220 | 0.0400 |
| 200 MHz | 0.0070 | * | * | * |
| *Clock pin upset voltage* | | | | |
| 1.2 MHz | 0.0008 | 0.0008 | 0.00044 | 0.00035 |
| 5 MHz | 0.0008 | 0.0008 | 0.0015 | 0.0012 |
| 10 MHz | 0.00085 | 0.00085 | 0.0028 | 0.0040 |
| 50 MHz | 0.0022 | 0.0020 | 0.0170 | 0.0200 |
| 100 MHz | 0.0044 | 0.0040 | 0.0200 | 0.0440 |
| 200 MHz | 0.0070 | 0.0055 | * | * |

* Denotes that no upset could be induced.

performed. The test involves varying the high- and low-input voltage level at different input pulsewidths and recording the energy at which an upset occurs.

### 5.3.3   Analog Video and Radio Frequency Circuit Noise and Immunity

#### 5.3.3.1   Thermal Noise

Analog, video, and RF circuits generate noncoherent noise that is caused by thermal noise in resistors, semiconductors, and ICs as well as $1/f$ noise in semiconductors. Analog-circuit noise may be modeled as an input noise current and an input noise voltage and is typically specified by manufacturers at 1 kHz in units of nV/$\sqrt{\text{MHz}}$ and pA/$\sqrt{\text{MHz}}$. Thermal rms noise voltage in resistors is given by the following equation:

$$E_{nt} = \sqrt{4kTR\text{BW}}$$

where

$k = 1.38 \times 10^{-23}$

$T$ = temperature in degrees Kelvin.

BW = bandwidth of the signal chain in which the resistor is used

R = value of the reisistor [$\Omega$]

The noise at the output of an amplifier may be found by computing the total resistance of the resistors included in each input and calculating the thermal-noise voltage developed by

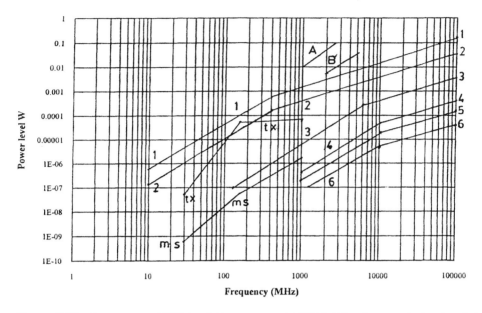

**Figure 5.63** A = measured TTL level, B = measured ECL10 K level, 1 = calculated TTL $f_T$ = 350 MHz, 2 = calculated ECL $f_T$ = 350 MHz, 3 = calculated VHSIC I TTL $f_T$ = 350 MHz, 4 = calculated VHSIC I TTL $f_T$ = 5 GHz, 5 = calculated VHSIC I ECL $f_T$ = 5 GHz, 6 = calculated VHSIC II ECL $f_T$ = 5 GHz, tx = induced power into 18-cm transmission PCB, $E$ = 3 V/m, ms = induced power into 18-cm microstrip PCB, $E$ = 3 V/m.

the resistors, the noise voltage developed across the resistors by the specified maximum amplifier input noise current, and the specified input noise voltage. Figure 5.64 illustrates the noise sources referenced to the input of an amplifier. The noise sources are considered noncoherent, and the rms sum of the noise sources is found by taking the square root of the sum of the squares of the individual noise sources. The noise is referenced to the input of the amplifier, so the output noise is the input noise times the gain of the amplifier.

In cascaded (i.e., a chain of) amplifiers, the maximum output noise may be computed by finding the noise per stage of amplification and multiplying by the remaining gain of the chain. The sum of the noise sources at the output, $N_t$, is again the square root of the sum of the squares:

$$N_t = \sqrt{(N_1 G_1 G_2)^2 + (N_2 G_2)^2}$$

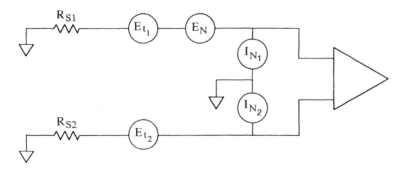

**Figure 5.64** Noise sources referenced to the input of an amplifier.

where

$N_1$ = total noise of the first stage referenced to the input

$G_1$ = gain of the first stage

$G_2$ = gain of the second stage

$N_2$ = noise of the second stage referenced to the input

The $1/f$ noise in operational amplifiers is significant only at very low frequency, because it reduces to the thermal-noise level at typically 6 Hz and, worst case, 30 Hz.

The rms noise voltage is less commonly used by RF engineers to characterize the intrinsic, component, induced noise in RF circuits; more often, noise power is used. In calculating the total noise power developed by an RF circuit or receiver, it is only necessary to add noise powers algebraically, multiplied by the power gain of amplifier stages and divided by losses, where applicable. One measure of noise is the noise factor, defined as

$$F = \frac{N_o/N_i}{P_o/P_i}$$

where

$P_o/P_i$ = available power gain of the receiver

$N_o/N_i$ = ratio of available noise output power and noise input power

The *available* power is that power density developed by a source into a conjugate load. When the ratio $N_o/N_i$ equals the gain, the receiver contributes no noise and $F = 1$.

In measurements of noise factor, very often a noise generator is used as a source and the noise factor is given by

$$F = \frac{P_{a(1/1)}}{kT_o BW}$$

where

$P_{a(1/1)}$ = available power required from the noise generator to produce a power

carrier-to-noise ratio (power in/noise in) of 1

BW = bandwidth [Hz]

$T_o$ = approximately 290 K($\mathring{A}$ 17°C)

$k$ = Boltzmann's constant = $1.38 \times 10^{-23}$

The noise figure of a circuit is either the noise factor expressed in decibels (thus, $F$ (db) = 10 log $F$) or synonymous with the noise factor, as defined by the IEEE.

An alternative measurement of RF circuit and receiver noise is noise temperature. The noise temperature of a receiver, in kelvin, is given by

$$T_r = FT_o - T_o$$

Therefore the noise factor is

$$F = 1 + \frac{T_r}{T_o}$$

The limiting sensitivity of a receiving system is determined by the ratio of received power to noise power that allows acceptable reception of information. In satellite communications a major source of noise is external atmospheric, cosmic, and ground radiation. The antenna is then considered a noise source, due largely to the noise flux incident on the antenna.

The total noise temperature of a receiving system is made up of the antenna noise temperature and the noise temperature of the receiving system up to the demodulator.

Passive components such as connectors, cables, and filters introduce a loss into the receiving system and increase the effective noise temperature. The effective noise temperature due to loss $L$ is

$$T_e = \left[\left(1 - \frac{1}{L}\right)L\right]T_o$$

and the output noise is

$$N_o = \left(1 - \frac{1}{L}\right)T_o$$

A figure of merit used to describe satellite communication systems is gain over temperature $(G/T)$, which is defined as

$$\frac{G}{T} = \frac{(P_o - N_o)/(P_{in} - N_{in})}{FT_o - T_o}$$

Once the antenna noise temperature is known, the receiver noise factor and input losses may be used to obtain $G/T$. In a cascaded receiver, $G/T$ is constant regardless of the position in the cascade at which $G/T$ is calculated.

As an example, we will calculate $G/T$ of a system after the antenna connecting cable and input filter and at the input of the receiver. Assume an antenna noise temperature $T_a$ of 50 K, a cable loss of 0.15 dB (1.035), a filter loss of 0.3 dB (1.072), and a receiver noise factor of 0.5 dB (1.122). The total loss is 0.45 dB (1.109) and the noise temperature at the receiver input due to antenna and losses is

$$T_i = \frac{T_a}{L} + \left(1 - \frac{1}{L}\right)T_o$$

$$= \frac{50}{1.109} + (1 - 0.902)290 = 45 \text{ K} + 28.42 \text{ K} = 73.42 \text{ K}$$

The noise temperature due to the receiver is

$$T_r = (1.122 \times 290) - 290 = 35.38 \text{ K}$$

Thus, the total noise temperature is $73.42 + 35.38 = 108.8$ K

The gain of the antenna is 30 dBi, from which the filter and cable losses are subtracted, resulting in a gain of 30 dB − 0.45 dB = 29.55 dB (901). $G/T$ is therefore 901/108.8 = 8.28/ K = 18.3 dB/K.

Very often the design engineer makes a careful worst-case calculation of the maximum intrinsic noise of an amplifier at the highest operating temperature but ignores induced noise from the electromagnetic ambient in which the circuit must function. For example, when the circuit is located in an enclosure that contains digital circuits or is powered by a switching-power supply, the induced noise may be many orders of magnitude higher than the intrinsic noise.

### 5.3.3.2　Coupling Modes to Analog/Video and Radio Frequency Circuits

When analog/video/RF circuitry share a board with digital circuits, the probability of EMI is high. Chapter 11, Case Study 11.1, examines coupling modes between digital logic and analog circuits contained on the same board. When the noise of an analog circuit must be restricted to the intrinsic noise level, careful design and layout are imperative. Ideally, power supplies to analog circuits should be dedicated to the analog circuit and not shared with digital circuits. Linear supplies should be used and not switching-power supplies. Where the main power comes from a switching supply used for a number of different circuits, the most susceptible analog circuits should be supplied by a dedicated linear supply. For example, when the switching supply provides $\pm 15$ V, this may be used as the input of a $\pm 10$–$12$-V linear supply. Radiated coupling between the input and output of the linear supply must be controlled in order to reduce high-frequency noise on the output of the supply. It is, however, possible to use a switching-power supply for the most sensitive circuits with the correct secondary power-line filter.

As an alternative to a large and inefficient linear supply, bulkhead-mounted and PCB-mount filters have been designed to allow the use of switching-power supplies for critical applications. These filters will reduce switching-power supply C/M and D/M noise to typically 30 $\mu$V of low-frequency ripple. With this type of filter, noise levels lower than the average linear supply can be achieved. The potential for crosstalk between wires, cables, and PCB tracks, as discussed in Chapter 4, should be examined.

In a very harsh EM environment, compartmentalization of the analog and digital circuits with the use of feedthrough filters may be required. The calculation of noise effects on analog circuits might be considered simply a function of circuit gain and bandwidth, and this is true of noise at in-band frequencies.

The potential for EMI at higher frequencies (i.e., at which the open loop gain of the amplifier is zero) should not be ignored. Nonlinearities due to input diodes, transistor junctions, or saturation of an input stage due to very high levels of RF can result in demodulation of AM or FM RF. Where the source of RF is unmodulated, DC offsets or gain may be affected.

Analog circuits are not immune to RF noise that is far out of band. For example, the output level of a 1.5-GHz high-power amplifier (HPA) was controlled by the output of a detector. The detector was contained in a shielded enclosure, but an op amp with an $f_T$ of 100 kHz was located outside of the enclosure. The unmodulated field generated by the HPA was rectified by the op amp's input circuit, which changed the output power level. The solution to the problem was to include a ground plane under the op amp and to connect surface-mount microwave capacitors with a low impedance at 1.5 GHz between each of the op amp's input pins and the ground plane. This solution has been effective in a number of similar EMI cases.

As an example of analog-circuit susceptibility to RF, a 741 op amp configured with a gain of 20 and with RF injected directly into either the inverting or noninverting input first exhibits a change in output level at an input power of $-20$ dBm (22.5 mV) and is saturated (i.e., close to the rail voltage) at 0 dBm (225 mV).

A common practical example of demodulation EMI is when a CB radio is operated in close proximity to a phonograph. Often the amplitude-modulated RF of the CB communication is demodulated by the input stage of the phonograph preamplifier, and the communication may be heard by the listener. The Nonlinear Circuit Analysis Program (NCAP) has been used to predict the demodulation EMI effects in bipolar, JFET-bipolar, and MOSFET-bipolar ICs (Ref. 7).

The addition of capacitors on analog or video circuit pins has the effect of reducing the impedance of the load at the problem frequencies. In conjunction with series resistance or inductance, an L-type filter can be formed and the attenuation is higher, especially when the noise source impedance is low. Very often the addition of a capacitor does not reduce the EMI response

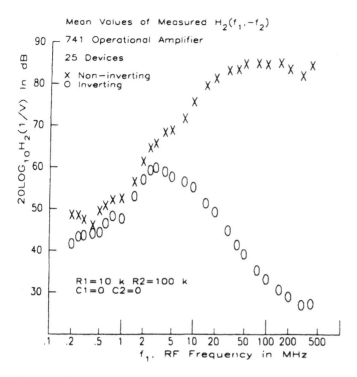

**Figure 5.65** Comparison of 20 log $1/x$ for inverting and noninverting inputs of the 741. $x$ is the op amp response and is dimensionless. (© 1991, IEEE.)

of the circuit. There are a number of possible reasons for this. The most common are: a different component or part of the circuit is susceptible; the capacitor is not the right value or type for the problem; the noise source impedance is too low; the capacitor is too far away from the rectifying semiconductor junction (it must be connected directly to the pin/s of the device), or the grounding around the device/s is not a sufficiently low impedance. The capacitor is most effective when its resonance frequency is at, or close to, the problem frequency.

Different analog devices exhibit different levels of immunity, and a comparison can be made based on measured data. The analog devices tested were the 741, OP27, LM10, LF355, and CA081 op amps. These devices were connected as an inverting amplifier with a gain of 10. The value of the input resistor used was 10 kΩ and that of the feedback resistor was 100 kΩ. Above some frequency the input capacitance of the device will reduce the level of RF, due to the low impedance of the capacitance, or when a series resistance is placed between the RF source and the input pin of the device the resistance/input capacitance will act as a filter.

Figure 5.65 (from Ref. 11) compares the response of the inverting input to the noninverting input of a 741 device. Figure 5.65 plots $1/x$, where $x$ is the circuit response. Therefore the higher the value in decibels, the lower the circuit response. Figures 5.65, 5.68, and 5.69 are provided for comparison purposes. Adding additional capacitors around the op amp, as shown in Figure 5.66, will decrease the response of the circuit to RF. An alternative filter for both C/M and D/M noise is shown in Figure 5.67.

Figure 5.68 compares the demodulated output of the device with different values of capacitor at locations $C_1$ and $C_2$.

The lower the value of the response shown in the figures, the more effective the filtering. Figure 5.69 compares the input RF voltage for a fixed-amplitude demodulated response for

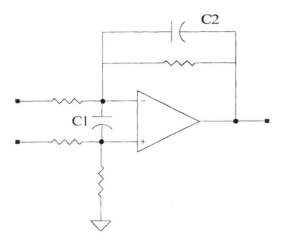

**Figure 5.66**   Additional capacitors around an op amp circuit.

different devices, and here the higher the level of RF voltage for a standard response, the more immune is the op amp type.

Most wireless designs have a performance specification that includes phase noise, FM, and spurious response. For example, a commercial aeronautical system requirement specifies that all fixed spurs (those that do not change with the carrier frequency) and all spurs remaining within 42 kHz of the carrier frequency must be at least 70 dB below the carrier ($-70$ dBc). Any spurs outside of the 42-kHz band that move with the carrier frequency are limited to $-55$ dBc. Other communication requirements (such as those for EW simulators) place the spurious limit at approximately $-60$ dBc. The EIA/TIA-250C specification on the ratio of video signal to periodic noise from 300 Hz to 4.2 MHz is 67 dB for short-haul links, 63 dB for satellite links, and 58 dB for long-haul links.

The use of digital receivers and digital signal processing (DSP) and switching-power supplies for the RF circuits makes the requirements increasingly difficult to meet. Generally speaking, if spurs close to the carrier must be 30–40 dB below the carrier ($-30$ to $-40$ dBc) and the RF circuit is in close proximity to sources of emission such as switching-power supplies and PCBs containing logic, then care must be taken in the design of the wireless for compatibility. In the case where the specification requires spurs to be at $-50$ to $-70$ dBc, extreme care must

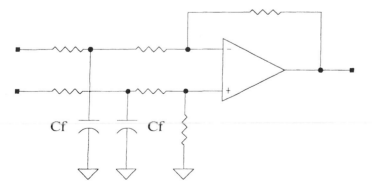

**Figure 5.67**   A C/M and D/M filter at the inverting input of an op amp.

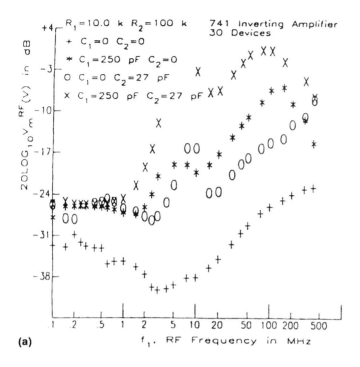

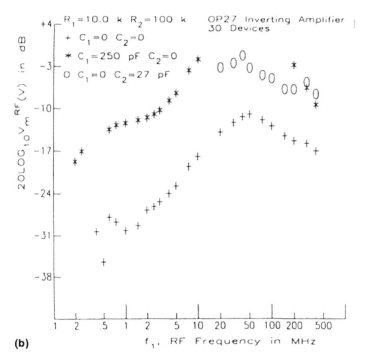

**Figure 5.68** (a) Comparison of the demodulated output of a 741 with and without additional capacitors. (b) Demodulated response at the output of an op 27 with and without additional capacitors. (© 1991, IEEE).

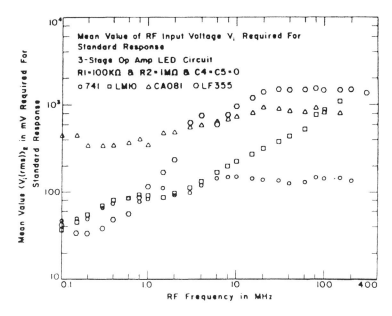

**Figure 5.69** Input voltage required for a standard output response for different devices. (© 1991, IEEE)

be taken, especially if signal levels are low. The following guidelines are based on achieving compatibility for RF circuits that can be located as close as 2 cm to sources of high-level emissions and where spurs must be at levels between −30 and −40 dBc.

High level in band spurs, typically generated by clock and data frequencies, will often make communications on the chosen channel unreliable to the point of being unusable (a denied channel). If other channels are available, then they can be selected. The alternative is to move the clock frequency out of band or to wiggle, sweep, or hop the frequency, as described later. Another problem caused by broadband noise sourced by switching-power supplies or data with a high-frequency component at a low repetition rate is the desensitization of the receiver. In one example, fire trucks and ambulances fitted with a new data terminal found that voice communication by radio to the base station was impossible when the vehicle was in the mountains. However, with the data terminal switched off, normal voice communication resumed. The problem is that in the mountains the radio's received level was low and the broadband emissions from the data terminal, picked up by the vehicle-mounted antenna, caused the radio to desense, resulting in loss of communication.

With PCMCIA wireless cards, the inclusion of the card in a PC can result in receiver desensitization due to broadband noise generated by the PC. Motorola has developed a tool that, unlike spectrum analysis measurements, provides an accurate measurement for wireless devices that accounts for the effects of the air protocol, the PC card antenna, and the interaction between the card and the host, typically a laptop computer. The system measures how much RF level is required to produce successful massaging. It emulates wireless wide-area network operation using packet data to determine the receiver sensitivity of the wireless card while inserted into the host PC card slot. As with system coverage, if the host computer is noisy, more RF signal is required to receive a message. This would translate to reduced network coverage. Thus PC manufacturers would have a competitive advantage if the PC emissions were significantly lower than the Class B requirements over the wireless frequency bands in use. A manufacturer of a dedicated data terminal with an imbedded wireless has designed the computer, imbedded wireless, and external antenna with extremely low coupling between these components, and this terminal has an even higher advantage in terms of increased network coverage and fewer denied channels.

There are a number of coupling mechanisms between noise sources and the RF circuit, and these include: conducted, radiated coupling, crosstalk, and common ground impedance coupling. These potential sources and techniques required to reduce the coupling are described throughout this book.

### 5.3.3.3  Noise Sources and Levels

In an ideal world, the actual noise sources that will be surrounding the RF PCB would be measured for emissions before the circuit was designed. In reality this can seldom be accomplished; instead, the design should be based on typical average levels of magnetic and electric field and typical conducted noise levels measured on equipment that contains digital circuits and switching-power supplies. Of particular interest are those levels that have resulted in EMI on similar equipment containing an imbedded wireless.

The magnetic and electric fields measured 3 cm distance from a reasonably well-laid-out digital PCB containing short (2.54 cm) microstrip used for a 40-MHz clock is shown in Table 5.14.

Chapter 11 discusses PCB layout in detail, but generally speaking longer traces will radiate more effectively than short traces and higher frequencies more effectively than lower frequency (see Chapter 11 for the limits of frequency and trace length). For example, a 5-cm-long microstrip with a 10-MHz clock radiated at a lower level than the 2.54-cm track with the 40-MHz clock. Compared to the microstrip, use of a stripline will reduce the level of the emissions by from 15 to 25 dB over the 20–600-MHz frequency range. In practice, where a stripline PCB layout is used with the upper and lower ground plane stitched together around the board with vias 3 mm apart, the radiation from the PCB is typically dominated by emissions from the ICs. If the PCB located close to the wireless PCB has yet to be designed and is under the control of the wireless manufacturer, the ''good'' PCB layout described in Chapter 11 should be followed. One additional tip is to add a low-value decoupling capacitor in parallel with all of the existing $0.01-1$-$\mu$F capacitors in the digital section of a PCB or adjacent digital PCB. The additional decoupling capacitors should be chosen to resonate at the center frequency of the wireless. Digital sections on the wireless PCB, such as DSP, should be laid out using the following considerations.

1. Keep oscillators, clock and data bus traces, and high-speed LSI chips as far away from the RF section as possible. Locate ICs so that the interconnects are as short as possible. Locate oscillators close to ICs that use the clock. Keep LSI devices,

**Table 5.14**  Levels of E and H Fields Close to a ''Typical'' Digital PCB

| $f$ (MHz) | $E_\theta$ (mV/m) | $H_\phi$ (mA/m) |
|---|---|---|
| Emissions 3 cm from a PCB microstrip layout, 40-MHz clock, 2.54-cm length | | |
| 100 | 25 | 0.23 |
| 360 | 83 | 1.0 |
| 950 | 8.5 | — |
| 5 cm of ribbon cable with 10-MHz clock | | |
| 360 | 301 | 3.6 |

which have fast data buses and clocks, and switching-power supplies as far away as possible from signal *and* power connectors and ribbon cable.

2. Use stripline over the digital section. If that is not possible, use localized stripline to imbed any digital and power interface traces routed to the wireless section.

3. Follow the grounding rules in Section 11.6, and avoid slots in the ground plane located directly under signal and clock traces, as shown in Figure 5.70.

4. Keep traces between ICs in the DSP section as short as possible. Minimize path lengths between sources and loads and between loads sharing a common clock or bus.

5. Keep "hot" signals as close to the signal return plane as feasible, and do not "weave" these traces through the ground plane from layer to layer except for short and necessary sections at ICs.

6. Reduce the drive current on tracks by minimizing loading, typically by the use of buffers at the load end or series resistance at the source end.

7. Reduce the number of signal tracks that change state simultaneously.

8. Design logic and software such that data is flowing on data buses intermittently, with as low a duty cycle as possible, and not continuously. If data is transmitted and received in packets with considerable gaps, then, if possible, cease all unnecessary computer operations during transmissions.

9. If digital clocks are designed to wiggle in frequency, sweep or hop then the occurrence of in-band spurs can be reduced. Alternatively if a clock is derived from a DCO, VCO, or phase-locked loop and the tuned frequency of the wireless is known, choose the clock frequency to generate spurs out of the wireless band or tuned channel.

10. Use an overall shield over the RF section, with a separate "enclosure" ground plane as described in Section 5.3.3.5 and Chapter 11.

### 5.3.3.4   Radio Frequency and Wireless

The levels described in Table 5.14 can create a problem with an RF circuit, this was illustrated by a receiver that exhibited spurs at 70 kHz and at 28 MHz (with a repetition rate of 70 kHz). The source of these spurs was a switching-power supply that generated a magnetic field incident on the RF section of the board of 13 mA/m at 70 kHz and 0.2 mA/m at 28 MHz. The problem

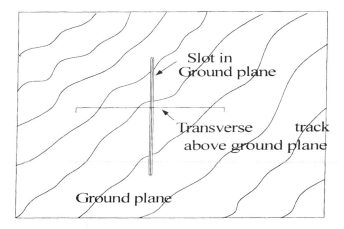

**Figure 5.70**   Ground plane with slot, formed by a row of vias, directly under data bus or clock.

was a 2-cm by 2-cm loop of PCB trace on the RF PCB that connected the oscillator signal to a mixer. The current flow in this 2-cm loop was calculated at approximately 30 μA at 70 kHz and 17 μA at 28 MHz. Shielding of the mixer and oscillator signal traces was only partially successful, and not at all at 70 kHz. The potential solution to this and other susceptibility problems is to keep PCB signal traces close together, to minimize loop area, or better still, to imbed the oscillator traces in stripline. The direct coupling into the loop at 70 kHz can be reduced by capacitively coupling the oscillator so that the capacitor presents a low impedance to RF and a high impedance to audio frequencies and to the switching-power supply fundamental frequencies and low harmonic frequencies. Although the susceptibility of specific RF circuits to incident fields is hard to predict, using all the "good" PCB layout techniques and localized shielding that RF engineers commonly employ to avoid self-compatibility problems, e.g., coupling from a high-level RF signal into a low-level signal, will also increase the immunity to externally sourced fields.

One technique that is effective at reducing conducted noise on digital data and clock lines that interface to an RF circuit is the inclusion of a bifilar wound balun on the digital lines and return/s. The balun will reduce C/M currents but will not degrade the operating speed of the digital lines. Nor is the balun affected by the DC supply current, for, as both lines are threaded through the balun, the DC magnetic field induced in the ferrite core is close to zero. To reduce low-frequency currents, the ferrite should be wound with as many turns of the signal/power and their returns as possible. For low-frequency NRZ data noise and switching-power supply noise, a high-permeability ferrite of typically 10,000 $\mu_o$ (where $\mu_o$ is the permeability of free space) is recommended. For high-frequency applications (>5 MHz), a lower-permeability material will typically achieve a greater attenuation.

When the parasitic capacitance between the turns of the balun results in a resonance at the problem frequency, the highest attenuation is achieved, although most ferrites are very low Q (lossy) and this is not a very significant effect.

If digital control and data lines are connected via a D subconnector to RF components or subassemblies contained in a conductive enclosure, the use of a filtered connector or filter adapter can reduce both C/M and D/M noise.

AC power at 50–400 Hz and their harmonics can also result in spurs, typically diagnosed as FM in a carrier. No effective shielding exists at these frequencies, short of high-permeability steel or mu-metal, and the best approach is to move the RF circuit as far away from transformers, AC fans, and AC power wiring as possible. All power wiring should be twisted to cancel the magnetic fields generated. As an extreme measure, the AC field may be canceled or bucked by the use of a coil through which the AC power current flows. The coil is then orientated to cancel the AC field incident on a susceptible component or circuit. Use of coupling capacitors that have a low impedance at RF and a high impedance at the power-line frequency should also be used. In one case the AC power current picked up on a loop formed by a semirigid cable, and an RF circuit mounted in a metal enclosure was shunted to chassis by a clamp placed over the semirigid cable. This measure reduced the AC power current flow into the sma connector mounted on the enclosure.

## 5.3.3.5  Shielding

Past experience has shown that when RF circuits are in very close proximity to the fields generated by logic and switching-power supplies, EMI can and often does exist. Shielding can, when implemented in conjunction with signal and power frequency filtering, be very effective. However, when signal and power connections are routed from the "noisy" external ambient into the shielded RF section of the circuit without filtering, the results may be no increase in immunity at all! Thus it is not enough simply to shield a circuit, it invariably requires additional filtering at the shield interface, as described in the filtering section.

All digital logic that switches nonreturn to zero data will radiate fields with a low-audio-frequency component and thin, small, shielded enclosures will provide 0 dB attenuation at these frequencies. The only effective technique is "good" RF PCB layout and the use of low-value capacitive coupling between stages of the RF. Chapter 11 describes shielding a PCB in detail; however, in general the shielding effectiveness of small, thin enclosures may not be adequate at switching-power supply frequencies of around 100 kHz. The weak link in any shield is at the seam between the separate components of the shield. In the PCB-mounted shield, this is the electrical connection between the shielded enclosure and the ground plane on the opposite side of the board. When the enclosure is constructed using a shield fence into which lids are clipped and the connection to the ground plane relies on the tabs in the fence material, the shielding effectiveness is compromised. The best solution to this problem is to include an upper "enclosure" ground plane, in the form of a loop approximately 3–5 mm wide on the fence material side of the board. This upper loop is not connected to any other (digital/RF/analog) ground plane, but is connected around the circumference of the loop to the lower PCB ground plane, which completes the enclosure. Vias are used to tack the upper loop down to the lower ground plane, and these should be located at intervals of approximately 3 mm. Then if the tabs on the shield fence do not provide a low-enough "contact impedance" with the lower ground plane, which forms the full enclosure, then the fence can be soldered to the upper loop. This configuration is shown in Figure 5.71.

Some of the types of material commonly used for the shield are 0.1–0.25-mm-thick beryllium copper or 0.2–0.5-mm-thick tin-plated steel. The magnetic-field shielding effectiveness of these types of shield with a low contact impedance seam (described earlier) is shown in Table 5.15. However, Table 5.15 applies only to an enclosure that is constructed completely of the materials shown. In the case of many PCB-mount enclosures, the loop that forms the enclosure is completed by an enclosure PCB ground plane. It is not, however, generally necessary to go to the expense of "buried vias" for this lower "enclosure" ground plane, for up to 1 GHz the ground plane can contain many small (2 mm) apertures without unduly reducing shielding effectiveness. At all costs, avoid long slots (2 cm) in this enclosure ground plane, for this will begin to further reduce the shielding effectiveness of the enclosure. However, in one instance

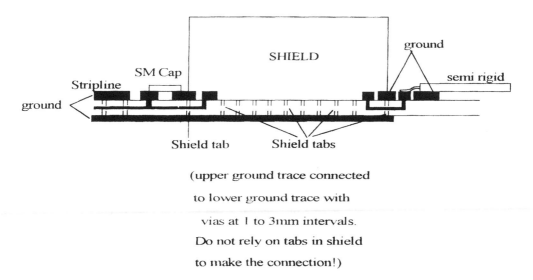

(upper ground trace connected

to lower ground trace with

vias at 1 to 3mm intervals.

Do not rely on tabs in shield

to make the connection!)

**Figure 5.71** Location of filter capacitor to the enclosure, and connection of RF signals through the shield into the RF section.

**Table 5.15** Magnetic-Field Shielding Effectiveness of 0.127-m ×
0.101-m × 0.01-m Enclosure

| $f$ | 0.5-mm-thick tin-plated steel | 0.2-mm-thick tin-plated steel | 0.2-mm-thick beryllium copper | 1 oz copper PCB material |
|---|---|---|---|---|
| 4 kHz | 6 | 4 | 0 | 0 |
| 10 kHz | 14 | 14 | 1 | 0 |
| 100 kHz | 33 | 33 | 25 | 10 |
| 1 MHz | 47 | 47 | 46 | 34 |
| 10 MHz | 54 | 54 | 54 | 52 |
| 100 MHz | 56 | 56 | 56 | 56 |

small apertures, added for thermal equalization, around the solder tabs of a coaxial connector attached to the printed circuit resulted in out-of-band spurs. In this case, a $-50$-dBm RF signal at 1.9 GHz was brought into the RF section via an external coaxial connector. When the small apertures around the connector tabs were filled with copper shim soldered in place, the spurs reduced to within specification. Some of the manufacturers of this type of shield are: Boldt Metal Industries A.K. Stamping with laminates made by Flexlan, Orion and Insul-Fab.

When RF current flows at gigahertz frequencies on, typically, semirigid cable, data cables, or well-shielded enclosures, where the current flow is a problem and additional shielding is not possible the use of absorber material over the enclosures or cables can be very effective. Gigahertz frequency absorber is available in rubber sheets that can be wrapped or glued or held in place by Velcro tape. As an alternative, potable epoxies may be used to coat assemblies.

### 5.3.3.6   Radio Frequency Grounding

Chapter 11 discusses PCB grounding in detail. But in general, the main goal in grounding an RF circuit is to avoid power and signal RF currents generated in the digital section or C/M currents on signal interfaces from flowing on the RF section power/signal ground plane. Two options exist: One is to ensure that no currents can flow through the RF ground plane, either by choosing the locations of the different grounding points or by covering the RF circuit and ground on both sides with a shield that is terminated either around the periphery to the RF ground or at one or two points only. A second grounding scheme is to isolate the digital and RF grounds. Either solution will work when implemented with care, although isolating the grounds is typically more difficult.

The goal in the isolated grounding scheme is to isolate all of the signal and power connections referenced to the digital ground from the RF ground. Radio frequency signals carried on coaxial cables must remain isolated from digital return as they exit the board or the equipment. This grounding scheme will provide DC isolation between the digital ground and the RF ground and allow the RF ground to be connected to chassis. That RF ground is connected to chassis somewhere is usually unavoidable. This is often due to the connection of shields of coaxial cables to chassis or building ground when they interface to the outside world. For example, the shield of an antenna cable located outside of a building and brought into an equipment rack should be grounded to the rack at the entry point. This provides a connection for secondary lightning currents and ensures that RF currents picked up on the antenna cable shield do not enter the equipment enclosure.

The isolated grounding scheme used in a typical wireless PCB is illustrated in Figure 5.72. The power supply must be of the isolating type. If it is a switching-power supply, the ripple and spikes at the output that will be attenuated by the filters shown in Figure 5.72. RF

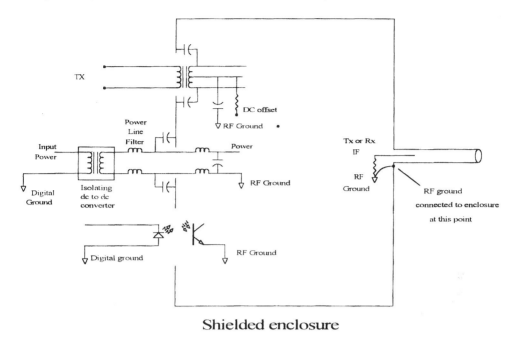

**Shielded enclosure**

**Figure 5.72** Grounding scheme with isolation between digital and RF ground.

C/M currents will also flow in the interwinding capacitance of the TX $I$ and $Q$ transformers. The solution is to add capacitors, as illustrated in Figure 5.72.

If these capacitors cannot be added, one technique to reduce coupling is to add a capacitor between the digital ground at the location of the entry of the $I$ and $Q$ signals into the enclosure, between the digital ground and the enclosure, which is in turn connected to chassis. This capacitor may be a ceramic having a 3–6-kV voltage rating if a hi-pot test is required. If a hi-pot test is applicable, then the isolating switching-power supply must also have an adequate voltage rating between primary to secondary, and this includes primary power return to secondary power return, and so must the $I$ and $Q$ transformers. Ensure that opto isolators are located at the shield and not inside the shielded enclosure as the traces to the LEDs can radiate.

The second grounding scheme must be used if one of the signal or control interfaces that reference digital ground or a power-supply ground cannot be isolated from the RF ground. In this case the RF ground is connected to the digital ground, and both will be connected to the chassis of the equipment and thereby to safety ground.

In this scheme the physical location of this common ground connection is important. In the equipment under consideration, the digital ground is connected to the equipment chassis ground and to RF ground at the location where the digital and $I$ and $Q$ signals interface to the RF section. The result is that RF currents set up by C/M noise voltages and by radiated emissions tend to flow back to the source via the chassis and not through the RF ground plane. In the instances where a signal return cannot be connected directly to chassis, isolation or filtering is required. If a DC connection cannot be made to the equipment chassis ground, then one or more high-quality capacitors can be used to provide an RF ground.

In the fully isolated grounding scheme, the PCB plane, which is used as part of the enclosure, and the RF ground plane should if possible be separate and should be isolated as far as possible from each other. In practice, a connection must be made from the enclosure to the shield of shielded cables that enter the RF section—in the example, the TX IF O/P and the RX

IF. It would be counterproductive to bring shields into the RF section without a connection to the enclosure, because any RF currents flowing on the shield, due to electromagnetic coupling from the digital section and the switching-power supply, in the equipment and from external sources such as AM and FM transmitters will reradiate inside the ''quiet'' shielded RF section. The reason to isolate the RF ground plane from the enclosure is that RF currents are set up in the enclosure by the electromagnetic ambient. In a perfect enclosure, these currents would be confined to the outside surface. However, in a practical enclosure a voltage is set up over the contact impedance of the seams, and this results in an internal current across the ground plane. Even in the scheme shown in Figure 5.73, in which the digital and RF grounds are connected through a balun, the RF ground should be connected to the enclosure at one point only.

In the grounding scheme where RF and digital ground are tied together at one point only, it is important that the RF ground plane section not be in the path of power return currents from the digital section back to the power supply via a connector. If the RF ground plane is in the path between the digital section and a connector used for the power and return, the RF ground plane should be isolated from this path. An alternative is to ground the RF ground around the periphery to the all-encompassing shield enclosure, in which case the RF noise currents flow through the shield back to the source. As mentioned in the PCB layout section in Chapter 11, the locations of the various sections of the board and the power and signal connectors play a vital role in achieving EMC.

Some of the latest RF circuits are differential. One of these devices is the HELA-10 manufactured by Mini-Circuits® This device has a 50–1000-MHz bandwidth, a gain of 10–13 dB, an NF of 3.5–4.7 dB and a 26–30 dBm output. To use this device with an unbalanced input and to provide an unbalanced output, balun transformers with either 50- or 75-$\Omega$ impedances are used at the device input and output. These baluns are also available from Mini-Circuits and can be used to provide C/M noise rejection and RF isolation, at least at certain frequencies.

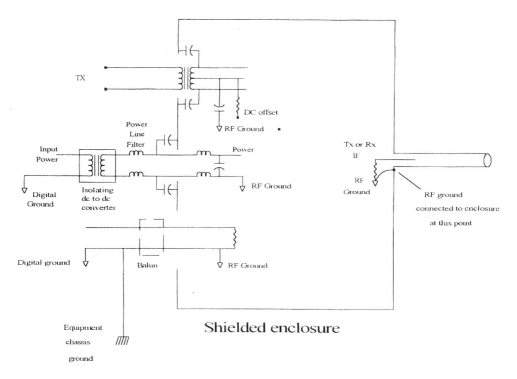

**Figure 5.73** Grounding scheme with RF and digital ground tied together through a balun.

If there is a golden rule to grounding it is to build some flexibility into the grounding scheme, especially in the breadboard, prototype, or first production models. For example, if a chassis plane is included in a motherboard or back plane. The PCB should be laid out to allow a direct connection or disconnection of the chassis plane from any analog/RF/digital ground and permit connection of the chassis plane through a low-impedance direct connection or capacitors to any or all of the remaining grounds at several locations around the board.

### 5.3.3.7   Filtering

Using either the isolated grounding scheme or the common grounding scheme, every connection between RF circuits inside the shielded enclosure and external circuits must have some level of filtering or C/M isolation. In the case where high-frequency signals enter the RF section, a number of options exist. For balanced transformer-coupled signals, in our example the TX, $I$, and $Q$, the parasitic coupling is almost certainly C/M via the interwinding capacitance of the transformer. If a transformer with an internal electrostatic shield is available, this should be used. Failing that, provision should be made for a low-value capacitor to be placed between the upper winding of the secondary and the enclosure, and a second capacitor between the lower winding and the enclosure. These can be removed if found unnecessary or if they interfere with the transformer coupling or balance.

Digital signals should preferably be connected via optoisolators when the digital and RF grounds are isolated, and these should be located outside of, but tight up against, the shield. For return to zero data, a transformer or a differential-input or quasi-differential-input amplifier may be used.

If digital signals cannot be connected through optoisolators, transformers, or differential amplifiers, then they can, in common with other single-ended signals, be routed through a ferrite balun. The signal return must always be routed through the balun along with the signal/s, and this almost certainly requires that the digital and RF ground plane be split as shown in Figure 5.74. One technique is to have slots in the PCB either side of the traces. A split balun can then be placed on either side of the board if this is found to be a requirement. The balun is shown in Figure 5.75.

In addition to presenting an impedance to C/M RF currents, the balun acts as a receiving and a transmitting antenna, and the best location is outside the shield, unless some source of high-level emission exists in close proximity to the balun, in which case it should be located just inside the shield, but still maintaining the split between digital and RF ground, as shown in Figure 5.74.

Although a signal is designated as an output, C/M currents can flow on the output connection and reradiate inside the shield or result in RF currents on the ground plane. Therefore, filtering, isolation, or a balun should also be placed on outputs. In our example, when digital

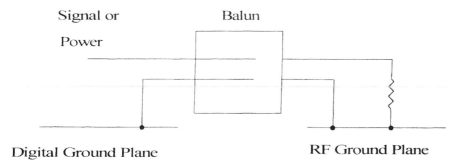

**Figure 5.74**   Connection of ground through the balun.

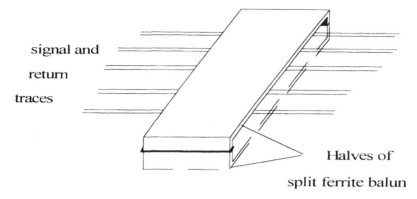

**Figure 5.75**  Split ferrite balun PCB mounting technique.

ground and RF ground are tied together, the RX1 and RX2 connections to the primary of the transformers should be routed through baluns, with the ground planes connected through the balun as shown in Figure 5.74. A better solution is to connect the shield of the RX1 and RX2 to the enclosure, in which case the balun is not required. If the RX1 and RX2 grounds are isolated from RF ground, then the transformer will reduce the flow of C/M RF current, with the addition of an RF (capacitive) chassis ground connection to digital ground, if required.

As described in Section 5.1.10.6, PCB-mounted switching-power supply regulators and filters can, when properly designed, reduce switching-power supply noise, which typically exhibits 400-mV short-duration spikes and 100-mV ripple, down to approximately 5-mV ripple with negligible HF component. Where extremely low ripple is required, typically satellite communication, radar simulators, and laser power supplies, small shielded PCB-mount filters have been designed, built, and tested that reduced ripple down to between 30 and 300 μV. Large-value D/M SM inductors should be located just outside of the shielded enclosure, followed by capacitors connected directly to the shielded enclosure, as shown in Figure 5.71. A C/M ferrite should also be located outside of the shielded enclosure over all of the power and return lines. However, if the C/M ferrite is located close to a high-level noise source, then RF currents will be induced into the ferrite. In this case, shield the ferrite or move it just inside the shield. The components inside the enclosure should be the smaller D/M capacitors and the linear regulators. With this configuration, the shield is used to minimize radiated coupling between the input components of the filter and the output components.

The RX IF input coax should have the coaxial shield tied to chassis as it enters the equipment and then tied to the shielded enclosure, as shown in Figure 5.71. A connection must also be made to the RF ground plane at that location. The RX IF input filter should attenuate a broad range of noise, including 60 Hz and its harmonics. This is typically achieved at power-line and switching-power supply frequencies by simply by using a coupling capacitor with a low series impedance from 400–900 MHz and a high impedance at low frequency. For this to work, the input filter and amplifier after the coupling capacitor must also maintain a low-, ≈50-Ω, input impedance at low frequency. The TX IF O/P connection must also contain a similar filter.

If the preceding guidelines are followed, then equipment has a very high probability of achieving the goal of EMC. Some of the additional components may be unnecessary, such as the baluns; however, if provision is made for them, they can easily be removed. Alternative techniques for grounding, shielding, isolation, and several combinations of power filtering can be designed to achieve the same goal, and these can be discussed. The most common problems usually occur in the practical realization of the design, for this reason the schematic and PCB should be reviewed prior to commitment to a PCB layout.

### 5.3.5   Radiated Emission Reduction

Printed circuit board emission is very dependent on the layout and use of ground planes or transmission line principles. The reduction techniques using PCB layout and circuit design are discussed in Chapter 11. The reduction in radiated emissions from wire and cables is discussed in Chapter 7.


## 5.4   TRANSIENT PROTECTION

Protection circuits and techniques exist for protection against high-level transients caused by a direct electrostatic discharge (ESD), the electrical fast transient (EFT), lightning strike, or the electromagnetic pulse (EMP) ambient. Protection may also be required against the lower-level secondary effects of lightning, EMP, and power-line-conducted transients. AC power may be protected as it enters the building or at outlets. Power strips and power-line filters may be purchased with built-in transient protection. A power-line filter with active tracking, designed for spike and transient suppression, with the trademark Islatrol is manufactured by Control Concepts. Protection against transients with voltages less than 1 kV, a rise time of 10 $\mu$s or longer, and pulse duration of 1000 $\mu$S may be provided on data lines by devices manufactured by MCG electronics. The data lines that may be protected include modems, RS232, RS422, RS423, 20-mA loop, telephone, video, and high-speed lines.

Unipolar and bipolar silicon transient voltage suppressors (TVSs) are available in DIP, surface-mount, radial- or axial-lead, and modular forms. Surface-mount devices are available rated at 1500 W of peak power and clamping voltages from 9.2 V to 189 V from Protek devices. Leaded devices, from General Instrument, can take 5000 W of peak power at 1 ms, can switch as fast as 1.0 ps, with clamping voltages of from 6.4 V to 122 V. Modular TVSs from Protek can take up to 15,000 W of peak pulse power dissipation at the 1-ms pulsewidth and clamp in subnanosecond times. California Micro Devices manufactures 17 and 18 channel ESD protection arrays which divert the ESD current pulse to either the positive or negative supply via diodes.

The Sidactor is a transient surge suppressor in a modified TO-220, TO-92, and surface-mount DO-214AA package, which, according to the manufacturer, has the following features: bidirectional protection, breakdown voltages from 27 to 720 V, clamping speed of nanoseconds, and surge current capability up to 500 A. The AVX Transguard® TVS is available in surface-mount (0603–1210) or pill configurations, with a peak current of 150 A for the 1206 chip. The metal oxide varistor (MOV) is capable of high peak currents (up to 70,000 A in a disc type, formerly from Siemens, for a very short duration) but it is slower than a TVS, at less than 15 ns. The lowest clamping voltage is higher than that of the TVS and for the Siemens disk type covers a range of 36 V to 1815 V, and so it has the possible advantage of a much higher maximum clamping voltage than the TVS. The peak clamping current for a 8/20-$\mu$s pulse is 2500 A for the 1100-V MOV, but the continuous current must be limited to achieve the 0.4-W average power dissipation. The average power must be limited to between 0.01 W and 0.6 W, dependent on the size of the varistor. The Sidactor and TVS and many diodes and zeners are fast enough to protect against the ESD and EFT test levels; however, the varistor will not be fast enough unless a filter is used between the source of the fast pulse and the circuit to be protected by the varistor. Thyristors and triacs have also been used in circuits designed to clamp transients to a very low level; however, when compared to the TVS, these devices are typically relatively slow. ABB HAFO manufactures a solid-state transient surge suppressor based on the thyristor diode structure. It is designed for lightning protection on telephone transmission lines. The smallest devices are available in a DIP or SOIC-16 package and are capable of 150-A pulse, a breakdown voltage of 60–80 V, and a nanosecond response time. ABB also manufactures a silicon unidirectional transient suppressor consisting of a PNPN thyristor diode integrated with a fast zener that has a short-circuit failure mode when overloaded. The peak current for a 8/20-μS pulse is 500 A, the static breakover voltage is 75 V, and the dynamic is 95 V at $dV/dt$ = 1.5 kV/μs. The on-state voltage is 3 V at 2 A, and the turn-on time is 20 ns, which is ok for lightning protection but too slow for EFT and ESD protection. A higher-current device from ABB can take 1000 A for the 8/20-μs pulse, has an on-state voltage of 6 V at 20 A, and has a breakover voltage of 280 V and a dynamic breakover voltage of 375 V with $dV/dt$ = 1.5 kV/μs.

### 5.4.1  Lightning Protection

A direct lightning strike can be modeled as a constant-current source with typical peak current of 20,000 A. However, approximately 10% of strikes have peak currents of 60,000 A or greater. If a 200,000-A lightning strike occurs to a metal aircraft frame, any electronic components mounted external to the frame can be exposed to magnetic fields as high as 318,000 A/m close to the surface and E fields as high as 250,000 V/m, assuming the strike does not occur directly to the externally mounted component. The preferred approach in protection is to avert a direct strike to cables, antennas, equipment, or equipment huts. Lightning rods or wires strung between poles create a so-called *zone of protection* beneath the rod or wire that extends in an arc from the top of the conductor to the ground. This form of protection is described in Section 8.6.1.

When antenna, control, AC or DC power, telephone wires, and other cables cannot be protected by lightning rods, then primary lightning strike arrestor devices are available, which are designed to protect the input of equipment. Varistors and gas discharge arrestors are capable of handling currents up to 60,000 A. At these high currents, the protection is usually one shot, and the arrestor fails safe (i.e., with a permanent short). This means the arrestor device must then be replaced. However, the cost of lightning arrestors is relatively low when compared to the cost of repairing receiving/transmitting equipment. The surge current handling capability of common protection devices versus the voltage developed across the device during a strike are shown in Figure 5.76. The striking voltage of a gas discharge arrestor is dependent on the rise time of the transient. Figure 5.77 illustrates the breakdown voltage versus the transient gradient for very fast gas gaps with DC breakdown voltages from 75 V to 7500 V. These fast arrestors are used by Fischer Custom Communications in a number of different products, both with leads and mounted with coaxial connectors, as described later.

A very low level of current flows in the gas discharge arrestor until the striking voltage is reached, after which the current increases to the milliamp region and the voltage drops to the glow voltage. As the current continues to increase, an arc occurs and the voltage drops to the arc voltage (approximately 20 V). On cessation of the strike, the current through the arrestor decreases until the arrestor extinguishes, assuming the supply current is less than the arc current and the supply voltage is less than the glow voltage. When used on an AC supply, the gas discharge device may not extinguish until the AC voltage waveform changes phase. On a DC supply, the arrestor may not extinguish until the supply is disconnected or the short causes a contactor, cutout, or fuse to open. One solution to the problem of not extinguishing on a 20–40-V supply is to place two arrestors in series across the supply. The disadvantage here is that the voltage spike generated across the arrestors, until the striking voltage is reached, and the arc voltage will be twice as high as for a single arrestor.

A new range of gas discharge arrestors designed for the much faster rise time of an EMP pulse may be used for lightning protection, with a resultant shorter striking time and strike voltage. Many lightning protection devices are manufactured specifically for use with coaxial or twisted-pair cables and in a bulkhead panel for mounting at entry points into buildings. The coaxial type will typically have a specified characteristic impedance, such as 50 Ω or 75 Ω

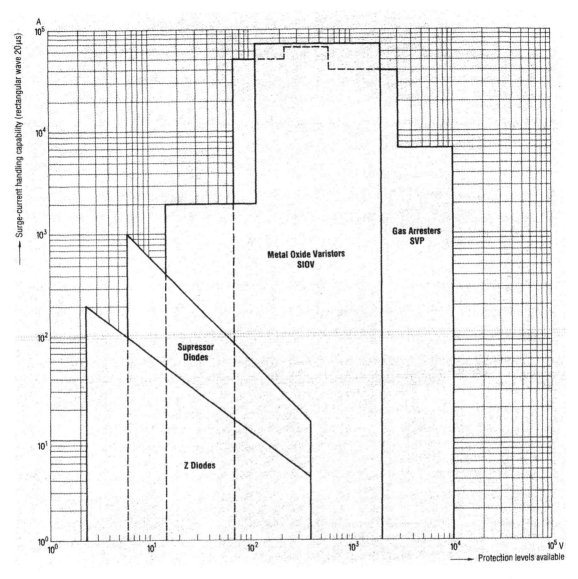

**Figure 5.76** Surge current handling capabilities versus the voltage protection levels available. (Reproduced courtesy of Siemens Ltd.)

over the frequency range of interest. Two basic types exist. Fischer Custom Communication manufactures a range of ''Spikeguard'' nanosecond transient protection designed for lightning and EMP. They exhibit fast response through the UHF region and are constructed from transmission-line, gaseous discharge components and silicon components.

The Spikeguard Lightning arrestors are available in metal enclosures designed for insertion in antenna, control, or telephone cables. The enclosures have attached grounding lugs and are designed for mounting on an antenna mast or ground rod. Typical VSWR characteristics for the Spikeguard and type N, UHF, and C coaxial connectors are: 100 MHz 1.2:1, 200 MHz 1.4:1, 300 MHz, 1.6:1, and 400 MHz 1.8:1.

# TRANSIENT VOLTAGE BREAKDOWN CHARACTERISTICS OF GAS GAPS

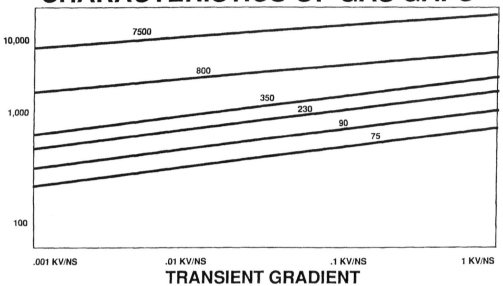

**Figure 5.77** Striking voltage versus risetime for gas discharge arrestors with DC breakdowns of from 75V to 7500V. Reproduced by kind permission of Fisher Custom Communications Inc.

The FCC-350 series is rated at 20 kA peak for a 10-μS duration and with breakdown voltages from 90 V to 20 kV. For DC breakdown voltage of 1 kV or higher, the overshoot voltages of transients having a risetime of 1 kV/ns is minimal, since the breakdown voltage is achieved in approximately 2 ns. The insertion loss to RF is approximately 0.2 dB. The FCC 450 series Spikeguards have been designed to provide transient protection for receivers and transmitters up to 100 W of output power and are capable of clamping in a few nanoseconds and dissipating 15 kW of power for transients having a pulse width of 10 μs and 1.5 kW for 1-ms transients. Other versions of the Spikeguard are available to protect HF transmitters to 1 kW and 2 kW. The FCC-550-10-BNC hybrid suppressors are designed for transient protection of signal and control circuits that are digital or analog with operating voltages of 6 V or less. Transients will be clamped to between 10 V and 20 V for the duration of transients with rise times of the order of 100 kV/μs and 1 MV/μs. Series impedance is 18 Ω, with a shunt impedance equal to or greater than 5 MΩ in parallel with no more than 10 pF of capacitance. These units come with BNC connectors.

Other versions of the FCC-550 suppressor wire connections are designed for 1-V control, analog, and digital signals and will limit the transient to 8.5 V for the duration of transients that are several thousand volts. On these suppressors, the series impedance is 80 Ω. The shunt impedance is several kilo-ohms in parallel with no more than 500 pF, except during transient suppression, when the impedance approaches zero ohms. The 550 series may be specified with clamping voltages between 8.5 V and 200 V.

Fischer also has Spikeguards designed for 60–400-Hz AC power-line protection.

Arrestors and transient filters made by the Polyphaser Corporation, and referred to as "protectors," are available for use in receiving or transmitting antenna cables and antenna control cables as well as in protection of twisted-pair, combiner, and AC power. The protectors are selected for use over a specific frequency range, e.g., for baseband (DC: 50 MHz), HF/VHF/

UHF 50–550-MHz, 450–900-MHz and other much narrower frequency band protectors, such as 800 MHz to 980 MHz for cellular/paging combiner protection. The 800-MHz and up, DC-blocked, non-gas-tube type are referred to as *microwave filters*. Polyphaser also manufactures: telephone semiconductor protectors, motor control, GPS, and PCS tower-top preamp protectors for 1.2–2-GHz, DC injection/DC path protectors, IBM twinax series protectors, and a range of GPS/PCS/cellular/pager/microwave protectors. The protectors either block DC or allow DC through, and come in weather-resistant, for outdoor mounting, and nonweather-resistant, for indoor mounting, versions and are for either transmit and/or receive-only applications. One Polyphaser high-power coaxial protector has a DC to 220-MHz frequency range and a 25-kW transmit power rating.

A Polyphaser protector designed for wireless local loop protection is the LSX series shown in Figure 5.78. The LSX series provides surge protection for a 20-kA waveform with a 8-µs rise time and a 20-µS pulsewidth (8/20 µs) at the 50% amplitude level. The throughput energy with a 3-kA @ 8/20 µs waveform is less than or equal to 0.5 µJ and the let-through voltage is ±3 V for a 3-kA @ 8/20 µs waveform. The frequency range is 4.2–6 GHz with an insertion loss of 0.1 dB typical, and the transmit power rating is 10 W continuous. The protector includes DC blocking and has a temperature range of −40°C to +85°C.

Another one of the many available Polyphaser protectors of this type is the PSX-D series, which provides surge protection against a 30-kA waveform with an 8-µs rise time and a 20-µs pulsewidth (8/20 µs) at the 50% amplitude level. The throughput energy with a 3-kA @ 8/20 µs waveform is less than or equal to 0.5 µJ, and the let-through voltage is ±3 V for a 3-kA @ 8/20 µs waveform. The frequency range is 1.7–2.3 GHz, with an insertion loss of 0.1 dB typical, and the transmit power rating is 500 W continuous. The protector includes DC blocking and has a temperature range of −40°C to +85°C. Polyphaser also manufactures a range of grounding and cable grounding systems and cable entrance panels, as well as providing courses and publications covering lighting protection and grounding solutions for communications sites.

Epcos (Siemens) also makes a range of gas discharge tubes in addition to MOVs for lighting protection. C. P. Clare manufactures mini- to small gas discharge surge arrestors, and data and application notes can be found at *www.cpclare.com*.

The advantage of gas discharge arrestors is the low self-capacitance, which allows their use at such high frequencies. Due to the shunting effect of lightning arrestors, it is imperative that a low-impedance ground connection be made to the case of the arrestor. For example, a 16-mm-diameter rod will develop, due to its self-inductance of 1.22 µH/m, 12 kV/m length during an average strike with a 2-µs rising edge and a peak current of 20,000 A. Despite a low-impedance ground, it is inevitable that some current will flow on the shields of cables connected to a lightning arrestor or on cables connected to an antenna protected by a lightning rod, due to potential gradients along the ground.

Case Study 8.3 describes a grounding scheme that reduces cable-induced currents during a lightning strike. A large ferrite core through which one to four turns of the interconnection cable/s are wrapped will increase the impedance of the cables and thereby reduce the magnitude of the cable current flow. Where possible, the core should be located close to the equipment to be protected, but certainly after the lightning arrestor, where used.

The voltage spike generated before the arrestor breaks down and clamps the voltage may damage transmitters, receivers, and telephone equipment. Semiconductor devices or varistors may be used to clamp the spike from the gas discharge arrestor or the voltage induced as a secondary effect of a strike to safe levels.

Some combinations of primary arrestor and secondary transient protection devices are shown in Figure 5.79. When secondary protection devices, such as zener diodes, diodes, and transient absorber zener diodes, are mounted on PCB, the ground path on the PCB back to the enclosure should be a low impedance and must be capable of carrying the high transient current.

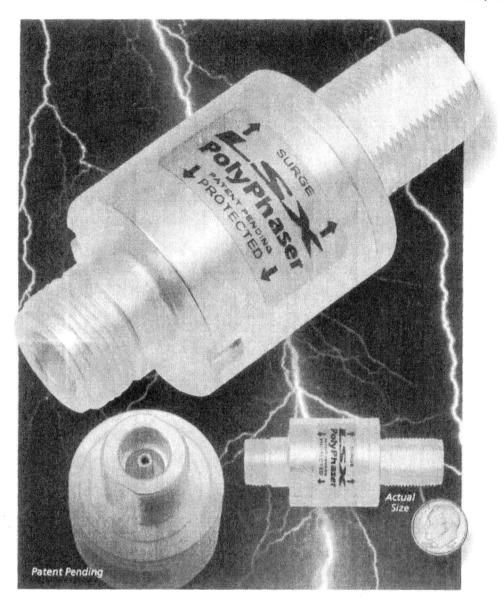

**Figure 5.78**   Arrestor manufactured by the Polyphaser Corporation.

The location of the protection device in the circuit is very important. When the device is placed after capacitors and inductors, the voltage developed across the device may be higher than the impressed transient and in the form of a damped sinusoid. The voltage developed across transient protection zeners, when tested using a generator capable of producing the MIL-STD-461 CS06-specified spike, can exhibit forward voltage drops two or three times higher than specified for the device, even when the current through the device is below the maximum rated.

The disadvantage in the use of both semiconductors and varistors at the input of high-frequency receivers or interface circuits is their high intrinsic capacitance. This capacitance is in parallel with the input and may degrade the signal. When the use of these devices is precluded,

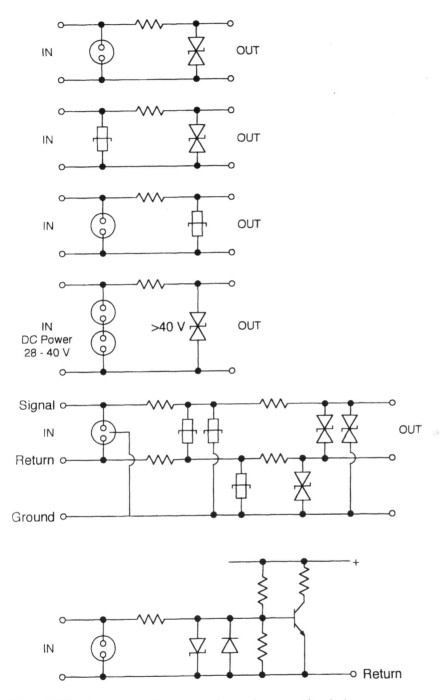

**Figure 5.79**  Combination of primary and secondary protection devices.

due to the high capacitance, the alternative is the use of a low-value blocking capacitor. Generally the receiver input impedance is 50–75 Ω, to which the antenna is connected via a shielded cable with a characteristic impedance of 50–75 Ω. With a capacitor rated at 1000 V or greater in series with the input to the receiver, the RF signal is virtually unattenuated, whereas the transient is considerably reduced in amplitude. As an example, consider a receiver operating at 40 MHz or greater with a 50-Ω input impedance. With a 1000-pF series capacitor, the voltage developed across the receiver input during the typical 2-μs rise time of the strike, and assuming a 1000-V spike, is reduced to 25 V. For additional protection, two low-capacitance diodes may be placed after the capacitor, with one diode connected between input and return and the second between input and supply. The advantage of the capacitor solution compared to a parallel protection device is that no high currents are shunted to ground. However, cables and PCB tracks must be capable of sustaining voltages of 1000 V or greater. With the diodes incorporated, higher currents flow, but the magnitude is limited by the time constant of the capacitor-and-diode combination.

The RTCA/DO-160D requirements for airborne equipment contains, in section 23, tests for direct lightning effects and in section 22 lightning-induced transient susceptibility. The tests in section 22 are designed to simulate the current and voltages induced in cables contained within aircraft or helicopters. Different test levels apply to cables that are protected by airframes composed of metal and composite skin panels and equipment in carbon fiber composite airframes whose major surface areas have been protected with metal meshes or foil. For this type of aircraft, the apertures are the main source of induced transients. In other aircraft (i.e., carbon fiber composite), the structural resistance of the airframe is also a significant source of induced transients. DO-160D describes both direct pin injection tests and induced cable bundle tests. Five different test levels are applicable to the pin injection tests and five different levels for the cable bundle tests. Both pulse and damped sinusoid test waveforms are specified. All pulse waveforms have a double exponential shape. Current waveform 1 is a waveform with a rise time to peak current of 6.4 μs and a fall time, to 50% amplitude, in 69 μs. Voltage waveform 2 has a rise time of maximum 100 ns and a fall time, to 50% amplitude, of 6.4 μs. Voltage/current waveform 3 is a damped sinusoid. Voltage waveform 4 has a rise time of 6.4 μs and a fall time, to 50% amplitude, in 69 μs. Current/voltage waveform 5a has a risetime of 40 μs and 5b 50 μs, and waveform 5a has a fall time, to 50% amplitude, of 120 μs and 5b 500 μs. The voltage waveforms are specified as open circuit and the current into a short circuit. The voltage and current at other loads can be calculated.

Table 5.16 shows the level 1 and level 5 voltage/current test levels for pin injection and Table 5.17 for the cable bundle injection.

In numerous measurements and analysis, the currents and voltages induced in shielded cables based on the cable-induced test levels can be relatively high. For example, with waveform 1 (6.4 μs/70-μs pulse) at test level 4, the test level is 750 V open circuit and 1500 A short circuit. In tests on a 9-m-long RG108A/U, twisted-shielded-pair cable, the internal (center conductor) open-circuit voltage at the load end of the cable was 264 V and the short-circuit current

**Table 5.16**  Lowest and Highest Pin Injection Test Levels

| | Waveforms | | |
|---|---|---|---|
| Level | Waveform 3: damped sinusoid $V_{oc}/I_{sc}$ | Waveform 4: 6.4 μs/69 μs $V_{oc}/I_{sc}$ | Waveform 5A: 40 μs/120 μs $V_{oc}/I_{sc}$ |
| 1 | 100/4 | 50/10 | 50/50 |
| 5 | 3200/128 | 1600/320 | 1600/1600 |

**Table 5.17**  Lowest and Highest Cable Bundle Injection Test Levels

| Level | Waveform | | | | |
|---|---|---|---|---|---|
| | Waveform 1: 6.4 μs/69 μs $V_L/I_T$ | Waveform 2: 100 ns/6.4 μs $V_T/I_L$ | Waveform 3: damped sinusoid | Waveform 4: 6.4 μs/69 μs $V_T/I_L$ | Waveform 5A: 40 μs/120 μs $V_L/I_T$ |
| 1 | 50/100 | 50/100 | 100/20 | 50/100 | 50/150 |
| 5 | 1600/3200 | 1600/3200 | 3200/640 | 1600/3200 | 1600/5000 |

825 A. The current is C/M, and if a number of circuits are connected to the conductors in the cable bundle, the current will divide between the circuits, with a magnitude dependent on the impedance of each circuit to chassis or safety ground. Traditional miniature secondary transient suppressors will either not carry the test current, or the voltage will rise across them to levels approaching the 264-V level. Some series impedance is necessary on each line before the transient suppressor. The minimum value for this series impedance depends on the test level, how the current is shared between the circuits, and the peak current-handling capabilities of the suppressor. The most common series impedance would be the resistor, although high-value C/M inductors are also useful, and for pulse and audio circuits, an isolating transformer will present a high C/M impedance. If a transformer is used, standard silicon or Schottky diodes may provide sufficient protection due to the low transient current coupled, primarily through the interwinding capacitance, of the transformer.

The level of voltage protection required depends on the maximum rating of the circuit to be protected. In some instances, Schottky diodes are used to limit the voltage to safe levels. The protection shown in Figures 5.80, 5.81, and 5.82 is primarily designed to protect the circuit, and some disruption of data is to be expected during application of the lighting test level. Transformer coupling is the best technique if data integrity is important; however, this can only be used on "return to zero" type of data communications.

### 5.4.2  Electrostatic Protection

It is a myth to assume that once an integrated circuit is mounted on a PCB, it is immune to an electrostatic discharge even when the IC input connections are not brought out to a connector. There is no better approach to protection than to avoid an ESD event by correct procedures, such as the use of a ground strap, antistatic bags, grounded mats, and deionizers. Integrated circuits mounted on PCBs where the input pins are brought out to an edge connector are just as sensitive to ESD as non-board-mounted ICs. Additional protection for CMOS logic and FET input op amps is relatively easy to achieve by inclusion of series resistance between the input and the device and either resistance or diodes/zener diodes between the input and ground.

Some of the devices that have been designed to protect against the fast rise time and high energy of EMP will ensure that effective protection against the lower-energy ESD event is achieved. The ESD pulse rise time is typically 15 ns, although the IEC 1000-4-2-specified ESD generator must provide a current spike into a low impedance with a risetime of 0.7–1 nS and a pulsewidth of 30–60 nS, and the turn-on time of the majority of EMP devices is 1–2.5 ns. Zener diodes, multilayer varistors, and diode surge arrestors are available with breakdown voltages as low as 6 V and as high as 440 V, with response times from picoseconds to nanoseconds. Connectors are also available with built-in ESD transient suppressors. A new suppressor array manufactured by Pulse-Guard® is designed to provide protection from direct pin injection ESD threats. The Pulse-Guard is a substrate that pushes on the front or rear of standard D-Sub and MIL-C-24308 I/O connectors.

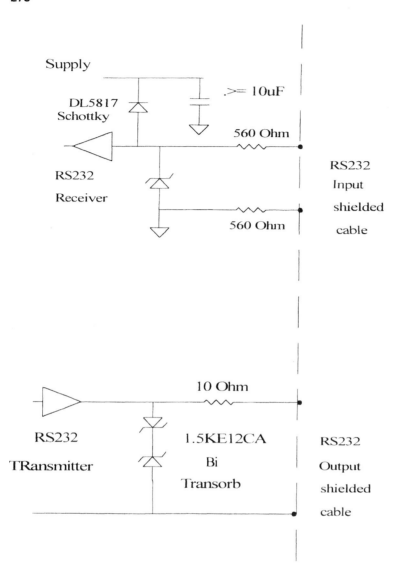

**Figure 5.80**   Secondary lightning protection of a single-ended (RS232 type) signal interface.


The European EMC requirements contained in EN50082-1 includes an 8-kV air discharge, and in EN50082-2 a 4-kV contact and an 8-kV air discharge, ESD test. IEC 1000-4-2 describes the ESD generator, which has a 150 pF charged up to the test voltage and which is discharged, as a single event, through a 330-$\Omega$ resistor. In an air discharge, the probe tip is brought close to the EUT as quickly as possible without causing physical damage, and the resistance of the discharge path is the 330-$\Omega$ resistor plus the resistance of the resultant plasma. The IEC 1000-4-2 test procedure has the EUT mounted on a conductive tabletop, described as a horizontal coupling plane (HCP) under the EUT. The nonconductive table is located above a ground plane (reference plane). The EUT is insulated from the HCP by a 0.5-mm-thick insulator. The EUT is located 0.1 m from a vertical coupling plane (VCP), which is also insulted from the HCP. The HCP is connected to the reference plane by a cable, with a 470-k$\Omega$ resistor located at each end, to prevent a buildup of charge. The VCP is also connected separately with an identical

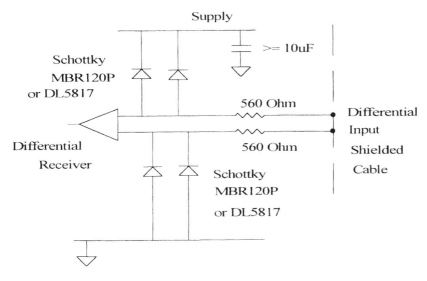

**Figure 5.81** Secondary lightning protection of differential input data circuit where the C/M input voltage must be limited to close to the supply voltage and to ground.

cable. The discharge return cable of the ESD generator shall be connected to the reference plane (ground plane under the table) at a point close to the EUT. Figure 9.31 shows this test set-up. During an air or direct-contact-ESD discharge to the HCP or VCP, the resultant current generates an E field that couples to the EUT. If the EUT enclosure is completely insulating with no exposed metal, e.g., fasteners, in either a direct or air discharge to the EUT, the charge is equalized around the nonconductive enclosure, but no noticeable discharge current is detected. In discharges to a metal enclosure, a rapid charging current flow, and in the air discharge a plasma, is seen at the probe tip. The most common problem occurs with discharges close to apertures in a metal

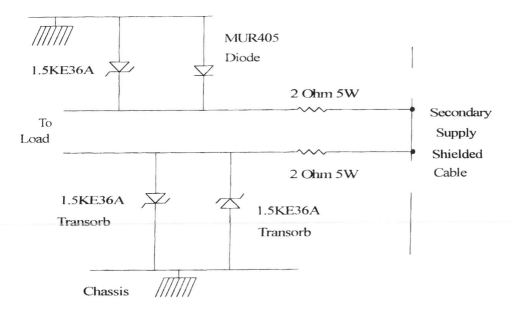

**Figure 5.82** Protection of secondary power (power provided by one equipment to another equipment).

enclosure filled with nonconductive material/components, for example, LCD displays, LEDs, touch buttons, keyboards. The field generated by the discharge can couple inside to PCB components through the aperture. Discharges to metal fasteners used to mount nonconductive components may also result in a susceptibility problem. Likewise, discharges to the pins of a connector or even the case of the connector, when it is isolated from the enclosure, are also problem areas. The current flow is predominantly C/M, caused either by the discharge or by the E field resulting from the discharge. Capacitors across the input pins of digital or analog devices can be very effective at increasing immunity, even when a relatively low level. If the capacitor is located between the PCB and chassis, then it needs to be a relatively high value to be effective at reducing the induced voltage. For although the 330-$\Omega$ resistor in the generator slows down the charging time for the capacitor, the peak instantaneous charging current is very high and the voltage may be high. For example, in a 4-kV direct discharge, a 10 pF will charge to 3750 V, a 100 pF to 2400 V, a 1000 pF to 520 V, an 0.01 $\mu$F to 59 V, and a 0.1 $\mu$F to 5.9 V.

If the capacitor is not connected between the PCB ground and chassis but is used between the input/output pins of a device, the D/M voltage developed due to C/M current flow or the incident E field can be drastically reduced. As with increasing immunity to RF, the capacitor/s should be located as close as possible to the junction/s to be protected. Another use for a capacitor is to slow down the voltage appearing across a slow-acting transient protection device and to give it a chance to clamp the voltage.

One device designed to protect in input circuits of receivers or the output circuits of transmitters from antenna-coupled ESD events is the AntennaGuard™ manufactured by AVX. These antenna protector chips have capacitance values less than or equal to 12 pF in the 0603 chip size and less than or equal to 3 pF in the 0402 chip size. Antenna protector chips are especially useful in the ESD protection and EMI filtering for high-gain FETs typically used in wireless products. The AntennaGuard may also be used with an inductor to make a low-pass filter. The varistor in the AntennaGuard can turn on in 300 ps to 700 ps and provide typical transient suppression of 15-kV air discharge ESD events, reducing the level to survival levels for most FET input preamplifiers.

### 5.4.3  Electromagnetic Pulse (EMP) Protection

The primary effect of EMP in shielded cables located outside of buildings, ships, and vehicles is a high current flow in the shields of cables and a high voltage developed across the inputs of devices connected to these cables. One form of protection is the use of cable shield ground adapters that shunt the high current to ground, for example, as the cable enters the metal wall of the building. The signal or power carried by the shielded cables may then be protected by gas discharge arrestors with fast response times of approximately 1 ns. Secondary protection on low-frequency signal lines and power may be achieved by fast-response semiconductor devices similar to those used for lightning protection. The Polyphaser Corporation provides an EMP or lightning protection device for DC remote/telephone line and Tone remote/dedicated line protection. Reliance Comm/Tec manufactures a building entrance terminal that incorporates EMP/lightning protection for 6, 12, or 25 pair tip and ring circuits.

The protection of high-frequency receivers, transmitters, and preamps against EMP is more difficult than for lightning protection, due to the fast rise time of the EMP event. One method is the use of protection tees made up of a shorted quarter-wave section in parallel with the input. The shorted quarter-wave section provides a low impedance to ground at frequencies other than the signal for which the tee is designed. Figure 5.83 shows the schematic of the tee, combined with a decoupling line, and the construction of a 900-MHz tee. The decoupling line is a concentric coaxial line in series with the signal of length equal to a quarter-wave at the signal frequency and may be used without the shorted quarter-wave section for protection. The frequency limits of the tee and decoupling line are 25 MHz to 18 GHz.

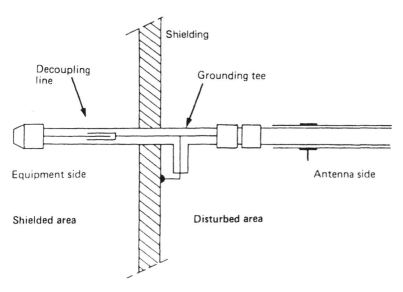

**Figure 5.83** Tee and decoupling line schematic and construction. (Reproduced with kind permission from Les Cables de Lyon Alcatel.)

The current induced by an EMP event on internal cables, which are shielded by the skin of a vehicle, ship, aircraft, or building, has been characterized in MIL-STD-461C as a damped sinusoid with a frequency component extending from 10 kHz to 100 MHz. The maximum common-mode current flow is 10 A, which is injected directly into pins and terminals (test CS10) or into cables (test CS11). Protection against CS10 and CS11 test levels may be achieved by filters and/or the semiconductor transient suppressors described for secondary lightning protection.

## ADDRESSES OF MANUFACTURERS MENTIONED IN THIS CHAPTER

ABB HAFO (thyristor/diode)
P.O. Box 520
S-175 26 Jarfalla
Sweden

AVX
P.O. Box 867
Myrtle Beach, SC 29578
(843) 448-9411

CAL Corporation (now EMS)
1725 Woodward Drive
Ottawa K2C 0P9
Canada

California Micro Devices
215 Topaz Street
Milpitas, California 95035
(408) 263-3214

CORCOM
844 E Rockland Rd.
Libertyville, IL 60048
(847) 680-7400

Erie Technological Products Inc.
2200 Lake Park Avenue
Smyrna, GA 30080

Fair-Rite Products Corp.
P.O. Box J
One Commercial Row
Wallkill, NY 12589-0288
888 FAIR-RITE/(914) 895-2055

Fischer Custom Communications
2917 W. Lomita Boulevard
Torrance, CA 90505
(310) 891-0635

Les Cables de Lyon (now Alcatel)
Department Haute Frequence et Electronique
35 Rue Jean Jaures
B.P. 20-95871
Bezons Cedex
France

MCG Electronics Inc. (USA)
12 Burt Drive
Deer Park, NY 11792
(516) 586-5125

Micrometals
5615 E. La Palma Avenue
Anaheim, CA 92807-2109
(714) 970-9400

Polyphaser Corporation (USA)
2225 Park Place
P.O. Box 9000
Minden, NV 89423
(775) 782-2511

Protek
P.O. Box 3129
Tempe, AZ 85282-3129
(602) 431-8101

Pulse-Guard
G&H Technology Inc.
750 W. Ventura Blvd.
Camarillo, CA 93010
(805) 389-5760

Reliance Comm/Tec (USA)
11333 Addison Street
Franklin Park, IL 60131
(312) 455-8010

Siemens (Epcos)
186 Wood Avenue So.
Iselin, NJ 08830
(732) 906-4300

Steward Manufacturing Co.
P.O. Box 510
Chattanooga, TN 37401-0510
(423) 867-4100

Tecor Electronics Inc. (Sidactor)
1801 Hurd Drive
Ewing, TX 75038-4385
(214) 580-1515

## REFERENCES

1. R. K. Keenan. Digital Design for Interference Specifications. The Keenan Corporation, Vienna, VA, 1983.
2. E. Kann. Design Guide for Electromagnetic Interference (EMI) Reduction in Power Supplies, MIL-HDBK-241B. Power Electronics Branch, Naval Electronic Systems Command, Department of Defense, Washington, DC, 1983.
3. H. M. Schlicke. Electromagnetic Compossibility. Marcel Dekker, New York, 1982.
4. R. B. Cowdell. Don't experiment with ferrite beads. Electronic Design, Issue 12, June 7, 1969.
5. Schematics of high immunity interface circuits, PLR2, PLD2, HD-1, HR-1, HAD-1, and HAR-1. Reproduced by kind permission of CAL Corporation.
6. D. J. Kenneally RF upset susceptibilities of CMOS and low power Schottky D-type, flip flops. IEEE International Symposium on Electromagnetic Compatibility. May 23–25, 1989, Denver, CO.
7. T. F. Fang and J. J. Whalen. Application of the Nonlinear Circuit Analysis Program (NCAP) to predict RFI effects in linear bipolar integrated circuits. Proc. 3rd Symp. Tech. Exhibition on Electromag. Compat. Rotterdam, May 1–3, 1979.
8. D. S. Britt, D. M. Hockanson, Fei Sha, J. L. Drewniak, T. H. Hubing and T. P. Van Doren. Effects of gapped groundplanes and guard traces on radiated EMI. IEEE International Symposium on Electromagnetic Compatibility. August 18–22, 1997, Austin, Texas.
9. F. B. J. Leferink, M. J. C. M. van Doorn. Inductance of printed circuit board ground plates. IEEE International Symposium on Electromagnetic Compatibility. August 9–13, 1993, Dallas, Texas.
10. L. O. Johansson. EMC Fundamentals PASSIVE and their EMC Characteristics. Compliance Engineering, July/August 1998.
11. H. G. Ghadamabadi, J. J. Whalen. Parasitic capacitance can cause demodulation RFI to differ for inverting and non-inverting operational amplifier circuits.

# 6

# Electromagnetic Shielding

## 6.1 REFLECTION, ABSORPTION, AND SHIELDING EFFECTIVENESS

### 6.1.1 Reflection in a Perfect Conductor

When an electromagnetic wave impinges on a sheet of metal, the electric field component induces a current flow in the sheet. Some electromagnetic theory of shielding states that the current flow has a current density of $2H_o$ or $2E_o/Z_w$, whereas other electromagnetic theories state that the current is equal to $H_o$. Figure 6.1 shows the incident electromagnetic wave and the induced current, from which it can be seen that the magnetic intensity produced by the induced current is such as to annihilate the magnetic intensity behind the sheet and to double the intensity in front of the sheet, assuming a perfectly conducting sheet. A standing wave is generated in front of the sheet, and this can be seen in Figure 6.1.

No metal is a perfect conductor (i.e., exhibits zero impedance) and thus complete reflection of an electromagnetic wave is not possible.

### 6.1.2 Transmission-Line Theory Applied to Shielding

Although wave theory may be used to predict the level of shielding obtained by a conductive sheet (1), the concept of wave impedance introduced by Schelkunoff in 1938, and the application of transmission-line theory to shielding, results in a considerably simpler solution. Using transmission-line theory, the reflection is explained by the impedance difference between the incident wave and the metal impedance.

Figure 6.2 illustrates the incident field with impedance $Z_w$ impinging on the near side of the metal shield, which exhibits a metal impedance $Z_m$ or a barrier impedance $Z_b$, as described in following sections. When the wave impedance is higher than the metal impedance, some of the incident field is reflected from the near side of the shield, due to the impedance mismatch. As the remaining wave propagates through the shield, it is attenuated. A fraction of the field incident on the far side of the interior of the shield is rereflected, due to the metal- to air-impedance mismatch. The remaining field propagates from the far side of the shield into space. A second rereflection of the internal field occurs at the near-side metal-to-air interface, and some small fraction of the return trip field leaves the far side of the shield. This small fraction is insignificant when the absorption through the shield is greater than numeric 5. The re-rereflected field as it leaves the metal is considered to be in phase with, and therefore adds to, the field that has propagated through the metal.

As discussed in Ref. 2, some authors state that the wave from the shield on the far side of the incident wave is either 377 $\Omega$ or the same impedance as the incident field. Other authors state that there is no reflection coefficient at the far side of the shield: the impedance of the wave is the same as the impedance of the barrier, and the wave impedance approaches 377 $\Omega$ only at a distance of $\lambda/2\pi$ in meters from the shield. In the measurements on aluminized mylar,

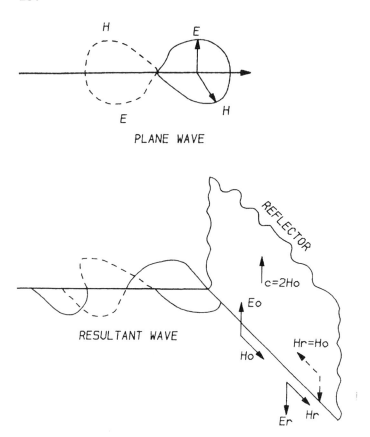

**Figure 6.1** Plane wave incident on a perfect conductor.

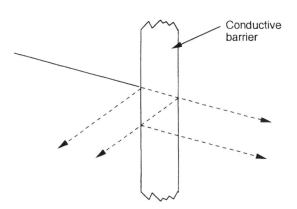

**Figure 6.2** Transmission losses at a conductive sheet.

discussed later, the results indicated that the current in the barrier material was consistent with the value of the H field and that the impedance of the wave emanating from the barrier was the same as the impedance of the barrier, or, in the case of the aluminized mylar, 2 $\Omega$.

### 6.1.3 Metal Impedance, Skin Depth, Barrier Impedance

The impedance of a metal is made up of resistance and inductance and is a function of its conductivity, permeability, thickness, and the frequency of the current flow in it. The impedance of a metal when its thickness is much greater than the skin depth is given by

$$Z_m = 369 \sqrt{\frac{u_r f}{G_r}} \quad [\mu\Omega/\text{sq}] \tag{6.1}$$

where

$\mu_r$ = permeability relative to copper

$G_r$ = conductivity relative to copper

$f$ = frequency in megahertz

We use the term $\Omega/\text{sq}$ to describe surface impedance or resistance. The question is often asked, per square what? Is it inches, centimeters, meters? The equation for the resistance of a conductor is

$$R = \frac{\rho l}{wt} \tag{6.2}$$

where

$l$ = length of the conductor

$\rho$ = resistivity

$w$ = width of the conductor

$t$ = thickness of the conductor

From Eq. (6.2) we see that, for a constant thickness and assuming $l = w$, the resistance is constant, regardless of the magnitude of $l$ and of $w$. Thus $w$ and $l$ may be 1 mm or 1 m and the resistance is the same. When the $\Omega/\text{sq}$ term is used to describe surface resistance, the thickness should be specified. When used to describe surface impedance and when the thickness is much greater than the skin depth at all frequencies of interest, the thickness of the conductor may be omitted, for as the frequency of current in a conductor increases, the current density does not remain constant throughout the depth of the metal but is greater at the surface. The skin depth of a metal is defined as the surface depth at a specific frequency at which 63.2% of the current flows. Equation (6.3) shows that the skin depth for any metal decreases with increasing frequency, and is lower for metals with high conductivity and permeability:

$$\delta = \frac{1}{\sqrt{\pi f \mu G}} \quad [\text{m}] \tag{6.3}$$

where

$\mu$ = permeability of the metal = $\mu_o \mu_r$

$G$ = conductivity in mhos/meter

$f$ = frequency in hertz

For copper, where $\mu_r = G_r = 1$,

$$\delta = \frac{0.066}{\sqrt{f}} \quad \text{[mm]}$$

where $f$ = frequency in megahertz. For any metal,

$$\delta = \frac{0.066}{\sqrt{f\mu_r G_r}} \quad \text{[mm]}$$

where $f$ = frequency in megahertz. Table 6.1 shows the skin depth of some of the common metals at several frequencies.

Due to the skin depth effect, the barrier impedance of the metal instead of the metal impedance may be required in the evaluation of shielding effectiveness. The barrier impedance of a metal is a function of the ratio of metal thickness to skin depth, in addition to the frequency dependency of the impedance. The barrier impedance of any metal is given by

$$Z_b = 369 \sqrt{\frac{\mu_r f}{G_r}} \frac{1}{(1 - e^{-t/\delta})} \quad \text{[}\mu\Omega/\text{sq]}\qquad(6.4)$$

where $f$ = frequency in megahertz and $t$ and $\delta$ are in the same units.

When $t/\delta \ll 1$,

$$Z_b = \frac{1.414}{0.058 G_r t} \quad \text{[}\mu\Omega/\text{sq]}\qquad(6.5)$$

where

$\qquad t$ = thickness [mm]

$\qquad G_r$ = relative conductivity [mhos/m]

Because no metal is a perfect conductor, the current flow in the metal sheet shown in Figure 6.1 is not confined to its surface but extends into the thickness of the metal. The impedance of a pure conductor is generally lower than that of the incident wave, with the possible exception of a current loop in very close proximity to a shield.

## 6.1.4   Reflection in Practical Conductors

From the explanation of reflection in Section 6.1.1 it may appear that for magnetic fields, where the electric field component is less than that of the magnetic, the reflection will be less complete than for an electric field, and this is supported by transmission-line theory. The impedance of a field changes with distance and frequency, and the metal impedance of the reflecting sheet is a function of frequency, permeability, and conductivity, as shown in Eq. (6.1). The equation for the reflection loss in decibels in an electric field when the metal thickness $t \gg \delta$ is given by

$$R_e = 353.6 + 10 \log\left(\frac{G_r}{f^3 \mu_r 2.54\ r_1^2}\right) \quad \text{[dB]}\qquad(6.6)$$

where

$\qquad f$ = frequency [Hz]

$\qquad r_1$ = distance from an electric field source [cm]

**Table 6.1** Skin Depth for Some Common Conductors

| Conductor | Conductivity [$\times 10^7$ mhos/m]† | Relative conductivity ($G_r$) | Relative permeability ($\mu_r$*) | Skin depth with frequency (mm) | | | | |
|---|---|---|---|---|---|---|---|---|
| | | | | 1 kHz | 1 MHz | 10 MHz | 100 MHz [$\times 10^{-3}$]† | 1 GHz [$\times 10^{-3}$]† |
| Aluminum | 3.54 | 0.6 | 1 | 2.7 | 0.085 | 0.027 | 8.5 | 2.7 |
| Copper | 5.8 | 1 | 1 | 2.0 | 0.066 | 0.02 | 6.6 | 2.0 |
| Gold | 4.5 | 0.7 | 1 | 2.5 | 0.079 | 0.025 | 7.9 | 2.5 |
| Tin | 0.87 | 0.15 | 1 | 5.4 | 0.170 | 0.054 | 17 | 5.4 |
| Nickel | 1.3 | 0.23 | 100 | 0.43 | 0.014 | 0.0045 | 1.4 | 0.44 |
| 4% silicon iron | 0.16 | 0.029 | 500 | 0.55 | 0.017 | 0.0055 | 1.7 | 0.55 |
| Hot rolled silicon steel | 0.22 | 0.038 | 1500 | 0.27 | 0.0087 | 0.0027 | 0.87 | 0.27 |
| Mumetal | 0.16 | 0.029 | 20,000 | 0.087 | 0.0027 | 0.00087 | 0.27 | 0.087 |

* At 10 kHz and B = 0.002 weber/m².

† Expressions in square brackets represent the exponent portion of the number e.g. $3.54 \times 10^7$ mhos/m or $3.54 \times 10^{-5}$ mhos/cm.

The equation for the reflection loss in a magnetic field is given by

$$R_m = 20 \log \left[ \frac{0.181}{r_1} \sqrt{\frac{\mu_r}{G_r f}} + 0.053 r_1 \sqrt{\frac{G_r f}{\mu_r}} + 0.354 \right] \quad \text{[dB]} \tag{6.7}$$

where

$f$ = frequency [Hz]

$r_1$ = distance from source [cm]

The equation for the reflection of a plane wave is given by

$$R_p = 168.2 + 10 \log \left[ \frac{G_r}{\mu_r f} \right] \quad \text{[dB]} \tag{6.8}$$

When the source of emission is neither a low-impedance current loop nor a high-impedance voltage source but a source that exhibits a characteristic impedance, $Z_c$ (e.g., a transmission line), we can use the circuit impedance to calculate the reflection in the near field. The equation for reflection loss using circuit impedance is given by

$$R_c = 20 \log \left[ \frac{Z_c}{(1.48 \times 10^{-3}) \sqrt{\dfrac{\mu_r f}{G_r \times 10^6}}} \right] \quad \text{[dB]} \tag{6.9}$$

where $f$ = frequency, in hertz.

### 6.1.5  Absorption

The variation in current density through a thick metal sheet is shown in Figure 6.3. The radiation from the far surface into space is proportional to the current flow on that surface. Therefore, due to the reduction in current flow, an absorption of the incident wave effectively occurs in

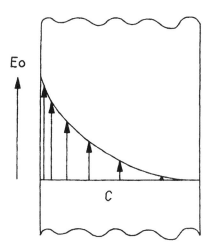

**Figure 6.3**  Variation in current density through a thick metal sheet. (From Ref. 2.)

the metal. The equation for absorption is independent of wave impedance, proportional to thickness, and inversely proportional to skin depth and is given by

$$A = (3.338 \times 10^{-3})t\sqrt{\mu_r fG_r} \tag{6.10}$$

where

$A$ = absorption loss [dB]

$t$ = thickness of shield [mils] (1 mil = 1/1000″ = 0.0254 mm)

$f$ = frequency [Hz]

### 6.1.6 Rereflection Correction

When the absorption loss in the thickness of the metal is not much greater than 5, the rereflection within the metal is appreciable and the rereflection gain should be included in the total shielding effectiveness equation. Often the term *rereflection loss* or *correction factor** is used to describe the effect of rereflections, but, since this is always a negative number, it subtracts from the total shielding effectiveness and is effectively a gain.

Re-reflection loss, which is always negative, is given as a magnitude in decibels:

$$R_r = 20 \log\left\{1 - \left[\left(\frac{Z_m - Z_w}{Z_m + Z_w}\right)^2 (10^{-0.1A}(e^{-j0.227A}))\right]\right\} \tag{6.11}$$

where

$Z_m$ = metal impedance

$A$ = absorption loss [dB]

$Z_w = 377\dfrac{2\pi r}{\lambda}$ when $r < \dfrac{\lambda}{2\pi}$ for a magnetic field

$Z_w = 377\dfrac{\lambda}{2\pi r}$ when $r < \dfrac{\lambda}{2\pi}$ for an electric field

$Z_w = 377\ \Omega$ for a plane wave

$r$ = distance from source [m]

The rereflection loss is a function of the ratio between the wave impedance $Z_w$ and the metal impedance $Z_m$ and of the absorption loss. For thin metals, where the thickness is less than the skin depth, $Z_b$ should be used instead of $Z_m$. When $Z_w$ is much greater than $Z_m$ or $Z_b$, the rereflection loss becomes an inverse function of the absorption loss only.

## 6.2 SHIELDING EFFECTIVENESS

*Shielding effectiveness* (SE) is defined as the ratio between the field strength at a given distance from the source without the shield interposed and the field strength with the shield interposed. The total shielding effectiveness of a conductive barrier in decibels is the sum of the reflective losses expressed in decibels, the absorption in decibels, and the rereflection loss in decibels; thus SE is given by

$$SE = R + A + R_r \quad [\text{dB}] \tag{6.12}$$

where $R$, $A$, and $R_r$ are in decibels.

* As discussed in Section 6.1.2 in measurements on aluminized mylar, 2 Ω, no reflection coefficient was seen.

The magnetic field reflection loss versus frequency, distance, and ratios of relative conductivity to relative permeability $(G_r/\mu_r)$ is shown in the nomogram of Figure 6.4.

The electric field reflection loss versus frequency, distance, and ratios of $(G_r/\mu_r)$ is shown in the nomogram of Figure 6.5. The plane-wave reflection loss versus frequency, distance, and ratios of $G_r/\mu_r$ is shown in the nomogram of Figure 6.6. The absorption loss versus frequency, thickness, and product $G_r^*\mu_r$ is shown in the nomogram of Figure 6.7. All of these nomograms

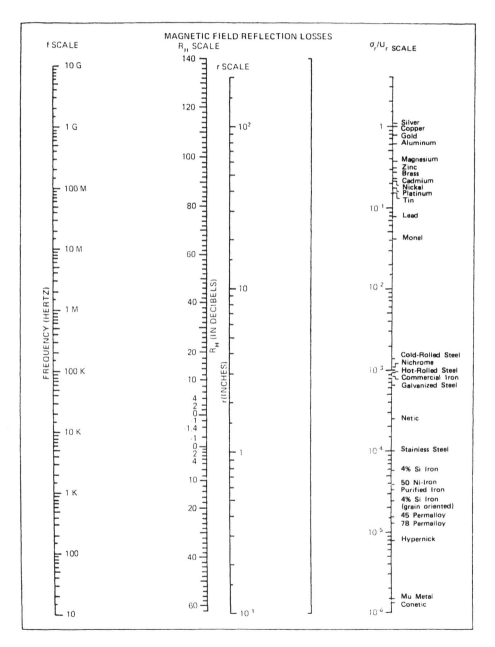

**Figure 6.4** Magnetic field reflection loss. (From Ref. 2. Reprinted from EDN, September 1972. © Cahners Publishing Company, Division of Reed Publishers, USA.)

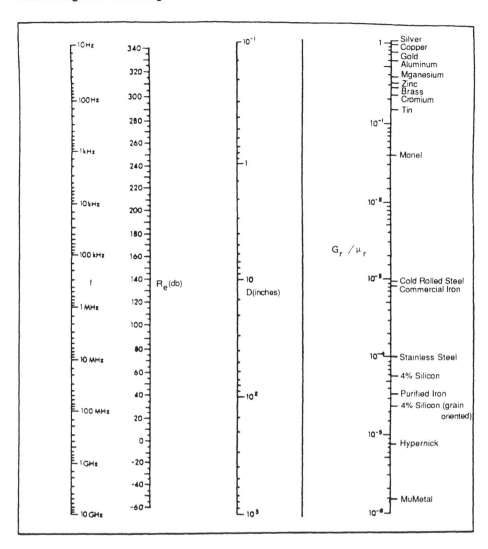

**Figure 6.5** Electric field reflection loss. (From Ref. 2. Reprinted from EDN, September 1972. © Cahners Publishing Company, Division of Reed Publishers, USA.)

are reproduced from Ref. 3. The nomograms in Figures 6.4–6.6 provide for a minimum ratio of $G_r/\mu_r$ of $1 \times 10^6$, so the shielding effectiveness of very-low-conductivity materials with a low permeability, such as graphite, may be found.

In choosing the location and type of shield, the following rules are useful, though not infallible:

Absorption losses increase with increase in frequency, barrier thickness, barrier permeability, and conductivity, all of which increase the ratio of barrier thickness to skin depth.

Assuming that the shield thickness is greater than the skin depth, the following are valid:

Reflection losses, above 10 kHz, generally increase with an increase in conductivity and a decrease in permeability (i.e., with a decrease in metal impedance).

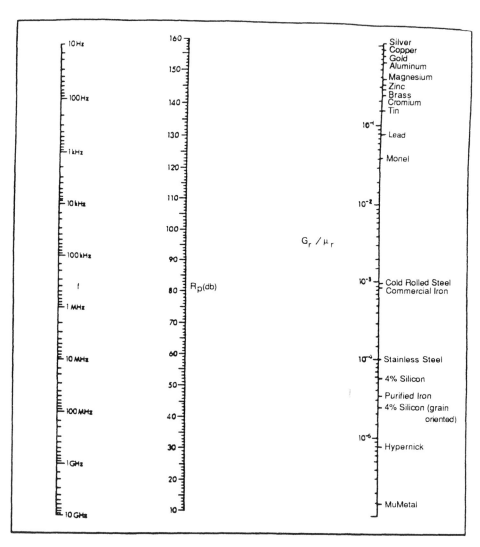

**Figure 6.6**  Plane-wave reflection loss. (From Ref. 2. Reprinted from EDN, September 1972. © Cahners Publishing Company, Division of Reed Publishers, USA.)

Reflection, E field, increases with a decrease in frequency and a decrease in distance from the source, both of which increase the ratio of wave impedance to metal impedance.

Reflection, H field, increases with an increase in frequency and an increase in distance from the source, due to a corresponding increase in wave impedance.

Reflection, plane wave, increases with a decrease in frequency as the metal impedance decreases.

One of the weaknesses of the transmission-line theory when applied to shielding is that the shielding effectiveness is dependent, in the near field, on the distance between the source of the field and the shield, due to the variation in wave impedance with distance. As discussed in Ref. 4, this leads to a contradiction with the basic law of reciprocity. Consider a transmitting loop antenna located at a fixed distance from a second receiving loop antenna. A shield is placed

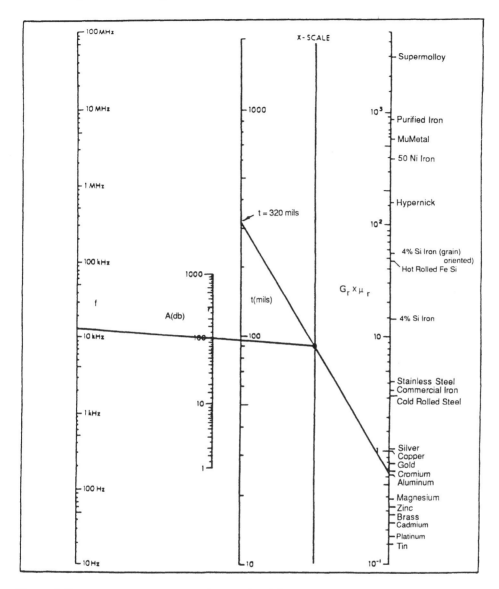

**Figure 6.7**  Absorption loss. (From Ref. 2. Reprinted from EDN, September 1972. © Cahners Publishing Company, Division of Reed Publishers, USA.)

at some distance, but not halfway, between the two antennas. The law of reciprocity states that the transmission must be equal either way when the transmitting and receiving source and load impedances are equal. Therefore the transmitting antenna may be used as the receiving antenna, and vice versa, with no change in the received signal. In contrast, the transmission-line theory of shielding makes the shielding effectiveness dependent on the location of the shield between the two loops. Lyle E. McBride of California State University, in a private communication, observed that in experiments the Schelkunoff assumptions are correct when the receiving loop in a shielding effectiveness measurement is open circuit but that the shielding effectiveness is independent of shield location when both transmitting and receiving loop impedances are matched. It therefore appears that the measurement of shielding effectiveness is not an intrinsic electromagnetic parameter but is dependent on the test setup, as discussed further in Section 6.5.2. Reference 1 compares the prediction of magnetic field shielding for copper, aluminum, and steel shields placed between circular loop antennas, and notes in the measurements a small but persistent dependence of the shielding effectiveness on the spacing between the two loops. In common with the author of Ref. 5, the predictions of shielding effectiveness in Ref. 1, using transmission-line theory, were made with the shield virtually displaced to coincide with the location of the receiving loop.

The author of Ref. 5 computes the wave impedance of a magnetic field generated by a magnetic dipole, which is approximately equal to that of a current loop if the radius is small, at one-third the wave impedance used in Eq. (6.7) for H field reflection and obtained from Eq. (2.29).

From Ref. 5, the wave impedance for an electric dipole is three times the wave impedance used in Eq. (6.6) for E field reflection and obtained from Eq. (2.30). Therefore we may expect the loss for the E field computed from Eq. (6.6) to be low and the loss for the H field computed from Eq. (6.7) to be high.

In practice, the H field is the most difficult to shield against, even here, thin aluminum foil, such as cooking foil, invariably provides sufficient shielding unless the source is a low-frequency high-magnitude H field. The greatest contributor to shielding degradation is typically the presence of joints and apertures and not the enclosure, unless the material has low conductivity.

Perhaps the question of greatest interest to the practicing engineer is how accurate the predicted shielding effectiveness is using the equations and nomographs presented here, especially for the shielding of H fields. The preceding SE equations were based on a shield of infinite size, whereas practical enclosures are limited in size.

Table 6.2 illustrates that the predicted shielding effectiveness of copper, aluminum, and steel shields of $\frac{1}{16}$-inch and $\frac{1}{8}$-inch thickness, as obtained from Eqs. (6.7) and (6.9) through (6.12), when compared to the measured H field attenuation given in Ref. 1, results in a reasonable correlation. Note that in the prediction, the distance between the transmitting loop and the shield was the same distance as used between the transmitting and receiving loops in obtaining the measured data.

The accuracy of the prediction is adequate for all but the most exact analysis. A more rigorous approach, but one mathematically more complex and requiring a computer, is to use the exact solution of the vector wave equation described in Ref. 1.

Shielding against a low-frequency (i.e., less than 1 kHz) magnetic field is often quoted as possible only by the use of a material exhibiting high permeability. It is true that the absorption achieved in material located close to the source is the primary loss mechanism and that appreciable attenuation is possible only by the use of high-permeability material. However, the total shielding effectiveness (absorption and reflection) of a 30-mil-thick copper sheet ($\mu_r = 1$, $G_r = 1$) is higher at frequencies below 3 kHz than a 30-mil silicon steel sheet [$\mu_r = 1500$, $G_r = 0.038$ ($\mu_r \times G_r = 57$)] at a distance of 38 cm from a magnetic field source.

The reason is that the reflection from the copper sheet, at 38-cm distance, is higher than that from the steel sheet. A thin, 20–50-micron (20 microns = 0.0008 inches = 0.02 mm, 50 microns = 0.002 inches = 0.05 mm) sheet with a relative permeability of 200 and a plating of either copper, zinc, nickel, tin, or chrome is available from the Iron-Shield Company, Barrington, IL 60010-0751. This material combines the advantages of high surface conductivity and moderate permeability and may be laminated to various nonconductive materials. The shielding effectiveness is shown in Figure 6.8a.

**Table 6.2** Predicted and Measured Shielding Effectiveness

| $f$ | $d$[cm] | Shielding effectiveness [dB] | |
| --- | --- | --- | --- |
| | | Predicted | Measured |
| Aluminum $^1/_{16}$-inch thick $G_r = 0.35$, $\mu_r = 1$ | | | |
| 100 Hz | 10 | 7.5 | 5 |
| 20 kHz | 10 | 50 | 42 |
| 2 kHz | 20 | 34 | 25 |
| 200 Hz | 35 | 21 | 12 |
| Aluminum $^1/_8$-inch thick, $G_r = 0.35$, $\mu_r = 1$ | | | |
| 1 kHz | 3 | 18 | 15 |
| 30 kHz | 3 | 66.9 | 65 |
| 100 Hz | 30 | 19 | 10 |
| 1 kHz | 30 | 36.6 | 28 |
| Copper $^1/_8$-inch thick, $G_r = 1$, $\mu_r = 1$ | | | |
| 100 Hz | 3 | 9.2 | 8 |
| 30 kHz | 3 | 101 | 98 |
| 100 Hz | 20 | 22 | 17 |
| 1 kHz | 20 | 43 | 38 |
| Steel $^1/_{16}$-inch thick, $G_r = 0.17$, $\mu_r = 112$ | | | |
| 100 Hz | 3 | 12 | 16 |
| 10 kHz | 3 | 91 | 102 |
| 1 kHz | 20 | 33 | 32 |

The reflection loss to a magnetic, electric, field or plane wave is dependent on the surface conductivity. When aluminum is left untreated, aluminum oxide builds up on the surface and the conductivity and reflection loss are reduced. Alodine, Iridite, and Oakite are high-conductivity passivation finishes and do not appreciably reduce the reflection loss, whereas anodizing results in a finish with very low conductivity. Figure 6.8b illustrates the degradation in reflection loss due to these types of finish. The reference 0-dB line represents clean, unpassivated aluminum, the reflection loss of which varies with frequency but is approximately 80-dB. The reflection loss degradation is specified for an E field source at low frequency and a plane wave at high frequency.

Coating a metal surface with an insulating finish such as the majority of lacquers and varnishes does not have an adverse effect, whereas certain types of paint containing carbon, graphite or conductive particles will reduce the reflection loss.

High-permeability sheet and stock designed to shield against DC to 60 Hz magnetic fields, typically for use around transformers and CRT tubes, is available from Eagle Magnetics Co. Ltd., P.O. Box 24283, Indianapolis, IN 46224, Magnetic Shield Corporation, 740 North Thomas Drive, Berrenville, Illinois 60106-1643 and Mμ Shield, P.O. Box 439, Goffstown, New Hampshire 03045. Such materials exhibit relative permeabilities from 20,000 to 300,000 and resistivities of 12–60 μΩ/cm cube.

Table 6.3 shows the shielding effectiveness and permeability versus incident magnetic field at DC and 60 Hz of high-permeability material. The equations are useful in calculating the shielding effectiveness, and an example is shown for a 3″ × 3″ × 3″ five-sided box around a source of magnetic field. Table 6.3 and the equations are reproduced by courtesy of Eagle Magnetic Company. Material with a permeability higher than 1000 tends to be sensitive to frequency and magnetic flux density within the material. Saturation of the material occurs at high flux densities, and the permeability decreases with increasing frequency. When shielding

**ATTENUATION OF AC MAGNETIC FIELD**

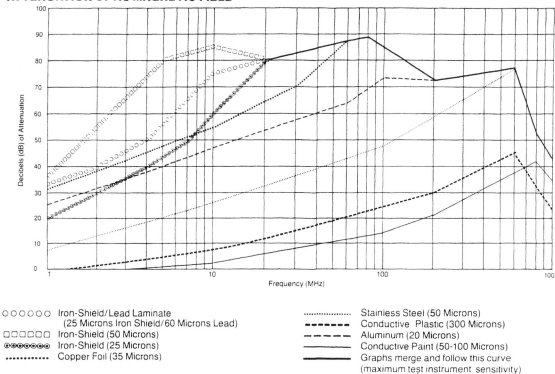

○○○○○○ Iron-Shield/Lead Laminate
                 (25 Microns Iron Shield/60 Microns Lead)
□□□□□□ Iron-Shield (50 Microns)
⊚⊚⊚⊚⊚⊚ Iron-Shield (25 Microns)
•••••••••••• Copper Foil (35 Microns)

················ Stainless Steel (50 Microns)
▬ ▬ ▬ ▬ ▬ Conductive Plastic (300 Microns)
▬ ▬ ▬ ▬ Aluminum (20 Microns)
▬▬▬▬▬ Conductive Paint (50-100 Microns)
━━━━━ Graphs merge and follow this curve
                 (maximum test instrument sensitivity)

**Figure 6.8a**  Examples of electric and magnetic field attenuation for various types of shielding. (Reproduced courtesy of Iron-Shield Company.)

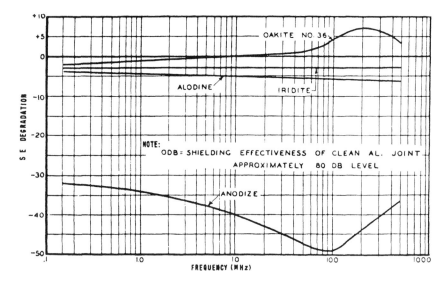

**Figure 6.8b**  Reflection loss degradation due to surface finishes on aluminum.

**Table 6.3** Shielding Effectiveness and Permeability versus External Field Strength, Frequency, and Thickness

EAGLE AAA .025″ (Thickness)  EAGLE AAA .025″ (Thickness)

| $H_1$ (Oerstead) | B (Gauss) | DC | | | | 60 Hertz | | | |
|---|---|---|---|---|---|---|---|---|---|
| | | μ(EFF) | G(AttRat) | SE($D_b$) | $H_o$ | μ(EFF) | G(AttRat) | SE($D_b$) | $H_o$ |
| 1 | 150 | 90,000 | 375 | 52 | .0028 | 30,000 | 125 | 41 | .008 |
| 5 | 750 | 140,000 | 565 | 55 | .008 | 40,000 | 167 | 44 | .030 |
| 10 | 1500 | 200,000 | 835 | 58 | .012 | 45,000 | 188 | 46 | .053 |
| 25 | 3750 | 300,000 | 1250 | 63 | .020 | 30,000 | 125 | 41 | .200 |
| 50 | 7500 | TOO CLOSE TO SATURATION | | | | TOO CLOSE TO SATURATION | | | |

EAGLE AAA .031″ (Thickness)  EAGLE AAA .031″ (Thickness)

| $H_1$ (Oerstead) | B (Gauss) | DC | | | | 60 Hertz | | | |
|---|---|---|---|---|---|---|---|---|---|
| | | μ(EFF) | G(AttRat) | SE($D_b$) | $H_o$ | μ(EFF) | G(AttRat) | SE($D_b$) | $H_o$ |
| 1 | 121 | 90,000 | 465 | 53 | .0021 | 25,000 | 127 | 42 | .0079 |
| 5 | 605 | 140,000 | 710 | 57 | .0071 | 34,000 | 173 | 45 | .028 |
| 10 | 1210 | 200,000 | 1020 | 60 | .0099 | 36,000 | 183 | 46 | .055 |
| 25 | 3025 | 300,000 | 1520 | 62 | .0165 | 26,000 | 132 | 43 | .190 |
| 50 | 7500 | TOO CLOSE TO SATURATION | | | | TOO CLOSE TO SATURATION | | | |

Equation #1 $B = \dfrac{2.5\,D\,H_1}{2T} = \dfrac{2.5\,(3)(1)}{2(.025)} = 150$ Gauss

Equation #2 $G = \dfrac{\mu T}{2L} = \dfrac{90,000(.025)}{2(3)} = 375$

Equation #3 $SE = 20\ \text{Log } G = 20\ \text{Log }(375) = 52\ dB_0$

Equation #4 $H_o = \dfrac{H_1}{G} = \dfrac{1}{375} = .0028$ Oerstead

$H_1$ = Internal Field (Oerstead)
$B$ = Flux That Shield Sees
$\mu$ = Effective Permeability
$C$ = Attentuation Ratio
$SE$ = Shielding Effectiveness ($D_b$)
$H_o$ = External Field (Oerstead)
$T$ = Thickness (Inches)

against low-frequency high magnetic fields, the use of a material such as steel followed by a high-permeability material is recommended. The first barrier of steel is designed to absorb the magnetic field, without saturation, to a level at which the second high-permeability barrier does not saturate. To illustrate the importance of choosing the correct shielding material for the job, we offer the following example: A shielded room was constructed, at great expense, of mumetal, with the aim of achieving a high level of shielding. Unfortunately, because the permeability of mumetal decreases rapidly with increasing frequency and the material exhibits a low conductivity, the shielding effectiveness of the room at frequencies in the megahertz region was less than that achieved with a much cheaper room constructed of cold-rolled steel.

## 6.3 NEW SHIELDING MATERIALS: CONDUCTIVE PAINTS AND THERMOPLASTICS, PLASTIC COATINGS, AND GLUE

With an increase in the use of nonconductive materials such as plastics and resins for the enclosures of computers, peripheral equipment, and the like, increased use is also made of conductive coatings and conductive moldable materials for the purpose of shielding. Conductive thermoplastics are available with a base material of nylon, polypropylene, polyethylene, thermoplastic rubber, SMA, ABS, polycarbonate, PES, polystyrene, acetal, etc. The filler may be carbon fibers, carbon black, metallized glass, nickel-coated carbon fibers, or stainless steel fibers. The higher the ratio of filler to base material, the higher the conductivity, up to a limit, but the greater the change in the physical properties of the base material. The most common change is that the material becomes brittle and will break more readily when dropped. The conductivity of conductive thermoplastics is typically lower than that of conductive coatings, with the use of carbon-filled plastic confined to electrostatic protection. The surface resistivity of conductive thermoplastics ranges from 10 $\Omega$/sq to 1 M$\Omega$/sq, and the volume resistivity from 1 $\Omega$/cm cube to 1 M$\Omega$/cm cube. One manufacturer of conductive plastics is Electrafil, Stoney Creek, Ontario L8E 2L9, Canada. Conductive coatings may be painted on or applied by flame spray, thermospray, or plasma flame spray. The surface resistance of a nickel-based paint versus thickness is shown in Figure 6.9, from which we see that the conductivity increases with increasing thickness in the thickness range of 1.5–3 mm. The manufacturer of the paint, along with other conductive coatings, is Achesons Colloids Company, Post Huron, MI 48060.

The shielding effectiveness of a material is often measured using the MIL-STD-285 test method, in which a transmitting antenna is located within a small, shielded enclosure that is in turn placed inside a shielded room. One face of the small, shielded enclosure is removed and covered by a panel of the material under test. A receiving antenna is placed inside the shielded room at a distance of 1 m from the panel under test. The *shielding effectiveness* is defined as the ratio of the measured field without the panel to the measured field with the panel closing the aperture. When a loop antenna is used as the transmitting and receiving antennas, the magnetic field shielding effectiveness is measured, whereas when monopole, dipole, or broadband antennas are used the measurement is of the E field or plane-wave shielding effectiveness, dependent on the frequency.

A shielding material that exhibits a high surface resistance (low conductivity) has a poor shielding effectiveness, often 0 dB, against a low-impedance H field and a very much higher shielding effectiveness against high-impedance E fields. In Ref. 2, tests are described on alumi-

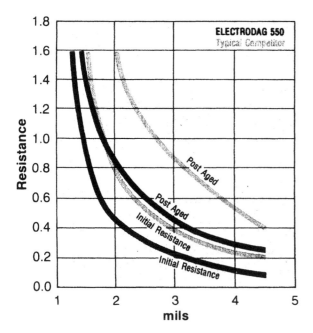

**Figure 6.9** Resistance vs. film thickness. (Reproduced courtesy of Acheson Colloids Company.)

nized Mylar with a 1.4-Ω/sq surface resistance and a thickness of approximately $2 \times 10^{-8}$ m (5 microns), which is 0.008 skin depths at 10 MHz. The size of the sample was 59 cm × 59 cm in a 2.44-m × 3-m × 2.44-m-high enclosure. Both E and H field shielding effectiveness was tested. The H field shielding effectiveness was measured with both an E field source (monopole antenna) and an H field source (loop antenna). With either source, the H field shielding effectiveness was 0 dB from 50 kHz to 10 MHz. The E field shielding effectiveness using the dipole as the source was 90 dB at 100 kHz, 72 dB at 500 kHz, 65 dB at 1 MHz, 51 dB at 5 MHz, and 43 dB at 10 MHz, as shown in upcoming Figure 6.11a. The E field shielding effectiveness using the loop antenna as the source was 0 dB from 100 kHz to 400 kHz, 9 dB at 1 MHz, and 28 dB at 10 MHz, as shown in upcoming Figure 6.11b.

Manufacturers of conductive coatings, paint, and conductively loaded (composite) materials often publish a predicted or measured shielding effectiveness without mentioning that the data is valid only for an E field. Figure 6.10 shows the predicted shielding effectiveness of a thin transparent conductive coating with a surface resistance of 5 and 20 Ω/sq, which is typically used as an EMI window in front of a display or CRT. The theoretical values of shielding effectiveness shown in Figure 6.10 for the lower-resistance 5-Ω/sq coating is much higher than seen in the measured E field shielding data for the 1.4-Ω/sq aluminized Mylar material shown in Figures 6.11a and 6.11b. The theoretical values in Figure 6.10 ignore the effect of the termination of the material at the surrounding metal, which is included in the measured values, and assumes a high-impedance E field.

Reference 6 describes a test on a small enclosure that contained a battery-powered 10-MHz oscillator driving two TTL gates. This oscillator noise source produced harmonics up to 200 MHz and beyond. Twelve different composite and plated plastic enclosures were manufactured and then tested with the noise source inside the enclosure and the E field receiving antenna

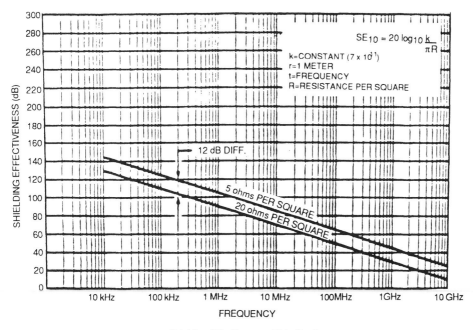

Shielding Effectiveness, Thin Coatings
(t << 2.5 MICRONS)

**Figure 6.10**   Theoretical shielding effectiveness of a thin (≪ 2.5 microns) coating.

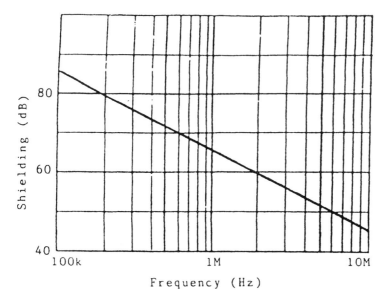

**Figure 6.11a** E field shielding effectiveness of 1.4-ohm aluminized Mylar using an electric dipole source. (© IEEE, 1992.)

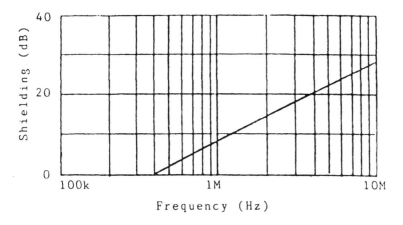

**Figure 6.11b** E field shielding effectiveness of 1.4-ohm aluminized Mylar using a loop antenna source. (© IEEE 1992.)

located 3 m from the enclosure. The measurements were comparative with the original polycarbonate nonconductive enclosure used as the control. Table 6.4, from Ref. 6, may be used to compare the different materials and coatings.

As so often happens in the application of an EMC improvement or an EMI fix, implementing an apparent improvement sometimes results in a worsening of the situation. Thus, applying a conductive coating to an enclosure may increase the level of radiated emissions. Because the field is to some extent contained by the enclosure, the common-mode current across the PCB increases. The common mode current on any unshielded cable connected to the PCB will also increase, resulting in increased radiation, especially when the cable length is resonant. Apart from the probability that emissions from unshielded cables connected to the enclosure will increase when the equipment is placed in a conductive enclosure, the radiation from the enclosure

**Table 6.4** Material Evaluation

| Sample | Material | Shielding effectiveness (70–300 MHz) [dB] | Relative cost |
|--------|----------|----------------------------------------|---------------|
| 1 | Polycarbonate glass filled | 0 (control) | 1.0 |
| 2 | Polycarbonate with copper paint | 25–30 | 2.25 |
| 3 | Polycarbonate with 10% SS/10% glass | 20–30 | 5.0 |
| 4 | Polycarbonate with 10% NiGraphite | 20–30 | 4.1 |
| 5 | Polyvinylchloride with 30% NiC fibers | 10–20 | 3.9 |
| 6 | Polycarbonate with 7% SS fibers | 25–35 | 4.3 |
| 7 | Polycarbonate with vacuum-deposited aluminum | 25–35 | 2.5 |
| 8 | Mica | 19–25 | 4.6 |
| 9 | Polycarbonate with 7% SS fibers | 21–31 | 3.1 |
| 10 | Polycarbonate with Ni paint | 19–27 | 2.25 |
| 11 | Polycarbonate with 10% SS fibers | 20–32 | 5.0 |
| 12 | Polycarbonate with special carbon | 12–21 | 2.9 |

SS = stainless steel.
*Source*: Ref. 6.

itself may increase at certain frequencies. As an example, consider a PCB and switching-power supply mounted on a metal base to which more than one power and signal ground connection is made. The base is surrounded by a plastic enclosure that is coated on the inside with graphite. The graphite is electrically connected to the metal base and, with the doors of the enclosure closed, an electrical connection exists between the graphite on the two halves of the doors. The equipment must meet FCC part 15 requirements placed on type A (commercial) equipment.

Measurements were made of the radiated E field emissions from the enclosure with the doors open and closed. The graphite acts as an absorber of radiation from sources located on the PCB; thus, some small level of attenuation may be expected with the doors of the enclosure closed. However, the absorption is offset at certain frequencies by radiation from the enclosure as a result of current flow in the graphite coating. The current flow in the graphite is a result of noise current flow in the metal base of the enclosure, due to multiple ground connections, some of which takes the alternative path through the graphite coating. The measured effect of opening or closing the doors was an increase in radiated emissions of 4.5 dB at 48 MHz and 4.7 dB at 107 MHz with the door open and an increase of 6 dB at 88 MHz with the door closed. The level of radiated emissions in this example were brought below the FCC limit by a combination of modifications. The radiation from unshielded cables connected to the equipment were reduced by the addition of ferrite baluns (beads) as the cables entered the enclosure. The radiation from both the enclosure and the cables was reduced by a modification in the grounding scheme on the PCB, including a reduction in the number of PCB power ground to metal base connections, and the inclusion of an additional decoupling capacitor. The graphite coating was retained as much for ESD protection as for its shielding effectiveness.

Emerson and Cuming manufacture a range of conductive surface coating materials. Some examples follow.

Eccocoat CC 2 is a silver lacquer with a surface resistivity of 0.1 $\Omega$/sq maximum (0.025-mm coating thickness).

Eccocoat CC 40A is an elastomeric silver-filled conductive coating with 0.001 Ω/sq to 0.05 Ω/sq for an 0.025-mm (1 mil) thickness.

Eccocoat CC 35U is an air-dry waterborne, nickel-filled acrylic coating with a surface resistivity of 1 Ω/sq for a 2-mil (0.05-mm)-thick coat.

Evershield of West Lafayette, IN, also manufactures coatings. These include:

A nickel-filled two-part urethane–based coating at 1.5 Ω/sq (2-mil thickness)
Copper acrylic, 0.3 Ω/sq (1-mil thickness)
Graphite acrylic, 7.5 Ω/sq
Silver PVC, 0.01 Ω/sq (1-mil thickness)
Silver acrylic, 0.04 Ω/sq (1-mil thickness)

The conductivity may be found from surface resistance as follows:

$$G = \frac{1}{R/sq \left[ \dfrac{2.54t}{1000} \right]} \quad \text{[mhos/cm]} \tag{6.13}$$

where $t$ = the thickness of the shield [mils].

Manufacturers' shielding effectiveness curves for these materials are invariably for electric field attenuation, which is the easiest type of field to shield against. MIL-STD-285 type measurements on silver- and nickel-loaded paints have shown that the shielding effectiveness is an inverse function of surface resistivity. Because the silver paint tested had a surface resistivity one-tenth that of the nickel, the measured shielding effectiveness was 20 dB greater. Increasing the thickness of the paint from 2 to 4 mils provided an additional 6 dB of shielding, but 6 mils of nickel produced 1–2 dB less shielding. More information on the test results may be obtained from Chomerics, Woburn, MA.

The shielding effectiveness of a flame-sprayed 22-gauge copper mesh panel compared to solid brass and against a magnetic field, measured in a TEM cell, is shown in Figure 6.12 (from Ref. 7). The surface resistance of flame-sprayed copper is 0.0006–0.001 Ω/sq for an unspecified thickness. The panel under test was mounted on an 0.6-m × 0.6-m × 0.2-m-deep enclosure mounted on the floor of a transverse electromagnetic (TEM) cell. The cell when driven by a signal source causes a current flow on the outside of the test enclosure and panel. A magnetic-field sensor was located on the outside of the enclosure under test and a second cell inside the enclosure. The magnetic-field attenuation of the panel was defined as the ratio of external field to internal field.

Due to the relatively high surface resistance and thinness of most conductive coatings, neither their shielding effectiveness nor their current-carrying capability is as high as that of pure metal. Thus, although useful in meeting FCC, EN55022/11, CS108, or similar commercial radiated emission requirements, they may not be suitable for the more stringent military standard requirements and certainly not for use where EMP or indirect lightning strikes must be protected against. Conductive polyester cloth and paper is available. Figure 6.13 (shielding effectiveness for a number of fabrics) is shown by courtesy of TW Trading International. The up to 26 dB of variation in shielding effectiveness seen in the graph is less a measure of the material characteristics and more a function of the MIL-STD-285 test method. In the test method, the test chamber and shielded room exhibits room resonances, and the reflections in the room cause large variations in the field incident on the test antenna. The material with the lowest shielding effectiveness (T 1500 NS) is a black sheer material with a polyester substrate plated with nickel (14.8% by weight). The material with the highest shielding effectiveness (T 2200 C) is a glossy

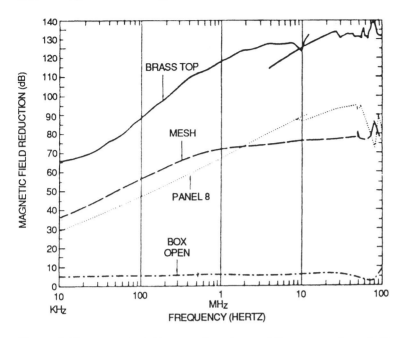

**Figure 6.12**   Magnetic field attenuation of a good flame-sprayed panel, 22-gauge copper mesh and a solid brass cover. (From Ref. 7. © 1985, IEEE).

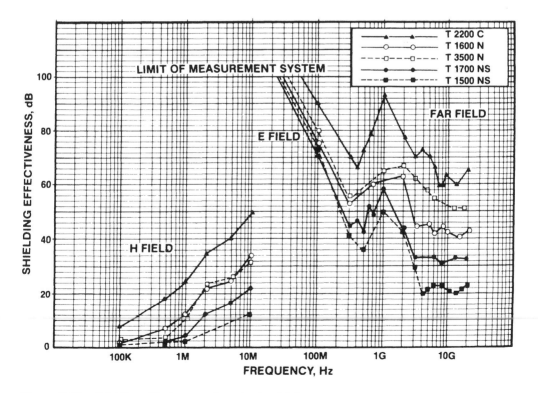

**Figure 6.13**   Shielding effectiveness of conductive fabrics.

material with a polyester substrate plated with copper (30% by weight). EMC wallpaper, louvre blind, carpet, and vinyl material are all available, so the construction of a well-furnished shielded room should be feasible!

Conductive adhesives with silicone or epoxy bases filled with silver, silver-plated copper, or copper particles are manufactured by Emerson and Cumings, Tecknit, Chomerics, etc.

The adhesives require that pressure be applied after both surfaces have been mated. One use for adhesives is therefore between the panels of an enclosure that are bolted together. When a sealant must be applied after a joint is made, a conductive caulking is available that may be applied to seal gaps around an enclosure or applied to the outside of a joint. Conductively loaded lubricants are available that may be used between moving surfaces, such as contacts, a rotating shaft, or on a piano hinge, to ensure a low-resistance connection.

In some EMI cases, especially at very high frequencies, shielding by reflection from a very conductive surface is not the best solution, because the incident field is simply reflected somewhere else. The alternative is the use of an absorbing material that converts the incident electromagnetic field into thermal energy inside the material. Microwave-absorber material is available in flexible foam, rigid sheet and stock, castable resin, cloth, plastic sheet, and ferrite tiles.

The only material with appreciable attenuation in the frequency range of 50 MHz to 50 GHz is the ferrite tile. The attenuation achieved by other materials is predominantly in the 3–20-GHz frequency range. In one instance of radiation from slots in a shield on a PCB, the source of emissions was a 2.4-GHz high-current driver. High-level emissions were seen at 2.4 GHz, 4.8 GHz, 7.2 GHz, and 9.6 GHz. Adding a number of magnetic and dielectric lossy material at the apertures or across the PCB did not measurably reduce emissions. However, adding low-frequency ferrite slabs at the apertures reduced emissions at all frequencies by typically 20 dB. In another instance, where surface currents at 10 GHz flowing on the control cable to a digital attenuator resulted in EMI, the placing of rubber-loaded microwave absorber over the cable and attenuator housing resulted in a significant reduction in the induced EMI level. From measurements it has been seen that at gigahertz frequencies, waves are launched both vertically and horizontally from a source, which may be one reason that laying absorber on the surface of a PCB may not be effective.

One application of a ferrite tile is to place it behind a wire mesh or honeycomb EMI air vent filter, with a sufficient gap to allow free air flow. At frequencies above that at which the air vent filter becomes an ineffective shield, typically 1 GHz, the ferrite tile with effectively absorb any radiation through the filter. Microwave absorbers are available from:

Advanced Absorber Products (Keene) Amesbury, MA 01913
Electronautics, Littleton, MA 01460
Plessey, San Diego, CA 92123-9214
Emerson and Cuming, Canton, MA 02021

## 6.4 SEAMS, JOINTS, VENTILATION, AND OTHER APERTURES

Inadequate shielding is due rarely to the properties of a conductive material but typically to the presence of seams and apertures. Thus when measuring radiation from a ''quick fix'' shield constructed of aluminum foil joined by conductive adhesive copper tape, the predominant source of any radiation is not from the foil but at the copper tape seam. Where an aperture exists in an enclosure manufactured from a conductive material, the shielding effectiveness of the enclosure is adversely affected.

Apertures are either deliberately included in an enclosure for the sake of ventilation, for the ingress of cables or for viewing displays or are unintentional, such as gaps in seams or joints

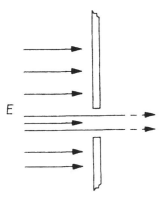

**Figure 6.14** E field coupling through an aperture.

around enclosure panels, doors, and connectors. The proximity of the source of radiation to the aperture and whether the source is an E field or H field determine the prediction technique used in assessing the attenuation of the enclosure with apertures and the steps required to improve the shielding effectiveness. When the source is physically small inside a large enclosure, such as a cabinet, the coupling through the aperture may be computed from the E field component of the wave generated by the source. The coupling of an E field through an aperture is illustrated in Figure 6.14. When the enclosure is small, the impedance of the inside of the enclosure is relatively low, and current flow on the inside of the enclosure may result in H field coupling through the aperture, as illustrated in Figure 6.15. The current flow on the inside of the enclosure may result from multiple connections of power or signal ground to the enclosure into which noise currents are directly injected, or via magnetic field induction via PCB tracks or unshielded cables in close proximity to the enclosure. When apertures exist, the level of current flow on the enclosure interior surface should be minimized by use of a single-point enclosure ground, the use of ground planes on PCBs (which are connected to signal or power ground and thus reduce coupling to the enclosure), and the use of shielded cables within the enclosure.

### 6.4.1 E Field Coupling Through a Thin Material

When the area of the aperture is much less than $\lambda/\pi$, the diameter of the aperture is much greater than the thickness of the material, and the enclosure, if it exists, is very large, the shielding effectiveness of the aperture may be found from the following equations. The dimensions and

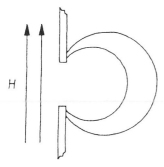

**Figure 6.15** H field coupling through an aperture.

distance from the source to the shield, $D_1$, and from the shield to the measuring point $d$ are shown in Figure 6.16. The source of radiation is considered to be a small loop, the E field from which reduces as a function of $1/r^2$ until a distance of $\lambda/2\pi$, after which the reduction is $1/r$ (where $r$ is the distance from the loop to the measuring point).

The *total shielding effectiveness* is defined as the ratio of the E field without shield and aperture to the E field with shield and aperture. When the source-to-aperture distance $D_1$ is much less than the aperture-to-measuring-point distance, $D$, point $d$ is in the far field, and if the aperture area is much less than $\lambda/\pi$, then the shielding effectiveness is given, from Ref. 8, by

$$SE = \frac{E \text{ at } d \text{ without shield and aperture}}{E \text{ at } d \text{ with shield and aperture}} = \frac{\lambda D_1}{A} \qquad (6.14)$$

where

$\quad A =$ aperture area [m$^2$]

$\quad d =$ measuring point

Assuming location $d$ is in the far field, the electric field is given by

$$E = E_o \frac{A}{D\lambda} \qquad (6.15)$$

where $E_o$ is the field incident on the aperture.

When $E_o$ is not known, the E field at location $d$ may be computed as the electric field from a loop using Eq. (2.26). The distance from the source is given by $r = D_1 + D$. Equation (2.26) is valid when "$r$" is in the near or far field. The field "$E$" at location "$d$," with shield and aperture, is computed from the field without shield and aperture reduced by the SE given by Eq. (6.14).

Equation (6.14) is valid for an electric field source contained within a small enclosure that contains an aperture, when measuring point $d$ is in the far field. E-field-radiated measurements were made in a shielded room from an E field source with and without a shield that contained an aperture. A solid-wall copper conduit located inside the shielded room was con-

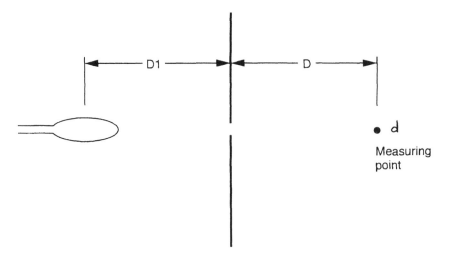

**Figure 6.16**   Source-to-aperture and aperture-to-measuring-point dimensions.

nected to an adapter mounted on the shielded-room wall. The conduit was used to shield a coaxial cable connected outside of the shielded room to a signal generator. The end of the conduit inside the shielded room was soldered to an 8-cm-diameter ground plane, to which the shield of the coaxial cable was also terminated. The center conductor of the coaxial cable was connected to a 5-cm-long rod which, with the circular ground plane, formed a monopole antenna. Measurements were made of the far field, inside the shielded room, radiated by the monopole antenna from 100 MHz to 4 GHz. A copper tube 7.5 cm long with a 7.5-cm-diameter and a 2.54-cm-diameter aperture was constructed. The tube had a lid soldered on at the one end and open, with beryllium copper finger stock soldered around the periphery, at the other end. The finger stock was connected under pressure to the monopole ground plane, so the monopole rod was effectively shielded by the can, except for the 2.54-cm aperture. The far-field measurements were repeated with the tube in place, from 100 MHz to 4 GHz. The measured shielding effectiveness showed the variation expected from measurements made in an undamped room; however, the average measured shielding effectiveness was within 5 dB of the predicted at 100 MHz, using Eq. (6.14), 15dB at 400 MHz, and 5 dB at 2 GHz. At 3 GHz and 3.6 GHz, negative shielding effectiveness was seen as a result of enclosure resonance.

When ''$d$'' is in the near field, i.e.,

$$d < \frac{\lambda}{2\pi} \tag{6.16}$$

$$E = \frac{E_o A}{D^2 \lambda} \tag{6.17}$$

The foregoing equations apply to rectangular apertures but are applicable to circular apertures, with an accuracy of 1 dB, when the apertures are small relative to a wavelength.

The effect of multiple apertures is dependent on the distance between the apertures and the measuring distance. When the distance $D$ to the measuring point $d$ from the apertures is larger than the area of the apertures and the source-to-aperture distance $D_1$ is much less than $D$, the radiation is considered to add coherently and the shielding effectiveness is

$$\text{SE} = \frac{\lambda D_1}{NA} \tag{6.18}$$

where $N$ is the number of apertures.

In close proximity, in contrast, the radiation from the apertures may be considered to add noncoherently and the shielding effectiveness is approximately

$$\text{SE} = \frac{\lambda D_1}{0.316NA} \tag{6.19}$$

Shielding effectiveness measurements were made with an E field source contained in a 0.39-m × 0.31-m enclosure with a depth of 0.185 m. The face of the enclosure contained 225 × 4-mm-diameter apertures. The measured shielding effectiveness was 9 dB higher than predicted, by Eq. (6.19), at 100 MHz, 10 dB higher at 300 MHz, and 0 dB at 400 MHz. The shielding effectiveness is inversely proportional to the product of the number of apertures and the aperture area. Thus separating the aperture with a very fine conductor is not very effective in increasing E field shielding. For example, the 4-mm-diameter apertures were covered in fine wire mesh with 0.7-mm apertures. The measured increase in shielding effectiveness was from 2 to 7 dB, which is close to that predicted using Eq. (6.19).

Due to reciprocity, Eq. (6.14) is also applicable to the case where a plane wave is incident on a small enclosure with an aperture, in this case the levels of the plane-wave H field component shielding and E-field component shielding effectiveness are generally different. One approach, which is described in Ref. 9, is to consider the enclosure as a waveguide and to assume a single $TE_{10}$ mode of propagation. Analysis both above and below the cutoff frequency for this mode and both E field and H field shielding are calculated as functions of frequency, aperture dimension, multiple apertures, enclosure dimension, wall thickness, and position within the enclosure. In addition, the damping effect and change in resonant frequency as a result of adding electronic assemblies and PCBs inside the enclosure are examined.

A rectangular aperture in an empty rectangular enclosure is represented by an impedance. The radiating source is represented by a voltage $V_o$ with an impedance $Z_o \approx 377\ \Omega$. The enclosure is represented by a shorted waveguide whose characteristic impedance and propagation constant are $Z_g$ and $K_g$, respectively. Figure 6.17 shows the dimensions of the enclosure and the equivalent circuit.

To validate the approach, sensors were placed within a range of enclosures, with apertures cut into one wall of the enclosure. A shielded loop antenna was used to measure the H field and a monopole antenna to measure the E field.

The enclosure was located inside a shielded room, and a stripline antenna, which radiates a field with an impedance of $\approx 377\ \Omega$ over the complete operating frequency range, a log periodic

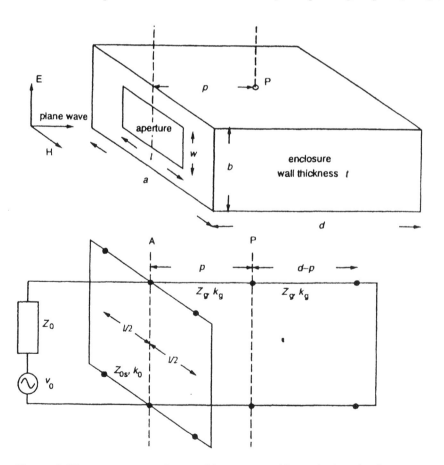

**Figure 6.17**  Rectangular enclosure with aperture and its equivalent circuit.

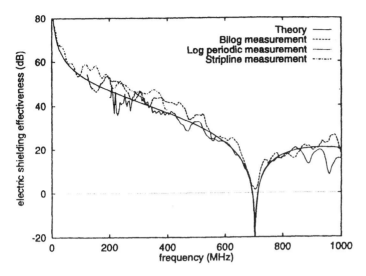

**Figure 6.18**   Calculated and measured $S_E$ at the center of a $300 \times 120 \times 300$-mm enclosure with $100 \times 5$-mm aperture. (© IEEE, 1998.)

antenna, or Bilog antenna, was used to generate the incident field. Figure 6.18 shows the calculated and measured electric field shielding effectiveness, $S_E$, at the center of a $300 \times 120 \times 300$-mm enclosure with a $100$-mm $\times$ 5-mm aperture. The correlation between theory and measurement is very good, with variation in the measured $S_E$ with the use of different source antennas and almost certainly because the source is located inside a poorly damped shielded room. Using Eq. (6.14) over the 50–500-MHz frequency range, the predicted shielding effectiveness is compared to the data, contained in Figures 6.18 and 6.19; this is shown in Table 6.5. Above 400 MHz, the shielding effectiveness, using Eq. (6.14), is higher than shown in the figures. This is because the enclosure is approaching resonance above 400 MHz.

The shielding effectiveness reduces with increasing frequency and, since the E field level is higher closer to the aperture, the shielding effectiveness increases with distance from the aperture. Therefore, locate susceptible circuits on PCBs as far away from the aperture as possible. The 300-mm $\times$ 120-mm $\times$ 300-mm enclosure resonates at approximately 700 MHz, and the shielding effectiveness is negative, i.e., field enhancement occurs. Figure 6.19 shows the same enclosure but with the aperture increased in size to 200 mm $\times$ 30 mm. The resonance and negative shielding can clearly be seen in the measured results. Figure 6.20 shows the $S_E$ for a 222-mm $\times$ 55-mm $\times$146-mm enclosure with $100 \times 5$-mm aperture. Adding PCBs to an enclosure both changes the resonant frequency and damps the enclosure, as shown in Figure 6.21. However, the PCBs must be mounted close to the inside walls of the enclosure to be effective. If the PCBs are attached to alternative walls of an enclosure, the effective path length increases and the resonant frequency decreases! Try to use the PCBs to divide the enclosure into a number of smaller compartments. The 300-mm $\times$ 300-mm $\times$ 120-mm enclosure was measured with one, two, and three $160 \times 4$-mm apertures. For these measurements, the box with a single aperture was used as the calibration standard, which helped iron out the shielded-room resonances in the shielding effectiveness curve. Increasing the number of apertures was found to reduce the shielding effectiveness. Table 6.6 shows the calculated reduction in $S_E$ at 400 MHz compared to measurements over the 200–600-MHz frequency range.

**Table 6.5** Comparison Between the Simple Eq. (6.14) and the Relatively Simple Approach Used in Ref. 9

| $f$ (MHz) | Shielding effectiveness (dB) Ref. 9 | Shielding effectiveness (dB) Eq. (6.14) | $\Delta$(dB) |
|---|---|---|---|
| Figure 7, 300 × 120 × 300-mm enclosure with 100-mm × 5-mm aperture | | | |
| 50 | 60 | 65 | 5 |
| 100 | 55 | 59 | 4 |
| 200 | 49 | 53 | 4 |
| 400 | 38 | 47 | 9 |
| 500 | 32 | 45 | 13 |
| Figure 8, 300-mm × 120-mm × 300-mm enclosure with 200-mm × 30-mm aperture | | | |
| 50 | 47 | 43.5 | 3.4 |
| 100 | 40 | 37.5 | 2.5 |
| 200 | 30 | 31.4 | 1.4 |
| 400 | 20 | 25.5 | 5.5 |
| 500 | 12 | 23 | 11 |
| Figure 10, 483-mm × 120-mm × 483-mm enclosure with 100-mm × 5-mm aperture | | | |
| 50 | 60 | 47.6 | 12 |
| 100 | 56 | 63.1 | 7.2 |
| 200 | 50 | 57 | 7 |
| 300 | 42 | 53.6 | 11.6 |

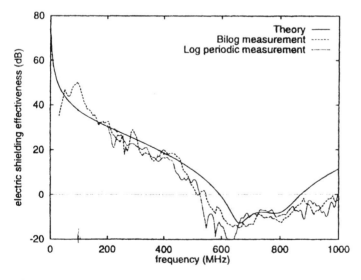

**Figure 6.19** Calculated and measured $S_E$ at the center of a 300 × 120 × 300-mm enclosure with 200 × 30-mm aperture. (© IEEE, 1998.)

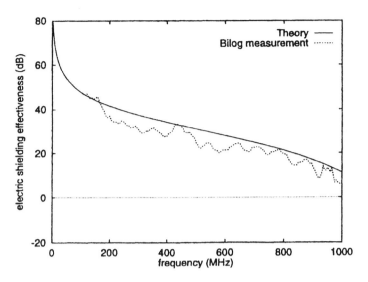

**Figure 6.20** Calculated and measured $S_E$ at the center of a 222 × 55 × 146-mm enclosure with 100 × 5-mm aperture. (© IEEE, 1998.)

The analytical solution predicts that both $S_E$ and the magnetic field shielding effectiveness $S_H$ are increased by dividing an aperture up into smaller apertures; thus the total area is kept the same and the number of apertures increases. Figure 6.22 shows the measured $S_E$ at the center of the 300 × 300 × 120-mm enclosure, with the area divided up into one, two, four, and nine apertures. In each case the total area was 6000 mm². As predicted, having more, but smaller, apertures improves the shielding. The practical application of the analysis and measurements are discussed in Ref. 9 and are, for the E field component:

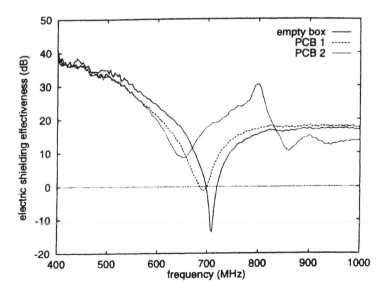

**Figure 6.21** Effect of adding PCBs to an enclosure. (© IEEE, 1998.)

**Table 6.6** Reduction in $S_E$ Due to Number
of Apertures, Compared to one, with Each
Aperture Size Kept Constant

| Number of apertures | $S_E$(dB) | |
| | Theory (400 MHz) | Measured (200–600 MHz) |
| --- | --- | --- |
| 1 | 0 | 0 |
| 2 | 5.6 | 3.7–4.4 |
| 3 | 8.8 | 6.6–7.7 |

Long, thin apertures are worse than round or square apertures, because, for a typical-size
enclosure, the theory predicts that doubling the length of the slot reduces $S_E$ by 12
dB while doubling the width reduces $S_E$ by only about 2 dB.

Doubling the number of apertures reduces $S_E$ by about 6 dB.

Dividing a slot into two shorter slots increases $S_E$ by about 6 dB.

At subresonant frequencies, doubling the enclosure dimensions while keeping the aperture
size and number constant is predicted to increase $S_E$ by about 6 dB, and larger-size
enclosures are recommended.

Doubling both the enclosure size and aperture is predicted to reduce $S_E$ by about 6 dB.

Doubling the enclosure size halves the resonant frequency, and, to avoid negative shielding
effectiveness, a smaller enclosure, or one that is full of electronic components, is
preferable.

Reference 9 uses a relatively simple approach to the problem of the shielding effectiveness
of rectangular enclosures with rectangular apertures. Reference 23 describes a rigorous full-
wave combined electric and magnetic field boundary simulator using a method of moment electrical field integral equation (MOM EFIE). This technique also shows a very good correlation
with measurements made in both a full anechoic room with ferrite tiles and a semianechoic
chamber in which absorbing cones are used everywhere except for the conducting floor. Despite
the good correlation between predicted and measured results, the shielding effectivenesses de-

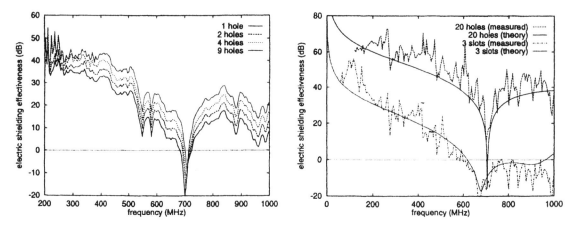

**Figure 6.22** Measured $S_E$ at the center of a 300-mm $\times$ 300-mm $\times$ 120-mm enclosure with one, two,
four, and nine apertures, with the total area held constant, i.e., smaller apertures.

**Table 6.7** Comparison of Shielding Effectiveness of a 500-mm × 500-mm × 500-mm Enclosure with a 50-mm × 200-mm Aperture

| $f$ (MHz) | MOM EFIE (Ref. 23) (dB) | Waveguide (Ref. 9) (dB) | Eq. (6.14) (dB) |
|---|---|---|---|
| 50 | 24 | 17 | 43 |
| 100 | 24 | 17 | 37.5 |
| 200 | 26 | 18 | 31.5 |
| 300 | 28 | 23 | 28 |

scribed in Ref. 23 are approximately 5–8 dB higher than obtained by the simple technique described in Ref. 9. The enclosure used for the comparison had the dimensions 50-mm × 500-mm × 500-mm with a 50-mm × 200-mm aperture. Table 6.7 compares the two techniques and Eq. (6.14) for this relatively large enclosure.

For E field shielding, the thickness of the material into which the aperture is placed does affect shielding effectiveness, unlike H field shielding. In practice it is not typically convenient to use thick material, and so the same effect can be achieved by bending a thin material over to form a flange at apertures or at seams.

In Ref. 24, a finite difference time domain (FDTD) analysis of shielding effectiveness is presented using a wire current source placed close to apertures. With a single 25-mm circular aperture and by increasing the thickness of the material from 1 mm to 12 mm, the analysis predicted an increase of 24 dB in shielding effectiveness. This is approximately 12 dB higher than predicted using the waveguide below cutoff effect shown in Eq. (6.21).

When the largest aperture dimension is greater than $1/4$ wavelength, the aperture may function as a horn or slot antenna and radiate efficiently. For example, a large 6-foot by 6-foot enclosure contained back panels connected to the frame of the enclosure by screws at 6-inch intervals. Both the frame and the panel were painted, and the electrical connection was tenuous, made by some of the screw heads that bit through the paint. The E field measured at 1-m distance, with the antenna pointing at the slot between the screws, was higher than when the panels on all four sides of the enclosure were removed. Thus the enclosure and slots functioned as an antenna with gain.

The shielding effectiveness of an aperture is a function not so much of the width of the gap at high frequencies but of the length of the gap and its depth, depending on the polarization of the wave or the direction of current flow across the gap. Because we are seldom able to determine either the wave polarization or the direction of current flow, we use the largest transverse dimension when estimating worst-case aperture attenuation. A metal enclosure with a removable lid is not an EMI-tight box unless the surfaces are optically flat and no distortion occurs between the lid and the side flanges. In practice, gaps occur between the lid and the side flanges that often are as much as one-half the distance between the fasteners when they are as close as 2 cm apart, and more than half the distance between fasteners when they are much more than 2 cm apart.

## 6.4.2 Waveguide Below Cutoff

The preceding section examined the shielding effectiveness of an aperture in which the diameter was much greater than the thickness of the material. This section examines the case in which the thickness of the material approaches or is greater than the gap dimensions.

In a waveguide, the dimensions are chosen such that the signal frequency is above the cutoff frequency and the attenuation in the waveguide is negligible. When the frequency of a plane wave incident on a waveguide is below the cutoff frequency of the waveguide, the plane wave is attenuated as it passes down the waveguide. It is just this effect that is used to describe the additional attenuation achieved in an aperture in a thick material.

The additional attenuation of a gap for a plane wave and an E field is given approximately by

$$A = 0.0018fl \sqrt{\left(\frac{f_c}{f}\right)^2 - 1} \quad [\text{dB}] \tag{6.20}$$

*or*

$$A = 0.0018fl \left(\frac{f_c}{f}\right) \quad [\text{dB}] \qquad \text{when} \frac{f_c}{f} > 1$$

where

$f$ = operating frequency [MHz]

$f_c$ = cutoff frequency [MHZ]

$l$ = gap depth for overlapping members of the thickness of the material for butting members (cm)

For a rectangular gap,

$$f_c = \frac{14980}{g}$$

where $g$ is the largest gap's transverse dimensions, in centimeters. For a round hole,

$$f_c = \frac{17526}{g} \tag{6.21}$$

When $fc \gg f$,

$$A = \begin{cases} 27\dfrac{l}{g} \quad [\text{dB}] \qquad \text{for a rectangular gap} \\[2ex] 32\dfrac{l}{g} \quad [\text{dB}] \qquad \text{for a circular gap} \end{cases}$$

When $l \ll g$, the attenuation tends to zero using the waveguide equation, and the attenuation will be that due to an aperture in a thin material, as described in Eq. (6.14). Equation (6.20) therefore does not provide the total shielding effectiveness of the metal but the increased attenuation due to the waveguide below the cutoff effect.

For a square array of holes, such as used for ventilation of a panel, the equation, when $f_c \gg f$, is

$$A = 20 \log\left(\frac{c^2 l}{d^3}\right) + \left(\frac{32t}{d}\right) + 3.8 \qquad [dB] \tag{6.22}$$

For a rectangular array $l_1$ by $l_2$, use

$$l = \sqrt{l_1 \times l_2}$$

Figure 6.23 shows the dimensions $c$, $d$, and $l$. The attenuation $A$ in this instance is defined as the increased attenuation of the hole pattern compared to the attenuation achieved if the total area $l \times l$ had been removed. The attenuation for multiple holes exhibits, according to Eq. (6.22), a higher value than the attenuation of a single aperture to the incident field. This apparent anomaly is explained as follows: The field strength developed across the larger area covered by the multiple holes, expressed in V/m, is higher than the field strength across the smaller area of a single hole, assuming a constant hole diameter, as shown in Eq. (6.15).

Equation (6.22), in this or similar form, is commonly found in the EMC literature and indicates that a constant attenuation with frequency exists as long as $f_c \gg f$. However, measurements of the attenuation of multiple holes in a solid 2-cm-thick sheet of metal, described in Ref. 10, show a measured attenuation that varies as much as 30 dB with frequency and may be 15 dB below that predicted for E fields and plane waves. Equations (6.20) and (6.22) are not valid for the attenuation of H fields.

In addition, the theoretical compensation for multiple apertures contained in Eq. (6.22) is not substantiated by test results, which show that the attenuation of multiple apertures is approximated by the $32t/d$ term (i.e., is equal to the attenuation of a single aperture).

Again the area of field incident on a number of apertures is greater than for a single aperture, and thus the field strength appearing on the far side of the apertures will be greater than that of a single aperture.

The waveguide below cutoff effect is used to advantage in the design of honeycomb shielded air vent panels and in the connection of a fiber-optic communication cable through a tube into a shielded enclosure. Figure 6.24 plots the theoretical attenuation of a ¹/₄-in.-wide rectangular honeycomb for several ratios of thickness to width, $w$.

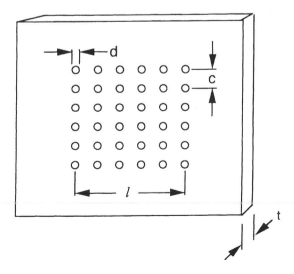

**Figure 6.23** Dimensions of array of holes in a panel.

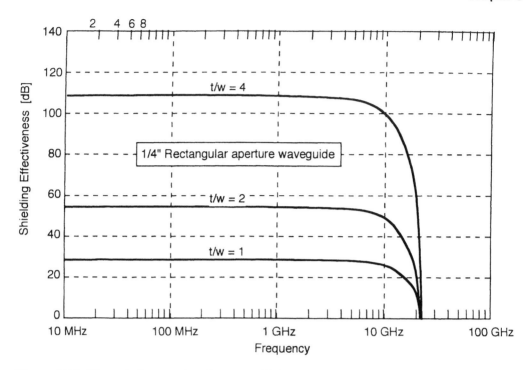

**Figure 6.24**  Theoretical attenuation of a waveguide below cutoff for a number of ratios of $t/w$.

### 6.4.3    Attenuation of a Thin Material with Joints and Apertures to an H Field

When the incident field is magnetic, a different mechanism predominates, especially when the thickness of the material is less than the width of the aperture.

MIL-STD-285 describes a technique for measuring the magnetic field shielding effectiveness of an enclosure, in which small transmitting and receiving loops are oriented such that the magnetic field is tangential to the surface of the enclosure. Consider the case where an aperture exists in a large enclosure and the frequency is high enough that diffusion through the enclosure material is insignificant. The distances between the transmitting loop and the aperture, $D_1$, and between the aperture and the receiving loop, $D$, and the orientation of the loops are shown in Figure 6.16.

The magnetic field at the measuring point d, distance $D$ from the shield with aperture, is given, from Ref. 20, by

$$H_\theta = \frac{\alpha_m}{2\pi} \left(\frac{1}{D}\right)^3 H_o \cos \phi \sin \theta$$

where

$\alpha_m$ = magnetic polarizability of an aperture, which equals $4(a^3/3)$ for a
         circular aperture

$a$ = radius of the aperture [m]

$\theta$ = angle between the measured magnetic field and the axis of the measurement current loop

$\phi$ = angle from the centerline of the aperture (90° from the surface of the enclosure) and the measuring location

Thus, the $H_\theta$ measured field is at a maximum when the field cuts the measuring loop at an angle of 90° and the loop is located on the centerline of the aperture (i.e., $\phi = 0°$). Assuming maximum field coupling to the receiving loop, the shielding effectiveness of the shield with aperture is given by

$$SE = \left(\frac{3\pi}{4}\right)\left(\frac{D_1}{D_1 + D}\right)^3 \left(\frac{D}{a}\right)^3$$

If $D_1 = D = \dfrac{D_1 + D}{2}$ then

$$SE = 0.295 \left(\frac{D}{a}\right)^3$$

Equation (6.18) for the shielding effectiveness of a shield with aperture to the E field generated by a small loop and the equation for the magnetic field attenuation of a shield with aperture are both functions of distance. This would appear to contradict the law of reciprocity, in which under certain conditions the transmitting and receiving loops may be interchanged and the received power will remain constant. Hoeft et al. remark, in Ref. 20, that ''the shielding effectiveness is a strong function (cubed dependence) of the distance between the shield and measuring location. Thus the shielding effectiveness is not an intrinsic electromagnetic parameter.'' This means the shielding effectiveness is dependent on the test setup, which includes the distances between the transmitting loop and the shield and the receiving loop and the shield. In contrast, the measurement of the transfer impedance of a shielded cable is an intrinsic function of the cable. For as long as the shield current is measured or known and the transferred voltage is measured, the transfer impedance of the cable may be computed and should be independent of the test method, unlike measurement of the transfer admittance of a cable, which is dependent on the source of electric field and the geometry of the test fixture and is not an intrinsic electromagnetic parameter.

The predicted shielding effectiveness using the equation has been compared, in Ref. 20, to measured values. In the measurement, the distance $D$ was fixed at 0.406 m (16″), with the aperture radius varied from 0.0076 m (0.3″) to 0.05 m (2″). The correlation between predicted and measured SE is remarkably good.

The attenuation of enclosures, joints, and apertures to H fields, based on Ref. 11, is considered as follows: We assume that the magnetic field (H out) impinging on the outside of an enclosure results in an H field (H in) inside the enclosure:

The attenuation of the enclosure is given by $\dfrac{H_{out}}{H_{in}}$ (6.23)

The resistance of the enclosure wall or the impedance of the defect (e.g., a joint, aperture, or gasket) is represented by $Z_s$ in Figure 6.25. The inside of the enclosure can be modeled as the inductance of a wide single-turn inductor, given by

$$L = \frac{\mu_o A}{l_1} \tag{6.24}$$

where

$\quad$ $A$ = area of the inside of the box = height × depth

$\quad$ $= l_2 \times d$

$\quad$ $l_1$ = width of the box

The inductive reactance of the inside of the box $(2\pi f L)$ is represented in Figure 6.25 as $Z_{in}$. The H field incident on the box causes a current to flow on the outside of the box. At low frequency where the thickness of the enclosure is less than a skin depth, the principal coupling mode is due to the current diffusion through the enclosure wall. $Z_s$ at low frequency is thus predominantly the AC resistance of the enclosure wall ($R_s$), given by

$$R_s = \frac{12e^{-t/\delta}}{G_t l_1} \quad [\Omega] \tag{6.25}$$

where

$\quad$ $t$ = thickness of the enclosure [m]

$\quad$ $G$ = conductivity [mhos/m]

$\quad$ $\delta$ = skin depth [m]

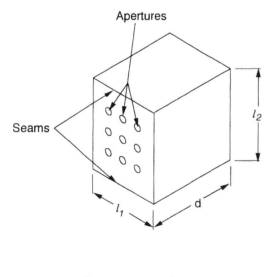

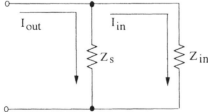

**Figure 6.25** Enclosure with defects, and equivalent circuit.

The conductivity is given by

$$G = G_o G_r$$

where $G_o = 5.8 \times 10^7$ mhos/m. The skin depth is computed from

$$\delta = \frac{1}{\sqrt{\pi f \mu G}} \quad \text{[m]}$$

where $\mu$ = permeability [H/m] and is given by

$$\mu = \mu_o \mu_r$$

where $\mu_o = 4\pi \times 10^{-7}$ [H/m].

When the metal contains a number of apertures, the conductivity of the metal is reduced by the apertures. The conductivity is then

$$G \times \frac{\text{Area of panel } (l_1 \times l_2) - \text{Total area of apertures}}{\text{Area of panel}} \tag{6.26}$$

The additional components of $Z_s$ when seams and joints are present are the following:

1. The DC resistance of the joint $R$, which is typically 15–30 $\mu\Omega$ for a riveted joint but may be as high as 1.5 m$\Omega$ for unplated metal.
2. The contact impedance $R_c$, which has the characteristics of an AC resistance and contains the term $\sqrt{f}$. The contact impedance is dependent on surface smoothness and finish. A typical value of $R_c$ for a riveted joint is $0.06\sqrt{f}$ $\mu\Omega$ and may be as high as $0.25\sqrt{f}$ $\mu\Omega$ for unplated metal.
3. Another component of a joint that contains a gap is the inductance of the gap $L$, which may have a typical value of $0.5 \times 10^{-12}$ H.

In addition, screws used to fasten the walls of an enclosure exhibit inductance. Although a complex impedance, the total impedance of a joint may be modeled by a simple series equivalent circuit and is given approximately by

$$R_s + lN(R + R_c\sqrt{f} + 2\pi fL) \tag{6.27}$$

where

$$l = \text{length of the joint}$$

$$N = \text{number of joints in the current path}$$

When the joint is gasketed, the expression after $R_s$ is replaced by the transfer impedance of the gasket at the frequency of interest.

The attenuation of the enclosure has been expressed as $H_{out}/H_{in}$. This is equivalent to the ratio $I_{out}/I_{in}$ which is given by

$$\frac{j\dfrac{2\pi f \mu_o A}{l_1}}{R_s + lN(R + R_c\sqrt{f} + 2\pi fL)} \tag{6.28}$$

where

$$A = \text{area of enclosure} = l_2 \times d$$

$$l = \text{length of joint (joint or slot in the current path)}$$

$$R = \text{DC resistance of joint}$$

$R_s$ = resistance of an enclosure wall as a function of skin depth
$R_c$ = contact resistance
$L$ = equivalent series inductance of slot
$N$ = number of joints in the current path

The magnitude of the internal $H$ field can be computed from Eq. (6.28) when $H_{out}$ is known. The limit in attenuation of an enclosure, with apertures, to an H field is given by

$$\frac{H_{out}}{H_{in}} = 0.445 \frac{A}{N\alpha l_1} \tag{6.29}$$

where

$N$ = number of apertures
$\alpha$ = magnetic polarizability of the aperture

The worst-case magnetic polarizability of different shapes of aperture is provided in Table 6.8, from Ref. 25. These values assume that the current flows over the longest dimension, that the apertures are less than $\frac{1}{6}\lambda$, and that the material is infinitely thin.

An approximate thickness correction, in decibels, for round apertures and magnetic fields is $32t/d$ and for square apertures $27.3t/l$, where $t$ is the thickness, $d$ is the diameter, and $l$ is the longest dimension.

As frequency increases, the attenuation of the enclosure, computed from Eq. (6.28), increases until the attenuation is limited by the apertures, Eq. (6.29), and is then constant with frequency.

Equation (6.29) is applicable to the attenuation of a single-layer wire mesh by assuming that the wire screen behaves like a sheet with square perforations. Measurements of wire have shown that the calculated values are all from 4 to 6 dB lower than measured values.

One type of shield that can be used effectively to shield H field emissions from a video display is very fine blackened wire. This type of wire mesh has typically 100 openings per inch (OPI) and, although the display is visible, will reduce the brightness of the display and may result in moiré patterns.

**Table 6.8**  Worst-Case Values of Magnetic Polarizability for Different-Shape Apertures in an Infinitely Thin Wall

| Shape of aperture | Magnetic polarizability $\alpha$ |
| --- | --- |
| Circle of diameter $d$ or radius $r$ | $\dfrac{d^3}{6}$ or $4\left(\dfrac{r^3}{3}\right)$ |
| Long, narrow ellipse of major axis $a$ and minor axis $b$ | $\dfrac{\pi}{3} \dfrac{a^3}{l_n(4a/b) - 1}$ |
| Square of side $l$ | $0.259l^3$ |
| Rectangles of length $l$ and width $w$ | |
| $l = \dfrac{4}{3}w$ | $0.2096l^3$ |
| $l = 2w$ | $0.1575l^3$ |
| $l = 5w$ | $0.0906l^3$ |
| $l = 10w$ | $0.0645l^3$ |

The main weak link is at the contact between the enclosure and the wire mesh, due to the contact impedance. The measured surface resistivity of a copper-plated stainless steel mesh with 100 OPI was 1.5 m$\Omega$/sq and that for copper mesh with 100 OPI was 3.3 m$\Omega$/sq.

### 6.4.3.1   Examples of H Field Attenuation of Enclosures

The application of an insulating compound to a joint may be a requirement for sealing against moisture or to increase the thermal conductivity of the joint. Figure 6.26 illustrates the joint impedance of a test sample made from two pieces of 12 $\times$ 24-inch aluminum joined by a 2-inch-wide backing strip riveted on 3-inch centers. A sealant was applied to the bond before the strip was riveted, and the configuration is referred to as the ''full-bond'' sample. The joint resistance of the ''joints-only'' sample, in which the sealant was applied to the exposed joints after the joint was riveted, is shown in Figure 6.27. Figures 6.27 and 6.28 are reproduced from Ref. 22, which describes experimental and theoretical analysis of the magnetic field attenuation of enclosures. The DC resistance of the full-bond sample was a measured 35 $\mu\Omega$ and of the joints-only sample it was 15 $\mu\Omega$.

The magnetic field attenuation of the two samples when mounted in a 0.6-m $\times$ 0.6-m $\times$ 0.2-m brass box with a thickness of 1.58 mm and a relative conductivity of 0.6 is shown in Figure 6.28.

The equivalent impedances of the two samples that fit the curves when used in Eq. (6.28) are as follows:

|             | $R$ <br> [$\mu\Omega$] | $R_c$ <br> [$\mu\Omega$] | $L$ <br> [$10^{-12}$ H] |
| ----------- | ---------------------- | ------------------------ | ----------------------- |
| Joints only | 40                     | $0.03\sqrt{f}$           | 0.1                     |
| Full bond   | 50                     | $0.10\sqrt{f}$           | 0.5                     |

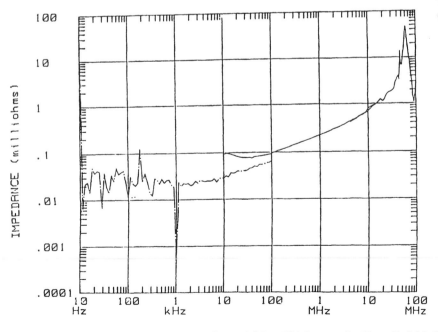

**Figure 6.26**   Measured joint impedance of the ''full bond'' joint sample. (From Ref. 22. © 1986, IEEE.)

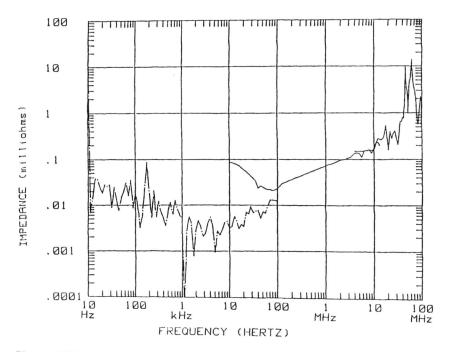

**Figure 6.27** Measured joint impedance of the "seams-only" joint sample. (From Ref. 22. © 1986, IEEE.)

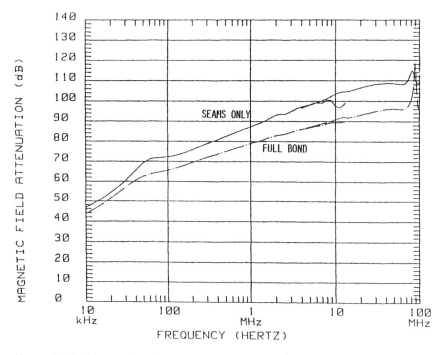

**Figure 6.28** Measured surface magnetic field reduction of the "full-bond" and "seams-only" joint. (From Ref. 22. © 1986, IEEE.)

The effect of the spacing of the fastener on transfer impedance is illustrated in Figure 6.29. In this example, the number of transverse screws in a large cableway were varied from no screws to one screw per 2-inch spacing. The transfer impedance is measured in a TEM cell in a test configuration similar to that used for magnetic field attenuation measurements. The transfer impedance is a measure of the voltage developed across the seam inside the test enclosure mounted on the floor of the TEM cell, divided by the current flow on the outside of the enclosure and across the joint under test.

The contact impedance and DC resistance of a joint is dependent on the surface finish, as illustrated in Figure 6.30. The unplated aluminum exhibits the lowest level of H field attenuation, whereas tin plating is almost as effective as a conductive gasket. Tin plating is an excellent choice because tin is malleable and tends to flow, filling up the small gaps in the joint.

The components of the equivalent impedance of the seams with different finishes are

| | $R$ $[\mu\Omega]$ | $R_c$ $[\mu\Omega]$ | $L$ $[10^{-12}\,H]$ |
|---|---|---|---|
| Tin plated | 30 | $0.015\sqrt{f}$ | 0.5 |
| Unplated | 1500 | $0.250\sqrt{f}$ | 0.5 |
| Gasketted and tin plated | 30 | $0.001\sqrt{f}$ | Negligible |

The effect of apertures on magnetic field attenuation is illustrated in Figure 6.31. We see that a hole as large as 4 inches in a $24'' \times 24'' \times 8''$ enclosure still results in a magnetic field attenuation of 50 dB. This explains why, as often seen in practice, a thin metal enclosure with large ventilation apertures, typically 1 inch, in close proximity to an H field source, such as a

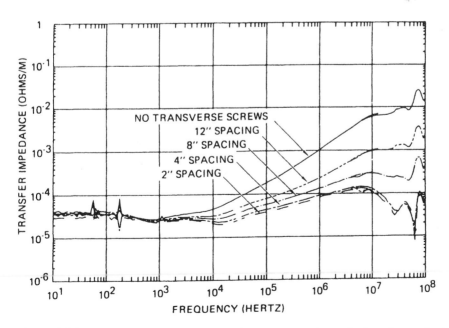

**Figure 6.29** Effect of transverse screw spacing on measured transfer impedance of a large cableway. (From Ref. 20. © 1986, IEEE.)

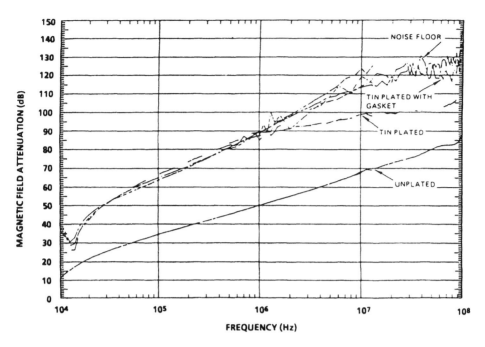

**Figure 6.30**   Magnetic field reduction of a deep drawn aluminum equipment case showing effects of tin plating and a gasket. (© 1986, IEEE.)

switching power supply/converter, will provide significant shielding. The level of shielding has been determined in practice by field measurements with the enclosure in place and removed.

A large number of small apertures will provide higher attenuation at high frequency compared to a single aperture of the same area. This is illustrated in Figure 6.32. The conductivity of the 0.6-m × 0.6-m × 0.2-m aluminum sheet containing 36,100 ¹/₆″ apertures is lower than the sheet with a single 4″ aperture; therefore, at frequencies below 30 kHz the multiple-aperture sample shows a lower attenuation.

Methods for reducing the coupling through ventilation apertures include wire mesh and honeycomb air vent filters. The published attenuation curves for these filters are often theoretical, and the attainable attenuation may be lower. The best type of honeycomb air vent filter is constructed with the metal segments, which make up the honeycomb, welded or soldered together. A more common method has the segments electrically connected via a bead or single spot of conductive adhesive. In addition, the edges of the metal segments are often connected to the frame in which the honeycomb is contained by spring pressure only. In these samples the contact impedance may be initially high and increases further with time, due to corrosion. The magnetic field attenuation performance of many commercially available honeycomb filters is very poor due to the low conductivity and, therefore, the appreciable skin depth of the honeycomb. It is not uncommon for equipment to fail MIL-STD-461 RE02 or similar radiated emission tests due to magnetic field coupling through poor honeycomb air vent panels, especially when the H field source is in close proximity to the panel.

In one sample of a honeycomb, filter cells were made from aluminum that had been chromated and were bonded by nonconductive adhesive. The electrical contact was made via the relatively high resistance of the chromate finish. The measured shielding effectiveness (SE) of the honeycomb was 15 dB at 100 kHz and 3 dB at 14 MHz. With a cadmium plating, the honeycomb achieved an SE of 20 dB at 100 kHz and 55 dB at 14 MHz. A much higher magnetic

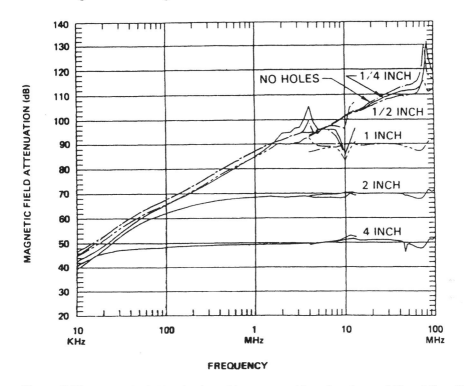

**Figure 6.31** Magnetic field reduction with apertures (size of enclosure 24″ × 24″ × 8″). (© 1986, IEEE.)

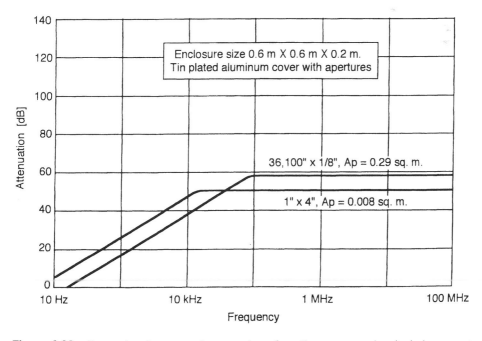

**Figure 6.32** Comparison between a large number of small apertures and a single large aperture with the same area.

field shielding effectiveness can be achieved with a steel or brass honeycomb. For optimum SE, steel cells are spot-welded together and then tin-plated or brass cells are soldered together. An alternative to a honeycomb, when high attenuation of magnetic fields is required, is the wire mesh screen air filter. The mesh must be passivated with a low-resistance finish in order to reduce the skin depth and therefore optimize the SE of the screen.

The impedance of a joint is dependent on the fastening method. The lowest-impedance seam is welded with an impedance close to the base metal when the welding is carefully controlled. The next preferred method is soldering. Riveted seams provide a low-impedance path at the rivet, for the head of the rivet flows into the surface of the metal. Screws exhibit an inductance of 1–3 nH, depending on the length of the thread from the head to the tapped metal. The impedance of pop rivets is no lower and in some samples higher than that of screws.

The H field shielding effectiveness of an enclosure with aperture changes dramatically at the resonant frequency of the enclosure, even when the aperture is not a resonant length. The analysis described for the E field shielding effectiveness of an enclosure when the source is an external plane wave, described in Ref. 9, is also applicable to the H field component shielding. As predicted in Eq. (6.29), the shielding effectiveness of an enclosure with aperture to an H field source, above some low frequency, is almost constant with frequency. This is shown in Figures 6.31 and 6.32 and in the theoretical curve in Figure 6.33, up to 400 MHz, and in the theoretical and measured curves in Figure 6.34, up to 500 MHz. However, at the resonant and antiresonant frequencies of the enclosure, the shielding either increases or becomes negative (i.e., a field enhancement occurs). These effects are shown in Figures 6.33 and 6.34. Figure 6.34 shows that the magnetic field shielding effectiveness increases (i.e., the magnitude of the H field decreases) the further away the measurement point is from the aperture. Therefore, locate susceptible circuits on a PCB as far away from the aperture as possible.

The practical application of the analysis and measurements are discussed in Ref. 9 and are, for the H field:

Long, thin apertures are worse than round or square apertures, because, for a typical-size enclosure, theory predicts that doubling the length of the slot reduces $S_H$ by 12 dB,

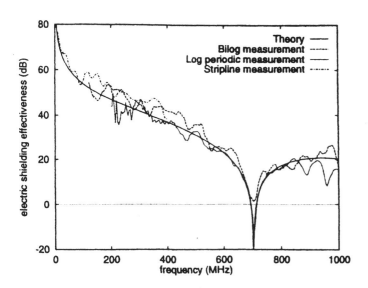

**Figure 6.33**   Calculated and measured $S_H$ at center in a 300 × 120 × 300-mm enclosure with a 100 × 5-mm aperture.

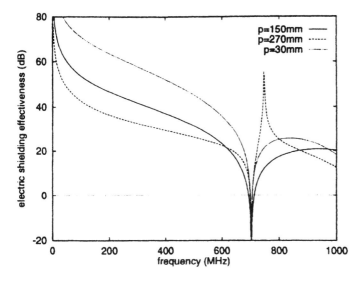

**Figure 6.34** Calculated $S_H$ at three positions in a $300 \times 120 \times 300$-mm enclosure with a $100 \times 5$-mm aperture, and measured $S_H$ at the center of the enclosure.

whereas doubling the width reduces $S_H$ by only about 2 dB. For magnetic field sources, using the magnetic polarizability of rectangles in Table 6.8, doubling the width reduces $S_H$ by 3–4 dB.

Doubling the number of apertures reduces $S_H$ by about 6 dB, and this is also seen at low frequency for magnetic field sources, as shown in Eq. (6.29).

Dividing a slot into two shorter slots increases $S_H$ by about 6 dB according to this prediction; however, for magnetic field sources, Eq. (6.29) predicts that the increase is closer to 12 dB.

At subresonant frequencies, doubling the enclosure dimensions while keeping the aperture size and number constant is predicted to increase $S_H$ by about 13 dB. For magnetic field shielding, larger-size enclosures are highly recommended.

Doubling the enclosure size halves the resonant frequency. To avoid negative shielding effectiveness, a smaller enclosure, or one that is full of electronic components, is preferable.

### 6.4.3.2 Measured Attenuation of a Magnetic Field Source Within an Enclosure

To ensure the method of calculating magnetic field attenuation is accurate for the case when the source is inside the enclosure, the following experiment was conducted.

A magnetic field was generated by a rectangular loop placed inside an enclosure with apertures. The size of the enclosure was 36 cm × 25.4 cm × 5 cm deep and 21 apertures of 1.5 cm × 1.5 cm and one aperture of 8 cm × 2.7 cm were punched out of the enclosure.

In commercial and military EMC requirements, the maximum level of radiated E field is often specified. As discussed in Chapter 2, a loop antenna generates both E and H field components, and either may be predicted and measured. As a wave couples through the defects in an enclosure, the magnetic and electric field components are not always attenuated to the same extent, and a change in the wave impedance occurs. In this case study, the loop is only 5 mm from the inside of the enclosure, and the wave impedance will be low both inside and outside of the enclosure. Thus we assume that the wave impedance of the field radiated from the outside of the enclosure is the same as the field from the loop when located outside of the enclosure.

A monopole antenna was located at a distance of 1 m from the loop antenna and measurements were made of the E field with the loop both inside and outside of the enclosure. The antenna factor of the monopole, which is calibrated assuming the far-field impedance of 377 $\Omega$, was corrected for the predicted wave impedance of the field 1 m from the loop antenna. The measured E field from the loop located outside of the enclosure was within 6 dB of that predicted, using Eq. (2.9).

The current flow induced on the inside of the enclosure by the loop antenna, based on the antenna location, was across the largest dimension, 8 cm, and it was therefore valid to use this dimension in Eq. (6.29). The upper limit of the attenuation when computed from Eq. (6.29) was dominated by the single 8-cm $\times$ 2.7-cm aperture. The 21 1.5-cm apertures placed the upper limit of attenuation at 65 dB, whereas the single large aperture placed it at 42 dB. The measured attenuation was computed from the ratio of E field with the loop outside of the enclosure to the measured E field with the loop inside the enclosure. A comparison of the predicted and measured attenuations shows an uncharacteristically good correlation:

| Frequency [MHz] | Attenuation [dB] | |
| --- | --- | --- |
| | Predicted | Measured |
| 12.0 | 42 | 45 |
| 17.0 | 42 | 40 |
| 24.0 | 42 | 40 |
| 27.6 | 42 | 48 |

## 6.5 GASKETING THEORY, GASKET TRANSFER IMPEDANCE, GASKET TYPE, AND SURFACE FINISH

### 6.5.1 Gasket Theory

One method of closing seams is to use conductive gasket material. In instances where the seam must be environmentally sealed, some form of gasket material is mandatory.

Figure 6.35 shows the current density through a metal plate with a seam closed by a gasket that has a lower conductivity or permeability than the metal plate. The voltage developed across the gasket may result in a current flow and the generation of a predominantly H field in the enclosure. Alternatively, when the impedance of the enclosure is high, which is more often true of large enclosures at high frequency, a predominantly E field may be generated by the voltage across the gasket.

The impedance of the gasket has components of resistance, contact impedance, and inductance, as does the impedance of the ungasketed joints discussed in the preceding sections. An additional component of gasket material that may become dominant is the capacitance between conductive particles or mesh used in the majority of gaskets.

### 6.5.2 Gasket Test Methods

To evaluate a gasket's performance under the wide range of wave impedances found in practical applications, radiated testing techniques using E fields, H fields, and plane waves are used.

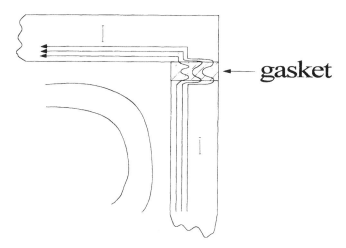

**Figure 6.35** Current density through an aperture closed by a gasket.

These techniques typically employ one or more metal cavities, with a source of field on one side of the barrier and a measuring antenna/probe on the other side of the barrier. The energy is coupled through the barrier via a seam that contains the gasket under test. The gasket is usually placed under a cover plate that is placed over an aperture in the barrier. The fields coupled through the aperture are measured with and without the cover plate, with and without the gasket under the cover plate, or in some cases with the conductive gasket replaced by a nonconductive gasket and nonconductive fasteners. The presence of conductive fasteners and conductive compression stops can dramatically change the gasket measurement, as discussed later. Also, the position of the antenna in these tests strongly influences the test signal coupled through the antenna. It is highly recommended that the test antenna be moved relative to the aperture and to the surrounding enclosure.

At least ten test methods exist for measuring the effectiveness of the addition of a gasket to a joint in reducing or preventing electromagnetic leakage through the joint. Of these ten test methods, only four are considered standardized, although considerable effort has been spent in developing the at-present nonstandard reverberation chamber test methods. These are discussed as alternative techniques. The IEEE Std 1302-1998 ''IEEE Guide for the Electromagnetic Characterization of Conductive Gaskets in the Frequency Range of DC to 18 GHz'' describes in detail the four standardized measurement techniques as well as five alternative techniques.

Table 6.9 compares the standardized gasket measurement techniques and also shows the repeatability for each of the tests. The ARP 1173–1998, Def Stan 59–103, and MIL-G-83528B show a typical repeatability of between 6 and 20 dB, which indicates that even comparative tests on gasket material using these techniques should be made at the same location and using the same relative antenna locations. The ARP 1173–1998, DEF Stan 59–103, and MIL-G-8352B are measurements of the relative transmission through an aperture, and ARP 1705-1981 is a transfer impedance test. ARP 1173–1988 uses metal spacers between the cover plate and the flange to set up an additional reference measurement. The difference between this reference measurement and the measurement with the gasketed cover in place is referred to as *shielding increase*. Similar modifications to the MIL-G-83528B measurement procedure are in use. In the IEEE Std 299–1997 measurement setup, the reference measurement is made in free space, without the test enclosure, and the comparison is made to the enclosure with the aperture covered by the gasketed plate. This method results in a very high measured shielding effectiveness. In

**Table 6.9**  Comparison of Standardized Gasket Measurement Techniques

| | Measurement technique | | | |
|---|---|---|---|---|
| | ARP 1705–1981 | ARP 1173–1988 | Def Stan 59-103 (17-Sep-93) | MIL-G-83528B (1993) |
| Measurement principle | Current injection | Aperture attenuation | Aperture attenuation | Aperture attenuation |
| Parameter measured | Transfer impedance | E/H/Plane-wave attenuation | E/H/Plane-wave attenuation | E/H/Plane-wave attenuation |
| Unit of measure | dB/$\Omega$-m | dB | dB | dB |
| Test sample configuration | Circular | Rectangular | Circular | Rectangular |
| Sample size | 150 mm diameter | 300 × 300 mm | 409 mm diameter | 610 × 610 mm |
| Frequency range | dc to 2 GHz | 400 Hz to 10 GHz | 10 kHz to 18 GHz | 20 MHz to 10 GHz |
| Typical dynamic range | 60–150 dB | 60–100 dB | 60–120 dB | 60–120 dB |
| Typical repeatability | ± 2 dB | ± (6–20) dB | ± (6–20) dB | ± (6–33) dB |

*Source:* © IEEE 1998

MIL-G-835528B, the reference level is determined by placing the transmitting and receiving antennas in line with the open aperture in the test chamber. This results in a lower reference level and lower shielding effectiveness compared to IEEE Std 299–1997, but the shielding effectiveness is higher than the ARP 1173–1998 "shielding increase." The ARP 1173–1988 test setup is illustrated in Figure 6.36.

One of the alternative test techniques is to use two reverberation chambers or mode-stirred chambers (MSCs). One of the problems with chamber measurements, seen in all other tests of the transmission through an aperture, are chamber resonances. These dramatically affect the measured shielding effectiveness. In the MSC technique, a paddle wheel is used in each chamber to randomize the field so that the gasket under test is exposed to fields of all angles and polarizations. By selecting the rotation rate of one paddle wheel to be high compared to that of the second, the measurement of maximum coupled fields inside one chamber is a function of the other chamber's tuner position. Shielding effectiveness is defined as the ratio of the power that is transferred into the second chamber with only a cover plate over the aperture to the power coupled through the aperture with the gasket under test inserted between the cover plate and the chamber wall. By using two large chambers, testing down to 200 MHz is possible, whereas in the nested stirred mode method, shown in Figure 6.37, the lower frequency is restricted to 500 MHz because of the size of the smaller test chamber.

One of the major concerns in selection of gasket material is in the interpretation of the manufacturers' test data. Can the results obtained using the various test methods be compared? Reference 26 describes tests made on four different gasket materials using the MIL-G-83528, MSC, transfer impedance, and modified ASTM D 4935–89 test methods. Despite attempts to standardize the tests and to linearize the test data, differences in the shielding performance of the four different gasket materials were up to 50 dB using the different techniques! No two test

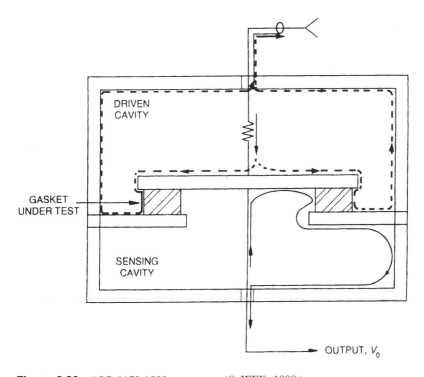

**Figure 6.36** ARP 1173-1998 test setup. (© IEEE, 1998.)

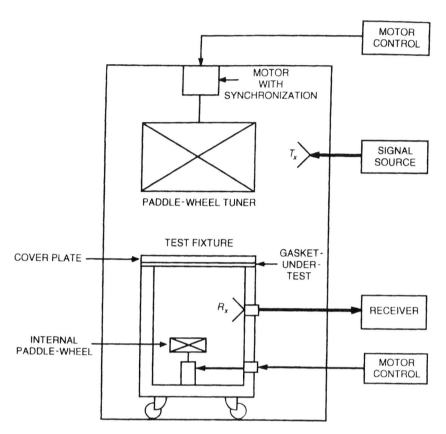

**Figure 6.37**  Nested stirred-mode method for characterizing EMI gaskets. (© IEEE, 1998.)

techniques showed a good correlation, and the shielding performances of the four different gasket materials also were not consistently higher or lower when measured using the four different test techniques. It therefore seems impossible to evaluate the relative performance of gaskets using different test methods. IEEE Std 1302-1998 also states: ''Any technique is unlikely to replicate the actual gasket application of interest. Therefore, each of the techniques should be considered as a 'platform' from which minor modifications and adjustments can be made to more closely approximate the application in question.''

Transfer impedance does provide an excellent and reproducible indication of the relative quality of the gasket; however, unless the test fixture represents the materials and dimensions of the equipment of concern, the test results can be misleading, as shown in the comparison of test methods.

The test setup for transfer impedance in accordance with ARP 1705–1981, Figure 6.36, shows a fixture with two cavities. The gap between the two cavities is closed by a metal plate sealed by the gasket under test. A current is caused to flow through an impedance across one face of the metal plate, through the gasketing material, returning on the walls of the driven cavity. A voltage output is taken from the opposite surface of the metal plate, and the ratio of measured voltage to drive current is described as the transfer impedance $Z_t$. In the test setup described here, the closing pressure was achieved by means of compressed air, so the effect of the impedance of the fasteners was not measured. A number of factors determine the value of

$Z_t$, including the composition of the gasket material, its thickness and width, the impedance of the mating shield joint surface, and the pressure on the gasket.

### 6.5.3 Relative Performance of Gasket Materials

Figures 6.38a, b, c, and d show the transfer impedance for a number of commercially available gasket materials at different closing pressures and frequencies, from Ref. 14. The purpose of these graphs is to illustrate the importance of closing pressure on transfer impedance: the higher the pressure, the lower the transfer impedance. Also, sample 2 shows an increase in transfer impedance with increasing frequency, which is indicative of a predominantly inductive transfer impedance. The "porcupine" type of wire mesh gasket, in which wires are imbedded vertically in neoprene foam, exhibit inductance. The gasket in Figure 6.38 is a graphite-loaded gasket and has a much higher DC resistance and at least 1000 times the transfer impedance of the other gaskets from 0.1 to 1 MHz. A graphite-loaded gasket seldom provides any additional level of shielding, especially when conductive fasteners at distances of at least 5 cm are used. The decrease in transfer impedance shown in Figure 6.38d is characteristic of a gasket with a predominantly capacitive transfer impedance, which is common with conductive particles imbedded in an elastomer. However, even above 10 MHz, the graphite-loaded gasket has a transfer impedance 15 times higher than the inductive monel wire mesh shown in Figure 6.38a.

The dimensions of the gasket material are important in determining the impedance presented to current flow; therefore the gasket transfer impedance is expressed in ohms per meter.

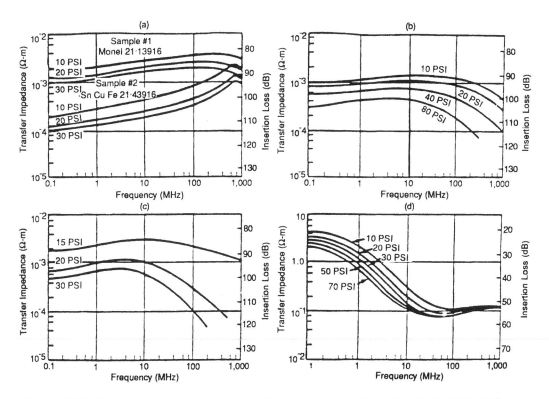

**Figure 6.38**  Transfer impedance for a number of gasket materials. (From Ref. 14. © 1982, IEEE.)

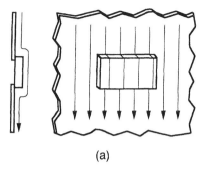
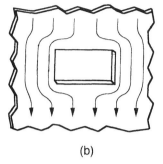

(a)                                                                                        (b)

**Figure 6.39**   Current flow around an aperture and through a gasketed metal plate.

Figure 6.39b shows the current flow around an aperture, Figure 6.39a shows the aperture closed by a gasketed metal plate.

The current density through the plate is considered to be a uniform 1 A/m. The impedance encountered by the current flow is dependent on the length of material presented to the current flow multiplied by 2, because two lengths of gasketing are in the current path. For example, for gasket material with a transfer impedance of 0.00018 $\Omega$/m, at 100 kHz and a length of 5 cm, the total impedance is

$$\frac{100 \text{ cm}}{5 \text{ cm}} \times 2 \times 0.00018 \ \Omega/\text{m} = 0.0072 \ \Omega$$

In practice, many of the manufacturers of gasket material use one of the relative aperture test techniques, based on and including MIL-STD-285. As seen in the attenuation measurement curves on conductive fabric in Figure 6.13, this type of measurement is subject to large variations and, due to dependence on antenna type, location, and size of both the test enclosure and the shielded room, to lack of reproducibility. Transfer impedance, in contrast, is a most useful parameter for evaluation of gasket shielding performance, due to the high level of repeatability achievable in the measurement.

Experience with the use of gasket material soon teaches that the inclusion of a conductive gasket may unexpectedly reduce the attenuation of an enclosure to external fields and increase the level of radiated emissions from the enclosure. When enclosures are constructed from machined metal passivated or plated with a low-resistance finish at least $\frac{1}{8}''$ thick, with approximately $\frac{1}{4}''$ overlap at the joint, and fasteners are approximately $1''$ apart, very few gasket materials will improve the performance of the joint and many will degrade it. One gasket material that may improve the performance of the machined enclosure, as measured by the DC resistance of the joint, is thin silver-loaded elastomer when inserted into a groove in the material. The resistance of the joint is at its lowest when the silver conductive gasket is in the form of an O-ring set into a groove in one of the mating surfaces. However, shelf-life measurements on silver-filled silicon elastomer have reported increased resistance with time. In these measurements, in which the material was stored for a year in uncontrolled humidity and temperature, a massive increase of 5000 times the DC resistance and an increase in transfer impedance of 1200 are recorded. Reference 14 presents the measured transfer impedance of a machined enclosure joint at 80 $\mu\Omega$/m at 50 MHz for the metal-to-metal joint and 7 $\mu\Omega$/m at 50 MHz for the joint when loaded with a $\frac{1}{4}$-inch-wide, $\frac{1}{16}$-inch-thick silver-loaded gasket. At 1 MHz, however, the silver-loaded gasket transfer impedance is slightly higher than the metal-to-metal impedance. Reference 27 describes degradation in EMI shielding and mating flange materials after environmental exposure. The test method described in Ref. 27 is a relative aperture technique similar to MIL-G-83528B. Two types of gasket material were tested: (1) a silver-loaded copper-filled silicone gasket (Ag-Cu gasket) and a silver-plated aluminum-filled silicone gasket (Ag-Al gasket). Stainless steel bolts and lock washers at a spacing of either 2.4″ (0.98 cm) or 2.667″ (1.05 cm) were used to provide compression, with either conductive or nonconductive compression stops to limit the compression on the gasket. One piece of interesting data obtained from the measurements is that the level of shielding effectiveness with the Ag-Cu gasket in place, using conductive compression stops, or with the gasket removed, leaving apertures between the fasteners, is the same up to 10 GHz! This is another illustration that the addition of a gasket material may not be beneficial and may be required only for environmental sealing and not for shielding. However, if the conductive compression stops are replaced by nonconductive stops, the relative level of shielding reduces by up to 60 dB.

Adding two narrow strips of silver-loaded elastomer to a joint in a sheet metal enclosure increased the transfer impedance at 50 MHz from 200 $\mu\Omega$ for the metal-to-metal joint to over 2 m$\Omega$ for the gasketed seam. The gasket in the thin sheet metal enclosure was not adequately compressed and provided a small surface area in contact with the metal, which may account for the higher impedance with the gasket than without. However, it is invariably a bad practice to seal a seam between thin metal, especially when the fasteners are more than 1 inch apart, with an incompressible material such as silicone elastomer, because the transfer impedance may be no lower and in fact may be higher than that attained with no gasket.

Gaskets are available that have been shaped to fit between the flange of a bulkhead connector and the enclosure wall. One experience with a connector gasket is related as follows.

The gasket was made from a foam silicone material with vertical monel wires (porcupine gasket material). The wires bite into the metal surface of the enclosure and the connector flange. In the example, the connector was mounted in a large rack. The connector was a MIL-STD type with a conductive cadmium-plated finish. When not in use, the connector was covered by a screw on a cadmium-plated cap. With the cap in place, the radiation from the connector and cap was measured, using a small magnetic field probe. Removing the gasket and relying on the metal-to-metal contact between connector flange and enclosure reduced the level of radiation by 7–13 dB. The most likely cause of the magnetic field was common-mode noise current flowing through the capacitance of the connector pin to case capacitance and diffusing through the impedance presented by the small number of vertical wires in the gasket material. Connecting a cable to the connector and measuring the E field radiation from the cable over the 30–100-MHz range showed an increase of between 6 and 10 dB with the gasket, compared to no gasket material. In this case a marked reduction in radiation was achieved by adding a shield over the wires inside the enclosure. The shield was connected to the inside of the enclosure and increased the wire-to-enclosure capacitance, which effectively shunted the noise currents to the enclosure and reduced current flow in the connector. In other experiments, the increase in radiation from a cable with gasket material was attributed to the voltage drop across the gasket as a result of current flow on the inside of the shield of the cable and the inside of the enclosure.

Further illustration of the importance of choosing the right gasket for the job, or avoiding the use of a gasket, is shown from following curves of Figures 6.40a, b, c, d, e, and f, from Ref. 16. The test panel was 20 cm square covering a 10-cm-square aperture in the transfer impedance test fixture. Eight #10/32 screws were used to apply pressure to the gasket under test. In addition to the gasket material, a nonconductive neoprene sheet, test numbers 101/201 and 102/202, was tested. In Figure 6.40b the effect of fastener spacing can be seen in tests with the neoprene spacer. As expected, the higher the number of fasteners, the lower the transfer impedance. With only two screws in the flange, a slot resonance was formed at approximately 380 MHz, and the measured transfer impedance increased to 1000 $\Omega$/cm. The brass plate has

| TEST NUMBER | | GASKET MATERIAL | WIDTH (cm) | THICKNESS (cm) | SQUARE SHAPE | CIRCULAR SHAPE |
|---|---|---|---|---|---|---|
| WITH BRASS | WITH IRIDITED ALUMINUM | | | | | |
| PANEL | PANEL | | | | | |
| 1 | · | Solid Brass Plate | · | 0.32 | · | · |
| · | 2 | Solid (Iridited) Aluminum Place | · | 0.25 | · | · |
| 100 | 200 | No Gasket Used | · | · | · | · |
| 101 | 201 | Neoprene (Nonconductive) Sheet | 0.64 | 0.24 | x | |
| 102 | 202 | Neoprene (Nonconductive) Sheet | 0.64 | 0.48 | x | |
| 103 | 203 | Monel Mesh - Compressed | 0.64 | 0.48 | x | |
| 104 | 204 | Monel Mesh - Compressed | 0.64 | 0.32 | x | |
| 105 | 205 | Monel Mesh over Silicon Foam Core | 0.64 | 0.32 | | x |
| 106 | 206 | Stainless Steal Spiral | 0.36 | 0.36 | | x |
| 107 | 207 | Tin - Placed Beryllium Copper Spiral | 0.36 | 0.36 | | x |
| 108 | · | Two Tin-Placed Beryllium Copper Spiral | 0.72 | 0.36 | | x |
| 109 | 209 | Expanded Monel Sheat | 1.0 | 0.36 | x | |
| 110 | 210 | Silver Filled Rubber | 2.5 | 0.16 | x | |
| · | 300 | 0.32 cm Aperture Aluminum Honeycomb | 15.0 | 1.25 | x | |

**Figure 6.40a**   Gasket test samples. (From Ref. 16. © 1976, IEEE.)

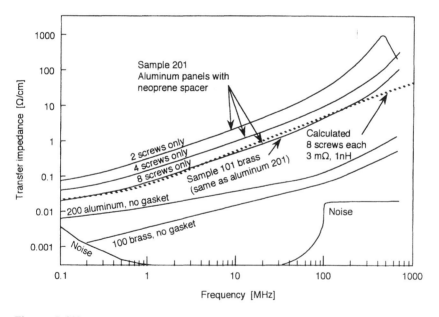

**Figure 6.40b**   Ungasketed and neoprene spacer transfer impedances. (From Ref. 16. © 1976, IEEE.)

a lower resistance than the aluminum, so the transfer impedance with the brass plate and no gasket or spacer is lower than with the aluminum. In Figure 6.40c the monel 195, 203, and 204 gaskets are not much better than the neoprene spacer up to 1 MHz, and neither is as good as the bare aluminum-to-aluminum joint up to 100 MHz. In figure 6.40d, the silver-filled rubber Xecom and porcupine gaskets are worse than the bare aluminum-to-aluminum joint from 0.1

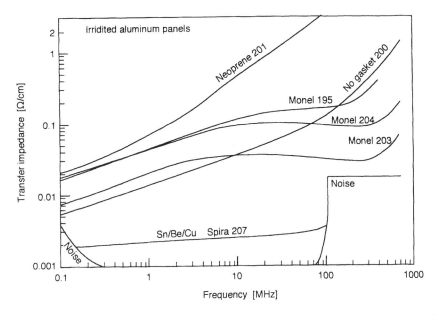

**Figure 6.40c** Transfer impedances with an aluminum panel. (From Ref. 16. © 1976, IEEE.)

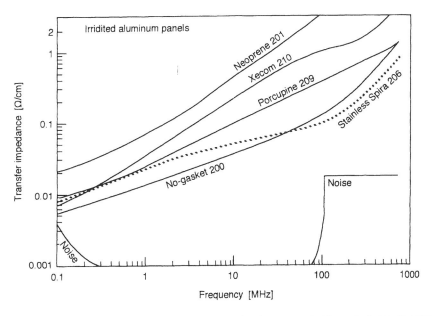

**Figure 6.40d** Transfer impedances with an aluminum panel. (From Ref. 16. © 1976, IEEE.)

MHz to 1 GHz. The transfer impedance of the stainless steel spira is almost the same as the no-gasket transfer impedance, so none of these gasket materials can be recommended over a bare metal-to-metal contact. From Fig. 6.40e the Xecom silver-filled rubber gasket used with a brass plate is no better than no gasket until above 60 MHz, and at 650 MHz the transfer impedance is 27% lower with than no gasket at all. The only gasket that shows a significant reduction in transfer impedance compared to no gasket is the Sn/Be/Cu spira in either a single-

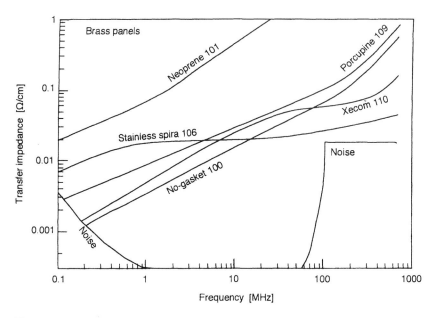

**Figure 6.40e**   Transfer impedances with a brass panel. (From Ref. 16. © 1976, IEEE.)

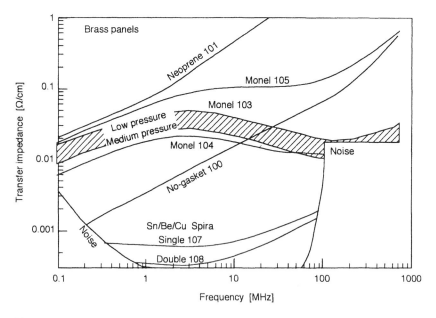

**Figure 6.40f**   Transfer impedances with a brass panel. (From Ref. 16. © 1976, IEEE.)

or double-row configuration as shown in Figure 6.40f. With either an aluminum plate or a brass plate, the Sn/Be/Cu spira gasket has a transfer impedance of 1.7% of the no-gasket brass plate at 100 MHz and 0.3% of the aluminum no-gasket plate at 100 MHz. Both the spira and Be/ Cu finger stock material is the gasket of choice when a gap must be filled. Even where environ-mental sealing is required, both types are available with a rubber or foam strip attached down the length of the gasket.

In life measurements, the increase in transfer impedance for stainless steel spira after storage for one year was minimal, and although a tin-plated beryllium spira gasket increased transfer impedance by up to a factor of 8, the transfer impedance after one year was still 1/100 of the transfer impedance of the stainless steel spira material. Figure 6.30 illustrates a very strong relationship between the surface resistance of the mating surfaces of an enclosure without a gasket and the magnetic field attenuation. Whereas Ref. 28 found that using a shield and cover plate with a surface preparation of either iridite, tin plate, cadmium plate, or nickel plate and aged for one year, a change in the contact resistance of the surface has very little impact on the transfer impedance, and therefore on the attenuation, when gaskets are used. Also, the gasket material is a very critical consideration in improving shielding effectiveness with beryllium copper, as seen in the other studies, significantly better than the other materials tested.

Reference 29 introduces the concept of the *effective transmission width* (ETW) of the gasketed seam. The ETW can be obtained from the transfer impedance of the gasketed aperture. Thus the shielding effectiveness of a gasketed joint in an enclosure can be obtained assuming the power loss through the gasket aperture is small relative to the wall losses of the enclosure.

Equation (6.30) provides the shielding effectiveness (SE) of an enclosure with a gasketed seam. The shielding effectiveness is directly proportional to the volume of the enclosure, which is true for the magnetic field shielding effectiveness of the enclosure shown in Eq. (6.28). The shielding effectiveness is inversely proportional to the Q of the enclosure, where the Q is the ratio of the energy stored in all resonant modes in the enclosure to that which would be stored with no losses. The Q of the enclosure is reduced when it is filled with electronics compared to an empty (high-Q) enclosure. The shielding effectiveness is also inversely proportional to the length of the gasket and the square of the magnitude of the transfer impedance of the gasket.

In addition, as we have seen, the shielding effectiveness is also dependent on the location where the field is measured within the enclosure:

$$SE = \frac{3Z_w V}{4l_g} \times Q|Z_t|^2 \tag{6.30}$$

where

$Z_w$ = wave impedance (m)

$V$ = volume of the enclosure (m)

$l_g$ = length of the gasket (m)

$Q$ = ratio of energy stored in all resonant modes in the enclosure to what would be stored with no losses

Some of the factors important in the choice of the correct gasket material follow.

*Level of E, H, or plane-wave attenuation required.*

*Maximum current-carrying requirement, resistance of material.* In addition to shielding, one use of gasket material is to provide a low-impedance path for noise current flow. Only certain materials can carry the high currents encountered in a lightning strike or EMP event (EMP typical current density is 3 A/mm).

*Gap tolerances.* When the gap to be filled does not exhibit a constant depth, a compressible material, with a first choice of finger stock, followed by wire mesh and finally conductive-cloth-over-foam type of gasket.

*Frequency of opening the aperture.* When a door or removable plate is opened often, then finger stock material, as used on shielded-room doors, is an obvious choice. A silicone foam with wire mesh may also be used. The foam strip is glued to one of the mating surfaces. Obvious though this may appear, experience has shown the importance of ensuring that the adhesive is not used between the wire mesh portion of the gasket and the enclosure!

*Fastener sizes and spaces.* The distance between fasteners must be small enough to avoid any buckling in sheet metal material, which will form gaps between the gasket and the enclosure. Fasteners at close distances used with a highly conductive coating on a thick material may eliminate the need for conductive gaskets!

*Mounting method.* Finger stock gasket material may be soldered or held in place by a metal strip fastened by screws or rivets or clipped over the edge of a frame or door (often small indentations are included in the finger stock clip that mate with small holes around the edge of the frame). When the spring finger of the finger stock touches both of the mating surfaces, the base of the finger stock may be glued in place. Conductive silicone materials may be glued, using conductive adhesive, or held in place until the fasteners are tightened. Conductive silicone may also be formed in place to fill slots in an enclosure. Wire mesh gaskets are available with attached neoprene foam that may be glued to the surface, with the foam acting as an environmental seal.

*Compression set.*

*Compression deflection.* Nonfoam silicone material is incompressible but will deform under pressure. Wire mesh gaskets are usually compressible, with the exception of sintered metal fiber, mesh in silicone, and wire screen types.

*Stress relaxation.* Load silicone material to 125% of rated, after which the material will relax to the rated load.

*Tensile strength.*

*Elongation.* This can increase the resistivity of silicone-loaded material and subsequent reduction in compression. Therefore, limit it to 10%, by the use of compression stops or containing the gasket material in grooves.

*Temperature range.* Most silicones are good from 55°C to 125°C.

*Outgassing.* This is usually achieved during the outgassing test on the material. For silicone conductive gaskets, some typical values, when tested for 24 hours at $10^{-6}$ torr and a temperature of 125°C, are: total mass loss 1% or less, collected volatile condensable 0.1%.

*Gas permeability.* For silicone-loaded material, this is usually less than for the base material.

*Flammability.* Most gasket material will support a flame but are self-extinguishing.

*Fungus resistance.* Most gasket materials will support fungus growth, but no internal change will occur.

*Cost.*

*Corrosion, oxidization, metal compatibility.* Dissimilar metals in contact will result in corrosion. The extent is dependent on the amount of moisture in the atmosphere and the incidence of condensation.

### 6.5.4 Waveguide Gaskets

A gasket material is often included in between the flanges of a waveguide. The reasons to use a conductive elastomer type of gasket are:

1.  To pressurize the waveguide to increase its peak power-handling capability
2.  To exclude the external environment, dust, or moisture

3. Usually incidentally, to seal the joint electromagnetically and reduce electromagnetic radiation

Two basic types of conductive elastomer exist. One is a thin, flat gasket, which is placed as a shim between the waveguide flange surfaces, the second is either an O- or D-shaped gasket, which is placed in a groove in one of the flange surfaces. With conductive elastomer gaskets, both a conductive metal-to-metal connection between the flange surfaces and a connection between the gasket material and the flange are made. The O- and D-shaped gaskets are suitable for choke-flange and grooved-contact-flange applications. If a thin, flat elastomer type of gasket is placed between the flange surfaces, which are then torqued up, the gasket material will cold-flow out of the joint. This results in the material's entering the waveguide opening and eventually results in a loose joint that will neither pressure-seal nor electromagnetically seal. Instead of using a solid elastomer, the cover-flange and flat contact-flange gaskets are die-cut from sheet stock containing an expanded metal reinforcement to eliminate cold-flowing of the elastomer material. This type of gasket can be supplied with a slightly raised lip around the iris opening for high-pressure, high-power applications. Standard waveguide gaskets are available that fit the standard UG, CPR, and CMR flanges, and custom gaskets can be made to cover the WR10 to WR23000 range of waveguides. The use of a silver-plated brass mesh gasket can be added to the groove of a flange choke. One reported problem with the fine wire mesh gasket is that the gasket can burn under heavy current, and the power-handling capability of this type of gasket remains unproven. None of the conductive elastomer gasket materials exhibit gasket contact impedances lower than predicted for a metal-to-metal flange impedance, in one case the gasket contact impedance was 390 times higher. However, none of the available waveguide gaskets use the elastomer material alone to seal the joint; an additional metal-to-metal contact is made. Tech-etch manufactures neoprene or silicone with aluminum and silicone in monel gaskets.

Parker seals manufactures a molded nonconductive gasket imbedded in a metal retainer. The main purpose of these gaskets is to allow pressurization of the waveguide or to environmentally seal the waveguide. Based on the Parker measurements, the RF flange resistance over the 2.6–3.6G-Hz frequency range was highest for pure conductive plastics, just lower for an impregnated wire mesh gasket, with the metal-to-metal flange gasket 1/9 of the impedance of the conductive plastic. Adding a knitted wire mesh gasket to an O-ring groove resulted in a very slightly lower impedance than with the metal-to-metal flange alone. The lowest impedance was a Parker knurled-aluminum gasket with a chemical film CL.3. The RF impedance of this gasket was 1/4 to 1/7 of the value for a bare metal-to-metal flange, so this would result in the lowest level of leakage from the waveguide joint. The most likely reason the serrated type of metal gasket is so successful at reducing the RF resistance is that the ridges are under tremendous pressure, due to the low surface area of the serration, and possibly cold-flow or weld into the adjacent flange material. The purpose of the nonconductive elastomer part of the Parker seal is for environmental or pressure sealing; if these are not required, then the use of the flange seal without elastomer is likely to be even better. The finish on the ridged gasket also affects the contact resistivity and therefore the RF impedance. The highest contact resistivity is achieved with a transverse ridged brass gasket with a Chem. Film CL3, the next lower resistivity is no finish (bare brass), followed by cadmium plate, gold, and silver, in that order. If a waveguide seam leaks around the edge of the flange, with or without a gasket, the gaps can be filled with conductive caulking, conductive adhesive, or silver-loaded grease. The addition of silver-loaded epoxy to the outside of the joint in a waveguide flange has been reported to reduce emissions by up to 20 dB, despite the epoxy's relatively high volume resistivity of approximately 2 m$\Omega$/ cm. The most probable reason for the silver-loaded epoxy's effectiveness is that the material enters the microscopic gaps in the two surfaces on the outside of the flange faces and, despite the poor conductivity, the skin depth of the epoxy in the 2–12-GHz range is small enough that virtually all of the current flows on the inside surface of the epoxy. One additional technique, applicable only to waveguide sections that contain standing waves, is to position waveguide flanges or components at locations of minimum current flow down the length of the waveguide. The voltage drop across the joint impedance is thereby reduced, as is the emission. The presence of transverse currents in the wall of the waveguide does not result in a voltage across the joint.

### 6.5.3 Conductive Finishes, DC Resistance, and Corrosion Effects on Gasket Materials

The DC resistance of materials and finishes is often increased by moisture, which increases corrosion, or by oxidization. Table 6.10, from Ref. 13, illustrates the initial DC resistance and the effect of 400 hr and 1000 hr at 95% relative humidity.

The DC resistance of a joint without gaps is a good indicator of impedance, for the inductance will be low and the contact impedance, from measurements, appears to track with resistance. Low values of DC resistance cannot be measured with any repeatability by the two-wire measuring technique, available with the majority of multimeters, due to the contact resistance of the probes. Instead, the four-wire technique described in Section 8.7.4 on bonding must be used. Here, a current is caused to flow across the two metal plates, which are joined together. The injection should be via two wide contacts; here, strips of conductive adhesive copper foil may be used. The voltage measuring contacts are then placed across the joint or seam, and the DC resistance is the ratio of measured voltage to injected current. Because the voltage-measuring contacts do not supply the current, the effect of contact impedance is greatly reduced.

**Table 6.10**  Resistance Measurements on Materials

|  |  | Resistance (milliohms) | | |
|  |  |  | At 400 hr | At 1000 hr |
| Material | Finish | Initial | 95% RH | 95% RH |
|---|---|---|---|---|
| Alum |  |  |  |  |
| 2024 | clad/clad | 1.3 | 1.1 | 2.0 |
| 2024 | clean only/clean only | 0.11 | 5.0 | 30.0 |
| 6061 | clean only/clean only | 0.02 | 7.0 | 13.0 |
| 2024 | light chromate conversion/same | 0.40 | 14.0 | 51.0 |
| 6061 | light chromate conversion/same | 0.55 | 11.5 | 12.0 |
| 2024 | heavy chromate conversion/same | 1.9 | 82.0 | 100.0 |
| 6061 | heavy chromate conversion/same | 0.42 | 3.2 | 5.8 |
| Steel |  |  |  |  |
| 1010 | cadmium/cadmium | 1.8 | 2.8 | 3.0 |
| 1010 | cadmium-chromate/same | 0.7 | 1.2 | 2.5 |
| 1010 | silver/silver | 0.05 | 1.2 | 1.2 |
| 1010 | tin/tin | 0.01 | 0.01 | 0.01 |
| Copper | clean only/clean only | 0.05 | 1.9 | 8.1 |
| Copper | cadmium/cadmium | 1.4 | 3.1 | 2.7 |
| Copper | cadmium-chromate/same | 0.02 | 0.4 | 2.0 |
| Copper | silver/silver | 0.01 | 0.8 | 1.3 |
| Copper | tin/tin | 0.01 | 0.01 | 0.01 |

Measurements of resistance across joints that are passivated by iridite are in the 10–100-μΩ region, whereas the joint impedance of two zinc dichromate passivated panels held together by pop rivets demonstrated a DC resistance of 1.5–2 mΩ. Anodizing, either hard black or clear, must not be used over contact areas over joints, seams, connector housings, panels, etc., for this finish is virtually an insulator. The surface resistivity of clear anodizing over a conductive surface was measured at a mean value of $5 \times 10^{11}$ Ω/cm, which is not even adequate for electrostatic discharge. When an enclosure must be anodized, mask those areas where joints are made, panels or doors are located, or connector housings mate. After anodizing, brush iridite on these areas. The brush iridite process is electroplating, with one electrode in the form of a brush used to locally apply the iridite.

Figure 6.30 illustrates the importance of low joint-contact resistance, for we see that the magnetic field attenuation of panels with 1.5 mΩ of DC seam resistance is up to 50 dB less than a panel with a 30-μΩ DC resistance.

The area, surface resistivity, and pressure on a joint affect the contact resistance. Table 6.11 from Ref. 30, shows the contact resistance of either bare or coated aluminum alloys. Five types of proprietary conversion coating with three types of plating and two joint compounds were measured. A summary of the findings is: Platings reduce the contact resistance of the joint when compared to bare aluminum, whereas chromate conversion coatings generally increase the contact resistance. Chromate-coating contact resistance varied from test to test and sample to sample over a wide range. Pressure cycling affects the contact resistance from cycle to cycle, with platings remaining reasonably stable and coatings showing random results. Conversion coating in contact with a plating shows a lower contact resistance than a joint with two conversion coatings. The typical reduction with pressure for a nickel bath plating over 6061/6061 with a surface roughness of 32 rms is shown in Figure 6.41.

A lower contact resistance than shown in Table 6.11 for iridite has been measured, of 60 μΩ for a 5.4-cm² area. Different formulations are available, so always specify the lowest-resistance version. Contact resistance of copper tape and epoxies are shown in Table 6.12. Note: much of this data is presented with different joint areas, and the pressure is unknown. However, those materials with a very high contact resistance can be identified.

Magnesium die castings have become popular for structural reasons, and the high-purity magnesium casting alloys AZ91D and AZ91E have demonstrated superior resistance to salt water corrosion, when compared to earlier commercial alloys. Data for different conductive treatments on AZ91D high-pressure die-cast magnesium in μΩ for 1-square-inch surfaces after exposure to humidity is shown in Table 6.13. #1, #20, and #21 are all acid chrome based. Neutral chromates, such as dichromate and dilute chrome pickle, have resistivities in the tens of thousands of micro-ohms and should be avoided. Zinc dichromate on aluminum also results in a significantly higher resistivity than electroplated zinc.

The closer the metal is in the galvanic series of metals, the lower the probability of corrosion. Figure 6.42 shows the standard potential of common metals used for shielding. The corrosion potential between silver and aluminum is expected to be high due to the large distance between the metals in the galvanic series. From measurements, the corrosion in a salt spray atmosphere of a silver-loaded silicone elastomer between aluminum plates was not excessive, possibly because the gasket material has a much higher resistance than pure silver, also because the gasket will seal the joint against the ingress of moisture. However, the measured transfer impedance of the silver-loaded silicone material when in contact with iridited aluminum increased from approximately 10 mΩ/m to 5Ω/m after a period of one year. In equipment designed for space application, negligible change is expected, due to the lack of an atmosphere. However, in ground-based applications, the potentially large increase in transfer impedance should be considered when making the choice of gasket material.

**Table 6.11**  Contact Resistance of 6061-T6 and 7075-T6 at 300 PSI (Milliohms) (First Pressure Cycle)

| Coating and surface roughness | 6061-T6/6061-T6 samples, joint area in in.$^2$ | | | 6061-T6/7075-T6 sample, joint area in in.$^2$ | | |
|---|---|---|---|---|---|---|
| | 10 | 5 | 2.5 | 10 | 5 | 2.5 |
| Bare aluminum—mill | 0.443 | 0.531 | 0.305 | 0.288 | 0.264 | 1.122 |
| Bare aluminum—32 RMS | — | 0.078 | — | — | 0.072 | — |
| Bare Aluminum—18 RMS | — | 0.126 | — | — | 0.072 | — |
| Alodine 1200—mill | 4.55 | 37.1 | 17.9 | 6.47 | 24.7 | 63.8 |
| Alodine 1200—32 RMS | — | 24.3 | — | — | 34.7 | — |
| Alodine 1200—64 RMS | — | 36.8 | — | — | 8.9 | — |
| Alodine 1500—mill | 0.639 | 5.80 | 23.4 | 5.89 | 5.06 | 4.82 |
| Alodine 1500—32 RMS | — | 2.14 | — | — | 14.7 | — |
| Alodine 1500—64 RMS | — | 13.1 | — | — | 7.97 | — |
| Alodine 600—mill | 5.87 | 21.6 | 12.4 | 8.4 | 14.9 | 24.6 |
| Alodine 600—32 RMS | — | 10.4 | — | — | 13.9 | — |
| Alodine 600—64 RMS | — | 10.8 | — | — | 7.9 | — |
| Tin—mill | 0.008 | 0.010 | 0.013 | 0.001 | 0.003 | 0.001 |
| Tin—32 RMS | — | 0.009 | — | — | 0.001 | — |
| Tin—64 RMS | — | 0.012 | — | — | 0.001 | — |
| Oakite—mill | 20.0 | 14.5 | 14.8 | 20.2 | 32.1 | 44.8 |
| Iridite—mill | 1.12 | 25.6 | 74.8 | 28.5 | 13.8 | 35.2 |
| Chromium—mill | 0.019 | 0.016 | 0.050 | — | — | — |
| Chromium—32 RMS | — | 0.017 | — | — | — | — |
| Alodine 1200/tin—mill | 0.332 | 0.249 | 0.539 | 0.606 | 1.01 | 0.715 |
| Alodine 1200/tin—32 RMS | — | 0.336 | — | — | — | — |
| Alodine 1200/tin—64 RMS | — | 0.090 | — | — | — | — |
| Nickel bath—mill | 0.009 | 0.023 | 0.035 | 0.023 | 0.006 | 0.016 |
| Nickel bath—32 RMS | — | 0.027 | — | — | 0.006 | — |
| Nickel bath—64 RMS | — | 0.029 | — | — | 0.008 | — |
| Nickel brush—mill | 0.010 | 0.011 | 0.008 | — | — | — |
| Nickel brush—32 RMS | — | 0.013 | — | — | — | — |
| Nickel brush—64 RMS | — | 0.009 | — | — | — | — |
| EJC | — | 0.193 | — | — | — | — |
| ALNOX | — | 0.018 | — | — | — | — |

*Source*: Ref. 30.

Tin-plated, stainless steel spiral, and monel mesh gaskets in contact with iridited, cadmium-plated, and nickel-plated panels exhibit a much lower increase in transfer impedance over a period of one year.

In Ref. 27, the shielding effectiveness of the Ag-Cu gasket when placed in contact with a stabilized copper-coated aluminum flange did not significantly change when exposed to a sulfur dioxide salt fog for 192 hours. Similarly, the shielding effectiveness of the silver-plated aluminum-filled fluorosilicone (Ag-Al) gasket, mated with the stabilized copper-coated aluminum flange, did not degrade by more than 14 dB when exposed. The volume resistivity of the gasket increased from 0.008 ohm-cm before exposure to 0.014 ohm-cm after. However, the shielding effectiveness of the nickel-coated graphite-filled fluorosilicone (NiGr) gasket material did degrade by up to 36 dB after exposure to $SO_2$ for 192 hours, so at some frequencies the shielding effectiveness of the NiGr gasket was only 24 dB higher than that of the flange without

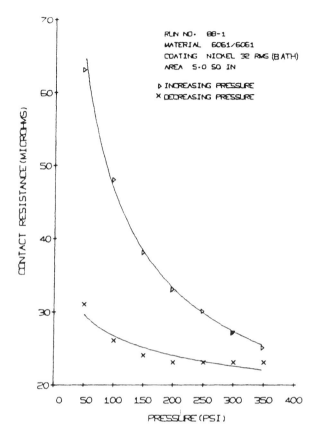

**Figure 6.41** Decrease in contact resistance with pressure for nickel bath plating.

**Table 6.12** Typical Contact Resistance of Coatings, Gaskets, Tapes, and Adhesives

| Material | Notes | Contact resistance | Area |
|---|---|---|---|
| Eccobond solder 57C | Cured at room temperature | 201–218 m$\Omega$ | 1 cm$^2$ |
| Chomerics 584–29 silver-loaded epoxy | Cured at room temperature | 106–268 m$\Omega$ | 1 cm$^2$ |
| Copper tape ribbed, nonconductive adhesive | Moderate finger pressure | 13 m$\Omega$ | 19 cm$^2$ |
| | After 24 hr | 9.4 m$\Omega$ | 19 cm$^2$ |
| | Full pressure | 37 m$\Omega$ | 19 cm$^2$ |
| | Low pressure | 11.7 m$\Omega$ | 19 cm$^2$ |
| Spraylat 599-A8219-1 copper conductive coating | Thickness = 2 mils | 16.8 m$\Omega$ | 8.82 cm$^2$ |
| Scotch 3M tape with conductive adhesive | | 11.7 m$\Omega$ | 19 cm$^2$ |
| Conductive fabric over neoprene gasket | Compressed to 2 mm | 550 m$\Omega$ | 1 cm$^2$ |
| Beryllium copper finger stock gasket | Compressed to 2 mm | 159 m$\Omega$ | 1 cm$^2$ |
| Multi-layers of very close wire mesh over neoprene | Compressed to 2 mm | 105 m$\Omega$ | 1 cm$^2$ |
| Tin-plated spring steel finger stock | Compressed to 2 mm | 8 m$\Omega$ | 1 cm$^2$ |

**Table 6.13** Contact Resistance of Conductive Treatments on AZ91D High Pressure, in μΩ, for 1-in.² Surfaces

| | Surface resistance, μΩ/days of cyclic humidity exposure | | |
| --- | --- | --- | --- |
| Treatment | Initial | 21 days | 50 days |
| Abraded | 70–140 | 140–340 | Erratic |
| #1 Chrome pickle MIL-M-3171, ASTM D 1732 | 140–200 | 90–180 | 90–130 |
| #18 Phospate | no cond. | no cond. | no cond. |
| #20 Modified chrome pickle | 150–220 | 130–170 | 90–150 |
| #23 Stannate immersion | 90–160 | 120–320 | 70–118 |
| #21 Ferric nitrate bright pickle | 50–130 | 90–120 | 10–50 |

gasket and with nonconductive compression stops, i.e., a seam with gaps. The volume resistivity of the NiGr gasket was highly variable, ranging from 0.085 ohm-cm to 0.372 ohm-cm.

The copper coating used in the tests was a two-part urethane coating containing a stabilized-copper filler, copper corrosion inhibitors, or aluminum corrosion inhibitor and an inorganic pigment, which was added over a MIL-C-5541 coating. This type of copper coating showed an initial SE 10–20 dB higher than a sample coated only with a conversion coating per MIL-C-5541, Class 3. Both types of coating with the Ag-Al gasket showed some degradation in SE after 1000 hours of salt spray.

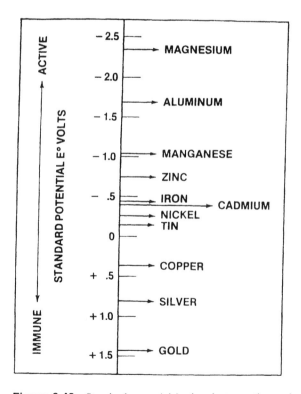

**Figure 6.42** Standard potential in the electromotive series for common materials.

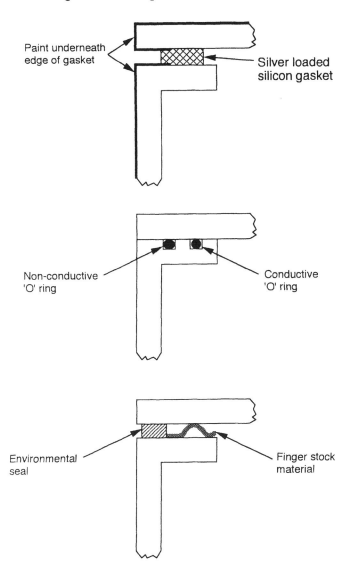

**Figure 6.43** Methods of excluding moisture from the conductive part of gaskets, thus reducing the possibility of corrosion.

Figure 6.43 illustrates methods of excluding moisture from the conductive part of gaskets thus reducing the probability of corrosion. Some of the many manufacturers of gasket material are as follows: Information on gasket material is available from Chomerics Ltd., 77 Dragon Court, Woburn, MA 01801. Chomerics produces an EMI/RFI gasket design manual and a gasket catalog. Tech-etch Ltd. of 45 Aldrin Road, Plymouth, MA, produces in addition to elastomer and metal mesh gaskets a range of beryllium copper finger material. Tecknit also manufactures a large range of gasket materials and supplies, a useful EMI shielding design guide, and a catalog, available from Tecknit, 129 Dermody Street, Cranford, NJ 07016. Instrument Specialities is another supplier of finger stock material. Spira, at 12721 Saticoy St. So., Unit B, No. Hollywood, CA 91605, manufactures the stainless steel and tin-plated beryllium copper spiral gasket material.

A manufacturer of conductive silver metallized nylon ripstop fabric bonded to foam is Schlegel, P.O. Box 23197, Rochester, NY 14692. The manufacturer of waveguide gaskets, as well as O-ring and molded gaskets is Parker Seals, 10567 Jefferson Blvd., Culver City, CA 90230. Two manufacturers of circuit board shields are Leader Tech, 14100 McCormick Drive, Tampa, FL 33626, and Boldt Metronics International, 345 North Erie Drive, Palatine, IL 60067.

## 6.6  PRACTICAL SHIELDING AND LIMITATION ON EFFECTIVENESS

Our discussion thus far has assumed the use of a totally enclosed box, cabinet, etc. Often, especially between printed circuit boards or to break an enclosure up into compartments, thin metal plates are used that are not totally connected around their peripheries to the inside of the enclosure. Figure 6.44 shows a metal plate inserted between a PCB that is a source of emission and a PCB containing susceptible circuits. The source of emission is likely to be due to switching current flowing around a current loop and thus a source of magnetic field. The current flow induced into the metal plate cannot be constant over the surface of the plate; instead it must reduce to zero at the edges of the plate. This edge effect means that the incident magnetic field is not canceled but reradiates from the edges of the plate, as shown in Figure 6.44.

Where the receptor circuit or PCB track to be shielded has a small area and is located close to the center of the PCB, the method of shielding described may be adequate. However, should the metal plate be grounded, either via a ground track on the receptor board or via a length of wire to the enclosure, the EMI situation may be exacerbated by either increased noise currents in the receptor PCB ground or radiation from the ground wire. If the distance between the two boards is kept constant and the metal plate is connected to signal ground on the receptor board, the coupling to the metal plate will be greater due to closer proximity to the source and the potential for EMI may again be increased.

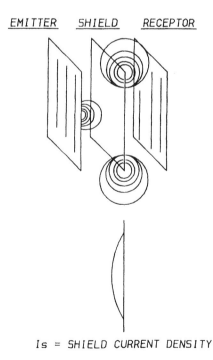

*EMITTER     SHIELD     RECEPTOR*

*Is = SHIELD CURRENT DENSITY*

**Figure 6.44**   Metal plate shield placed between two PCBs.

If the metal plate cannot be connected around its periphery to the inside of the enclosure, then there is no reason to ground it at all. Increasing the distance between the source and the receptor board with or without the metal shield may be an adequate solution.

Placing susceptible circuits in small, shielded enclosures on PCBs or in enclosures with shielded input, output, and power cables may be required for low-level high-input-impedance circuits. For example, a high-input-impedance op amp circuit with a gain of 1000 was used to increase the "pink noise" level from a noise source diode. The output of the amplifier should have been noncoherent noise, whereas the measured signal contained coherent amplitude-modulated signals from local AM transmitters. Placing a small enclosure made of copper foil laminated to cardboard over the circuit and connecting the shield to power ground reduced the interference. However, it was not until a ground plane was added under the enclosure on the track side of the board and electrically connected to the enclosure on the component side, thus completing the shield, that the interference was eliminated.

## 6.7 COMPARTMENTALIZATION

Placing susceptible circuits or noise sources in a compartment within an enclosure is an option that may be mandatory when dealing with high-level noise sources or low-level analog, video, or RF circuits. Just as with shielded enclosures, the effectiveness of the compartment may be compromised by bringing in power or signal lines that contain noise voltages. Feedthrough filters should be designed into the compartment. The need for the filters may then be assessed by removal, and those filters found unnecessary may be replaced with nonfilter-type feedthroughs.

When apertures exist in the compartment, the levels of E and H fields generated from noise sources should be calculated from the equations of Chapter 2, the coupling through the aperture may be predicted from equations presented in this chapter and the response of the receptor circuit may be predicted using the techniques described in Chapter 2.

## 6.8 SHIELDING EFFECTIVENESS OF BUILDINGS

Shielded rooms are designed to achieve specific levels of attenuation and are used principally for EMI measurements and to contain classified electronic information or restricted transmissions. The characteristics of shielded rooms used in EMI measurements are discussed in Chapter 9. In this section we examine the shielding effectiveness of commercial buildings and residential houses. The reduction in the signal strength of a radio transmission as it propagates inside a building is undesirable when reception is required. Most of us have noticed the reduction in signal strength as a portable radio is moved from a position close to a window to one further inside the building. However, building attenuation is a desired attribute when the source of the field is either high-level broadcast transmissions, lightning, or radar, which have the potential for interference.

In an EMC/EMI investigation that includes a building in the radiated propagation path, data on the average attenuation for different types of buildings is required. When susceptible equipment is contained in the building, the magnitude of the field incident on the building may be predicted from the distance and source characteristics. When the building is used to shield a receptor from a source, the diffraction of the wave around the building must be accounted for, in addition to attenuation through the building. Diffraction and coupling around a structure are discussed in Chapter 10.

As noted, in enclosure shielding the E field and H field components may not be equally attenuated by a material, and the wave impedance may change after transmission through a medium.

In Ref. 17, both the E and H fields of ambient broadcast signals were measured inside and outside seven different types of buildings:

1. Single-family detached residence of the split-level ranch type. *Materials*: wood-framed with wood and brick siding, aluminum foil–covered insulation in the ceiling.
2. Single-family detached residence of the wood-framed, raised ranch type. *Materials*: concrete block walls on the lower levels and aluminum siding covering the upper levels; brick veneer front on the main level.
3. Single-story concrete block. *Materials*: concrete blocks with a large storefront window; structure framed with steel columns and a steel bar joist–supported roof; interior divided into cubicles with demountable steel and glass. An extensive array of ducts, pipes, and cables occupied the space above a suspended ceiling.
4. Single-story concrete block. *Materials*: concrete block, steel-framed building constructed on a reinforced-concrete slab. Interior description similar to 3.
5. Four-story office. *Materials*: Steel-framed with preformed-concrete exterior wall panels; one side covered with metal exterior wall panels. The floors are corrugated steel covered with poured concrete. Interior description similar to 3.
6. Four-story office. *Materials*: steel frame with brick exterior walls; corrugated steel floors covered with poured concrete. Interior description similar to 3.
7. Twenty-story office. *Materials*: Steel frame with marble exterior wall panels; interior steel columns spaced approximately 7 m apart; corrugated steel floors covered with poured concrete. The interior is uncluttered, with few interior walls and partitions.

The level of attenuation depends on the interior location at which the measurement was made. Thus measurements were made at a number of different locations in each building. Table 6.14 provides a summary of the fitted mean, minimum, and maximum measured attenuations of the seven buildings.

The attenuation of UHF radio signals by buildings has been measured in Ref. 18. Table 6.15 gives a summary of the effects of frequency and building type on the attenuation. The average attenuation from measured data is 6.3 dB. The table shows correction factors that must be added to or subtracted from 6.3 dB to account for frequency, room location, and building materials used in the construction.

When equipment housed in a building is found to be susceptible to a source of high-power radiated emissions, such as a transmitter, and it can be proven that the EMI problem is marginal, then a few decibels of additional attenuation may be all that is required to eliminate EMI. The first step in such an investigation is to prove that the coupling path is truly radiated to the equipment inside the building and not radiated to or conducted on external power or signal lines. The next step is to ensure that a modicum of additional shielding is all that is required.

One method of enhancing the building attenuation is to paint it with conductive paint. In Ref. 19, a powdered nickel acrylic base conductive paint was used on concrete or concrete block walls. The measured attenuation after painting increased by 20 dB over the 100–350-MHz frequency range and by 16 dB for the pulsed field from lightning. Additional shielding at windows may be achieved by the use of wire mesh in a frame electrically connected to the conductive paint. A transparent plastic film coated with copper or gold may also provide sufficient shielding at windows, but is a more costly solution. Care must be taken to ensure that the conductive side of the film contacts the paint either directly or via a metal window frame.

When a higher level of shielding than achievable with conductive paint is required, the use of sheets of metal foil soldered or taped together on the inside walls and ceiling is an alternative. Earth is a good absorber of electromagnetic radiation; therefore, installing equipment

**Table 6.14** Building Attenuation

| Building | Frequency | H field mean [dB] | H field limits [dB] | E field mean [dB] | E field limits [dB] |
|---|---|---|---|---|---|
| 1 | 20 kHz | 0 | — | 32 | — |
|   | 1 MHz | 0 | — | 30 | — |
|   | 500 MHz | 0 | −5 to 3 | 0 | −5 to 8 |
| 2 | 20 kHz | 3 | — | 22 | 12 to 33 |
|   | 1 MHz | 0 | −5 to 2 | 9 | −8 to 34 |
|   | 500 MHz | 10 | 2 to 20 | 12 | 7 to 20 |
| 3 | 20 kHz | −1 | — | — | — |
|   | 1 MHz | 3 | — | 18 | — |
|   | 500 MHz | 8 | −3 to 19 | 8 | −3 to 15 |
| 4 | 20 kHz | −3 | | 28 | |
|   | 1 MHz | 12 | 5 to 25 | 28 | 12 to 28 |
|   | 500 MHz | 8 | — | 10 | — |
| 5 | 20 kHz | 2[a] | — | — | — |
|   | 1 MHz | 3[a] | — | 22[a] | — |
|   | 500 MHz | 6[a] | — | 3[a] | — |
| 6 | 20 kHz | 3[a] | — | 32[a] | — |
|   | 1 MHz | 8[a] | — | 28[a] | — |
|   | 500 MHz | 10[a] | — | 10[a] | — |
| 7 | 20 kHz | 20[a] | — | 40[a] | — |
|   | 1 MHz | 20[a] | — | 35[a] | — |
|   | 500 MHz | 10[a] | — | 10[a] | 0 to 25[a] |

[a] Measured 1 m from the outer wall of the buildings. Measurements made 15 m from the outer wall showed attenuation of up to 35 dB for magnetic fields and up to 50 dB for electric fields. Mobile homes provide an average of 28 dB of attenuation over the 20-kHz to 500-MHz frequency range.

**Table 6.15** Attenuation Correction Factors

| | | Correction factor[a] [dB] |
|---|---|---|
| Frequency | 2569 MHz | 1.160 |
|  | 1550 MHz | 0.390 |
|  | 860 MHz (V) | −0.169 |
|  | 860 MHz (H) | 0.140 |
| Construction | Wood siding | −0.58 |
|  | Brick veneer | 0.58 |
| Insulation | Blown in ceiling | −0.8 |
|  | Ceiling and walls | 0.8 |
| Room position | Exposed walls | −0.3 |
|  | No exposed walls | 0.3 |

[a] Average attenuation from which the correction factors are added/subtracted is 6.3 dB.

below ground level and/or building up a 2-m-thick wall of earth around the room housing the equipment is another potential solution.

## 6.9  COMPUTER PROGRAM FOR EVALUATING SHIELDING EFFECTIVENESS

```
'Program for evaluating shielding effectiveness
'Function for logarithms to base 10
DEF FNLOG (x)
FNLOG = LOG(x)/LOG(10)
END DEF
'Initialize constants for PI and speed of light as well as other
'parameters used in the program
C = 3E + 08
PI = 3.14159
f = 100: Gr = .35: Ur = 1: R1 = 8.78: t = 62.5: ANS$ = "H"
GOTO ComputeAttenuation:
AttenuationInputs:
CLS
PRINT "COMPUTATION OF SHIELDING ATTENUATION"
PRINT ""
PRINT "INPUTS:"
PRINT "[A] Compute H or E-field reflection", ANS$
PRINT "[F] Frequency, F",, f; "Hz"
PRINT "[D] Distance from conductor, R1", R1; "inches"
PRINT "[T] Thickness of shield, t", t; "mils"
PRINT "[G] Relative conductivity, Gr", Gr
PRINT "[U] Relative permeability, Ur", Ur
PRINT "[X] Exit program"
PRINT "Please select variable you wish to change (A,B,etc)"
PRINT ""
PRINT "OUTPUTS:"
GOSUB Outputs1:
MonopoleLoop:
SEL$ = INKEY$
IF LEN(SEL$) > 0 THEN LOCATE 11, 1
IF LEN(SEL$) > 0 THEN PRINT SPACE$(70)
LOCATE 11, 3
IF (SEL$ = "A") OR (SEL$ = "a") THEN INPUT "Enter H-field(H)/E-field(E) ", ANS$
IF (SEL$ = "F") OR (SEL$ = "f") THEN INPUT "Enter Frequency ", f
IF (SEL$ = "G") OR (SEL$ = "g") THEN INPUT "Enter relative conductivity ", Gr
IF (SEL$ = "U") OR (SEL$ = "u") THEN INPUT "Enter relative permeability ", Ur
IF (SEL$ = "D") OR (SEL$ = "d") THEN INPUT "Enter distance from conductor ", R1
IF (SEL$ = "T") OR (SEL$ = "t") THEN INPUT "Enter thickness of shield ", t
IF (SEL$ = "X") OR (SEL$ = "x") THEN END
IF LEN(SEL$) > 0 THEN GOTO ComputeAttenuation:
GOTO MonopoleLoop:
ComputeAttenuation:
IF (ANS$ = "E") OR (ANS$ = "e") THEN ANS$ = "ELECTRIC (E-field)"
```

IF (ANS$ = "H") OR (ANS$ = "h") THEN ANS$ = "MAGNETIC (H-field)"
LAMBDA = (C / f) * 39.37 'LAMBDA = wavelength in inches
Tcm = t * .00254 'Shield thickness in cm for calculation
      'of skin depth
'Reflection of electric field
Re = 353.6 + (10 * FNLOG(Gr/(f ^ 3 * Ur * R1 ^ 2)))
 'Reflection of magnetic field
Rm = 20 * FNLOG(((.462/R1) * SQR(Ur / (Gr * f))) + ((.136 * R1) / SQR(Ur / (f * Gr)))
+ .354)
'Absorption by shield
A = .003334 * t * SQR(Ur * f * Gr)
'Metal impedance
Zm = 369 * SQR((Ur * (f / 1000000!)) / Gr) * .000001
'Wave impedance
GOSUB ComputeZw: 'Zw is different depending on whether one is looking
      'at magnetic or electric fields
K = ((Zm − Zw) / (Zm + Zw)) ^ 2
'K is a function of Zw and is different depending on the type of field as well
IF K < 1E-09 THEN K = 1E-09'Program sets minimum K at 1.0E-05
'The re-reflection, Rr, will be different depending on type of field since
'it is a function of K
Rr = 20 * FNLOG(1 − (K * 10 ^ (−.1 * A) * EXP(−.227 * A)))
SEe = Rr + Re + A
SEm = Rr + Rm + A
GOTO AttenuationInputs:
Outputs1:
IF ANS$ = "ELECTRIC (E-field)" THEN GOSUB ElectricFieldOutputs:
IF ANS$ = "MAGNETIC (H-field)" THEN GOSUB MagneticFieldOutputs:
RETURN
ElectricFieldOutputs:
PRINT "Shielding Attenuation",, SEe; "dB"
PRINT "Wave Impedance",,Zw; "Ohms"
PRINT "Re-reflection attenuation, Rr",Rr; "dB"
PRINT "Reflection of electric field, Re",Re; "dB"
PRINT "Absorption loss, A",, A;"dB"
PRINT
RETURN
MagneticFieldOutputs:
SkinDepth = .0066 / SQR(Ur * Gr * (f / 1000000!))
Zb = 369 * SQR((Ur * (f / 1000000!)) / Gr) * (1 / (1 − EXP(−Tcm / SkinDepth)))
Rrb = 20 * FNLOG(1 − ((((Zb * .000001) − Zw) / ((Zb * .000001) + Zw)) ^ 2) * 10 ^
(−.1 * A) * EXP(−.227 * A))
PRINT "Shielding Attenuation",,SEm; "dB"
PRINT "Wave Impedance",,Zw; "Ohms"
PRINT "Re-reflection attenuation, Rr", Rr; "dB"
PRINT "Re-reflection based on Zb, Rrb",Rrb; "dB"
PRINT "Reflection of magnetic field, Rm",Rm; "dB (Based on Zm)"
PRINT "Absorption loss, A",,A;"dB"
IF A > 10 THEN PRINT : PRINT "NOTE: For A > 10: A = 10 in calculating Rr"

```
RETURN
ComputeZw:
IF (ANS$ = "MAGNETIC (H-field)") AND (R1 < LAMBDA / (2 * PI)) THEN
   Zw = (377 * 2 * PI * R1) / LAMDDA
END IF
IF (ANS$ = "ELECTRIC (E-field)") AND (R1 < LAMBDA / (2 * PI)) THEN
   Zw = (377 * LAMBDA) / (2 * PI * R1)
END IF
IF R1 > = LAMBDA / (2 * PI) THEN Zw = 377
RETURN
```

## REFERENCES

1. J. R. Moser. Low frequency shielding of a circular loop electromagnetic field source. IEEE Trans. on Electromag. Compat.. Vol. EMC9, No. 1, March 1967.
2. A. Broaddus, G. Kunkel. Shielding Effectiveness Test results of Aluminized Mylar. IEEE EMC Symposium Record, 1992.
3. R. B. Cowdell. Nomograms simplify calculations of magnetic shielding effectiveness. EDN p. 44, September 1972.
4. T. Sjoegren. Shielding effectiveness and wave impedance. EMC Technology, July/August, 1989.
5. A. C. D. Whitehouse. Screening: new wave impedance for the transmission line analogy. Proc. IEEE, Vol. 116, No. 7, July 1969.
6. D. R. Bush. A simple way of evaluating the shielding effectiveness of small enclosures. IEEE EMC Symposium Record.
7. L. O. Hoeft, J. W. Millard, J. S. Hofstra Measured magnetic field reduction of copper sprayed panels. IEEE EMC Symposium Record, Sept. 16–18, 1985, San Diego, CA.
8. R. K. Keenan. Digital Design for interference specifications. The Keenan Corporation.
9. M. P. Robinson, T. M. Benson, C. Chrisopoulus, J. F. Dawson, M. D. Ganley, A. C. Marvin, S. J. Porter, D. W. P. Thomas. IEEE Transactions on Electromagnetic Compatibility, Vol. 40, No. 3, August 1998.
10. H. Bloks. NEMP/EMI Shielding. EMC Technology, Nov./Dec. 1986.
11. L. O. Hoeft, J. S. Hofstra. Experimental and theoretical analysis of the magnetic field attenuation of enclosures. IEEE Trans. Vol. 30, No. 3, August 1988.
12. L. O. Hoeft. How big a hole is allowable in a shield: theory and experiment. 1986 IEEE EMC Symposium Record. Sept. 16–18, 1986, San Diego, CA.
13. E. Groshart. Corrosion control in EMI design. 2nd Symposium and Technical Exhibition on Electromagnetic Compatibility. Montreux, Switzerland, June 28–30, 1977.
14. A. N. Faught. An introduction to shield joint evaluation using EMI gasket transfer impedances. IEEE International Symposium on EMC. Sept. 8–10, 1982, New York.
15. R. J. Mohr. Evaluation Techniques for EMI seams. IEEE International Symposium on EMC. Aug. 25–27, Atlanta, GA.
16. P. J. Madle. Transfer impedance and transfer admittance measurement on gasketed panel assemblies and honeycomb air vent assemblies. IEEE International Symposium on EMC, 1976.
17. A. A. Smith. Attenuation of electric and magnetic fields by buildings. IEEE Trans. on EMC. Vol. EMC20, No. 3, Aug. 1978.
18. P. I. Wells. IEEE Transactions on vehicular technology. Vol. VT 26, No. 4, November 1977.
19. H. E. Coonce. AT&T Bell Labs, G. E. Marco, AT&T Technology Inc. IEEE National Symposium on EMC, Apr. 24–26, 1984.
20. L. O. Hoeft, T. M. Sales, J. S. Hofstra. Predicted shielding effectiveness of apertures in large enclosures as measured by MIL-STD-285 and other methods. IEEE National Symposium on EMC. May 23–25, 1989, Denver, CO.

21. L. O. Hoeft, J. W. Millard, J. S. Hofstra. Measured magnetic field reduction of copper sprayed panels. IEEE EMC Symposium. Aug. 20–22, 1985, Wakefield, MA.

22. L. O. Hoeft. The case for identifying contact impedance as the major electromagnetic hardness degradation factor. IEEE EMC Symposium Record. Sept. 16–18, 1986, San Diego, CA.

23. F. Olyslager, E. Laermans, D. De Zutter, S. Criel, R. D. Smedt, N. Lietaert, A. De Clercq. IEEE Transactions on Electromagnetic Compatibility, Vol 41, No 3, August 1999.

24. B. Archamneault, C Brench. Shielded air vent design guidelines from EMI modeling. IEEE International Symposium Record, 1993.

25. J. P. Quine. Theoretical formulas for calculating the shielding effectiveness of perforated sheets and wire mesh screens.

26. G. J. Freyer, J. Rowan, M. O. Hatfield. Gasket shielding performance measurements obtained from four test techniques. IEEE International Symposium on EMC, 1994.

27. P. Lessner, D Inman. Quantitive measurement of the degradation of EMI shielding and mating flange materials after environmental exposure. IEEE International Symposium on EMC, 1993.

28. G. Kunkel. Corrosion effects on field penetration through apertures. IEEE Electromagnetic Compatibility Symposium Record, 1978.

29. IEEE Guide for the Electromagnetic Characterization of Conductive Gaskets in the Frequency Range of DC to 18 GHz. IEEE Std 1302-1998.

30. B. Kountanis. Electric contact resistance of conductive coatings on aluminum. IEEE Electromagnetic Compatibility Symposium Record, 1970.

# 7

# Cable Shielding, Coupling from E and H Fields, and Cable Emissions

## 7.1  INTRODUCTION TO CABLE COUPLING AND EMISSIONS

Cables are a major source of radiated emissions and receptors in electromagnetic coupling. Commercial equipment that is designed for EMC and includes a well-thought-out grounding scheme and wiring and PCB layout can meet commercial EMC requirements with a number of unshielded cables connected to the equipment. When the equipment enclosure is also unshielded, the sources of emission are likely to be both the enclosure and the cables, although in some cases emissions from cables are higher when the enclosure is shielded!

For equipment that must meet the more stringent MIL-STD-461 RE02 or similar requirements, the use of unshielded cables is the exception. Assuming equipment does not generate noise voltages, which is typical of analog, video, or low-level RF circuits with linear power supplies, the level of noise current on unshielded cables may be low enough to ensure meeting the radiated emission requirements. However, the same low-signal-level equipment is likely to be susceptible to noise voltage induced in unshielded cables during exposure to the E field generated in the radiated susceptibility RS03 test. In many cases, even equipment with shielded cables fails these tests. This chapter describes some of the reasons why this is possible.

Apart from EMC requirements, the severity of the EM environment may necessitate the use of shielded cables. Also, some standard interfaces, such as GPIB, MIL-STD 1553, and Ethernet, require the use of shielded cable.

The use of fiber-optic cables may appear to eliminate the need for concern about cable radiation and coupling. However, with the present high cost of fiber-optic links and the need for power interconnections, the use of shielded cables is not quite obsolescete.

## 7.2  CABLE SHIELDING EFFECTIVENESS/TRANSFER IMPEDANCE

The shielding effectiveness of a shielded cable is dependent on a number of factors in addition to the cable characteristics: cable length relative to the wavelength of the field incident on the cable or the wavelength of the current flow on the cable core (where this current flow is the source of emission), the wave impedance (i.e., a predominantly H field, E field, or a plane wave), and the termination of the shield. One of the most common questions is how to terminate the shield of a cable, whether at both ends or at one end only, and if so which end. We shall see indeed that how the shield of a cable is terminated has an effect on its shielding effectiveness. For all types of shielded cable, the shielding effectiveness is not constant with frequency but can be characterized over approximately three frequency ranges:

60 Hz–100 kHz
100 kHz–30 MHz
30 MHz–10 GHz

The physical characteristics of a cable play a large role in its shielding effectiveness. Types of shielded cable are single braid, double braid, triple braid, two types of foil, braid and foil, conduit, semirigid and flexible corrugated conduit. Many cable manufacturers publish curves of attenuation plotted against frequency. The method of measurement is of importance in the evaluation of these curves. For example, the induction or near-field leakage information is found using probes, current loops, test fixtures, etc., whereas in far-field measurements large broadband antennas are often used. Some of these test methods are described as follows:

*TEM Cell*: The TEM cell is a transmission line used either to generate transverse electromagnetic waves or to measure the radiation from a current-carrying cable. A cable may generate radial or axial fields as well as TE and TM modes, and ideally all these should be measured. However, it is believed that the TEM mode is the predominant one.

*Absorbing Clamp*: The clamp inductively detects emissions from a cable in the frequency range 30–1000 MHz. The clamp fits over the cable under test, and the radiation from the shielded cable is compared to the radiation of an unshielded cable carrying the same current. The difference between the two radiated values is the shielding effectiveness of the cable.

*Antenna Site*: The open field site is most often used to measure the emissions from a current-carrying cable and is of particular use when the cable interconnects two pieces of equipment. Broadband antennas, a spectrum analyzer with a preamplifier or an EMI receiver with peak hold capability are required for this test setup. This test method is often limited, especially at high frequency, by the ambient noise level; it is, however, an effective method of measuring the radiation from a cable, especially in the far field. The height of the cable above the ground plane and the cable length are two important parameters affecting the results obtained using this method. A source of error in this test is radiation from the signal source and its power lead.

Due to the number of variables, the reproducibility obtained with this test method is relatively poor. When the actual configuration in which the cable will be used, i.e., connected to equipment, vehicles, or close to structures and ground, is being simulated, this is the method of choice.

*Shielded-Room Test Method*: This test method minimizes the coupling between the signal source and the receiving antenna by enclosing the receiving antenna inside a shielded room, with the cable entering and exiting the room via feedthrough fixtures connected to the wall of the room. This test method suffers from the same room resonance and reflection errors inherent in the MIL-STD-285 test method.

*Reverberation Chamber or Mode-Stirred Chamber Test Method*: The reverberating chamber or mode-stirred chamber is well suited to immunity testing and for cables and connectors, because the method is not sensitive to cable layout. Testing on components has been performed up to 40 GHz.

*Test Fixtures*: Many test fixtures have been designed and built to measure the transfer impedance of cables using the methods described in Section 7.2.2. The use of an outer solid tube, with the coaxial cable under test placed concentrically inside the tube, the whole forming a triaxial transmission line, is a common test jig. The triaxial assembly is terminated either in its characteristic impedance or in a short circuit, and a current is caused to flow in the cable under test. Such a test fixture is capable of transfer impedance measurement up to 3 GHz. However, at these high frequencies, great care is required in the correct termination of the cable, which tends to dominate the transfer impedance, especially when very short lengths of cable are used. Alternatively, any error due to an impedance mismatch must be well understood. In addition, a difference in the propagation delay in the shielded cable under test, which will have a core insulation with a relative permittivity greater than 1, and the air cored test fixture will introduce long-line effects.

One simple transfer impedance test method has been adopted in the European International Electrotechnical Commission Publication IEC 96-1, Radio Frequency Cables. In this method a current is injected into a shielded cable by an injection cable that forms a transmission line in parallel with the shielded cable under test. One possible criticism of test methods confined to current injection on the shield of the test cable is the omission of electric field coupling through the transfer admittance of the cable. Generally we find the transfer impedance the most useful measure of the shielding effectiveness of a cable in EMI predictions.

The concept of transfer impedance was introduced in Section 6.5, which described gasket transfer impedance. Cable surface transfer impedance is specified in milliohms or ohms per meter length of cable. To obtain the overall shielding picture, we must include the shield termination technique and its transfer impedance. Where the shield connection is via the connector, the transfer impedance of the backshell plus that of the two mating halves of the connector, as well as the receptacle-to-bulkhead transfer impedance, must be included.

The transfer impedance is of use when the current flow on the cable sheath is known. The current flow on the shield may be due to a wave incident on the cable or to the signal current when the shield is the return path for the core current or to common-mode noise between the chassis of two units of equipment connected together by the cable shield.

There often is a relationship between the specified shielding effectiveness of a cable and its transfer impedance. One definition of cable shielding effectiveness is the ratio of the current flow on the shield of a cable ($I_s$) to the current flow on the core/s ($I_c$), usually expressed in decibels:

$$SE = 20 \log\left(\frac{I_S}{I_c}\right) \quad [\text{dB}] \tag{7.1}$$

The core current $I_c = V_{oc}/2R_o$, where $R_o$ is the termination resistance at each end of the shielded cable. When $R_o$ is not given, the assumption that $R_o$ equals the characteristic impedance of the cable will be correct for the majority of the test methods used. $V_{oc}$ is the open-circuit voltage on the core, and for electrically short lengths of cable,

$$V_{oc} = Z_t I_S l \tag{7.2}$$

where $l$ = length of cable [m].

For samples that are electrically short, one can obtain the shielding effectiveness in terms of the transfer impedance and the termination resistance:

$$SE = 20 \log\left(\frac{2R_o}{Z_t} l\right) \tag{7.3}$$

Due to the different definitions of shielding effectiveness and the various test methods, the conversion of a manufacturer's shielding effectiveness figure to transfer impedance may not be without error. However, when the transfer impedance is not otherwise available, and assuming the cable is terminated in its characteristic impedance:

$$Z_t = \frac{2R_o}{10^{SE/20}} \tag{7.4}$$

### 7.2.1  Frequency Dependency 60 Hz–100 kHz

For the majority of flexible coaxial cables, the wall thickness of the shield does not approach one skin depth until approximately 100 kHz. The skin depth is defined as the surface thickness of a metal in which 63% of the current is flowing. For copper, skin depth $= 0.0066\,f$ (MHz) [cm]. The transfer impedance below 100 kHz, except for thick-walled shields such as conduit, is approximately equal to the shield resistance $R_S$.

Other factors important in assessing the shielding effectiveness of a coaxial cable at low frequency are cable inductance and the eccentricity of the inner conductor/s relative to the outer shield. A coaxial cable exhibits an inductance per unit length of

$$L_c = 0.14 \, \log\left(\frac{d_S}{d_i}\right) \qquad [\mu H/ft] \tag{7.5}$$

where $d_S$ is the diameter of the shield and $d_i$ is the diameter of the center core. For a coaxial cable above a ground plane, the inductances of inner and outer conductors are

$$L_i = 0.14 \, \log\left(\frac{4h}{d_i}\right) \qquad [\mu H/ft] \tag{7.6}$$

$$L_S = 0.14 \, \log\left(\frac{4h}{d_S}\right) \qquad [\mu H/ft] \tag{7.7}$$

where $h$ is the height above the ground plane. The mutual inductance between the inner and outer conductors is given by

$$L_m = 0.14 \, \log\left(\frac{4h}{d_S}\right) \qquad [\mu H/ft] \tag{7.8}$$

It can be seen that Eq. (7.7) equals Eq. (7.8), so the mutual inductance is identically equal to the shield inductance. This can be visualized by considering that all the flux produced by the shield current encircles the center conductor. The mutual inductance is independent of the position of the center conductor relative to the shield. Further, the cable inductance $L_c = L_i - L_S$.

In transmission-line theory, the mutual inductance of a transmission line such as a coaxial cable is normally ignored, because the assumption is made that equal and opposite currents flow and therefore the mutual inductance between shield and center conductor is effectively zero. Figure 7.1a shows the schematic of a coaxial cable connected at both ends to a ground plane with a signal current flow on the cable. This is a typical shield connection and is effective at shielding against both the ingress and the egress of magnetic radiation, above a given frequency, as well as electric fields. As we shall see later, if either or both ends of the cable shield were removed from the ground plane, then the magnetic field shielding is improved. Figure 7.1b is the equivalent circuit of the cable and Figure 7.1c is the equivalent open-wire line producing the same external magnetic field as the shielded-wire line produces.

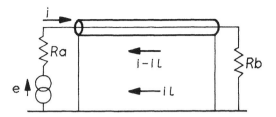

**Figure 7.1a** Schematic of a coaxial cable with the shield connected to a ground plane at both ends. (© 1967, IEEE.)

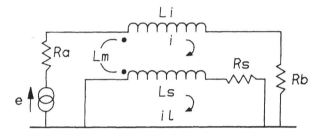

**Figure 7.1b** Equivalent circuit of the cable. (© 1967, IEEE.)

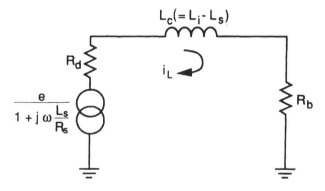

**Figure 7.1c** Equivalent open-wire line producing the same external field as a shielded-wire line produces. (© 1967, IEEE.)

How the attenuation of magnetic fields is achieved is described as follows. The leakage current flow in the ground plane $I_L$ is the current $i$ attenuated by the factor

$$\frac{1}{1 + j\omega \dfrac{L_S}{R_S}} \tag{7.9}$$

From Eq. (7.9) it can be seen that at low frequency the cable is ineffective at reducing the magnetic field from the cable because $I_L$ is large. At high frequencies the magnitude of $I_L$ reduces, the current in the shield approaches the center conductor current due to the mutual inductance between the center conductor and the shield, and the external magnetic field tends to cancel. The position of the center conductor relative to the shield plays a role in the external magnetic field produced by the cable. With equal shield and center conductor current and a perfectly concentric cable, the external magnetic field is exactly canceled.

One application of the effect of mutual inductance between pairs, or a number, of wires in close proximity (bundled) is to provide a return wire in the bundle with the supply or signal wires, even when the return is connected to ground at both ends. Then, at higher frequencies, a large proportion of the supply or signal return current will flow in the return wire even though the DC resistance of the alternative ground path is lower than that of the return wire. The attenuation coefficient $a$ is plotted versus $(fL_S/R_S)$ in Figure 7.2. For the low-frequency case, where a current flow is either induced into the shield of a cable due to an incident magnetic field or caused by a common-mode voltage between signal return and the ground plane, the voltage appearing across the shield $V_S$ is equal to

$$V_S = -j\omega M I_S + j\omega L_S I_S + I_S R_S \tag{7.10}$$

Since $L_S = M$, then $V_S = R_S I_S$. Therefore, $R_S$ at low frequency equals the transfer impedance. In Figure 7.3a the current path is shown for common-mode noise, and it can be seen that $V_S$ adds to the signal voltage $V$ and appears as noise in the signal.

The shielding effectiveness of the coaxial cable against a magnetic field may be seen by comparing the shielded case to an unshielded cable. The unshielded case is shown in Figure 7.3b. Here the induced current due to the magnetic field flows through the signal and source impedances and generates a noise voltage across the load $R_L$. With the shield connected, as shown in Figure 7.3c, the current flow in the shield generates a noise voltage across the resistance of the shield in the same way as the common-mode current did in Figure 7.3a, and the noise voltage $V_S$ is again equal to $R_S I_S$. If we assume that the magnetic-field-induced current flow in the loop is the same for the shielded and unshielded cables and that all of the signal current

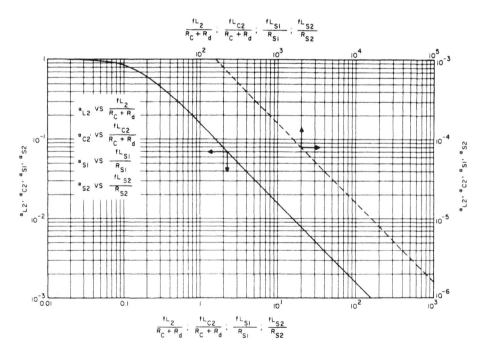

**Figure 7.2**  Attenuation coefficient $a_s$ versus $fL_s/R_s$. (© 1967, IEEE.)

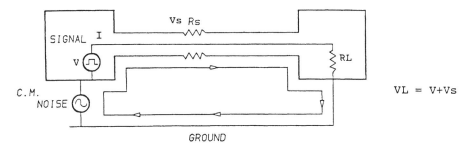

**Figure 7.3a**  Current path for common-mode induced current in a shielded cable.

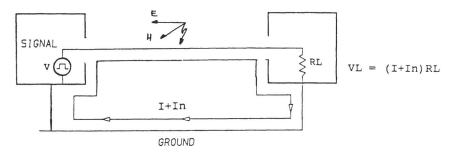

**Figure 7.3b**  Unshielded case for magnetic waves.

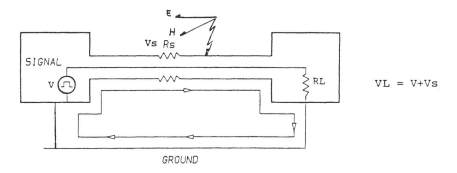

**Figure 7.3c**  Shielded case for magnetic waves.

returns on the inside of the shield, then the shielding effectiveness of the cable is given by the ratio of the load resistance to the shield resistance. Thus with a load resistance of 1000 $\Omega$ and a total shield resistance of 20 m$\Omega$, the shielding effectiveness is $20 \log 1000/20 \times 10^{-3} = 94$ dB. The lower the DC resistance of the shield, the higher the shielding effectiveness of a cable at low frequency. A braided-wire shield will have a lower DC resistance than a foil-type shield; and the thicker the braid or the more braids in the shield, the lower the DC resistance.

As we can see by comparing Figures 7.3b and 7.3c, a shielded cable with a tin copper braid shield does exhibit magnetic field shielding effectiveness when both enclosures are connected to ground, even at DC. However, the tin copper braid shield has a relative permeability of 1; if a shield with a higher permeability is used, increased magnetic field shielding will be achieved. This improvement is greatest at DC and at power-line frequencies from 50 Hz to 400 Hz. The

Magnetic Shield Corporation, of 740 North Thomas Drive, Bensenville, IL 60106-1643, manufactures a shielded four-center conductor cable in which the braid of the shield is manufactured from CO-NETIC AA wire that has a relative permeability of 30,000. Placing cables in a seamless galvanized cold rolled steel conduit, which has a relative permeability of close to 200, will provide approximately 20 dB of attenuation against 60-Hz magnetic fields.

### 7.2.2 Frequency Dependency 100 kHz–22 GHz

From the discussion on shielding effectiveness at low frequency, the transfer impedance of a shielded cable was seen to equal the DC resistance of the shield. At some frequency where the current density through the shield is no longer uniform, $Z_t$ no longer equals $R_S$.

When a current flow is induced by an incident field, the current density is greater on the outside of the shield. When the shield is used as a signal return path or the current is the result of common-mode noise on the center conductor and the shield, the current density is greater on the inside of the cable shield. The transfer impedance of several types of cable are shown in Figure 7.4, from which can be seen that for a solid copper shield of wall thickness 0.89 mm, the transfer impedance relative to the DC resistance begins to decrease at frequencies as low as 20 kHz. Transfer impedance is defined as

$$Z_t = \frac{V_t}{I_S} l \qquad [\Omega] \tag{7.11}$$

$$V_t = Z_t I_S l \qquad [V] \tag{7.12}$$

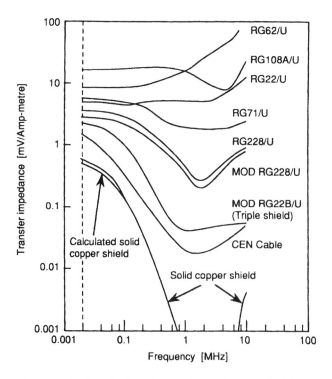

**Figure 7.4** Transfer impedance of several types of cable.

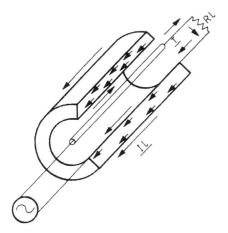

**Figure 7.5** Shield current flow (transfer impedance).

where

$V_t$ = transferred noise voltage [V]

$I_s$ = shield current [A]

$l$ = length of the cable

Figure 7.5 shows a shield current flow due, for example, to an incident electromagnetic field. Only a percentage of the shield current flows on the inner surface of the shield, and it is this current flow that generates a voltage between the inner surface of the shield and the center conductor. Two basic methods are used to measure the transfer impedance, these are shown in Figures 7.6 and 7.7.

In the test configuration of Figure 7.7, the current flow is provided by a generator connected to both ends of the shield, which simulates the situation where current flows on the outside of the shield, either due to an incident electromagnetic wave or due to a common-mode current.

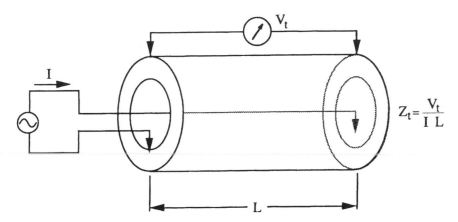

**Figure 7.6** Transfer impedance test method 1.

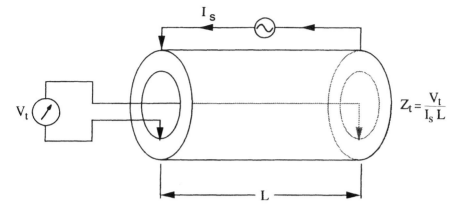

**Figure 7.7** Transfer impedance test method 2.

As the frequency is increased, the voltage developed across the shield increases due to its inductance; however, this is not the transfer voltage measured at one end of the cable between the center conductor and the inner surface of the shield.

The voltage developed from one end of the center conductor to the other end is equal to the shield voltage, with a difference equal to the transferred voltage $V_t$. It is not necessary to make a physical electrical connection to the inside of the shield when measuring $V_t$; a connection to the end of the shield, inside or outside, will suffice.

In Figure 7.6 the current flow is between the center conductor and the shield and $V_t$ is measured across the ends of the shields.

Some of the test fixtures used to measure surface transfer impedance are the MIL-C-85485 and the IEC 96-1A triaxial fixtures the quadraxial and the quintaxial. The generic triaxial line looks like a coaxial cable in which the center conductor is the shielded cable under test. The drive voltage is connected to the shield of the cable under test, and the outer cylinder of the test fixture is used as the current return path. The triaxial fixture has the internal line, the cable under test, terminated in its characteristic impedance at the drive end, and the outer line, the return current line, either terminated in its characteristic impedance or a short circuit at the far end. A detector that has the same characteristic impedance as the cable is used at the far end of the cable under test to monitor the transferred voltage. When the outside line is terminated, the upper test frequency is extended, as cable resonances are reduced; however, the long-line effects at high frequency are not eliminated, since the inner-line and outer-line phase velocities are usually different. This difference is due to the higher core dielectric constant of the cable under test compared to the air core outer line. One difficulty with the triaxial fixture is ground loop problems, which interfere with measurements at low frequency. These problems can be reduced by the addition of ferrite baluns on each end of the cable under test. The quadraxial test fixture will typically have all three lines correctly terminated, and the current in the drive line is relatively independent of frequency. However, due to the extra exterior line, the outer diameter is larger than that of the equivalent triaxial test fixture and non-TEM waves can occur at a lower frequency than with a triaxial fixture. Reference 1 describes two triaxial test methods and one quadraxial test method in more detail. Figure 7.8 shows the measured transfer impedance of an RG-58C/U coaxial cable measured using the quadraxial and two triaxial test methods.

Reference 1 describes the predominant high-frequency coupling as ''porpoising'' coupling in this sample rather than aperture coupling. Thus the surface electric field makes only a minor contribution to the measurement. At low frequency, the triaxial measurements are higher than

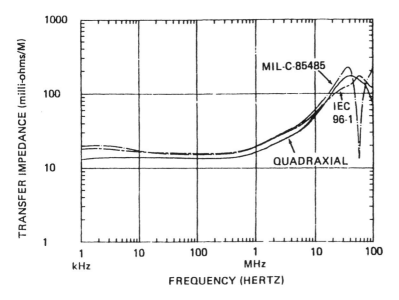

**Figure 7.8** Surface transfer impedance of RG-58C/u as measured by quadraxial and triaxial (IEC 96-1A and MIL-C-85485) test fixtures.

the quadraxial due to the ground loop problems. At high frequency, the triaxial response is smoother than that of the MIL-C-85485, which has a dielectric constant of 2.6, and the resonance occurs at 50 MHz. The IEC 96-1A fixture, which has a dielectric constant of 1, has a resonance above 100 MHz, this is not seen in Figure 7.8.

A test method that in effect makes an easy-to-construct triaxial fixture is the pull-on braid method. This is illustrated in Figure 7.9. The cable under test has the shield and center conductor short-circuited at the drive end, and this is connected to a pin in a connector. An additional braid is "milked" on over the insulation of the shielded cable under test. The center conductor of the cable under test is connected to a pin in the connector at the signal-detector end. The milked-on braid shield is then connected to the signal-source connector case and the signal-detector connector case. The far-end connector pin connects to a detector/measurement device that has the same impedance as the cable under test, and the source signal is driven through an impedance equal to the cable impedance. In this triaxial fixture, the cable under test is terminated correctly at both ends and the outer cylinder to cable shield is short-circuited. If the cable under test is electrically short, then the measurement error is low. At high frequencies, where the cable is electrically long, a correction factor must be made for resonances. The input impedance of the outer coaxial circuit can be measured by use of a network analyzer and the current computed, but this correction factor can introduce severe errors.

Other test methods include the IEC 96-1 line injection test method, in which an injection wire that is typically a flat copper braid is taped to the cable under test. The great advantage in the injection line test method is that it is capable of measuring surface transfer impedance to very high frequencies (above 1 GHz).

The construction of the test fixture is simple and requires only a coaxial cable, injection wire, signal generator, and measuring device. Figures 7.10 and 7.11 show the test setup. The transition from the coaxial injection cable to the injection wire and the transition back can easily be matched to the impedance of the coaxial injection line up to very high frequencies. With a 50-cm coupling length, the phase velocities in the injection circuit and in the cable under test

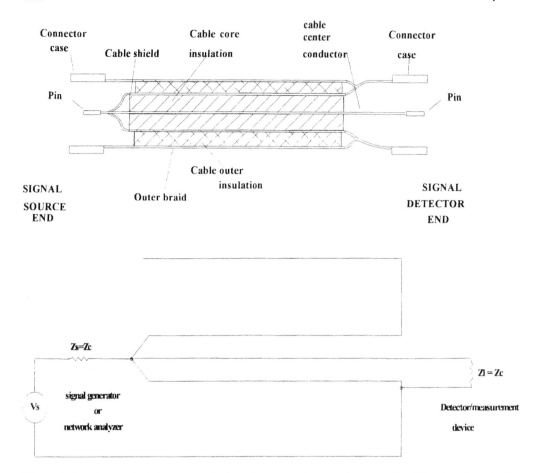

**Figure 7.9** Simple triaxial-type transfer impedance test fixture using a "milked"-on braid, and schematic.

are sufficiently well matched to permit measurements up to 3 GHz. With a 10-cm coupling length, measurements up to 20 GHz are feasible; however, at frequencies above even 1 GHz, extreme care must be taken at the transitions and in matching the source and load ends of the cables. Ferrite baluns on the near and far ends of the cable under test and on the near and far ends of the injection lines are recommended to reduce common-mode current caused by line radiation. And it is recommended that the signal source be placed outside of a shielded room, with the cable under test shield terminated at the shielded-room wall. A coupling transfer function between the excitation current and the near- and far-end coupling are contained in the IEC 96-1 standard. One potential source of error is that this method excites only a portion of the circumference of the cable shield. The question of whether this yields the same test results as with other methods is addressed in Ref. 2. Here the test results obtained with the line injection method were compared with the test results using a quadraxial fixture. It was found that in the frequency range up to 50 MHz the results were reasonably close and the results up to 1 GHz were more credible using the line injection method.

One other simple test method is to place the cable under test at a fixed height above a ground plane, which forms a two-wire transmission line with the coaxial cable and its image in the ground. If an RG58 cable is placed with its outer insulation in contact with the ground plane, the characteristic impedance of the transmission line is approximately 50 Ω; and at a

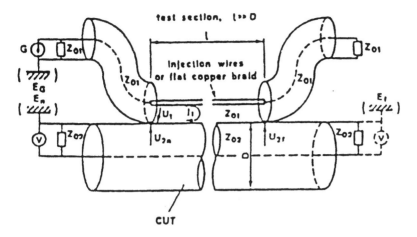

**Figure 7.10**   Schematic diagram of the line injection test setup. (© 1998 IEEE.)

height of 5 cm above the ground plane it is approximately 317 Ω. The shield of the transmission line can be excited either by an injection current probe or directly by a signal source with a series impedance equal to the transmission-line impedance. The most common configuration has the shield of the cable under test short-circuited at the far end. In this configuration, shown in Figure 7.12a, a current probe is used to monitor current. When the cable is electrically short (less than 1/10 of a wavelength), the current probe measurement can be used directly with the measurement of the transferred voltage to determine transfer impedance. At higher frequencies a correction factor is required and errors can be introduced. A very short cable length can be used to increase the frequency range; however, the influence of the connector-to-connector and connector-to-bulk head transfer impedances become significant. Measurements made on a 50-cm-long cable 5 cm above the ground plane have shown a good correlation between a transmission-line method (with direct injection, a short circuit at the far end, and a current probe monitor) to line injection and triaxial methods up to 1 GHz.

A second transmission-line test setup, shown in Figure 7.12b, uses direct injection and terminates the shield of the cable at the far end in the characteristic impedance of the transmission line, which reduces the long-line effects. Placing the cable insulation on the ground plane means that the velocity of propagation in the cable-to-ground plane transmission line is approximately equal to that of the shielded cable. This would not be true with a predominantly air

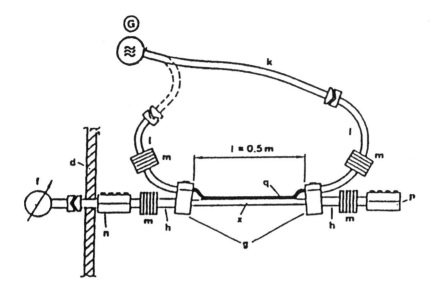

**Complete installation for practical transfer impedance measurements**

| | |
|---|---|
| X | Cable under test (CUT) |
| d | Screened room wall |
| G | Generator (synthesizer or tracking generator, etc.) |
| f | Test receiver (spectrum analyzer, network analyzer, etc.) |
| g | Launcher to injection wire |
| h | Brass tube for additional screening for CUT |
| i | Feeding cables for injection wire (low loss, approximatively 0.5 m) |
| k | Feeding cable from generator |
| m | Ferrite rings (length approximatively 100 mm) |
| n | Additional screening for connection between screened room and CUT |
| p | Additional screening for terminating resistance of CUT |
| q | Injection wire |

**Figure 7.11** Pictorial diagram of the setup used in the line injection test. (© 1998 IEEE.)

space between the cable and the ground. A spacer may be used between the cable and the ground plane that has a similar permittivity to the cable-core insulation material to minimize this effect. In this test setup, the addition of the measuring device will present a complex impedance between the shield of the cable under test and ground, in parallel with the terminating resistor. To control this impedance, many ferrite baluns should be placed on the shielded cable connecting the measuring device to the termination point. Even with the baluns in place, it is recommended that a network analyzer be used to measure the transmission-line impedance and to detect any major deviation from the required impedance.

In all, five different types of coupling to cables have been identified, as described in Refs. 3 and 4. These include *axial transfer impedance* induced by a magnetic field parallel to the cable, such as generated down the axis of a solenoid or when the magnetic component of an

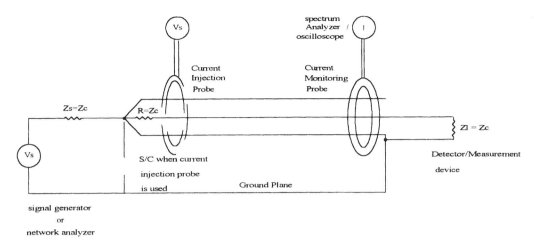

**Figure 7.12a**  Transmission-line transfer impedance test setup with S/C termination and current probe monitoring.

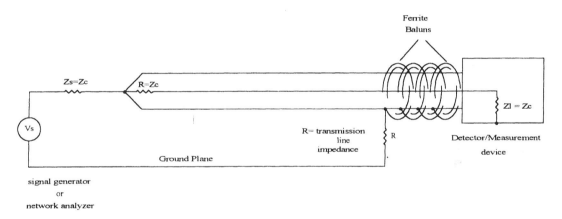

**Figure 7.12b**  Transmission-line transfer impedance test setup with the transmission line terminated in its characteristic impedance.

electromagnetic field is parallel to the cable, which sets up circumferential currents. It also includes *parallel electric field coupling*, which would occur if the cable were placed between parallel plates at different RF potentials. A third type is *parallel transfer impedance*, caused by a magnetic field loop that penetrates the shield. All of these new couplings disappear for a typical coaxial cable with a single conductor that is concentric. It has been shown that for multiconductor cables, the new coupling modes can induce both common-mode and differential-mode signals in the cable. The circumferential current flow caused by the axial magnetic will not introduce an EMI voltage in a multiconductor cable in which the conductors are straight, but it will introduce a voltage in twisted conductors, such as contained in the twisted-shielded-pair cable.

At low frequency, in practical cables a magnetic field is generated by the cable, due to eccentricity between the center conductor and the shield. This magnetic field is given by

$$H = \frac{I \Delta r}{2\pi R^2} \tag{7.13}$$

where

$R$ = distance between the core and the measurement point

$\Delta r$ = eccentricity between the shield and core

Equation (7.13) assumes equal and opposite currents in the shield and core. For a cable where the transfer impedance decreases with frequency, the current flow on the outer surface, assuming the test method shown in Figure 7.7, decreases, as does the external magnetic field generated by the cable.

From Figure 7.4 we see that the transfer impedance of the solid copper shield continuously reduces with increasing frequency, unlike the braided flexible-shield cable, for which it typically begins to increase above 2 MHz. This effect is due primarily to a leakage of the external magnetic field through the apertures in the braid, although electric coupling due to an external electric field can also penetrate the holes in the braid. A second coupling mechanism is referred to as *porpoising coupling*. Porpoising coupling results from current that is pulled into the shield on the strands of wires, or carriers, of which the shield is constructed. This form of coupling has been attributed to the contact impedance between the carriers, which forces some of the current flow to remain on the carrier instead of flowing into the next carrier. If this mechanism is valid, then the porpoising coupling of a cable is expected to increase with an increase in contact impedance with use, due to flexing of the cable that results in loosening of the carriers and due to corrosion. A second cause for porpoising coupling has been proposed: an inductive effect due to the twist in the carriers. Reference 5 says that most cable samples have a surface transfer impedance signature that was indicative of porpoising coupling. Reference 6 notes that the aperture and porpoising coupling should be out of phase and describes a cable with an optimized shield in which wires have been removed and then degraded until the aperture and porpoising coupling components of the transfer impedance tend to cancel.

In open-weave cables, the aperture coupling is often predominant above 1 MHz and becomes more so in the majority of cables above 30 MHz. When aperture coupling is predominant, the transfer impedance of a cable is

$$Z_t = j\omega MA \tag{7.14}$$

where MA is the mutual inductance due to shield apertures which has a nominal value of $3 \times 10^{-10}$H/m.

The electric field coupling through apertures is modeled by a transfer admittance, which may not be accounted for in measured transfer impedance, especially when the test fixture injects a current flow on the shield of the cable under test. A correction may be obtained for transfer admittance by multiplying $Z_t$ by $(1 + A_e/A_m)$, where $A_e$ is the electric polarizability of the apertures and $A_m$ is the magnetic polarizability of the apertures. From Ref. 9, the ratio $A_e/A_m$ for circular apertures is 0.5, which corresponds to an approximate braid weave angle of 40°. Reference 10 provides additional values for elliptical apertures at a number of braid weave angles. It should be noted that the transfer admittance correction is strictly valid only when the shield is terminated to ground at both ends by the characteristic impedance of the cable above the ground. Sections 7.2.3.2 and 7.2.3.3 also discuss transfer admittance.

With the advent of foil-type cables, one might expect them to achieve the low transfer impedances of a solid copper shield; however, this is not the case. In the construction of early-type foil shields, a spiral wrap was used; later a longitudinal edge contact was used. At first the longitudinal join was insulated, forming a slot. However, most foil braids of recent manufacture are folded at the longitudinal edges. Figure 7.13a shows the transfer impedances of combination braid- and foil-construction shields. Above approximately 50 MHz, the transfer impedance

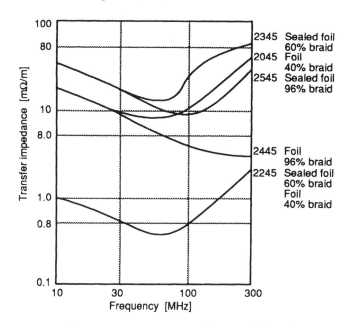

**Figure 7.13a** Foil and braid type cable transfer impedance.

of the combination shield tends to increase. This is due to the AC resistance of the longitudinal joint in the foil. As with the transfer impedance of gasket material, the current is not confined to the outer surface of the foil. But because of an impedance change at the joint, the current diffuses into the center of the foil at the joint.

As the combination cable is flexed, the transfer impedance rises. Figure 7.13b, from Ref. 15, shows a large increase from 0.004 $\Omega$ to approximately 0.6 $\Omega$ for the worst case, unsealed foil-and-braid cable after 49,000 flexures.

Transfer impedance data is available on both military- and RG-type cables. Figure 7.14a reproduces transfer impedance curves for coaxial cables and triaxial cables and Figure 7.14b for twinaxial cables, by permission of Belden Wire and Cable. It has been found that cables of the same type from different manufacturers may have very different transfer impedances, so the data in Figures 7.14a and 7.14b is valid for the cables manufactured by Belden. RG 58 is a very common 50-$\Omega$ coaxial cable, and we see a very large variation of up to 20 dB$\Omega$ (factor of 10) between cables from different manufacturers, as shown in Figure 7.15, from Ref. 11. Reference 11 describes one of the transmission-line techniques for the measurement of transfer impedance. For this reason it is advisable either to obtain the transfer impedance from the cable manufacturer or to use one of the simple test methods, such as the transmission-line or line injection methods, to measure the transfer impedance of the cable of choice.

The shield construction of cables shown in Figure 7.14a are

| Belden cable number | Shield type |
| --- | --- |
| 9259, 8254, 9555, 9269, 9268, 9862, 9228 | Bare copper braid, 95% coverage |
| 9889 | Duofoil (TM) with 4/24 AWG drain wires |
| 9888 | Triax, Two bare copper braid, 95% coverage |

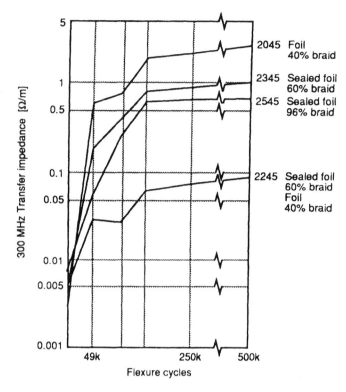

**Figure 7.13b**  Increase of transfer impedance with flexure. (From Ref. 15. © 1979, IEEE.)

The shield construction of cables shown in Figure 7.14b are

| Belden cable number | Shield type |
|---|---|
| 9271 | Beldfoil[™] with stranded copper drain wire |
| 9272 | Tinned copper braid, 93% coverage |
| 9851 | Foil with shorting fold and stranded copper drain wire |
| 9207 | Tinned copper braid, 95% coverage |
| 9463 | Beldfoil with 57% coverage tinned copper braid |
| 8227 | Tinned copper braid, 85% coverage |
| 9860 | Duofoil (™) with 92% coverage tinned copper braid |

The terms *single-, double-, and triple-braid shielded cable* refer to the number of braids in a shield that are, typically, in electrical contact.

A comparison between the transfer impedances shown in Figure 7.4, 7.14a, and 7.14b indicates that the cables with multiple-braid shields exhibit a lower transfer impedance. The upper curve of the RG22B/U in Figure 7.4 is for a double-braid cable, whereas the lower curve for the modified RG22B/U is that for a triple-shield cable.

The term *triaxial* refers to a cable with a center conductor and two shields that are electrically isolated from each other. Triaxial connectors exist that allow the center shield connection

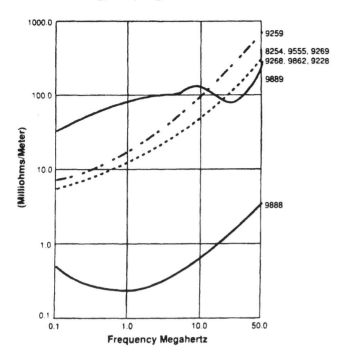

**Figure 7.14a** Transfer impedance of Belden coaxial and triaxial cable. (Reproduced by courtesy of Belden Wire and Cable.)

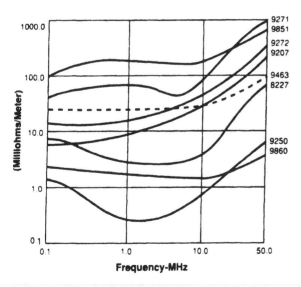

**Figure 7.14b** Transfer impedance of Belden twinaxial cable. (Reproduced by courtesy of Belden Wire and Cable.)

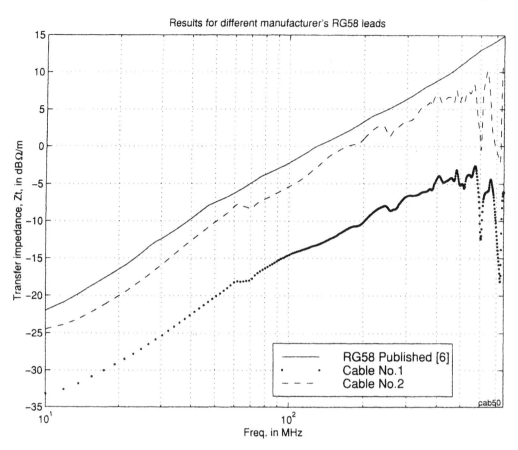

**Figure 7.15**  Transfer impedance of two samples of RG58 from different manufacturers. (From Ref. 11. © IEEE, 1998.)

to continue through the connector and remain isolated from the enclosure and in which the outer shield is connected to the case of the connector. The triaxial configuration allows the signal return currents to flow on the center shield and the outer shield to be connected to the equipment enclosure. Thus the transfer voltage does not appear directly in the signal path. The remaining coupling between the inner and outer isolated shields is via electric field coupling through the transfer admittance. This advantage of the triaxial cable over a double-braid cable is considerably reduced if the signal ground is connected to the enclosure at both ends. One example of the correct use of a triaxial cable is shown in the HAR-1 circuit in Figure 5.62, in which the signal is transformer coupled at the receiver end. At high frequencies it becomes increasingly difficult to isolate signal grounds from the enclosure, due to stray capacitances, so the transfer impedance of triaxial cable in which the shields are connected together at both ends is of interest. The transfer impedances of isolated, single-, double- (triaxial), and triple-braided cables with the braids shorted are shown in Figure 7.16 (from Ref. 8). We see that at frequencies below that at which the cable length is equal to $\lambda/2$, the transfer impedance is lower for the multibraided cables; however, above this frequency the difference is negligible due to intersheath resonances or different propagation delays down the shields.

Twinax cable is constructed of two center conductors with an overall shield, and quadrax is a two-center-conductor cable with two isolated shields. A typical use for these cables is in

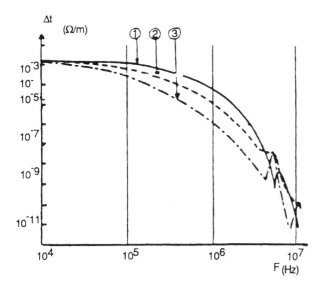

**Figure 7.16** Comparison of the transfer impedances of isolated, single-, double-, and triple-braided cables, in which the braids are shorted together at both ends. (From Ref. 8. © IEEE, 1967.)

fully balanced or single-ended driver/differential input circuits. Typical shield connections are shown in Figure 7.17a and b. The shielded-cable transfer voltage is common mode; that is, it appears equally on all the center conductors within a shielded cable. Thus to achieve maximum immunity, the input circuit must exhibit a sufficiently high common-mode noise rejection, and any imbalance in the input impedances of the circuit should be minimized. The quadrax cable will exhibit a lower transfer impedance than the twinax by approximately the same magnitude as shown for triax cable. The options for shield connections in Figure 7.17 show the outer shield connected to the enclosure at both ends, for reasons discussed in Section 7.5 on shield termination.

Reference 12 describes a comparison of five cables with shields constructed of copper braid and aluminized plastic foil, with a sixth cable shield constructed of braid alone. These cables were examined for use with 1-Gb/s signals. In most of the cables the braid was on the outside and provided the turn-to-turn contact for the foil shield. In two cables the braid was inside the foil but still provided the turn-to-turn contact (Figure 7.18b). The core four of the cables was a balanced quad (one pair transmits and one pair receives, with no individual shielding of the pair. In two cables the pair was individual shielded. In one cable the metal of the internal shield faced the inside, and turn-to-turn contact was made with a drain wire (Figure 7.18d). In another cable the metal of the internal foil faced the outside and the braid made the turn–turn contact (Figure 7.18c). The highest transfer impedance is that of the shielded twisted pair, or inner shield. The measurements showed that the aluminized foil plastic tape did not provide a good electromagnetic shield by itself, particularly at high frequencies. This is consistent with experience with inexpensive computer serial and parallel cables, where the shield is an external foil alone with an internal drain wire. Although this type of cable is better than an unshielded type, its shielding effectiveness is not as high as a braid type of cable.

The lowest transfer impedance is with two layers of shielding, one the outside layer and the second the individual shield over the pairs of conductors. Where the plastic foil insulation is between the braid shield and the aluminized layer, resonances occur due to the difference in

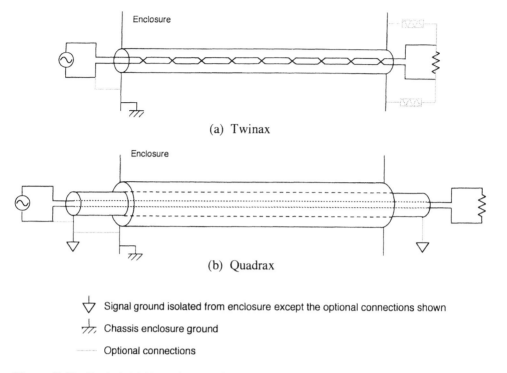

(a) Twinax

(b) Quadrax

▽ Signal ground isolated from enclosure except the optional connections shown

⏚ Chassis enclosure ground

........... Optional connections

**Figure 7.17**   Typical shield terminations for a twinax and quadrax cable. (© IEEE, 1980.)

propagation delay as a result of the plastic between the shielding layers. This effect is shown in Figure 7.16 for insulated shields. Reference 12 concludes with:

> The measured transfer impedance of six cable samples that used combinations of braid and foil showed that these cables do not have the classic $R + j\omega M_{12}$ frequency dependence. Above a few MHz, they exhibited a frequency dependence that was approximately proportional to

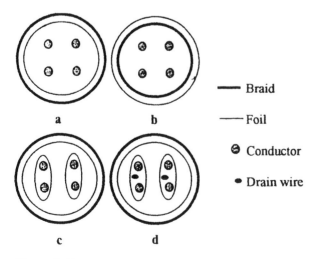

— Braid

— Foil

⊘ Conductor

• Drain wire

**Figure 7.18**   Cross sections of the cable construction used in the 1-Gb/s cables.

the square root of frequency. This suggests that the coupling mechanism is due to contact resistance somewhere in the shield. The two cables with individually shielded pairs had lower transfer impedances than those that used an overall shield over a balanced quad core. The transfer resistance of the 6 cables ranged from 9 to 21 m$\Omega$. This is in the range of single-braid cable shield and is appropriate for 1-Gb/s interconnect cables. At 500 MHz, the transfer impedance ranged from a little more than 10 m$\Omega$/m to 135 m$\Omega$/m. The cables with individually shielded pairs were best (12.4 and 21.8 $\Omega$/m).

Figure 7.19 compares the transfer impedances at 500 MHz.

Thus far we have examined the voltage induced in shielded cables when the length of the cable is less than 0.5$\lambda$, where $\lambda$ is the wavelength of the current flow on the shield of the cable. Equation (7.12), which gives the magnitude of the induced voltage, is no longer valid when the cable is longer than 0.5$\lambda$, due to so-called long-line effects.

Long-line effects are considered in Section 7.2.3 (30 MHz–10 GHz), although for long cables they may occur in the frequency range from 100 kHz to 30 MHz.

Should an unshielded or a shielded cable be the culprit in a susceptibility/immunity problem or result in excessive radiation, the addition of an overbraid may solve the problem. Tin-plated copper braid is sold as shielding and bonding cables with IDs of 3.18 mm to over 25.4 mm. These braids can be either attached to a circular connector by use of a hose clamp, soldered to a brass connector, or clamped under a strain relief on a none EMI backshell, as described in

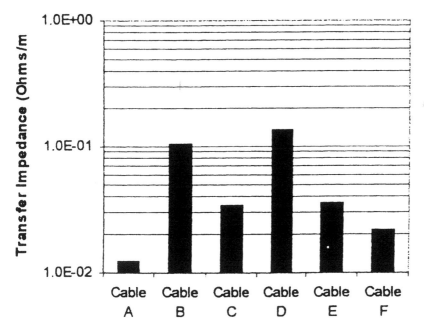

**Figure 7.19** Transfer impedance (extrapolated) of the six combination braid/foil shields at 500 MHz. (Ref. 12 © IEEE 1998.)

    A = figure 1d, Individually shielded pair with the metal of the foil faced inside (drain wire used for turn-to-turn contact) plus overall shield

    B = figure 1a, Overall shield, no individual shielding of the pair

    C = figure 1b, Braid inside the foil, no individual shielding of the pair

    D = figure 1b, Braid inside the foil, no individual shielding of the pair

    E = figure 1a, Overall shield, no individual shielding of the pair

    F = figure 1c, Individual shielded pair with an overall shield.

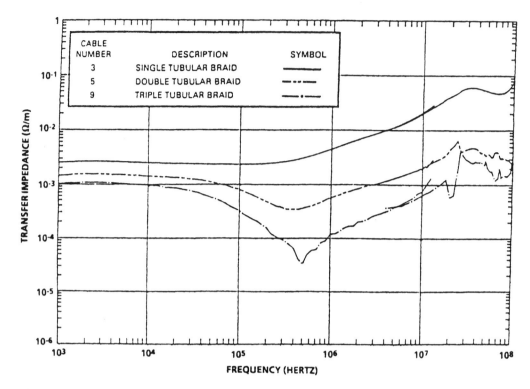

**Figure 7.20**  Measured surface transfer impedance of 1-m-long tin-plated copper tubular braid shields. (© IEEE, 1988.)

the section on connectors. The measured transfer impedance of 1-m length of single tubular braid, double tubular braid, or triple tubular braid is shown in Figure 7.20.

Above 30 MHz the leakage effects are more pronounced, because the holes in the braid of a shielded cable become more effective antennas with increasing frequency. The method of testing cables in the gigahertz region becomes difficult because the geometries must be more rigorously controlled. Forward and backward waves can be generated in the test fixture, and the magnetic field induced in the cable is confined to the TEM mode. Therefore the test fixture may be far removed from the real-life use of shielded cables where radial, axial, and circumferential fields can be generated by a cable at very high frequencies. Using the line injection test method, the transfer impedances for single- and double-braid cables up to 22 GHz are shown in Figure 7.21 (from Ref. 13). The transfer impedance increases almost constantly up to 20 GHz. Figure 7.22 shows the near- and far-end coupling for a single-braided cable. The far-end transfer impedance increases rapidly above 10 GHz. Reference 13 attributes this to radiation loss through the shield and makes the point that cables used at these frequencies are in some way antennas. Much earlier data on the measured transfer impedance in the gigahertz region is shown in Figure 7.23a for single-braid cable and in Figure 7.23b for double-braid (from Ref. 21). The transfer impedance of the single-braid RG58C/U is 400 mΩ/cm, or 40 Ω/m at 4 GHz, which is similar to that of the single-braid cable in Figure 7.21, which has a transfer impedance of approximately 30 Ω/m at 6 GHz. The double-braid cable GR cables in Figure 7.23b have an impedance of 40 mΩ/cm, or 4Ω/m, which is much higher than the double-braid shielded cable in Figure 7.21 with a transfer impedance of 0.2 Ω/m at 6 GHz. However, as we have seen, even samples of the same type of cable show a surprising variation in transfer impedance.

$Z_t$

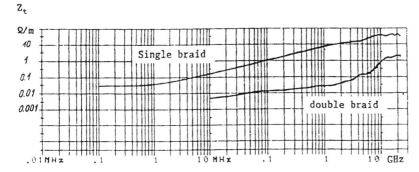

**Figure 7.21** Single- and double-braid shielded cable transfer impedance up to 22 GHz using the line injection test method. (From Ref. 13. © IEEE, 1992.)

$Z_t$

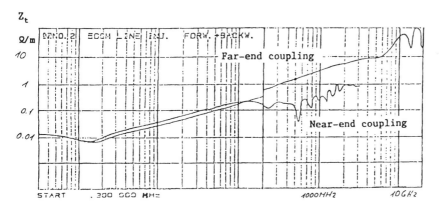

**Figure 7.22** Single-braided cable showing near- and far-end transfer impedance. (From Ref. 13. © IEEE, 1992.)

Modern triple braid stainless steel armored cables have a shielding effectiveness of approximately 80 dB which, using Eq. (7.3), corresponds to a transfer impedance of 10 mΩ/m.

### 7.2.2.1 Semirigid Cable

Semirigid cable is constructed of a solid metal outer sheath, normally copper. The inner insulation is either solid or air with spacers used to support the inner conductor. The cable either is hand-malleable or must be bent by machine. Connectors are soldered directly to the ends of the cable or crimped on.

This type of cable is the best available for the prevention of EMI at high frequency. When properly soldered at the cable-to-connector interface, the only source of ingress or egress of radiation is at the connector mating, due to the transfer impedance of the connector-to-bulkhead interface. Manufacturers offer a quick-connect type of semirigid connector, in which the cable is crimped to the connector, thus adding an additional transfer impedance. In common with all shielded cables the low-frequency performance is dependent on the thickness and conductivity of the shield and the concentricity of the cable. A new form of solid shield is the "semiflexible" cable, which is a braided shield in which the shield braids have been soldered together. The cable is much more flexible than the semirigid variety, with almost the same shielding effectiveness, but is less flexible than a small-diameter braided cable.

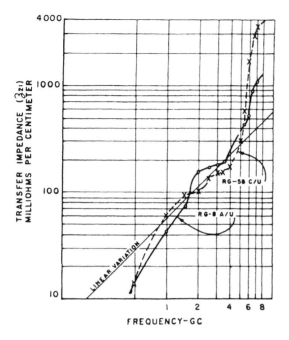

**Figure 7.23a**  Single-braid coaxial transfer impedance up to 8 GHz.

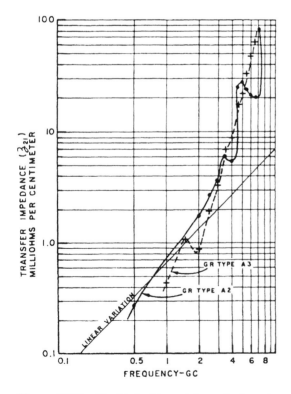

**Figure 7.23b**  Double-braid coaxial transfer impedance up to 8 GHz.

### 7.2.2.2 Long-Line Effects

From Eq. (7.14) it is seen that when the transfer impedance is dominated by aperture coupling, transfer impedance increases monotonically with increasing frequency. From Eq. (7.12) it appears that the transferred voltage is proportional to the transfer impedance, the shield current, and the length of the line and, where Eq. (7.14) applies, to frequency; however, this is not true when the line length is greater than $0.5\lambda$. Such a cable is termed *electrically long*. The wavelength of a wave in air is equal to $300/f$ (MHz), whereas the wavelength of the same wave in a material with a relative permittivity greater than 1 is $300/(f \text{ (MHz)} \times \sqrt{\varepsilon_r})$. When the shield of a cable is terminated to ground at both ends, the cable is resonant at frequencies for which the line length equals $(k\lambda/2)$, where $k$ is an integral multiple, i.e., 1, 2, 3,. . . .

The transfer voltage for an electrically long cable in which the shield is terminated to ground at both ends with the characteristic impedance of the cable above ground, derived from Ref. 7, is given by

$$V_t = I_s Z_t \frac{\sin \theta}{\theta} \qquad \text{[V]} \tag{7.15}$$

where

$Z_t$ = transfer impedance $[\Omega/\text{m}]$

$I_s$ = shield current [A]

and

$$\theta = \frac{2\pi f}{300}(\sqrt{\varepsilon_r} + 1)\frac{l}{2} \qquad \text{[radians]}$$

where

$f$ = frequency [MHz]

$\varepsilon_r$ = relative permittivity of the cable sheath

$l$ = length of cable [m]

Figure 7.24 plots the ratio of $\sqrt{(\sin\theta/\theta)^2}$, in decibels, for a 20-m-long cable with a relative permittivity of 2.2. From Figure 7.24 an envelope reduction of 6 dB per octave or 20 dB per

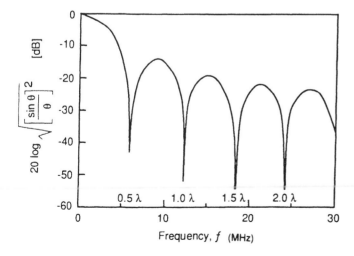

**Figure 7.24** Plot of $20 \log/R((\sin \theta/\theta)^2)$ for a 20-m cable with a sheath relative permittivity of 2.2.

decade is seen, i.e., a monotonic decrease with increasing frequency above the first resonance frequency of the cable. Equation (7.14) predicts a monotonic increase in transfer impedance with increasing frequency, so the envelope of the transfer voltage, above the first resonance frequency, should be constant with frequency. One definition of the shielding effectiveness of a cable is the ratio of the shield current $I_s$ to the core (center conductor/s) current $I_c$ with the cable terminated, between core and shield, in its characteristic impedance.

The measured voltage, in decibels referenced to the test fixture voltage of a 1.18-m-long RG/58A cable from 1 MHz to 1 GHz is shown in Figure 7.25. The center reference graticule is $-75$ dB, which corresponds to a shielding effectiveness of 69 dB. The graticule is graduated 10 dB per division. As the measured voltage increases by 10 dB above the reference graticule, the shielding effectiveness decreases by 10 dB. Thus, at 300 MHz, the shielding effectiveness from Figure 7.25 is 59 dB.

Figure 7.26 plots the measured voltage, in decibels, referenced to the test fixture input for a Raychem shielded twisted pair 10595-24-2-9 of 1.2-m length. The center graticule is $-70$ dB, which is equivalent to a shielding effectiveness of 51 dB.

### 7.2.2.3  Transfer Admittance

Sections 7.2.1 and 7.2.2 briefly discussed the transfer admittance of shielded cables. In most cases the transfer admittance is a secondary phenomenon that can be ignored. When a cable is electrically long, a current is set up on the cable regardless of whether the shield is terminated to ground or not. For an electrically short cable with the shield connected to ground at both ends, an incident field will, depending on the angle of incidence, set up a current flow on the cable. The worst-case figure in these cases is the transfer impedance alone. However, for an electrically short cable on which a high-impedance field is incident, the cable current may be low and the coupling may be via the charge that appears on the cable. This also applies to an electrically short cable connected to an enclosure that is disconnected from ground. Another situation would

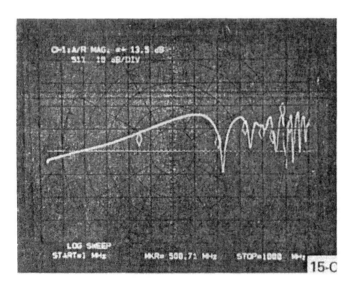

**Figure 7.25**  Measured voltage for a 1.18-m length of RG/58A cable, in decibels, referenced to the test fixture input voltage, as a function of frequency on a log scale. The center reference graticule is at $-75$ dB, corresponding to a shielding effectiveness of 69 dB. The first marker is at 10 MHz, all other markers are at 100-MHz steps.

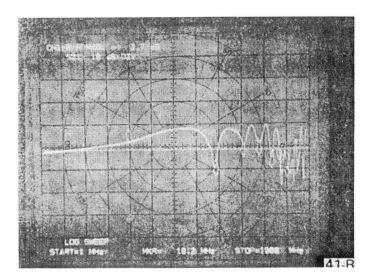

**Figure 7.26** Measured voltage, in decibels, for a 1.2-m length of Raychem shielded twisted-pair cable 10595-24-2-9. The center reference graticule is 70 dB, which is equivalent to a shielding effectiveness of 51 dB.

be an electrically long cable over which a field in the gigahertz frequency range is incident over a small area of the cable. This would in practice occur when a radar beam is incident on a section of cable. At gigahertz frequencies, the predominant coupling may be of the localized field through the apertures in the shield.

Transfer admittance is not an intrinsic electromagnetic parameter, for it is a characteristic of the cable and its surroundings (the test setup in the case of measurements). The through-elastance $K_T$, which is a property of the cable alone, can be used to characterize coaxial cables, and it can be related to the transfer admittance $Y_T$ by

$$K_T = \frac{Y_T}{j\omega C_1 C_2}$$

where $C_1$ is the per-unit capacitance of the outer circuit (test fixture to cable shield) and $C_2$ is the per-unit length capacitance between the two conductors of the coaxial cable. Obviously $Y_T$ is dependent on the measurement setup, because $C_1$ will differ from one setup to the other.

Reference 14 describes through-elastance measurements and provides the following conclusions: On coaxial cables of any length and matched at both ends, i.e., the cable matched to its characteristic impedance between the center conductor and shield at both ends, the transfer admittance can be neglected up to 6 MHz and cannot even be measured. However, a cable connected at both ends to a high impedance and submitted to a high-impedance field would be more susceptible to transfer admittance coupling.

In the measurements described in Reference 14, the $K_T$ measurements are very close to the $Z_T$ measurements, and the transfer impedance coupling may be the only one of concern.

## 7.3 SHIELD TERMINATION EFFECTS ON TRANSFERRED VOLTAGE

One of the most common errors made in the use of shielded cables is to connect the shield to the backshell of the connector or to the enclosure by a "pigtail" (either a length of braid or, more

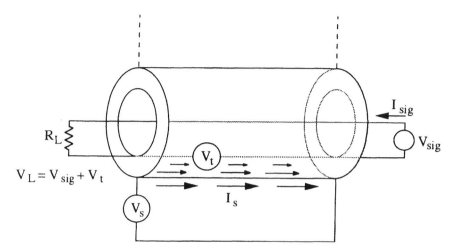

**Figure 7.27**  Common-mode current inducing a noise voltage ($V_t$) into the signal.

commonly, a length of wire). Alternatively, the pigtail may be connected through a connector pin and then to the inside of the equipment enclosure/chassis. As we shall see in Section 7.6 on radiated emissions from cables, the use of a pigtail to connect the shield at both ends of the cable results in emissions close to an unshielded cable. Likewise, the EMI voltage induced in a shielded cable with pigtail connections approaches that of a two-wire unshielded cable. The noise voltage developed across the transfer impedance by the shield current flow $I_s$ is effectively in series with the signal, $V_{sig}$, in the configuration of Figure 7.27.

   If the shield is terminated at the backshell of the connectors and via the connector to the chassis, then the shield current flow $I_s$ is effectively isolated from the signal path, with the exception of current that diffuses through the cable, couples through the cable apertures, and diffuses through the transfer impedances at the connector. However, if the shield is connected

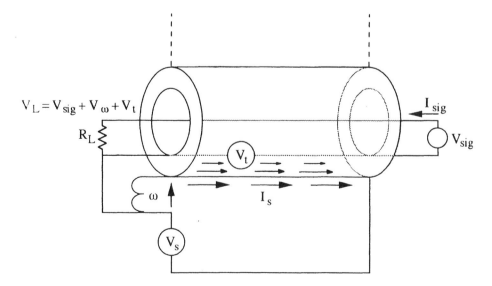

**Figure 7.28**  Shield connection extended into the enclosure.

**Table 7.1** Noise Voltages $V_t$ and $V_w$, Cable Transfer Impedance and Inductive Reactance of the Wire versus Frequency

| Frequency [MHz] | $V_t$ [μV] | $V_w$ [mV] | $Z_t$ [mΩ] | $Z_w$ [Ω] |
|---|---|---|---|---|
| 0.1 | 70 | 0.035 | 70 | 0.035 |
| 1 | 50 | 0.350 | 50 | 0.350 |
| 10 | 20 | 3.5 | 20 | 3.5 |
| 100 | 20 | 35 | 20 | 35 |
| 300 | 70 | 350 | 70 | 350 |

via a length of wire to the chassis and the signal path continues from that point to the load resistor $R_L$, as shown in Figure 7.28, then the shield current $I_s$ develops a noise voltage across the inductance of the length of wire $V_w$. Assume a shield current flow of 1 mA (this current may be due to the common-mode voltage source as shown, or to an incident electromagnetic wave) and a 26-gauge two-inch length of wire connecting the shield via a pin in the connector to chassis. Assume also that a moderately effective foil/braid shield is used and that the 26-gauge wire is 1 inch above the chassis and therefore has an inductance of 0.056 μH. Table 7.1 shows the two noise voltages $V_t$ and $V_w$, the cable transfer impedance, and the inductive reactance of the wire versus frequency.

Table 7.1 illustrates how important it is to terminate the shield to the backshell of the connector. Had a second wire been attached between the shield and the load resistor, then $V_w$ would be effectively isolated from the signal; however, the 2-inch length of wire carrying $I_s$ is still capable of radiating noise to internal circuitry. Also, when the second connection between the load resistor and the shield exists and the load resistor is connected to a signal ground, which is connected at some other location to the enclosure or to chassis ground, then some fraction of the noise current is injected into the signal ground. The optimum connection of a shield is at 360° around the circumference of the braid to a metal backshell. EMI backshells achieve the desired low-impedance connection, typically by clamping the end of the shield around the circumference to the backshell. Backshells are available from Glenaire that allow the connection of the shields of a number of cables, using the same connector, to the backshell. Where a connector must be used that does not allow the clamping of the shield to the backshell, the braid may be terminated by one of the following techniques: soldering the braid to the backshell, use of conductive adhesive (note that the conductive adhesive cannot be used as strain relief; use instead a cable clamp or a thin bead of nonconductive epoxy), the insertion of a canted coil spring between the cable clamp nut and the bared braid, hose clamp. One supplier of a suitable spring is Bal Seal Co., Santa Ana, California. An alternative backshell is a shielded heat-shrink boot, which contacts the connector and the cable shield as the boot is heat-shrunk. The supplier and transfer impedance of a heat shrink boot is provided in Section 7.8.

## 7.4 COUPLING FROM E AND H FIELDS

The coupling into a shielded or unshielded wire disconnected from ground and far removed from a ground plane was examined in Section 2.5.4. Chapter 2 also described the receiving properties of a loop far removed from a ground plane (Section 2.2.5). In this section we shall examine the current flow induced in either an unshielded wire/cable or a shielded wire/cable located above a ground plane. Provided that the height of the wire above the ground plane, $h$, is much less than the length, $l$, and $h \ll \lambda$, then the single wire can be considered a two-wire transmission line formed by the wire and its electromagnetic image in the ground plane. The terminating impedances of the transmission line are $Z_0$ and $Z_1$ and may be close to 0 $\Omega$ when the line is a shielded cable with both ends connected to the ground plane. The characteristic impedance of the line is given by

$$Z_c = \sqrt{\frac{(Z_i + j\omega l_e)\omega l_e}{jk^2}} \tag{7.16}$$

where

    $Z_i$ = distributed series resistance of the two wire line [$\Omega$/m]

    $k = \omega\sqrt{\mu_o e_o} = \dfrac{2\pi}{\lambda}$

    $l_e = \dfrac{\mu_o}{\pi} \ln \dfrac{2h}{a}$

    $\mu_o = 4\pi \times 10^{-7}$ [H/m]

    $a$ = diameter of wire [m]

When $l \gg h$ and $h \ll \lambda$, transmission-line theory is expected to be accurate. However, when $hl \ll \lambda$, circuit theory using the inductance of a rectangular loop of wire will provide a more accurate solution. The impedance of the loop (with image), including termination resistances, is given by

$$Z = \sqrt{2 \ (Z_0 + Z_1)^2 + (j\omega L)^2} \tag{7.17}$$

The inductance of a rectangular loop of wire $L$ with image is

$$L = l \ln \frac{4hl}{a(l + d)} + 2h \ln \frac{4hl}{a(2h + d)} + 2d - \frac{7}{4}(l + 2h) \tag{7.18}$$

where $d = \sqrt{(2h)^2 + l^2}$. When $l \gg h$ and $h \gg a$, then $L = l_e l$

Equation (7.19) multiplies the physical height $h$ above the ground plane by 2 to account for the electromagnetic image and may be used to find the inductance of a wire loop by changing $4h$ to $2h$ and $2h$ to $h$. The current in the termination impedance $Z_0 = I_0$ and $Z_1 = I_1$. These currents are

$$I_0 = I_1 = \frac{-j\omega 4hB_{in}}{\sqrt{2(Z_0 + Z_1)^2 + (\omega L)^2}} \tag{7.19}$$

$B_{in}$ is the incident magnetic flux density, and the illumination is considered as causing worst-case maximum coupling. Maximum coupling occurs for a plane wave when the plane of incidence is coincident with the plane of the loop, with the magnetic field perpendicular to the plane of incidence, as shown in Figure 7.29.

The magnetic flux density $B_{in}$ is related to the magnetic field $H$ by the permeability of free space $\mu_o$ ($4\pi \times 10^{-7}$ [H/m]); i.e.,

    $B = \mu_0 H$

When $Z_0 = Z_1 = 0$ $\Omega$, which is true for a shielded cable, and where the transmission-line equation applies, then

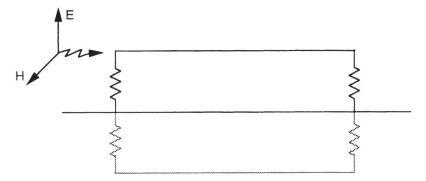

**Figure 7.29** Magnetic field induction in a transmission line formed by a wire over a ground plane.

$$I_0 = I_1 = \frac{4E_o h}{Z_c} \tag{7.20}$$

$E_o$ may be found from $H$ by the relationship $E = H \times Z_w$. The angle of incidence of the electromagnetic wave does not affect the magnitude of the induced current; however, the magnetic field vector must be perpendicular to the plane of the loop for the assumption of maximum coupling to remain valid. Figure 7.30 compares the results of calculations using transmission-line theory and from a computer program (The Numerical Electromagnetic Code, NEC).

Resonance can be seen at 7.5 MHz and 15 MHz, and these correspond to the frequencies where $\lambda/2$ and $\lambda$ equal the line length of 20 m. Other resonances occur at $l = k\lambda/2$, where $k$ is an integral multiple. Transmission-line theory does not predict these resonances, though they do exist. The resonances may be calculated based on the termination impedance of the transmission line. In practical shielded cables, the impedance is predominantly the impedance of the vertical section of cable as it terminates to the enclosure. Less typically, the cable does not bend up or down as it connects to the enclosure, in which case the termination impedance is the impedance of the enclosure to ground plane connection, which may be confined to the impedance of the enclosure wall. This impedance may be found from the impedance of a ground plane as

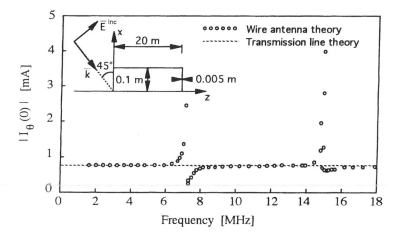

**Figure 7.30** Termination current for a rectangular loop oriented normal to a perfect ground plane and illuminated by a plane wave. (© 1987, IEEE.)

described in Section 5.1.2. The impedance of the vertical section of the transmission line is given by

$$Z_v = 60\left(\ln\frac{2h}{a} - 1\right) \tag{7.21}$$

where $a$ is the radius of the vertical section and $h$ is the height.

If the incident plane wave has a vertical E field component, then some coupling to the end sections will occur. However, because we assume that the length of the transmission line is much greater than the height of the line above the ground plane, we ignore any contribution from coupling to the vertical section of the line. Theoretically, when the termination impedance is zero, the termination current is infinitely high. In practice, some terminating impedance will always exist. Figure 7.31 (from Ref. 16) shows the predicted cable current amplitude for a 1-V/m E field incident on a 20-m-long line of radius 5 mm and at a height of 50 cm above the ground plane. The cable current was predicted by use of method-of-moments antenna theory and by a modification to transmission-line theory described in Ref. 16. We see that as the termination impedance approaches the characteristic impedance of the cable above a ground plane, the maximum current reduces to that predicted by the transmission-line approach. Often in radiated susceptibility tests, such as the MIL-STD-461 RS03 test where the interconnection cables are located at a height of 5 cm above a ground plane, equipment is found to be susceptible at frequencies corresponding to the resonant frequency of the interconnection cables. This is especially true for shielded cables with a low-impedance connection of the shield at one or both ends to the ground plane. A simple test, which may be used to verify shielded-cable resonance susceptibility, is to short the shields of the interconnection cables to the ground plane at some distance along the length of the cable and thus shift the resonance frequency. The use of ferrite beads or baluns at the ends of shielded cables will often increase the termination impedance and reduce the cable resonance currents sufficiently to achieve EMC. When this is impractical, the type of shield or number of braids may be changed to increase shielding effectiveness. As described in Section 7.2 on transfer impedance, there is a limit to the level of shielding achievable in a flexible cable, and in some instances the only solution is to increase the immunity of the equipment.

Transmission-line theory or the inductance of a loop may be used to find the shield current flow on a shielded cable. The transfer voltage is then found by use of Eq. (7.15). The (sin $\theta$)/$\theta$ function of Eq. (7.15) is valid for a cable with a matched shield termination only. Therefore, when an unmatched termination exists, the transferred voltage should be multiplied by a factor that accounts for the increase in shield current at resonance. When the cable shield is terminated at both ends to the enclosure, a nominal multiplication factor of 5 may be used. To more accurately account for resonance and termination effects, software such as GEMACS, NEC, MININEC, and a commercially available Radiated Immunity software form SpectraSoft can be used to model cables above a ground plane, typically as a two-wire transmission line formed by the cable and its electromagnetic image in the ground plane. In one example, a 2.5-m length of cable located 5 cm above a ground plane was modeled using GEMACS and the SpectraSoft Radiated Immunity software. Also, the simple transmission-line equation (Eq. 7.20) was used to predict the current flow, but this does not account for cable resonances, which are significant when the termination impedances are low. The termination impedances at both ends of the cable, which represent a typical short circuit, were set at 50 nH in series with 2.5 mΩ. This would be the case for a shielded cable with the shield terminated to the ground plane at each end. The transmission-line and termination impedances were modeled in GEMACS using wire segments and a MOM analysis, with a plane wave incident on the structure. The current flow on the cable with a horizontally polarized field broadside onto the cable was analyzed, as well as a vertically

(a)

(b)

(c)

(d)

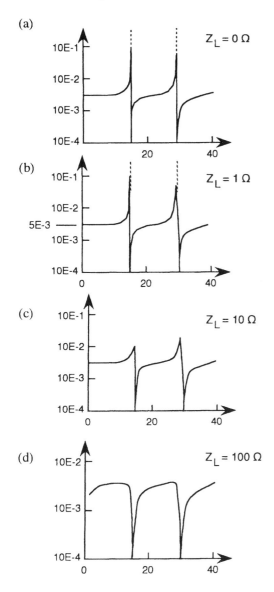

**Figure 7.31** : Cable current with 0-, 1-, 10-, and 100-Ω termination impedances. (From Ref. 16. © 1988, IEEE.)

polarized sidefire field, i.e., illuminating the vertical end section. In either case it was the maximum current in any of the wire segments down the length of the cable that was recorded. The same analysis was performed using the SpectraSoft Radiated Immunity program, which provided a plot of cable current in the cable versus frequency so that the current at resonant and antiresonant frequencies could be seen. It was not clear in the use of this program if the predicted current was an average value or a peak value. The peak current at the first resonant frequencies of 60 MHz and 180 MHz and an incident field of 40 V/m were analyzed.

In addition, cable currents at 1-GHz and 4-GHz and an E field of 3500 V/m, which would be representative of illumination by radar, were analyzed.

The current flow using Eq. (7.20) for the horizontal, broadside on, field was also predicted.

The height of the cable above the ground plane must be much less than λ for this equation to remain valid, so the prediction at 4 GHz was omitted. Also, Eq. (7.20) does not account for current flow at resonance, and this is significant at 60 and 180 MHz, where the termination impedances were lower than the transmission-line impedance. However, at 1 GHz the termination impedances are high and the transmission line equation predicts a higher current flow than was calculated by GEMACS or the SpectraSoft Radiated Immunity program. The results of the predictions are shown in Table 7.2, with a good correlation between the GEMACS program and the SpectraSoft program.

At gigahertz frequencies it may be more appropriate to conduct a shielding effectiveness test on cables in the configuration in which they are to be used. One test was made of the coupling through a cable when exposed to a local field from 1 to 2 GHz. A capacitive injection clamp was used to inject a high-level E field into the cable, as shown in Figures 7.32 and 7.33. A 20-W TWT amplifier was used to generate the field, and the capacitive injection probe was designed to have a VSWR of less than 3:1 to avoid damage to the TWT. Other TWTs require an even lower VSWR. A 2.4-m length of RG58 coaxial cable was set up on 5-cm blocks above a ground plane, with the capacitive injection clamp placed around the cable. A number of locations down the length of the cable were tried for the injection probe. It was found that a large number of locations down the length of the cable resulted in the same cable current and transferred voltage; however, the maximum induced values were highly dependent on moving the injection probe a centimeter or so relative to the cable at any of these locations. The far end of the RG58 was terminated with 50 Ω, while at the opposite end the center conductor led to a pin in an MIL 38999 connector. The reference to a ''38999 connector'' implies the combination of a Matrix D38999/24WF35SA female connector, a Bendix JD38999/26WF35PA male connector (mating half), and a Glenair M85049/19-19W06 backshell (mounted on the Bendix connector). The direct termination of the RG58 cable shield to the EMI backshell is not the ideal 360° termination around the periphery of the cable, but it is representative of the type of connection used in the cable harness. An aperture exists between the cable shield and the backshell, as can be seen in upcoming Figure 7.35, and coupling from currents flowing on the shield, through this aperture, is almost certainly one reason why the attenuation of the direct connection is so poor. In Ref. 17 a conductive elastomer is described that is used to plug the hole in the backshell and to make a connection around the periphery of the shield Figure 7.38 shows an improvement at some frequencies with the conductive elastomer compared to an unshielded cable at frequencies from 0.2 to 1.2 GHz.

The 38999 connector was mounted on the end of a metal box, that also had an N-type

**Table 7.2** Comparison of Cable Currents Induced by a Plane Wave Incident on a Two-Wire Transmission Line, Using GEMACS, Commercial Radiated Immunity Software, and Eq. (7.20)

| $f$ (MHz) | E field (V/m) | Field orientation | Current using TX line Eq. (7.20) (mA) | GEMACS peak current (mA) | Radiated immunity software (mA) |
|---|---|---|---|---|---|
| 60 | 40 | Vertical (sidefire) | — | 6.1 | 7 |
| 60 | 40 | Horizontal (broadside) | 43 | 320 | 150 |
| 180 | 40 | Vertical (sidefire) | — | 7.4 | 5.6 |
| 180 | 40 | Horizontal (broadside) | 43 | 90 | 65 |
| 1000 | 3500 | Vertical (sidefire) | — | 860 | 500 |
| 1000 | 3500 | Horizontal (broadside) | 3760 | 1800 | 1200 |
| 4000 | 3500 | Vertical (sidefire) | — | 600 | 110 |
| 4000 | 3500 | Horizontal (broadside) | — | 630 | 300 |

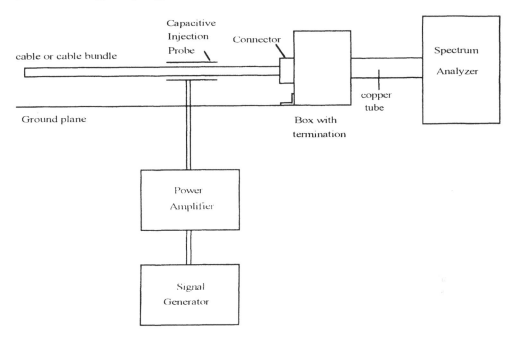

**Figure 7.32** Capacitive injection clamp setup.

**Figure 7.33** Injection clamp.

connector mounted on the opposite end. Inside, a short length of RG213 coax led from the N-type connector to the pin that connected to the RG58 cable. The unshielded length of RG213 cable was kept to a minimum. The box setup can be seen in Figure 7.34. The box lid was fastened with EMI gasket electromagnetically sealing the box, as shown in Figure 7.35. The signal from the N-type connector was taken via a coaxial cable covered in a solid copper conduit to a spectrum analyzer located outside of the shielded room where the measurements were made. The injected signal was swept from 910 MHz to 2 GHz, and the level was recorded on the spectrum analyzer. This test was performed three times in three different configurations. The first was to connect the RG58 shield directly to the pin in the Matrix connector to simulate an unshielded wire, and the level from 910 MHz to 2 GHz was recorded. The test was then repeated with the center conductor connected to the pin and the coax shield terminated directly at the backshell of the connector. Once again the test was repeated, this time with the shield terminated via a 1″ pigtail. These two levels were subtracted from the unshielded-wire reference level to obtain the shielding effectiveness of the cable.

The results of this test give the shielding effectiveness achieved with a direct shield termination as well as with a 1″ pigtail termination using the 38999 connector. The results are shown in Figure 7.36. As shown in the graph, the attenuation achieved with a good shield termination is higher than that achieved with 1″ pigtails at most frequencies. Above 1400 MHz the coupling may have been primarily aperture coupling and not transfer impedance coupling due to the long-line limit on transfer impedance. Why the direct termination and the pigtail is better than an unshielded cable from 1700 MHz to 2000 MHz is unclear. However, the test setup is representative of the actual cable, connector, and backshell configuration.

The next test was on a cable with an ''ARINC'' connector with a provisional copper backshell. This backshell was open, and the copper surround was used to solder the cable shields via the 1″ pigtail. Because the makeshift backshell was wide open at the back, each conductor in the cable had short unshielded lengths of wire leading from where the shield ended to the pins in the ARINC connector. Radiated coupling to these lengths of wire from the currents set

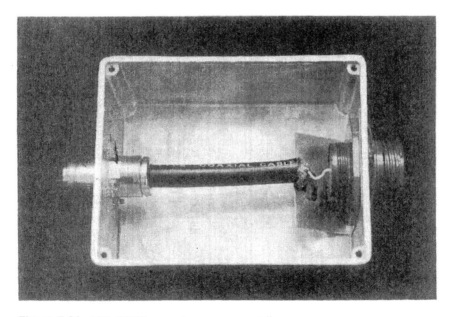

**Figure 7.34**   MIL-38999 connector measurement jig.

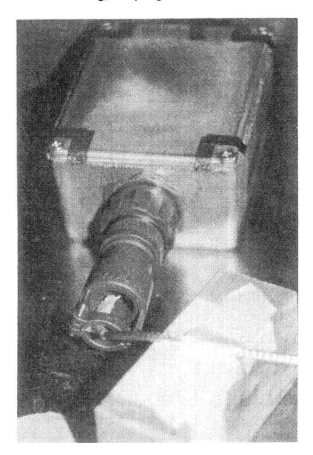

**Figure 7.35**   MIL-38999 connector measurement jig with lid fastened.

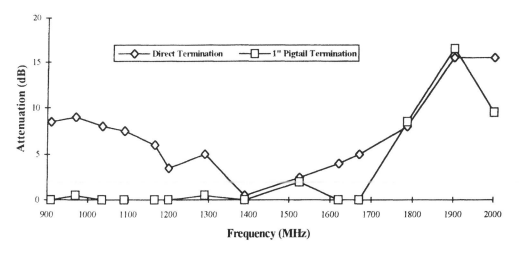

**Figure 7.36**   RG58 shielding effectiveness.

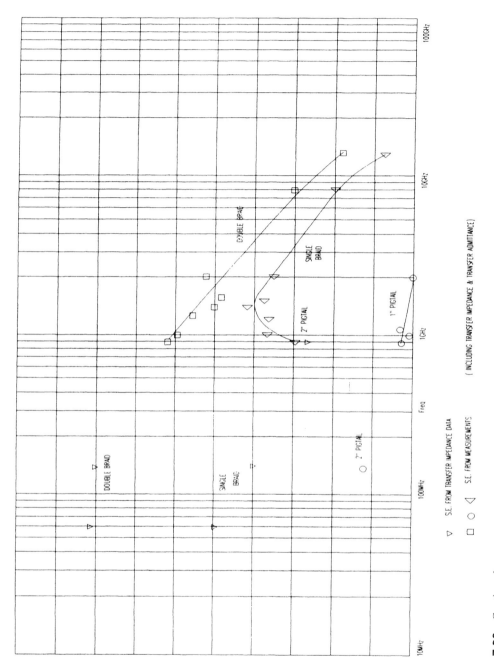

**Figure 7.36** Continued

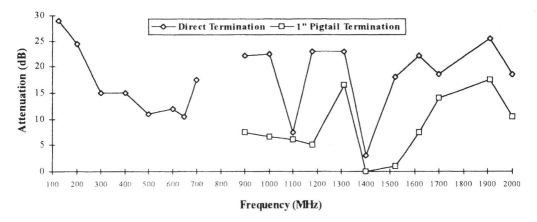

**Figure 7.37** Shielding effectiveness of a representative cable harness with ARINC connector and provisional ''open'' backshell. Direct connection and 1″ pigtail.

up on the cable shield is very probable. Figure 7.37 shows the attenuation with a 1″ pigtail and a direct connection of the braid of the shielded cable to the provisional backshell.

### 7.4.1 Polarization and Angle of Incidence

In the following discussion we refer to a wave with *horizontal* polarization as one in which the electric field component is parallel to the ground plane and a *vertically* polarized wave as one in which the electric field is perpendicular to the ground plane.

In many cases we do not know the polarization or the angle of incidence of the wave,

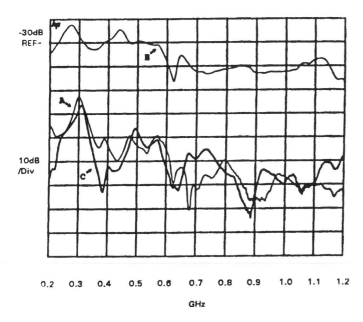

**Figure 7.38** Voltage induced in: (A) shielded cable with a standard EMI backshell; (B) unshielded cable; (C) shielded cable with conductive elastomer backshell. (© IEEE, 1995.)

and a worst-case coupling of either the magnetic or electric field component must be assumed. One example in which the assumption of worst-case coupling is valid is the RS03 test setup, in which, above 30 MHz, both vertical and horizontal orientations of the E-field-generating antenna must be used. The antenna is located 1 m from the edge of the ground plane. Thus for frequencies below 50 MHz the field is near field and exhibits curvature and horizontal, vertical, and radial field vectors. When the RS03 test is conducted in a shielded room, the electromagnetic image of the antenna is formed in the ceiling and the angle of incidence of the wave on the interconnection cables is both directly from the antenna and that of the reflected wave from the ceiling. Thus, especially at low frequency, the assumption of worst-case coupling in the RS03 test setup is perfectly valid. In some instances both the polarization of the wave and the angle of incidence are known. One example is when the orientation and locations of a transmitting antenna that couples to an above-ground signal or power cable are known. The angle of incidence and polarization determine the directivity of the configuration that is used to correct the predicted open-circuit voltage induced in a transmission line above ground. Reference 10 contains directivity patterns for vertical polarization, for which the directivity may be as high as 3.7, and for horizontal polarization, which may have a directivity as high as 2. Reference 10 also provides information on the compensation in the coupling prediction when the transmission line is located above a ground plane with less than perfect conductivity.

## 7.5  SHIELD TERMINATION

Thus far we have assumed that the shield of a cable is connected to ground at both ends (symmetrical connection) via the metal walls of an enclosure. We have seen that this configuration provides shielding against an incident magnetic field but that this shielding effectiveness is limited at low frequency. When one or both of the enclosures can be disconnected from ground, greatly increased shielding against a magnetic field may be achieved. However, when it is the shield of the cable that is disconnected from an enclosure at one end and the two enclosures remain connected to the ground plane, the configuration is effectively the same as shown in Figure 7.3b with minimum attenuation of magnetic-field-induced voltages. When a shielded cable is used to connect a component such as a potentiometer mounted on a front panel, thermistor, strain gauge, or heating element in which the electrical connections are isolated from chassis ground, then the shield should be connected as shown in Figure 7.39. Here the shield is isolated from ground at the component end and maximum shielding against both $E$ and $H$ fields is achieved. The capacitance between the component (and associated connections) and enclosure must be minimized for this to remain valid at high frequencies. As with all guidelines, exceptions exist; one in which a shield is deliberately disconnected from the enclosure at low frequencies

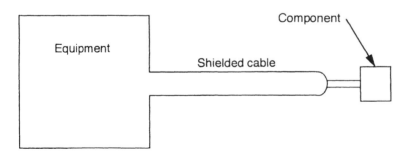

**Figure 7.39**   Shield connection for components with low capacitance to ground (chassis).

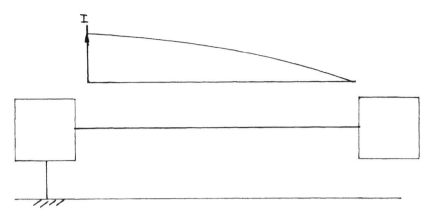

**Figure 7.40a**   Current flow in a near-resonant length of shielded cable in which one enclosure is connected to ground (asymmetrical connection).

at one end but effectively connected at RF is described later in this section. In the following discussions we assume that the shield is connected to the enclosures at both ends of the cable, and isolation of the cable at one end (asymmetrical connection) means that the enclosure is disconnected from ground.

Assuming the conditions for use of transmission-line theory are met and that the length of the cable is close to or greater than $\lambda/4$, then Eq. (7.20) may still be used to determine the cable current flow for the asymmetrical connection. Figure 7.40a illustrates the current distribution on the shield due to an incident electromagnetic wave for the asymmetrical connection and where the cable length is a significant fraction of a wavelength but not greater than $\lambda/4$. The effect of removing one end of the ground connection (asymmetrical connection) is merely to move the frequency of resonances; the magnitude of the shield current at nonresonant frequencies is of the same order. The resonances for the asymmetrical connection occur when

$$l = \frac{(2k + 1)\lambda}{4} \tag{7.22}$$

where $k = 0, 1, 2, 3, \ldots$. Thus resonances for the asymmetrical connection occur for line lengths that equal $0.25\lambda$, $0.75\lambda$, $1.25\lambda$, etc., and for the symmetrical connection when the line length equals $0.5\lambda$, $\lambda$, $1.5\lambda$, etc.

Figure 7.40b illustrates that a current does flow on a shield when both ends are discon-

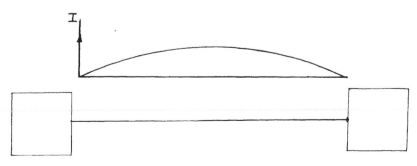

**Figure 7.40b**   Current flow in a cable and enclosures disconnected from ground (or for the symmetrical connection).

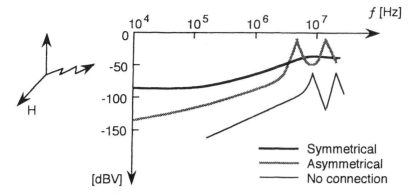

**Figure 7.41a**   EMI voltage induced in coaxial cable with asymmetrical, symmetrical, and no connection of the shield and enclosures to ground. (From Ref. 18.)

nected from ground. The current distribution for the symmetrical connection is the same as the distribution shown in Figure 7.40b for the totally disconnected configuration.

Figure 7.41a, from Ref. 18, illustrates the voltage induced in a leaky coaxial cable, one in which the transfer impedance increases linearly with frequency, i.e.,

$$Z_t = R_s + j\omega L_t \tag{7.23}$$

and for which $L_t = 16$ nH/m and $R_s = 9$ mΩ/m. The incident electromagnetic wave is a 1 = V/m plane wave, and the cable has a diameter of 8 mm, a characteristic impedance of 50 Ω, and a length of 15 m and is located 30 cm above a ground plane. The angle of incidence of the plane wave ensures worst-case coupling for all three shield terminations, which are symmetrical, asymmetrical, and no connection. Below 1 MHz, the symmetrical connection exhibits a lower level of shielding than either the asymmetrical or no connection. At 5 MHz, where the cable length is approximately equal to $\lambda/2$, the symmetrical connection achieves the highest level of shielding. In Figure 7.41b experimental results are shown compared to theoretical, in which the difference in the asymmetrical and symmetrical connections is no greater than 25 dB at 10 kHz in the measured values. This is attributed to additional E field coupling through the transfer

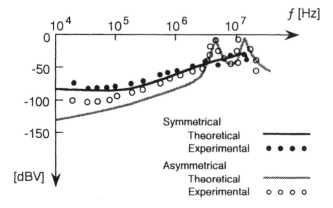

**Figure 7.41b**   Comparison of calculated and measured EMI voltages. (From Ref. 18.)

admittance of the shield, which increases the measured EMI voltage, especially for the asymmetrical connection.

In practice a lower transfer impedance than indicated is achieved above 1 MHz for the majority of cables, and therefore the shielding effectiveness will typically be higher. When the incident electromagnetic wave is not a plane wave but predominantly an electric or magnetic excitation, then at frequencies where the line length is less than the wavelength, the symmetrical or asymmetrical connection of shields results in different values for the shielding effectiveness of the cable.

For a magnetic field incident on the same cable configuration (i.e., 15-m cable length etc.), the asymmetric connection achieves a higher level of shielding effectiveness below 1 MHz but a lower level above 5 MHz. For an electric field, the symmetrical connection achieves a higher level of shielding than the asymmetrical at frequencies below 5 MHz, for the 15-m cable length, with little to choose between the two at higher frequencies. Again, measured values of shielding show less difference than predicted for the two configurations and for E and H fields, due to the coupling through the transfer admittance of the cable.

In practice it is difficult to achieve a true asymmetric connection unless the inner conductors of the shielded cable terminate on a physically small component, such as a transducer, at the end where the shield is floated.

Where considerable circuitry exists, either capacitance to ground or a ground connection via a power supply and the line cord safety ground may negate the true asymmetrical connection. The configuration we have considered, of cables running close to a ground plane, is normally achieved only on spacecraft, in vehicles, aircraft, and ships, and in the RS03 test setup.

When the cabinets or cases of two pieces of equipment are connected together via shielded cable and conductive connector backshells, the second connection is often made via additional interface cables or via the two equipment line cords to the safety ground. Where the incident electromagnetic field induces current flow on the shields of two or more adjacent interface cables, transmission-line theory may be used to calculate the magnitude of the current flow. In contrast, when a second case-to-case connection is made via safety grounds, a high impedance due to inductance is included in the loop and circuit theory is applicable. These two configurations are shown in Figure 7.42.

It should be emphasized that in neither of the situations in Figure 7.42 is the cable shield disconnected from the enclosure; only the enclosure is disconnected from the ground plane. This is also true for Figures 7.40a, and b. Where the cable shield is disconnected from the enclosure at one end, the attenuation of the configuration to a magnetic field is negligible. With the shield disconnected, some attenuation to an E field is achieved, especially at low frequency and where circuit impedances are high or where the cable length is much greater than the wavelength, in which case a low level of attenuation to a plane wave is achieved. The level of plane-wave attenuation for a leaky cable will be approximately the same as for the asymmetrical connection, shown in Figure 7.41a, at 5 MHz and above.

One example where the shield of a cable was deliberately disconnected from an enclosure is as follows. The shielded cable was kilometers long and in proximity to a 60-Hz magnetic field. Connecting the shield of the cable to the enclosure at both ends would result in very high levels of 60-Hz current flow on the shield of the cable, which in turn would be coupled via mutual inductance to the center conductors of the cable. Disconnecting the shield of the cable meant that the center conductors of the cable were exposed to the 60-Hz field with virtually no shielding. The signals were differential and terminated in a high impedance at the receiver, so the 60-Hz current was limited by the high input impedance of the receiver. The 60-Hz common-mode voltage appearing at the receiver input was rejected by the common-mode noise rejection of the receiver. In addition to 60 Hz, the long cable carried currents induced by FM and AM radio

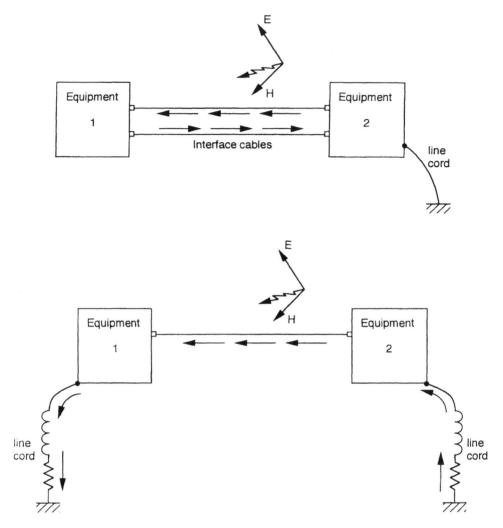

**Figure 7.42**  Loop area and common-mode current induced by an electromagnetic wave.

transmissions. The cable was electrically long and achieved the level of attenuation predicted in Eq. (7.15), which is independent of the type of shield termination (asymmetrical or symmetrical), as shown in Figure 7.41a. Thus the type of shield termination appears ideal for the application. The problem is that the shield at the end of the unterminated cable enters the enclosure and, due to a high level of noise voltage between the shield and the enclosure, radiates inside the equipment. One solution is to add an RF capacitor, with a sufficiently low impedance connection, of approximately 100-pF value, between the end of the shield and the enclosure. The capacitor maintains the isolation of the shield at 60 Hz but effectively connects the shield to ground at RF. The correct location for the capacitor is outside of the enclosure to ensure that the noise currents do not flow on the inside of the enclosure. Not only was the location of the capacitor outside of the enclosure not feasible, but the connection of the physically small capacitor between the shield of a large cable and chassis was extremely difficult. A potential solution was to use the high permittivity of the cable insulation and, by either bringing the cable through a metal tube or wrapping the cable with metal foil, to form a 100-pF capacitance between the shield of the cable and the enclosure.

## 7.6 EMISSIONS FROM CABLES AND WIRES

Many of the considerations we have given to coupling from E and H fields into cables apply to the generation of E and H fields by cables. One of the more useful axioms to be remembered in EMC work is that where a time-varying current exists, so does an electromagnetic field; conversely, where a time-varying field impinges on a conductor, a current flows. The path and, where possible, the magnitude of the current flow must first be identified in predicting the resultant radiation.

Where differential currents flow on a multiconductor cable and the conductors are close together, the amount of radiation from the cable is often less than for a lower-magnitude common-mode current that flows in a loop of large area or on an isolated cable. The practicing engineer with little exposure to EMC will often dismiss the potential for radiation from a signal or power interconnection cable on the grounds that the signal- or differential-noise voltage is at a low frequency and, thereby, ignore the common-mode contribution. The EMC engineer, on the other hand, may incorrectly ignore differential-mode currents due to the common experience that the source is so often common-mode current flow.

A practical instance of this is a digital data bus connecting two pieces of equipment in which common-mode noise voltage exists between the PCBs in the equipment and the equipment case. A current flow may then exist between the case-to-PCB connection and the data bus returning via a ground connection in the far enclosure, as illustrated in Figure 7.43. In the case of shielded cables, the differential current flow on the center conductors induces only a low level of current on the inside of the shield, whereas any common-mode voltage between the center conductors and the shield results in a typically higher current flow on the inside of the shield. Even without a connection of any one of the conductors to the far-end enclosure, a current still flows on the inside of the shield due to the impedance of the transmission line formed by the center conductors and the shield.

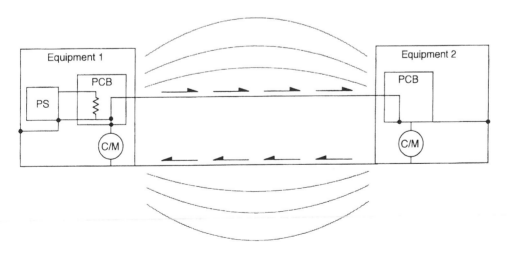

**Figure 7.43** Radiation as a result of common-mode current generated by noise voltages within equipment.

When a piece of equipment is connected to ground, for example, via the AC safety ground, it is often common-mode currents on the AC power cord that is the major contributor to radiation. As described in Ref. 22, experiments with a battery-powered broadband noise source connected via a length of cable to a load in which no ground connection existed still resulted in the detection of common-mode current on the interconnection cable. The most likely explanation for this surprising result is displacement current flow via capacitance between the circuit and grounded structures. When a ground connection is made to a similar setup with a radiating loop, a common-mode current flows on the ground connection that may be the principal source of radiation, as discussed later and in Chapter 11, Sections 11.1 and 11.6.2, on PCB radiation.

The most accurate method of obtaining the common-mode current flow is measurement by use of a current probe around the shield of a cable or around all of the conductors in an unshielded cable. Alternatively, circuit theory may be used to calculate current flow on cables based on measured or predicted values of common-mode voltage and circuit or transmission-line impedance. When the current flow on the inside of the shield is known, the current flow on the outer surface of the shield of a cable may be derived from the methods used to calculate cable-to-cable coupling below 200 kHz and from the transfer impedance of the cable above 200 kHz. The field radiated by the cable may then be calculated by use of the monopole, electric current element, or current loop equations, whichever are applicable to the cable geometry, contained in Chapter 2. Alternatively, moment method techniques may be used. In the example shown in Figure 7.43, the geometry of the current path must be known in order to arrive at the loop area.

## 7.6.1  Emission from Loops

For loops of any shape but that do not fall under the definition of a transmission line, the equations of Section 2.2.3 may be used. These equations are valid when the load impedance at the end of the loop is less than or equal to the transmission-line impedance of the loop, as discussed later. For calculations of the electromagnetic wave in the far field, the equations may be simplified to Eq. (7.24) with identical results:

$$E = \frac{Z_c I_L \beta^2 A}{4 \check{s} R} \tag{7.24}$$

where

$Z_c = 377 \ \Omega$

$I_L = $ maximum loop current

$\beta = \dfrac{2\pi}{\lambda}$

$A = $ area of loop

## 7.6.2  Emission from Transmission-Line Geometries

To calculate the radiation from a transmission line made up of either two cables or a cable above a ground plane, use is made of the concept of the *radiation resistance* of the line. Radiation resistance of an antenna or a transmission line is used to describe that part of the conductor resistance that converts a fraction of the power delivered to the load into radiated power. In an efficient antenna, the radiation resistance is designed to be high and the resistance, which converts the input power into heat, is low. In the transmission line, the opposite is true. The radiation resistance for a resonant section of two-wire line of length $\lambda/2$ (or integral multiples of $\lambda/2$)

when the line is short-circuited (symmetrical connection), or $\lambda/4$ and other multiples shown in Eq. (7.22) when the line is open-circuited or terminated in a load higher than $Z_c$ (asymmetrical connection), is

$$30\beta^2 b^2 \qquad (7.25)$$

where $b$ is either the distance between the two-conductor line or twice the height of the cable above a ground plane. The radiated power for the resonant line is given by

$$30\beta^2 b^2 I^2 \qquad (7.26)$$

where $I$ is the current flow on the line, either measured or calculated. For the majority of EMI conditions, the radiated noise covers a wide range of frequencies, and resonant line conditions can be expected. For example, when the source of current is converter or digital logic noise, the harmonics may range from kilohertz up to 500 MHz.

Where a single frequency with negligible harmonics is the source of current in the line and the cable length is electrically long, i.e., $l > 0.25\lambda$, but either the line is not a resonant length at the frequency of interest or the line is terminated in its characteristic impedance, the radiation resistance equals

$$30\beta b^2 \qquad (7.27)$$

and the power is

$$30\beta b^2 I^2 \qquad (7.28)$$

The magnetic field some distance $R$ from the line is

$$H = \sqrt{\frac{Pk}{4\pi R^2 Z_w}} \qquad (7.29)$$

where

$$P = \text{radiated power}$$
$$Z_w = \text{wave impedance}$$
$$k = \text{directivity, which is approximately 1.5 for a current loop and resonant}$$
$$\text{line and 1.0 for a nonresonant line}$$

and the electric field is

$$E = \sqrt{\frac{Z_w Pk}{4\pi R^2}} \qquad (7.30)$$

In the near field, the wave impedance is close to the characteristic impedance $Z_c$ of the transmission line, which may be calculated from Eq. (7.16). The wave impedance then changes linearly until the near-field/far-field interface, at which $Z_w = 377\ \Omega$. The E field radiated by the current loop may be obtained from $E = H \times Z_w$. The equation for calculating the wave impedance in the near field is

$$Z_w = \frac{\lambda/2\pi - R}{\lambda/2\pi}(Z_c - 377) + 377 \qquad (Z_w < 377\ \Omega)$$

where $R$ is the distance from the radiation source, in meters.

Figures 7.44 a, b, and c compare the calculated and measured magnetic field from a transmission line 2 m long suspended 5 cm above a ground. In the near field, the calculated H field using transmission-line theory produces results closer to the measured values, whereas large errors are obtained for the use of current loop equations. In the far field, the two calculated values tend to converge and the maximum error between calculated and measured is approximately 6 dB.

The criteria for assessing when circuit theory or transmission-line theory is applicable are the same as described in Section 7.4 on coupling into wires. The directivity factor of 1.5 does not take into account the height of the antenna with reference to the transmission line. Even more important when measurements are made in a shielded room, as were those shown in Figure 7.44, are the reflections of the transmission line from the ceiling and wall of the room. Compensation for reflections is discussed in Section 9.3.2.

The transmission-line model is applicable to the RE02 and DO-160 test setups, in which cables are located 5 cm above a ground plane and 10 cm from the edge of the ground plane for a distance of 2 m. When shielded cables are used, the shield is often terminated at the equipment enclosure, which is bonded to the ground plane, at one end and connected to the ground plane, after the 2-m length, at the other end. If we assume that the cable length between the ground connections is 3 m and the relative permittivity of the dielectric is 2, then the cables first resonant frequency is 35 MHz, at which $\lambda/2 = 3$ m. A typical RE02 narrowband limit at 35 MHz is 22 dBµ V/m measured at a distance of 1 m from the edge of the ground plane. The current flow on the shield of the cable must be below 9 µA at 35 MHz in order to meet the specification limit using Eq. (7.28), from which the predicted H field at a distance of 1.05 m from the cable is 3.47E-8 A/m. The characteristic impedance of the transmission line is approximately 317 Ω, and thus the wave impedance 1.05 m from the transmission line is approximately 377 Ω. The predicted E field is therefore 3.47 E-8 A/m $\times$ 377 = 13 µV/m = 22 dBµ V/m.

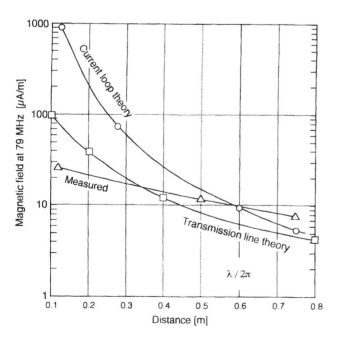

**Figure 7.44a** Comparison of calculated and measured fields from a transmission line with distance at 79 MHz.

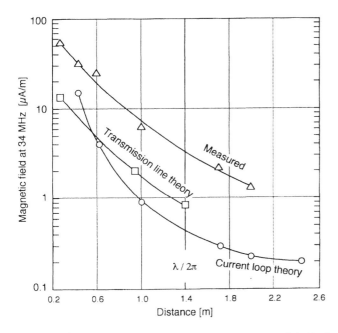

**Figure 7.44b** Comparison of calculated and measured fields from a transmission line with distance at 34 MHz.

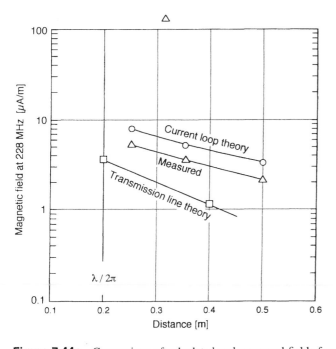

**Figure 7.44c** Comparison of calculated and measured fields from a transmission line with distance at 228 MHz.

In Figure 7.44b the measured H field was approximately 12 dB above the predicted value at a frequency of 34 MHz and a distance of 1 m, based on the measured cable current. Some of the difference may be accounted for by reflections from the ceiling and wall of the shielded room, plus errors in antenna calibration and the damping effect of the current probe on the cable when the cable current is measured. Thus engineers must set a design goal of less than 1 μA of common-mode current on a cable at 20–35 MHz to ensure meeting RE02 limits.

Measurements were also made in a well-damped chamber that, over the frequency range of the measurements, showed a correlation of between 0 and 4 dB to measurements on an OATS. A 2-m-long insulated #16AWG wire was raised 5 cm above and 10 cm from the edge of a ground plane and terminated either in a short circuit, in an open circuit, or with 317 Ω. The current on the line was monitored by use of a current probe. The current on the cable when terminated with the characteristic impedance of the line is constant down the length of the line. For the short-circuit and open-circuit termination, the current probe was moved up and down the length of the line to intercept the highest level of current. The measuring antenna was located at a distance of 1 m from the edge of the ground plane and raised and lowered in height (not a requirement in MIL-STD-462 or DO-160) to measure the highest E field from the test setup. The antenna orientation was changed to intercept both the horizontally and the vertically polarized field. Either Eq. (7.26) or Eq. (7.27) was used to calculate the power, and Eq. (7.30) was used to calculate the E field.

Table 7.3 provides a comparison of the measured and predicted E fields. The accuracy of the prediction is typically acceptable, with some error in the current measurement introduced by the presence of the current probe on the cable. The type and size of antenna used affected the measurement, with a significant difference in the data using a log periodic/biconical versus a reference dipole antenna. A smaller contribution to the error is from the chamber.

### 7.6.3  Emissions from Loops With and Without Attached Cables

The situation arises where the radiation from a loop that either is disconnected from any ground connection or more commonly is connected to ground is required. One example is wiring or PCB tracks inside an unshielded enclosure to which a ground connection is made.

One practical example of such a loop connected to a square wave generator is as follows. A 20-cm × 5-cm rectangular loop is connected to a square-wave generator that drives the loop via a 51-Ω resistor with a 2.5-V 1-MHz square wave. The 1-m-long cable driving the loop is a double-braid type wrapped four times around a ferrite balun to reduce radiation from the cable. The pulse generator is shielded and the AC power line is shielded at the generator enclosure. Thus the prediction is that a low level of current will flow on the cables. The majority of EMI sources are pulses or square waves, and in meeting EMI specifications we are interested in the frequency domain spectral density. Thus in an EMC prediction we must often convert from the time domain to the frequency domain. The maximum current flow in our example loop at the fundamental 1-MHz frequency is 2.5 V/51 Ω = 49 mA. Using the equations of Figure 3.1, the amplitude at the 21st harmonic (21 MHz) is 62 dBμA.

Using this current in the computer program that computes the E field from a loop results in a predicted E field of 0.28 mV/m, with no compensation for reflections from ceiling and floor. A very approximate correction is to add 6 dB for reflections, resulting in a predicted E field of 0.56 mV/m. The E field from the loop measured by a monopole antenna located 1 m from the loop is shown in Figure 7.45. The level measured by the monopole is −57 dBm at 21.7 MHz. A 26-dB preamplifier is used, so the output level from the monopole is −57 dBm − 26 dB = −83 dBm. This converts to a voltage of 16 μV. The antenna factor of the monopole is 40, and so the measured E field is 0.64 mV/m. Removing the loop from the end of the cable

**Table 7.3**  Predicted and Measured E Field From a Wire above a Ground
Plane in an MIL-STD-462/DO-160 Test Setup

| Frequency (MHz) | Calculated level (dBμV/m) | Measured level (dBμV/m) | Delta (dB) |
|---|---|---|---|
| **Short-circuit termination, horizontally polarized field** | | | |
| 83 | 91.1 | 85.5 | 5.6 |
| 140 | 95.6 | 91.0 | 4.6 |
| 200 | 96.7 | 88 | 8.7 |
| 260 | 96.0 | 85.0 | 11 |
| **Short-circuit termination, vertically polarized field** | | | |
| 77 | 90.5 | 89.0 | 10.5 |
| 146 | 96.0 | 89.0 | 7.0 |
| 200 | 96.7 | 90.5 | 6.2 |
| 252 | 96.7 | 83.5 | 13.7 |
| **Open-circuit termination, horizontally polarized field** | | | |
| 55 | 85.5 | 85.5 | 0.0 |
| 118 | 93.1 | 86.0 | 7.1 |
| 166 | 96.1 | 96.0 | 0.1 |
| 228 | 94.9 | 85.0 | 9.9 |
| 293 | 89.0 | 87.0 | 2.0 |
| **Open-circuit termination, vertically polarized field** | | | |
| 48 | 84.4 | 86.0 | 1.6 |
| 110 | 92.6 | 90.0 | 2.6 |
| 172 | 95.4 | 93.5 | 1.9 |
| 225 | 94.8 | 94.9 | 0.8 |
| 297 | 89.2 | 87.0 | 2.2 |
| **Line terminated in its characteristic impedance of 317 Ω, horizontally polarized field** | | | |
| 35 | 76.2 | 77.0 | 0.8 |
| 140 | 78.2 | 80.0 | 1.8 |
| 233 | 81.4 | 77.0 | 4.4 |
| 45 | 76.3 | 77.0 | 0.7 |
| 77 | 77.6 | 81.0 | 3.4 |
| 112 | 80.3 | 75.0 | 5.3 |
| 151 | 78.5 | 81.5 | 3.0 |
| 185 | 80.4 | 85.5 | 5.1 |
| 216 | 80.1 | 84.5 | 4.4 |
| 250 | 80.7 | 81.0 | 0.3 |

and replacing it with a near-perfect short is expected to reduce emissions considerably. However, as shown in Figure 7.46, this is not the case. A measurement of the current flow on the AC power-line cable was made using a current probe, with the results as shown in Figure 7.47. The level at 21.7 MHz is $-58$ dBm, which, accounting for the 26-dB preamplifier, results in a probe output level of $-84$ dBm $= 14$ μV. The transfer impedance of the current probe at 21 MHz is 0.89 Ω, so the cable current is 15.7 μA. Using this current in the computer program for E field from an electric current element results in a predicted E field of 0.65 mV/m. This example illustrates very nicely the importance of common-mode current on cables as the predominant source of radiation, for approximately the same E field is generated by the 15.7 μA on the AC cable as by a current of 1.26 mA in the loop. In subsequent measurements the loop was located in the shielded room close to a tube through which the double-shielded cable was fed, and the

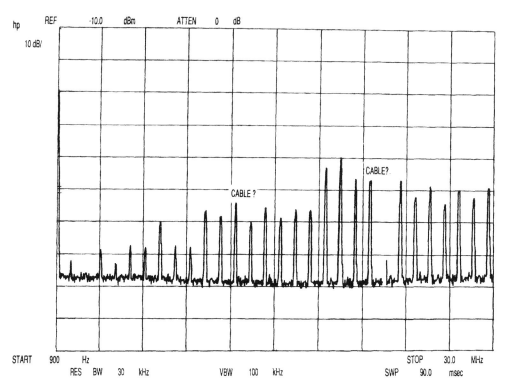

**Figure 7.45**   Output voltage from a monopole antenna with the loop in circuit.

generator and power cable were located outside of the room. With this configuration, no emissions were measured with the end of the generator cable shorted.

### 7.6.3.1   Loop Termination

The loop may be terminated in a low-impedance load, representative of logic gates, or a high-impedance load, representative of analog circuits, or, less commonly, an open circuit. The equations for radiation from a current loop assumes a short-circuited loop. Therefore an important question is how applicable the equations are when, as is so often the case, the loop is terminated in an impedance.

The currents flowing in a loop terminated in a load impedance are composed of the uniform current flowing into the load and a nonuniform displacement current that decreases to zero at the end of the loop, as shown in Figure 7.48. The magnitude of the nonuniform current is determined by the input impedance of a short open-circuited transmission line, which, from Ref. 10, is given by

$$Z_{\text{in}} = \frac{(377 \text{ or } Z_o)\lambda}{2\pi l} \tag{7.31}$$

where $l$ is the length of the transmission line, in meters.

The uniform loop current is given by $V_{\text{in}}/R_l$ and the nonuniform current by $V_{\text{in}}/Z_{\text{in}}$. The E field radiated by the nonuniform current may be modeled by two electric current elements separated by a distance $h$, which is the distance between the two conductors of the loop, in meters. The direction of current flow in the two elements is out of phase, and so are the E fields

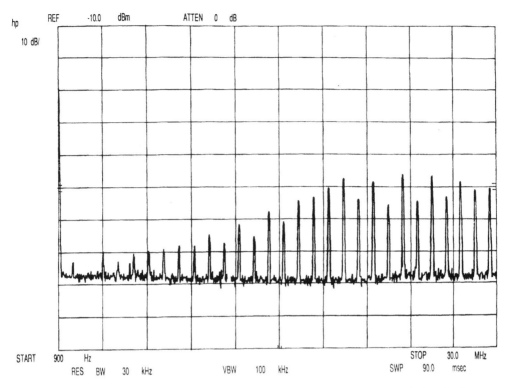

**Figure 7.46**   Output voltage from the monopole antenna with a short circuit replacing the loop.

generated by the elements. At some measuring distance in the plane of the loop the measured E field is the difference between the fields generated by the two elements. As an example, consider the 20-cm-long by 5-cm-wide loop open-circuited at the end. The input impedance of the loop at 30 MHz, from Eq. (7.31), is 3 k$\Omega$. The voltage applied to the loop is 0.55 V, so the value of nonuniform current is 0.183 mA. The E field at a distance of 1 m from the closest conductor in the loop is required. Using Eq. (2.12) for the E field at 1 m from an electric current element results in 7.6 mV/m. The E field at a distance of 1.05 m from the furthest conductor is 6.555 mV/m. The difference between the E fields is 1 mV/m and is the magnitude of the predicted field at 1-m distance.

In the test setup shown in Figure 7.49, the loop was either unterminated or terminated with a short circuit, a 50-$\Omega$, 1-k$\Omega$, or 10-k$\Omega$ resistor. The E field resulting from either the nonuniform or the uniform currents for each termination was calculated and measured by a monopole antenna at a distance of 1 m from the loop over a frequency range of 1–30 MHz. Table 7.4 provides a summary of the predicted and measured E fields at 5.8 MHz and 30 MHz. With a short circuit or a 50-$\Omega$ termination, it is the E field from the uniform loop current that dominates, with minimal contribution from the current element sources. With a termination impedance greater than 377 $\Omega$, the dominant source is the nonuniform current in the loop modeled by the electric current elements. With a fixed loop current, the measured E field is constant with a loop termination from a short circuit to a 50-$\Omega$ load; for a fixed input voltage, the E field is constant from a 1-k$\Omega$ load to an open circuit.

When a long wire is connected to the loop, the nonuniform current flows on the long wire as shown in Figure 7.50. The E field with the wire attached both grounded and ungrounded is

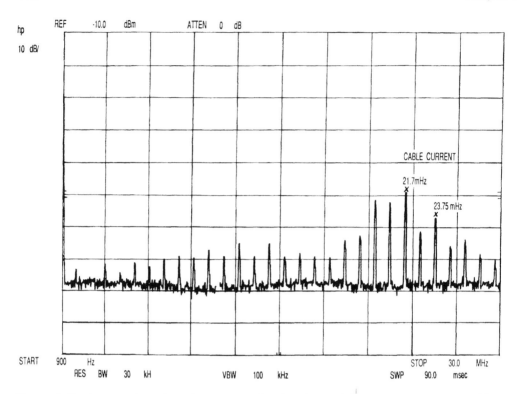

**Figure 7.47**  Output voltage from the current probe placed around the generator AC power cable.

reproduced in Figure 7.51, from Ref. 19, from which we see that the field is almost independent of the current flowing in the loop. Here is yet another example of the common-mode current on a wire acting as the dominant source of the field. It is no wonder that engineers with EMC experience tend to neglect differential-mode current flow on cables. The danger in doing so is when not all of the differential-mode current returns on the cable, for when current returns on the ground, a large loop is formed that may become the major radiation source.

The current flow on the long cable is difficult to predict without the use of the moment method model, as discussed in Ref. 19. The solution when the equipment exists or a breadboard model can be constructed is to measure the current flow on the long cable by use of the current

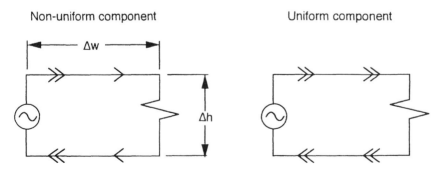

**Figure 7.48**  Uniform and nonuniform current flow in a loop.

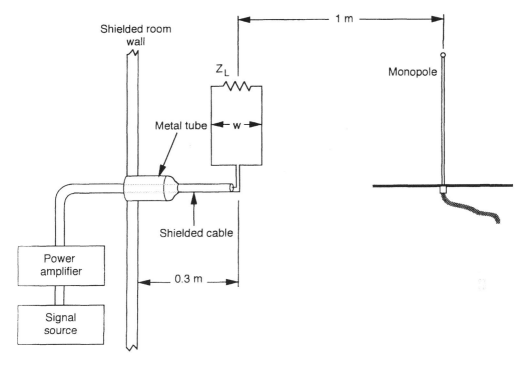

**Figure 7.49** Loop measurement test setup.

probe. Once the current is known, the electric current element equation may be used to predict the E field from the cable, as previously described. The electric current element model should be valid only for infinitesimally short lengths of wire. However, when the model is compensated for long lengths of wire, the predicted field is invariably higher than seen in near-field measurements and the unmodified model is found to be accurate.

Although a complete shield around the loop will completely eliminate the induction of common-mode current from the nonuniform loop current source, an incomplete shield may actually increase the common-mode current on the long cable. Reference 19 shows that the

**Table 7.4** Predicted and Measured E Field from a Loop with Different Terminations

| Frequency [MHz] | Termination | Current loop [mV/m] | Electric current elements [mV/m] | Measured [mV/m] |
|---|---|---|---|---|
| 5.8 | S/C | 0.40 | 0 | 0.49 |
| 5.8 | 50 Ω | 0.40 | 0.2 | 0.49 |
| 5.8 | 1 k Ω | 0.02 | 0.2 | 0.39 |
| 5.8 | O/C | 0 | 0.2 | 0.39 |
| 30 | S/C | 2.4 | 0 | 0.90 |
| 30 | 50 Ω | 2.4 | 1 | 0.90 |
| 30 | 1 k Ω | 0.122 | 1 | 0.41 |
| 30 | O/C | 0 | 1 | 0.41 |

S/C = short circuit; O/C = open circuit.

NON-UNIFORM
COMPONENT

**Figure 7.50** Nonuniform current flow on a loop connected to a long wire.

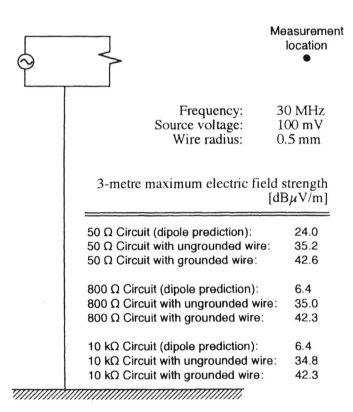

Measurement
location
●

| Frequency: | 30 MHz |
| Source voltage: | 100 mV |
| Wire radius: | 0.5 mm |

3-metre maximum electric field strength
[dB$\mu$V/m]

| | |
|---|---|
| 50 Ω Circuit (dipole prediction): | 24.0 |
| 50 Ω Circuit with ungrounded wire: | 35.2 |
| 50 Ω Circuit with grounded wire: | 42.6 |
| | |
| 800 Ω Circuit (dipole prediction): | 6.4 |
| 800 Ω Circuit with ungrounded wire: | 35.0 |
| 800 Ω Circuit with grounded wire: | 42.3 |
| | |
| 10 kΩ Circuit (dipole prediction): | 6.4 |
| 10 kΩ Circuit with ungrounded wire: | 34.8 |
| 10 kΩ Circuit with grounded wire: | 42.3 |

**Figure 7.51** The E field from a loop connected to a long wire. (From Ref. 19. © 1989, IEEE.)

impedance of the long wire close to the loop is high, approximately 3 kΩ, 10 cm below the source, and therefore the inclusion of a ferrite bead on the line reduces the current only slightly.

Although the common-mode current flow on interconnection cables is more often the result of common-mode voltage between ground and the conductors or shield of the cable, we see from the foregoing that differential-mode currents in the cable can induce a common-mode current on an attached wire. This is also shown for PCBs in Sections 11.2 and 11.3.

## 7.7  REDUCTION IN THE EMISSION OF E AND H FIELDS FROM CABLES

The principal EMI reduction technique is to reduce at source. Where the source of current flow is common-mode noise voltage, RF capacitors between the signal ground and chassis ground or the use of filter connectors may be effective. Other techniques, such as the use of beads and baluns, are dealt with in Sections 5.1.10.3 and 5.1.10.4.

The introduction of an impedance (i.e., inductor or resistor in the return current path) is often an effective reduction method. When the current flow on a cable is reduced in both amplitude and frequency content to the extent compatible with normal operation of the interface/power line, the remaining reduction techniques are

> The use of shielded cable exhibiting a low transfer impedance at the frequencies of interest, possibly by use of double-, triple-, or quad-braided shielded cable
>
> The use of connectors and backshells with a low transfer impedance
>
> Where used, removal of a "poor" connector bulkhead gasket, such as graphite-loaded, porcupine, wire-screen types, and replacement or installation of an effective gasket, such as thin-sheet or O-ring silver-loaded elastomer
>
> Isolation of the signal return current from the shield by use of triaxial or twinaxial cable
>
> Reduction in the loop area of the cable (achievable by decreasing the height of a cable above the ground plane or bundling together the cables that carry a differential-mode current)

When a shielded cable cannot be used, any spare conductors in the cable should be connected to a "clean" ground. If the cable enters a metal enclosure, current flow on the outside of the enclosure is likely to be minimal, whereas current flow on the inside of the enclosure may be considerable. Thus connection of the spare conductors or the shield of the cable to the outside of the enclosure is likely to be effective, whereas connection of spare conductors or the shield of a cable to the inside of the enclosure or to a noisy signal ground may increase the level of radiation!

## 7.8  SHIELDED CONNECTORS, BACKSHELLS, AND OTHER SHIELD TERMINATION TECHNIQUES

The effectiveness of a shielded cable is often compromised by use of a connector that exhibits a high transfer impedance or low shielding effectiveness (i.e., high leakage).

Connectors specifically designed for use in harsh EM environments usually include metal fingers that make electrical contact between the mating halves of the connector and reduce the high transfer impedance inherent in a purely metal-to-metal interface that is not under pressure.

In addition, the bulkhead mounting section of the connector often contains a groove for the inclusion of an O-ring type of EMI gasket. Where the surface is flat, a rectangular gasket may be used of the correct material, as discussed in Chapter 6.5. Use of the wrong material, as previously noted, may decrease the shielding effectiveness. EMI backshells are designed to connect the shields concentrically around the backshell with a low-impedance connection, with another low impedance at the backshell-to-connector interface. Backshells designed for less harsh EM environments but that provide some EM protection, for example, to meet FCC levels of emission, are often constructed of nonconductive material coated with a thin layer of deposited conductive material. It should be emphasized that such backshells are not suitable for the conduction of high-level transient current such as that indirectly induced in cables by EMP or a lightning strike. The shielding effectiveness of various types of backshells designed for D-type connectors is plotted in Figures 7.52 and 7.53, from Ref. 20. The test setup is shown in Figure 7.54, in

KEY

| | | | |
|---|---|---|---|
| 1 | Unshielded Cable | 6 | Cast Zinc (AMP) |
| 2 | Pigtail Connection | 7 | Cast Zinc (Northern Technologie |
| 3 | Copper Tape (Chomerics) | 8 | Drawn Metal (Malco Microdot) |
| 4 | Copper Foil | 9 | Drawn Metal (Kern) |
| 5 | Metalized Plastic (AMP) | 10 | Quad Shielded Cable |

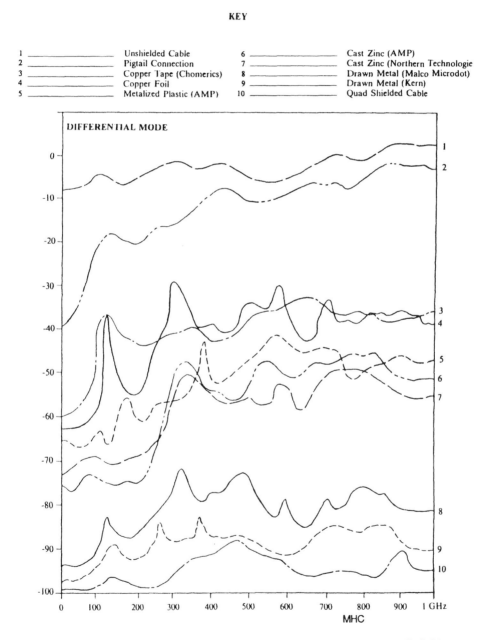

**Figure 7.52**  Differential-mode noise, backshell shielding effectiveness. (From Ref. 20.)

KEY

| | | | | |
|---|---|---|---|---|
| 1 | Unshielded Cable | 6 | Soldered Copper Backshell |
| 2 | Pigtail Connection | 7 | Cast Zinc (Northern Technologies) |
| 3 | Metalized Plastic (AMP) | 8 | Drawn Metal (Malco Microdot) |
| 4 | Copper Tape (Chomerics) | 9 | Drawn Metal (Kern) |
| 5 | Cast Zinc (AMP) | 10 | Quad Shielded Cable |

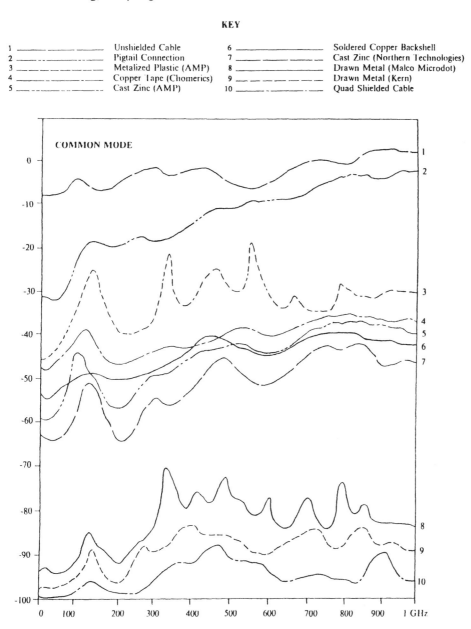

**Figure 7.53** Common-mode noise, backshell shielding effectiveness. (From Ref. 20.)

which a 3-foot welded-aluminum cube enclosure contained the test antenna and the cable under test. The cable was brought into the enclosure via feedthrough connectors and was terminated in the characteristic impedance of the system. Antenna reflections were calculated and removed from the final data; however, some of the artifacts of the system are still present in the curves. The cable was used as the radiation source, and the test antenna was used to measure the relative magnitude of the field generated. The zero- or close-to-zero-dB reference is the field measured from the unshielded cable, with shielded cable radiation at minus-dB levels referenced to 0 dB.

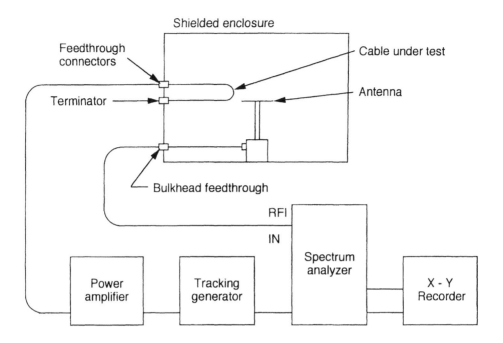

**Figure 7.54**   Backshell shielding effectiveness test setup.

The maximum attenuation is achieved by the quad shielded cable, which was solidly attached to the walls of the test chamber. The same cable was used in the connection to the various backshells. The connector used with all of the backshells was a 25-pin D type with solder-plated metal parts and EMI controlling dimples that are used to ensure good contact.

In the differential-mode test setup, the shield of the cable carries no signal current, because the signal conductor and its return are isolated from the shield. In the common-mode setup, the signal return is made via the shield of the cable. The results of the common-mode test indicate shielding effectiveness 5–15 dB lower than for the differential-mode test.

The metallized plastic backshell in the common-mode test (curve 3) exhibited the lowest level of shielding for a correctly terminated shield, that is, one not connected via a wire pigtail. One important factor found by experiment is that a degradation of as much as 30 dB in the best readings was seen when the backshell was not installed or terminated correctly.

Curve 2 is a clear example of the importance of terminating a shield correctly; here a (2-inch) pigtail was used to connect the shield to the backshell. As discussed in Section 7.3 on coupling to shielded cables, the termination of the shield with a pigtail results in almost-negligible shielding, especially at high frequencies. Curve 1 is the unshielded reference cable. The curves may best be used by first calculating or measuring the potential radiated emission from an unshielded cable and comparing the predicted electromagnetic field at a specified distance, either 1 m, 3 m, or 30 m (depending on the EMC requirements) to the maximum allowable level contained in the requirements. The reduction due to the use of the quad shielded cable and different types of backshell may then be assessed and the most cost-effective type of backshell selected for the application. It should be noted that the quad shielded cable used during the measurements exhibits a high level of shielding; use of a different type of shielded cable will almost certainly reduce the overall level of shielding.

When MIL-STD-461 or similar radiated susceptibility tests are a requirement or the worst-case electromagnetic ambient is known, coupling to the cable with the different types of back-shells may be calculated as follows. The combined transfer impedance of the cable and connector may be found very approximately from the shielding effectiveness of the combination and the characteristic impedance of the test setup using Eq. (7.4). The current flow on the shield may be calculated using the methods described in this chapter, and the transfer voltage is then found from the transfer impedance.

As a rough guideline, any of the backshells tested, when correctly terminated to the shield, are likely to meet FCC emission limits. For MIL-STD-461 and TEMPEST applications, only the solid metal backshells are likely to suffice. Figure 7.55 show cables terminated via very effective "Glenaire" backshells and by use of the ubiquitous pigtail. Above 1 GHz, connectors and backshells are common sources of leakage. Figure 7.56a shows the leakage characteristics of four types of HF connector (from Ref. 21), and Figures 7.56b and 7.56c are the leakage characteristics translated into transfer impedance. Even the best backshells may be limited by loose-threaded interfaces. In one instance, military equipment failed a radiated emission test, and it was found that the connector backshell was loose. As the backshell was tightened, the emissions could be seen to reduce until they disappeared into the noise level and the equipment passed the emission requirements with a significant safety margin. When the MIL-C-38999

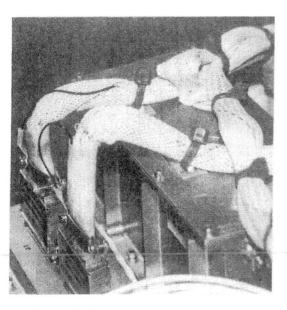

**Figure 7.55**   Left: cable shields terminated in a pigtail. Right: cable shields terminated in a "Glenaire" EMI backshell.

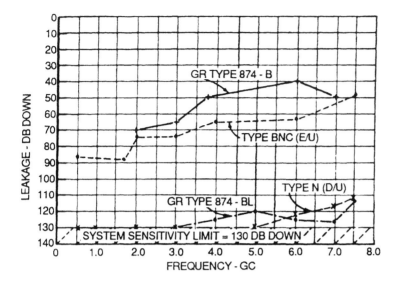

**Figure 7.56a**  Leakage characteristics of the type BNC, type N, and General Radio Type 874B and Type 874BL connectors. (From Ref. 21.)

Series III/IV connector–to-backshell interface was loose, the measured transfer impedance was 0.5 Ω at 100 MHz; as the connection was tightened, the transfer impedance was reduced, as shown in Table 7.5.

Figure 7.56d shows the combined transfer impedance of the interfaces of the braid of a shielded cable to the EMI backshell, the backshell to the connector, the two halves of the connector, and the connector to the bulkhead. The connector is a D (non-EMI) type, either gold-plated over brass or zinc-plated (ASTM-B633) over steel. The backshell is manufactured by GlenAire. Adding Technit Consil A silver aluminum–filled silicon gasket material between the connector and the bulkhead filled up the gap at the interface and reduced the transfer impedance above 1 GHz, as shown in Figure 7.56d.

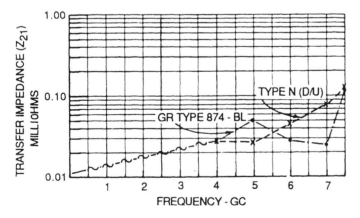

**Figure 7.56b**   Surface transfer impedance of the type N and General Radio Type 874 BL coaxial connectors.

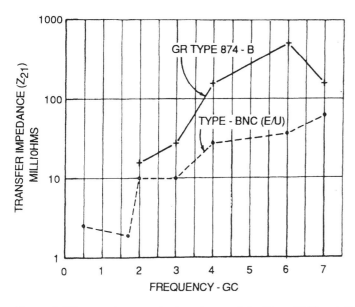

**Figure 7.56c** Surface transfer impedance of the Type BNC and General Radio Type 874 B coaxial connectors.

An alternative to a solid backshell is a shielded heat-shrink boot. The interior surface of the boot is covered with a conductive coating. As the boot is heat-shrunk, the coating makes electrical contact with the metal of the connector housing and the braid of the shielded cable. A shielded heat-shrink boot for D-type connectors is available from Raychem Canada Ltd., 113 Lindsay Avenue, Dorval, Quebec, H9P 2S6. The transfer impedance of the Raychem boot is 38 m$\Omega$ at 10 kHz, increasing almost linearly to approximately 70 m$\Omega$ at 100 MHz. As expected, the transfer impedance of the heat-shrink boot is higher than that of the solid cast GlenAire backshell, and therefore the shielding effectiveness will be lower.

One way of terminating a number of shields to a backshell is via a compressible disk of conducting elastomer material with wire-insertion holes, as described in Ref. 17. This conductive elastomer is pressed against the entry of the backshell and is used to terminate the individual shields. Figure 7.38 shows the attenuation compared to an unshielded cable and a standard backshell.

**Table 7.5** Effect of Torque on Connector and Backshell Transfer Impedance at 10 MHz and 100 MHz

| Torque (in.-lb) | Transfer impedance at 10 MHz (m$\Omega$) | Transfer impedance at 100 MHz (m$\Omega$) |
|---|---|---|
| 25 | 300 | 500 |
| 50 | 15 | 9 |
| 100 | 3 | 3 |
| 150 | 2 | 2 |
| 200 | 1 | 1.5 |

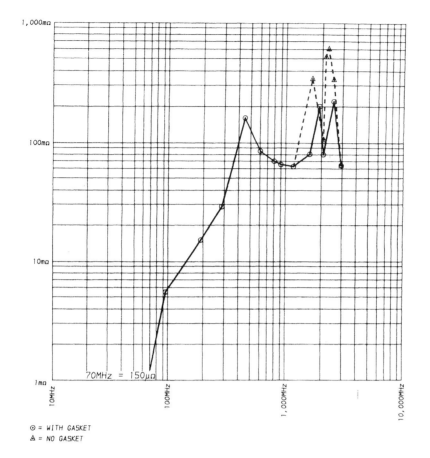

⊙ = WITH GASKET
△ = NO GASKET

**Figure 7.56d**  Surface transfer impedance of the two halves of a 25-pin D-type connector with ''Glen-Aire'' backshell.

When a tin-plated copper braid is used over either an unshielded or a shielded cable to improve the shielding effectiveness, one technique for terminating the braid to the connector case is by clamping under a hose clamp. The transfer impedance of this type of termination is shown in Figure 7.57 from Ref. 23.

Additional data on the transfer impedance of a 15-pin D-type, N, UHF, BNC, and SMA connectors are shown in Figure 7.58 from Ref. 24. N1 is the N-type connector attached to an RG214 cable, and N2 is the N-type attached to a solid tube. This illustrates that it was not the N-type connector transfer impedance that was dominant but rather that of the cable. The SMA connector was attached to a semirigid cable, and the characteristic dip in transfer impedance of a solid-walled cable is seen at 10 MHz. However, above 10 MHz the connector transfer impedance predominates, and it rises to 1.5 mΩ at 1 GHz. Data on a gold-plated SMA connector soldered to a semirigid cable states an attenuation of 100 dB at 1 GHz, which can be converted to a transfer impedance of 0.5 mΩ. It is important to make a solder connection of the SMA connector to a semirigid or a semiflex cable, for a crimped termination exhibits a higher transfer impedance.

The dips around 800 MHz for the N- and D-type connectors are caused by the differences

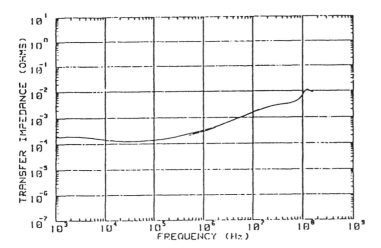

**Figure 7.57** Transfer impedance of braid shield terminated to the connector case with a hose clamp. (From Ref. 23; © 1995 IEEE.)

in wave velocities at the inside and outside of the connector under test, which becomes important because of the larger length of these connectors.

Additional data on a 15-position D-type connector shows a transfer impedance of 40 m$\Omega$ at 100 MHz, 400 m$\Omega$ at 1 GHz, and 3 $\Omega$ at 3 GHz. These values are much higher than the values in Figure 7.56d but lower than the values in Figure 7.58 at 100 MHz and almost the same at 1 GHz. The problem with transfer impedance measurements on D-type connectors is the typically poor contact area between the connector case and the mounting surface.

Although connectors are the most common method for terminating shields to an enclosure, bulkhead, or ground, other methods may be more suitable. For example when signal and return conductors in shielded cable must be connected to a terminal block, the most common shield termination method is the ubiquitous "pigtail." A much better method is the cable clamp, typically constructed from a brass block, or tin-plated metals, as shown in Fig. 7.59. The insulation is peeled back from the end of the cable to expose the braid shield. The safest and most effective way of removing the insulation without damaging the shield is to cut around the insulation using nylon fishing line. The shield is placed in the clamp, and a solid metal strip is tightened over the shield, as shown in the figure.

If a cable clamp cannot be used, the braid of the shield can be soldered directly to a solder lug, which is bolted to the chassis, as shown in Figure 7.60, keeping the braid length very short.

These shield termination techniques are effective up to 1 GHz. But increasingly the shield attenuation is degraded due to the radiated coupling from the shield to the unshielded sections of cable. If a cable must terminate at a bulkhead and a connector is not desirable, the bared shield can be soldered to a brass plate connected to a slot cut or milled into the top of the bulkhead, as shown in Figure 7.61. If the cable must be soldered at either end to a circuit, the cable and brass plate can be manufactured as a cable assembly.

EMI transits are used for the penetration of cables and waveguides and to achieve a conductive bond to bulkheads and to the decks of ships. Multiplug transits (NELSON) are available

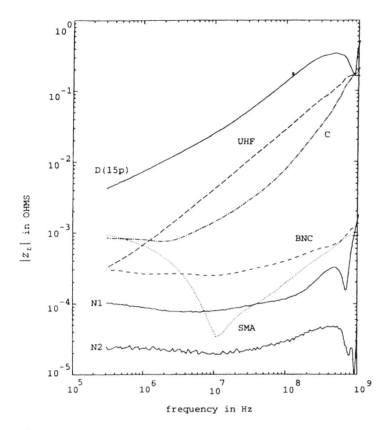

**Figure 7.58** Transfer impedances of N, SMA, BNC, C, UHF and 15-pin high density $D$ connector. From Ref. 24; © 1998 IEEE.

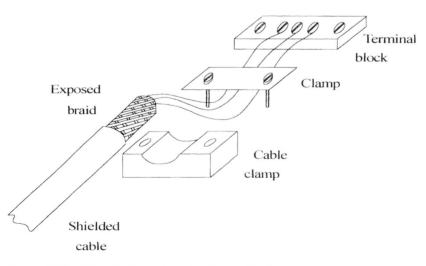

**Figure 7.59** Cable shield terminated under a cable clamp.

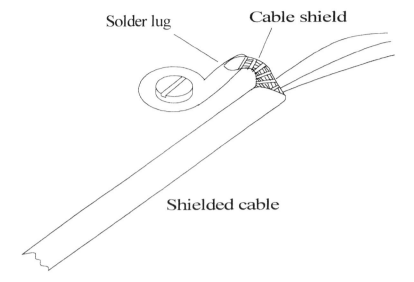

**Figure 7.60**  Cable shield terminated at a solder lug.

that can be inserted into pipes with diameters of 2–8 inches; by the use of Nelson insert blocks, these can accommodate a wide range of cables and waveguides.

A poured transit is another technique for bonding the shield of cables or waveguides to a deck or bulkhead. In a poured transit, a conductive compound is used to make electrical contact at the penetration of the cable or waveguide; either sealing compound or a swell strip and impregnation fluid is used to seal the conductive compound from moisture, thereby minimizing

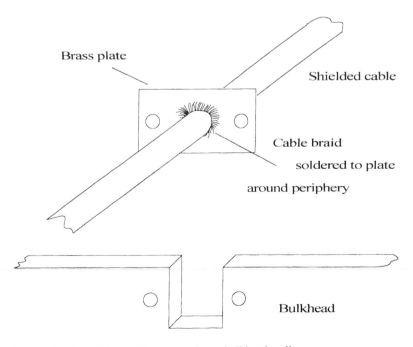

**Figure 7.61**  Cable shield connected to a bulkhead wall.

corrosion. The major disadvantage of the poured transit is the difficulty of adding or removing a cable.

Another type of transit is the nonpoured rubber block transit, which is constructed of upper and lower module halves of conductive rubber that clamp the cable and are mounted into a frame. By peeling off the inner sheets of the module, it can be modified to fit a number of different cable diameters. The insulation is removed from the cable, exposing the shield, and this exposed area is clamped between the module halves. To make a good electrical contact between the cable shield and the metal frame, a strip of copper tape is placed on the inner surface of each half of the module prior to clamping. The modules are then assembled in rows in the frame. Once the packing space in the frame is full of modules, some of which may be empty and others of which may contain cables, the modules are compressed via a compression plate and the upper space filled with a top packing piece. These transits can be fire resistant, gas-tight, and watertight if required. Roxtec and Brandshuttechnik (bst) manufacture transits; more details can be obtained from Jastram, 188 Bunting Rd., Unit 310, St Catherines, Ontario, Canada, L2M 3Y1.

The computer programs at the end of Chapter 2 can be used to calculate the field coupling to a loop, transmission line, and a short wire terminated on a ground plane (monopole antenna), as can the emissions from a loop, short wire/cable, or transmission line. Chapter 11 contains case studies of coupling to wires and cable shielding against 2–31,000-V/m fields over the frequency range from 14 kHz to 40 GHz. These techniques can be used to calculate the current flowing through the transfer impedance of the connector and the transferred voltage.

## 7.9 PRACTICAL LEVEL OF CABLE SHIELDING REQUIRED TO MEET MIL-STD/DO-160C OR COMMERCIAL RADIATED EMISSIONS REQUIREMENTS

The practicing electronic design engineer who is not a specialist in EMC often requires an idea of the level of cable shielding and the type of cable shield termination required to meet an EMI requirement. To help provide some of this information, a cable shielding effectiveness test was made with a number of different test configurations and cable shield termination techniques. The source of RF current flow on the cable was a battery-supplied clock and driver contained within a well-shielded enclosure that generated the fundamental frequency and harmonics up to 1000 MHz. With the source powered up in the enclosure and without a cable attached, the emissions from the RF source were measured and found to be below the noise floor of the preamplifier and spectrum analyzer over the 40–1000-MHz frequency range. Thus the source of emissions is confined to the cable alone. A log periodic biconical antenna was used to measure radiated emissions over the 40–1000-MHz frequency range. The clock frequency was 40 MHz, and the driver output was 3.8 V peak, with a rise time of 5 ns and a fall time of 5 ns. An MIL-STD D38999 connector was mounted on the enclosure and used for the clock signal. A Glenair M85049 non-EMI backshell with a cable strain relief was used on the external connector. A 2.8-m-long RG58 50-Ω cable was used to connect the signal to a 50-Ω load at the far end of the cable. The 50-Ω load was contained in a brass enclosure that was soldered on all sides, and the braid of the RG 58 cable was soldered to the case around 360°, as the cable entered the brass enclosure. The RG58 shield was thus used as the signal return, and the full signal current flowed on the inside of the shield.

Figure 7.62a shows the connection of the source, shielded cable, and load for the case where the shield braid is connected to the strain relief clamp on the backshell. Figure 7.62b shows the source and load enclosure connected to the ground plane in the MIL-STD/DO-160 test setup. Figure 7.62c is the typical commercial test set-up for radiated emissions contained

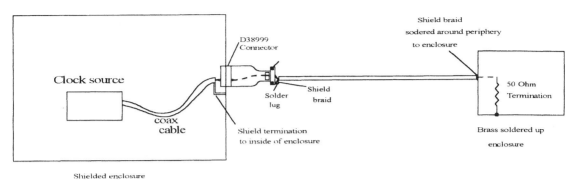

**Figure 7.62a** Enclosure, source, and load connection.

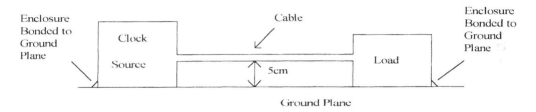

**Figure 7.62b** MIL-STD-462/DO-160 radiated emission test setup.

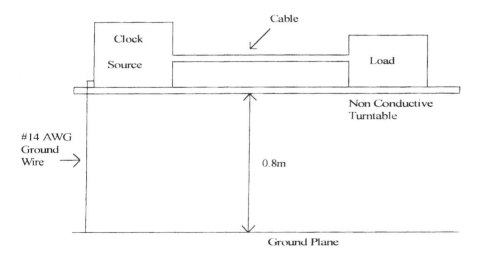

**Figure 7.62c** ANSI C63.4 Radiated emission test setup.

in ANSI C63.4. The cable shield was terminated at the source end in one of the following four methods:

1. Shield taken through a pin in the D38999 connector and then connected to the inside of the enclosure, using a pigtail in the external connector and inside the enclosure
2. Shield connected using a 1-inch pigtail to a solder lug fastened beneath one of the screws in the backshell strain relief

3. Shield braid connected directly to a solder lug fastened beneath one of the screws in the backshell strain relief
4. Large overbraid clamped to the shield of the RG58 cable using two hose clamps and to the Glenaire backshell using a large hose clamp (this termination technique produces results close to an EMI backshell with 360° shield termination)

The first test setup was in accordance with MIL-STD-462 or DO-160, with the two enclosures bonded to a ground plane. The cable was routed 5 cm above and 10 cm from the front edge of a ground plane located 1 m above the conductive floor in a semi-anechoic chamber. The tip of the measurement antenna was located 1 m from the cable. Radiated emission measurements were made at the fundamental and the harmonic frequencies of the 40-MHz clock up to 1000 MHz. The radiated emissions were narrowband in nature, and the measured level was almost independent of the measurement bandwidth.

Figure 7.63 compares the emissions with the four shield termination techniques to the MIL STD 461C Part 2 narrowband limits for aircraft showing curve 1 for the Army and curve 2 for the Air Force and Navy. As the figure shows, only the full overbraid shield termination, #4, comes close to meeting the requirements, and even this "best" method results in emission that are 5 dB above curve 1 at 120 MHz and 5 dB below curve 2 at 120 MHz. The weak link is not the #4 shield termination technique but the high level of current flowing at the second and third harmonic (80 MHz and 120 MHz) and the limited shielding effectiveness of the RG58 at these frequencies. Although the shielding effectiveness of the cable continues to decrease with increasing frequency, it does reach a limit, due to the long-line effect, whereas the current flow on the cable at these upper harmonics continues to reduce. Another reason for reduced radiation from the cable is that it becomes electrically long and a less effective antenna. The net effect is that the radiated emissions above 120 MHz are well below either the #1 or #2 curve when the full overbraid termination technique is used. For the braid-to-backshell shield termination, #3, the limitation is the cable shielding effectiveness at 120 MHz. Above 120 MHz the braid termination and the aperture in the backshell at the strain relief are the weak links,

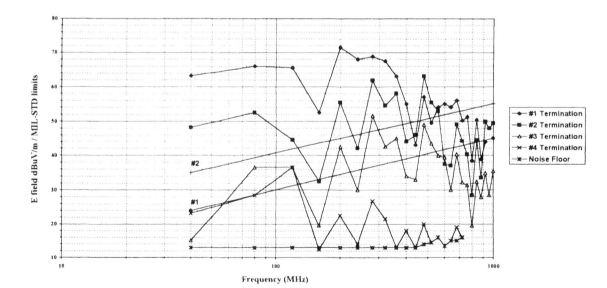

**Figure 7.63**  Radiated emissions compared to MIL-STD-461C Part 2 with the four types of shield termination.

and this means that even the #3 termination is above curve 1 at a number of frequencies and even above curve 2 at 280 MHz. The 1-inch pigtail is, as we would expect, less effective than a short braid termination, and bringing the shield through a pin in the connector and attaching this to the inside of the enclosure results in the highest level of emissions, i.e., is the least effective. One reason is the increased length of the shield connection caused by the external pigtail in the backshell, the pins of the connector and in the internal pigtail.

Another effect, which can be demonstrated very effectively, is RF current that flows on the inside of the enclosure, which is caused by internal fields and C/M ground connections. This current will be conducted out on the cable shield. To reduce emissions, the use of a double-braid shielded cable will be required to meet the curve 1 requirements. Alternatively, a twisted shielded pair can be used in which the signal return current flows on one of the pair and not on the shield. An even better method is to use an overall braid and to use the RG58 shield as a signal return path. The shield is connected to signal ground inside the enclosure using a pin in the connector. The overall braid shield is connected at both ends of the cable to an EMI backshell. This overall braid is not used to carry signal return current, and thus the radiation, due to the limited shielding effectiveness in the cable and due to the shield termination, is dramatically reduced. This type of overbraid is ideal for the typically multipin D38999 connector, for the cable bundle inside the shield can contain a mix of unshielded wires, individually shielded wires, and multiconductor shielded cables.

The same RF source cable and enclosure were also measured in the ANSI C63.4 radiated emission test setup on a 3-m open area test site (OATS). In this test setup, the enclosure containing the source and the enclosure containing the load were mounted on nonconductive turntable located 0.8 m above the OATS ground plane. The 2.8-m-long cable was laid out with a horizontal section and also looped down at the back of the table. A 14-AWG wire connected the source enclosure to the OATS ground plane. Measurements were made with the antenna oriented both horizontally and vertically and scanned in height to measure the highest emission levels. The turntable was also slowly rotated, again to measure the highest level of emissions. Measurements were made using the quasi-peak detector and a 120-kHz bandwidth.

Figure 7.64 shows the maximum level of emissions with the antenna either vertically or horizontally oriented using the braid shield termination #3 and the full overbraid #4. The radiated emissions are compared to the FCC Part 15 Class A and Class B limits in Figure 7.64. The shield terminations with #3 and with #4 are above the Class B limit, and the best shield termination, #4, is just at the Class A limit at 120 MHz. The problem here is again the signal current return on the shield and the limiting shielding effectiveness of the cable. We know that the #1 and #2 shield termination methods are less effective, and we know by how much, so these were omitted from the OATS tests. It may be surprising that the FCC limits are exceeded, for the measurement distance is 3 m instead of the MIL-STD 1m. Also, MIL-STD-461 curve #1 is 9 dB below the FCC Class B limit at 120 MHz. The reason that radiated emissions are increased in the ANSI C63.4 test setup is that the cable is 0.8 m above a ground plane and the MIL-STD-462 distance is only 5 cm; the closer a cable is routed to ground, the lower the level of radiation.

The measurements in Figures 7.63 and 7.64 are valid for a 40-MHz 3.8-V clock into a 50-$\Omega$ load and a 2.8-m-long cable. The results may be scaled to your specific signal level and the level of emissions corrected. For example, the emissions may be adjusted based on signal current, which in the test setup is equal to the shield current. Thus, for a 40-MHz clock with a 1-V signal into a 100-$\Omega$ load and the same rise and fall times as the 40-MHz clock, the reduction in emissions will be approximately 20 log [(3.8/50)/(1/100)] = 17 dB at any of the harmonics. For different clock frequencies, a Fourier analysis of the current at the different harmonics can be made and compared to a Fourier analysis of the current at the harmonics of the 40-MHz clock pulse used in the test. Even though the harmonics of the two clock pulses

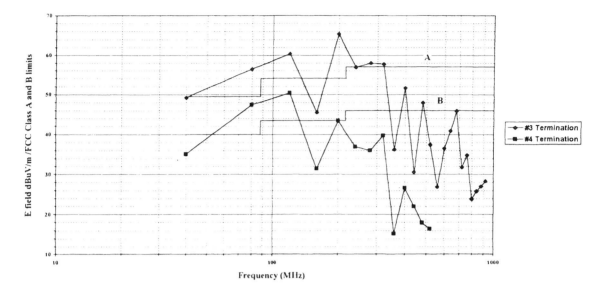

**Figure 7.64**   Radiated emissions measured on a 3-m OATS with two types of shield termination and plotted against the FCC Class A and FCC Class B limits.

do not exactly coincide in frequency, an approximate correction to the emissions shown in Figures 7.63 and 7.64 can be made.

## REFERENCES

1.  J.S. Hofstra, L.O. Hoeft. Measurement of surface transfer impedance of multi-wire cables, connectors and cable assemblies. Record of the IEEE International Symposium on EMC, 1992.
2.  L.O. Hoeft, T.M. Sala, W.D. Prather. Experimental and theoretical comparison of the line injection and cylindrical test fixture methods for measuring surface transfer impedance of cables. IEEE 1998 Symposium record.
3.  F. Boyde, E. Clavelier. Comparison of coupling mechanisms on multiconductor cables. IEEE transactions on electromagnetic compatibility, Vol 35, No 4, November 1993.
4.  F. Broyde, E. Clevalier. Definition, relevance and measurement of the parallel and axial transfer impedance. IEEE Symposium record 1995.
5.  J.S. Hofstra, M.A. Dinallo, L.O. Hoeft. Measured transfer impedance of braid and convoluted shields. Record of the IEEE International Symposium on EMC 482–488, 1982.
6.  L.O. Hoeft, J.S. Hofstra. Experimental evidence for porpoising coupling and optimization in braided cable. 8th International Zurich Symposium and Technical Exhibition on EMC. March 7–9, 1989.
7.  K.L. Smith. Analysis and measurement of CATV drop cable RF leakage. IEEE Trans. on Cable Television, Vol. CATV-4, No. 4, October 1979.
8.  B. Demoulin, P. Degauque. Shielding performance of triply shielded coaxial cables. IEEE Trans. on Electromag. Compat. Vol. EMC-22, No. 3, August 1980.
9.  K. Casey. Low frequency electromagnetic penetration of loaded apertures. IEEE Trans. on Electromag. Compat. Vol. EMC-23, No. 4, November 1981.
10. E.F. Vance. *Coupling to Shielded Cables.* Robert E. Krieger, Malabar, FL, 1987.
11. A.P.C. Fourie, O. Givati, A.R. Clark. Simple technique for the measurement of the transfer impedance of variable length coaxial interconnecting leads. IEEE Transaction on EMC, Vol. 40, No. 2, May 1998.
12. L.O. Hoeft, J.L. Knighten. Measured surface transfer impedance of cable shields that use combinations of braid and foil and are used for 1-Gb/s Data Transfer. IEEE EMC Symposium, 1998.

13. L.O. Hoeft. Measured electromagnetic shielding performance of commonly used cables and connectors. IEEE Transactions on EMC, Vol. 30, No. 3, August 1988.

14. F. Broyde, E. Clavelier, D Givord, Pascal Vallet. Discussion of the relevance of transfer admittance and some through elastance measurement results. IEEE Trans on EMC, Vol. 35, No. 11, November 1993.

15. M.A. Dinallo, L.O. Hoeft, J.S. Hofstra, D. Thomas. Shielding effectiveness of typical cables from 1 MHz to 1000 MHz. IEEE International Symposium on Electromagnetic Compatibility, 1982.

16. P. Degauque, A. Zeddam. Remarks on the transmission line approach to determining the current induced on above-ground cables. IEEE Trans. on Electromag. Compat., Vol. 30, No. 1, February 1988.

17. J.C. Santamaria, Louis J. Haller. A comparison of unshielded wire to shielded wire with shields terminated using pigtail and shields terminated through a conducting elastomer. IEEE EMC Symposium Record, 1995.

18. G. Chandesris. Effect of ground connection on the coupling of disturbing signals to a coaxial line. 4th International Zurich Symposium and Technical Exhibition on Electromagnetic Compatibility, March 10–12, 1981.

19. T. Hubing, J.F. Kaufman. Modeling the electromagnetic radiation from electrically small table top products. IEEE Trans. on Electromag. Compat., Vol. 31, No. 1, February 1989.

20. D. Fernald. Comparison of shielding effectiveness of various backshell configurations. ITEM 1984.

21. J. Zorzy, R.F. Muehlberger. RF leakage characteristics of popular coaxial cables and connectors, 500 MHz to 7.5 GHz. The Microwave Journal, Nov. 1961.

22. C.R. Paul, D.R. Bush. Radiated emissions from common mode currents. Proceedings of the IEEE International Symposium on Electromagnetic Compatibility, 1982.

23. L.O. Hoeft, J.S. Hofstra. Transfer impedance of overlapping braided cable shields and hose clamp shield terminations. IEEE International Symposium record 1995.

24. F.B.M. van Horck, A.P.J. van Deursen, P.C.T. van der Laan, P.R. Bruins, B.L.F. Paagmans. A rapid method for measuring the transfer impedance of connectors. IEEE transactions on electromagnetic compatibility, Vol 40, No 3, August 1998.

# 8
# Grounding and Bonding

## 8.1   INTRODUCTION TO GROUNDING

Perhaps one of the most difficult topics is that of grounding. Even the definition of the term *ground* is not universally accepted, for example, "One of two conductors used to carry a single-ended signal." Or the signal or power return may be termed a *ground*. In the majority of cases the ground is an electrical connection either to the earth or to a structure that serves a similar function to the earth.

However, we also use the term *floating ground*, which, as its name implies, is either disconnected from the enclosure, the earth, or any other structure or connected via a high impedance. Grounds often serve different functions, and it is common to encounter a number of grounds in equipment, subsystems, or systems. Throughout this book, the function of the ground, e.g., safety ground or signal ground, is usually described.

In this chapter we use the term *signal ground* to describe a reference plane that is as close to an equipotential plane as is feasible. To emphasize the definition, the term *reference ground* is often used. The term *safety ground* refers to a conductor (green wire in America) that has the purpose of conducting fault current and operating a circuit breaker or fuse and in limiting the voltage to ground during a fault. The purpose of the green wire is to protect personnel and, secondarily, the equipment. The definition of *return* used here is "A conductor that has the primary function of returning signal or power back to the source." The problem with this definition is that very often the ground and the return are the same. Thus an analog signal return is often connected to a ground plane to which the analog supply return is connected.

The term *earth* or *earth ground* is used to describe a low-resistance connection to earth, often made via rods, plates, or grids. The earth ground may be used to conduct fault and lightning current or to tie a reference ground as close to ground potential as possible. It should not be used for both functions! The term *bond* is used to describe the mechanical connection holding two structures together. Depending on the purpose of the bond, a maximum value of resistance is often specified for the bond.

One of the problems in discussing grounding is the difficulty in describing the geometry of the connection. A schematic may show a ground connection but not be able to describe the physical realization. As described in Chapter 5, Section 5.1.2, above a relatively low frequency (typically kilohertz) the DC resistance no longer determines the impedance of a connection; it is the total impedance, including AC resistance, and inductive and capacitive reactance. Both parallel and series resonance circuits are formed and are determined by the geometry of the ground connection.

The implementation of the ground connection is extremely important. In Chapter 7, in the section on cable shielding, it was seen that the addition of a 2-inch pigtail shield connection to ground reduced the shielding effectiveness of a cable to almost that of an unshielded cable. Likewise, the implementation and location of a signal ground to chassis or enclosure ground connection can make the difference between the equipment's passing or failing EMI requirements or in exhibiting EMI.

It is advisable to consider the low-frequency and high-frequency grounding schemes in equipment and systems separately. It is the low-frequency scheme that is shown in power and grounding diagrams and in grounding schematics.

## 8.2 SAFETY GROUNDS, EARTH GROUNDS, AND LARGE-SYSTEM GROUNDING

Some of the reasons for grounding are

Compliance with safety requirements
Lightning protection
Reduction of noise in receiving antennas/equipment
Reduction of common-mode noise between equipment cases/cabinets
Reduction in radiated emissions?
Protection against electrostatic buildup
Line-to-ground filter component return
Reduced crosstalk

The question mark after reduction in radiated emissions serves to remind us that radiated emissions may increase with a second low-impedance ground connection when a primary path to ground exists, as described in Chapter 7. Where possible, the connection of the AC line safety ground should be separated from any conductor that connects equipment grounds together or to a common earth ground.

It is imperative that a grounding system designed to protect against direct lightning strikes be isolated from all other grounds, except for a single connection at the earth ground electrode. The lightning ground must be capable of carrying the lightning strike current without fusing, thereby maintaining the conductive path.

Two examples of large-system grounding schemes are given in Figures 8.1 and 8.2. Figure 8.1 is an ideal large-system grounding scheme in which the steel girders or rebars in the building are bonded together. In some steel-framed buildings, the steel columns are used as part of a lightning protection scheme. Even when the steel frame of a building is not intentionally used as a lightning ground, lightning may punch through a wall of the building and current may flow on the structure. The typical 20,000 amps to more than 250,000 amps of primary lightning strike current then travels down the steel column, using the easiest path, to earth ground.

If the building is constructed of reinforced concrete, then the current will flow through the rebars to ground. In reinforced concrete, wood, or masonry buildings it is advisable to provide down-conductors on the outside of the building for lightning protection. It should be remembered that for all practical gauge of conductors, the conductor will exhibit an impedance, due to self-inductance, higher than that of a metal plate. Consequently, if the conductor is in close proximity to a large structure or sheet of metal, then the lightning current may well flash over and continue to earth ground in the lower-impedance metal.

In the ideal building ground scheme of Figure 8.1, the steel girders or rebars are welded together, whereas in the typical grounding scheme no steps are taken to ensure a low-resistance bond between structure members. When a lightning strike occurs to the typical building structure, voltages appear across the ends of the poorly connected metal structure, and an E field is generated. The voltage may be high enough to result in a sideflash to adequately grounded structures, such as cables and equipment enclosures.

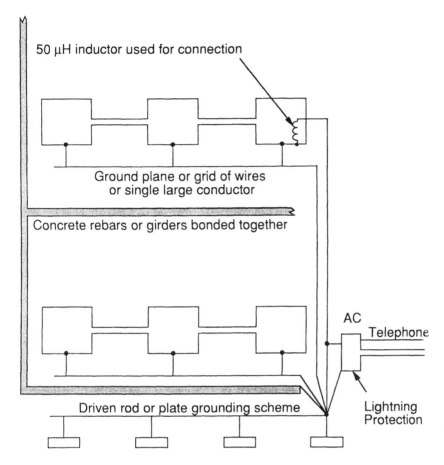

50 μH inductor used for connection

Ground plane or grid of wires
or single large conductor

Concrete rebars or girders bonded together

AC

Telephone

Driven rod or plate grounding scheme

Lightning
Protection

**Figure 8.1** Ideal large-system grounding scheme.

The AC safety ground is a common source of RF voltage induced in the power wiring by AM/FM transmitters or by equipment that is a major source of conducted power-line noise, as described in Chapter 1. The technique, shown in Figure 8.1, which reduces safety ground RF current flow, is to connect the safety ground to a low-impedance earth ground at the entry point of the power into the building. The safety code takes precedence over all other grounding rules, but it is possible to tailor the grounding scheme and comply with the code. Most electrical safety codes require connection of the safety ground and neutral to earth ground at the main service panel but do not specify a low-impedance connection.

RF voltages induced on the safety ground by equipment within the building may be reduced by the inclusion of one or more inductors in series. The inductors must be capable of carrying a fault current and thus allow the fuse or circuit breaker to function. The inductors must also exhibit a low DC resistance and low impedance at 50/60 Hz in order to maintain a low touch potential in the event of a short circuit of AC hot or neutral to the enclosure. In the Figure 8.1 grounding scheme it would be correct to assume that the major portion of the fault current would flow into the ground plane and via the #2 or lower gauge wire or plate connecting the ground plane to the earth ground. An obvious question is why have the safety ground connection via the green wire? Why not rely on the grounding scheme to provide the safety ground connection? The reason the green wire must be retained is that the enclosure ground connection,

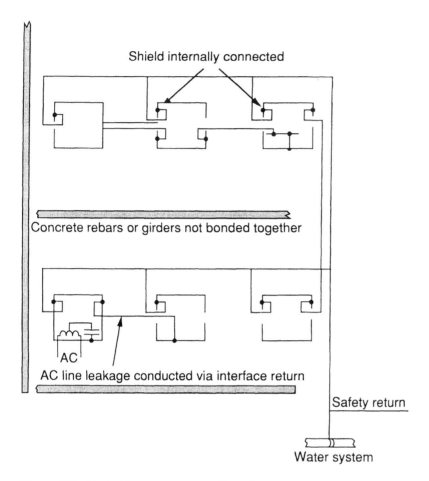

**Figure 8.2**   Typical large-system grounding scheme.

shown in Figure 8.1, may be inadvertently disconnected, whereas the "green wire" is normally disconnected only when the power connector is removed from the outlet. It is for this reason that safety codes often make the green wire mandatory.

The scheme of Figure 8.1 assumes that the equipment is located on separate floors inside the building and, to achieve as low an impedance ground as possible, either a #2 (or lower) gauge, wire or a wide plate a minimum 1/4 inch thick connects the ground planes on the two floors together. A lower impedance may be obtained in a building with steel girders by using the girders as a parallel path to the wire/plate. The problem is that the girders may carry lightning strike currents and will almost certainly carry AC power leakage currents from equipment such as rotating machinery, the housings of which are often electrically connected to the structure. In addition, the frame of the building is an effective antenna at AM broadcast frequencies and thus carries RF currents. For these reasons it is inadvisable to use the building structure in a grounding scheme.

In Figure 8.2, the typical ground scheme, it is the safety ground and the shields of cables and conductors within the cable that connect the enclosures together. One connection may be made via a signal return in which RF currents and 50–180-Hz leakage currents may flow. The typical large-system ground may be adequate for digital equipment, but for analog/instrumentation equipment, the grounding may result in EMI, as discussed in Section 8.3.2 on multipoint grounding.

The voltages appearing between the earth ground–to-neutral connection within a building and earth ground outside of a building will contain AC power and RF frequencies. For example, in a three-story office building the measured voltage at a distance of 100 feet between the two earth grounds was 3 V at 60 Hz and approximately 350 mV at 1.1 MHz. When earth grounds are much further apart, a concomitant increase in voltage and current flow in conductors connecting the grounds together may be expected.

Thus far the assumption has been made that the large system is powered from the same service. When this is not the case, the configuration shown in Figure 8.3a may apply. The AC neutral is connected to safety ground and earth ground in both services, and thus both the safety ground and earth ground carry some of the neutral return current. The safety grounds are connected to the metal enclosures, as often required by the safety code. A problem exists when a signal return is also connected to the enclosure in both pieces of equipment. The connection may not be immediately obvious, especially when the signal return is routed via a PCB to power return, which is in turn connected via a power supply to the enclosure. Whatever the route, some small percentage of the AC neutral return current will flow in the signal return, with the potential for EMI. One solution to the problem is to use two feeders from a service to the two pieces of equipment, in which case the only safety ground–to-neutral connection is at the feeder, as shown in Figure 8.3b. Disconnecting one enclosure from earth ground will reduce earth-borne leakage current.

The paramount concern in any power distribution is to comply with the applicable electrical safety code. The use of isolation transformers on the input side of both of the services shown in Figure 8.3a will interrupt the neutral current flow in the signal return conductor. However, the solution is expensive and may not comply with safety regulations. When the equipment is

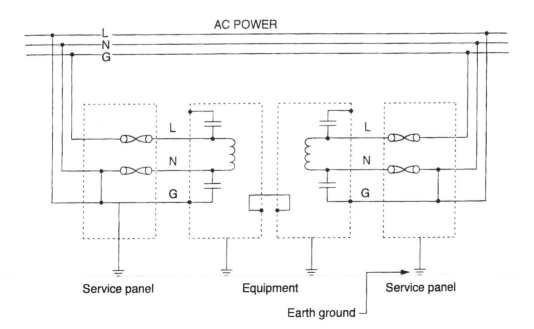

**Figure 8.3a**  Equipment connected to separate service panels in which neutral current flows in a signal return.

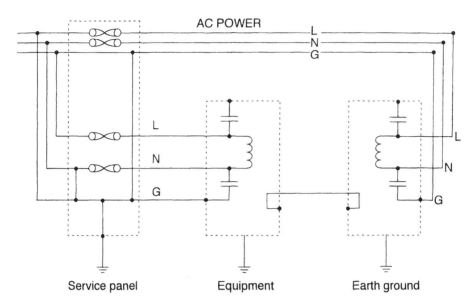

**Figure 8.3b** Equipment supplied by separate feeders from a single service panel and with a single structure-to–earth ground connection.

part of a system and is located some distance from other equipment in the system, it is preferable to use one location as the source of power and to feed the outlying equipment from the single source. Assuming an adequately sized ''green'' wire is taken to the outlying equipment, no additional earth ground is required and the problem of low-frequency current flow in single-ended signal returns is reduced. Two situations where the isolation of equipment, such as an antenna, from earth grounds is inadvisable is when an earth ground exists for the purpose of lightning protection and is located close to the antenna structure. To avoid flash-over, the antenna structure should be connected *by a single connection* to the earth ground. The second reason to rethink isolation of equipment from earth is when the equipment is in the proximity of a source of high-power E field (e.g., a transmitting antenna). The connection of the equipment structure to a low-impedance earth ground will typically reduce the RF current flow on interconnection cables, thereby reducing the probability of EMI.

The sources of electrical codes seldom consider the requirements of instrumentation systems; however, grounding schemes may be designed that improve signal grounding and that comply with safety regulations.

## 8.2.1  Earth Ground

Grounding to earth is required for safety, to minimize touch potentials due to AC line leakage or equipment breakdown or direct or indirect lightning strikes, or where a receiving or transmitting antenna uses the earth as a counterfoil (ground plane beneath an antenna that forms part of the antenna design). In practice, a low-impedance connection to earth has been found to reduce the incidence of failures in equipment, as reported in Ref. 1, pp. 81–107.

### 8.2.1.1  Earth Resistance

Requirements on the maximum value of earth resistance imposed by insurance, power, and oil companies vary from 1 to 5 $\Omega$. In a survey of USA Federal Aviation Administration (FAA)

**Table 8.1** Resistivities of Different Soils

| Soil | Resistivity, ohm-cm | | |
|---|---|---|---|
| | Average | Min. | Max. |
| Fills—ashes, cinders, brine wastes | 2,370 | 590 | 7,000 |
| Clay, shale, gumbo, loam | 4,060 | 340 | 16,300 |
| Same—with varying proportions of sand and gravel | 15,800 | 1,020 | 135,000 |
| Gravel, sand, stones, with little clay or loam | 94,000 | 59,000 | 458,000 |

*Source*: U.S. Bureau of Standards Technical Report 108.

| Soil | Resistivity, ohm-cm (Range) | | |
|---|---|---|---|
| Surface soils, loam, etc. | 100 | — | 5,000 |
| Clay | 200 | — | 10,000 |
| Sand and gravel | 5,000 | — | 100,000 |
| Surface limestone | 10,000 | — | 1,000,000 |
| Limestones | 500 | — | 400,000 |
| Shales | 500 | — | 10,000 |
| Sandstone | 2,000 | — | 200,000 |
| Granites, basalts, etc. | | 100,000 | |
| Decomposed gneisses | 5,000 | — | 50,000 |
| States, etc. | 1,000 | — | 10,000 |

*Source*: Evershed & Vignoles Bulletin 245.

Air Route Control Centers, Air Traffic Control Towers, and Long-Range Radar Sites, contained in Ref. 1, pp. 135–147, the measured earth resistance varied, depending on the site, from 0.1 $\Omega$ to 12.8 $\Omega$.

The resistivity of earth, expressed in ohms per centimeter, is surprisingly high. Table 8.1 shows typical values for different soils. The effect of moisture is shown in Table 8.2, salt content in Table 8.3, and temperature in Table 8.4. The conductivity of the soil is primarily electrolytic, so the higher the concentration of moisture and sodium ions, the lower the resistance. The value

**Table 8.2** Effect of Moisture Content on Earth Resistivity

| Moisture content, % by weight | Resistivity, ohm-cm | |
|---|---|---|
| | Top soil | Sandy loam |
| 0 | $1,000 \times 10^6$ | $1,000 \times 10^6$ |
| 2.5 | 250,000 | 150,000 |
| 5 | 165,000 | 43,000 |
| 10 | 53,000 | 22,000 |
| 15 | 21,000 | 13,000 |
| 20 | 12,000 | 10,000 |
| 30 | 10,000 | 8,000 |

*Source*: ''An Investigation of Earthing Resistance'', by P. J. Higgs, I.E.E. Jour., vol. 68, p. 736, February 1930.

**Table 8.3** Effect of Salt
Content on Earth Resistivity

| Added salt, % by weight of moisture | Resistivity, ohm-cm |
|---|---|
| 0 | 10,700 |
| 0.1 | 1,800 |
| 1.0 | 460 |
| 5 | 190 |
| 10 | 130 |
| 20 | 100 |

For sandy loam—moisture content,
15% by weight; temperature, 17°C
(63°F).

of earth resistance is dependent on the area, quantity, and depth of ground rods/plates. A low resistance is achieved because of the large area of the ground shell surrounding the earth electrode, as shown in Figure 8.4.

The effect of depth of a driven rod on earth resistance is shown in Figure 8.5. As the depth of a ground rod increases, the area of the shell around the rod increases. At a depth of approximately 6 feet, the resistance begins to level out. To decrease the resistance further, use multiple rods, treat the soil with salt, or increase the moisture content. The effect of increasing the number of ground rods is shown in Figure 8.6. A useful nomograph relating the basic factors affecting earth resistance is shown Figure 8.7

The foregoing information is reproduced by courtesy of AVO/Biddle Instruments, 4651 S. Westmoreland, Dallas, TX 75237, which manufactures a range of earth testers. The earth tester is used in a three-terminal, fall-of-potential earth resistance test. Two current terminals are provided on the tester and two voltage. A current is caused to flow between the ground electrode connected to the $C_1$ terminal and a small electrode placed approximately 50 feet away and connected to the $C_2$ terminal. The resistance between the small electrode and earth is unimportant, for the requirement is only that sufficient current flow in the ground to develop a potential, and so it is not necessary to bury the electrode more than an inch or two into the soil.

The $P_1$ (potential) terminal of the Megger is connected to the $C_1$ (current) terminal, and the $P_2$ terminal to a second small electrode located 62% of the distance between the two current

**Table 8.4** Effect of Temperature on Earth
Resistivity[a]

| Temperature | | Resistivity, ohm-cm |
|---|---|---|
| C | F | |
| 20 | 68 | 7,200 |
| 10 | 50 | 9,900 |
| 0 | 32 (water) | 13,800 |
| 0 | 32 (ice) | 30,000 |
| −5 | 23 | 79,000 |
| −15 | 14 | 330,000 |

[a] For sandy loam, 15.2% moisture.

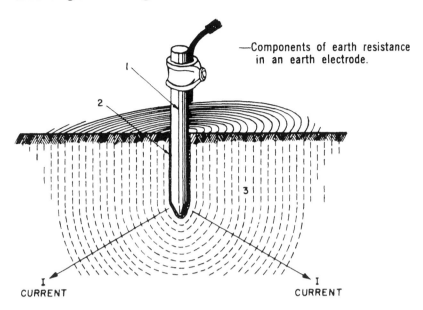

—Components of earth resistance
in an earth electrode.

CURRENT    CURRENT

**Figure 8.4**  Ground shell surrounding earth electrode.

electrodes (i.e., at approximately 31 feet from the electrode under test). The Megger measures the potential difference between the earth electrode, and the $P_2$ and computes the earth resistance from the current flowing between the two current terminals. The earth resistance test setup is shown in Figure 8.8.

Tables 8.1–8.4 and Figures 8.4–8.8 are reproduced courtesy of AVO/Biddle, Dallas TX 75237.

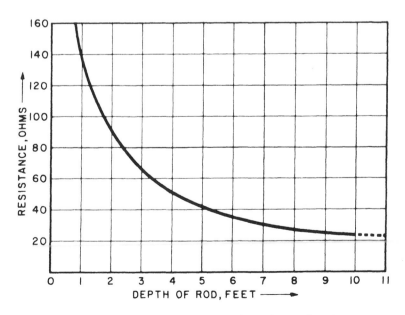

**Figure 8.5**  Effect of depth of electrode on decreasing earth resistance.

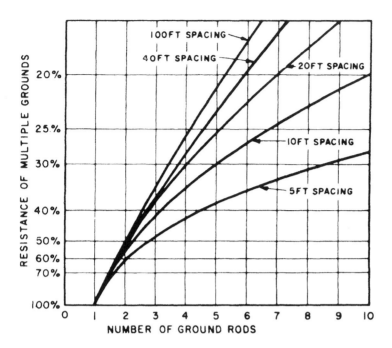

**Figure 8.6**  Effect of increasing the number of ground rods on earth resistance.

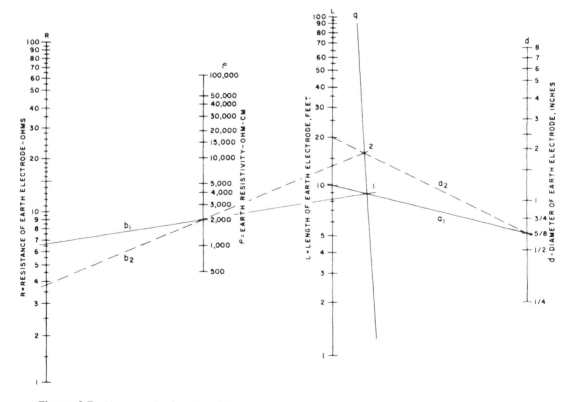

**Figure 8.7**  Nomograph of earth resistance.

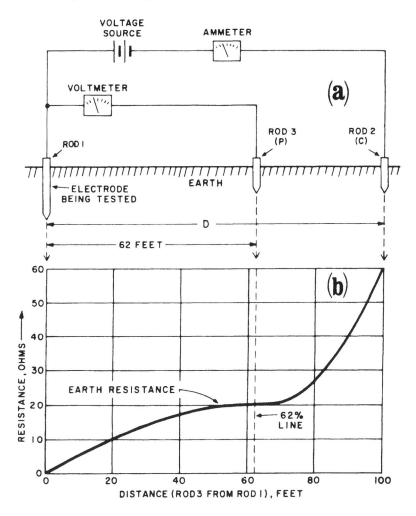

**Figure 8.8**   Earth resistance measurement test setup.

## 8.3   SIGNAL GROUND AND POWER GROUND

### 8.3.1   Signal Ground

An *ideal signal ground* is defined as an equipotential plane that serves as a reference point. In practice, a signal ground is used as a (hopefully) low-impedance path for current to return to the source. All conductors exhibit impedance, so when current flows in a ground conductor, a voltage drop occurs. A good signal grounding scheme limits the magnitude of the current flow and the impedance of the path. The major cause of impulsive current flow is signal switching current and changes in IC supply current as the device changes state and these currents flow back to the source on the power return, which is often the signal return. As we saw in Chapter 4, the closer the proximity of a conductor that carries RF current (signal or power-line noise) to a conductive ground plane, the lower the inductive and capacitive crosstalk. The plane does not need to be grounded or connected to either the signal or power return to be effective. However, ungrounded ground planes will have voltages developed between them and other ground planes and other grounded structures, which in turn can result in radiated coupling or crosstalk.

Chapter 11, Section 11.6, describes PCB grounding, with Figure 11.19 illustrating common-mode voltages developed across a ground plane. Section 11.6.2 describes "Good" and "Bad" PCB ground planes and Section 11.6.3 describes grounding a PCB within an enclosure. In Chapter 5, the isolation or connection of an RF ground plane to digital/analog ground is discussed.

### 8.3.2   Grounding Philosophy

One of a number of different grounding schemes may be selected; however, very often the designer/systems engineer does not have a choice. The choice may be limited when the grounding scheme is dictated by either a larger system into which equipment or subsystem must fit or by commercially available peripherals to which equipment must connect.

Where a designer may have some control over the grounding scheme is within equipment. However, when standard, RS232, GPIB, or similar interfaces are used, signal-to–chassis ground connections are likely in the equipment in which the interfaces terminate, and the use of a single-point scheme is not feasible. The basic grounding schemes are

1. Single point
2. Single-point star
3. Multipoint
4. Single-point DC, multipoint RF
5. Floating ground

In practice, schemes 1 and 2 become scheme 4 above some frequency, and scheme 5 becomes scheme 3, due to parasitic stray capacitance between circuit board tracks/signal wiring and nearby grounded structure and between signal conductors in cables and shields. Examples of the first four schemes are shown in Figures 8.9, 8.11, 8.12, and 8.13.

The aim of a single-point ground at both the board level and at the system level is to reduce DC and low-frequency current flow in the structure. The upper frequency beyond which the single-point ground scheme is no longer effective is determined by the magnitude of capacitance between the signal ground and structure or chassis. The single-point DC, multipoint RF ground scheme recognizes the limitation of the single-point scheme and deliberately introduces an RF ground via one or more capacitors in parallel. This scheme has the advantage of reducing common-mode noise voltage on the signal ground at a designated location. One preferred location is at the point of entry of interface signals to equipment. We have seen in Chapter 7, on cable radiation, that the predominant source is common-mode voltage on both shielded and

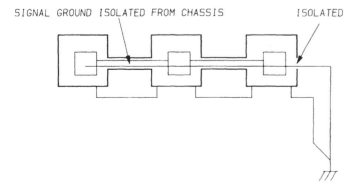

**Figure 8.9**   Single-point ground taken from one piece of equipment to another.

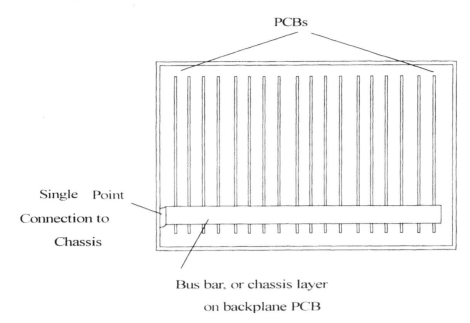

**Figure 8.10** Single-point ground connecting together PCBs within an enclosure.

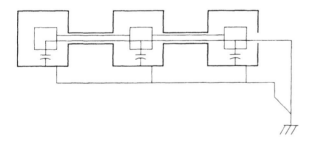

**Figure 8.11** Single-point DC, multipoint RF ground.

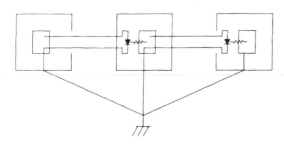

**Figure 8.12** Single-point star.

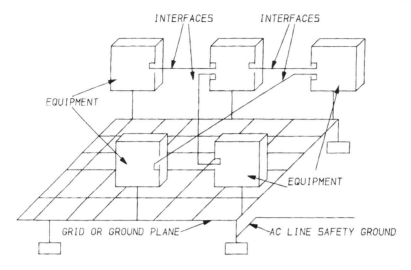

**Figure 8.13** Multipoint ground.

unshielded cables. A second location of choice for the RF (capacitor) ground within equipment is at a common ground point to which a number of different types of ground are connected. The disadvantage of the multipoint RF grounding scheme is that RF current flow in signal grounds is typically higher than in the single-point ground. Assume that the potential source of common-mode noise current is known, for example, the return of a supply that generates common-mode noise between the return and case/chassis. The introduction of an inductor in the supply return may be effective in reducing the current flow on the signal return or ground plane connected to the return and reduce the voltage developed across the RF ground. Some examples of this technique are provided later. Whenever an inductor and capacitor are used in combination, a resonant circuit is formed and the inclusion of a damping resistor is advisable.

The single-point star ground scheme is a variation of the single-point scheme that reduces common-mode noise coupling from one piece of equipment to another by the use of isolated interfaces, such as the opto-isolator shown, or signal transformers, or differential-input receivers.

In a similar scheme, the interfaces are isolated and the ground is taken from equipment to equipment as shown in Figure 8.9. This scheme may appear as a great waste of the circuitry required for isolation. However, the signal ground comes as close as possible to the ideal reference ground, which it is often termed, for neither signal return currents nor power supply return currents flow on the reference ground. Figure 8.10 shows a single-point ground for different PCBs within an enclosure, typically with the use of a bus bar, or a chassis layer on a backplane PCB, to connect the PCB grounds together and to the enclosure. Although at first glance this may not seem to be a single-point ground, the impedance of the ground connection from PCB to PCB and from the bus/backplane to chassis can be kept very low, and so at lower frequencies this is in effect a single-point ground. In very large systems, such as the space station, a single-point ground scheme becomes unwieldy and a so-called layered single-point scheme is used in which the chassis ground connection is made at a subsystem or logical EMC element (i.e., a group of equipment pieces in fairly close proximity). The use of of the term *single point* for this layered scheme is not strictly accurate, however, for the scheme is much different from the multipoint ground.

The multipoint ground philosophy recognizes that the manufacturer of bought-out equipment is obliged to connect safety ground to the chassis and that the signal and secondary power ground is also often connected to the enclosure. In the multipoint scheme, the signal interfaces are usually single ended, with the signal return connected to the enclosure in both pieces of equipment. The scheme by its nature introduces ground loops between the safety grounds and the signal returns. As discussed in Section 8.2, AC safety grounds connected when the power cord is plugged into the receptacle are required by most electrical safety codes.

Many systems made up of computer equipment function correctly with multiple ground loops. Exceptions occur when the neutral return current flows in the signal return, as discussed in Section 8.2, or large common-mode noise appears between the safety grounds of equipment. Large common-mode voltages are encountered when equipment pieces are located some distance apart and connected to different earth grounds, especially when sources of power-line noise, such as welders, milling machines, elevators, and control and automation equipment, are connected to one of the earth grounds. With the use of single-ended analog/instrumentation interfaces in which the signal return is connected between the two earth grounds, the potential for EMI is high. The use of shielded cables on single-ended circuits or where the shield is the return path for signal currents will not reduce the potential for EMI, because the problem is conducted noise. Multipoint grounds between equipment should ideally be less than $0.1 \lambda$ in length. If longer, the resonance and antiresonance effects seen with electrically long conductors in Chapter 5 may occur, with resultant high currents or very high impedance.

Decreasing the impedance of the ground connection between equipment pieces located far apart is difficult due to the typically low-impedance source of the noise current and the correspondingly large size of conductor required. The solution may be high common-mode noise immunity differential-interface circuits, as described in Chapter 5.

The multipoint ground scheme illustrated in Figure 8.13 is the same as used in Figure 8.1 for ideal large-system grounding. This scheme reduces the common-mode noise voltage developed between the enclosures of the equipment but does not eliminate ground loops. The interconnections shown between equipment enclosures represent single-ended signal grounds connected to the enclosures at both ends.

Conventional wisdom dictates that multipoint grounding with very short length for the ground conductors be used above 10 MHz, and multipoint grounding is common in equipment containing RF circuits.

At the PCB level, multipoint grounding may not be required, and some advantages from a single-point connection between the PCB ground plane and the enclosure exist. However, single-point grounding is seldom used. More common is the connection of the PCB ground plane of an RF circuit at more than one location around the edge of the PCB to an enclosure provided to shield the circuit. Radio frequency engineers have followed this practice for years. One possible reason for multipoint grounding at the PCB level is that the PCB ground plane does not provide a sufficiently low impedance due to crosshatching or interruption of the ground plane by tracks. In this case the use of a solid ground plane formed by the enclosure may improve the performance of an RF circuit or reduce the radiated emissions from the board.

When other circuits reference the enclosure as ground or when seams and gaps occur in the enclosure, the goal should be to reduce the RF current flow on the inside of the enclosure to the greatest extent. This can be achieved by reducing leakage currents from the circuit and by reducing the radiated coupling to the enclosure. The leakage currents can be reduced by grounding at a single point or several points in close proximity (at the same time reducing the capacitance between the enclosure and the RF ground). Leakage current and radiated emissions can be reduced by good RF design, such as the use of transmission lines formed by the signal and return tracks or by the signal tracks in close proximity to an adequate PCB ground plane. When a signal track is in close proximity to a ground plane, the RF return current tends to concentrate in the ground plane under the track, resulting in a minimum impedance in the return path. Thus, good RF design practice tends to reduce RF leakage current in the enclosure, reduce radiated emissions, and minimize common impedance coupling and conducted EMI problems.

Decreasing the capacitance between RF semiconductors and the enclosure to which they are heatsunk can also reduce RF current flow in the enclosure.

Measurements on the shielding effectiveness of small enclosures surrounding RF circuits have shown the importance of the grounding scheme, although the results have not always been as predicted. For example, shielding effectiveness measurements were made on enclosures containing a PCB on which several loads were mounted. An RF signal was connected to the loads via a coaxial connector mounted through the enclosure. Either a single connection from the PCB ground plane to the enclosure was made via the braid of the shielded cable inside the enclosure or the ground plane was connected around the periphery to the enclosure. One measurement was taken on an enclosure made up of an extruded section covered by a lid and capped at the top and bottom by plates. The dimensions of the enclosure were 26.5 cm $\times$ 22.5 cm $\times$ 4 cm, and the enclosure contained seams but no apertures. The PCB was either mounted in the center of the enclosure for the case when the ground plane was isolated or mounted close to the solid base of the enclosure when the ground plane was connected around the periphery to the enclosure. A comparison of the shielding effectiveness of the single-point connection and the multipoint connection of the ground plane follows with the single point ground slightly better at most frequencies.

| Frequency [MHz] | Shielding effectiveness | |
|---|---|---|
| | Single-point ground [dB] | Multipoint ground [dB] |
| 100 | 73.5 | 76.0 |
| 160 | 74.0 | 70.0 |
| 195 | 64.0 | 57.0 |

Another enclosure used to shield the same PCB had the dimensions 23.5 cm $\times$ 22.5 cm $\times$ 1.8 cm. The enclosure was constructed of a solid base plate to which a cover was attached. The defects in the enclosure were the seams plus 38 $\times$ 8 mm $\times$ 2.2 mm slots in the cover. In this measurement the ground plane was either connected around the periphery to the enclosure or mounted on 1.8-mm-thick nylon washers by nylon screws. The total capacitance between the ground plane and the enclosure with the PCB mounted on the insulating washers was measured at 125 pF. A secondary effect of mounting the PCB on the insulating washers was to move the PCB closer to the cover. The shielding effectiveness of the enclosure with the PCB either single-point or multipoint grounded follows:

| Frequency [MHz] | Shielding effectiveness | |
|---|---|---|
| | Single-point ground [dB] | Multipoint ground [dB] |
| 100 | 66 | 78 |
| 131 | 39 | 59 |
| 195 | 64 | 64 |
| 250 | 86 | 86 |
| 297 | 84 | 76 |

One potential reason for the degradation in shielding effectiveness at some frequencies with the single-point ground scheme is the potential for increased radiated coupling to the cover as the PCB is moved in closer proximity, due to the 1.8-mm thickness of the nylon washers.

One lesson that can be learned from the shielding effectiveness measurements is that the grounding scheme should be chosen on a case-by-case basis. Although multipoint grounding is the method of choice for the interconnection of PCBs, modules, and equipment that contains RF circuits, even at this level exceptions to the rule do exist.

Floating ground schemes are difficult to realize, test, and maintain. In practice, leakage paths exist with megaohms of resistance and the floating ground floats to some undefined potential. The inclusion of a high-value resistor between the ground and chassis will reduce the static potential of the ground and is recommended for static discharge protection. Connections of signal interfaces and peripheral equipment is likely to invalidate the floating ground scheme when signal-to-chassis connections are made in the external equipment. Test devices must have high-impedance differential inputs. The advantage of the floating ground is the very low value of low-frequency currents that flow in the ground. Above some frequency, the ground becomes a multipoint system due to parasitic capacitance. Shielding and grounding RF and digital circuits are discussed in 5.3.3.5 and 5.3.3.6 respectively.

## 8.3.3  Single-Point Ground

The single-point ground scheme is often described as the ideal. How and where the single-point ground is made are vitally important, and in some instances it is better to make more than one connection even though this does technically violate the single-point philosophy.

When secondary power (e.g., +5 V, +/−15 V) isolated from the input power return is provided within equipment, the point of connection of the secondary power return to chassis is considered the single-point ground for the equipment. When DC primary power is used directly by equipment, the point after the power-line input filter might conveniently be considered the single point for the connection of the different signal grounds within the equipment.

An alternative location for the single-point ground may be chosen, e.g., close to an RF component that, by its construction, connects power return and signal return to the conductive enclosure of the component, or at a signal interface where the signal return is connected to chassis, or at an A/D converter where analog and digital grounds are connected together.

It is neither mandatory nor, often, beneficial to make the single-point ground at a power supply within the equipment, although the advantage of this location is that common-mode voltage generated by the supply is shorted by the ground connection. The disadvantage is that common-mode voltages appearing on circuit boards and at the point of entry of interface signals cannot be shorted out by connection of the return to chassis without violating the grounding scheme. One solution when the supply return is isolated from the chassis is to reduce both differential- and common-mode noise from the supply with an L- or T-type filter connected as shown in Figure 8.14. The chassis ground may then be made at the interface circuit. In this configuration, noise currents generated on circuit boards and flowing into the power supply on the boards' ground plane or signal return tracks are reduced by the series inductance of the L- or T-type filter.

Using a nonisolated supply or adding a second chassis ground connection in Figure 8.14 at the supply return will result in the flow of noise currents in the enclosure. These noise currents may be reduced by inclusion of an inductor in the return. When apertures are present in the enclosure and the internal enclosure current is increased, increased aperture coupling must be expected. When the shielding integrity of the enclosure is high, the increased current flow in the enclosure may not significantly increase radiation from the enclosure.

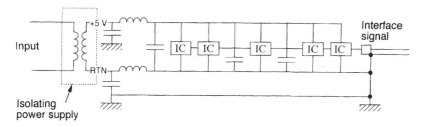

**Figure 8.14**  Reduction in power supply noise voltage by an *LC* filter, thus allowing a ground chassis connection at the interface.

The signal grounds within a unit should be divided into the following, where applicable and feasible:

1. Analog/video ground
2. Digital ground
3. RF ground
4. Control signal ground

These four grounds should be connected to the equipment single-point ground only, and shall be isolated from each other to the greatest extent possible, including capacitive coupling between them.

It is easier to design the separation into the system and then interconnect, when necessary, than to attempt to separate after the fact. For example, a PCB layout in which analog and digital grounds are separated may very easily be connected together; in fact, the provision of terminal posts for the purpose is recommended. However, separation of grounds after the PCB is made may be out of the question. Where an A/D converter or similar device requires an analog ground connection and a digital ground, the analog ground single-point connection shall be made at the device in question. Even where device manufacturers supply separate digital and analog ground pins, very often the grounds are either connected within the device or a very low maximum voltage difference between the grounds is recommended.

Common impedance coupling is a frequent source of noise problems in instrumentation, analog, and video circuits. Figure 8.15 illustrates the mechanism of common impedance coupling. The return current from the fiber-optic driver flows into the ground impedance of the U1

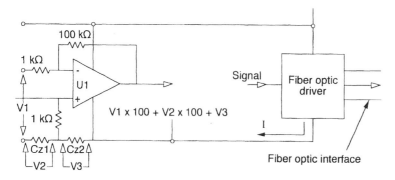

**Figure 8.15**  Common impedance coupling.

analog circuit resulting in the noise voltages $V_2$ and $V_3$. Typically the sources of ground current are digital or RF circuits.

### 8.3.3.1 Floating the Analog Ground on the Digital Ground

One technique that has proven very effective in reducing the current flow in analog ground is to float the analog ground on the digital. This scheme is shown in Figure 8.16. The noise currents generated by the logic as it changes state flows back on the digital supply return, with little current flow in the analog ground. The analog circuit supply in Figure 8.16 is an isolating DC-to-DC converter, which is required to achieve maximum isolation between the analog supply return and the chassis and minimum digital supply current in the analog ground. When a supply is used in which the return is connected to the enclosure, the floating analog grounding scheme is still partially effective, due to some level of isolation provided by the impedance of the inductor in the analog supply return, shown in Figure 8.16.

In the floating analog scheme, large common-mode voltage appears between the enclosure and the analog ground, which often causes concern when measured on an oscilloscope. However, on looking at the output bits of the A/D converter with a nonvarying input voltage, such as zero volts, no change in the status of the bits is seen; thus a differential noise voltage lower than the weighting of the least significant bit (LSB) is present. The location of the digital ground to chassis ground is deliberately at the point where a digital interface signal leaves the board. In a practical system, the chassis connection may be made at a connector when the interface is external to the equipment. The reason for this choice of location is to reduce the common-mode voltage on the interface signal and return conductors, thus reducing radiation from the cable. If this is not a consideration and the major concern is to reduce noise at the input of the A/D, then the chassis ground connection should be made at the A/D converter. This is often difficult unless the PCB is directly above chassis, in which case a metal standoff can be used to bring the chassis up to the PCB at that point. Alternatively, if the motherboard/backplane has a chassis

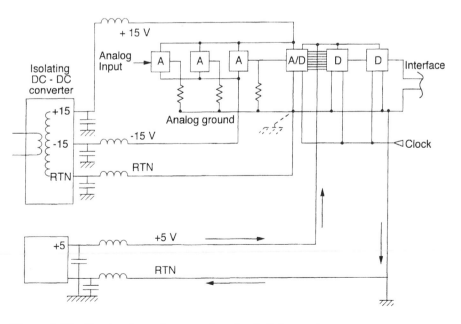

**Figure 8.16** Floating the analog ground on the digital ground.

connection, this should be brought through as many pins are available on the PCB edge connector, and the A/D should be located at the edge of the board close to the connector to maintain a short connection length to chassis.

### 8.3.4 Modified Differential Op Amp Circuit

Figure 8.17a illustrates a common type of circuit used as an analog/video/RF amplifier. The output voltage equals the signal voltage times $R_2$ divided by $R_1$, plus the contribution of the voltages developed across the ground impedances $Z_{g1}$ and $Z_{g2}$.

By addition of a single resistor, as shown in Figure 8.17b, the circuit will reduce the magnitude of the ground current and virtually eliminate the noise voltage generated by $V_{com}$. The signal output voltage is the same as achieved with the circuit in Figure 8.17a for the same differential input voltage, regardless of whether the noninverting input voltage is at the output ground potential or at a positive or negative potential referenced to ground.

When $R_1 = R_2$ and $R_3 = R_4$, the circuit has a gain of unity. The input to $R_3$ is shown in Figure 8.17b connected to a signal return, in which case the circuit inverts the input signal. Alternatively, the input may be true differential or the input to $R_1$ may be connected to a signal return and $R_3$ used as the signal input in which case the circuit is noninverting.

This circuit is invaluable when rejection of common-mode input voltages is required, hence the name *modified differential*. When a higher isolation is required between the two grounds, a true differential circuit, in which the input signals are applied typically to FET input, voltage follower op amps followed by the differential amplifier, shown in Figure 8.17b, may be used.

The limitations on rejection is the common-mode rejection ratio of the op amp and the match in resistor values.

The modified differential circuit may be used as an instrumentation receiver or used to separate analog and digital grounds on a PCB, in which case the input is referenced to analog ground and the output voltage is referenced to digital ground, or to separate system/subsystem grounds.

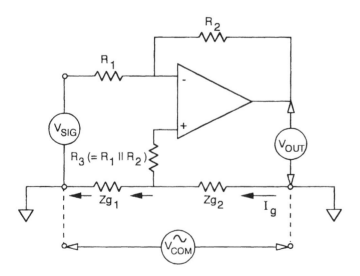

**Figure 8.17a**   Typical op amp gain stage or inverter.

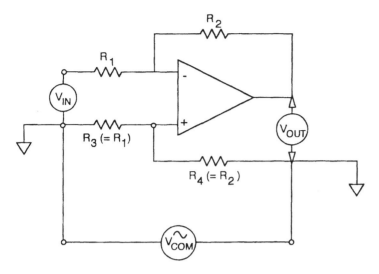

**Figure 8.17b** Modified differential circuit with common-mode noise rejection.

## 8.4 GUIDELINES FOR SIGNAL GROUNDING

Grounds between equipment shall, where feasible, be DC or low-frequency isolated, normally at the input (receiver) end of the interface. This can be achieved by capacitors, transformers, high-input-impedance amplifiers, and opto-isolators.

High-power stages of a circuit shall be connected the closest to the single-point ground, with the most sensitive circuit the furthest from the single point. This ensures that power return currents do not flow in the ground connection or track used by sensitive circuits and thus minimizes the risk of interference and LF frequency instability, sometimes referred to as *motorboating* in audio circuits.

Always return decoupling capacitors to the appropriate ground. That is, if +10 V is used as the digital circuit supply, do not return a decoupling capacitor connected to +10 V to analog or other ground.

Where possible, separate grounds. It is always possible to connect later.

Always build flexibility into the grounding scheme. Use 0-$\Omega$ resistors or jumpers to connect grounds, because these may be removed as desired. Place ground pads on PCBs that can be connected via ground straps to ungrounded piggy back PCBs/memory modules/PCMCIA cards, frames, conductive structures, etc. that are mounted on the PCB.

Always use a solid low-impedance signal ground connection between circuits, backplanes, and PCB, PCB-to-PCB, and different grounds located on the same PCB, when digital or transient signal interfaces exist between these circuits. Keep the signals and ground/s as close together as possible, and in PCB layouts keep the ground directly under the signal traces without any interruption in the ground plane, such as slots.

The grounding philosophy should be decided on a system-by-system basis. The manufacturer of bought-out equipment used in a system often connects digital and analog/video grounds to chassis and also, for safety reasons, the AC safety ground to chassis. Thus a single-point grounding scheme may be impractical. In other units/systems, multipoint grounding is the preferred method.

The physical location and lengths of ground connections shall be included in mechanical/ layout diagrams.

When ground connections are much longer than 1/10 of the wavelength of the noise frequency or the ground is in close proximity to chassis, the equivalent circuit of the ground connection must be used in an EMI/EMC evaluation. Section 11.2.1 includes Case study 11.1, which provides an example of modeling the ground structure.

Also see Section 5.3.3.6 on RF circuit grounding and Section 11.6 on PCB grounding

## 8.5   POWER AND GROUNDING DIAGRAMS

A power and grounding diagram should be constructed showing the connections of primary and secondary returns, as well as the location of filters and the use of shielded or twisted shielded wires within equipment, subsystems, or systems. The power and grounding diagram allows an overview of the realization of the grounding philosophy and, when problems occur, a comparison of the actual implementation to the desired scheme. Very often the two are different! A number of examples of power and grounding for both a single-point ground system comprising a large number of pieces of equipment as well as a single-point DC and multipoint RF system are illustrated in the succeeding pages.

The following power and grounding diagrams are reproduced by kind permission of Canadian Astronautics Ltd. and the Canadian Space Agency.

## 8.6   GROUNDING FOR LIGHTNING PROTECTION

### 8.6.1   Lightning Conductors

The purpose of a grounding scheme for lightning protection is to carry a direct lightning strike to earth ground with no, or at least minimal, damage to the equipment or structure to be protected. For example, lightning protection may be required for antennas mounted on a mast, buildings, or power/signal lines entering a building.

Buildings may be protected by a mesh of lightning conductors with short vertical rods attached to the top of the building. The mesh is connected to earth ground by vertical conductors. If the building framework is steel, then the steel columns may be used as the vertical conductors. The purpose of lightning conductors is not to discharge a thunder cloud but to divert a lightning strike from the structure to be protected to the lightning protection scheme. The protection scheme typically provides a higher conductivity to ground than the building, and the scheme extends in height above the building. Thus the probability of a strike at the building location may be higher than if the protection were not there. Therefore, install lightning protection only if it is capable of handling a direct strike of at least 25,000 amps and preferably one of 300,000 amps. Not installing the protection is likely, at least, to reduce the probability of a strike.

The intent of the first pointed lightning protection rods was to dissipate the charge around the structure so that electric fields associated with the charge would not cause electrical breakdown of the air. It was soon found out that the opposite occurred, and lighting strikes to these pointed rods increased. The majority of lighting protection schemes are of this type. As the use of electronics has increased dramatically, the incidence of malfunction or destruction in sensitive equipment due to lightning has increased, and the idea of keeping the very high currents induced in a lightning strike away from the sensitive electronics is appealing. This should be possible by the use of a dissipation array systems (DAS) or a charge transfer system (CTS). The theory behind these systems is that they should cause a current to flow from the grounding system due to the ionization of the air around the system. The more current that is flowing from the system into the air and the longer the duration of this current flow, the higher the space charge. An increased space charge will weaken the electric field and so the probability of an air breakdown is reduced. Early analysis showed the size of the canopy or umbrella or the quantity of sharpened points would be impractically large. However, these systems have been designed and are in use. Little information on their effectiveness is available, although in Ref. 4 the lightning occurrence probability before and after installation of a DAS remained the same. This was attributed to the fact that the DAS was not installed to the manufacturer's recommendation, caused by building construction conditions in Japan. Another lightning grounding scheme, made up of a gold-plated ball mounted at the maximum height and connected to a coaxial lightning conductor, was commercially available. This system was computationally modeled and a small physical scale model of the transmission line was constructed and tested with a lighting generator. In neither the computer model nor the scale model was any improvement seen over a solid ground wire.

Alternatively, a structure may be protected by erecting a mast, rods, or wire above the structure at a sufficient height that the structure falls under an arc of protection, as shown in Figure 8.18. Some lightning protection recommendations show a cone of protection in which a line is drawn at an angle of 45° from the vertical to the ground. Any structure within the cone is considered to be protected from a direct lightning stroke. In practice, lightning strikes have occurred to antennas that are side mounted on a mast within the 45° cone of protection. The National Fire Protection Association (NFPA) recommends instead the arc of protection. Here the assumption is made that a 150-foot-radius circle touches the top of the structure and the ground at 360° points around the structure and that equipment within this volume is protected. Other more conservative models for the arc of protection are used. A lightning strike may be modeled as a constant-current source that must find a path either from a cloud to ground or from ground to a cloud. In practice, a number of strikes occur one right after another. A conductor that carries the strike current must do so without fusing. The majority of lightning strikes generate currents below 75 kA, although strikes as high as 300 kA have been recorded (Ref. 2). The minimum conductor size capable of conducting a 75-kA strike with a half-amplitude time of 100 μs without fusing is No. 12 AWG. For a 300-kA, 100-μs half-amplitude strike, the minimum conductor size is No. 6 AWG.

Perhaps more important than the current-carrying capability of the conductor are the DC resistance of the joints and its self-inductance. It is these parameters that determine the voltage appearing across the conductor and across the joints. Sideflashes occur when the lightning con-

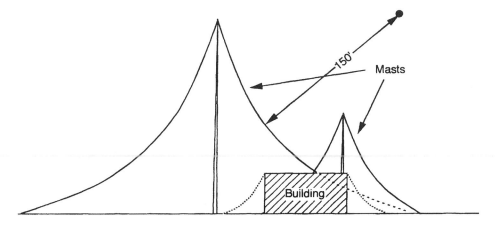

**Figure 8.18** Arc of protection under masts and buildings.

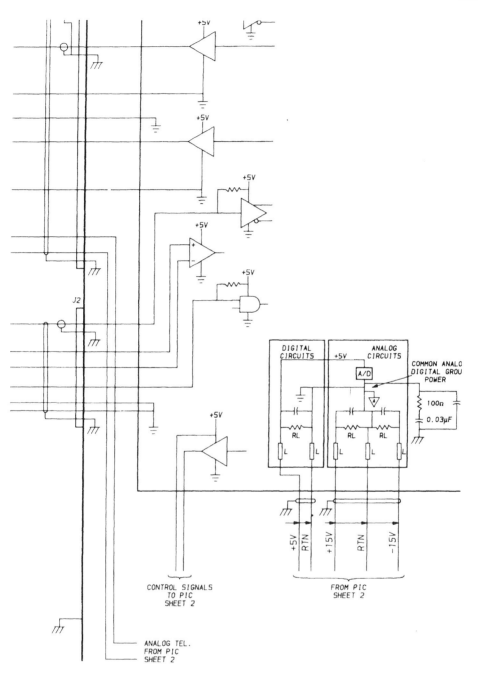

**Scheme 1**   Single-point DC, multipoint AC grounding scheme. Power and signal isolated from chassis by at least 10 MΩ. Maximum capacitance or equivalent impedance to chassis is 0.5 μF. Single signal ground connection from unit to unit. Single-point DC, multipoint AC grounding scheme. Power and signal ground isolated from chassis by at least 10 MΩ. Maximum capacitance or equivalent impedance to chassis is 0.5 μF. Single reference signal ground connection from equipment to equipment.

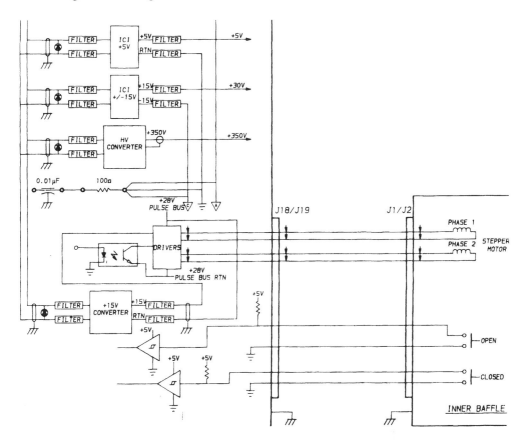

**Scheme 1** Continued

ductor is in close proximity to the structure and the voltage differential is too high. When joint resistance is high, arcing may occur across the joint, and local heating therefrom causes the joint either to fuse apart or to weld together. The major cause of high joint resistance is corrosion. To reduce corrosion, the use of dissimilar metals in the lightning conductor scheme should be avoided, and after the joint is made it should be painted over. Copper and aluminum are the preferred conductor materials, with galvanized steel a third choice. Conductors, other than those buried in the ground or intended to be the first contact point with the air, should be protected, for example, by PVC coating or paint. The inductance of a wide, thin strip should be used where feasible, because its inductance is lower than that of a round conductor when its cross sectional area is larger, as described in Section 5.1.2. When the vertical conductors are a meter or more apart, the total inductance from the top of the building to the ground is the inductances of the vertical conductors in parallel. It is this total inductance that determines the voltage drop between the air contact point and the ground. If we assume a single 100-foot-high grounding conductor of 0.62-inch diameter, then the inductance of the conductor, from Eq. (4) of Section 5.1.2, is 46 μH. We also assume a lightning current waveform with a 2-μs rise time and a peak current of 25 kA. The voltage developed between the top of the conductor and the ground is approximately 575 kV, from

$$V = L \frac{di}{dt}$$ (8.1)

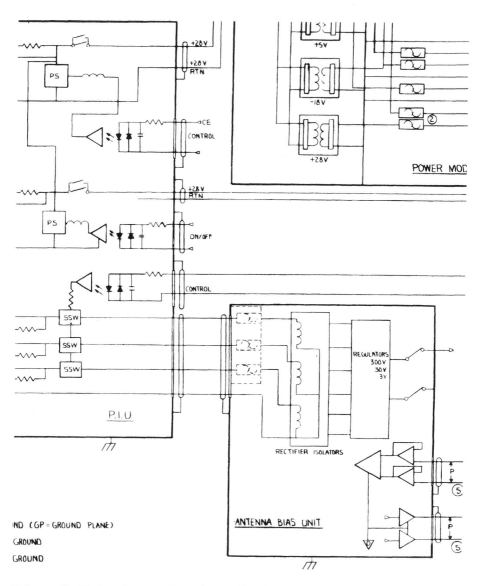

**Scheme 2**  Single-point grounding scheme. Signal ground connected to chassis at one point. Power ground isolated from signal ground and chassis by at least 1 MΩ. Equipment-to-equipment signal grounds isolate single-point grounding scheme.

The total inductance of the scheme may be reduced by adding more conductors in parallel. However, when conductors are in close proximity and share the current, the total inductance is not as low as when the conductors are far apart. For, in close proximity, the field from one conductor interacts with the adjacent conductor and results in an increase in inductance. The conductors should be spaced at least 1 m apart to minimize inductance and a maximum of 10 m apart to maximize the protection of the building.

Right-angle bends increase the inductance of the conductor; therefore add a radius to any bend. When the building has a metal-clad exterior or is of reinforced concrete or steel girder

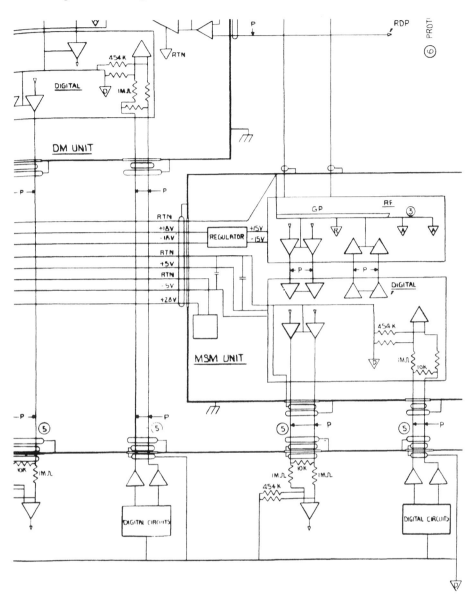

**Scheme 2** Continued

construction, the inductance of the down-conductors is reduced by proximity to the conductive structure, as described in Section 5.1.2.

Further information on the inductance of antenna masts and the application of protection devices is obtainable from Ref. 3.

The minimum separation distance between any of the upper components of the protection scheme and the building may be established from the potential difference between them. The impulse breakdown of air is dependent on its ionization and on the geometry of the electrodes. Figure 8.19 plots the minimum clearance for a given impulse voltage for any shape of electrode but assumes nonionization of the air. The source of Figure 8.19 is the International Electrotechni-

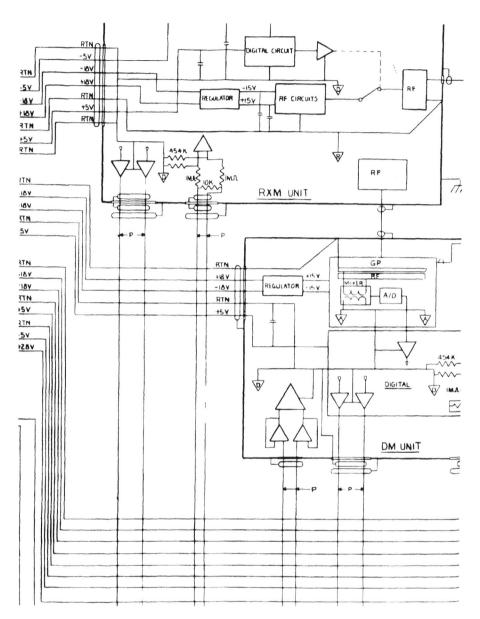

**Scheme 2**  Continued

cal Commission Document 28A, ''Insulation Coordination of Low-Voltage Systems and Equipment.'' Bends should be made with as large a radius as possible, and an increased minimum
clearance should be used at the bend to reduce the probability of breakdown.

Large mechanical forces are generated on conductors that carry high currents; therefore
all conductors should be mechanically bonded together and anchored firmly to the structure.

Figure 8.20 illustrates a typical grounding scheme for rooftop-mounted cellular or PCS
base station antennas. The antenna coaxial cable is brought into a metal cableway B, and at

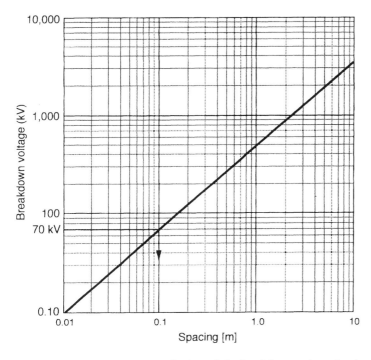

**Figure 8.19** Curve for determination of the breakdown voltage in air as a function of spacing.

this point the shield of the cable is bonded to the ground strap/braid/wire C. Also at this point, a combined primary lightning surge arrestor and filter A is inserted in the coaxial cable. The coaxial cable is then taken inside the cableway until the point of entry into the building D. The cableway must be bonded by a bond strap on either side of the seams in the cableway. At least two ground straps/braid or wires are run alongside the cableway and taken down the *outside* of the building to ground rods or, depending on the resistivity of the soil, a ground network, D. If lightning were to strike any of the antennas, the current would flow on the antenna coaxial cable until the junction with the ground wire. The current would then continue flowing on both the ground wire and the outside of the cableway and via the ground straps on the outside of the building into the soil. The voltage developed on the center conductor of the antenna cable is shunted to the shield by the surge protector A, which if it also contains a filter will attenuate the voltage further in amplitude to protect the cellular/PCS equipment. One major mistake that is made is to bond the ground system to a metal vent pipe or the metal of heating/ventilation equipment on the roof and to omit the external ground wires. Metal vent pipes are often connected together inside the building by the use of rubber gaskets, and the lightning strike can then flash over to sensitive equipment within the building and possibly start a fire. If the connection is made to an AC safety ground that penetrates the building, the same flash-over or fusing can occur, again with the possibility of fire. Therefore never bring a lightning protection ground scheme inside a building. The external ground wire/braid can be covered with a plastic cover that has a color similar to the building's, e.g., white or brown, if the aesthetics of the grounding scheme is a problem.

The IEC 1312-1, May 1995, document ''Protection Against Lightning Electromagnetic Impulse'' provides further information, as does MIL-HDBK-419A, Chapter 3, and Reference 3.

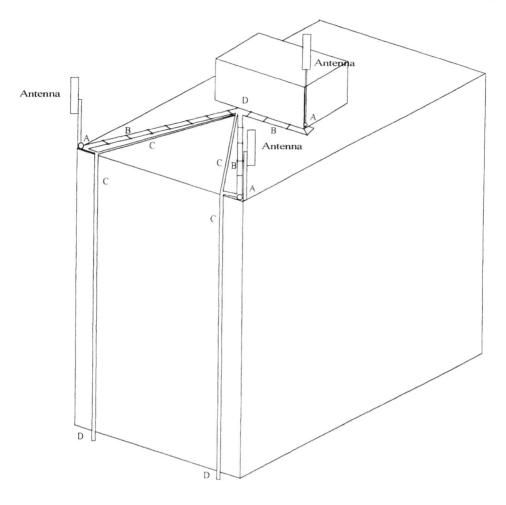

A = Primary lightning surge protection and filter

B = Metal cableway bonded at seams

C = Ground strap, braid, or wire

D = Ground rod

**Figure 8.20**  Typical lightning protection scheme for building-mounted cellular phone/PCS base station antennas.

## 8.6.2 Ground Potential

A potential gradient appears across the ground, with equipotential circles spreading out from the point of contact. A lightning strike at an area of the earth's surface that has a high resistivity tends to charge the area to a potential closer to the cloud's potential and a high potential gradient exists proceeding away from the immediate area of the strike. The purpose of the lightning protection scheme is to reduce the potential gradient between structures and equipment. A typical ground scheme for buildings and an antenna installation is shown in the following case study.

### 8.6.3 Case Study 8.1: Grounding for Lightning Protection at a Communication Site

The antenna structure shown in Figure 8.21 lies under the arc of protection of the lightning conductor rod, which is constructed of 1.6-cm-diameter material with a length of 4.3 m. The minimum distance between the rod and the antenna is 20 cm. Assuming the rod moves in a high wind such that the minimum distance reduces to 15 cm, the impulse breakdown voltage, from Figure 8.19, is 90 kV. Modeling the inductance of the rod and its resistance in parallel with that of the support structure to which it is bonded and using a rise time of 2 μs, the maximum lightning strike current protected by the rod is 60 kA. Thus the antenna protection scheme is adequate for approximately 80% of lightning strikes.

The lightning conductor is connected to a #2 AWG wire mesh that is also connected to rebars in the concrete base of the antenna and to an adjacent metal fence.

An X-IT rod is used to provide a low-resistance ground connection. The X-IT rod is a metal tube that contains a chemical that leaches out through apertures in the tube and permeates the surrounding soil. The moisture in the air is used to form an electrode with the chemical, which results in a low resistance. The use of a metal rod immersed in ground that has been soaked in a sodium solution is an alternative way of achieving a low earth resistance. X-IT rods are manufactured by Lyncole, Torrance, CA 90502.

The buildings at the site are clad with cold-rolled steel with a thickness of at least 1 mm. The cladding is constructed of a number of sheets that are bonded together. The DC resistance of the building from the center of the roof to the surrounding ground ring is approximately 5 mΩ. The buildings are thus adequately protected from lightning strikes. The maximum touch potential within the building during a 25-kA strike has been calculated at 125 V.

An approximately 25-m-long No. 2 AWG grounding wire connects the buildings to the base of the antenna and is run in a bermed PVC duct along with the antenna interconnection cables.

Several ground conductors within cables and the shields of cables will share the current

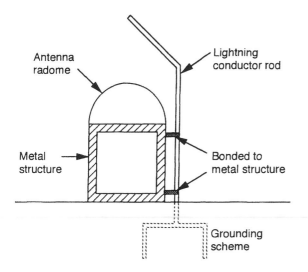

**Figure 8.21**  Antenna structure with lightning conductor rod.

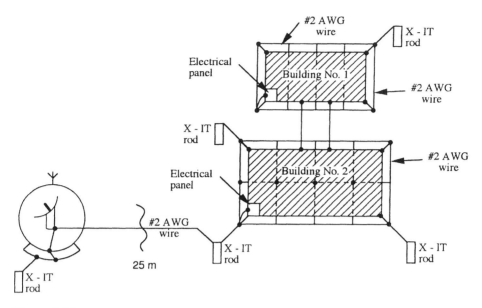

**Figure 8.22**   Communication site grounding scheme.

with the No. 2 AWG grounding wire during a lightning strike. The No. 2 AWG grounding wire ensures that the full lightning current does not flow in the interconnection cables.

During the fast initial transition of the lightning current, the inductance of the cables will limit the current flow. When the earth resistance is 50 Ω or less, the division of the current between the cables and of the ground between the antenna and the buildings becomes significant, especially during the fast initial transition.

The shields of cables are connected to the metal building at the entry point of the cables in order to shunt shield currents to ground. Large ferrite baluns are used on cables after the shield termination point in order to reduce remaining common-mode current flow. The differential-mode voltage induced into the center conductors of cables is reduced by one or more of the methods of transient suppression described in Section 5.4.1.

A reduction in the induced current in interconnection cables would be achieved by running the cables in a metal conduit, welded at the joints, instead of the PVC pipe. The No. 2 AWG wire should remain outside of the metal conduit, or it may be eliminated, since the metal conduit functions as the ground conductor.

At sites where the electromagnetic ambient was in the volts-per-meter region, the welded metal conduit was required to eliminate EMI, and the improvement in lightning protection was a secondary consideration.

The grounding system at buildings No. 1 and No. 2, as shown in Figure 8.22 can be described as follows. Three lengths of No. 2 AWG bare copper wire positioned equidistant beneath the two buildings, cad-welded to the ground ring system (which comprises No. 2 AWG wire around both buildings) and the center leg between the buildings. The wires are buried beneath the earth to maximize the contact area with the earth. The two buildings are bonded together via two No. 2 AWG wires of approximately 1.4-m length and spaced 6 m apart. Four X-IT rods are used to connect the ring to earth via low-resistance paths.

## 8.7 BONDING

### 8.7.1 General

*Electrical bonding* is a term used for the process of connecting metal structures together in order to achieve a low-resistance contact. Bonding is used in lightning protection schemes to ensure that lightning strike currents may be carried between structures, or structure and conductors, such as lightning arrestors and earth grounding rods. Other reasons for bonding are to provide static protection and in the realization of a ground reference plane. Antenna components also require low-impedance bonds, for currents flow on the junctions between these components. Adequate bonding is vital in minimizing passive intermodulation in antennas. Each of these bonds should be made so that the mechanical and electrical properties of the path are determined by the connected members and not by the interconnection junction. Further, the joint must maintain its properties over an extended period of time, to prevent progressive degradation of the degree of performance initially established by the interconnection. Bonding is concerned with those techniques and procedures necessary to achieve a mechanically strong, low-impedance interconnection between metal objects and to prevent the path thus established from subsequent deterioration through corrosion or mechanical looseness. The resistance of the bond and surface area over which the contact must be made are dependent on the purpose for which the bond is made.

### 8.7.2 MIL-B-5087 and MIL-STD-464

MIL-B-5087 has been cancelled, but it did define a number of different bonds and the requirements for each type, and these requirements have been used on military equipment that is still in the field as well as a number of different space programs. Although MIL-B-5087 has been cancelled and is now contained in MIL-STD-464, MIL-STD-464 still references the bonding classes of MIL-B-5087.

These bond types from MIL-B-5087 are as follows:

*Class A for antenna installations*—no bonding resistance specified, but a negligible impedance over the operating frequency was required.

*Class C for current return path*—impedance limited by maximum allowable voltage drop as shown in Table 8.5.

*Class H for shock hazard* < 0.1 ohm dc.

*Class L for lightning protection*—control internal vehicle voltages to 500 V.

*Class R for RF potential* < 2.5 milliohms from electronic units to structure.

*Class S for static discharge* < 1 ohm.

**Table 8.5** Maximum Allowable Voltage Drop for Class C Bonds

| Nominal system voltage | Maximum allowable voltage drop equipment operation | |
|---|---|---|
| | Continuous | Intermittent |
| 28 | 1 | 2 |
| 115 | 4 | 8 |
| 200 | 7 | 14 |

There were also less obvious requirements in MIL-B-5087, such as 2.5-milliohms requirement on connector shells. MIL-STD-464 points out that there is no scientific basis for the 2.5-milliohms requirement for Class R bonds. And although the equipment case–to-structure class R requirement is not important in most instances, the 2.5 milliohms is still a good number for several electrical bonds, such as terminating shields to connectors and bonding connector to equipment case. It is also a good design value where a good bond is required for other purposes. The other bonding values of MIL-B-5087, for shock protection, current return paths, and static charge, are still valid today.

MIL-STD-464 states that for lightning protection, metallic structural members (aluminum, steel, titanium, and so forth) provide the best opportunity to achieve an electrical bond on the order of 2.5 milliohms. A bond of this level will limit the induced voltage on system cabling to 500 volts from lightning strike attachments (200 kA) to system structure.

Bonding resistance is specified and measured, not bonding impedance. The bonding resistance is not necessarily an indicator of the impedance of the bond, which in EMI reduction is often of paramount importance. The impedance of ground planes and rectangular and circular conductors, which may be used as bonding straps or jumpers, are described in Chapter 5, Section 5.1. As discussed in Chapter 5, stray capacitance, self-inductance, and conductor length result in both series and parallel resonant circuits, and these must be accounted for when considering the bond at RF frequencies.

A poor electrical bond when exposed to two or more high-power fields may result in passive intermodulation and reradiate intermodulation products, as described in Chapter 10.3.

### 8.7.2.1   MIL-STD-464 Methods of Electrical Bonding

An electrical bond may be made by welding, brazing, soldering, riveting, bolting, conductive adhesive, or conductive grease or by spring finger stock material. The preferred method of bonding is welding, followed by brazing and soldering. A bolt or rivet is used to connect two conductive surfaces together with sufficient force to provide a low-resistance bond. The thread of a screw or bolt must not be relied upon to provide either a DC current or RF bond.

Conductive grease or spring finger stock material may be used to provide a bond between moving surfaces, for example, conductive grease between spring-loaded contacts, piano hinges, or rotating shafts. When the contact is not under pressure, the bond should, if possible, be backed up by a flexible bonding strap. The bond achieved by a circular spring finger stock around a shaft or conductive grease between the shaft and bearing walls will provide a lower impedance than a bonding strap. When conventional bonding methods, such as bolts, rivets, and screws, cannot be used, then the use of conductive adhesive may be allowed.

### 8.7.2.2   MIL-HDBK-419A Bonding Practices

MIL-HDB-419A describes in brief the following grounding practices: Equipment emission and susceptibility requirements for proper system operation should be accomplished with the most cost-effective combination of interference reduction techniques. Bonding is an essential element of the interference control effort. MIL-HDBK-419A presents design and construction guidelines to aid in the implementation of effective bonding of equipment circuits, equipment enclosures, and cabling. These guidelines are not intended as step-by-step procedures for meeting EMC specifications. Rather, they are aimed at focusing attention on those principles and techniques that lead to increased compatibility between circuits, assemblies, and equipments.

   a.   Welded seams should be used wherever possible, because they are permanent, offer a low-impedance bond, and achieve the highest degree of RF tightness.

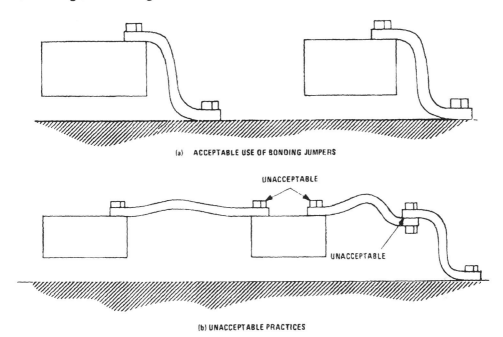

**Figure 8.23a**   Bonding jumper: acceptable and unacceptable uses.

b.   Spot welds may be used where RF tightness is not necessary. Spot welding is less desirable than continuous welding because of the tendency for buckling and the possibility of corrosion between welds.

c.   Soldering should not be used where high mechanical strength is required; the solder should be supplemented with fasteners, such as screws or bolts.

d.   Solder must not be used to form bonds that may be reasonably expected to carry large currents, such as those produced by power-line faults or lightning currents.

e.   Fasteners such as bolts, rivets, and screws should not be relied upon to provide the primary current path through a joint.

f.   Rivets should be used primarily to provide mechanical strength to soldered bonds.

g.   Sheet metal screws should be used only for the fastening of dust covers on equipment

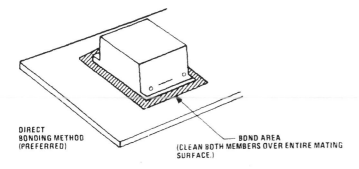

**Figure 8.23b**   Bonding of subassemblies to equipment chassis.

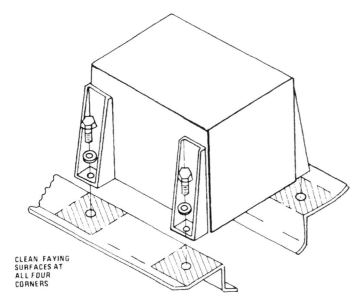

CLEAN FAYING
SURFACES AT
ALL FOUR
CORNERS

**Figure 8.23c**   Bonding of equipment to mounting surface.

or for the attachment of covers, to discourage unauthorized access by untrained personnel.
h.   Bonds that cannot be made through direct metal-to-metal contact must use auxiliary straps or jumpers. The following precautions should be observed when employing bonding straps or jumpers (see Figure 8.23a and b).
   (1).   Jumpers should be bonded directly to the basic structure rather than through an adjacent part, as shown in figure 8.23a.
   (2).   Jumpers should not be installed two or more in series.
   (3).   Jumpers should be as short as possible.

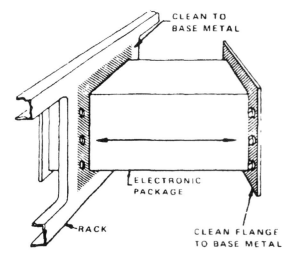

CLEAN TO
BASE METAL

ELECTRONIC
PACKAGE

RACK

CLEAN FLANGE
TO BASE METAL

**Figure 8.23d**   Typical method of bonding equipment flanges to frame or rack.

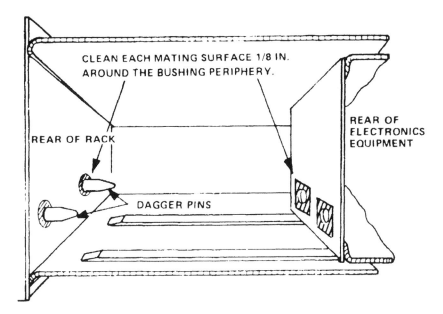

CLEAN EACH MATING SURFACE 1/8 IN. AROUND THE BUSHING PERIPHERY.

REAR OF ELECTRONICS EQUIPMENT

REAR OF RACK

DAGGER PINS

**Figure 8.23e**   Bonding of rack-mounted equipment employing dagger pins.

(4). Jumpers should not be fastened with self-tapping screws.

(5). Jumpers should be installed so that vibration or motion will not affect the impedance of the bonding path.

(6). Jumpers should be made of tinned copper, cadmium-plated phosphor bronze, aluminum, or cadmium-plated steel.

(7). Mating metals should be selected to offer maximum galvanic compatability (see Section 8.8.2).

i.  Where electrical continuity across the shock mounts is necessary, bonding jumpers should be installed across each shock mount. Jumpers for this application should have a maximum thickness of 0.06 cm (0.025 in.) so that the damping efficiency of the mount is not impaired. In severe shock and vibration environments, solid straps may be corrugated or flexible, coarse wire braid may be used.

j.  Where RF tightness is required and welded joints cannot be used, the bond surfaces must be machined smooth to establish a high degree of surface contact throughout the joint area. Fasteners must be positioned to maintain uniform pressure throughout the bond area.

k.  Chassis-mounted subassemblies should utilize the full mounting area for the bond, as illustrated in Figures 8.23b and 8.23c. Separate jumpers should not be used for this purpose.

l.  Equipment attached to frames or racks by means of flange-mounted quick-disconnect fasteners must be bonded about the entire flange periphery, as shown in Figure 8.23d. Both the flange surface and the mating rack surface must be cleaned over the entire contact area.

m.  Rack-mounted packages employing one or more dagger pins should be bonded as shown in Figure 8.23e.

n.  The recommended practices for effective bonding of equipment racks are shown in Figure 8.23f. Bonding between the equipment chassis and the rack is achieved

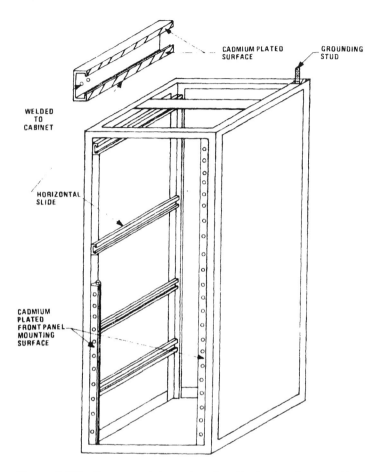

**Figure 8.23f**  Recommended practices for effective bonding in cabinets.

through contact between the equipment front panel and the rack front brackets. These brackets are bonded to the horizontal slide, which in turn is welded to the rack frame. The ground stud at the top of the rack is used to connect the rack structure to the facility ground system.

o.   Where hinges are used, establish an alternate electrical path through the use of thin, flexible straps across the hinges, as shown in Figure 8.23g.

p.   Standard MS-type connectors and coaxial connectors must be bonded to their respective panels over the entire mating surfaces, as illustrated in Figure 8.23h. Panel surfaces must be cleaned to the base metal for no less than 0.32 cm (1/8 in.) beyond the periphery of the mating connector.

q.   In ideal situations, cable shields should be bonded to the connector shell completely around the periphery of the shield with either compression or, preferably, soldered bonds.

r.   When an RF-tight joint is required at seams, access covers, removable partitions, or other shield discontinuities, conductive gaskets should be used. They may also be used to improve the bond between irregular or rough bonding surfaces. Gaskets should be sufficiently resilient to allow for frequent opening and closing of the joint

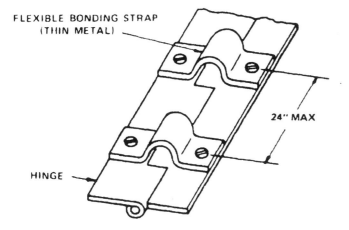

**Figure 8.23g** Method of bonding across hinges.

   and yet be stiff enough to penetrate any nonconductive films on surfaces.
s. Gaskets should be firmly affixed to one of the bond members with screws or conductive cement or by any other means that does not interfere with their operation. The gaskets may be placed in a milled slot to prevent lateral movement.
t. All bonds that are not in readily accessible areas must be protected from corrosion and mechanical deterioration. Corrosion protection should be provided by ensuring galvanic compatibility of metals and by sealing the bonded joint against moisture.

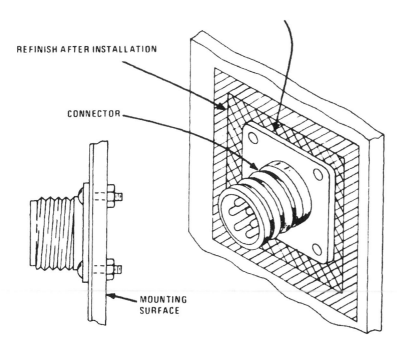

**Figure 8.23h** Bonding of connector to mounting surface.

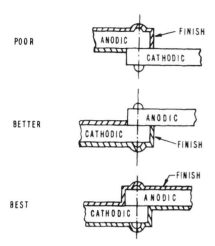

**Figure 8.24**  Finishing over dissimilar metals.

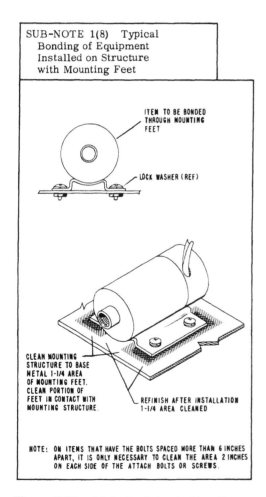

**SUB-NOTE 1(8)**   Typical Bonding of Equipment Installed on Structure with Mounting Feet

ITEM TO BE BONDED THROUGH MOUNTING FEET

LOCK WASHER (REF)

CLEAN MOUNTING STRUCTURE TO BASE METAL 1-1/4 AREA OF MOUNTING FEET. CLEAN PORTION OF FEET IN CONTACT WITH MOUNTING STRUCTURE.

REFINISH AFTER INSTALLATION 1-1/4 AREA CLEANED

NOTE: ON ITEMS THAT HAVE THE BOLTS SPACED MORE THAN 6 INCHES APART, IT IS ONLY NECESSARY TO CLEAN THE AREA 2 INCHES ON EACH SIDE OF THE ATTACH BOLTS OR SCREWS.

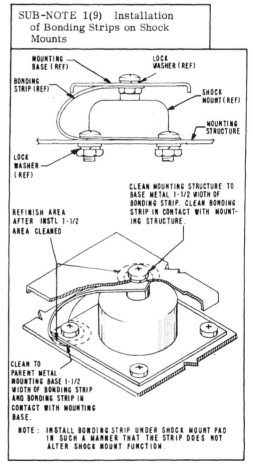

**SUB-NOTE 1(9)**   Installation of Bonding Strips on Shock Mounts

MOUNTING BASE (REF)

LOCK WASHER (REF)

BONDING STRIP (REF)

SHOCK MOUNT (REF)

MOUNTING STRUCTURE

LOCK WASHER (REF)

CLEAN MOUNTING STRUCTURE TO BASE METAL 1-1/2 WIDTH OF BONDING STRIP. CLEAN BONDING STRIP IN CONTACT WITH MOUNTING STRUCTURE.

REFINISH AREA AFTER INSTL 1-1/2 AREA CLEANED

CLEAN TO PARENT METAL MOUNTING BASE 1-1/2 WIDTH OF BONDING STRIP AND BONDING STRIP IN CONTACT WITH MOUNTING BASE.

NOTE: INSTALL BONDING STRIP UNDER SHOCK MOUNT PAD IN SUCH A MANNER THAT THE STRIP DOES NOT ALTER SHOCK MOUNT FUNCTION.

**Figure 8.25a**  Methods of electrical bonding.

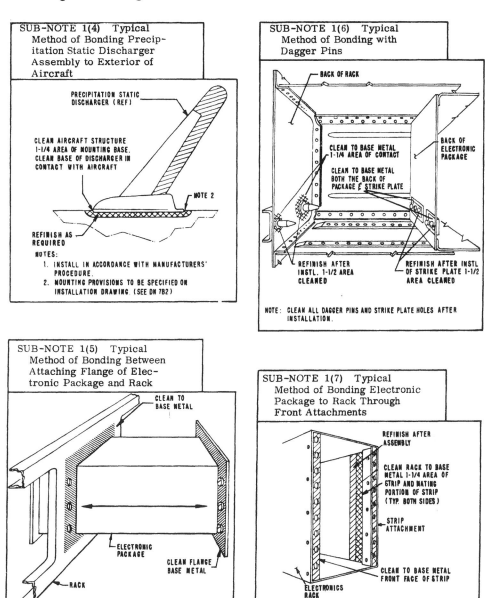

**Figure 8.25b** Methods of electrical bonding.

### 8.7.3 Corrosion, Dissimilar Metals, and Oxidization

When the contact is between dissimilar metals in the presence of moisture, corrosion can occur. The further the metals are apart in the electrochemical series, the greater the electrolytic reaction (corrosion). One method that may be used to reduce the reaction is to insert a metal that is intermediate in the electrochemical series between the two dissimilar metals to be joined. The intermediate metal may be a surface coating or plating, or it may be a thin piece of metal, such as a washer. In the contact of dissimilar metals, one metal acts as a cathode and the other as the anode. Because a cathode is the source of electron flow, the smaller the cathode relative to

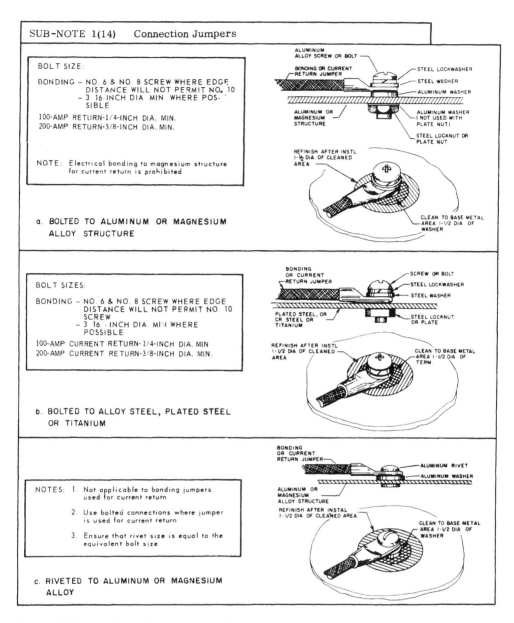

**Figure 8.25c** Methods of electrical bonding.

the anode the lower the electron flow and the lower the corrosion. The most effective methods of minimizing electrolytic corrosion are either to use similar metals or to reduce the moisture between dissimilar metals.

Moisture may be excluded by finishing the surfaces after the joint has been made, typically by paint or by plating or caulking. Figure 8.24 shows that it is better to apply the finish over both of the surfaces to be joined. Where this is not possible, finishing the cathodic metal is better than finishing the anodic metal only.

Metals such as aluminum and steel will oxidize and build up a high-resistance coating.

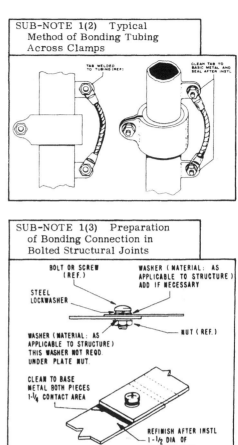

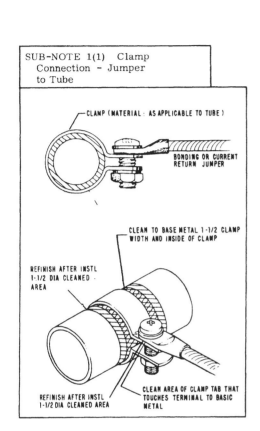

**Figure 8.25d**  Methods of electrical bonding.

The exclusion of air by passivation of the two surfaces to be joined will inhibit the oxide buildup. First the surfaces must be cleaned and degreased, and in some cases it is advisable to etch the surface. After cleaning, the surfaces may be coated with a conductive passivation, such as oaktite #36, alodine #1000, iridite #14, or iridite #18P. A chromate finish may be used but does not provide as low a resistance as the preceding finishes. Zinc, tin, or gold plating may also be used to achieve a high level of passivation. Anodizing is usually nonconductive and should be avoided. Cleaning or etching the surfaces and finishing with paint after joining, although not as effective as passivation, will minimize the oxidization at the joint.

Some methods of bonding, reproduced from the Air Force Systems Command Design Handbook DH 1-4 Electromagnetic Compatibility, are shown in Figures 8.25a, b, c, d.

### 8.7.4  Bonding Test Methods

The use of a two-terminal resistance meter to measure the resistance of a joint will lead to wide fluctuations in the reading, ranging from an open circuit to a fairly low resistance, depending on the pressure used to contact the probes to the joint to be measured. The problem with the

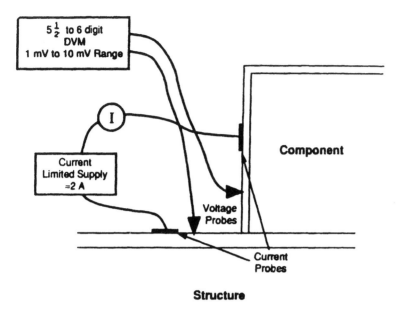

**Figure 8.26**  Four-terminal bonding resistance test method.

two-terminal method is that the current is injected into the surface of the material by the same probe as used to measure the voltage, and in reality it is the probe-to-metal contact impedances of the two probes that are measured. The four-terminal technique injects the current by two terminals and measures the voltage by use of another two terminals, and the measurement is much less sensitive to contact pressure. The resistance is found from Ohm's law by the measurement of the current and the voltage. One method of applying the current and voltage terminals is to solder leads to conductive adhesive copper tape and to apply the tape to the two surfaces. The current terminals should be placed some distance from the joint to be measured, whereas the voltage terminals should be placed close to the joint. The voltage terminals must be placed in the direct path between the two current terminals to minimize measurement errors. The four-terminal measurement technique may be used to make comparative measurements of the surface resistance of metals and finishes. Here the width of the current and voltage contacts should be constant from measurement to measurement and may be most readily obtained by use of conductive adhesive copper tape of a fixed width for all four terminals and from measurement to measurement. The four-terminal test method is illustrated in Figure 8.26.

## REFERENCES

1.  FAA/Georgia Tech. Workshop on the grounding of electronic systems. March 1974. Distributed by National Information Service U.S. Department of Commerce.
2.  AFSC Design Handbook Electromagnetic Compatibility. DH-1-4. Department of the Air Force, Wright Patterson Air Force Base, Ohio 45433, 1971.
3.  Lightning Protection and Grounding Solutions for Communication Sites. Polyphaser Corporation, P.O. Box 9000, Minden, NY 89423-9000 (775) 782-2511.
4.  Nobuo Kuwabara, Tetsuya Tominaga, Masaru Kanazawa, Shoich Kuramoto. Probability Occurrence of Estimated Lightning Surge Current at Lightning Rod Before and After Installing Dissipation Array System (DAS). NTT Multimedia Networks Laboratories 3-9-11 Midori-cho, Musashino-shi, Tokyo 180-8585, Japan, 1998, IEEE.

# 9

# EMI Measurements, Control Requirements, and Test Methods

## 9.1  INTRODUCTION

EMI measurements may be divided into those imposed by EMI requirements, which also impose a test method, and measurements useful in finding the source of EMI problems. These problems include lack of self-compatibility or interference with other systems. Equipment manufacturers may not possess specialized EMI test equipment and therefore wish to make measurements with standard electronic test equipment, such as the oscilloscope. This chapter deals with the use of standard and specialized test equipment and perhaps most importantly with common sources of error in their use. It has been said that one measurement is worth a hundred predictions; however, EMI measurements may significantly affect the parameter measured, resulting in error. The location of antennas may affect the calibration curve, for example, errors greater than 30 dB are common in undamped shielded-room measurements of radiated emissions. A measurement with an oscilloscope may corrupt the measured signal as a result of either the additional ground connection or by radiated pickup on the oscilloscope test leads. Thus great care has to be exercised in the selection and application of the measurement technique, otherwise the prediction may be more accurate than the measurement. A design should not be based solely on measurements of performance, such as filter attenuation and shielding effectiveness; instead, the measurement should be used to validate the design. When large discrepancies exist between theory and measurement, both should be examined for error. It is incorrect to assume that measurement data is intrinsically more accurate than predicted values. The measured parameter may not be the intended; for example, small loop antennas may intercept the radial magnetic field and not the intended horizontal when the loop orientation is incorrect. The incorrect test equipment may be used for the measurements, leading to error. For example, a manufacturer of equipment that failed a commercial radiated emission test was using an AM radio with a rod antenna to measure a relative reduction in field after EMI fixes were implemented. The criterion used was the very subjective one of reduction in audio noise. A potential source of error was the radio's automatic gain control (AGC), which may have a 20-dB dynamic range. Thus a reduction of 20 dB or greater in field strength may be achieved with no perceptible change in measured noise level. Had the AGC voltage been measured by a DVM or small panel meter, a more accurate measurement might have been achieved. However, even this technique has the limitation that the AGC is operative only over a limited range of signal strength.

This chapter describes the use of commonly available or easily manufactured and relatively inexpensive equipment that may be used for simple diagnostic measurements as well as the use of specialized equipment.

## 9.2 TEST EQUIPMENT

### 9.2.1 Oscilloscope

The oscilloscope is one of the most common pieces of test equipment found in electronic laboratories. The oscilloscope has both a number of major limitations when used for EMI tests and
a number of advantages. A major limitation is that the majority of EMI test requirements place
limits specified in the frequency domain. Thus any oscilloscope measurement of a complex
waveform must be converted into the frequency domain in order to compare measurements with
limits. For measurements of single frequencies with low harmonic distortion, the oscilloscope
may be useful, assuming its sensitivity is adequate. When a number of frequencies are present,
measurements of a low-amplitude high frequency superimposed on a high-amplitude low frequency may be impossible. In this case a frequency-domain measurement taken by using a
spectrum analyzer is the preferred method. When measurements of transient noise in the time
domain are required, an oscilloscope is the correct instrument to use. Another limitation is the
conversion of common-mode noise into differential across the oscilloscope's high-impedance
single-ended input. A simple test for the presence of common-mode noise is to connect the
probe ground clip to the signal ground and at the same ground point the probe tip. Although
the input to the oscilloscope is thereby shorted out, any common-mode current flow on the
shield and center conductor develops a differential voltage across the input impedance of the
scope, and the magnitude of the common-mode contribution to the measured noise may be
judged. A voltage is developed across the impedance of the shield due to a shield current flow,
and it is this voltage that appears across the input of the oscilloscope. In this test some of the
voltage measured may result from radiated pickup by the loop formed between the probe and
the ground wire. The spectrum analyzer also has a single-ended input, but the 50–75 $\Omega$ input
impedance develops a lower differential voltage from a common-mode current flow than the
$1–10$-M$\Omega$ impedance of the oscilloscope, or the voltage drop across the impedance of the shield
is reduced because the input impedance is across the shield impedance. One disadvantage the
spectrum analyzer has is that the low input impedance may load the signal, especially when the
source impedance is high; also, the input is DC coupled, and a decoupling capacitor must be
used when measuring signals/noise existing on high DC voltages. Oscilloscope and spectrum
analyzer single-ended inputs connect the signal return to the enclosure of the oscilloscope or
spectrum analyzer and thus to the AC safety ground. This additional ground connection may
drastically alter measured noise levels and susceptibility characteristics of the equipment. The
single-ended input and chassis ground connection characteristic of the oscilloscope may be useful when these are required to simulate the input of a piece of equipment with the same characteristics. Thus the single-ended output signal of equipment under test may be correctly terminated
by the use of an oscilloscope. One additional effect is that the inductance of the probe ground
wire and the input capacitance of the probe form a resonant circuit that changes the wave shape
or amplitude of a measured signal or noise. The resonant circuit may cause ringing on a transient,
which may be incorrectly attributed to a circuit under test resonance. The LC circuit forms a
low-pass filter and may attenuate the amplitude of high-frequency signals. The inductance of
the ground wire should be reduced to a minimum by the use of as short a connection as feasible
and by locating it close to the probe. Alternatively, a field effect transistor (FET) input probe
with a low input capacitance may be used. When the oscilloscope single-ended input characteristics are undesirable, the use of a differential input is a solution. Differential plug-ins for oscilloscopes do not necessarily provide the highest level of common-mode noise rejection because
of a poor balance between the two inputs caused in part by the use of separate test leads. A
higher level of common-mode noise rejection is achieved by the use of a differential scope
probe, in which the differential input is located at the probe tip. This type of probe has the

advantage of lower pickup on the test lead from the radiated ambient and lower loading and resonance effects. When a differential input is unavailable, the two channels, A and B, of some oscilloscopes may be used in an A − B mode or an inverted A-plus-B mode, which effectively measures differential voltage when the input to one channel is connected to the voltage and the second to the return. When the ground clips are connected together at the probe but not connected to either the ground or the enclosure of the equipment under test, the ground loop problem inherent in single-ended inputs is eliminated. The common-mode voltage is not totally removed from the measurement using this technique, due to the poor balance between the two inputs. The level of common-mode contribution to the measurement may be checked by connecting the tips of both channel A and channel B probes to the signal ground. Some compensation for imbalance may be achieved by adjusting one or both of the compensation trimmer capacitors in the probes for minimum measured level. Radiated pickup on the test probes may be reduced by twisting the probe leads together, which reduces the pickup loop area and tends to cancel the field-induced currents. Another method of differentiating between common-mode and differential-mode noise sources, and of isolating the oscilloscope ground from the signal ground, is to measure the noise current by use of a current probe. In some instances, especially when the circuit impedances are low, the level of noise voltage is low, and it is the noise currents that result in radiation and should be measured (this is often true for current flow in chassis ground). The oscilloscope may be used in relative measurements of electric field by attaching a length of wire to the probe tip. When an oscilloscope has a 50-$\Omega$ terminated input or when a 50-$\Omega$ external termination is used, the small calibrated loop and other antennas described in Chapter 2, Section 2.6 may be used. Antenna or current probe output levels are typically too low for use with an oscilloscope, even on the most sensitive range, and the use of a preamplifier is almost mandatory. Another limitation of the oscilloscope, in addition to the loss of frequency component information, is the limited upper-frequency response of the average analog oscilloscope.

In the CS01 and CS06 series injection test on AC power lines, the oscilloscope may have 120–220-V common-mode voltage applied. Even with the use of the ×10 probe, many oscilloscopes provide erroneous measurements with this level of C/M voltage. The modern digital scope seems to be more susceptible than the older analog. One potential solution is to place the injection in the neutral line, which should have a much lower C/M voltage referenced to chassis/safety ground.

## 9.2.2 Spectrum Analyzer

One of the most useful measuring instruments for diagnostic tests, and one that is gaining acceptance for use in EMI testing requirements and for certification, is the spectrum analyzer. An EMI receiver and spectrum analyzer are similar in their basic functions. However, a number of important differences exist; these are expanded on further in this chapter. One major difference that makes the spectrum analyzer superior for diagnostic measurements is the CRT display. Some EMI receivers have the capability of connection to an oscilloscope or monitor that is used to display amplitude versus frequency, instead of the usual oscilloscope display of amplitude versus time. However, the display capability is often far inferior to that of the spectrum analyzer. Another feature of the spectrum analyzer that is different from the EMI receiver is its capability of displaying emissions in a short-duration sweeptime (i.e., short-duration changes in amplitude are more easily discernible). To enable a fast sweep rate, the IF filter components must be capable of fast charge and discharge, which results in a filter with a Gaussian shape. The receiver that typically does not allow a fast sweep rate has a rectangular-shaped filter. The advantage of the rectangular filter is greater selectivity due to the very much reduced bandwidth below the 3-dB down point. For example, in comparing the 3–60-dB down bandwidths of the EMI receiver and spectrum analyzer, the ratio in the EMI receiver is typically 1:2, whereas for the spectrum analyzer it is 1:14. Thus, assuming a 100-Hz 3-dB bandwidth, the 60-dB bandwidth for the receiver is typically 200 Hz, and for the spectrum analyzer it is 1.4 kHz.

A limit on sweeptime does exist, and where it is set too short most spectrum analyzers display an out-of-calibration condition. The spectrum analyzer can suffer from compression and overload, just as described in the section on the preamplifier, and the same methods can be used for detection and correction. The spectrum analyzer and EMI receiver have a built-in variable-input attenuator that may be used to check for compression as follows: By searching over the frequency range of the spectrum analyzer to find the maximum input level and adjusting the input attenuator so that the displayed amplitude is at maximum, or lower, input compression and overload can be avoided. If on adjusting the input attenuator the displayed magnitude changes, the spectrum analyzer is compressing. This is because the spectrum analyzer automatically adjusts the displayed level regardless of the attenuator setting. In compression the effective gain of the front end changes and so the displayed level does not remain constant. Hewlett-Packard publishes a useful booklet entitled "EMI Measurement Solutions Using the Spectrum Analyzer Receiver," and some of the figures and following information are adapted from that source by kind permission of Hewlett-Packard. A block diagram of the spectrum analyzer is shown in Figure 9.1.

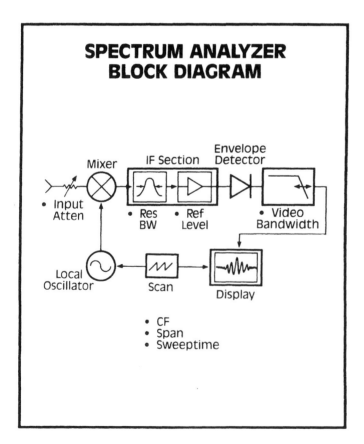

**Figure 9.1**  Spectrum analyzer block diagram. (Reproduced with kind permission from Hewlett-Packard.)

The spectrum analyzer is a swept tuned super-heterodyne receiver with an extremely wide input frequency range (typically 10 kHz to 1.3 GHz, but as wide as 100 Hz to 22 GHz). The incoming signal mixes with the local oscillator; when the mixed product equals the IF, the signal passes through to a peak detector. Some spectrum analyzers allow switching in a quasi-peak detector, whereas other spectrum analyzers allow the connection of an external quasi-peak detector. The front panel of the spectrum analyzer contains controls for the frequency span, sweeptime, resolution, and video bandwidths and reference levels. Additional functions on most spectrum analyzers include a marker that may be used to read off the frequency and amplitude of a selected component. The accuracy of the displayed frequency is a function of the resolution bandwidth and the frequency span. A typical specification for frequency readout accuracy is $\pm 3\%$ of span and of resolution bandwidth. Therefore a frequency measured with a 2-GHz span may be displayed with up to $\pm 60$-MHz error. For maximum accuracy, the minimum span and resolution bandwidth should be used. On some spectrum analyzers a counter is incorporated with the marker function, with which counter accuracy frequency measurements may be made. Maximum or peak hold and storage of one or more sweeps are also useful functions that enable capturing short-duration emissions and comparing before and after emission profiles. The resolution bandwidth control on the spectrum analyzer has a number of uses, one of which is in reducing the noise floor of the spectrum analyzer. The narrower the resolution bandwidth, the lower the noise level and the better the displayed signal-to-noise ratio.

Assume that broadband coherent noise and a single-frequency narrowband source are both present and are measured with the spectrum analyzer. As the resolution bandwidth is widened, more of the spectral lines generated by the broadband source, which are close together in frequency, will lie within the envelope of the resolution bandwidth, and the displayed signal level will increase. The narrowband signal level, assuming it is not swamped by the broadband signal, will not increase with an increase in bandwidth, for only one frequency is measured in the envelope of the bandwidth.

The resolution bandwidth, sometimes referred to as the IF bandwidth, also determines selectivity and sweeptime. Because the spectrum analyzer traces out its own IF filter shape as it tunes past the signal, the spectral emission lines are not displayed as infinitely narrow lines. Thus if the resolution bandwidth is changed, the width of the display changes. This is important in differentiating between adjacent signals (selectivity). Hewlett-Packard specifies a 3-dB bandwidth (i.e., the width over which the amplitude is no more than 3 dB down on the maximum amplitude). If two adjacent signals are closer together than the 3-dB bandwidth, they cannot be differentiated in the display. In contrast, if the signals are further apart than the 3-dB bandwidth, they may be differentiated.

One of the definitions of narrowband (NB) or broadband (BB) impulsive noise refers to the resolution bandwidth of the measuring instrument. When separate narrowband and broadband limits are specified, it is important that typical bandwidths for narrowband and broadband measurements be established and agreed upon by the procuring agency before commencement of tests. In some cases the agency that imposes EMI measurements will also define the measurement bandwidths. Where this is not true, upcoming Tables 9.20 and 9.21 provides typical NB and BB measurement bandwidths. As illustrated in this chapter, whether impulsive noise is displayed as narrowband or broadband depends on the resolution bandwidth. It is important to differentiate between the two types of noise for a number of reasons. The FCC allows a 13-dB relaxation in the conducted emission limit if the noise is determined to be broadband, and the VDE, early versions of MIL-STD-461, and DO-160 requirements specify different emission levels for broadband and narrowband noise. In diagnostic measurements, tracing the source of emissions may be facilitated by determination of the emission type. For example, a single high-intensity emission may exist in the multiple spectral emission lines from a broadband source.

If the single high-intensity emission is determined to be narrowband, then the source is not the broadband source, although this may excite the narrowband signal. The source is instead an oscillation, which may then be traced to a power supply, circuit instability, or a resonant circuit. Signals are characterized as narrowband with respect to the measuring instrument's bandwidth when there is only one spectral component of the signal contained within the filter bandpass. Each spectral component of a signal is individually resolved and displayed in the frequency domain. This means that the frequency of the emission may be read directly off of the $X$ coordinate of the spectrum analyzer display. Changing the frequency span of the display changes the spacing of the spectral emission lines. For example, for a frequency span of 20–200 MHz and 10 divisions, the frequency per division of display is 18 MHz. Thus two emissions 18 MHz apart will be spaced one division apart. Changing the frequency span to 20–110 MHz increases the spacing to two divisions. Changing the spectrum analyzer sweeptime will not change the spacing when the display is in the frequency domain. The spectral emissions lines may be either narrowband signals (i.e., sinewave sources) or impulsive noise, which is displayed as narrowband signals and which may be classified as narrowband when the spectrum analyzer bandwidth is set to a narrowband value. For impulsive noise sources displayed in the frequency domain, the pulse repetition frequency (PRF) may be determined by measuring the frequency spacing between the individual spectral lines. In our example, the PRF is 18 MHz. Figure 9.2 illustrates the narrowband display and the associated narrowband characteristics.

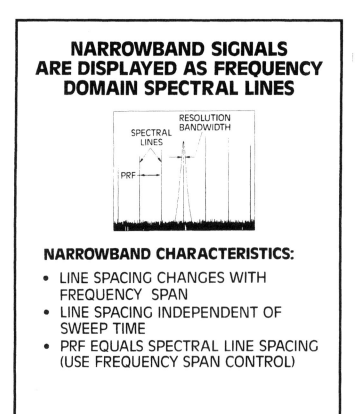

**Figure 9.2**  Narrowband display. (Reproduced with kind permission from Hewlett-Packard.)

If we now assume an impulsive noise source at a much lower repetition rate than 18 MHz, for example, 50 kHz, and the resolution bandwidth is increased to include more than one spectral line, then individual spectral lines are no longer resolved and the signals are added under the resolution bandwidth curve. The impulsive noise occurs in the time domain at a specific repetition rate, unlike a continuous wave. Thus impulsive noise sources, when more than one spectral emission line is covered by the resolution bandwidth, are displayed in the time domain. The amplitude of the display is now equal to the envelope of the $(\sin x)/x$ spectrum.

A scanning analyzer will therefore display a pulse every $1/PRF$ seconds, with an amplitude proportional to the spectrum envelope amplitude at the frequency to which the analyzer is tuned, and the signal will be displayed as a broadband signal. Due to the time-domain display of broadband noise, changing the frequency span does not alter the spacing between the pulses as was true of the narrowband display. Instead the spacing between the pulses is changed by changing the sweeptime. A typical characteristic of a broadband display is that the signals appear to "walk" across the CRT, because the analyzer sweeptime is generally not locked to the PRF of the signal. The PRF is found by taking the reciprocal of the sweeptime between the individual pulses. Thus for a sweeptime of 0.2 ms and a one-division spacing between the pulses, the sweeptime between pulses is 0.2 ms/10 = 20 μs and the PRF is 1/20 μs = 50 kHz. Note that, as described in Section 3.1.1, the PRF of a pulse with mark/space ratio of unity is often mistakenly calculated at half the actual PRF. This is because the pulses are at zero amplitude at the sidelobes. Figure 9.3 illustrates an impulsive signal, displayed as a broadband time-domain signal, and its associated broadband characteristics.

Differentiation between narrowband and broadband signals may also be achieved by changing the video bandwidth, which may produce an averaging of the high-frequency component at the output of the envelope detector. The averaging occurs because the video bandwidth filter is low pass and reduces the amplitude of the high-frequency components. The conditions under which this is achieved are as follows: (1) when the video filter bandwidth is narrower than the resolution filter bandwidth; (2) when the video bandwidth is less than the lowest PRF; (3) the frequency sweep must be slow enough to allow the filters to charge completely; and (4) the spectrum analyzer must be in the linear amplitude display mode. Average detection reduces the displayed amplitude of broadband signals but has no effect on the amplitude values of narrowband signals. One further test used to differentiate between narrowband and broadband signals is the tuning test. In the tuning test the spectrum analyzer is tuned one impulse bandwidth (in the case of VDE requirements) or two impulse bandwidths (in the case of the MIL-STD-462) on either side of the center frequency of the emission under investigation. The impulse bandwidth is defined as the 6-dB filter bandwidth, whereas some spectrum analyzer manufacturers, such as Hewlett-Packard, quote a 3-dB resolution bandwidth. A change in peak response of 3 dB or less indicates a broadband emission, while a change of greater than 3 dB indicates a narrowband emission. Table 9.1 illustrates the four methods used to differentiate between narrowband and broadband sources. That four methods exist indicates that the differentiation is not easy and that often more than one method must be tried before a conclusion can be made. If the results are inconclusive or contradictory, the tuning test is the preferred method used in arbitration. In some cases the decision must be made by the EMC engineering authority in charge of tests and equipment compliance or by the agency imposing the EMC requirements.

An illustration of the effect of resolution bandwidth on displayed amplitude is shown in Figure 9.4 for an impulsive signal with a 1-kHz PRF and a 260-kHz mainlobe width. At narrowband settings, the individual spectral lines are captured in the resolution bandwidth, resulting in a narrowband display. As the receiver bandwidth is increased, more of the spectral lines are captured, the amplitude increases, and the display is broadband. Finally, as the bandwidth is

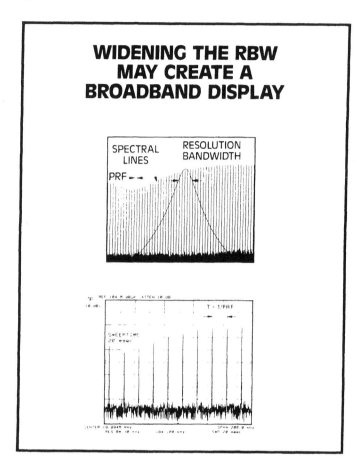

**Figure 9.3**   Broadband display. (Reproduced with kind permission from Hewlett-Packard.)

further increased, the full signal spectrum is captured by the resolution bandwidth, and any further increase does not result in an increase in amplitude, resulting in a quasi-narrowband display.

Some of the criticisms of the spectrum analyzer are its wide input frequency range, which leaves it prone to compression and overload, and its lack of sensitivity, which is to a large extent answered by the use of a preselector and amplifier. The Hewlett-Packard 85685A preselector has a separate 20-Hz to 50-MHz input and a 20-MHz to 2-GHz input. It contains protection against high-voltage transients at the low-frequency input, and the preselector filters include a number of low-pass and tuned bandpass filters. Both high- and low-frequency paths contain preamps with 20-dB gain, which improves the noise figure. The preselector also contains a comb generator that may be used to amplitude-calibrate the system to meet the $\pm 2$-dB CISPR specification. An additional advantage of the RF preselector is in measuring broadband noise that covers a very wide frequency range, for due to the very wide frequency range of the spectrum analyzer input, a high spectral density may be applied to the mixer, even though the resolution bandwidth filter reduces the spectral density applied to the peak detector. Thus the dynamic range available, limited by the maximum mixer input level, may not be adequate. By limiting the bandwidth applied to the mixer with the preselector filter, the dynamic range of the instrument may be greatly increased.

## BROADBAND SIGNALS ARE DISPLAYED AS TIME DOMAIN PULSE RESPONSES

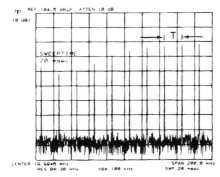

### BROADBAND CHARACTERISTICS:

- PULSE SPACING INDEPENDENT OF FREQUENCY SPAN
- PULSE SPACING CHANGES WITH SWEEP TIME
- PERIOD (T) EQUALS SPACING OF PULSE RESPONSES (USE SWEEP TIME CONTROL)
- PRF = 1/T

**Figure 9.3** Continued

**Table 9.1** Methods for NB and BB analysis. (Reproduced with kind permission from Hewlett-Packard.)

| METHODS | NB | BB | CRT RESPONSE |
|---|---|---|---|
| TUNING TEST 'TUNE' △ BW$_i$ | △ AMPL > 3 dB | △ AMPL < 3 dB | |
| PRF TEST △ SWEEPTIME | NO △ SPACING | △ SPACING | |
| PEAK VS. AVG DET △ VIDEO BW | NO △ AMPL | △ AMPL | |
| BANDWIDTH TEST △ RESOLUTION BW | NO △ AMPL | △ AMPL | |

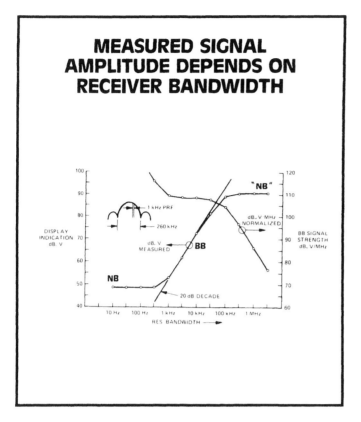

**Figure 9.4**  Narrowband and broadband display of impulsive noise with amplitude versus bandwidth. (Reproduced with kind permission from Hewlett-Packard.)

For European Union (EU), European Norm (EN), VDE, Industry Canada, and FCC commercial certification tests, the quasi-peak detector is specified; however, if the EUT passes the requirement using a peak detecting instrument, then the results are acceptable, for, compared to quasi-peak, a peak measurement is worst case. The quasi-peak detector was instituted to attempt to account for the annoyance factor of noise. The annoyance factor of low-repetition-rate noise is considered lower than that of high-repetition noise, and due to the charge and discharge time constant built into the quasi-peak detector, its response is less to low-repetition-rate noise sources. In practice, above a 10-kHz repetition rate, the quasi-peak and peak responses are very close; and this becomes increasingly true as the repetition rate increases above 10 kHz. As the quasi-peak detector incorporates both a charge and discharge time constant, an additional effect is to average out the peak amplitude measured in successive slow scans of a signal that changes amplitude after each scan. This type of variation is very typical of multiple noise sources, which can change from being in phase, and therefore additive, to out of phase, and therefore subtractive, with time. This averaging effect often results in a lower measurement using the quasi-peak detector compared to the peak detector, which has a very much faster charge time.

One important factor in the choice of spectrum analyzer, preamplifier, and EMI receivers are low production of intermodulation, spurious, and harmonic responses. Additional factors in the choice of the spectrum analyzer and EMI receiver are: flat ±2-dB gain vs. frequency; adequate frequency accuracy and stability; wide dynamic range; high sensitivity and low noise floor

(EMI receivers have often been quoted as exhibiting lower noise and higher sensitivity than the spectrum analyzer, but this is no longer true of the more expensive analyzer); outputs for IF, video display, plotter, LO (local oscillator), and audio; acceptance by governing agencies (some do not accept a spectrum analyzer); computer control capability. The spectrum analyzer uses a Gaussian shaped resolution bandwidth filter and the receiver uses an impulse bandwidth. In the measurement of broadband noise a correction factor of approximately 4dB has to be made to the spectrum analyzer measurement when compared to the receiver, whereas no correction factor is required for narrowband noise.

### 9.2.3 Preamplifiers

Preamplifiers are available with gains of 20–26 dB (numeric 10–20) and a typical noise figure of 0.08 μV when measured on a spectrum analyzer with a 10-kHz bandwidth. Thus it is possible to amplify a 1-μV signal level to between 10 and 20 μV, depending on the gain of the preamplifier, and to differentiate between the signal and noise, depending on the resolution bandwidth.

A typical broadband preamplifier will have a 10-kHz to 1-GHz frequency range with a flatness of 1 dB. The maximum output voltage from a typical preamplifier is 0 dBm (i.e., 225 mV). If the input voltage to the preamplifier is greater than 12 mV, assuming a gain of 26 dB, then compression occurs (i.e., the output of the preamplifier does not increase beyond 225 mV). The problem is that the gain of the preamplifier effectively reduces at all frequencies when compression is present.

As an example, assume a measurement is made with a preamplifier connected to the input of a spectrum analyzer and a signal is displayed at 10 MHz on a frequency span of 1–500 MHz. Assume the input level to the preamplifier at 10 MHz is 50 μV, at the same time a frequency of 10 kHz and a magnitude of 1 V are input to the preamplifier but not displayed on the spectrum analyzer. The 1-V level at 10 kHz causes compression in the preamplifier, and the output at 10 MHz is displayed as 200 μV instead of the correct 1000 μV.

One technique that can be used to detect compression, if the input signal of interest is above the noise floor of the spectrum analyzer, is to bypass the preamplifier (i.e., plug the input signal directly into the spectrum analyzer), and the displayed signal should reduce by the gain of the preamplifier, which in the example is 26 dB. When the input signal level is too low, a 3–12-dB attenuator may be connected to the input of the preamplifier, after which, in the absence of compression, the signal should be reduced by the appropriate level (3–12 dB). The probability of compression can also be determined by setting the frequency span of the spectrum analyzer to the frequency range of the preamplifier and noting high-level (i.e., close to 0 dBm) signals at any frequency. The solution to the compression problem is to include a filter, preferably tunable, at the input of the preamplifier, which attenuates the unwanted frequency but not the frequency of interest. Alternatively the 3–12 dB alternator may be adequate.

A second result of overload caused by an input signal that is too high is clipping and distortion at the output of the preamplifier. In the frequency domain the distortion results in an output that exhibits coherently related spectral emissions covering a wide range of frequencies, whereas the input signal may be a single frequency. In some instances overload can cause the preamplifier to oscillate. Constant attention to signs of, and testing for, compression and overload in preamplifiers, spectrum analyzers, and receivers must be made to ensure that equipment does not either pass EMI requirements due to compression or fail due to spurious response caused by overload. Preamplifiers are also available that cover the 1–18-GHz frequency range at gains of 9–40 dB. It is important to choose an amplifier with a noise figure referenced to the input at least 3 dB lower than the lowest signal to be amplified. Generally the narrower the frequency range of the amplifier, the lower the noise figure. Often the noise floor is specified in dB instead of (dBμV or μV)/bandwidth or nV/$\sqrt{\text{Hz}}$. The noise specified in dB is the noise figure described, with a typical test method, in Section 5.3.3. Often the noise figure is referenced to the noise generated by the preamplifier input termination resistor.

Low-noise preamplifiers are especially useful in making broadband measurements at low levels with a spectrum analyzer, for the preamplifier can drastically increase the signal-to-noise ratio obtainable with the spectrum analyzer alone.

### 9.2.4  EMI Receivers

Some of the advantages and disadvantages of the EMI receiver, sometimes referred to as an EMI meter, compared to the spectrum analyzer are discussed next.

All the advantages of the preselector are built into the EMI receiver, without the need for an add-on unit with its additional cost and space requirements. In EMI receivers, fundamental mixing is always used, whereas in some spectrum analyzers a combination of fundamental mixing and harmonic mixing is used. *Fundamental mixing* means that the local oscillator, which is used in a super-heterodyne receiver to down-convert the input signal to the IF, covers the entire input frequency range. Fundamental mixing results in the lowest conversion loss and the lowest noise figure or highest sensitivity. *Harmonic mixing*, on the other hand, results in lower sensitivity. In measuring impulsive noise, the charge time of the detector is important, for to capture the peak amplitude the charge time should be less than $1/(10 \times \text{IF BW})$. EMI receiver manufacturers claim that the receiver has a faster charge time than the spectrum analyzer. This may not be an intrinsic function of the receiver, and in choosing between the receiver and analyzer a comparison of the charge times should be made. Spectrum analyzers provide a narrowband continuous wave (CW) calibration signal, whereas most receivers contain impulse generators in addition to a CW source. Many receivers contain AM and FM detection, as do some spectrum analyzers. AM or FM detection is extremely useful in determining the source of a signal, especially in an ambient site survey, and also in determining the susceptibility of the system exposed to the signal. For example, a system may well be more susceptible to an audio frequency AM compared to a CW unmodulated signal. For completeness, phase demodulation should be, but seldom is, provided. In most radiated susceptibility measurements made in a shielded room, the measuring instrument monitoring the susceptibility test level is located outside of the shielded room and is thus not exposed to the test level. For the rare instances where the monitoring equipment is located inside the shielded room, the typically superior shielding of the EMI receiver will be an advantage and will reduce the spurious response of the instrument to the field in which it is immersed. Likewise, where diagnostic measurements are made of radiated or conducted emissions from an EUT with the measuring instrument located in the shielded room, the lower level of radiated emissions typical of an EMI receiver may be an advantage. As with all emission tests, the ambient inside the room, which includes radiation from the measuring instrument, with the EUT powered down should be monitored. One important feature of the spectrum analyzer or receiver is the capability of programming the instrument, which is of benefit in automating measurements. A typical program will allow entry of the calibration curves of antennas and probes, from which the true noise levels may be calculated and displayed against the specification limits.

The modern receiver combines all the advantages of the spectrum analyzer and the EMI receiver. For example, the Rohde and Schwarze ESMI receiver has a 20-Hz to 26.5-GHz frequency range, extendable to 110 GHz with external mixers.

It has a spectrum analyzer display and selectable spectrum analyzer (overview mode) 3-dB resolution bandwidth or an EMI receiver 6-dB resolution bandwidth. Correction factors can be entered into a transducer table to compensate for antenna factors and current probe transfer impedance. The ESMI includes an built-in selectable preamplifier with a gain of 10 dB from 100 Hz to 26.5 GHz. The 16 RF preselector filters are fixed bandpass from DC −9 kHz, 9–150 kHz, 150 kHz to 2 MHz, 2–10 MHz, 10–30 MHz, 30–50 MHz, 50–80 MHz, 80–100 MHz, 110–140 MHz, 140–260 MHz, 260–450 MHz, 450–700 MHz, 700–1000 MHz, 1–1.9 GHz, 1.9–5 GHz, and 4.9–26.5 GHz. The filter is a YIG type. AM or FM demodulation is provided, and the instrument can be controlled via a RS-232-C, IEEE 488, or parallel (Centronics) interface. Rohde and Schwarze also supply an EMI measurement software package to control the receiver.

### 9.2.5  Signal Generator and Power Amplifiers

The best type of signal source in a radiated or conducted susceptibility test is the analog sweep generator or a frequency synthesizer that can sweep continuously over the frequency range of interest. The disadvantage of the analog type of generator is that typically the sweep must be made manually. A frequency synthesizer–based generator, on the other hand, may usually be programmed via pushbuttons on the front panel or via a computer, using a computer interface such as the GPIB bus. The more modern frequency synthesizer sweep generator is similar to an analog generator and sweeps continuously, whereas the older or simpler types sweep by stepping the frequency. The susceptibility of an EUT often occurs at a single frequency or a limited number of frequencies. This selectivity is commonly a result of increased current flow in circuits or cables at resonance. Depending on the Q of the resonant circuit, the susceptibility may be seen only at a single frequency, with a sharply decreased response either side of that frequency. Use of a generator that must be stepped in frequency, despite a small step size, increases the possibility that a resonant frequency at which susceptibility occurs may be missed. A potential disadvantage of the frequency synthesizer generator is the typically high level of harmonics generated compared to an analog generator. Power amplifiers are required to generate the specified level of E field in a radiated susceptibility test and the specified voltage at the input of the EUT in a conducted susceptibility test. Some examples of the frequency range and typical output powers of amplifiers are provided in Table 9.2. An amplifier is available from Dressler that is rated at 75 W from 9 kHz to 250 MHz, and this is capable, with the majority of antennas and injection probes, of generating the EN 50081-2 C/M RF test level and the RF electromagnetic field of EN 50081-2 and EN 50082-1 up to 250 MHz.

The power required to generate a specified E field from an antenna is a function of the gain of the antenna and the distance between the antenna and the EUT. MIL-STD 462 and DO-160 specify a distance of 1 m. An equation that provides the minimum power required for a specified E field is:

$$P = \frac{E^2 4\pi r^2}{Z_w G}$$

where

$r$ = distance [m]

$Z_w$ = wave impedance at the distance $r$ [Ω]

$G$ = gain of the antenna

Some of the important factors in the choice of a power amplifier are unconditional stability from an open-circuit to a short-circuit load and any value of inductive or capacitive load. When a power amplifier exhibits a high gain (40 dB or greater), great care must be taken in routing and shielding of the input and output cables. If these cables are too close together or not ade-

**Table 9.2**  Typical Power Amplifiers

| Frequency range | Power output | Manufacturer/s |
| --- | --- | --- |
| 10 Hz–20 kHz | 1200W | Techtron |
| DC–1000 MHz | 50W | Amplifier Research |
| 10 kHz–100 MHz | 1250W | Amplifier Research |
| 10 kHz–75 MHz | 100W–2000W | IFI |
| 9 kHz–220 MHz | 75W | Dressler |
| 10 kHz–220 MHz | 200W | Amplifier Research |
| 10 MHz–1200 MHz | 10W | LCF |
| 100 MHz–1000 MHz | 500/1000W | Amplifier Research |
| 200 MHz–1000 MHz | 10W–400W | IFI |
| 10 kHz–1000 MHz | 10W–100W | IFI |
| 1 MHz–100 MHz | 5000W–15000W | IFI |
| 0.8 GHz–2G Hz | 50W | ITS Electronics Inc. |
| 1–4.2 GHz | 10W | Amplifier Research |
| 1–2 GHz | 200W | Amplifier Research |
| 2–4 GHz | 200W | Amplifier Research |
| 4–8 GHz | 200W | Amplifier Research |
| 8–18 GHz | 200W | Amplifier Research |
| 1–2 GHz | 20W | CPI (formerly Varian) |
| 2–4 GHz | 20W | CPI |
| 4–8 GHz | 20W | CPI |
| 8–18 GHz | 20W | CPI |

quately shielded, positive feedback may occur, with a potential for the generation of full output voltage across the load. When the load is an antenna, hazardous levels of E field may be generated, even though the input from the signal generator is set to a very low level or even disconnected. Routing the input cable too close to the antenna is a common source of positive feedback. In radiated susceptibility tests, locate the power amplifier outside the shielded room and use a short cable to connect the signal generator to the amplifier. A good power amplifier generates low levels of harmonics, spurious responses, and low levels of broadband noise. A front panel meter that indicates power output is a useful function, especially in avoiding input overload and in monitoring instability. Instability and positive feedback can destroy the power amplifier. But even more important, it can destroy the EUT due to the potentially high level of E field. When monitoring the output power of an amplifier by use of a power meter, EMI receiver, or spectrum analyzer, use a power attenuator at the input of the measuring instrument capable of dissipating the full output power of the amplifier or a bidirectional coupler. When using an attenuator, the level of attenuation depends on the input power rating of the measuring instrument. For example, a typical spectrum analyzer 50-$\Omega$ input is rated at 1 W. Consider that the output level of a 400-W power amplifier must be adjusted to a specified level, which is monitored on a spectrum analyzer. The maximum input level to the spectrum analyzer is 1 W, and the power amplifier may generate up to 25% more power than the rated value of 400 W (i.e., 500 W). The level of attenuation required is given by 10 log $(P_{in}/P_{out})$ = 10 log 500/1 = 27 dB. The voltage attenuation required is given using $V = \sqrt{50P}$ as follows:

$$20 \log \frac{\sqrt{50P_{in}}}{\sqrt{50P_{out}}} = 20 \log \frac{\sqrt{50 \times 500}}{\sqrt{50 \times 1}} = 27 \text{ dB}$$

The power and voltage attenuation are the same, because the attenuator input impedance and the spectrum analyzer input impedance are the same, 50 Ω. Thus, in our example a 26-dB attenuator, which is a common value, rated at 500 W would be used between the output of the power amplifier and the input of the spectrum analyzer. To be ultrasafe, an additional 3-dB attenuator may be added to ensure achieving at least 27 dB of attenuation.

## 9.2.6  Current Probes

Current probes used in EMC measurements are somewhat different from the current probe used with oscilloscopes. The major difference is that EMC-type probes usually have a larger-diameter opening, thus allowing measurements on larger-diameter cables. They do not measure DC currents but frequencies as low as 20 Hz and as high as 1 GHz. Current probes can accurately measure microamp levels of AC current in the presence of up to 200 A of DC current. When the current probe frequency response lies much above 60 Hz, the probe is capable of measuring low levels of AC current at high frequency in the presence of high levels of 60-Hz AC current. No single current probe can measure over the 20-Hz to 1-GHz range of frequencies; instead, a number of probes must be used. Two probes are the minimum required to cover this wide frequency range, with one probe typically covering a 20-Hz to 200-MHz frequency range and a second probe covering the range of 20 MHz to 1 GHz. The current probe does not contact the conductors over which it is placed; instead, a voltage is generated by the magnetic field surrounding the cable by transformer action. The current probe when placed over a cable that contains a signal line and return or many signal lines and returns measures the common-mode current but not the differential-mode currents. As we have seen, the major contributor to cable radiated emissions is the common-mode current flowing on cables or cable shields. The calibration of the probe is typically of transfer impedance, which is the voltage developed across a 50-Ω load divided by the current flow measured by the probe.

For example, when the transfer impedance ($Z_t$) at a specific frequency is specified as 0 dB Ω (i.e., 0 dB above 1 Ω), the transfer impedance is 1 Ω. An example of the use of $Z_t$ follows: If the current measured is 1 μA, then the voltage developed across the 50-Ω load is

$$V = IZ_t = 1 \ \mu A \times 1 \ \Omega = 1 \ \mu V$$

Such a low level of voltage is likely to be in the noise floor of the spectrum analyzer, and thus a preamplifier must be used. This type of current probe, regardless of whether it is used with an oscilloscope or a spectrum analyzer, is designed to be loaded with 50 Ω.

When the transfer impedance is provided in dBΩ, this is subtracted from the measurement in dBμV to obtain the current in dBμA. For example, if the transfer impedance is −20dBΩ and the measured level is 25dBμV, then the current is 45dBμA (25dBμV − −20dB = 45dBμA). If the transfer impedance is 5dBΩ and the measurement is 25dBμV, then the current is 20dBμA.

Although the current probe is typically shielded, it does respond to incident E fields. Thus any measurement using the current probe should commence with a measurement of the ambient (i.e., with the current probe lying close to the cable to be measured but not placed around it). When the coaxial cable connecting the probe to the measuring equipment is acting as an antenna and responding to radiated fields, the use of baluns on the cable may reduce the level of pickup. Another technique often effective in reducing pickup on the cable is to add an additional over-braid and thus increase the shielding effectiveness of the cable.

Some manufacturers provide a calibration in terms of the transducer factor $k$, which is equivalent to $1/Z_t$. The transducer factor in dB has to be added to the measured level in dBμV to obtain the current in dBμA.

### 9.2.7  Magnetic Field Antennas

The construction of these antennas is described in Section 2.6 and their use in Section 9.3.2. The construction of a 13.3-cm shielded loop antenna is described in MIL-STD-462 for RE01 measurements 30 Hz to 30 kHz and in MIL-STD-462D:1993 for RE101 measurements from 30 Hz to 100 kHz. Magnetic field antennas are used in a similar manner to the current probe with a preamplifier and spectrum analyzer. They are shielded or balanced to provide attenuation of the E-field-induced voltage, but the level of attenuation is limited. The magnetic field antenna may be used in close proximity to a shielded enclosure or cables, and the purpose is to ''sniff'' around apertures and seams for leakage, indicating a faulty joint or increased joint transfer impedance. The H field antenna is more sensitive to the level of electromagnetic ambient than is the current probe and should be used in as low an ambient as feasible.

### 9.2.8  Broadband Antennas

Broadband antennas used in EMI measurements differ from broadcast antennas in a number of aspects. High directivity in a broadcast antenna is often a desirable parameter, whereas in EMI measurement antennas with too high a directivity are a disadvantage. In radiated emission measurements the sources of emissions may be 2 m or more apart, which necessitates movement of the antenna and subsequent multiple scans when the antenna directivity is too high. Likewise, in radiated susceptibility tests the antenna should generate as uniform an E field over the EUT and interconnecting cables as possible with the antenna located as close as 1 m from the EUT. The antennas are designed to cover as wide a frequency range as possible in order to reduce

**Model 3301B Rod Antenna**

**Figure 9.5a**  EMCO Model 3301B active rod E field antenna (shown without counterpoise). (Reproduced with kind permission from the Electro-Mechanics Co. Ltd.)

the number of antennas required to cover the typical 14-kHz–18-GHz frequency span used in E field susceptibility tests. EMI measurement antennas are designed to both radiate and receive, with the exception of active antennas, which are designed to receive only. Thus, calibration curves of both the gain and the antenna factor (AF) of the antenna are provided by the manufacturer, to enable the calculation of the power level required to generate the specified E field and to convert the measured antenna output voltage to an E field. The gain and the antenna factor are either calculated for far-field conditions, measured on an open-field test site, free space antenna range and measured at distances of 1 m, 3 m, or 10 m from the source. Low-frequency, 10-kHz–30-MHz antennas are often calculated for far-field conditions, thus, when used at a distance of 1 m from a source, the AF may be very much different from the calculated value. The AF is also very sensitive to the incident wave impedance. Using antennas in a shielded room can drastically alter the calibration, due to loading effects caused by proximity to the floor and ceiling of the room as well as the EUT. In addition, room resonances and reflections change the apparent gain and AF of antennas, as discussed in Section 9.5 on shielded rooms. Another factor is the variable input impedance of the antenna with changing frequency. EMI measurement antennas are specified with a nominal input impedance of 50 Ω, whereas the actual impedance of the antenna may be much different, resulting in a high voltage–standing-wave ratio

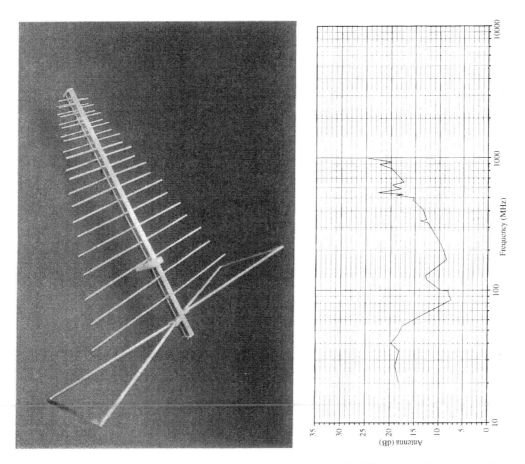

**Figure 9.5b**   Combined log periodic and biconical antenna, 20–1300 MHz, with typical antenna factor vs. frequency. (Reproduced with kind permission from EMC Consulting Inc.)

## Antenna Factor

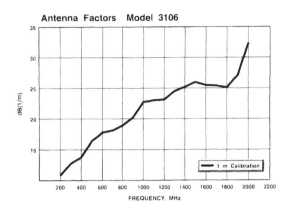

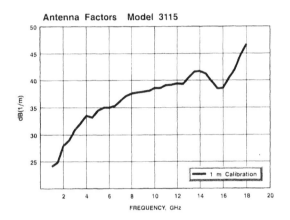

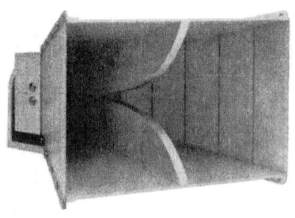

**Model 3115 Double-Ridged Waveguide Horn Antenna**

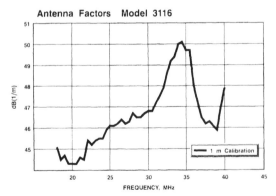

**Model 3116 Double-Ridged Waveguide Horn Antenna**

**Figure 9.5c**   Antenna factor vs. frequency curve for three models of double-ridged guide antennas. (Reproduced with kind permission from the Electro-Mechanics Co. Ltd.)

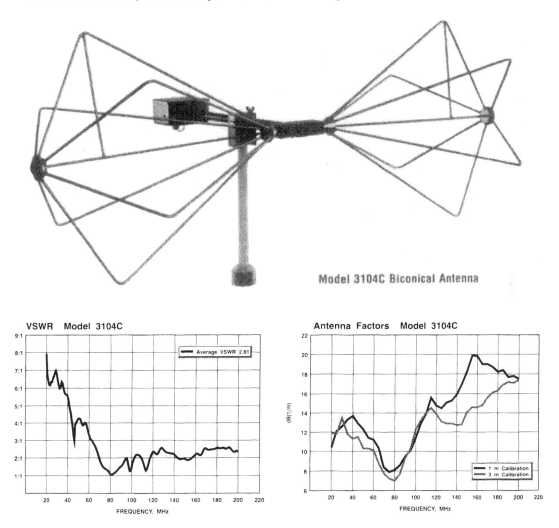

**Figure 9.5d** Antenna factor vs. frequency for a 20–200-MHz biconical antenna. (Reproduced with kind permission from the Electro-Mechanics Co. Ltd.)

(VSWR). A high VSWR affects the power amplifier driving the antenna in ways that are unique to the amplifier in use. When the antenna factor calibration is based on measured data, the antenna impedance is accounted for in the calibration, assuming the curve is not idealized (that is, when the sharp dips and peaks commonly seen in an antenna calibration are not smoothed out). Some figures and curves of antenna factor and gain for typical EMI measurement antennas are shown in Figures 9.5a, b, c, and d.

## 9.3 DIAGNOSTIC MEASUREMENTS

Diagnostic measurements are used to locate the source of emissions and the area of susceptibility. In addition, measurements on circuits and equipment may be used in an EMC prediction to provide the magnitude of fields, voltages, or currents on which an analysis is based. Measurements on internal wiring, PCBs, enclosures, and cables may be used in an equipment-level, subsystems-label, or system-level EMC prediction.

### 9.3.1   Radiated Measurement

The most useful antennas for diagnostic measurements are physically small and preferably directional. The small antenna is able to pinpoint the source of radiation to a cable, aperture, or seam in an enclosure and even down to a track or IC on a PCB, when the antenna is sufficiently small. Measurements that are close to the source, and are therefore predominantly near-field, may be made with the small antenna. The advantage of near-field measurements are that the measurements are often less affected by the electromagnetic ambient; however, even near-field measurements may not adequately discriminate against the ambient. For example, assuming equipment must meet the low-level emission requirements of MIL-STD-461, or similar, and the near-field measurements must be made in an unshielded room, it is common for currents induced on cables and equipment by the electromagnetic ambient and conducted on the power cord to the equipment enclosure and cables to result in radiation levels that exceed and mask the equipment emissions. It is therefore imperative, when measurements must be made outside of a shielded room, to compare the levels with the equipment powered down to those with the equipment powered up. It may be necessary either to move the equipment to a quieter location, to lower the ambient by switching off other equipment, or to surround the equipment with absorber foam (useful above 200 MHz), to obtain EUT emissions above ambient measurements.

### 9.3.2   Magnetic Field Measurements

Emission specifications typically place limits on the E field; however, because an H field is always present when the fields are time varying, H field measurements are indicators of the potential E field magnitude. The advantage of H field measurements, compared to E field measurements, are lower sensitivity to reflections and room resonances and lower sensitivity of the probe to metallic structure proximity effects.

Figure 9.6 illustrates the typical fields around a cable connected to a piece of equipment. The cable is located above a ground plane on a nonconductive table, which is a typical commercial test configuration. If we assume a common-mode current flow on the cable, then the resultant fields are a longitudinal E field and a circumferential H field. Additional fields are present due to displacement current flow.

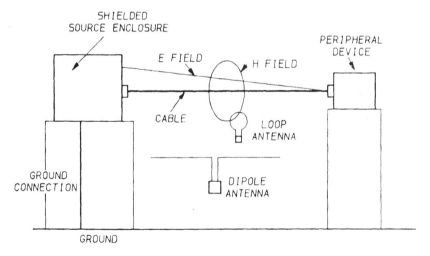

**Figure 9.6**   Typical fields around a cable connected to a piece of equipment placed on a nonconductive table.

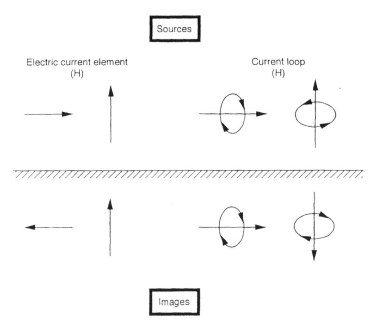

**Figure 9.7a** Orientation and relative phase of the electric current element and current loop magnetic field sources and images in a perfectly conducting ground plane.

Displacement current may flow between the cable and the floor, as well as the ceiling when measurements are made in a shielded room. The additional fields are then an electric field tangential to the cable and a magnetic field in the same plane as the cable. Numerous measurements have shown that it is the longitudinal E field and the circumferential H field that typically predominate. However, this is not always the case, and when making H field measurements, with the types of loop antenna described in Chapter 2, the loop should be oriented both horizontally and vertically, referenced to the cable under test. Note that the orientation of the plane of the loop must be 90° to the field to be measured. The circumferential H field above a perfectly conducting ground plane results in an image (reflection) in the ground plane that is out of phase with the source, whereas the image due to the horizontal (displacement current) field is in phase with the source. Figure 9.7a illustrates both the source and the image for the electric current element and the current loop. At distances from the source less than half a wavelength, the source- and image-contributed fields add vectorially, this addition is illustrated in Figure 9.7b.

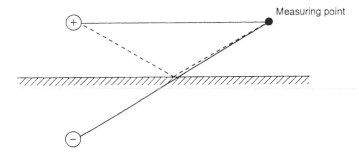

**Figure 9.7b** Direct path, reflected path, and image for a horizontal electric current element.

The total magnetic field is the sum of the source and image generated fields. The H field from a cable is approximately that of the electric current element (Eq. 2.10), for a loop it is given by Eq. (2.24) When $dD$ is the direct path length and $dR$ is the indirect path, the total field is given by $H(r = dD) + mH (r = dR)$, where $r$ is the distance used in Eq. (2.10) or Eq. (2.24), $m = \pm 1$, depending on the orientation of the electric current element or loop. For example, a horizontally oriented cable (which may be modeled as an electric current element) generates an image out of phase with the source and $m = -1$. For the displacement-current-generated field, the source and image are in phase and $m = +1$. In contrast, when an electric current element is vertical above a groundplane (the monopole antenna is a good example), the magnetic field is circumferential and the E field starts on the rod of the monopole and finishes on the ground plane. Thus the image is in phase with the source. The relative magnitudes of the vertical (displacement) E field and horizontal (electric current element) E fields may be measured with a small antenna, typically a bowtie from 20 to 250 MHz or a tunable monopole from 250 MHz to 1 GHz. From a measurement of the polarization of the predominant E field, the predominant H field is known. For example, when an antenna is oriented to intercept a horizontally polarized E field and measures a higher level than when vertically polarized, the E field is horizontal, the H field is circumferential, and the image subtracts. When the direct path length $dD$ or the indirect path length $dR$ is greater than half a wavelength, the relative phase of the two fields may be calculated, with the resultant total field the vector sum of these. For example, when the direct path length is different from the reflected path length by exactly half a wavelength, the direct and reflected fields at the measuring point are in phase for the horizontal current element, and out of phase for the vertical current element. This attribute is used in an antenna in which a small fixed frequency dipole is located horizontally 1/4 wavelength above a metal reflector. This ensures that the source and image fields in front of the dipole are in phase.

Another measurement technique useful in evaluating the E or H field radiated by a cable is to measure the current flow on the cable by use of a current probe and to use the measured current in Eq. (2.10) to obtain the field from a cable far removed from a ground plane or Eq. (7.28) for a cable above a ground plane. For greatest accuracy, the "typical" calibration curve provided by most probe manufacturers should be replaced by a calibration of the probe, using a setup representative of the cable under tests. At high frequencies, where the cable length may be greater than half a wavelength, the probe should be moved up and down the cable until a maximum reading is obtained. A source of inaccuracy that must be accepted is the loading effect of the current probe on the cable, which is likely to increase the cable current flow. Figure 9.8 illustrates the use of the current probe in evaluating the radiated emissions from an interface or power cable. A typical test method for use with the current probe is as follows.

### Current Probe Test Method

*Step 1*: The spectrum analyzer frequency sweep should be set to the frequency range of the probe in use, initially with a 100-kHz resolution bandwidth.

*Step 2*: A measurement of the ambient with the EUT switched off and the current probe placed on the nonconductive work surface should be made and a record taken using either a plotter or an oscilloscope camera.

*Step 3*: With the EUT switched on and the current probe placed over the cable, monitor the emissions displayed on the spectrum analyzer and move the current probe on the cable in order to maximize the amplitude displayed.

*Step 4*: Ensure that no compression has occurred in either the preamplifier or the spectrum analyzer, as described in Section 9.2.3.

*Step 5*: Where the amplitude displayed on the spectrum analyzer is in dBm, convert to voltage: $V = \sqrt{10^{dBm/10} * 1 \times 10^{-3} * 50}$.

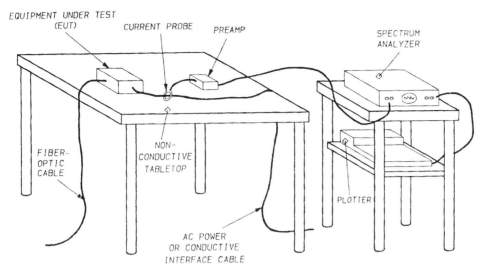

**Figure 9.8**  Test setup for measurement of cable current flow.

*Step 6*: Read the maximum amplitude displayed on the spectrum analyzer and note the frequency. From the current probe transfer impedance calibration curve, read the transfer impedance at the frequency of the emission under investigation and convert from voltage to current using

$$I = \frac{V}{Z_t}$$

*Step 7*: Convert the measured current to an E field at the specification distance using Eq. (2.10) or Eqs. (7.26, 7.28 and 7.30).

The magnitude of the H field from a cable above a ground plane is obtained by use of Eq. (7.28), from which the E field magnitude may be obtained by multiplying by the wave impedance. Compare the magnitude of the E field to the E field limit. The transfer impedance will be low at some frequencies, which means the current amplitude displayed at those frequencies will also be low. It is therefore important that any peak emissions at frequencies where $Z_t$ is low also be converted to current measurements, from which the E field is calculated and compared to the limit.

Repeat steps 1–7 using the remaining current probe/s required to cover the 14-kHz–1-GHz frequency range for equipment that must meet the MIL-STD requirements or the 30-MHz–1-GHz range for FCC requirements.

Using a combination of current probe measurements and small magnetic field and electric field antennas to conduct measurements in a shielded room, a correlation of, on average, 6 dB and, worst-case, 13 dB to the measurements obtained on one specific open-field test site measurements has been achieved. However, a number of different measurements are required to identify room resonances and reflection, these are described in detail later in this chapter. Choice of the optimal measuring distance and correct orientation of the antenna is of paramount importance in achieving maximum accuracy. In the near field, the predicted reduction in the H field from a loop antenna is a function of $1/r^3$, whereas the H field reduction for an electric current element (cable) is a function of $1/r^2$. From measurements made using a transmitting

**Table 9.3**  Reduction in Measured Magnetic Field Compared to Predicted.

| Freq. [MHz] | $\lambda/2\pi$ [m] | 0.1–0.2 m | | 0.2–0.3 m | | 0.3–0.5 m | |
|---|---|---|---|---|---|---|---|
| | | Meas. [dB] | Pred. [dB] | Meas. [dB] | Pred. [dB] | Meas. [dB] | Pred. [dB] |
| 20 | 2.380 | 18 | 18 | 7 | 10.5 | | |
| 100 | 0.470 | 15 | 18 | 4 | 10.5 | 5.3 | 13 |
| 220 | 0.217 | 9 | 18 | 0 | 3.5 | | |

loop and a receiving loop in a shielded room it has been found that the near-/far-field interface does not occur abruptly at $\lambda/2\pi$, rather, the $1/r^3$ law changes gradually through $1/r^2$ to $1/r$ in the far field.

Table 9.3 illustrates the reduction in measured magnetic field compared to the predicted, based on a $1/r^3$ reduction in the near field. The calculation used in obtaining the reduction is given in the following example: The initial measuring distance is 0.2 m and the loop is moved to 0.3 m; the reduction in dB is given by

$$20 \log \frac{1}{(r_1/r_2)^3} = 20 \log \frac{1}{(0.2 \text{ m}/0.3 \text{ m})^3} = 10.5 \text{ dB}$$

When near-field measurements are confined to finding a source of emission (e.g., cable or enclosure) or to measuring the effectiveness of EMI fixes, relative measurements are adequate. Often it is required to know how close the emissions from a piece of equipment are to the commercial specification limits, without removing it to an open-field test site or without removing a piece of equipment to a shielded room for MIL-STD-462–type measurements. Using small E and H field antennas we may measure the fields directly, or with the current probe we can measure current and predict the fields from equipment enclosures and cables. When H field measurements are made in the near field, we may extrapolate out to 1-m, 3-m, or 150-m distances, whichever the specification limit requires. The distance from emission source to measurement is measured and the near-/far-field interface distance (FFID) is calculated. From a cable the reduction in H field is assumed to be a function of $1/r^2$, and for magnetic field coupling due to current flow on the inside of the enclosure it is $1/r^3$ out to the FFID distance. For wire sources inside the enclosure, when the circuit impedance is less than 377 $\Omega$, the source is predominantly magnetic and $1/r^3$ should be used. For wires connected to circuits with impedances higher than 377 $\Omega$, the $1/r^2$ law should be used. When the specification distance is in the near field, the E field can be obtained from the magnitude of the H field calculated at the specification distance (SD) times the wave impedance. When SD is in the far field, to obtain the reduction from the FFID to the specification distance, use Eq. (9.1). When the measuring distance (MD) is greater than $\lambda/2\pi$, replace FFID in Eq. (9.1) with MD to obtain the reduction in H field in dB.

$$20 \log \frac{1}{\text{FFID}/\text{SD}} \tag{9.1}$$

The specification distance E field in dB$\mu$V/m from the FFID magnetic field is

$$E \text{ [dB}\mu\text{V/m]} = H_{\text{FFID}} \text{ [dB}\mu\text{A/m]} - \text{SD reduction [dB]} + 51.5 \text{ dB} \tag{9.2}$$

or numerically

$$E \ [\mu V/m] = H_{\text{FFID}} \ [uA/m] \times \frac{\text{FFID}}{\text{SD}} \ [\text{reduction numeric}] \times 377$$

In practice one or more emissions at the same frequency can emanate from different sources (e.g., more than one cable and aperture and seams in the enclosure). The far field is therefore a composite of these separate sources. The composite H field, $H_t$, may be calculated from

$$H_t = \sqrt{H_1^2 + H_2^2 + H_3^2} \tag{9.3}$$

where $H_1$, $H_2$, and $H_3$ are separate sources.

In practice, if the measurements are conducted over a long time period, the sources can add algebraically. In comparisons between measured far E fields and E fields obtained from near H field measurements, the prediction magnitude is usually lower than the measured. This may be the result of not adding the separate sources of fields, or it may be due to the use of the $1/r^3$ law out to the far-field interface distance, whereas the $1/r^3$ law should be modified close to the far-field interface distance. Conversely, near E field measurements tend to be high when extrapolated out to a far-field distance and compared to far-field measurements. The major reason for this is thought to be the room resonance and reflections inherent in a shielded-room measurement, described in Section 9.5.1. In addition, the electric current element model is strictly true only for distances of greater than half a wavelength from a ground conductor. Also these reductions are strictly valid for point sources.

In making H field measurements on equipment that must meet the low-level MIL-STD-461 specification limits, a preamplifier and a spectrum analyzer are invariably required. The sensitivity of an oscilloscope is typically not adequate except for measurement 0.5–0.2 m from a source on equipment required to meet the relatively high commercial specification limits.

### 9.3.3  Conducted Measurements

In this section we use the term *conducted measurements* to cover the direct measurement of voltage or current or the indirect measurement of current using the current probe. This type of measurement is useful in locating the source of an emission at a specific frequency and in ensuring that a circuit or supply is not oscillating. To infer that a high magnitude of measured voltage or current will necessarily result in a high level of radiation is not necessarily true, due to the importance of the circuit geometry in determining radiation levels. Nevertheless, conducted measurements, typically using the oscilloscope, are extremely useful in determining the presence of differential- and common-mode noise. When the equivalent circuit of a coupling path can be determined and the applied voltage or current is measured, the EMI voltage at the receptor may be calculated and the need for modifications to the coupling path can be determined.

### 9.3.4  Susceptibility/Immunity Tests

Diagnostic susceptibility/immunity testing is useful when a piece of equipment has failed a radiated susceptibility/immunity test requirement and the susceptible cable or enclosure side has to be found. Another use is when searching for an intermittent conducted susceptibility problem. If spikes or a sinewave are injected onto a power or signal interface, which is a likely candidate for examination, and the occurrence of the intermittent fault increases, the chances are that the source of the noise has been found. The injection of an additional noise level is best accomplished by use of an injection transformer wound on a ferrite toroid core. By placing only a few turns on its secondary, the additional low impedance in series with the power or signal line is less likely to affect the circuit under test. The primary should typically have more windings than the secondary and enough to represent a sufficiently high source impedance to the signal generator or power amplifier. Although injecting an additional noise level is indicative of susceptibility, it is no guarantee that the problem has been found. The next step is to measure the noise level and to note the frequencies at which susceptibility occurs. A filter may then be designed to achieve at least 20 dB of attenuation at the susceptibility frequencies, unless these are close to the signal frequencies. If, by the insertion of the filter, the intermittent fault disappears, then the susceptible circuit has been located. If the filter degrades a signal or cannot be permanently incorporated into the equipment, then the source of noise must be found. Clues as to the source are contained in the susceptibility test level and frequencies. In addition, the level and type of noise, differential or common mode, present without the additional injected noise helps in locating the source. If the noise is common mode, an RF capacitor between signal or power ground and chassis may suffice. Alternatively, on signal interfaces an isolation transformer with electrostatic shield between the primary and secondary or an optoisolation circuit may be used to reduce the common mode coupled into the receiver circuit. When the noise is differential, the noise is injected into the signal or power line at some stage, and the location and source must be found.

As described in detail in Chapters 4 and 5, the noise may be injected via radiated coupling, crosstalk, common ground coupling, or a noisy source signal. Radiated susceptibility may be determined by using a loop to generate a magnetic field. The loop is driven by either a source of wide-frequency-range noise, such as a squarewave, or a sweep generator that sweeps the frequency over the range at which susceptibility was either demonstrated or suspected. Above approximately 20 MHz, a simple one-turn loop or the balanced loop antenna described in Chapter 2, Section 2.6 may be used to generate the field. At low frequency, a multiturn loop should be used to ensure that an adequately high impedance is presented to the signal source. Some power amplifiers are specified as capable of driving a short circuit; however, based on the high mortality of power amplifiers, an impedance of at least a few ohms at the lowest frequency is recommended. Certification radiated susceptibility tests are typically E field, although in the MIL-STD-461 RS04 test an H field is used. Typically it does not matter whether the current flow on a cable or enclosure is caused by an H field or an E field source during a susceptibility test. If equipment is suspected of susceptibility to specifically a low-frequency E field, then a small wand, monopole, or bowtie antenna may be used; however, it is very difficult to generate adequate E fields with these antennas. Above approximately 250 MHz, small directional antennas, such as the conical log spiral or conical spiral, may be used to radiate a specific location and thus determine the susceptible cable or equipment. The antenna is used in close proximity and pointed at the location under investigation.

## 9.4  COMMERCIAL EMI REQUIREMENTS AND MEASUREMENT

Commercial requirements are typically confined to limits on conducted and radiated emissions. However, with the advent of the European Union (EU) EMC Directive, mandatory and routine immunity (susceptibility) test requirements were imposed for the first time on commercial equipment. The EMC requirements encompass equipment from RF arc welders to electric drills, household equipment, and digital devices. The definition of *digital devices* is often very broad. For example, the Federal Communications Commission (FCC) defines a digital device as any equipment containing a clock at a frequency above 9 kHz using digital techniques. Thus a pinball machine or similar device may qualify. The prime purpose of the FCC requirements contained in subpart B of part 15 of its rules is to reduce emissions to levels that do not interface with nearby radio and TV reception. In a great many countries, people own a radio or TV or both, and it is to reduce complaints from owners that the majority of commercial EMI requirements are imposed. In this section, only requirements and test methods imposed on digital devices are considered.

### 9.4.1  FCC Rules on Emissions from Digital Devices

Compliance with the FCC requirements is a prerequisite for legal importation and marketing of such apparatus in the United States. The technical requirements for limiting unwanted RF emissions from unintentional radiators, such as computers, videocassette recorders, and microwave ovens, and intentional radiators are contained in Part 15 of the FCC Rules (47 CFR Part 15). The rules for Industrial, Scientific, and Medical Equipment are contained in Part 18 (47 CFR Part 18).

Specific devices are exempt; these include: transportation vehicle, hydro power, digital device used exclusively as industrial, commercial, or medical *test* equipment or included in an appliance (microwave oven, dishwasher, clothes dryer, air conditioner (central or window), etc.), digital devices that have a power consumption not exceeding 6 nW, joystick/mouse, devices that use or generate a clock below 1.705 MHz *and* are not connected to AC power.

Devices that are connected only part of the time via a battery charger or that obtain AC power through another device are *not* exempt from conducted emission measurement. If one device is connected to the public utility AC power and supplies DC power to a second device (or a number of devices), then a conducted emission measurement is made on the AC power-line side, and the composite emissions from all devices, which include those on the DC power lines that are reflected or conducted through the AC to DC power supply, must meet the emission limits. Thus a wall-socket-mounted AC-DC power supply may have an FCC sticker that says that it meets the FCC requirements, and yet that power supply must be retested with a load device for conducted emissions.

As yet the FCC has not set immunity requirements, although section 302 of the Communications Act of 1934, as amended, gives the FCC authority to regulate the immunity of home entertainment products. The FCC also has the authority to regulate any device that in its operation is capable of causing harmful interference to radio communications, i.e., a broad power of regulation.

The FCC places different requirements on digital devices, depending on their intended use. Devices marketed for use in a commercial, industrial, or business environment, exclusive of a device marketed for use in the home or by the general public, are classified as Class A. Devices marketed for use in a residential environment, notwithstanding use in a commercial, business, or industrial environment, are classified as Class B. The rationale is that commercial equipment is typically located further away from TV receivers than domestic equipment, which may be used in the next room.

Examples of Class B devices include, but are not limited to, personal computers, calculators, and similar electronic devices that are marketed for use by the general public. The FCC may also classify a device that repeatedly causes harmful interference as a Class B device.

Class A emission limits are higher (less stringent) than Class B limits, which are lower (more stringent).

Tables 9.4a and b provide the radiated and conducted requirements for Class B equipment and Tables 9.5a and b for Class A equipment.

For an intentional radiator, the spectrum shall be investigated from the lowest radio frequency generated within the device, without going below 9 kHz, up to:

1.  If the intentional radiator operates below 10 GHz, to the tenth harmonic or up to 40 GHz

**Table 9.4a** FCC Class B
(Residential) Radiated Emission
Limits

| Frequency [MHz] | Distance [m] | Field strength [$\mu$V/m] |
|---|---|---|
| 30–88 | 3 | 100 |
| 88–216 | 3 | 150 |
| 216–960 | 3 | 200 |
| Above 960 | 3 | 500 |

Bandwidth of measuring instrument is not
less than 100 kHz.

**Table 9.4b** FCC Class B
(Residential) Conducted Emission
Limits

| Frequency [MHz] | Maximum RF line voltage [$\mu$V] |
|---|---|
| 0.45–30 | 250 |

Bandwidth of measuring instrument is not
less than 9 kHz.
CISPR quasi-peak function is used for both
conducted and radiated measurements.
FCC allows a 13-dB relaxation when the
conducted emission is determined as BB.

**Table 9.5a** FCC Class A (Commercial) Radiated
Emission Limits

| Frequency [MHz] | Distance [m] | Field strength [$\mu$V/m] |
|---|---|---|
| 30–88 | 10 | 90 |
| 88–216 | 10 | 150 |
| 216–960 | 10 | 210 |
| Above 960 | 10 | 300 |

Bandwidth of measuring instrument is not less than 100 kHz.

2. If the intentional radiator operates at or above 10 GHz and below 30 GHz, to the fifth harmonic or to 100 GHz, whichever is lower

3. If the intentional radiator operates at or above 30 GHz, to the fifth harmonic of the highest fundamental or to 200 GHz, whichever is lower

4. If the intentional radiator contains a digital device, to the upper frequency limit for the intentional radiator or up to the highest frequency applicable to the digital device, whichever is the higher frequency

**Table 9.5b**   FCC Class A
(Commercial) Conducted Emission
Limits

| Frequency [MHz] | Maximum RF line voltage [μV] |
|---|---|
| 0.45–1.705 | 1000 |
| 1.705–30 | 3000 |

Bandwidth of measuring instrument is not
less than 9 kHz.
CISPR quasi-peak function is used for both
conducted and radiated measurements.
FCC allows a 13-dB relaxation when the
conducted emission is determined as BB.

For an unintentional radiator, including a digital device, the lowest frequency measured shall not be below the lowest frequency for which a radiated emission limit is specified. The upper frequency limit to be measured, above 1 GHz, is:

| Highest frequency generated or used (MHz) | Upper frequency of measurement range (MHz) |
|---|---|
| 1.705–108 | 1000 |
| 108–500 | 2000 |
| 500–1000 | 5000 |
| Above 1000 | 5th harmonic of the highest frequency or 40 GHz, whichever is lower |

On any frequency or frequencies above 1000 MHz, the radiated limits are based on the use of an average detector function (see 47 CFR for further information on peak detector measurements). Unless otherwise specified, measurements above 1000 MHz shall be performed using a minimum resolution bandwidth of 1 MHz. Conducted AC power-line emission measurements are always performed using a CISPR quasi-peak detector.

Any equipment that comes under the definition of digital devices and that is used in the United States must have a representative sample tested. Most digital devices are subject to verification by the manufacturer, which is an equipment authorization process whereby the manufacturer, after testing for compliance, provides information to the user and labels the equipment with the statement required by 47 CFR 15.19.47. CFR 2.901 provides more information on this verification program. The FCC declaration of conformity (DOC) is a procedure where the responsible party makes measurements or takes other necessary steps to ensure that the equipment complies with the appropriate limits. The submittal of units or data is not required by the FCC. The DOC attaches to all items identical to that tested. If a product must be tested and authorized under a DOC, a compliance information statement shall be supplied with the product at the time of marketing or importation. The party responsible for a DOC must be located within the United States.

The description of verification does not differ in any substantive way from the declaration of conformity, but verification can be performed by a responsible party outside of the United States.

The single difference in the description is the replacement of "other" with "the" in the following: "makes measurements or takes *the* necessary steps to insure that the equipment complies." Verification shows that the device or product has been shown to be capable of compliance. A record of the measurement data and details of an appropriate test site that demonstrates compliance with the applicable regulations (ANSI C63.4), original design drawings, and description of equipment test, etc., must be made. These records shall be retained for two years after manufacture of the equipment has been permanently discontinued.

The FCC has the right to review the records and to request a sample of the equipment.

Notification is an equipment authorization issued by FCC. The applicant must make measurements and report that such measurements have been made, but neither a unit nor the data needs to be submitted to the FCC. Notification attaches to all items identical to that tested.

*Identical* means identical within the variation that can be expected to arise as a result of quantity production techniques.

Obviously any changes in software or hardware means the equipment is not identical. However, although it can affect emissions, a change in manufacturer of an IC with the same part number is allowed.

Type acceptance is a route not available for unintentional radiators but is confined to a station authorization. Type acceptance also applies only to units that are identical.

The routes to compliance for manufacturers outside of the United States are verification for Class A digital devices, peripherals, and external switching-power supplies and equipment, and the same is true for almost any Class B switching-power supplies, digital devices, and peripherals. However, if the equipment falls under the category of CB receiver, TV interface device, scanning receiver, superhet receiver, then the more onerous route of certification applies.

Certification is a bilateral equipment authorization required by the FCC for those pieces of equipment deemed to have a significant potential for causing radio interference (e.g., personal computers). The manufacturer must test its equipment for compliance and send a copy of the report of measurements, application fee, description equipment, photographs, etc., to the FCC for review and approval. The equipment must also be labeled. The FCC review process takes on average 30 days. The administrative rules are contained in 47 CFR 2.901.

The FCC label should state: "This device complies with part 15 of the FCC rules. Operation is subject to the following two conditions: (1) This device may not cause harmful interference and (2) this device must accept any interference received, including interference that may cause undesired operation."

If the Class A limits have been met, then the statement that the equipment has met the "Class A requirements of the FCC Part 15" should be included in the operating manual.

The definition of *digital device* is: an unintentional radiator that generates and uses timing signals and pulses in excess of 9,000 pulses (cycles) per second and uses digital techniques. This includes computers, data processing equipment, and other types of information technology equipment.

On 17 September 1993, the FCC modified Part 15 by allowing manufacturers of digital devices the option of using either CISPR Publication 22 or the technical standards in Part 15 to demonstrate compliance with Part 15. The EU has incorporated CISPR 22 as a normalized standard (EN 55022). There are several important caveats to this option: A slight difference in the limits exists between CISPR 22 (1993) and the Part 15 limits. Manufacturers must choose between CISPR 22 (1993) and Part 15, but cannot use a combination of both standards. Manufacturers who use CISPR 22 must use the measurement procedure C63.4, which in effect gives more detail than CISPR 22. CISPR 22 is in the process of change, and the reference to C63.4 may not be necessary in the future. The tests for line-conducted measurements only should show compliance with the U.S. voltages of 110 V/60 cycles. For digital devices with oscillators above 108 MHz, measurements must also be made on frequencies above 1000 MHz using an instrument with a peak and average detector. Both the FCC and CISPR limits below 1000 MHz are based on quasi-peak measurements. FCC measurements above 1000 MHz are based on the use of an average detector with a peak limitation of 20 dB above the average value. Measurements above 1000 MHz are required under 47 CFR 15.33.

CISPR 22 states that if the field strength at 10 m cannot be made because of high ambient noise levels or for other reasons, measurements may be made at a closer distance, e.g., 3 meters. An inverse proportionality factor of 20 dB per decade should be used to normalize the measured value to the specified distance for determining compliance. However, care should be taken in measurement of large test units at 3 meters and at frequencies near 30 MHz due to the problem of measuring close to the near field. A comparison between the Part 15 radiated and conducted emission limits and the CISPR 22/EN55022 limits (as well as other EN limits) is made in upcoming Tables 9.7, 9.8, and 9.9. The CISPR 22 Class A and Class B limits are identical to the EN 55022 Class A and Class B limits shown in the tables.

Any modification, however minor, to a circuit or software can change the emission profile. For example, a change in software, device function, or grounding shield termination or the addition of a signal interface can significantly change emissions. It is therefore recommended that the equipment be retested after any modification. The equipment verification in accordance with FCC CFR 47 applies to identical equipment. However, the FCC, in Public Notice 7, 1982, cautioned that

> The manufacturer is cautioned that many changes which on their face seem to be insignificant are, in fact, very significant. A change in the layout of a circuit board or the addition, removal or rerouting of a wire, or even a change in logic, will almost surely change the emission characteristics (both conducted and radiated) of the device. This is particularly true with a device housed in a nonmetallic enclosure. Whether this change in characteristics is enough to throw the product out of compliance can *best* be determined by retesting.

It is clear that retesting is the preferred route, but it is not mandatory, and the notice does not stipulate that this testing needs to be at verification level with plots, etc. Therefore, when the testing of every minor modification to a wide range of products becomes onerous and costly, the modifications may be reviewed by an EMI expert, or "quick look" testing can be performed. This "quick look" test for radiated emissions on an OATS and conducted emissions in a shielded room can be made without plots of the emissions and, depending on equipment complexity and functional operating modes, will typically take no longer than three hours, with the data presented in tabular form only.

The FCC routinely retests equipment to ensure continuing compliance and is empowered to fine or to seize equipment that is noncompliant. Lack of conformity in equipment is brought to the notice of the FCC through complaints of EMI, referral of competitors, monitoring of trade shows and literature, or the FCC's sampling program. Every fall the FCC investigators go to the COMDEX computer show and assess fines totaling $400,000 or more on hundreds of manufacturers that run afoul of the FCC marketing rules.

First the FCC can issue a "Marketing Citation," a letter informing the company that it has apparently violated FCC rules and perhaps requesting that a product sample be sent to the FCC laboratories. If the violations continue, or the equipment is found by the FCC to be in technical violation of the rules, a Notice of Apparent Liability will be issued. It can assess a fine of up to $10,000 for each occurrence and up to $75,000 per violator for multiple occurrences.

Historically, at this point Canadian manufacturers have usually opted to recall all devices that are available from their distributors in the United States, and apply EMI fixes.

Many compliance engineers in companies in Canada require their equipment to be 6 dB below the FCC limits, although this is not an FCC requirement, and furthermore require testing of samples in a production run to reduce the risk that any of the exported equipment will fail a retest.

If the "administrative" sanctions are ignored, then the FCC can take "judicial" measures, which are imposed only after the commission becomes convinced that the cited party will not cooperate. The penalty is: "any person who willfully or knowingly does . . . any act, manner, or a thing in the relevant sections (501, 502) of the Communications Act . . . shall, upon conviction thereof, be punished for such offense, by a fine of not more than $100,000 or by imprisonment for a term not exceeding one year." An even stronger deterrent for manufacturers outside of the United States is an order to stop the importing of the culprit equipment. But how this is covered by the North American Free Trade Agreement (NAFTA) is not known.

The test methods for verification are contained in ANSI C63.4-1992, although the FCC has accepted some alternative test methods, described later. The commission *encourages* the use of the ANSI C63.4 procedure for all testing. Any party using other procedures should ensure that such procedures can be relied on to produce measurement results compatible with the FCC measurement procedures. The description of the measurement procedure used in testing the equipment for compliance and a list of the test equipment actually employed shall be made part of an application for certification or included with the data required to be retained by the party responsible for devices authorized by a DOC, notification, or verification. The use of an FCC-recognized test facility is not a requirement, although one may assume that test data from a recognized facility is more readily acceptable to the FCC. The FCC has allowed the radiated emission measurement data using a GTEM cell that has been correlated, usually by use of a computer program, to an open area test site (OATS). However, the measurement from an OATS shall take precedence, and this is the method the FCC uses. ANSI C63.4 allows the use of a 3-m or 10-m site in an anechoic chamber, but this site must meet a more stringent normalized site attenuation (NSA) test than applicable to an OATS. Weather-proof OATS in which either the EUT or the measurement antenna and the EUT are contained in a totally nonconductive enclosure are accepted by the FCC at present, although this may change, because they have a concern about dirt buildup on the outside of the enclosure.

ANSI C63.4 replaces MP4. Section 11 of ANSI C63.4 provides specific information on the measurement of information technology equipment (ITE) and includes information on the operating conditions of hosts, peripherals, and visual display units. Additional information on the placement of tabletop systems, hosts, monitors and keyboards, external peripherals, interface cables, floor-standing equipment, combination tabletop and floor-standing equipment, conducted emission measurements, and radiated emission measurements is provided.

ANSI C63.4 specifies the use of a tunable dipole, or a dipole tuned only above 80 MHz and set to the 80-MHz tuned length when used between 30 and 80 MHz. Alternatively, broadband antennas may be used when correlatable to a dipole with an acceptable degree of accuracy. In practice, to achieve reasonable accuracy the antenna may have to be calibrated against a calibrated dipole using, typically, the three-antenna calibration method. ANSI C63.4 requires that all antennas be individually calibrated to NIST or an equivalent standards reference organization. Antennas calibrated to the methods of ANSI C63.5-1988 meet the traceability requirements. ANSI C63.5 describes two fundamental approaches to antenna calibration, both assuming a perfect reference. One approach assumes a perfect antenna whose characteristics are known to a high degree of accuracy by virtue of its standard construction. The "Roberts" tuned dipole antenna is the reference antenna; if the construction drawings are followed carefully, the antenna factor will match the predicted values to within 1 dB. A four-antenna set can be built to cover the 30–1000-MHz frequency range. The antenna factor of any other antenna may be derived by substitution with the reference antenna. In the substitution calibration, the separation distance between the transmitting antenna and the reference antenna or the antenna being calibrated is 10 m. The transmitting antenna can be any antenna, and it should be at least 2 m above the ground. The reference antenna is used as the receiving antenna, which is adjusted in height to be anywhere from 2.5 to 4 m above the ground. The height should be chosen to avoid a null by finding a location where the received signal is either maximum or is varying slowly. After the signal strength is noted with the reference antenna, it is replaced with the antenna being calibrated, located at exactly the same height and position as the reference dipole. (Scanning the receiving antennas in height while recording the maximum received signals will facilitate the measurements, especially at higher frequencies, where many peaks and nulls of the field are present). The AF of the unknown antenna can be modified by the presence of the ground, and this is minimized by the 2.5–4-m height chosen. But this is not a free-space calibration, nor does it compensate for the presence of the ground. The second approach assumes a perfect, or at least a nearly ideal, "standard site." The standard site will typically be located in a large, open field far from reflecting objects such as trees, buildings, overhead wiring, bumps, or hills and will almost certainly contain an "oversize" ground plane. It would seem that an NSA calibration using identical antennas on a "standard site" is all that is required to obtain their AF. However, ANSI C63.5 restricts the use of NSA measurements on an open-area test site that is also used for the calibration of the antennas. The argument is that site imperfections may have been erroneously assigned to the antenna factors. C63.5 allows an exception for a large open-area test site, i.e., the near-perfect site described earlier, but the antennas must be calibrated using horizontal polarizations on a propagation path independent of the path used for EMI, and the NSA, measurements. An alternative is to make two NSA measurements on two independent sites or on two independent paths on a large site, for example, two independent 10-m paths on a 30-m test site. The standard site method described in C63.5 requires three site attenuation measurements if three different antennas are used. These measurements are made, with the same distance between antennas, with the height of the fixed antenna the same and with the scan height of the second antenna the same for all three measurements.

ANSI C63.5 contains tables of the predicted maximum received ground wave field strength for the source antenna heights of 1 m and 2 m.

The three different antennas are calibrated in pairs, e.g., 1 and 2, 1 and 3, and 2 and 3.

Three equations that contain the measured site attenuations, the frequency and the maximum predicted ground wave field, from the tables, are solved simultaneously to obtain the antenna factors.

If two identical antennas are to be calibrated, then a single measurement is required and a single equation. If any one or two of the antennas are tuned dipoles and the NSA measurement is made at 3 m, then a correction, contained in ANSI C63.5, for one or both dipoles must be made.

When broadband antennas, such as the log periodic, are calibrated close to a ground plane, the presence of the ground plane can significantly change the current distribution on the antenna and its input impedance, which results in a very different antenna factor compared to a close-to-free-space calibration. Another way of explaining this effect is that the image of the antenna in the ground plane interferes with the antenna above the ground plane. This effect becomes especially noticeable when the antenna is a wavelength or multiple wavelengths from the ground plane. Thus, with an antenna 1 m above the ground plane, the antenna factor is substantially different from the free-space value at 300 MHz, although the effect is noticeable over the 200–400-MHz frequency range. In the standard site antenna factor, the search antenna height covers the 1–4-m range, so these proximity effects are accounted for in the calibration. A typical "close to free space" calibration, made at a 3-m separation, is achieved when both antennas are at a height of 4 m over a poorly conducting surface, i.e., no ground plane and the surface between the two antennas is covered in large blocks of foam absorber material or ferrite tiles.

The antenna calibration should be performed at the distance to be used; i.e., if the measurement distance is 3 m, then a calibration at 3 m is required and if measurements are also made at 10 m, then a 10-m calibration shall be used. The calibration should be made in accordance with ANSI-C63.5. If this is not the case, difficulty may be experienced in measuring the required open-area normalized site attenuation (NSA). The antennas used for measurements from 1 to 40 GHz should be calibrated and linearly polarized, such as: double-ridged guide horns, rectangular waveguide horns, pyramidal horns, optimum-gain horns, and standard-gain horns, although the use of a log-periodic dipole array is also allowed. The beam of the antenna must be large enough to encompass the EUT, or provision shall be made for scanning the EUT. The largest aperture dimension of a horn antenna ($D$) should be small enough so the measurement distance in meters is equal to or smaller than $D^2/2\lambda$. Standard-gain horn antennas need not be calibrated beyond whatever is provided by the manufacturer, unless they are damaged or deterioration is suspected. Standard-gain horn antennas have gains fixed by their dimensions and dimensional tolerances. Because they are invariably used in the far field and have a 50-$\Omega$ impedance, the simple relationship provided by Eq. (2.53) may be used to calculate the antenna factor from the gain.

Moving the measurement antenna over the surfaces of the four sides of the EUT or down the length of an interface cable or another method of scanning of the EUT is required when the EUT is larger than the beam width of the measuring antenna. It is preferred that 1–40-GHz measurements be performed on an open-area test site or in an absorber-lined room. However, measurements may also be performed where there is adequate clearance, considering the radiation pattern of the EUT, to ensure that reflections from any other objects in the vicinity do not affect the measurements. A conducting ground plane is not required, but one may be used for measurements over 1 GHz.

One of the many reasons for noncompliance in a retest of an EUT may be the use of a broadband antenna in the original test, which exhibits high voltage–standing-wave ratios (VSWRs) at low frequency and resultant large peaks and troughs in the calibration curve. Thus an appreciable discrepancy may exist between a measurement made with a dipole antenna and with a broadband antenna. The commonly achievable average VSWR for a 20–1200-MHz very broadband antenna is 2:1, excluding the effect of the proximity of a ground plane to the antenna. However, at low frequency the VSWR may be as high as 30 at 20 MHz. Radiated emission measurement distances under the FCC rules are 3 m for Class B equipment and 30 m for Class A equipment. ANSI C63.4 allows measurements at 3 m, 10 m, or 30 m. Tests shall be made with the antenna positioned in both the horizontal and vertical planes of polarization. The measurement antenna shall be varied in height above the conducting ground plane to obtain the maximum signal strength. For vertical polarizations, the minimum height shall be increased so that the lowest point of the antenna is at least 25 cm above the ground plane.

Tests shall be made in an open, flat area characteristic of cleared, level terrain. Such open-area test sites (OATS) shall be void of buildings, electric lines, fences, trees, underground cables, pipelines, etc. except as required to perform the tests. A suggested layout for the OATS is in the form of an ellipse with a major diameter of $2F$ and a minor diameter of $\sqrt{3}F$, where $F$ is the measurement distances of either 3 m, 10 m, or 30 m.

ANSI C63.4 requires that the OATS used for radiated emission measurements be validated by making horizontal and vertical NSA measurements. Antenna spacings used for making site attenuation measurements shall be the same as the spacings used for the EUT compliance tests at frequencies from 30 MHz to 1000 MHz. The measured NSA data shall be compared to that calculated for an ideal site and the measured NSA shall be within ±4 dB of the theoretical NSA. This ±4-dB tolerance includes instrumentation calibration errors, measurement technique errors, and errors due to site anomalies. C63.4 provides suggestions for actions to be taken if a site fails to meet requirements. The most common error is to make the attenuation without at least the minimum specified ground plane. Even damp soil with a high humus content is highly unlikely to meet requirements. Even with a conductive ground plane, such as hardware cloth, the area of the ground plane must be sufficiently large, with maximum dimension apertures of 3 cm and with the sections of the plane properly bonded together. Another source of error is in the calibration of the antenna factor of the transmitting and receiving antennas. It is possible to use the free-space "two identical antenna" calibration method or the "three-antenna method," in which two antennas are identical and the third is the unknown, but the preferred method is the ANSI-C63.5 standard site method. If the free-space AF calibration of a broadband antenna is used in the NSA calibration, then the proximity of the fixed-height antenna to the ground plane will almost certainly mean that the required NSA is not achievable at certain frequencies. One step that may help is to change from the 1-m fixed height to 1.5 m, or vice versa.

ANSI C63.4 says that measurements can be made at a location other than an open-area test site, such as a weather-protected site, an absorber-lined room, a dedicated laboratory, or a factory site, provided the alternative site meets the site attenuation requirements over the volume occupied by the EUT and the minimum conducting ground plane requirements are met.

As of publication of C63.4-1992, weather-protected sites are exempt from the multiple NSA measurements required of alternative test sites. In addition, the use of a GTEM cell with a factor programmed into the calibration of the GTEM to achieve correlation with the open-area site attenuation requirements is described in Section 9.5.2.2. For alternative test sites, C63.4 states that a single-point measurement of the NSA measurement is insufficient to pick up possible reflections from the construction and/or RF-absorbing material comprising the walls and ceiling of the facility. For these sites, a "test volume" is defined, and the transmit antenna may have to be placed at various points, up to 20, within the test volume. The points and test method are described in detail in C63.4. The smaller the test volume, the lower the number of calibration points. For example, for a volume no larger than 1 m in depth by 1.5 m in width by 1.5 m in height, a minimum of eight measurement positions may be required.

Weather-protected open-area test sites have been constructed from fiberglass girders and panels, with nylon or fiberglass nuts and bolts to secure the structure. An alternative is a pressurized rubber tent, plastic radome, or cloth tent with fiberglass poles placed over the turntable on which the EUT is located. With the antenna remaining outside the weatherproof structure, the height of the structure may be limited, typically to the height of the maximum EUT plus the 0.8-m table height. All of these sites can meet the NSA requirements, at least when first constructed and before potential degradation by conductive contaminants.

For a company that owns a totally wooden structure with a tiled roof made of asphalt or wood shingle or ceramic, but not steel or aluminum, a proposal has been made that enables the use of this type of building as a low-cost, covered open-area test site.

First a temporary, or permanent, open-field test site area is constructed, using the antenna, antenna mast, and turntable intended for use in the weatherproof building. This site must have a ground plane, typically constructed from hardware cloth, and this may also be subsequently used in the weatherproof building. A typical area for this open-field site is a car park, a field, or a backyard. The open-field test site area must be on even ground, with obstacles such as bushes or trees, and building outside of the "obstruction-free area for site with a turntable" shown in Figure 5 of ANSI C63.4. A NSA measurement is made on this open-area site that must conform to the ANSI C63.4 requirements. A broadband swept-frequency noise source (typically battery powered) that covers the full 30–1000-MHz frequency range without any gaps, contained in a conductive shielded enclosure, is constructed. Horizontal and vertical cables are attached to the broadband signal output level, and these cables cover the proposed test volume. The vertical cables are draped over the back of the table, with a configuration similar to that in Figure 9.7. The radiated emissions from this dummy EUT may be used to obtain a correlation between the two test sites. Alternatively, the emission test setup used for correlation of the GTEM to an open-area test site may be used. First the noise source is measured on the open-area test site with the antenna moved from 1 to 4 m both horizontally and vertically and the turntable rotated. The same measurement is made on the covered site, with the EUT and cables placed in exactly the same positions (to achieve this, the cables and source can be taped down to a wooden plane or table). By comparing the two sets of emission measurements, a correlation between the open-area site and the less-than-ideal weatherproof site can be obtained, and from this a correction table can be generated, without the need for the covered site to meet the NSA requirements. However, the covered site should be constructed of nonconductive materials and contain minimum conductive sources of reflections, such as AC power wiring, and contain at least the minimum-dimension ground plane. Although this proposed test site could theoretically achieve as good a correlation as that between a GTEM and an open-area test site, the FCC will not consider this proposition. Although it is only possible to guess at the reason why this approach is unacceptable, it may be that some modification in the proposed test site, e.g., addition of conductive objects, is more likely to occur than any modification in the GTEM cell. Another possible objection is that the volume of the EUT and the orientation of the cables may affect the calibration, because reflections from the test area will change. However, a similar error may occur in the way in which the GTEM is excited by the EUT in the standard three orthogonal calibration technique, described in Section 9.5.2.2. Although this type of facility is not acceptable for certification/verification testing, it may be used for comparative measurements when making modifications to a product to reduce radiated emissions.

The minimum ground plane dimensions for AC power conductive measurements are at least 2 m × 2 m in size extending at least 0.5 m beyond the vertical projection of the EUT. If the EUT normally does not make contact with the ground plane, the ground plane shall be covered with insulating materials between 3 and 12 mm thick.

A conducting ground plane is required at radiated emission sites that shall extend at least 1 m beyond the periphery of the EUT and the largest measuring antenna, and cover the entire area between the EUT and the antenna. It shall be made of metal, with no holes or gaps, with the largest dimension greater than 3 cm. A larger ground plane may be required for certain test sites and especially if the NSA requirements are not met.

The ambient radio noise and signals, both conducted and radiated, are required to be at least 6 dB below the allowable limit. In a true open-area test site this is invariably not possible due to a multiplicity of RF, VHF, and UHF transmitters and broadband noise sources. Use of an absorber-lined room or a GTEM cell are two of the measurement techniques that will meet the 6-dB requirement. The alternative recommendations, such as measurements at a closer distance and using a narrower measurement bandwidth for narrowband and signals masked by broadband noise, typically meet with limited success. The recommendation that measurements of critical frequency bands be made during hours when broadcast stations are off the air and at times when the ambient contribution from industrial equipment is reduced to less than the 6-dB level is rarely followed. In one case where this recommendation was followed, a relatively low-level ambient existed at a location at least 55 km from a large town or city and with low TV and radio signal strength. The measured signals from these sources were still well above the limit, even at a measuring distance of 3 m. In addition, a broadband ambient noise source was measured during the workday, and it was still present at 1 o'clock, 3 o'clock, and 6 o'clock in the morning! The noise source was a large fan used to dry silage on a farm that was in operation day and night for months. The realistic technique for minimizing the effect of the ambient on the EUT emissions is to go to a low-frequency span and reduced measurement bandwidth to differentiate between the ambient and the EUT emissions. This technique usually requires switching on and off the EUT to determine its contribution to the measured emissions. The frequency of peak emissions detected in the prescan will help in finding the EUT emissions. Because EUT emissions are notoriously unstable in frequency, one alternative technique to identify EUT emissions is to use a small antenna, such as the 20–1000-MHz bowtie, located close (typically 20 cm) to the EUT and cables. The measurements made with this antenna will typically lift EUT emissions well out of the ambient noise and allow identification of the emission in measurements at 3 m.

If an anechoic chamber or semianechoic chamber is available, then a prescan of the radiated emissions can be made, with the antenna typically at a distance of 1 m from the EUT. The prescan is used to identify radiated emissions from the EUT measured on the OATS. If any of the radiated emissions are masked by the ambient, an emission close in frequency but out of the ambient should be measured. This measurement on the OATS at 3 m or 10 m can be compared to the prescan measurement of the two frequencies. For example, if an emission at 102 MHz is 6 dB below the limit and a second emission at 102.5 MHz is masked by an FM radio station, then a comparison of these two emissions measured in the anechoic chamber should be made. If the emission at 102.5-MHz is 6 dB below the emission at 102 MHz, then the probability that the 102.5-MHz emission when measured on the OATS will also be at least 6 dB below the limit is high.

Testing at a manufacturer's location or at a user installation is permitted if the equipment cannot be set up on an open-area test site or alternative test site. In this case both the equipment and its location are considered the EUT. The radiated emission and conducted emission measurements are considered unique to the installation site. However, if three or more representative locations have been tested, the results may be considered representative of all sites with similar EUTs. The voltage probe, and not the LISN, shall be used for the conducted emission measurements, and the conducting ground plane shall not be installed for user's installation testing unless one or both are to be a permanent part of the installation.

The table height on which non-floor-standing equipment is placed must be 0.8 m for both AC line-conducted measurements and radiated emissions. Figures 9.9a and 9.9b show the test configurations for tabletop ITE equipment. Equipment that can be used as either tabletop or floor-standing equipment shall be tested only in a tabletop configuration. For conducted measurements in a shielded room, tabletop devices shall be placed on a platform of nominal size, 1 m × 1.5 m, raised 80 cm above the ground plane. The vertical conducting surface of the shielded room shall be located 40 cm to the rear of the EUT. Floor-standing devices shall be placed either directly on the conducting ground plane or on insulating material if the equipment is typically isolated. All other surfaces of tabletop or floor-standing equipment shall be at least 80 cm from any other grounded conducting surface, including the case or cases of one or more LISNs.

Excess interface cable length will be draped over the back edge of the tabletop for tabletop equipment. If any draped cable extends closer than 40 cm to the conducting ground plane, the excess shall be bundled in the center in a serpentine fashion using 30–40-cm lengths to maintain the 40-cm height. If the cables cannot be bundled due to bulk, length, or stiffness, they shall be draped over the back edge of the tabletop unbundled, but in such a way that all portions of the interface cable remain at least 40 cm from the horizontal conducting ground plane, as shown in Figures 9.9a and 9.9b.

The system shall be arranged in one typical equipment configuration for the test. In making

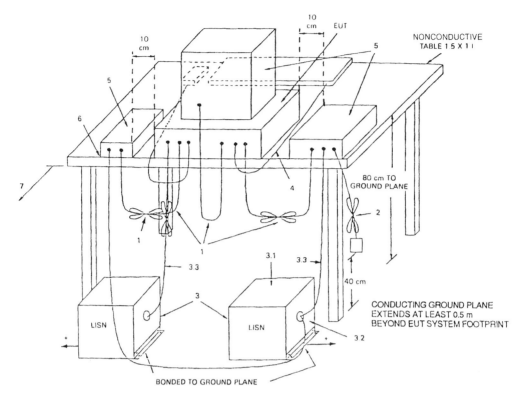

**Figure 9.9a**   Test configuration for tabletop equipment conducted emissions (© 1992 IEEE).

1.   Interconnecting cables that hang closer than 40 cm to the ground plane shall be folded back and forth forming a bundle 30 to 40 cm long, hanging approximately in the middle between ground plane and table.
2.   I/O cables that are connected to a peripheral shall be bundled in center. The end of the cable may be terminated if required using correct terminating impedance. The total length shall not exceed 1 m.
3.   EUT connected to one LISN. Unused LISN connectors shall be terminated in 50 Ω. LISN can be placed on top of, or immediately beneath, ground plane.
   3.1   All other equipment powered from second LISN.
   3.2   Multiple outlet strip can be used for multiple power cords of non-EUT equipment.
   3.3   LISN at least 80 cm from nearest part of EUT chassis.
4.   Cables of hand-operated devices, such as keyboards, mouses, etc., have to be placed as close as possible to the host.
5.   Non-EUT components being tested.
6.   Rear of EUT, including peripherals, shall be all aligned and flush with rear of tabletop.
7.   Rear of tabletop shall be 40 cm removed from a vertical conducting plane that is bonded to the floor ground plane (see 5.2).

any tests involving several pieces of tabletop equipment interconnected by cables or wires, it is essential to recognize that the measured levels may be critically dependent on the exact placement of the cables or wires. Thus preliminary tests shall be carried out while varying cable positions in order to determine the configuration for maximum or near-maximum emission. During manipulation, cables shall not be placed under or on top of the system test components unless such placement is required by the inherent equipment design. This preliminary test, often referred to as a "prescan," is a firm requirement and is often conducted in a shielded room. The cable configuration used during the test is crucial. In the past, the FCC spent approximately half an hour positioning the cable for maximum emissions in the preliminary prescan test, with the antenna positioned close to the equipment. The cables were then taped in position and the equipment taken to the open-field test site for a certification-type test. The prescan is used to determine the emission characteristics of the EUT, and the frequency of peak emissions is noted

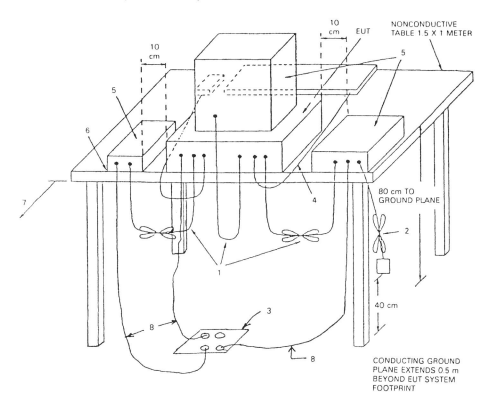

**LEGEND:**
1.  Interconnecting cables that hang closer than 40 cm to the ground plane shall be folded back and forth forming a bundle 30 to 40 cm long, hanging approximately in the middle between ground plane and table.
2.  I/O cables that are connected to a peripheral shall be bundled in center. The end of the cable may be terminated if required using correct terminating impedance. The total length shall not exceed 1 m.
3.  If LISNs are kept in the test setup for radiated emissions, it is preferred that they be installed under the ground plane with the receptacle flush with the ground plane.
4.  Cables of hand-operated devices, such as keyboards, mouses, etc., have to be placed as close as possible to the controller.
5.  Non-EUT components of EUT system being tested.
6.  The rear of all components of the system under test shall be located flush with the rear of the table.
7.  No vertical conducting wall used.
8.  Power cords drape to the floor and are routed over to receptacle.

**Figure 9.9b**   Test configuration for tabletop equipment radiated emissions (© 1992 IEEE).

and is used to help differentiate between the EUT and ambient emissions when measurement are made on the open these are conducted on an open area when these are made on the open-area test site. Cables are not normally manipulated for floor-standing equipment. Instead, the cables should be laid out as shown in Figures 9.11 and 9.12.

Each EUT current-carrying power lead, except the safety ground, shall be individually connected through a line impedance stabilization network (LISN) to the input power source. See section 7.2.1 of ANSI C63.4-1992 for more information on the use of the LISN. The LISN uses a combination of inductors and capacitors to present a standard power-line impedance to the EUT. ANSI C63.4-1992 describes two LISNs, both of which will achieve the characteristic impedance of Figure 9.13. One is usable from 10 kHz to 150 kHz, although this may be extended to 30 MHz; the second is usable from 0.15 MHz to 30 MHz. It is the second LISN that is most

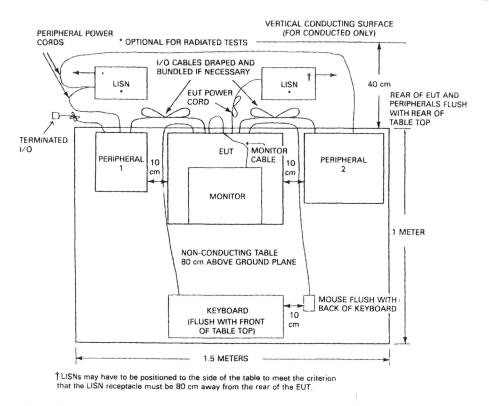

**Figure 9.10**   Test configuration for tabletop equipment—top view (© 1992 IEEE).

commonly used for testing to FCC requirements. The voltage measurement instrument (spectrum analyzer/receiver) with a 50-input termination, is connected to a port of one of the two LISNs to measure the noise voltage developed by the EUT across the impedance of the LISN. The ''measuring instrument port'' of the second LISN must be terminated in 50 Ω. The LISN serves a secondary function of isolating supply-generated noise from the test equipment. Often the LISN is not adequate and an additional power-line filter inserted between the power source and the LISN is required. As with all EMI measurements, a background measurement with the EUT powered down should be conducted prior to conducted emission tests. The use of a calibrated and correctly terminated current probe that fits around each of the current-carrying conductors, separately, may be used when permitted. A LISN should be inserted between the EUT conductors and the mains (power) outlet. The probe is placed between the EUT and the LISN as near the LISN as possible. If an appropriate LISN satisfying the current requirements of the EUT is not commercially available, the LISN may be eliminated and the current probe placed between the EUT and the mains outlet. Although the current probe measurement without LISN will not achieve good repeatability, the lack of a suitable LISN is one good reason to use the current probe technique. ANSI C63.4-1992, in Appendix F, describes two techniques for calibrating the LISN. One requires a signal generator and receiver/spectrum analyzer and the second a network analyzer. Neither method checks the LISN impedance under AC power load conditions.

The measurement of the impedance of the LISN with the maximum AC line current flowing through the LISN may be made at a number of frequencies by use of an impedance meter. The technique involves shorting the two phases of the LISN together (load side) and supplying the input via a variable low-voltage AC supply (e.g., a variac and isolating step-down trans-

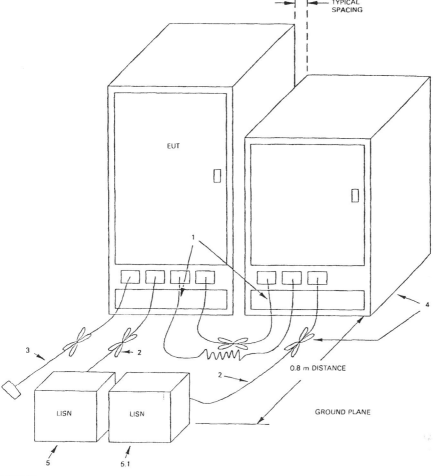

**Figure 9.11** Test configuration for floor-standing equipment conducted emissions (© 1992 IEEE).

LEGEND:
1. Excess I/O cables shall be bundled in center. If bundling is not possible, the cables shall be arranged in serpentine fashion. Bundling shall not exceed 40 cm in length.
2. Excess power cords shall be bundled in the center or shortened to appropriate length.
3. I/O cables that are not connected to a peripheral shall be bundled in the center. The end of the cable may be terminated if required using correct terminating impedance. If bundling is not possible, the cable shall be arranged in serpentine fashion.
4. EUT and all cables shall be insulated from ground plane by 3 to 12 mm of insulating material.
5. EUT connected to one LISN. LISN can be placed on top of, or immediately beneath, ground plane.
   5.1  All other equipment powered from second LISN.

former). The impedance meter may be connected between the case of the LISN and the shorted outputs of the LISN. It would be wise to test for the presence of a common-mode supply frequency voltage, which may damage the impedance meter before making the connection.

A modified conducted emission test setup that may be utilized at a user's installation using the voltage probe is shown in Figure 9.14. A voltage probe may also be used where the use of a LISN is impossible due to high current requirements of the EUT. In the test setup with the voltage probe, both the EUT and the installation environment are tested. The problem with the test method is that AC power-line conducted noise from locations outside of the jurisdiction of the user are also measured. The addition of a large ferrite balun with 2–10 turns on all the

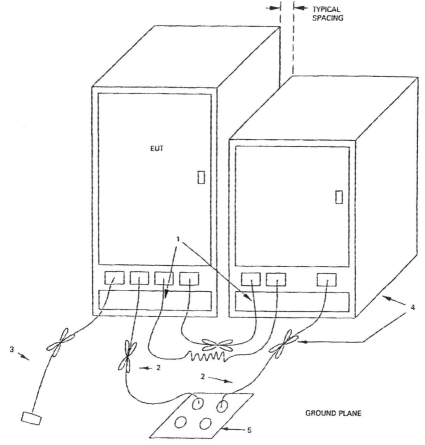

**Figure 9.12**   Test configuration for floor-standing equipment radiated emissions (©1992 IEEE).

power conductors, including safety ground, can help reduce the ambient noise. An alternative
test method that is not approved by the FCC but that has a number of advantages over Figure
9.14 is shown in Figure 9.15. One advantage of the test circuit compared to the LISN is that
it may be easily constructed. The inductor may be wound on a metal oxide or ferrite toroid,
and the 10-$\mu$F capacitor should, where feasible, be an RF feedthrough type. By the choice of
a large enough metal oxide toroid, a current capacity of up to 200 amps may be achieved. The
second advantage of the proposed setup is that the current probe measurement is isolated from
power-line-generated noise by the series inductor. The test circuit can provide an impedance
within the tolerance limits of Figure 9.13. Although the proposed setup is useful in diagnostic
and precertification measurements, only an FCC-type LISN or the line probe, at the user's site,
is likely to be accepted by the FCC for certification purposes.

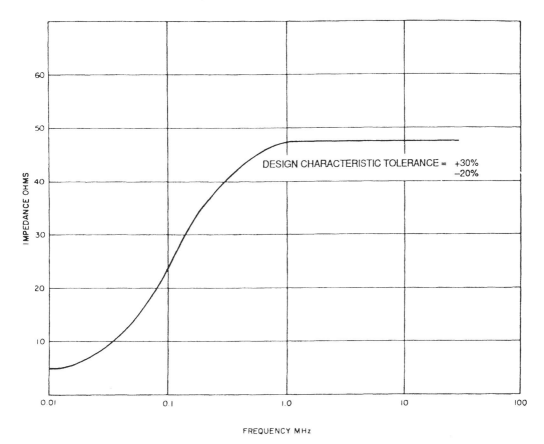

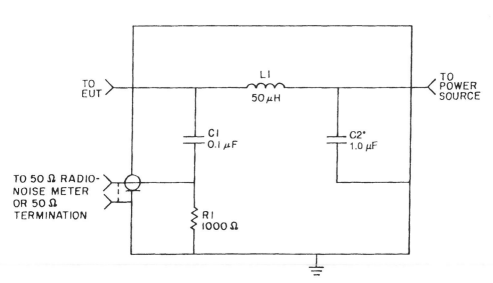

*IN SOME LISNs, A SERIES RESISTANCE IS INCLUDED IN SERIES WITH CAPACITOR C2,
E.G., CISPR PUBLICATION 16 (1987) [6].

**Figure 9.13**   FCC LISN schematic and impedance versus frequency characteristics (©1992 IEEE).

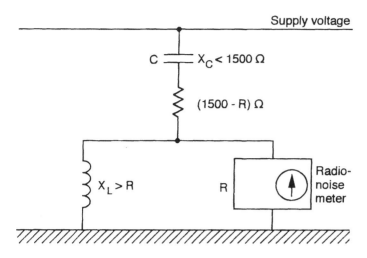

**Figure 9.14** Line probe for use at a user's installation.

Measurements of radio noise power using the absorbing clamp are allowed only for restricted frequency ranges and if specified in the individual equipment requirements and are not allowed for measurements to FCC requirements.

ANSI C63.4 mentions the use of an artificial hand if the EUT is normally operated in the hand, although this is understood not to be a requirement for FCC testing. The test site location should be in as quiet an electromagnetic environment as feasible. For example, locating a test site close to an airport may be convenient for transportation of equipment, but it is invariably a noisy environment.

No matter how apparently minor, changes in an equipment's software program or configuration can make large changes in the emission profile, both radiated and conducted. For example, a piece of equipment appeared to change overnight from within FCC radiated emission limits to outside. The reason was that prior to testing on the second day the designer had slipped in a new PROM with the updated version of software. The new software initialized interfaces in parallel instead of sequentially and increased emissions during the power-up sequence. Other

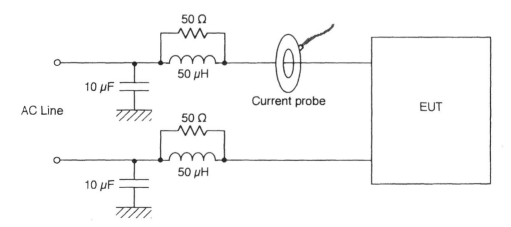

**Figure 9.15** Proposed conducted emission setup for field and diagnostic measurements.

causes of increased emissions have been traced to a change in the manufacturer of ICs and in a signal interface transformer that was changed from a shielded to an unshielded type.

### 9.4.1.1  Radiated emission test set-up and open area test site case study

A small piece of electronic equipment had been tested for radiated emissions on numerous occasions for the manufacturer, throughout the development of the product and on to the production version, and these tests were performed by at least three test personnel. The requirements were the FCC Part 15 and the EN55022 Class B. The software run by the manufacturer was considered ''worst case'' as it exercised the display, LED, audio, signal interface, and memory. The EUT was powered by a battery and a single cable which contained power and a serial signal interface was taken down to the battery located on the OATS ground plane and then brought back to the table top. The EUT was tested both with the battery-terminal connected and disconnected from the ground plane. The cable orientation was as close to that shown in 9(c) of ANSI C63.4 with the power/signal cable brought back up to the nonconductive table top and then looped back down again in a serpentine fashion, with the loops at least 40cm above the ground plane. ANSI C63.4 mentions that *In order to replicate emission measurements, it is important to carefully arrange, not only the system components, but also system cables, wires and ac power cord.* ANSI C63.4 also discusses cable orientation as follows: *It is essential to recognize that the measured levels may be critically dependent upon the exact placement of the cables and wiring. Thus preliminary tests may be carried out while varying cable positions in order to determine the maximum or near maximum emission level.* In the test site under discussion the tests were made using a turntable and the radiated emissions have been found to vary dramatically depending on the rotation of the table. After a maximum emission has been found the cables are then separated or moved within the confines of the recommended ANSI configuration. Again this manipulation of cables has been performed by at least three different personnel! This reorientation is also only made at one (worst case frequency) whereas different cable orientations will maximize the emissions at other frequencies. This approach is taken to limit the amount of time required for a test, but by limiting the basic cable orientation to the ANSI C63.4 recommendation the absolute worst case emissions may not be found.

The test site, which we shall designate A, was a 3m open area test site (OATS) which had been tested in accordance with ANSI C63.4 for normalized site attenuation (NSA). The NSA test had been conducted in 1997 and the measured site attenuation was within $+/-4$dB of the theoretical NSA provided in ANSI C63.4 with a safety margin i.e. the measured NSA was lower than the $+/-4$dB tolerance. In all of these tests on site A the equipment under test (EUT) met all of the radiated emission measurements with a safety margin of at least 8.8dB, below either the FCC or EN55022 limit.

In near field measurements, close to the EUT and the cable, on all of the different EUTs tested, the measured levels were also consistent with a low level emitter. The same type of product, but not the identical unit with the same serial number, was then tested in a 3m semi-anechoic chamber test site (test site C), which reported emissions up to 3.65dB above the limit i.e., 12.46dB higher than the 3m OATS measurements. At exactly the same frequency the difference in measurements could have been as high as 25dB!

At first different software and different cables were used in the test on site C. However, even when the EUT was retested with the same software and cable the EUT was above the limit. The EUT was then tested on a 10m OATS (test site B) and also in a GTEM (test site D). The measurements made on a 3m OATS and a 10m OATS often do not correlate to the far field inverse linear distance extrapolation factor of 20 log 10m/3m = 10.5dB. It is common that in moving the EUT location from 3m to 10m that the reduction at some frequency/ies is less than 10.5dB. Thus an EUT may be just within specification when measured at 3m and out of specification at 10m. However, this does not account for the 14dB difference measured between sites A and B. The correlation between a GTEM and an OATS has a typical mean value of 6dB with 11–15dB differences in individual measurements. Unfortunately OATS to OATS correlation can be similarly poor. In the GTEM measurements the test engineer made the cable "dormant" by the addition of three ferrites, and so if cables were the main source of emissions the GTEM measurements should have been the lowest. Also the test on site C was repeated using a short cable and with the battery placed on the top of the turntable with the same high levels of emissions and this again led us to believe that cables were not a factor. A comparison of the measurements made on the four test sites and compared to the limit is shown in Table 9.6. Thus −8.8dB means that the emission is 8.8dB below the spec and +5.4dB is above. A dash at any location in Table 9.6 means that on the specific test site that the emission was low and in some case negligible!

These test results did not look good for test site A and so extensive measurements were made on the site and test equipment. One possibility was that an EMI receiver was used on sites B and C and a spectrum analyzer (S/A) was used on site A. A variation in the measurement of broadband noise can be expected due to the different shape of the impulse bandwidth used in the receiver and the resolution bandwidth used in the S/A. The analyzer typically requires a 4dB correction for broadband noise. However, the measurements using the S/A did not vary significantly with change in RBW, as the emission was predominantly narrowband. Also, in the final measurements made on site E (a 3m OATS almost identical to A) a receiver was used and a good correlation was seen between sites A and E, not consistent with a 4dB systemic error.

The test site A measurements were made using two different spectrum analyzers which were calibrated and were checked against a calibrated signal source. Quasi-peak and peak measurements were the same and so the quasi-peak detectors were not a problem. Cable attenuation and preamplifier gain were checked and were OK. Before using the test site a "sanity check" is made of the antenna, cables, and preamplifier using the ambient from a number of FM radio stations (about the only advantage of the OATS, apart from accuracy) and the levels during all tests on the EUT were within 1dB of the values measured when the site was first constructed.

The only remaining possibility is the antenna calibration and the NSA characteristics of

**Table 9.6**  Worst Case Radiated Emission Measurements Compared to the Limit for the EUT When Measured on Four Different Test Sites (−dB is Below the Limit, +dB is Above the Limit)

| f(MHz) | Site A 3m OATS | Site B 10m OATS | Site C 3m Anechoic chamber | Site D GTEM |
|---|---|---|---|---|
| 197.66 | −7.7 | — | — | — |
| 204.2 | — | +2.6 | — | — |
| 206.7 | — | — | — | −3.8 |
| 208 | — | 0 | — | −5.9 |
| 212 | — | +4.6 | +3.2 | — |
| 220 | −8.8 | +4.4 | +3.2 | — |
| 228 | — | +5.4 | +3.65 | — |
| 235.88 | −20.8 | — | −6 | — |
| 244.3 | — | −1.3 | — | — |
| 249.5 | −18.8 | — | — | — |
| 252 | — | −2.1 | — | — |
| 279.06 | −16.3 | — | −12.9 | — |
| 368 | — | −7.6 | — | — |

**Table 9.7** Delta NSA Using Two
Log Periodic Biconical Antennas

| $f$(MHz) | $\Delta$NSA (dB) |
|----------|------------------|
| 200      | 0.82             |
| 250      | 1.68             |
| 300      | 2.1              |

the site. A log periodic/biconical antenna and a Roberts dipole were both used during measurements on the EUT and it seems inconceivable that both would be equally out of specification. Nevertheless two of the log periodic biconical antennas were tested on a free space range and the coupling between the antennas was exactly as predicted, based on the input signal level, their gain, and antenna factor. Also as expected, the free space measurement failed the NSA requirements with a difference of 5.7dB, due to the absence of the ground plane.

The equation for $A_N$ is:

$$A_N = V_{\text{Direct}} - V_{\text{Site}} - AF_T - AF_R \qquad \text{(all terms in dB)}$$

where $V_{\text{Direct}}$ is the direct measurement via the cables, $V_{\text{Site}}$ is the radiated measurement, using the transmit and receive antennas and the same cables, $AF_T$ is the transmit antenna factor and $AF_R$ is the receive antenna factor.

As the direct measurement is made with a calibrated output signal generator this measurement also tests that the spectrum analyzer is measuring accurately the amplitude of the frequencies of interest, with the expected cable attenuation. N.B. During radiated measurements the same spectrum analyzer reference level is used as for direct.

The radiated measurement uses the calibrated antenna factors, which are critical in meeting the NSA requirements, and so if we assume a very good OATS performance the NSA measurement is also a way of reconfirming the accuracy of the antenna calibration. This was further tested by replacing the biconical/log periodic receive antennas with two reference dipoles and repeating the NSA measurement.

The difference between measured and theoretical NSA over the frequency range of interest using two log periodic biconical antennas is shown in Table 9.7 and using two dipoles in Table 9.8. These measurements show that the site is at a worst case within 2.1dB of the predicted and at the critical frequency of the EUT (220MHz) is within approximately 1dB of the value for NSA, interpolated between 200 and 250MHz.

Even though site A met all of the requirements it was decided to test the EUT yet again on a 3m test site (E). These measurements showed the EUT passed the requirements, but by only 1.6dB, unlike the 8.8dB found on site A. However, one difference in the cable configuration used on test site E was that the cable was taken down from the center of the turntable, and not

**Table 9.8** Delta NSA Using Two
Dipole Antennas

| $f$(MHz) | $\Delta$NSA (dB) |
|----------|------------------|
| 200      | 0.31             |
| 250      | 1.69             |
| 300      | 1.9              |

**Table 9.9**  Comparison of Radiated Emissions Measured on Test Site A and Test Site E with the Same EUT, Software, and Cable Orientation

| f(MHz) | Test facility and antenna | Level (dBμV/m) | Limit (dBμV/m) | Δ(dB) |
|---|---|---|---|---|
| 212 | Site E log/bicon | 31.4 | 40 | −8.6 |
| 212 | Site A log/bicon | 31.6 | 40 | −8.4 |
| 220 | Site A dipole | 31.2 | 40 | −8.8 |
| 220 | Site E dipole | 35 | 40 | −5 |
| 228 | Site A log/bicon | 35.7 | 40 | −4.3 |
| 228 | Site E dipole | 38.4 | 40 | −1.6 |
| 236 | Site A log/bicon | 36.2 | 47.5 | −11.3 |
| 236 | Site E log/bicon | 37.1 | 47.5 | −10.4 |

the edge, and then brought back from the battery and coiled in the center of the turntable in a large loop.

The exact same cable configuration was tested at site A and the difference in emission measurements between site A and E was a worst case 3.8dB which is an acceptable difference based on antenna factor calibration errors, NSA errors, measuring instrument errors, and the difference in receiver to S/A bandwidth shape. This final test did show that cable orientation does play a role in the level of emissions, by as much as 2.9dB in this particular instance.

In the tests conducted by test site B and C which showed the highest levels of emissions, the length of cable after the battery was coiled on the ground plane. This cable orientation seems far away from the spirit of the ANSI recommended layout and may explain the large difference in measurements, although apparently the measurements were the same with the short cable and the battery placed up on the top of the turntable.

Table 9.9 shows the emissions in dBμV/m compared to the EN55022 Class B limit.

### 9.4.2  Canadian Requirements

Industry Canada places requirements regulating the sale, offering for sale, and use of digital apparatus in Canada. The requirements place limits on the permissible level of interference from digital apparatus manufactured in Canada or imported into Canada after January 31, 1989. The latest requirements, applicable after November 22, 1997, are contained in ICES-003, Issue 3. The requirements do *not* apply to digital apparatus used:

a. In a transportation vehicle
b. As an electronic control, either by a public utility or in an industrial plant
c. In a power system, either by a public utility or in an industrial plant
d. As test equipment, including an oscilloscope and a frequency counter, in an industrial, commercial, or medical environment
e. As a medical computing device, under the direction of a licensed health care practitioner
f. In machinery, apparatus, or equipment
   I. The primary function of which is to apply energy to a process or material through the action of an electric motor or a resistive heating element
   II. That draws a steady-state current that does not exceed:
     A. In the case of an electric motor, 20 A rms
     B. In the case of an electric heating element, used either alone or in conjunction with an electric motor, 50 A rms

     III.  That operates from an alternating current voltage supply that does not exceed 150 V rms

     IV.  Where the machinery, apparatus, or equipment is a portable tool and has an input power that does not exceed 2 kW

g.  In central office telephone equipment operated by a telecommunications common carrier in a central office

h.  In a device having a power consumption not exceeding 6 nW

i.  In a device in which both the highest frequency generated and the highest frequency used are less than 1.705 MHz and that neither operates from nor contains provision for operation while directly or indirectly connected to the AC power lines

j.  Solely for demonstration and exhibition purposes as a prototype unit

The requirements do not apply to units or models of digital apparatus for which the manufacturer, importer, or owner has been granted a special permission by the Minister.

The Minister may grant a special permission where:

a.  The manufacturer, importer, or owner has presented a written application giving:

     I.  The reasons for the request

     II.  An analysis based on sound engineering principles showing that the unit or model of digital apparatus will not pose a significant risk to radiocommunications

     III.  A guarantee of compliance with all the conditions the Minister may set in the special permission

b.  The Minister is satisfied that the unit or model will not pose a significant risk to radio communication

The special permission is valid only if:

a.  The unit bears a label stating that it is operating under special permission and setting out the conditions of that special permission and setting out the conditions of that special permission.

b.  The unit complies with all conditions set out in the special permission.

The Minister may revoke or amend the special permission at any time without notice.

*Digital apparatus* is defined as electronic apparatus that generates and uses timing signals at a rate in excess of 10,000 pulses per second and that utilizes radio frequency energy for the purpose of performing functions including computations, operations, transformations, recording, filing, sorting, storage, retrieval, and transfer, but does not include ISM equipment.

"Class A digital apparatus" means a model of digital apparatus for which, by virtue of its characteristics, it is highly unlikely that any units will be used in a residential environment, which includes a home business.

"Class B digital apparatus" means any digital apparatus that cannot qualify as Class A digital apparatus.

A representative type of model of each digital apparatus shall be tested in accordance with C108.8-M1983, "Electromagnetic Emissions from Data Processing Equipment and Electronic Office Machines," or alternatively CAN/CSA-CISPR 22-96, "Limits and Methods of Measurement of Radio Disturbance Characteristics of Information Technology Equipment" (the latter is an adoption without modification of CISPR 22:1993, second edition).

The procedural requirements are:

A written notice indicating compliance must accompany each unit of digital apparatus to the end user. A suggested text is:

> This Class A digital apparatus complies with all the requirements of the Canadian Interference Causing Equipment regulations.
>
>      Cet appereil numérique de le classe A respecte toutes les exigences du Reglement sur le matériel broulleur du Canada.

The notice shall be in the form of a label that is affixed to the device. Where, because of insufficient space or other restrictions, it is not feasible to affix a label to the apparatus, the notice may be in the form of a statement included in the user's manual.

Any suitably equipped laboratory or organization can perform the tests. There are no restrictions as long as the specified test methodology is followed. It is the responsibility of the manufacturer or the importer to ensure the validity of the test results.

The field intensity of radiated emissions may be measured at a distance other than that described in the table of limits but not less than 3 m. The measurement result shall be extrapolated to the prescribed distance as described in C108.8 or CSA-CISPR 22.

The description of the measurement site for both conducted and radiated emission measurements are the same in ANSI C63.4:1992 as published in C108.8-M1983. Although no open-field site calibration test is required by Industry Canada.

The conducted emission equipment limits set out in C108.8 are identical to the FCC requirements, and so are the radiated emission limits up to 1 GHz. The ICES-003 requirements limit radiated emission measurements to 1 GHz, whereas the FCC requires measurements above 1 GHz as required under 47 CFR 15.33. Some of the minor differences between MP4 and C108.8 have been removed in the adoption by the FCC of ANSI C63.4:1992. Both ANSI C63.4 and the C108.8 test method have the table height set at 0.8 m for conducted and radiated measurements. C108.8 allows the use of a dipole tuned to 80 MHz but used below 80 MHz with appropriate correction factors, which is especially useful for measurements made with vertical polarization. The FCC had discussed rejecting this technique on the grounds that the antenna factor and gain are meaningless under these conditions; however, this technique is allowed in ANSI C63.4. ANSI C63.4 and C108.8 allow measurement closer to the EUT than the specified distance, but not less than 3 m, when the ambient exceeds the limit. A correction factor is then added to the limit to account for the close in measurement. This provision may well be of assistance when the 30-m distance specified for Class A limits apply. In common with the FCC requirements, the preferred measurement instrument is a CISPR (Comité International Special Des Perturbations Radioélectriques) type. However, both the FCC and Industry Canada allow the use of a spectrum analyzer with the quasi-peak detector. Conducted narrowband emission limits are the same as the FCC limits, and the LISN is the same. Where the FCC allows a 13-dB relaxation for broadband noise, C108.8 publishes narrowband limits (and broadband limits that are 13 dB higher). ANSI C63.4 conducted emission test setup complies with that of C108.8 and requires that test site ambient noise level should be at least 6 dB below the FCC limits. C108.8 states that the ambient noise level shall preferably be at least 6 dB below the limit. If the ambient noise level is higher, it must be shown that it does not interfere with the measurements of the source emissions from the EUT. C108.8 also states that the ambient level should not exceed the Class A radiated emission limits. But if the ambient field is higher than the limit, then a "close in" measurement technique is described.

The author has retested equipment with FCC stickers indicating compliance and found emissions above the limits. The FCC routinely retests equipment and finds certified equipment above limits, and thus it is questionable how effective regulations are unless some policing by the regulating body is undertaken. The unofficial position of Industry Canada is "that computers do not result in EMI," and that may be the reason Industry Canada does not plan to independently retest equipment to ensure continuing compliance on production units.

However, should the operation of equipment result in interference, the radio investigator may elect to investigate whether the equipment meets the appropriate limits. The Department of Industry Radiocommunication Act provides for penalties upon summary conviction not to exceed a fine of $5,000 or one year imprisonment or both, in the case of an individual, or not exceeding $25,000 in the case of a corporation.

### 9.4.3 German Regulations

The EMC standards cover numbers 0838 to 0879 in the DIN-VDE series.

The standards committees of the VDE responsible for these standards were involved in the CISPR committees and in the EMC committee of the IEC, and so there is no significant discrepancies between the CISPR and IEC publications and the VDE specifications. Many European standards are included as harmonized national standards in the VDE series. The following is a condensed list of both European harmonized and national standards reproduced from Ref. 1.

| VDE No. | EN No. | Description |
|---|---|---|
| VDE 0838 | EN 60555 Parts 1–3 | Disturbances in supply systems caused by household appliances and similar electrical equipment |
| VDE 0839 | EN 50081 and EN 50082 | EMC: Generic emission and immunity standards |
| | EN 50160 | Characteristics of electricity supplied by public distribution systems |
| VDE 0843 | Parts of IEC-801 | EMC for industrial process measurement and control equipment (immunity) |
| VDE 0845 | — | Protection of information technology equipment against lighting, electrostatic discharges, and overvoltages |
| VDE 0846 | — | Measurement apparatus for the judgment of EMC. Deals mainly with instruments for measuring the effect of disturbances on the mains and for immunity tests |
| VDE 0847 | EN 61000—4 series | Mainly contains measurement methods for immunity testing |
| VDE 0848 | — | Hazards by electromagnetic fields. Methods for measurement and calculation |
| VDE 0855 | — | Antenna and cable-distribution systems for television and sound broadcast signals |
| VDE 0870 | — | Electromagnetic influence; definitions |
| VDE 0871 | CISPR23 | German translation of CISPR 23. ISM equipment is now covered by VDE 0875 Part 11 (EN 55011) and Information processing equipment is covered by VDE 0878 Part 3 (EN 55022) |
| VDE 0872 | EN 55013 | EMC of sound and television broadcast receivers (and video recorders) emissions |
| | EN 55020 | Immunity |

| VDE No. | EN No. | Description |
|---------|--------|-------------|
| VDE 0873 | CISPR18 | Measures against radio interference from electrical utility plants and electric traction systems; radio interference from systems of 10 kV and above. |
| VDE 0875 | | Suppression of radio disturbances caused by electrical appliances and systems |
| | EN 55011 | Part 11: Industrial, scientific, and medical (ISM) equipment |
| | EN 55014 | Part 14: Electric-motor operated and thermal appliances for household and similar purposes, electric tools, and similar electrical apparatus |
| | EN 55015 | Part 15: Fluorescent lamp and fluorescent lamp luminaries |
| VDE 0876 | — | Radio interference measuring apparatus (test receivers and accessories) |
| VDE 0877 | — | Measurement methods of radio interference |
| VDE 0878 | | Radio interference suppression for information and telecommunication systems and apparatus |
| | EN 55022 | Part 3: Limits and methods of measurement |
| | | Part 200, 242, 243, and 244: Immunity requirements |
| VDE 0879 | — | Radio interference suppression of vehicles, vehicle equipment, and internal combustion engines |
| | | Part 1: Suppression of interference to radio reception at a distance from the vehicle |
| | | Part 2: Suppression of interference to radio reception within the vehicle |
| | | Part 3: Measurements on vehicle equipment |

VDE 0879 Part 1/06.79: The RFI suppression of vehicles and of generators and internal combustion engines for radio reception in far distances, is harmonized with CISPR 12 and the EU council directive 72/245/EEC. An EMC directive explicitly for vehicles is 95/54/EC as a successor to Directive 72/245/EEC. The work is carried out by the ACEA (Association des Constructeurs Européens d'Automobiles) and includes both limits and test methods for emission and susceptibility measurements. In the United States, the SAE (Society of Automotive Engineers) has issued SAE J551 for vehicles and J1113 for components, which correspond to the CISPR and ISO standards.

VDE 0875 and EN 55015:1993 cover frequencies from 9 kHz to 400 GHz for electrical lighting and similar electrical equipment and systems. The E field emissions are measured using a triple loop antenna set up around the device under test. The E field limits cover a 9-kHz to 30-MHz frequency. These standards also cover magnetic field emissions measured with a 2-m-diameter loop antenna from 9 kHz to 30 MHz. In one case, conducted emissions are not measured directly; instead, the attenuation of the ballasting and starting arrangement is measured from 0.15 to 1.6 MHz. In two additional cases, the interference voltage is measured at the terminals of the lamps.

VDE 0872 Part 13/08.91 and EN 55013 provide limits and test methods for radio interference characteristics of broadcast receivers and attached equipment. In addition to E field measurements at 3-m distance and the typical 0.15–30-MHz power-line measurement, it imposes voltage limits at the antenna connector and power measurements from 30 to 300 MHz on all cables.

VDE 0875 Part 14/12.95 and EN 55014:1993 provide limits and test methods for radio interference from equipment using electric motors and electric heating equipment for domestic and similar uses, electric tools, and similar equipment. The standard covers the frequency range of 9 kHz to 400 GHz. The limits are for radio interference voltage (148.5 kHz to 30 MHz) and radio interference power (30–300 MHz) on connecting cables, measured with absorbing clamp transducers. There are significant relaxations for interferers that only emit clicks, such as automatically operated switches in laundry machines, refrigerators, thermostatically controlled irons, and ironing machines.

VDE 0875 Part 11/07.92 and EN 55011:1991 cover radio interference voltage 0.15–30 MHz, radio interference field 150 kHz to 1 GHz, and radiated interference power 1–18 GHz. Limits are under consideration for the ranges of 9–150 kHz and 18–400 GHz. EN 55011:1991 states that radiated emissions from 0.15 MHz to 30 MHz are under consideration and, under radiated interference power 1–18 GHz, limits emissions to 57dB(pW) erp referred to a half-wave dipole in the frequency band 11.7–12.7 GHz.

The VDE 0876 ''Radio interference measuring apparatus'' and VDE 0877 ''Measurement methods of radio interference'' documents are the German equivalent of CISPR 16. There are no plans to harmonize these standards by issuing an EN document. VDE 0876 specifies the test receiver with both quasi-peak and average detectors and the click-rate analyzer for disturbances caused by switching operations.

The LISN is specified and the impedance characteristic and the equivalent circuit on which it is based are binding, whereas the circuit diagrams shown in the standards are only examples. Reference 1 continues with the demonstrably true statement that

> Simply reproducing the circuits will not, in our experience, lead to a fully compliant LISN, because the high-power inductors have to be measured using an impedance analyzer to ensure that they also have the necessary RF characteristics. Particular care must be taken that they do not resonate within the working frequency range.

Conducted voltage measurements are not always possible and a current transducer in the form of a clamp can be used. VDE 0876 specifies an accuracy of 1 dB but does not give specific sizes for RF current clamps. It does provide extensive data on the absorbing clamps, which are typically used for the measurement of radio interference power on cables.

The use of three different voltage probes is proposed for: measurement of interference voltages on power and control lines; measurement on very short lines; and measurement on receiving antennas.

VDE 0876 suggests the use of electrically shortened dipoles or monopoles for measuring field strength below 30 MHz (although neither VDE nor EN standards require this). They should be no longer than 1 m for measurement at distances up to 10 m, and so the MIL-STD 1-m rod antenna may be suitable. For frequencies above 30 MHz, tuned dipoles or broadband antennas are required. Each rod of the dipole for the 30–80-MHz frequency range must have a length corresponding to a $\lambda/4$ resonance at 80 MHz. The VSWR of all antennas must be less than 2. The VSWR requirement is unlikely to be met by the dipole measuring at 30 MHz and tuned to 80 MHz. Also, the biconical antenna and the ultrabroadband biconical/log-periodic antenna, which have VSWRs of up to 30 at 30 MHz, will not comply. It is difficult to know what antenna can be used from 30 MHz to 80 MHz that will fulfill the requirement.

VDE0877 describes the measurement of radio-interference voltages, measurement of radio-interference fields and measurement of radio-interference power on cables. It describes the use of the LISN, probes, and the artificial hand. It describes a suitable radiated emission test site, including the use of the normalized field attenuation curves to check the suitability of the test site. It describes the use of the Meyer De Stadelhofen (MDS) absorbing clamp for power measurements on cables. For smaller devices with a side length up to 1 m, the absorbing clamp replaces the much more complex field strength measurement. For this reason, this technique is the only alternative given in the VDE 0875 Standard Part 1 (household appliances, electric tools) for measurement of radiated emissions.

Figure 9.16a compares the early radiated VDE limits and the FCC limits, converted to a 1m measuring distance, to the MIL-STD-461B narrowband (NB) and broadband limits (BB), specified at 1m. This comparison is not totally accurate due to the different measuring bandwidths and the simple $1/r$ ratio used for the distance conversions. However it does show that the MIL-STD limits are significantly lower than even the FCC Class B limit.

Figure 9.16b compares the early VDE and FCC conducted emission limits to the MIL-STD-461B limits. Here the conversion is from a current flowing into a 10μF RF capacitor for the MIL-STD measuring technique to a voltage developed across a LISN for the commercial measuring technique. Again some error is inherent in converting from a current to voltage.

### 9.4.4  Japanese EMI Requirements on Computing Devices

The control of emission on EDP equipment in Japan is undertaken on a voluntary basis under the supervision of the Voluntary Control Council for Interference (VCCI). Although a voluntary

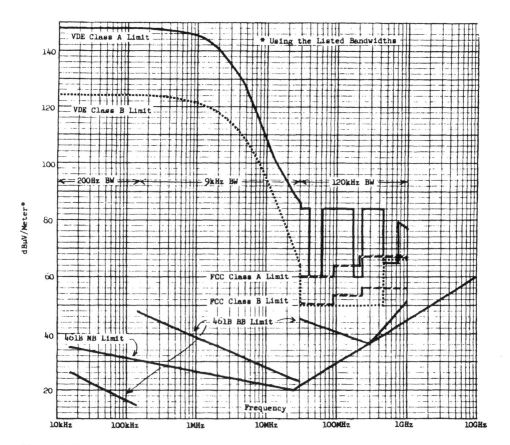

**Figure 9.16a**   Comparison of commercial and military radiated emissions. © 1986 IEEE.

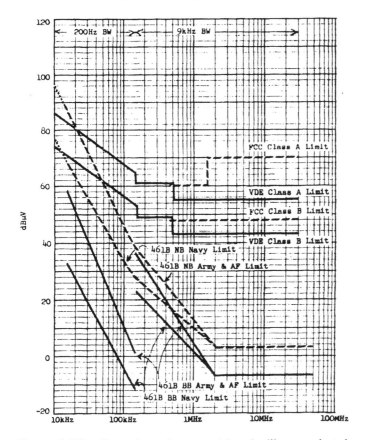

**Figure 9.16b** Comparison of commercial amd military conducted emission limits. © 1986 IEEE.

requirement, many Japanese manufacturers comply as a sign of product integrity. Also, equipment manufactured in Japan for export to the United States must meet FCC requirements and is therefore likely to meet VCCI requirements. Equipment imported to Japan is not required to meet VCCI requirements; however, membership in the VCCI is allowed to foreign manufacturers and may well be of value in the marketing of equipment in Japan. The VCCI requirements are modeled on and use the same limits as contained in CISPR 22. Tests on equipment are performed at a VCCI-approved facility, and the test report is submitted to the VCCI, which issues a certificate of compliance. Compliant equipment is then labeled with a VCCI mark. The VCCI routinely retests certified equipment and, should it fail, requires the manufacturer to modify the equipment. The members agree to abide by the VCCI decisions, which may include revocation of the certificate on noncompliant equipment. Detailed VCCI regulations are available from Compliance Engineering (508-264-4208) or Dash Straus and Goodhue (508-263-2662).

## 9.4.5 European Economic Community Directive 89/336/EEC

The purpose of the directive is to allow free movement of goods between member states of the community. This is accomplished by overcoming technical barriers to trade imposed by differing EMC requirements and procedures in different member states.

The proposal from the Commission of the European Community was adopted by the Council of European Communities, which issued the Council Directive of 3 May 1989 on the approximation of the laws of member states relating to electromagnetic compatibility (89/336/eec). The EMC directive has been passed into law in at least 14 of the 15 member states. The directive became active in 1992, with a 1-year transition to 1 January 1993, which under pressure from manufacturers was extended by three years to 1996. CE marking on compliant equipment is mandatory throughout the European Union (EU). From 1995, the so-called CE mark directive made it optional for most products to include the Low Voltage Directive on the declaration. From 1997 it became mandatory to do so. Council Directive 73/23/EEC is the low-voltage Council Directive.

The low-voltage directive applies to all electrical equipment designed for use with a voltage rating of between 50 and 1000 V for alternating current and between 75 and 1500 V for direct current. EN60950 covers the safety of ITE equipment, EN61010-1 covers electrical equipment for measurement, control, and laboratory use, and EN60065 contains the safety requirements for mains-operated electronics and related apparatus for household and similar use.

The Official Journal of the European Union contains the list of applicable EMC requirements. The number of different limits and requirements applicable to specific devices or that are generic is 102 at the moment.

If a specific type of equipment is not covered in one of the requirements, then one of the generic standards may be used.

The directive covers a vast range of apparatus, encompassing as broadly as possible all electrical appliances, systems, and installations whether or not they are connected to the mains (the public utility AC power supply). The directive includes devices connected to the electricity distribution network and the public telecommunications network. No upper or lower limit on the power output or selection of transmission frequencies is imposed. The directive directly covers several sectors of electrical and electronic engineering, in particular household appliances, consumer electronics, industrial manufacturing, information technology, radio communications, and telecommunications apparatus.

The devices that are exempt from the EMC directive are:

Passive devices, since these are not liable to cause or be susceptible to disturbance
Devices in which the emissions, without additional shielding or filtering, are intrinsically
    far below the most stringent limits of the relevant EMC standards and are inherently
    immune in the intended electromagnetic environment
Amateur radio equipment
Motor vehicles (a different EMC directive covers these)
Active implantable medical devices
Equipment intended for use in aircraft in flight (covered by a different regulation)
Marine equipment (covered by a different regulation)

A couple of partial exclusions, such as weighing machines and tractors, also exist.

The EMC directive applies to all devices that contain electric or electronic elements, with the exception of devices whose EMC is regulated through other directives. These directives are:

| | |
|---|---|
| Motor vehicles | (EC Directive 95/54/EC) |
| Tractors for agriculture and forestry | (EC Directive 72/322/EEC) |
| Active medical implants | (EC Directive 90/385/EEC) |
| Medical equipment | (EC Directive 93/42/EEC) |
| Personal protective equipment | (EC Directive 89/686/EEC) |
| Telecommunications terminal equipment | (EC Directive 91/263/EEC) |

These post-1987 directives are termed "New Approach Directives," which means primarily that they do not contain standards and restrict the technical content to the minimum. The standards have been and still are in the process of being developed independent of the directives. Equipment that is exempt from the EMC directive is electromagnetically passive equipment, such as equipment that does not contain oscillators or clocks or do any switching. One example would be a linear power supply connected to a passive resistive load. Some engineers extend this to include equipment that by its nature produces emissions far below the most stringent limits and is, by its nature, immune. However, for this type of equipment there is no risk in making a declaration of conformance (DOC). Equipment that is explicitly excluded is quartz wrist watches and filament lamps (bulbs). Any component that has a direct function is considered equivalent to an apparatus, and if it is to be placed on the market for final use, it must be CE marked. For example, computer cards and motherboards, computer disc drives, lift controls, electric motors (excluding induction motors), and PLCs have a direct function, whereas cables, resistors, transistors, and capacitors, outside of an electronic circuit, do not. A system made up of a number of finished products is considered a final apparatus and must comply with the EMC directive. Components placed on the market for distribution and use in which the direct function is available without further adjustment or connections other than simple ones that can be performed by anyone are considered equivalent to "apparatus" as defined by the EMC directive. Other components that perform a direct function, such as plug-in cards, smart cards, and input/output modules designed for incorporation into computers, are apparatus commonly found in retail outlets and are available to the general public. Once cards of this type are inserted into a PC, they perform a direct function for the user. They must therefore be considered as apparatus and are, consequently, subject to the provisions of the EMC directive. This does not mean that they must necessarily be intrinsically compliant from the EMC point of view in all cases, if this is either impossible or impractical. However, in such cases they must be designed in such a way that they become fully EMC compliant (emissions and immunity) when they are installed as intended in the apparatus, in any of its possible variants and configurations, without exceptions, and used in the electromagnetic environment determined by the manufacturer. The instructions accompanying the component must clearly indicate these requirements, the pertinent limitations of use, and how to comply without resorting to an EMC specialist (such components are available to non-EMC specialists, for a wide range of applications). The manufacturer has the ultimate responsibility for this decision. These EMI mitigating components may include a shielded cable or the use of a ferrite balun on interconnection cables. It is advisable to include these EMI components with the product if economically viable. If each of the parts of the system bears the CE mark, then the system does not, although the system manufacturer has to provide the system with an EU DOC and therefore should be satisfied that the complete system meets the requirements.

Manufacturers may gain a competitive edge by illegally adding a CE mark to equipment for which it is unnecessary. For example, the author has seen a number of cables and directional couplers, which are passive components, mounted to a metal plate to which a CE mark has been added.

The paths to compliance are contained in articles 10.1 and 10.2 of the directive for all but radio transmitters, which are covered in article 10.5. Article 10.1 is the self-declaration route, which, in the author's experience, is the most popular. The technical construction file (TCF) route, contained in article 10.2, is the most difficult.

Article 10.1 states:

In the case of apparatus for which the manufacturer has applied the standards referred to in article 7(1), the conformity of apparatus with this Directive shall be certified by the manufacturer or his authorized representative established within the community.

Despite the phrase "established within the community," neither the manufacturer nor the organization making the declaration need be resident within the EU. One of the normalized standards specific to the class of product or one of the generic standards has to be applied for the self-declaration route. However, there is some disagreement as to whether "applied" means testing or not. The directive does not specify testing to an EN standard but does specify compliance. If a manufacturer believes that the product is in compliance, then it can make a declaration. However, test results may be required to prove the validity of the declaration, and a manufacturer may be taking a risk in making the declaration without a test, for in Chapter 11 it is demonstrated how difficult it is to lay out a compliant unshielded PCB that contains high-speed devices. And in Section 7.6 it was shown how very low RF current flow on cables can cause equipment to fail radiated emission requirements.

Despite the difficulty of meeting both emission and immunity requirements, and therefore the risk involved in not testing, some, typically smaller, companies who wish to save money have made the declaration without a test. Some companies have put forward the argument that they have no competitors who make similar equipment in Europe and who can retest the equipment to ensure that it is compliant, and therefore of low risk. The member states of the EU could have followed the approach taken by the FCC and implemented policing of the requirements by retesting products. Alternatively, they could have followed the approach taken by Industry Canada, which waits until a complaint of interference by a product occurs. Reference 2, which is a very good source of information on standards and compliance issues, provides the information that the Swedish enforcement program took three frequency converters off the market in 1997 and that the German regulatory authorities have fined 19 companies so far. The German authority verifies conformity declarations regarding their formal correctness and logical plausibility and performs sampling tests of products sold in Germany. During the first six months of 1999, 25,384 items were inspected; of these, 3582 products were subjected to EMC testing. Of the tested items, 23.7% (849) did not comply. So it obvious that the policing approach has been taken by the EU.

The standards to which the declaration is made must be European norms that have been published in the Official Journal (OJ) of the European Commission and that have been adopted by at least one of the member states. For example, some standards, such as the ENVs, are drafts, and the prEN's are prestandards and have not been published in the OJ as yet.

The directive requires that the DOC be signed by "someone empowered to enter into commitments on behalf of the manufacturer or his authorized representative established within the community." Who satisfies this criterion is still debated. Some companies use the signature of the EMC/compliance engineer, although it would appear that unless this person is a director, he may not be empowered to enter into legal commitments.

The question that is most often asked is, since this is a legal document with potentially legal and commercial penalties, who takes the responsibility? Initially it was thought that the individual signing the document was not responsible but was merely signing on behalf of the company. Later this was considered incorrect. The present position appears to be that the signatory takes the responsibility.

A good source of additional information is: Guidelines on the application of Council Directive 89/336/EEC of 3 May 1989 on the approximation of the laws of the Member States relating to electromagnetic compatibility 30-8-93, No. L220/21. Yet more information is available on the Internet at http://europa.eu.int.

The recommended documentation to accompany the equipment is as follows.

For equipment that meets class A of the EN55022 requirements, the following warning must be included in the instructions for use:

**Warning**

This is a Class A product. In a domestic environment this product may cause radio interference, in which case the user may be required to take adequate measures.

An additional certificate does not need to accompany the product, although most manufacturers include it and it is highly recommended, since this is the declaration that customs officials will review. A CE mark alone may result in an investigation. An *example* of the declaration is provided nearby. If the low-voltage directive does not apply, it is better to say so and to state why than to leave it out. The theory is that it is better to add the title of documents with an explanation than to leave out what may appear to be an applicable document. Make sure that the standards to which the declaration is made are the latest version!

---

**Declaration of Conformity**

**Application of Council Directive**          **EMC Directive 89/336/EEC**
                                      **Low Voltage Directive 73/23/EEC**

**Standards to which Conformity is Declared**         EN55022 1995 ,
                                              EN50082-1 1998
                                              EN 61000-4-3 1997
                                              etc.

**Manufacturers Name**

**Manufacturers Address**

**Importers name**                                **(if available)**

**Importers address**                           **(if available)**

**Type of equipment**

**Model No.**

**Serial No. (Optional)**              **Year of Manufacture (Optional)**

**I, the undersigned, hereby declare that the equipment specified above conforms to the above Directives and Standards.**

**Signed on behalf of the Manufacturer and Importer**

**Place**                         **Signature**

**Date**                           **Full name**

                                    **Position**

Example of Declaration of Conformity

The CE conformity marking that is to be attached to the equipment shall take the following form:

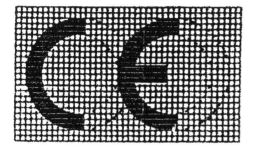

If the CE marking is reduced or enlarged, the proportions shown in the drawing must be respected. The various components of the CE marking must have substantially the same vertical dimension, which may not be less than 5 mm.

In the case of apparatus for which the manufacturer has not applied, or has applied in part, the standards referred to in article 7(1) or failing the existence of such standards, the manufacturer or its authorized representative established within the community shall hold at the disposal of the relevant competent authorities, as soon as the apparatus is placed on the market, a technical construction file (TCF). This file shall describe the apparatus, set out the procedures used to ensure conformity of the apparatus with the protection requirements referred to in article 4, and include a technical report or certificate, one or other obtained from a competent body (CB).

The technical construction file is applicable where no relevant standards are available or where the product is too large to be tested in a laboratory and the manufacturer does not see fit to make a declaration without test. The manufacturer may also want to declare for a product range based on testing one or more products in that range. In this case a justification as to why these tests represent a coverage of the product range should be made. Another situation in which to use the TCF is where products are certified to existing EMC standards not recognized by the EU but that are possibly in compliance with some or all of the requirements of the directive or where harmonized standards have been applied in part only.

Unlike the self-declaration route, the TCF must be approved by a third party, the competent body.

If testing has been done to nonharmonized standards, then the TCF may make a justification as to why these standards are relevant. How the apparatus is described, what procedures are used to ensure conformity, and the relationship between the file itself and the report of certificate, written by the CB, are not described and are therefore open to interpretation. A draft guidance document, but which has no legal standing, was produced by the NAMAS EMC working group and was reviewed by the German Group of Competent Bodies and the authorities. This document is intended as a starting point for information on how to prepare the TCF. The TCF will typically have the following contents: description of the apparatus, identification of the apparatus, technical description of the apparatus, technical rationale, detail of significant aspects, test data, report of certificate from competent body. Because it is mandatory to use a CB for the report or certificate, the CB should also be used to provide guidance in the preparation of the TCF. For example, the role of the TCF is tailored based on the reason for choosing the TCF route.

The task of the CB is described in German law in paragraph 2.8 of the EMVG as follows: The CB issues technical construction files and conformity declarations after positively confirming the protective requirements of the Directive. The CB must demonstrate: competence in EMC, independence, and professionalism. In addition, EN 45011/CENELEC/Sept 89 General Criteria for Certification Bodies operating Product Certification is applicable. An association of competent bodies was formed that obtained the support of the Commission of the European Community by agreeing to comply with the definition for the forming of notified bodies contained in CERTIF 94/6.

Using the TCF approach without confirming EMI testing is novel and not without difficulty, for as we have seen throughout this book, a design may be adequate in one application but fail in another. An example is a power-line filter that may achieve adequate attenuation when used with one piece of equipment but may exhibit insertion gain when used with a second piece of equipment. Assuming the design approach is adopted, then the need for accurate EMC predictions and competent and experienced CBs to interpret them will increase. Should many manufacturers opt to incorporate EMC into their designs and follow the TCF route, the market for a powerful and comprehensive EMC prediction computer program may justify the expense of developing such a tool. An EMC prediction program would aid in standardizing the evaluation of the design if the same program were used by manufacturers in all member states.

The CB may require the test lab (TL) to be accredited to EN 45001. However, when testing to harmonized EN standards, nonaccredited third-party test labs can be used. Because only harmonized standards can be applied for the DOC route, it stands to reason that nonaccredited labs can be used for the DOC, as they routinely are. The Commission of the EU has indicated to interested parties in the United States that the EU will accept the test data from an independent laboratory outside of an EU country to support the declaration of conformity made by the manufacturer of the equipment. The EU has not to date ruled on accreditation of test facilities outside of EU countries for the use of CBs. The TL must, however, comply with the IEC 1000-4-5 testing and measurement techniques, which include the open-area test site attenuation calibration and compliance with the field uniformity requirements for radiated immunity testing. Up until 1997 the CB could operate only in EU countries, although in 1997 the EU was developing and signing agreements with non-EU countries, allowing them to appoint CBs nationally that will have the same status as CBs in Europe. The requirements for apparatus designed for the transmission of radio communications are that it shall be certified by an EC declaration of conformity. But in addition, an EC type of examination certificate concerning this apparatus (type approval) shall be provided by a competent body. In Germany the competent body is the BZT.

Table 9.10 describes the more common normalized standards for EMC, their applicability, and the types of equipment.

Product standards are special limits for emission and immunity for a specific category of devices. They are never less stringent than the generic standards over which they always take precedence when they are available for the specific product family. Special product standards are spoken of when the EMC requirements for a product are embedded in another standard for the product.

Generic standards are applicable to a particular electromagnetic environment. They specify requirements and tests that are used by all products and systems placed in the specified environment, unless superseded by a specific product standard, which takes precedence. The two environments are (1) industry, residential, commercial, and light industry, and (2) industry.

Basic standards define and describe the EMC problem, measurement and test methods, principle measuring instruments, and test setup. They do not provide limits and are typically referenced by the generic and product standards.

| | |
|---|---|
| EN 50091-2 1995 | UPS Part 2 EMC requirements |
| EN 60730-2-18 1999 | Automatic electrical controls for household and similar use |

**Table 9.10** Normalized Standards for EMC

| | Type of compatibility | | | |
| --- | --- | --- | --- | --- |
| | Emissions | | | Immunity, all aspects |
| Product family | Power-line harmonics | Voltage fluctuations | RF emissions | |
| Electric household appliances and electric tools | EN 61000-3-2 | EN61000-3-3 | EN 55014 | EN 50082-1, future EN 55104 |
| Lighting equipment | EN 60555-2 | | EN 55015 | EN 50082-1 |
| Radio receivers and audio equipment | EN 60555-2 | | EN 55013 | |
| Data processing, information technology | EN 60-555-2 republished as 61000 (part 3, section 2, available as draft) | | EN 55022 | EN 50082-1 (used until superseded by prEN 55024) |
| Signaling in the public low-voltage power | EN 50081-1 | EN 50081-1 | EN 50065-1 | EN 50082-1 |
| ISM equipment | EN 50081-2 | EN 50081-2 | EN 55011 | EN 50082-2 |

Part 2

EN 60947-5-3 1999    Low voltage switchgear and control gear. Part 5-3
EN 60947-6-2 1993    A1 = 1997 Low voltage switchgear and control gear Part 6-2
EN 61000-6-2 1999    Electromagnetic compatibility (EMC) Part 6-2
EN 61543 1995        residual current operated protective devices (RCDs) for household and other use
EN 61812 1996        A11 = 1996 Specified time relays for industrial use Part 1

The following are the basic standards published up to 2000;

CISPR 16-1/1993 First part of the new edition: Specification for radio disturbance and immunity measuring apparatus and methods. Part 1 Radio disturbance and immunity measuring apparatus. There is no EN planned to replace CISPR 16.

The EN 61000-x-x standards 1–6 sections are derived from the IEC 801-2–6 standards, which were in turn converted to the IEC 1000-4-x standards. The IEC 1000-4-x standards are:

IEC 1000-4-1     Overview of the immunity tests; basic EMC publication
IEC 1000-4-2     Electrostatic discharge (ESD) immunity test
IEC 1000-4-3     Radiated radio frequency electromagnetic field immunity test
IEC 1000-4-4     Electrical fast transient/burst immunity test
IEC 1000-4-5     Surge immunity test
IEC 1000-4-6     Immunity to conducted disturbances, induced by radio frequency fields
IEC 1000-4-7     General guide on harmonics and interharmonics measurements and instrumentation, for power supply systems and equipment connected thereto
IEC 1000-4-8     Power frequency magnetic field immunity test
IEC 1000-4-9     Pulse magnetic field immunity test
IEC 1000-4-10    Damped oscillatory magnetic field immunity test
IEC 1000-4-11    Voltage dips, short interruptions, and voltage variations immunity test

The following are the harmonized standards published up to 1999 that are specifically EMC, or contain EMC requirements (among others), or contain voltage or current ripple requirements. © European Communities 1995–2000

| | |
|---|---|
| EN 50065-1-1991 | Amendment A1 = 1992, A2 = 1995, A3 = 1996. Emissions limit for signaling on low-voltage electrical installations |
| EN 50082-1 1992 | Generic emission standard for residential, commercial, and light industrial environments |
| EN 50082-2 1993 | Generic emissions standard for the heavy industrial environment |
| EN 50082-1 1997 | Generic immunity standard for residential, commercial, and light industrial environments |
| EN 50082-2 1995 | Generic immunity standard for the heavy industrial environment |
| EN 50083-2 1995 | A1 = 1997. Cabled distribution systems for TV, sound, and interactive media |
| EN 50090-2-2 1996 | Home and building electronic systems (HBES) Part 2-2. |
| EN 50091-2 1995 | UPS Part 2 EMC requirements |
| EN 500130-4 1995 | A1 = 1998. Alarm systems, fire, intruder, and social |
| EN 50148 1995 | Electronic taximeters |
| EN 50199 1995 | Arc welding equipment |
| EN 50227 1997 | Control circuits and switching elements, proximity sensors, DC interface for proximity sensors, and switching amplifiers |
| EN 55011 1998 | A1 = 1999. Emission limits for Industrial, Scientific, and Medical (ISM) RF equipment (supersedes EN 55011 1991) |
| EN 55013 1990 | A12 = 1994, A13 = 1996. A14 = 1999. Emissions limit for broadcast receivers and associated equipment. Amendment 11t and Amendment 12 |
| EN 55014 1993 | A1 = 1997, A2 = 1999. Emission limits for electrical motor operated and thermal appliances for household and similar purposes, electrical tools and similar apparatus Part 1 (supersedes EN 55014 1987) |
| EN 5014-2 1997 | Emission limits for electrical motor operated and thermal appliances for household and similar purposes, electrical tools and similar apparatus Part 2 |
| EN 55015 1996 | A1 = 1997, A2 = 1999. Radio disturbance characteristics of electrical lighting and similar equipment (supersedes EN 55015 1986) |
| EN 55020 1994 | A11 = 1996, A12 = 1999, A13 = 1999, A14 = 1999. Immunity requirements for broadcast receivers and associated equipment (supersedes EN 55020 1987) |
| EN 55022 1998 | Emission limits for information technology equipment (ITE), based on CISPR 22 1993 |
| EN 55024 1998 | Information technology equipment. Immunity characteristics |
| EN 55103-1 1996 | Audio, video, audiovisual, and entertainment lighting control apparatus for professional use. Part 1 emissions |
| EN 55103-2 1996 | Audio, video, audiovisual, and entertainment lighting control apparatus for professional use. Part 2 immunity |
| EN 55104 1995 | Immunity requirements for household appliances, tools and similar apparatus |

| | |
|---|---|
| EN 60269-1 1993 | Amendment 1. Low-voltage fuses (based on IEC 269-1) |
| EN 60282-1 1993 | High-voltage fuses (based on IEC 282-1) |
| EN 60439-1 1999 | A11 = 1996. Low-voltage switchgear |
| EN 60521 1995 | Class 0.5, 1, and 2 alternating-current watt-hour meters (based on IEC 521 1988) |
| EN 60555-2 1987 | Disturbances in supply systems caused by household appliances and similar electrical equipment. Part 2 Harmonics |
| EN 60555-3 1987 | A1 = 1991. Disturbances in supply systems caused by household appliances and similar electrical equipment. Part 3 Voltage fluctuations |
| EN 60669-2-1 1996 | A11 = 1997. Switches for household and similar fixed-electrical installations |
| EN 60669-2-2 1997 | Switches for household and similar fixed-electrical installations |
| EN 60669-2-3 1996 | Switches for household and similar fixed-electrical installations |
| EN 60945 1993 | Marine navigational equipment, general requirements (based on IEC 945) |
| EN 60947-1 1999 | Low-voltage switch gear and control gear. |
| EN 60730-1 1995 | A1 = 1996. Automatic electrical controls for household and similar use |
| EN 60730-2-5 1995 | Automatic electrical controls for household and similar use |
| EN 60730-2-6 1995 | A1 = 1997. Automatic electrical controls for household and similar use |
| EN 60730-2-7 1991 | A1 = 1997. Automatic electrical controls for household and similar use |
| EN 60730-2-8 1995 | A1 = 1997. Automatic electrical controls for household and similar use |
| EN 60730-2-9 1995 | A1 = 1996, A2 = 1997. Automatic electrical controls for household and similar use |
| EN 60730-2-11 1995 | A1 = 1997. Automatic electrical controls for household and similar use |
| EN 60730-2-14 1997 | Automatic electrical controls for household and similar use |
| EN 60730-2-18 1999 | Automatic electrical controls for household and similar use |
| EN 60870-2-1 1996 | Telecom equipment and systems |
| EN 60687 1992 | Alternating-current static watt-hour meters for active energy (classes 0.2S to 0.5S) |
| EN 60945 1997 | Maritime navigation and radio communications equipment and systems |
| EN 60947-1 1999 | Low-voltage switch gear and control gear. Part 1 |
| EN 60947-2 1996 | A1 = 1997. Low-voltage switch gear and control gear. Part 2 |
| EN 60947-3 1999 | Low-voltage switch gear and control gear. Part 3 |
| EN 60947-4-1 1996 | A2 = 1997. Low-voltage switch gear and control gear. Part 4 |
| EN 60947-4-2 1996 | A2 = 1998. Low-voltage switch gear and control gear. Part 4 |
| EN 60947-5-1 1991 | A12 = 1997. Low-voltage switch gear and control gear. Part 5 |
| EN 60947-5-2 1997 | Low-voltage switch gear and control gear. Part 5 |
| EN 60947-5-3 1999 | Automatic electrical controls for household and similar use. Part 2 |
| EN 60947-6-1 1991 | A2 = 1997. Low-voltage switch gear and control gear. Part 6 |
| EN 60947-6-2 1993 | A1 = 1997. Low-voltage switch gear and control gear. Part 6.2 |
| EN 61000-3-2 1995 | A1 = 1998, A2 = 1998. Limits for harmonic current emissions (equipment current < = 16 A per phase) |

| | |
|---|---|
| EN 61000-3-3 1995 | A1 = 1998. Limits for voltage fluctuations and flicker (equipment current < = 16 A per phase) |
| EN 61000-6-2 1999 | Electromagnetic Compatibility (EMC). Part 6-2 |
| EN 61008-1 1994 | A2 = 1995, A14 = 1998. Electrical accessories (RCCBs) |
| EN 61009-1 1994 | A2 = 1995, A14 = 1998. Electrical accessories (RCBOs) |
| EN 61036 1996 | Alternating-current electronic watt-hour meters for active energy (classes 1 and 2) |
| EN 61037 1992 | A1 = 1996, A2 = 1998. Electronic ripple control receivers for tariff and load control |
| EN 61038 1992 | A1 = 1996, A2 = 1998. Time switches for tariff and load control |
| EN 61131-2 1994 | A11 = 1996. Programmable controllers. Part 2: Equipment requirements and tests (based on IEC 1131-2 1992) |
| EN 61268 1996 | Alternating-current static var-hour meters for reactive energy (classes 2 and 3) |
| EN 61326 1997 | Electrical equipment for measurement control and laboratory use |
| EN 61543 1995 | Residual current operated protective devices (RCDs) for household and other use |
| EN 61547 1995 | Equipment for general lighting purposes |
| EN 61800-3 1996 | Adjustable-speed electrical power drive systems. Part 3 |
| EN 61812-1 1996 | A11 = 1996. Specified time relays for industrial use. Part 1 |
| EN 12015 1998 | Product family standard for lifts, escalators, and passenger conveyers. Emissions |
| EN 12016 1998 | Product family standard for lifts, escalators, and passenger conveyers. Immunity |
| EN ISO 14982 1998 | Agricultural and forestry machines |

The following draft standards or prestandards have been proposed. If any of these apply to your apparatus, it would be advisable to make sure the draft standard or prestandard has not been converted to a harmonized standard.

| | |
|---|---|
| prEN 50098 | Radiated emission testing of physically large telecommunication systems (prETS 300 127: 1991) |
| ENV 50102 | EMC requirements for ISDN terminal equipment<br>Part 1: Emission requirements<br>Part 2: Immunity requirements |
| ENV 50140 | Basic EMC standard: Immunity to RF fields 80–1000 MHz |
| ENV 50141 | Basic EMC standard: Immunity to conducted disturbances 0.15–80 MHz |
| ENV 50142 | Basic EMC standard: Immunity to surges |
| ENV 50147 | Anechoic chambers<br>Part 1: Shield attenuation measurement<br>Part 2: Alternative test site suitability with respect to site attenuation<br>Part 3: Shielded anechoic enclosure performance for immunity tests of radiated radio frequency electromagnetic fields |
| EN 50160 | EMC: Voltage characteristics of electricity supplied by public distribution system |
| EN 60601-1-2 | Medical electrical apparatus |

<table>
<tr><td></td><td>Part 1: General safety requirements, 2nd supplement EMC</td></tr>
<tr><td>prEN 61000-2-2</td><td>Compatibility levels for low-frequency conducted disturbances and signaling in public low-voltage power-supply systems. IEC 1000-2-2:1990 modified.</td></tr>
<tr><td></td><td>German standard: DIN VDE 0839 Part 2 Draft 11.92</td></tr>
<tr><td>prEN 50096</td><td>EMC requirements for ISDN equipment (prETS 300 126:1991</td></tr>
<tr><td>prEN 50098</td><td>Equipment engineering (EE): Radiated emission testing of physically large equipment</td></tr>
</table>

The European Telecommunications Standards Institute (ETSI) has published a range of product standards and drafts standards that cover radio equipment and systems. Details on these and updates to the preceding standards can be obtained from the Web at http://europa.eu.int/comm. (enterprise/newapproach/standardization/harmstds/reflist/emc.html)

*Radiated and Conducted Emissions*

The EN radiated and conducted emission limits are very close to the FCC limits. Table 9.11 lists the common EN and FCC radiated emissions at the specified distances; Table 9.12 shows the limits converted to 3 m using a far-field inverse linear distance extrapolation factor. NB: The FCC and EN55022 allow these extrapolated 3-m limits to be used for measurements at 3 m, whereas EN55011 requires that the 10-m limits also be applied to measurements at 3 m. Thus Table 9.8 is useful for a direct comparison of the different requirements as well as for comparison of measurements at 3 m to the FCC and EN55022 limits. NB: The FCC also requires measurements above 1000 MHz, under certain conditions in accordance with 47 CFR 15.33, and for these an average detector function is used.

Table 9.13 compares the conducted emission of the EN and FCC requirements above 0.15 MHz, and Figure 9.17 plots the conducted emission limits from 0.15 to 0.5 MHz for the EN 55011 group 1, class B, the EN 55022 class B, and the EN 50081-1 requirements.

These tables and plots can be obtained on laminated card from EMC Consulting Inc.

Generic emission requirements apply to electrical and electronic apparatus for which no dedicated product or product family emission standard exists. The meaning of class A and B equipment varies from one document to another. For example, in the generic emission requirement EN 50081-1, the definition of domestic, commercial, and light industrial locations is: locations that are characterized by being supplied directly at low voltage from the public mains.

EN 50081-2 covers generic emissions for equipment operating in industrial environments, which are defined as: where industrial, scientific, and medical (ISM) equipment is present; where heavy inductive or capacitive loads are frequently switched; and where currents and associated magnetic fields are high. EN 55011 covers all ISM equipment and separates them into groups and classes. Group 1 ISM equipment contains all equipment in which there is intentionally generated and/or used conductively coupled radio frequency energy that is necessary for the internal functioning of the equipment. Group 1 covers digitally controlled equipment in which the conductively coupled radio frequency is a by-product of the digital signals and clocks.

Group 2 ISM equipment includes all equipment in which radio frequency energy is intentionally generated and/or used in the form of electromagnet radiation for the treatment of material and spark erosion. The equipment is then divided into Class A equipment that is suitable for use in all establishments *other* than domestic, and that directly connected to a low-voltage power supply network that supplies buildings used for domestic purposes. Class B equipment is suitable for use in domestic establishments and in establishments directly connected to a low-voltage power supply network that supplies buildings used for domestic purposes. Unlike the preceding EN requirements, EN 55022, which covers ITE equipment, classifies such equipment

**Table 9.11** EN and FCC Radiated Emission Limits as Specified

| $f$ (MHz) | EN55011[a] Group 1, Class A / EN55022 Class A / EN50081-2 — Specified limit at 30 m (dBμV/m) | EN55011[a] Group 1, Class B / EN55022 Class B / EN50081-1 — Specified limit at 10 m (dBμV/m) | FCC Class A — Specified limit at 10 m (dBμV/m) | FCC Class B — Specified limit at 3 m (dBμV/m) |
|---|---|---|---|---|
| 30–88 | 30 | 30 | 39 | 40 |
| 88–216 | 30 | 30 | 43.5 | 43.5 |
| 216–230 | 30 | 30 | 46.5 | 46 |
| 230–1000 | 37 | 37 | 46.5–960 MHz | 46–960 MHz |
| | | | 49.5 > 960 MHz | 54 > 960 MHz |

| | |
|---|---|
| EN | Europaeische Norm (European Standard) |
| EN55011 | Industrial Scientific Medical: |
| | Group 1, Class A—Nondomestic or not connected to low-voltage supply |
| | Group 1, Class B—Domestic or connected to low-voltage supply |
| | ([a]No increase in the 10-m level allowed for measurements at 3 m) |
| EN55022 | Information Technology Equipment: |
| | Class A—Commercial |
| | Class B—Domestic |
| EN50081-1 | Generic Standard for Domestic, Commercial, or Light Industrial |
| EN50081-2 | Generic Standard for the Industrial Environment |
| FCC | Part 15, Subpart J: |
| | Class A Computing Device—Commercial |
| | Class B Computing Device—Residential |
| | Class A—Commercial |
| | Class B—Domestic |
| EN50081-1 | Generic Standard for Domestic, Commercial or Light Industrial |
| FCC | Part 15, Subpart J: |
| | Class A Computing Device—Commercial |
| | Class B Computing Device—Residential |

*Source*: Reproduced by kind permission of EMC Consulting Inc.

as Class B when intended primarily for use in the domestic environment, which is in turn defined as an environment where the use of broadcast radio and television receivers may be expected within a distance of 10 m of the apparatus concerned. Class A equipment is defined as a category of all other ITE that satisfies the Class A ITE limits but not the Class B limits. Such equipment should not be restricted in its sale, but the class A warning, described previously, should be included with the equipment.

### 9.4.5.2 Immunity Tests

Unlike the majority of commercial requirements, the EN requirements include immunity tests. The generic immunity requirement for residential, commercial, and light industry environment, EN 50082-1:1998, specifies the following tests:

*Radiated Immunity*: Radiated radio frequency energy from 80 to 1000 MHz at 3V/m unmodulated, with the test performed with a 1-kHz 80% am using the test setup described in EN 6100-4-3. Applicable to enclosure port.

**Table 9.12** EN and FCC Radiated Emission Limits Converted to 3 m

| f MHz | EN55022 Class A | EN55022 Class B | FCC Class A at 3 m | FCC Class B at 3 m |
|---|---|---|---|---|
| | EN50081-2 at 3 m | EN50081-1 at 3 m | | |
| | Corrected limit (30 m to 3 m) (dBµV/m) | Corrected limit (10 m to 3 m) (dBµV/m) | Corrected limit (30 m to 3 m) (dBµV/m) | Specified limit (no correction) (dBµV/m) |
| 30–88 | 50 | 40 | 49.5 | 40 |
| 88–216 | 50 | 40 | 54 | 43.5 |
| 216–230 | 50 | 40 | 57 | 46 |
| 230–1000 | 57 | 47 | 57–960 MHz | 46–960 MHz |
| | | | 67.5 > 960 MHz | 54 > 960 MHz |

| | |
|---|---|
| EN | Europaeische Norm (European Standard) |
| EN55011 | Industrial Scientific Medical: |
| |     Group 1, Class A—Nondomestic or not connected to low-voltage supply |
| |     Group 1, Class B—Domestic or connected to low-voltage supply |
| |     (EN 55011 does not allow a reduction in emission levels for measurement at 3 m and so the limits at |
| |         30 m and 10 m also apply to 3 m) |
| EN55022 | Information Technology Equipment: |
| |     Class A—Commercial |
| |     Class B—Domestic |
| EN50081-1 | Generic Standard for Domestic, Commercial, or Light Industrial |
| EN50081-2 | Generic Standard for the Industrial Environment |
| FCC | Part 15, Subpart J: |
| |     Class A Computing Device—Commercial |
| |     Class B Computing Device—Residential |

*Source*: Reproduced by kind permission of EMC Consulting Inc.

*Electrostatic Discharge*: Air discharge 8 kV and contact discharge of 4kV with test setup IEC EN 61000-4-2, also for enclosure port.

*Electrical Fast Transient (EFT)*: The requirements for ports for signal lines and control lines and functional earth ports is 0.5 kV, rise time 5 nS, and pulsewidth 50 nS at a repetition rate of 5 kHz with the test setup described in IEC 61000-4-4, capacitive clamp. The EFT pulse is injected common mode and for signal and control lines is only applicable to ports interfacing with cables whose total length according to the manufacturer's functional specification may exceed 3 m. The same EFT pulse is specified for DC power ports with no cable length restrictions; however, the requirement is not applicable to input ports intended for connection to a dedicated battery or a rechargeable battery that must be removed or disconnected from the apparatus for recharging. Apparatus with a DC power input port intended for use with an AC-DC power adapter shall be tested on the AC power input of the AC-DC power adapter specified by the manufacturer or, where none is so specified, using a typical AC-DC power adapter. The test is applicable to DC power input ports intended to be connected permanently to cables longer than 10 m. The EFT pulse for input and output AC power ports is 1 kV, 5-nS rise time, 50-nS pulsewidth at a repetition rate of 5 kHz.

*Surges*: The surge test is applicable to AC and DC power and in some cases to signals. The generator shall have a 1.2-µs rise time and a 50-µs pulse width at 50% amplitude into an open circuit and an 8-µs rise time, 20-µs pulsewidth into a short circuit. The

**Table 9.13** EN and FCC Conducted Emission Limits

| | E55011 Group 1, Class A | | EN55011 Group 1, Class B | | | |
| | EN55022 Class A | | EN55022 Class B | | | |
| | EN50081-2 | | EN50081-1 | | FCC Class A | FCC Class B |
| $f$ (MHz) | Quasi-peak (dBμV) | Average (dBμV) | Quasi-peak (dBμV) | Average (dBμV) | Quasi-peak (dBμV) | Quasi-peak (dBμV) |
|---|---|---|---|---|---|---|
| 0.150–0.450 | 79 | 66 | See Figure 9.16 | See Figure 9.16 | | |
| 0.450–0.500 | 79 | 66 | See Figure 9.16 | See Figure 9.16 | 60 | 48 |
| 0.500–1.600 | 73 | 60 | 56 | 46 | 60 | 48 |
| 1.600–5 | 73 | 60 | 56 | 46 | 69.5 | 48 |
| 5–30 | 73 | 60 | 60 | 50 | 69.5 | 48 |

| | |
|---|---|
| EN55011 | Industrial Scientific Medical: |
| | Group 1, Class A—Nondomestic or not connected to low-voltage supply |
| | Group 1, Class B—Domestic or connected to low-voltage supply |
| EN55022 | Information Technology Equipment: |
| | Class A—Commercial |
| | Class B—Domestic |
| EN50081-1 | Generic Standard for Domestic, Commercial, or Light Industrial |
| EN50081-2 | Generic Standard for the Industrial Environment |
| FCC | Part 15, Subpart J: |
| | Class A Computing Device—Commercial |
| | Class B Computing Device—Residential |

*Source*: Reproduced by kind permission of EMC Consulting Inc.

source impedance of the generator is 2 Ω. The charge (O/C) voltage for DC power is 0.5 kV line to ground and line to line. For AC power the charge voltage is 2 kV line to earth and 1 kV line to line. The line-to-line pulse is injected via an 18-μF capacitor and line-to-ground via a 9-μF capacitor in series with a 10-Ω resistor. The test procedure, including injection methods, is contained in EN 61000-4-5.

*Radio Frequency Common Mode*: Applicable to ports for signal lines and control lines as well as DC and AC power ports and functional earth ports. The RF is injected common mode at a level of 3-V rms unmodulated, but during the test this level is modulated at 1 kHz 80% am. The test methods, including injection methods and calibration method, are included in EN 61000-4-6.

*Voltage Dips and Voltage Interruptions*: These are applicable to AC power only and include a voltage shift at zero crossing, 30% reduction in amplitude for 10 ms for output ports, and a 60% reduction for 100 ms for input ports. A voltage interruption of greater than 95% of the applied voltage shall be applied for 5000 ms to input ports only. Voltage dip and voltage interruption test methods are described in EN 61000-4-11.

Apparatus shall not become dangerous or unsafe as a result of the application of the tests defined in EN 50082-1. A functional description and a definition of performance criteria, during or as a consequence of the EMC testing, shall be provided by the manufacturer and noted in the test report, based on criteria described in EN 50082-1. The criteria are basically (A) the apparatus shall continue to operate as intended during the test, (B) the apparatus shall continue

## EN55011 Group 1, Class B,
## EN55022 Class B
## and EN50081-1
## Conducted Emissions Requirements
## for 150kHz to 500kHz

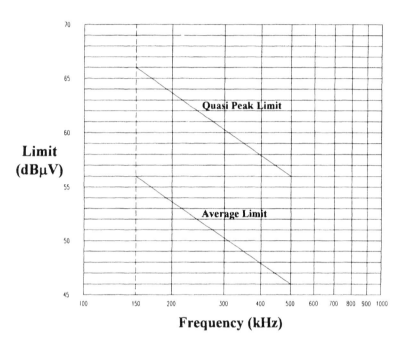

**Figure 9.17**   EN conducted emission limits below 500 kHz.

to operate as intended after the test, and (C) temporary loss of function is allowed, provided the function is self-recoverable or can be restored by the operation of controls. Which criterion applies is typically decided by the criticality of the apparatus. The tests shall be made in the most susceptible operating mode and shall be varied to achieve maximum susceptibility. If the apparatus is part of a system or can be connected to auxiliary apparatus, then the apparatus shall be tested while connected to the minimum configuration necessary to exercise the ports in accordance with EN 55022. EN 50082-1 also describes the selection of terminations, inappropriate tests, and immunity test requirements. The rule is that previously compliant equipment that is modified must remain compliant. If the modification changes the configuration and this was not covered in the original declaration, then a new declaration is required and any tests that have been added since the original declaration must be included. This is one reason why it makes sense to include any immunity tests that are still ''for information only'' if the product life is sufficiently long.

Because the 1.2-μs/50-μs surge test is an addition to the latest version of EN 50082-1, wall-mount or modular AC-to-DC power supplies may have a CE mark and yet not have been tested with the surge test.

The generic requirements for industrial environment, EN 50082-2, specifies for enclosure ports: the RF field at 10 V/m when unmodulated, but with 80% 1-kHz AM, 80–1000-MHz frequency range, at 900 MHz $\pm$ 5 MHz, a 10 V/m unmodulated, but modulated during the test with a 50% duty cycle pulse and 200-Hz repetition rate; the power frequency magnetic field at 30 A/m 50 Hz (applicable only to apparatus containing devices susceptible to magnetic fields); and ESD with 4 kV contact and 8 kV air discharge.

For signal lines and data buses not involved in process control etc., the requirements are: C/M injected RF, 0.15–80 MHz at 10 V rms, when unmodulated but modulated at 80% AM (1 kHz) with a source impedance of 150 $\Omega$; the EFT requirement is 1 kV peak, and both are applicable only to cables exceeding 3 m in length.

For ports for process, measurement, and control lines and long bus and control lines, the 10-V-rms RF C/M injection applies, as does the EFT, but at a peak voltage of 2 kV.

For DC input and output power ports, as well as AC input and output power ports, the same 10 V rms RF C/M injection and the 2-kV EFT pulse applies.

The informative annex in EN 50082-2 indicates the following tests that may be included in the standard: power frequency C/M injection of 10 V rms at 50 Hz for signal lines and data buses not involved in process control and 20 V rms for process, measurement, and control lines; 2 kV C/M and 1 kV D/M 1.2 µs/50 µs for process, measurement, and control lines, and 0.5 kV D/M and 0.5 kV C/M for DC power and 4 kV C/M and 2 kV D/M for AC power; voltage deviation and voltage variation for DC power and voltage dips, voltage interruptions, voltage fluctuation, and low-frequency harmonics for AC power.

Because the 1.2-µs/50-µs surge test has been included in the most recent update to EN 50082-1, it is very likely that it will be included in an update to EN 500882-2.

The modulated and unmodulated test levels are defined in ENV 50140, in which if the unmodulated specified test level is 1 V/m rms or 2.8 V p-p, then when amplitude modulated at 80%, the rms value 1.12 V and the peak-to-peak value is 5.1 V. This increase in level with modulation is important, for many signal generators provide the same output level modulated or unmodulated, and the capability of the power amplifier must include the peak-to-peak levels. It is also important to find out if any E field monitoring device is used, whether it measures peak or rms values and also what the displayed value is.

EN55024 1997 are the specific immunity requirements for information technology equipment. And these are identical to the latest EN 50082-1, with the *very important addition* of the 1.2-µs/50-µs surge voltage at 1 kV between ground and the cable on signal and telecommunications ports.

The emission and immunity tests described in the harmonized standards refer to test methods, test setups, and test location calibration criteria. For radiated emission measurements, the test area must typically be a calibrated open-area test site, described in ANSI C63.4 1992 and in Appendix A of EN 55022.

## 9.4.5.3 Emission Tests

In EN 55011, Class A ISM equipment may be measured either on a test site or in situ as determined by the manufacturer. Class B ISM equipment shall be measured on a test site. The test site shall meet an ambient requirement of at least 6 dB below the limit, which is almost impossible for an open-area test site. The test site for measurements from 9 kHz to 1 GHz shall have a ground plane for both conducted and radiated measurements with the minimum size and shape as depicted in Figure 9.18. Details on the validation of this test site await the third edition of CISPR 16, but it is likely to be the open-area attenuation calibration. Equipment shall, if possible, be placed on a turntable and rotated. If a turntable is not possible, the antenna shall be positioned at various points in azimuth for both vertical and horizontal polarizations.

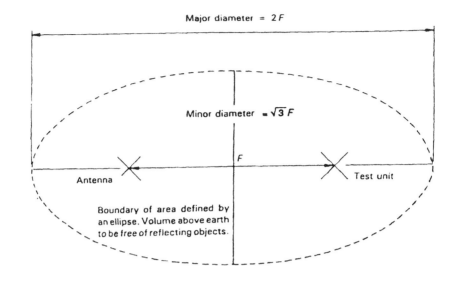

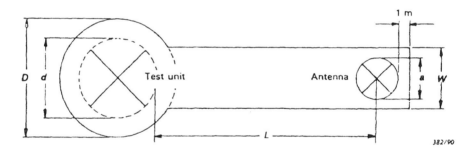

$D = (d + 2)$ m, where $d$ is the maximum test unit dimension
$W = (a + 1)$ m, where $a$ is the maximum antenna dimension
$L = 10$ m

**Figure 9.18**  Characteristics of radiated emissions test site and minimum ground plane for radiated emissions measurement. (Reproduced by permission of CENELEC, Brussels, Belgium, ©CENELEC.)

The configuration of the equipment shall be varied to maximize emissions. Interconnecting cables shall be of the type and length specified in the individual equipment requirements. If the length can be varied, the lengths shall be selected to produce maximum radiated emissions. For conducted measurements, excess lengths of cable shall be bundled at the approximate center of the cable, with bundles of 30–40 cm in length. Where multiple interface ports are all of the same type, connecting a cable to just one of those ports is sufficient, providing it can be shown that the additional cables would not significantly affect the results. This is difficult to achieve, for the C/M noise on two or more cables can drive against each other, just like the rods of a dipole, and the fields generated by the cables can add or subtract at the location of the antenna. A description of the cable and equipment orientation must be provided so that results can be repeated. If the equipment can perform separately any one of a number of functions, then the equipment shall be tested while performing each of these functions.

Fundamentally, EN 55011 requires that worst-case, maximized emissions be performed. EN 55011 also describes the measuring equipment requirements, antennas, connection to the electricity supply on a test site, "in situ" measurements disposition of medical equipment, types of medical, industrial, scientific, laboratory, and measuring equipment, microwave cooking appliances, and other equipment in the 1–18-GHz frequency band.

For conducted measurements, the 50-$\Omega$/50-$\mu$H V-network (LISN) shall be used, or the voltage probe when the LISN cannot be used. EN 55011 supplies additional information on connection of the EUT to the electricity supply network when performing measurements on a test site.

EN 55022:1994 is the emission document for information technology equipment (ITE), and it provides the same basic guidelines for EUT configuration, general measurement conditions, methods of measurement, etc., as provided in EN 55011, with some additional EUT configuration details specific to ITE equipment. The test site information is more specific, in as much as a site attenuation measurement requirement is imposed. In general the site shall be validated by making site attenuation measurements for both horizontal and vertical polarization fields in the frequency range of 30 MHz to 1000 MHz. The distance between the transmitting antenna and receiving antennas shall be the same as the distance used for radiated emission tests of the EUT. A measurement site shall be considered acceptable if the horizontal and vertical site attenuation measurements are within $\pm 4$ dB of the theoretical site attenuation of an ideal site (see also CISPR 16). The problems inherent in achieving the site attenuation requirement are discussed in Section 9.4.1. The test site shall characteristically be flat, free of overhead wires and nearby reflecting structures, and sufficiently large to permit antenna placement at the specified distance and provide adequate separation between antenna, EUT, and reflecting structures. *Reflecting structures* are defined as those whose construction material is primarily conductive. The test site shown in Figure 9.18 is presented in both EN 55022 and EN 55011, with a minimum alternative test site, shown in Figure 9.19, presented in EN 55022. A conducting ground plane shall extend at least 1 m beyond the periphery of the EUT and the largest measuring antenna and shall cover the entire area between the EUT and the antenna. It should be of metal, with no holes or gaps having dimensions larger than one-tenth of a wavelength at the highest frequency of measurement. The minimum-size ground plane is shown in Figure 9.18. A larger-size conducting ground plane may be required if the site attenuation requirements are not met. As an alternative, tests may be conducted on other test sites that do *not* have the physical characteristics shown in Figure 9.18 or 9.19. Evidence shall be obtained to show that such alternative sites will yield valid results. Annex A describes a method of testing an alternative site whereby a measurement antenna is moved within a volume in both horizontal and vertical polarizations. A second antenna is scanned in height from 1 m to 4 m. This alternative site test measurement method could be performed in a semianechoic chamber designed for at least a 3-m test area but would preclude the use of the GTEM cell. However, if the GTEM, or any other test fixture, can be proven to yield valid test results, then its use may be acceptable.

EN 55022 states that tests can be made on one appliance only or on a sample of appliances using a statistical approach, and the requirement then is that 80% of the mass-produced appliances must comply with the limits with at least 80% confidence. EN 55022 states that subsequent tests are necessary from time to time on appliances taken at random from production, especially when only one appliance has been tested. However, although this retest may be highly recommended, it is well to remember that the widely held view is that testing is not mandatory to make a declaration of conformity for an appliance.

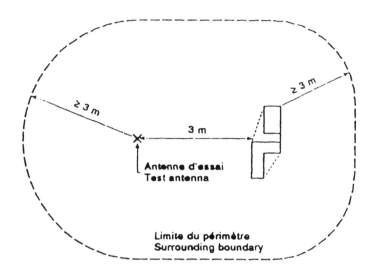

There shall be no reflecting object inside the volume defined on the ground by the line corresponding to this figure and defined in height by a horizontal plane ≥3 m above the highest element of either aerial or equipment under test.

**Figure 9.19**   Alternative minimum radiated emission test site (EN 55022:1994). (Reproduced by permission of CENELEC, Brussels, Belgium, ©CENELEC.)

*Immunity Testing*

One of the most common errors when performing immunity tests is the lack of an inadequate method of testing any or all the functions of the EUT and an effective method of flagging a failure during or after the test. Digital, analog, video, or RF should flow on all signal interfaces during the test. Digital data should be monitored either by a typical interface device that can flag an error in the data or by a computer that can perform error checking. An alternative to an external device is to loop back the data to the EUT at the end of the signal cable. However, an effective means of flagging the data on the EUT by an audible or visual warning is required. Video data should be displayed on an external monitor, which, if part of the EUT, must remain inside the test setup. If the monitor is used merely to test the video signal, then it may be located outside of the test area. However, adding long cables between a computer and a monitor in order to locate the monitor away from the test area will invariably result in EMI during a radiated immunity or C/M RF conducted immunity test. This typically takes the form of distortion of the image or bars in the image. In this case it is probably more realistic to have the monitor located close to the computer, and this is less likely to result in EMI. If a display must be viewed inside a semianechoic chamber during a radiated immunity test, then a door with a wire mesh screen may be used. More convenient is the use of a video camera located inside the room, with an external monitor. In some cases, where a large-diameter ''waveguide below cutoff'' port is installed in the chamber, a set of mirrors has been used to view the display outside of the chamber!

If a bar code reader or similar device is connected to the EUT, the bar code should be read and displayed or monitored by the EUT on a repetitive basis. This usually entails the writing of separate software for immunity testing, and this is often required to ensure that all the interface signals are exercised as well as disk drives, alarms, displays, wireless, etc. during the EMI test.

If a writing board or touch display requires contact on the surface to read information, this can easily be achieved by a weighted wooden structure containing a suitable pen.

It is not uncommon to be asked to perform immunity tests where no indication of a pass/ fail condition is available from the EUT; in this case there is no sense in performing the test!

It is common for radiated and conducted RF immunity tests that the EUT pass/fail performance criterion is A, which means that "the apparatus shall *continue to operate as intended*. No degradation of performance or loss of function is allowed below a performance level specified by the manufacturer, when the apparatus is used as intended. The performance level may be replaced by a permissible loss of performance. If the minimum performance level or permissible performance loss is not specified by the manufacturer, then either of these may be derived from the product description and documentation and what the user may reasonably expect, from the apparatus if used as intended." Thus in testing to the criterion A level, the functional/monitoring test equipment must also continue to function correctly during the EMI test.

For ESD, EFT, and surge tests on either AC/DC power or signal lines, the pass/fail criterion is B. Criterion B is described as "The apparatus shall continue to operate as intended *after* the test. No degradation of performance or loss of function is allowed below a performance level specified by the manufacturer, when the apparatus is used as intended. The performance level may be replaced by a permissible loss of performance. During the test, degradation of performance, however, is allowed. No change of actual operating state or stored data is allowed. If the minimum performance level or permissible performance loss is not specified by the manufacturer, then either of these may be derived from the product description and documentation and what the user may reasonably expect from the apparatus if used as intended." Thus in testing to the criterion B level, the functional/monitoring test equipment must continue to monitor or store the data, which must remain uncorrupted during the test. As described later, this may be very difficult to achieve, and in some cases a display may momentarily blank or the I/O card in the monitoring equipment may lose the data due to the impressed transient. In this case, as long as data and the correct monitoring of data resumes after the test, the equipment can be said to have passed at criterion B, which is the requirement, even though it is the monitoring device that is susceptible.

Before a radiated immunity test can be performed, the test facility field uniformity test described in IEC 6000-4-3 and ENV 50140 must be performed. If a usable test area of 1.5 m × 1.5 m is required, the field uniformity over this area and at a height of 0.8 m above the floor of the facility must be measured. The calibration is performed without the EUT present, and 12 of 16 points on a grid over the plane must meet the 0-to +6-dB field uniformity criterion referenced to 3 V/m or 10 V/m. For testing larger EUTs it is required that the intensity of the field at a height of 0.4 m above the ground plane, as well as over the full width and height of the largest EUT, be measured and reported in the test report. Figure 9.20 shows the calibration of the field.

The preferred test facility consists of an absorber-lined shielded enclosure, an example of which is shown in Figure 9.21, in which the anechoic lining material on walls and ceiling has been omitted for clarity. Both the type and construction of the test area and the type of transmitting antenna can lead to errors larger than 6 dB. For example, many anechoic chambers exhibit low-frequency resonances, which can result in a wide variation in E field over the test volume. Addition of absorber loads at strategic locations in the room or ferrite tiles at strategic locations on the walls or any other of the steps discussed in Section 9.51 (9) may be applied. If the transmitting antenna has too high a gain, then the field strength in the main lobe of the

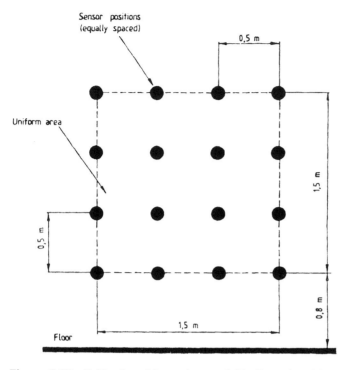

**Figure 9.20**  Calibration of immunity test field. (Reproduced by permission of CENELEC, Brussels, Belgium, ©CENELEC.)

antenna can vary by more than 6 dB. If the required test area is smaller than 1.5 m × 1.5 m, then the number of test points can be reduced.

One very important guideline, which should be applied to MIL-STD or any type of radiated susceptibility tests but which seldom is, is that the test field be calibrated on the empty test area before the test and the test then performed with the same output power to the field-generating antenna as during the calibration. Control of the field strength during the test is of no value, because the field will be distorted when the test object is present and the field may not be correctly measured.

The test sample shall not be placed closer than 1 m to the field-generating antenna, although a test distance of 3 m is preferred.

Annex A of ENV 50140, which is for information, discusses field strength to be expected at the biconical antenna from 20 to 300 MHz and the log-periodic from 80 to 1000 MHz. Circularly polarized antennas may be used only if the output power from the power amplifier is increased by 3 dB. It is recommended in semianechoic chambers (those with no absorber on the floor) that additional absorber be placed on the floor in the illumination path from antenna to EUT. Striplines, TEM, and (although it is not mentioned) the GTEM cells may be used only if the field homogeneity requirements are met and if the EUT and cable can be arranged as required by the standard. Additionally, wires/cables cannot exceed one-third of the dimension between septum and outer conductor. Thus the TEM, GTEM, and stripline are typically useful for small EUTs. A screened room with limited absorbing material may be used if the field homogeneity requirement and all other requirements are met and an open antenna range providing legal limits on radiating E fields is met and absorbing material is placed on the floor.

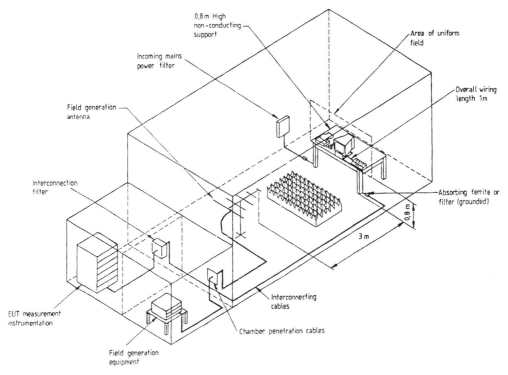

0,8 m High
non-conducting
support

Incoming mains
power filter

Field generation
antenna

Interconnection
filter

EUT measurement
instrumentation

Field generation
equipment

Area of uniform
field

Overall wiring
length 1m

Absorbing ferrite or
filter (grounded)

0,8 m

3 m

Interconnecting
cables

Chamber penetration cables

NOTE  Anechoic lining material on walls and ceiling has been omitted for clarity.

**Figure 9.21**  Example of semianechoic chamber suitable for radiated immunity tests. (Reproduced by permission of CENELEC, Brussels, Belgium, ©CENELEC.)

*Conducted Immunity C/M RF*

The generic conducted immunity (C/M injection), specified for equipment used in industrial environments, is covered by EN 50082-2, Part 2. This references ENV 50141 for the test method. The generic requirement for conducted immunity on equipment used in residential commercial and light industry is contained in EN 50082-1, Part 1, which reports that C/M injection is ''under consideration'' and which references EN 61000-4-6. The test frequency is 0.15 MHz to 80 MHz AM modulated at 80%. Three methods of injection are provided. The direct coupling is for tests on shielded or unshielded cables; the basic test setup for shielded cables is shown in Figure 9.22, and the calibration test setup for unshielded cables is shown in Figure 9.23. The value for R is N × 100 Ω where N = number of conductors in unshielded cable. An alternative coupling method is via a coupling/decoupling network (CDN). The injection may be made either directly via the CDN, as shown in Figure 9.24, or by an inductive injection clamp into the impedance of the CDN, as shown in Figure 9.25. If a CDN is not available, the method shown in Figure 9.25 can be used with the terminating resistors shown in Figure 9.23 as long as the impedance of the terminating resistors is in accordance with the requirements of ENV 50141 and EN 61000-4-6. EN 61000-4-6 describes a separate setup for level-setting. The injection test setups are described as follows.

**Equipment Checklist**

1. Signal generator 0.15–80 MHz with either an AM modulation capability of 1 kHz at 80% or an external modulation input.

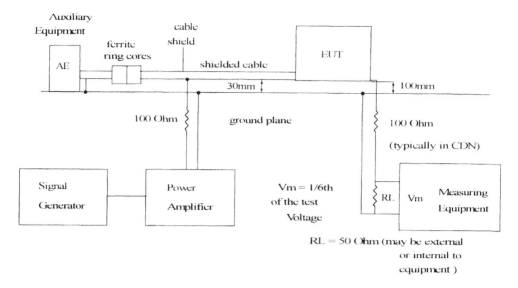

**Figure 9.22** C/M injection direct test method for screened (shielded) cables.

2. Signal generator for 1 kHz. If generator in (1) requires an external modulation source. If generator in (1) does not have an external modulation input, then a modulator is required between the signal generator output and the power amplifier in (3).

3. Power amplifier 0.15–80 MHz. Typically 10–75 W, depending on injection method, with 50-ohm source impedance.

4. Inductive injection clamp or coupling/decoupling networks (CDNs) or capacitor/resistor networks (see 6) for direct coupling.

5. Ferrite ring cores for the direct coupling into screened (shielded) cables test set.

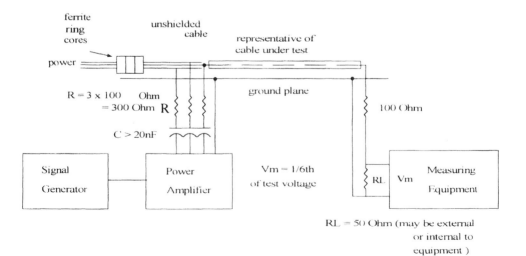

**Figure 9.23** Test setup for calibration, direct coupling to nonscreened (unshielded) mains and (power) cables.

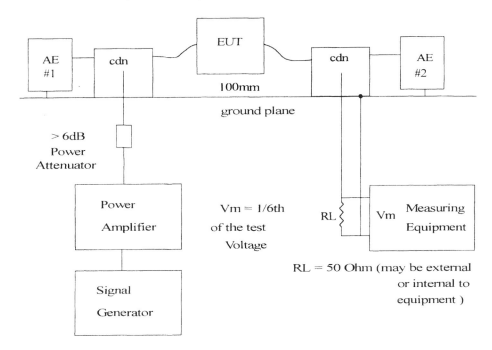

**Figure 9.24**   C/M injection directly via the CDN, with CDNs or resistive networks as the loads.

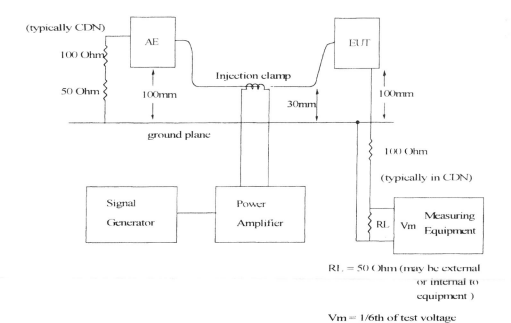

**Figure 9.25**   Test setup for injection via the inductive clamp, with resistive networks, or more typically CDNs, as the loads.

6. 100-ohm and 50-ohm series injection and load resistors and for direct coupling into nonscreened (unshielded) cables <20-nF coupling capacitors.
7. Oscilloscope with at least a 100-MHz bandwidth and a 50-$\Omega$ input impedance or a 50-$\Omega$ terminating resistor.
8. Spectrum analyzer or EMI receiver.
9. >6-dB power attenuator. Required for CDN injection only.

### Setup and Operations

For direct coupling into shielded cables and unshielded cables: Connect the equipment as shown in Figure 9.22 for shielded cables and Figure 9.23 for unshielded and ensure that the full 3 V or 10-V rms unmodulated can be developed across the 150-ohm load. Because the voltage is measured across only the 50-ohm part of the load and an impedance of 250 $\Omega$ is presented in series, the 10-V rms (28-V pk-pk) level is met when the voltage is one-sixth, or 1.67 V rms (4.71 pk-pk measured using the oscilloscope). With the carrier amplitude modulated at 80%, ensure that the pk-pk voltage increases. EN 61000-4-6 shows a test setup for level setting. Also measure the harmonics using the spectrum analyzer/receiver and ensure that they are at least 15 dB below the carrier level.

For unshielded cables during the test, the signal is injected in parallel into all the power conductors in the cable via a resistor and capacitor in series. The total value of all of the resistors in parallel (the C/M resistance) is 100 $\Omega$. Thus, if four lines are tested, 4 $\times$ 400 $\Omega$ resistors are used.

For the coupling directly via the CDN or via the inductive clamp with CDNs or resistor networks:

1. Ensure that the CDN has been calibrated as described in ENV 50141 and meets the $Z_{ce}$ common-mode impedance requirements (with all lines connected together while tested for the impedance) of 150 $\pm$ 20 $\Omega$ from 0.15 to 26 MHz and 150 +60/−45 $\Omega$ from 26 to 80 MHz when tested in accordance with Figure 7c of ENV 50141: 1993.
2. Connect the equipment as shown in Figure 9.24 for direct injection or as shown in Figure 9.25 for injection via the inductive clamp. Ensure that the full 10-V rms unmodulated can be developed across the 300-ohm load presented by the CDNs or CDN and resistor network. Because the voltage is measured across only the 50-ohm output of the CDN and a series impedance of 250 ohms is included in the two CDNs or resistor network, the voltage developed across the 50-$\Omega$ termination is one-sixth of the test voltage. Thus, with a 10-V rms (28-V pk-pk) level applied, the test is met when the measured voltage is 1.67 V rms (4.71 V pk-pk measured using the oscilloscope). EN 61000-4-6 shows a setup for level setting. With the carrier amplitude modulated at 80%, ensure that the pk-pk voltage increases across the 50-ohm load. Also measure the harmonics using the spectrum analyzer/receiver and ensure that they are at least 15 dB below the carrier level.

When the EUT does not have any conductive surface (such as a terminal contained within a plastic enclosure), a more realistic test setup is with the EUT completely disconnected from the ground plane, this is shown in Figure 9.26.

If a computer or other device is used to monitor the performance of the EUT during immunity testing, then this can be protected from the test level by locating it outside of the shielded room, by the use of ferrite baluns on the cable, or by a filtered connector at the shielded-

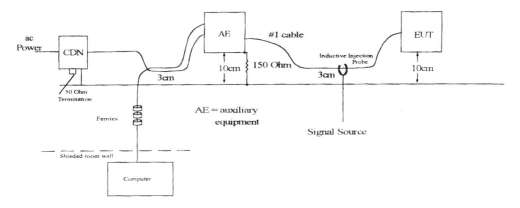

**Figure 9.26** Inductive clamp injection with the EUT disconnected from ground.

room entry. Figure 9.27 shows a similar test setup with direct injection of the C/M RF into the power lines.

For coupling via the EM-clamp, see Figure A.5.2 of ENV 50141:1993 or Annex A of EN 61000-4-6.

*Radiated Immunity H Field*

The power frequency H field immunity requirement is contained in EN 50082-2, with the test method described in EN 61000-4-8 (IEC 1000-4-8). One calibration test setup, using the MIL-STD 13.3-cm loop antenna available in many test facilities, is shown in Figure 9.28. The transmitting loop is either a 1-m × 1-m single square induction coil connected to a ground plane, a double square loop, 1 m per side and spaced 0.6 m apart, or a single rectangular coil 1 m × 2.6 m, all of which are described in EN 61000-4-8. The calibration method is as follows.

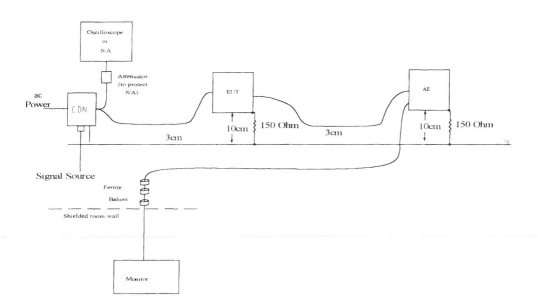

**Figure 9.27** C/M RF injection via the CDN into the power lines.

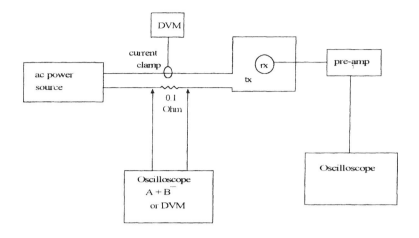

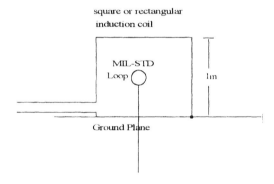

**Figure 9.28**  Calibration test setup for loop antenna.

## Equipment Checklist

1.  AC power source for continuous-mode operation with a current capability of 1–100 A divided by the coil factor (typically 0.833, resulting in a current of 1.2–120 A). The power source is typically the variac and step-down transformer used in the MIL-STD-462 RS02 test.

2.  AC power source for short duration, 1–3 s, with a current capability of 300–1000 A divided by the coil factor (typically 350–1200 A).

3.  AC power current clamp or the equipment in (7) and (8) or in (8) and (9).

4.  AC voltmeter for use with the current clamp.

5.  0.1-ohm resistor rated at $I^2 \times 0.1$, where $I$ is the test current. At 100 A the resistor may be constructed of a number of resistors with a total wattage rating of greater than 1 kW.

6.  An oscilloscope used for differential measurements (channel A + B or A − B) and an input voltage range of at least 50 V DC or the differential probe with ×10 attenuator, or an AC voltmeter.

7.  1-m × 1-m single square induction coil with ground plane or double 1-m-side, 0.6-m-spaced square induction coil or single rectangular coil 1 m × 2.6 m as described in EN 61000-4-8.

8. 13.3-cm shielded loop H field antenna, as described in MIL-STD-462 RE01, 30 Hz to 30 kHz, 1997, and MIL-STD-462D RE101, 30 Hz to 100 kHz, 1993.
9. Low-frequency preamplifier, if required, with MIL-STD loop in (8).

**Setup and Operations**

*For AC power source continuous current*:

1. Connect the variac to the AC power line, with the adjustable winding connected to the high-voltage winding of the step-down transformer.
2. Connect the 0.1-ohm resistor across the low-voltage winding of the step-down transformer with a sufficiently long connection to accommodate the jaws of the current clamp.
3. Connect the current clamp around the low-voltage connection of the step-down transformer and the clamp to the AC voltmeter.
4. Adjust the variac until the voltage measured across the voltmeter is correct for the specified test current.

*For AC power frequency test using the 0.1-ohm resistor and oscilloscope for continuous current*:

1. Connect the variac to the AC power line, with the adjustable winding connected to the high-voltage winding of the step-down transformer.
2. Connect the 0.1-ohm resistor across the low-voltage winding of the step-down transformer.
3. Connect the oscilloscope or AC voltmeter across the 0.1-ohm resistor.
4. Adjust the variac until the voltage measured across the voltmeter or measured by the oscilloscope is equal to the specified test current multiplied by 0.1.

*For AC power frequency test using the 0.1-ohm resistor and oscilloscope for short-duration current*:

1. Connect the short-duration current generator to the 0.1-ohm resistor.
2. Connect the oscilloscope differentially across the 0.1-ohm resistor.
3. Adjust the short-duration current generator until the voltage measured by the oscilloscope is equal to the specified test current multiplied by 0.1.

*Calibration of the loop for coil factor*:

1. Connect the current source to the induction coil via an 0.1-ohm resistor if the oscilloscope is used to measure current, or directly if a current clamp is used to measure current. The test setup is shown in Figure 9.26.
2. Connect the 13.3-cm loop to the oscilloscope, terminated in a 50-ohm resistor.
3. Adjust the current in the induction coil to a known value, typically 10–50 A.
4. Place the MIL-STD 13.3-cm loop in the center of the induction coil and orient the loop for a maximum voltage reading on the oscilloscope.
5. Use the conversion factor of the MIL-STD 13.3-cm loop or the manufacturer's calibration at the power-line frequency to obtain the H field in amps per meter. For example, assume that the voltage measured from the loop is 11 mV pk-pk, which converts to 3.9 mV rms = 72 dBμV. Assume the conversion factor for the loop is 68.5 dB at the power-line frequency; then the magnetic field in dBpT is 72 dBμV + 68.5 dB = 140.5 dBpT = $1.06 \times 10^7$ pT. The magnetic field in amps per meter is $1.06 \times 10^7$ pT $\times 7.936 \times 10^{-7} = 8.4$ A/m. The coil factor of the transmitting loop is given

by the magnetic field strength divided by the input current. With a transmitting induction loop current of 10 A, the coil factor of the transmitting loop is 8.4 A/m/10 A = 0.84.

*Conducted Immunity EFT*

The extrafast transient (EFT) immunity test is specified in EN 50082-1, EN 50082-2, and EN 55024, with the test setup described in IEC 1000-4-4 and EN 61000-4-4. The EFT generator has the following stringent requirements placed upon it: Up to 4 kV O/C and 2 kV when loaded with 50 Ω. The waveshape measured across 50 Ω is to be in accordance with Figure 3 of IEC-801-4 (5 ns ± 30% rise time to 90% amplitude and 50 ns ± 30% at 50% amplitude). The pulse repetition rate is to be 2.5–5 kHz, depending on output voltage, with a burst duration of 15 ms and a burst period of 300 ms. A gas discharge tube is shown in IEC 1000-4-4 as the device used to generate these very fast pulses. However, gas discharge tubes have a limited life and are often erratic in triggering. A patented semiconductor device has been manufactured by Behlke Electronic GmbH, Am Auernberg 4, D-61476 Kronberg, Germany, Tel +49(0) 6173-9229020, Fax +49 (0) 6173-929030. This is a very reliable and easy-to-use device for those who want to build their own EFT generator.

The EFT generator can best be calibrated using a calibration network/high-voltage attenuator. This network will load up to a 2.5-kV pulse with 50 Ω and divide down the voltage so that a fraction of the high-voltage pulse can be developed across the 50-Ω input of a very high-speed oscilloscope. This network can be calibrated to ensure that the waveshape of the EFT pulse measured by the oscilloscope is virtually identical to the EFT generator output pulse.

For the EFT test the following equipment is required.

**Equipment Checklist**

1. EFT generator of up to 4 kV O/C output and 2 kV when loaded with 50 Ω. The waveshape measured across 50 Ω to be in accordance with Figure 3 of IEC 1000-4-4 (5 ns ± 30% rise time to 90% amplitude and 50 ns ± 30% at 50% amplitude). The pulse repetition rate to be 2.5–5 kHz with a burst duration of 15 ms and a burst period of 300 ms. If a calibration of the generator cannot be made, ensure that the generator has a calibration certificate and that the calibration was conducted less than a year prior to the proposed test.
2. Coupling/decoupling network for AC/DC power supply lines in accordance with IEC 1000-4-4.
3. Capacitive coupling clamp for I/O circuits may also be used for AC/DC power when the coupling/decoupling network cannot be used. Capacitance between 50 and 200 pF.
4. Oscilloscope and coaxial cable with length of well-insulated wire connected to center conductor of the shielded cable.

**Capacitive Coupling Clamp or Coupling/Decoupling Network Test Setup**

Connect the capacitive clamp or coupling network to a ground plane with a minimum size of 1 m × 1 m, as shown in Figures 9.29a and 9.29b. Locate the signal/power cable under test through the capacitive clamp, and locate the EUT at a distance of less than 1 m between the end of the clamp and the EUT. The clamp has a coaxial connector at one end and a second coaxial connector at the other end. Always connect the output of the EFT generator to the connector closest to the EUT. The minimum distance between the EUT or the capacitive coupling clamp and all other conductive structures other than the ground plane shall be more than 0.5 m. The capacitive coupling clamp is the first choice for the test on I/O or communications

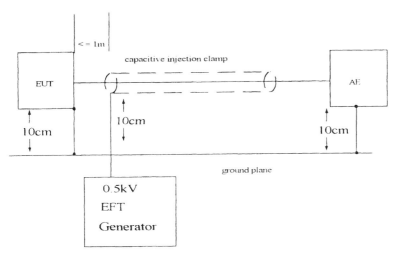

**Figure 9.29a**  Typical capacitive clamp injection of the EFT pulse with both EUT and AE tested.

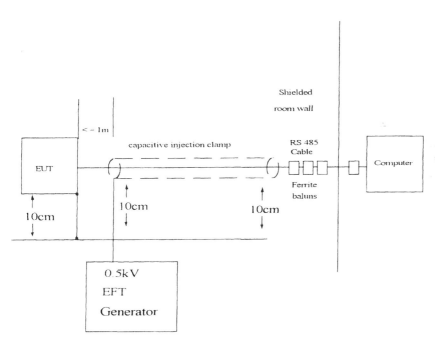

**Figure 9.29b**  Typical capacitive clamp injection of the EFT pulse with the monitoring equipment protected and outside the shielded room.

ports. If the clamp cannot be used due to mechanical problems (size, cable routing) in the cabling, it may be replaced by a tape or a conductive foil enveloping the lines under test. The capacitance of this coupling arrangement with foil or tape should be equivalent to that of the standard coupling clamp. As an alternative it might be useful to couple the generator output to the terminals of the lines under test via discrete 100-pF capacitors. If only one EUT is to be tested and not the AE, or if a computer or monitor is used to monitor the performance of the EUT during the test, it is advisable to locate this equipment outside of the shielded room. Ferrite baluns can be added on the interconnection cables or a transient suppressor/filtered connector at the entry of the cable into the shielded room, as shown in Figure 9.29b. Even with these precautions it is not uncommon for the I/O board on a computer to be destroyed during the EFT test, especially when coupling occurs within the EUT. In this case use a fiber-optic coupled I/O board.

A typical time period for calibration of the EFT generator is once a year, unless it has sustained some obvious damage, in which case an immediate recalibration is recommended. Figure 9.30 shows the CDN coupling of the EFT pulse.

*Conducted Immunity Electrostatic Discharge (ESD)*

The ESD generator is described in EN 1000-4-2 and has the following specifications:

| | |
|---|---|
| Energy storage capacitance | 150 pF ± 10% |
| Discharge resistance | 330 Ω ± 10% |
| Charging resistance | Between 50 and 100 MΩ |
| Output voltage | Up to 8 kV (nominal) for contact discharge and 15 kV (nominal) for air discharge |
| Tolerance of the output voltage indication | ±5% |
| Polarity of the output voltage | Positive and negative (switchable) |
| Holding time | At least 5 seconds |
| Discharge, mode of operation | Single discharge (time between successive discharges at least 1 s) |
| Waveshape of the discharge current | The peak current is reached with a rise time of 0.7–1 ns. At 30 ns, the current is at 53.3% of the peak, and at 60 ns at 26.6% of the peak. |

A discharge electrode with a round tip and having specific dimensions is shown for air discharges; a sharp-pointed tip with a specific angle for the tip is shown for contact discharges.

A current-sensing transducer, for use with a 1-GHz-bandwidth oscilloscope, is described for the calibration of the ESD generator. Dimensions for the transducer are provided in IEC 1000-4-2, and these transducers are commercially available. It is advisable to mount the transducer in the wall of a shielded room, with the ESD generator on the outside and the oscilloscope

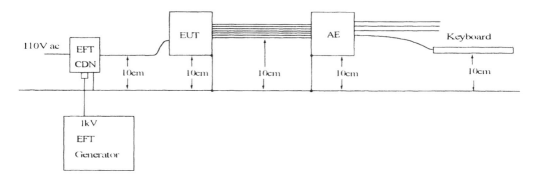

**Figure 9.30** Typical CDN coupling of the EFT pulse.

on the inside. Alternatively, a small faraday cage can be constructed, with the transducer mounted to a face made of solid sheet metal. The oscilloscope can be mounted inside the faraday cage and either viewed through a wire mesh or operated in a storage mode. A typical time period for calibration of the ESD generator is once a year, unless it has sustained some obvious damage, in which case an immediate recalibration is recommended.

The contact discharge should be made to conductive surfaces and to the horizontal coupling plane (HCP) and vertical coupling plane (VCP). Some typical conductive surfaces on an EUT contained in a nonconductive enclosure are screws and other metal fasteners, connector housings, pins in connectors, and metal covers. Some of the most potentially susceptible areas are pins in a connector to which no cable is attached, around a reset switch, and a plasma, LCD, or LED display. Air discharges should be made to insulating surfaces such as video display screens or keyboards. Some of the most potentially susceptible areas are displays and anywhere where a metal fastener, PCB, or other metal is close to the surface in a nonconductive enclosure.

A ground reference plane shall be provided on the floor of the test area. It shall be a metallic sheet (copper or aluminum) of 0.25-mm minimum thickness. Any other metallic materials may be used, but they shall have at least 0.65-mm minimum thickness. The minimum size of the ground reference plane shall be 1 m$^2$, and it shall project beyond the EUT or coupling plane by at least 0.5 m.

A distance of 1 m minimum shall be provided between the EUT and the wall of the test facility and any other metallic structure; this requirement may restrict the use of a small shielded room for this test. The EUT shall be connected to the grounding system, in accordance with its installation specifications. No additional grounding connections are allowed. If the EUT is contained in a nonconductive enclosure and DC power is derived via an isolated power supply connected to the AC outlet, no obvious ground connection is made. Often the equipment will interface to a computer or monitor, in which case the ground is often made via these devices, typically via a single-ended digital/analog or video signal interface. Just as with the EFT test, the ESD current can flow via the EUT into the peripheral equipment, resulting in damage. In one instance the equipment was connected inside a vehicle and the current path to ground was via the power return cable and via the battery terminal to the chassis of the vehicle. Analyzing the path for the ESD event current flow is very important when solving ESD susceptibility problems. The connection of the earth cable to the ground reference plane and all bondings shall be low impedance. Coupling planes are connected to the reference ground plane with cable that contains 470-k$\Omega$ resistors at each end of the cable. The ESD test setup shall consist of a wooden table 0.8 m high standing on the ground reference plane. The horizontal coupling plane is 1.6 m $\times$ 0.8 m and shall be placed on the wooden table. The EUT and cables are isolated from the HCP by an insulating support 0.5 mm thick. The ESD generator return is connected to the ground reference plane. The test programs and software, as with all immunity tests, shall be chosen to exercise all normal modes of operation of the EUT. The use of special exercising software is encouraged, but it is permitted only where it can be shown that the EUT is comprehensively exercised. For the ESD test, performance criterion B is required for a pass, so the correct functioning of the EUT after the test has been conducted must be demonstrated.

For direct application of discharges to the EUT, the ESD shall be applied to such points and surfaces of the EUT as are accessible to personnel during normal usage. The test shall be performed with single discharges. On preselected points, at least 10 single discharges (in the most sensitive polarity) shall be applied. Multiple discharges are recommended when attempting to find the source of ESD susceptibility or after modifications designed to harden against ESD.

The ESD generator is to be held perpendicular to the surface to which the discharge is applied. The discharge return cable is to be kept at a distance of at least 0.2 m from the EUT. For contact discharges, the tip of the discharge electrode shall touch the EUT, before the discharge switch is operated. In the case of air discharges, the round discharge tip of the electrode shall be approached as fast as possible (without causing mechanical damage) to touch the EUT. After each discharge, the discharge electrode shall be removed from the EUT. The generator is then charged for a new, single discharge. The discharge switch, which is used for contact discharge, shall remain closed. At least 10 single discharges shall be made to the HCP at points on each side of the EUT. At least 10 single discharges shall be made to the center of one vertical edge of the VCP. The VCP is 0.5 m × 0.5 m in size and is placed parallel to, and positioned at a distance of 0.1 m from the EUT. Figure 9.31 illustrates the ESD test setup.

*Conducted Immunity Surge Test*

The surge generator is referred to as a combination wave (hybrid) generator. The major characteristic of the combination wave generator is that it delivers a 1.2-μs/50-μs voltage surge under open-circuit (O/C) conditions and a 8-μs/20-μs current surge into a short-circuit (S/C) with an effective output impedance of 2 Ω.

Thus at a peak open-circuit voltage of 2 kV, the peak short-circuit current is 1000 A.

The characteristics of the generator, from IEC 1000-4-5, are as follows.

Open-circuit output voltage          At least as low as 0.5 kV to at least as high
                                     as 4.0 kV

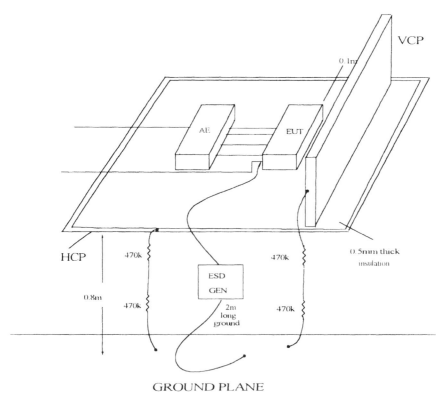

**Figure 9.31**  ESD test setup.

| Waveform of the surge voltage | Rise time of 1.2 μs ± 30% from 0.1 × Vpk to 0.95 × Vpk. Time from 0.1 × Vpk on rising edge to 0.5 × Vpk on falling edge = 50 μs ± 20% |
| Short-circuit output current | At least as low as 0.25 kV to at least as high as 2.0 kV |
| Waveform of the surge current | Rise time of 8 μs ± 20% from 0.1 × Ipk to 0.95 × Ipk. Time from 0.1 × Ipk on rising edge to 0.5 × Ipk on falling edge = 20 μs ± 20% |
| Polarity | Positive/negative |
| Phase shifting | In a range between 0° and 360° versus the AC line phase angle |
| Repetition rate | At least 1 per minute |

A generator with floating output shall be used.

Additional resistors of 10 Ω or 40 Ω shall be included to increase the required effective source impedance, as described later, and under these conditions the open-circuit voltage waveform and the short-circuit current waveform in combination with the coupling/decoupling network are no longer as specified.

Coupling/decoupling networks are used to isolate the power or signal source from the surge, and these should not significantly influence the parameters of the generators, e.g., open-circuit voltage, short-circuit current.

Capacitive coupling for power supplies allows the surge to be applied line to line or line to earth (ground) while the decoupling network is also connected.

The coupling capacitors are either 9 μF or 18 μF, and the inductance in the decoupling network is 1.5 mH.

The residual surge voltage on unsurged lines with the EUT disconnected shall not exceed 15% of the applied voltage, and shall not exceed 15% on the power supply inputs when the EUT and power supply are disconnected.

Figure 9.32 shows the capacitive coupling into AC/DC power line to line, and Figure 9.33 shows the capacitive injection line to earth.

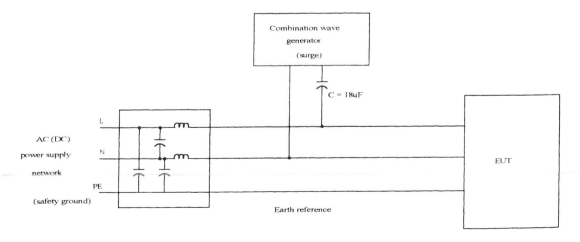

**Figure 9.32** Capacitive injection of the surge on AC/DC power line to line.

Capacitive coupling is the preferred method for unbalanced unshielded I/O circuits when there is no influence to the functional communication on that line. The application is via a 0.5-μF capacitor in series with a 40-Ω resistor, in accordance with Figure 10 of IEC 1000-4-5. Coupling via arrestors is the preferred coupling method for unshielded balanced circuits, as shown in Figure 12 of IEC 1000-4-5.

Manufacturers of commercial equipment are used to meeting FCC radiated and conducted emission requirements, and therefore the EN emission requirements do not generally present any more of a problem, whereas the immunity requirements may present a problem for certain types of equipment.

Military/aerospace equipment radiated susceptibility test levels contained in MIL-STD-461 are specified from 1 V/m and above. However, a very great difference exists between equipment designed to meet military/aerospace requirements and commercial equipment. For example, MIL-STD-461 type of equipment is typically contained in shielded metal enclosures with few if any apertures, use effective power-line filters, and are connected with shielded cables. Commercial equipment is often contained in plastic unshielded enclosures or metal enclosures that contain large slots, which often enhance emissions from that location, and use unshielded interface cables. Thus the incident E field in a radiated immunity test results in RF currents that flow on cables and on PCBs inside the enclosure. When equipment uses digital techniques and digital signal interfaces, it is itself a source of high levels of electromagnetic fields. It is self-compatible and therefore typically immune to the specified radiated susceptibility test levels and the resultant RF currents set up in the equipment. The problem occurs with telephone and audio equipment and their associated signal interfaces, in which a high incidence of susceptibility has been seen. In the early drafts of the EN documents, the C/M conducted immunity requirement appeared to apply to all classes of equipment, and much of the telephone, audio, and low-signal-level control equipment was tested with radiated immunity and C/M conducted immunity test levels and had a problem meeting the requirements.

Sections 5.1.10, 5.3.2 and 5.4 discuss how the immunity of equipment may be increased both in the design and after production.

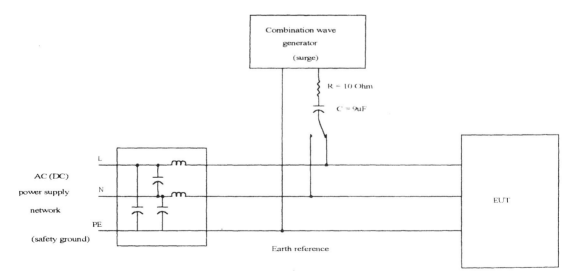

**Figure 9.33**  Capacitive injection of the surge on AC/DC power line to earth.

### 9.4.6   Taiwan EMC Law

At present, only RF emissions are regulated, although the proposed immunity standards may be adopted. The Ministry of Economic Affairs published the requirements in Commodity EMC Regulation. The Bureau of Commodity Inspection and Quarantine (BCIQ) is responsible for issuing EMC type approval certificates and for granting laboratory accreditation. By law, only Chinese National Standards (CNS) may be used as inspection standards. The emission standards are based on IEC and CISPR, as follows:

| CISPR Standard | Covers | Chinese National |
|---|---|---|
| CISPR 11 | ISM equipment | CNS 13306 |
| CISPR 13 | Radio/TVs | CNS 13439 |
| CISPR 14 | Appliances | Products covered by IEC 335 |
| CISPR 22 | ITE/Telecom | CNS 13438 |
| CISPR 16 + ANSI 63.4 | Methods | CNS 13306 |

The process to secure an EMC type approval certificate from the BCIQ and to allow equipment to be sold in Taiwan is to obtain an EMC test from a BCIQ-accredited test laboratory, submit a test report and other documentation to the BCIQ for issuance of the type approval certificate. The other documentation required includes:

Completed application form
Product sample (if requested)
Two copies of product catalog
Sufficient 4- by 5-inch color photos of the product to show its appearance and construction
        as well as component placement in the chassis and nature of the chassis assembly
Instruction manual and technical specifications
Block diagram showing all oscillator frequencies
Circuit diagram
Application fee

The BCIQ intends to audit products offered for sale in Taiwan to ensure that they conform to the type approval. Nonconforming products will be withdrawn from the market and their type approval certificates revoked. Aside from nonconformance, other reasons for the revocation of a type approval certificate include: false labeling that is not corrected immediately; failure promptly to provide a sample to the BCIQ when requested; failure to pay a processing fee in a timely manner, and fraudulent acquisition of a certificate or improper use of the same. At present, no BCIQ-accredited EMI laboratories exist outside of Taiwan.

## 9.5   SHIELDED ROOMS, ANECHOIC CHAMBERS, TRANSMISSION LINES, AND CELL ANTENNAE

The use of a shielded room for low-level electromagnetic measurements is often predicated on the high electromagnetic ambient outside of the room; also, the room may be required to contain susceptibility test levels. The use of a shielded room results in errors in the measurement of radiated emissions and in the generation of radiated susceptibility test fields. The cause of errors and some techniques available to reduce them are described in the following section.

### 9.5.1 Shielded-Enclosure Internal Fields and Antenna Errors

Shielded rooms are used primarily either to contain electromagnetic fields within them, typically for either Tempest requirements or during radiated susceptibility tests, or to reduce the ambient electromagnetic environment to low levels. Two principle types of shielded enclosures are those with low-loss, reflective interior walls and those with lossy interior walls (e.g., the anechoic chamber). When using a shielded enclosure with reflective walls for radiated emission or radiated susceptibility tests, the accuracy of measurements are adversely affected by

1. Enclosure internal reflection
2. Capacitive or inductive coupling of antennas to the conductive walls
3. Distortion, noncoherent and standing waves due to reflections from the interior surfaces
4. Enclosure excited in the TEM mode
5. Other resonance modes

Some characteristics that affect all of these factors are the shape and size of the enclosure, the position of the test setup, the size and gain of the antenna, and the presence and location of personnel and test equipment and the enclosure.

Errors of as high as $\pm 40$ dB are possible in radiated emission measurements made in a shielded room, with a change in field strength of approximately 15 dB due to a 1–2-cm change either in the spacing between the EUT and the antenna or in the location of measuring equipment or personnel. The continued use of nonanechoic chambers for measurements of electromagnetic waves is a puzzle to most antenna engineers. However, shielded rooms have been used for many years for the MIL-STD-462 and DO-160 type of tests.

Both DO-160 and MIL-STD-462, prior to MIL-STD-462D, require the use of a ground plane on which nonportable equipment is bonded. Most, but not all, procuring agencies using the foregoing specifications require that the ground plane be connected to a vertical conductive wall and located at a height of 1 m above a second conductive surface (shielded-enclosure floor); however, this does not rule out the use of an anechoic or semianechoic chamber (a chamber in which the floor is not covered with absorber). The resonances of a shielded room with and without equipment and the coupling between antennas located a meter apart compared to the coupling in free space are reproduced in Figures 9.34a and 9.34b, respectively, from Ref. 3. Figure 9.35 shows the principal reflection paths in a shielded room that cause a spatial interference pattern due to the phase difference between the primary and reflected wave. It is theoretically possible to measure either an infinitely high wave impedance (i.e., no magnetic field component) or zero wave impedance (i.e., no electric field component) due to the interference. The gain of a current element or short dipole antenna is affected by the proximity of the antenna to the ground plane. The gain of the current element over a ground plane may be found from $d$, the effective length, and $\beta = 2\pi/\lambda$. When $\beta d$ is small, the gain of the current element over the ground plane is approximately three times that for an element in isolation. When $\beta d = 2.9$, the gain is a maximum of 6.6 and the length of the antenna $d$ is $0.46\lambda$.

A knowledge of the gain is important when predicting E fields from an antenna located in a shielded room, for susceptibility measurements or for an EMC prediction of radiated emission from a current element close to a ground plane.

After almost 50 years of conducting MIL-STD radiated emission measurements in shielded rooms without damping, the requirement that RF absorber material be used in shielded rooms when performing radiated emission or radiated susceptibility measurements was introduced in MIL-STD-462D, 11 January 1993, although other, unspecified, test sites may be used.

The MIL-STD-462D/MIL-STD-461E absorber requirements, as well as other, often less

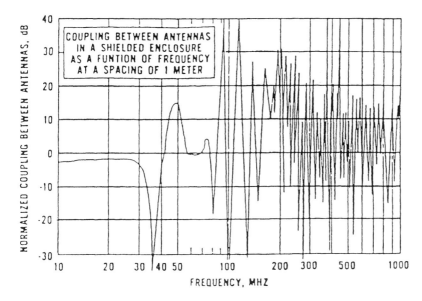

**Figure 9.34a**   Antenna coupling in a shielded room. (From Ref. 3.)

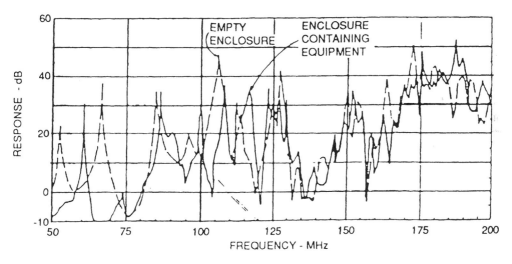

**Figure 9.34b**   Shielded-enclosure resonance effects versus frequency for constant drive with the enclosure empty and with equipment in it. (From Ref. 3.)

expensive, methods used to reduce errors inherent in shielded-room measurements are discussed next.

  1.  *MIL-STD-462D/MIL-STD-461E absorber requirements.* The RF absorber may be carbon-impregnated foam pyramids or ferrite tiles or a combination of these types of absorber material. The RF absorber shall be placed above, behind, and on both sides of the EUT and behind the radiating or receiving antenna, as shown in Figure 9.36. Minimum performance of the material shall be as specified in Table 9.14. The manufacturer's certification of their RF absorber material (basic material only, not installed) is acceptable.

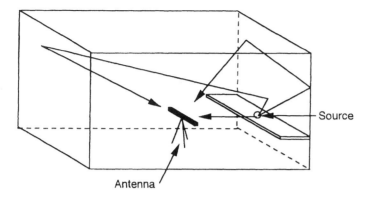

**Figure 9.35**  Multiple-path reflections in a shielded room.

2.  *Measurements made in the shielded room at a specific location with antennas calibrated at the same location.* As with all antenna factor measurements, these should be made in the far field and above the frequency where a TEM mode may occur in the room. This technique is most effective when used with high-gain antennas, such as horns, at frequencies above 500 MHz and relies on the fact that the principal resonances in the room are compensated for, to a degree, in the AF calibration, as are reflections. Although less effective, the technique may be applied to broadband antennas at frequencies below 500 MHz. The transmitting and receiving antennas should be identical, and the receiving antenna should be located at a specific location in the room. The AF calibration is then valid only for that specific location. The AF is derived using the two-antenna test method described in Section 2.5.3. The accuracy of the AF obtained by use of this method is sensitive to the location of the transmitting antenna, to the location of equipment in the room, and to the exact location of the receiving antenna. Interpretation of the resultant AF is often difficult, due to the sharp dips and peaks in the curve from room effects. Despite these sources of error, the resultant AF is more accurate than that published by the manufacturer, and at the very least room resonances may be identified by examination of the measured AF.

3.  *Stirred mode or reverberation room.* The reverberation room contains a large paddle that moves and changes the resonance frequencies of the room. Figure 9.37a illustrates the setup. The fields inside a reverberation room can be accurately described as isotropic and noncoherent and exhibit a constant, average, uniform field in the large inside volume of the room. The room can be characterized by separating the time and spatial variations, and this can be achieved by filtering the revolution frequency of the moving paddle from the change in frequency caused by the stirred mode. Thus a calibration curve of the room that compares the measured field strength to the free-space-equivalent field strength can be constructed. The accuracy of the correlation is typically ±6 dB from 45 MHz to 1 GHz and ±0.5 dB from 1 to 18 GHz. Although a fairly recent development, the use of reverberation chambers has been looked at by CISPR (International Special Commission for Radio Interference) and the U.S. Navy. MIL-STD-1377 describes a tuned-mode test setup that is useful above 200 MHz and is an improvement on MIL-STD-285 for the test of enclosure shielding effectiveness. The tuned-mode test setup is shown in Figure 9.37b. An alternative to the MIL-STD-462 radiated susceptibility test setup (RS03), using a stirred-mode reverberation room useful over the frequency range from 200 MHz to 1 GHz is described as follows: The components used inside the room are shown in Figure 9.37. The room contains a tuner stirrer. Either it is stepped and stopped, with a measurement made during the stopped interval, or a sample-hold technique may be used, with the stirrer in continu-

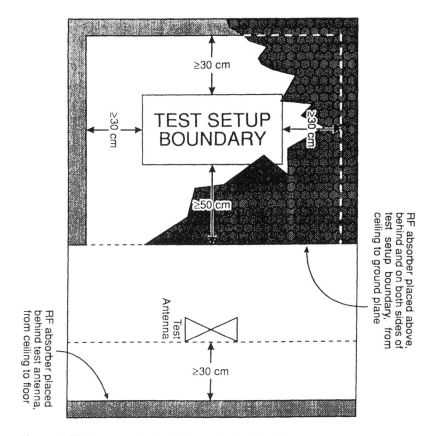

**Figure 9.36**   MIL-STD-462D RF absorber loading diagram.

ous stepped motion. The sampling is then accomplished during the stopped time of the stirrer. The stirrer is rotated through 360° in a minimum of 100 steps. Where fewer than 100 steps is used, a derating correction is required, as described later. The step rate must be slow enough to ensure that the EUT has time to respond to the changing electromagnetic environment. A power meter is used to monitor the output of the signal generator/power amplifier applied to the transmitting antenna. The transmitting antenna may be a matched long-wire antenna, up to the resonant frequency of the long wire. Isolators may be required to protect the signal source and power meter from large reflection coefficients at the test power frequencies and levels. The

**Table 9.14**   Absorption at Normal Incidence of MIL-STD-462/MIL-STD-461E Specified Material

| Frequency | Minimum absorption |
|-----------|--------------------|
| 80–250 MHz | 6 dB |
| Above 250 MHz | 10 dB |

Shielded enclosure

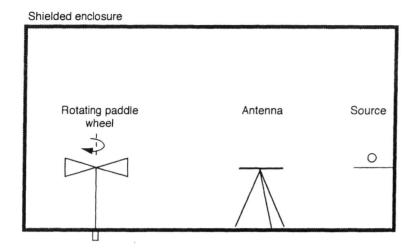

**Figure 9.37a**   Stirred-mode perturbation test configuration.

room is calibrated at a given calibration input power $P(\text{cal})$ that generates a given average calibration E field $E(\text{cal})$. The E field is measured by a receiving antenna and receiver or spectrum analyzer. To obtain the required input power, $P_{in}(\text{test})$, to develop the specified test field, $E(\text{test})$, the following equation is used:

$$P_{in}(\text{test}) = P_{in}(\text{cal}) \times \left(\frac{E(\text{test})}{E(\text{cal})}\right)^2 + 3 \text{ dB} + \text{step derating (dB)}$$

Shielded enclosure

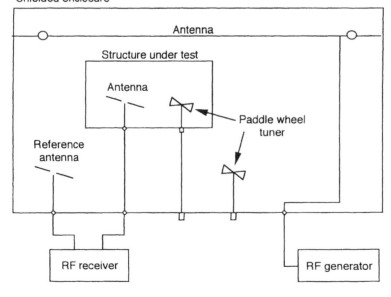

**Figure 9.37b**   Tuned-mode perturbation test configuration.

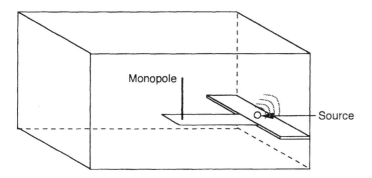

**Figure 9.38**   RE02 radiated emission shielded room configuration for frequencies below 30 MHz which may excite the room.

When the number of steps is less than 100, the step derating [given by $10 \log(100/N)$, where $N$ is the number of steps] is added. The test frequency is not swept in the stirred-mode test, unlike the standard RS03 test. Instead, the stirrer is swept through 360° at a number of fixed frequencies. Some recommended frequencies are 200, 250, 300, 350, 400, 500, 600, 800, and 1000 MHz.

Other frequencies at which the EUT may be susceptible should be used, if known. The time during which the stirrer is stopped is determined by the time required to complete a functional test on the EUT and to determine if a malfunction has occurred. Thus test times may be extremely long, especially if the 100 steps per rotation are adhered to. Another disadvantage is that the frequency is not swept, and susceptibility at a specific resonance frequency may be missed.

If the EUT is found to be susceptible, then the level of the E field generated should be closely monitored, using the stirrer position at which susceptibility occurred, if known, and adjusted to the specification level. If the EUT is susceptible at the specification level, the input power is reduced until the EUT functions as normal and the test E field is either measured or calculated. Although a more accurate method of controlling the test E field than the standard RS03 test method, the inaccuracies inherent in measuring the magnitude of the E field with a receiving antenna or probe inside the room remain.

4.   *Correction of a shielded room for electric and magnetic dipole excitation.* Below the first resonance frequency of a shielded room it (the shielded room) can be excited and function as a large TEM cell. The room is typically excited during radiated emission measurements, as per MIL-STD-462 method RE02, by the EUT, placed on the ground plane, or interconnection cables; this is illustrated in Figure 9.38. A test is described in Ref. 10 of the excitation of a shielded room by either an electric field or a magnetic field source in the frequency range of 10 kHz to 30 MHz. It was found that the position of the source of emission on the ground plane (i.e., proximity to the wall) affected the magnetic field source results by approximately 6 dB, whereas the field from an electric source varied by up to typically 25 dB. This relative insensitivity to reflections in the magnetic field has been seen in numerous measurements. One of the characteristics of an excitation mode in a room is that the measurement antenna may be moved from the measurement location to almost any location in the room with very little change in the measured E field. Once the source of emission is discovered (enclosure, aperture, or cable) a correction of the measured results based on the excitation of the TEM mode of the room may be made.

One correction technique that may be applicable is to change the effective dimensions of the room by adding metal structures. If the coupling to the room can be reduced, then the excitation of the room may be reduced. Moving the EUT and cables or even isolating the EUT from the ground plane may effectively reduce the coupling. Although in the radiated emission (RE02) type of tests it is a requirement that the EUT be bonded to the ground plane, some measure of the reduction in E field with the room nonresonant may be useful in applying for a waiver. Absorber material placed on the wall behind the EUT and on the ceiling above it, although not very effective at low frequency, can also reduce the coupling to the room.

5. *Long-wire antenna, multiwire transmission lines, and parallel-plate antennas.* The long-wire antenna technique uses the excitation of the room as a TEM cell for radiated susceptibility measurement up to 30 or 50 MHz. The typical configuration is shown in Figure 9.39a. The generation of test fields at low frequency from conventional antennas is difficult. The long-wire antenna, multiwire transmission line, and parallel-plate (stripline) antenna are all capable of generating high E fields, from 14 kH to 30 or 50 MHz for the long-wire/multiwire and up to 200 MHz for the parallel plate, with an E/H ratio of 377 $\Omega$, with the benefit of low cost. Because the wave impedance is 377 $\Omega$, only the E or H field need be measured to determine the unknown field. A transmission line, such as a coaxial cable or an air dielectric stripline will create an internal 377-$\Omega$ field when it is terminated by its characteristic impedance, even when the transmission-line impedance is far from 377 $\Omega$. When a transmission line is terminated in other than its characteristic impedance, the field impedance is directly proportional to the load impedance. The author is indebted to a private communication from Ref. 5 for a clearer understanding of the relationship between field impedance and circuit impedance ($E/H = V/I$) and that when $V/I$ is replaced by the characteristic impedance of the line, $Z_c$, the $E/H$ ratio reduces to 377 $\Omega$.

This has a very important implication in generating high E fields using a low-level power amplifier. If a step-up transformer is used between a power amplifier and a long-wire antenna or a stripline antenna, then a higher level of E field can be generated than possible without the transformer, and a closer match to the optimum load, for most power amplifiers, of 40–50 $\Omega$, is achieved. A suitable transformer is commercially available that has step-up ratios of $\times 2$, $\times 3$, $\times 4$, $\times 7$, and $\times 10$ and a frequency range of 10 kHz to 30 MHz. The transformer is heavy, but it is multipurpose because it is also designed to transform the DO-160-specified lightning pulse from 750 V, 70 μs, 1500 A to 1600 V/320 A, 70 μs, and the 750-V, 500-μs, 5000-A pulse to

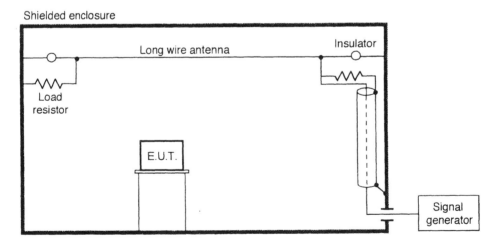

**Figure 9.39a** Long-wire antenna in a shielded-room configuration for radiated susceptibility tests in the typical configuration.

1600 V, 500 µs, 1600 A, and it can be used to increase the amplitude of a damped sinusoid waveform with a frequency of 10 kHz up to 30 MHz.

A small long-wire antenna can be designed with an impedance close to 400 Ω, modeled as a wire over a ground plane, by locating the wire 0.75 m above the tabletop ground plane. The long wire should be approximately 2 m below the ceiling of the room and at least 3 m from the nearest wall, which typically requires moving the table away from a wall. In this configuration using the ×3 step-up ratio of the transformer (impedance change of ×9) and with the long wire terminated in the characteristic impedance of the line, an E field as high as 200 V/m can be generated using a 75-W power amplifier! This configuration is shown in figure 9.39b. By comparison, a conventional E field generator requires 800 W to generate a 200-V/m field 1 m from the antenna. Adding a second wire 0.25 m from the first wire will improve the field uniformity of the line, but due to the lower impedance (approximately 280 Ω) the step-up ratio of the transformer is limited to ×2 and approximately 150 W is required to generate 200 V/m. With the long-wire antenna strung across a 2.43-m- (8′) high room, in the more typical configuration shown in Figure 9.39a, but with the use of a transformer, the E field can be approximately 50 V/m with a distance between wire and ceiling of 0.25 m, and 100 V/m with a distance of 1 m and the 75-W power amplifier. This is approximately 2.8 times the fields generated without the transformer. A multiwire transmission-line antenna is described in detail in Ref. 6. The multiwire antenna is constructed from metal tubes terminated in an array of high-power resistors connected to the chamber wall. The line is easy to assemble and dismantle and can generate 100-V/m fields up to 50 MHz, over a 4-m-deep × 6-m-wide and 2.8-m-high test area with a ±2-dB spatial field uniformity. The multiwire antenna has a much higher field uniformity than the long-wire antenna and can accommodate higher EUTs, but it is not nearly as efficient as the long-wire, or stripline, with step-up transformer, for it requires a power amplifier of 2 kW to generate 100 V/m.

A parallel-plate transmission line, often referred to as a stripline antenna, open TEM cell, or open-cage parallel-plate antenna, achieves high efficiency and a uniform field between the

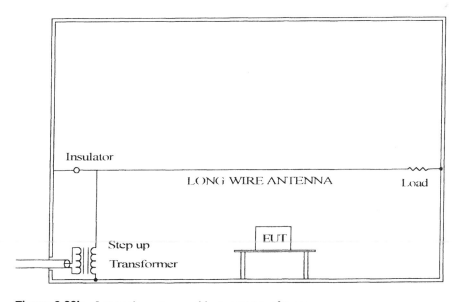

**Figure 9.39b**  Long-wire antenna with step-up transformer.

lines. It is terminated in the characteristic impedance of the line and will generate a TEM 377-Ω impedance field between the plates with the E field vertically polarized. If the upper surface of the EUT is equal to or less than one-half the width of the lines, then the upper surface of the EUT can be as close as 3 cm to the upper line when used with an upper frequency of 30 MHz, although this may result in multimoding; i.e., more than the TEM is generated. The recommended maximum EUT height is three-fourths of the distance between the lines. The upper useful frequency is given by $\lambda = 2h$, where $h$ is the distance between the plates. In practice this upper frequency is typically limited to 200 MHz for large lines and may be achieved only with great care in the manufacture of the terminating resistance and with a small EUT.

The length of the line is not limited as long as it is correctly terminated in its characteristic impedance.

Use of the parallel-plate line with a step-up transformer is not as effective as the long-wire antenna, due to the lower characteristic impedance of the line. However, a parallel-plate antenna with a 1-m distance between plates and an upper plate width of 0.4 m can be used with a ×2 step-up transformer, which matches the line to the power amplifier reasonably well and can generate an E field of up to 100 V/m with a 75-W power amplifier.

The parallel-plate or stripline antenna has been described in the Society of Automotive Engineers Aerospace Information Report AIR 1209. One of the parallel-plate antennas described has a 1-m-wide upper plate with 1/16″ slots at 1/2″ centers and an unslotted lower plate. The distance between the upper plate and the lower plate is 1 m, and the characteristic impedance of the line is 120 Ω. The termination impedance is three separated layers of 377 Ω/sq. conductive plastic film. The film distributes the impedance more evenly at the termination and reduces the coupling to the shielded-room wall from the termination by 36 dB. The recommended grounding of the lower plate is at the termination only over the complete width of the plate. The input to the line is described as a wave launcher rather than a matching section, which has an empirically determined shape.

The recommended frequency range of the line is 0 Hz to 30 MHz, and the maximum recommended EUT size is 0.75 times the strip separation.

The second line is a MIL-STD-462 type in which the height-to-width ratio is not unity. Typical characteristic impedance for these lines is anywhere from 80 Ω to 105 Ω, depending on the height-to-width ratio; the normally used frequency range is 14 kHz to 30 MHz. Again the recommended maximum EUT height is 0.75 times the distance between the plates, and the ground plane on the table may be used as the lower plate. It is recommended that the top plate of the line be at least 30 cm from the shielded-room wall. The feed point is tapered to match the 50-Ω power amplifier impedance to the impedance of the line. When used with a matching transformer, this tapered section is not required.

One variation of the stripline is the open TEM cell, in which the input section is tapered up from the drive point, located on the ground plane, to the upper plate and tapered down from the upper plate to the 50-Ω termination, located on the ground plane. An open TEM cell with the dimensions shown in Figure 9.40a is described in Ref. 7. This stripline has the great advantage of very high E fields with a low-power amplifier and a useful upper frequency of 400 MHz. The disadvantage with this antenna is that the maximum size of EUT is 0.5 m × 0.25 m × 0.06 m in height.

Because all of the plate/stripline antennas described in this section are open on the sides, they will radiate and must typically be used in a shielded room. One great disadvantage with using these antennas above 30 MHz is that the radiation from the antenna will result in room resonances that change the E field generated between the lines of the antenna. Reference 7 describes errors as high as 49 dB in the E field generated by the open TEM cell antenna due to room resonances. The techniques described in (9) to damp room resonances with ferrite tiles

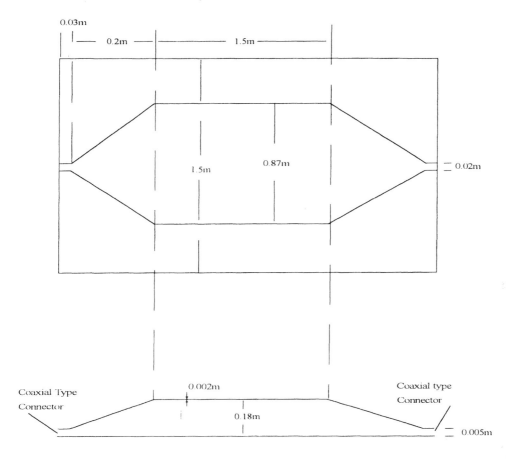

Maximum EUT dimensions = 0.5m x 0.25m x 0.06m height

**Figure 9.40a** Open TEM cell antenna usable up to 400 MHz.

and absorber loads may be employed. Reference 7 describes surrounding just the open TEM cell antenna with ferrite and small pyramid absorber, with a resultant reduction in the amount of absorber required. With this configuration, with the absorber experimentally optimized for location, the errors were reduced to within a ±3-dB range up to 400 MHz.

Figure 9.40b shows a stripline antenna with a 0.6 m height which has been calibrated to 220 MHz. Maximum EUT height is 0.4 m and the stripline is used in a damped room.

6. *Tuned resonator.* The shielded enclosure becomes a large rectangular resonant cavity at resonant frequencies. The resonant frequencies of an enclosure are given by

$$f \text{ [MHz]} = 150 \sqrt{\left(\frac{k}{l}\right)^2 + \left(\frac{m}{h}\right)^2 + \left(\frac{n}{w}\right)^2}$$

where

$l$ = length of enclosure in meters

**Figure 9.40b**   0.6m height stripline useful to 220 MHz.

$h$ = height of enclosure in meters

$w$ = width or depth of enclosure in meters

$k, m, n$ = a positive integer 0, 1, 2, 3, . . .

Note: Not more than one of $k$, $m$, or $n$ can be zero at the same time.

Reference 5 provides an approximation for the maximum number of resonant modes that can exist below a specific frequency:

$$N = 155f^3V$$

where

$f$ = frequency [GHz]

$V$ = volume of cavity [m³]

A 2.59-m-high, 3.2-m-wide, and 6.25-m-long enclosure theoretically exhibits 68 resonant modes over the 20–200-MHz frequency range. Measurements on the same-size enclosure described in Ref. 9 identified 25 resonant frequencies, with the possibility that some resonance peaks and troughs were masked by an overlapping resonance. The technique proposed in Ref. 9 to reduce the up to 48-dB variation in measured E field at resonance is to contain the field within a wave trap. A coaxial resonator having a helical inner conductor forms a high ''Q'' circuit of small physical dimensions. A 7.62-cm section of copper tubing of 5.7-cm inside diameter was constructed with field coupling to the resonator accomplished by an 20.3-cm-long antenna soldered to the high-potential end of the inner conductor. The construction enabled the resonator to be tuned manually. The resonator was positioned on a 91.4-cm² aluminum plate connected to the enclosure's metal framework. The antenna was oriented for maximum coupling to the vertical E field component. The exact location of the resonator was found to be unimportant as long as it was located further than 30.48 cm from the walls. Measurements were made using a monopole transmitting antenna and a dipole receiving antenna in a shielded room with and without the tuned resonator, and with the same test configuration using an open-field test site. The resonator was tuned to the frequency of the transmitted fields during the test.

Figure 9.41 illustrates a massive reduction of 38 dB at 52 MHz in the resonance field strength due to the tuned resonator, which results in a measurement only 3 dB above that of a specific open-field test site. The maximum deviation between the tuned resonator in the shielded-room measurement and the open-field measurement is 15 dB at 54 MHz.

It would appear that the tuned resonator technique holds the promise of a relatively easy and inexpensive solution to the problem of room resonance. An automatic test setup may be possible in which the scan of the EMI receiver or spectrum analyzer is synchronized to the tuning of the resonator, which may be driven by a stepper motor or similar device. A computer could be used to control the spectrum analyzer sweep and a servo system, used to drive the resonator and sense its position.

7. *Computing near-field antenna coupling, including reflections, in a shielded room.* Above approximately 50 MHz, near-field coupling between antennas that are physically close (0.25–1 meter) can be represented by a mutual impedance. At frequencies up to 150 MHz the coupling is predominantly capacitive, and above 200 MHz it is predominantly the radiation mode coupling between the antennas. With the reflections in the shielded-room walls described as images of the antennas, possessing mutual and self-impedance, Ref. 10 describes a method of compensating for up to two reflectors (shielded-room surfaces). Alternatively, a computer program such as GEMACS, described in Section 12.4.3.2, may be used to model the antenna as a system of wires and, where applicable, plates, and the surfaces of the shielded room as

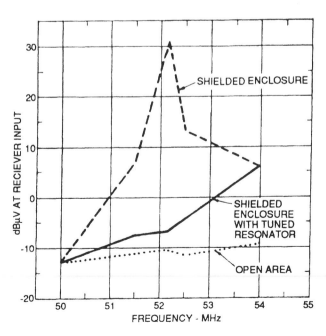

**Figure 9.41** Comparison of measurements made in a shielded room with and without a tuned resonator and on an open-field test site. (©1970, IEEE.)

ground planes. The composite E field at some distance from the transmitting antenna or the current flow in the load of the receiving antenna may then be computed. It should be pointed out that use of the GEMACS program in this type of application is not without cost in time and effort.

8. *Optoisolated, physically small antennas.* A source of perturbation of the field within a shielded room during radiated emission measurement is the antenna and the metallic connection of the antenna to the measuring equipment. By using small antennas in close proximity to the source of emission, similar to the technique described in Section 9.2.7 but with the use of electrically isolated electric field antennas, the perturbation of the electric field close to the EUT is reduced and the ratio of direct to reflected waves incident on the antenna is increased. Figure 9.42 illustrates the optoisolated antenna setup described in Ref. 1.

9. *Use of ferrite tiles and absorber loads placed at strategically located positions to damp room resonances.* The advantage of this technique is that only a small number of ferrite tiles and small amount of foam absorber material is required, which results in a less expensive modification to the room.

The problem with the standard foam absorber material is that it is placed on the conducting surfaces of the shielded room. At the surface of the room the E field reduces to zero, and long, typically 36″, absorber cones are required to obtain any attenuation at frequencies down to 200 MHz. The correct location for the foam absorber is closer into the room, and that is exactly the location where blocks of absorber material (absorber loads) are placed. The magnetic field is at a maximum on the surface of the room, and a ferrite tile, which is predominantly an H field absorber, is ideally suited for mounting on the shielded-room wall. Shielded rooms (semianechoic chambers) have been constructed in which all walls and ceiling are covered in ferrite tiles, often with absorber cones placed on top, and although this approach is effective, it is very costly. When retrofitting a self-supporting room with ferrite tiles it is advisable to check with a structural engineer or the supplier of the room to provide the loading capability, because ferrite

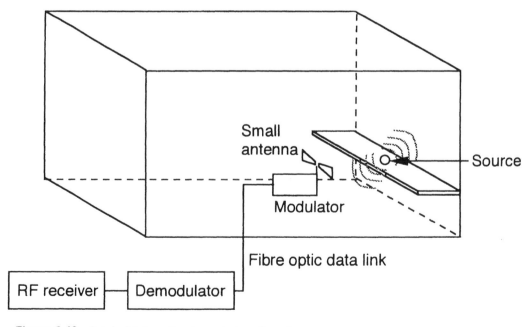

**Figure 9.42**  Optoisolated small antenna test configuration.

tiles are extremely heavy and can add approximately three times to the weight of a panel. Most rooms are self-supporting with spans of up to 10 feet. If ferrite tiles are added to small rooms, then stiffeners on the walls or ceilings may be all that is required. For rooms with spans larger than 10 feet, an external support made of columns and girders, from which the ceiling panels are suspended, may be required. For the small amount of additional ferrite described in this section, the only precaution may be to ensure that personnel do not climb on top of the self-supporting room.

Reference 11 describes the addition of an absorber load to damp resonances in a room, and Ref. 12 describes the addition of a very limited number of ferrite tiles. The absorber load alone reduced resonance affected from $\pm 30$ dB down to $\pm 15$ dB over the frequency range from 50 MHz to 200 MHz. The effect of ferrite tiles was to reduce resonance effects to 10–12 dB over the frequency range from 50 MHz to 230 MHz.

A 12′ by 12′ by 8′-high chamber was loaded with panels 2′ by 2′ covered with ferrite tiles. One panel was placed on each side wall, two on the end wall where the table was located, one panel on the rear wall, which contained the door, and three panels on the ceiling. In addition, two movable loads constructed of a wooden container 2′ by 1.3′ and 3.25′ high were filled with absorber foam. Figure 9.43 illustrates one of the wall panels loaded with ferrite and an absorber load. The 12″ absorber cones are constructed of microwave absorber, which will have no appreciable absorbing properties below 1 GHz and so are not expected to alter the test results below

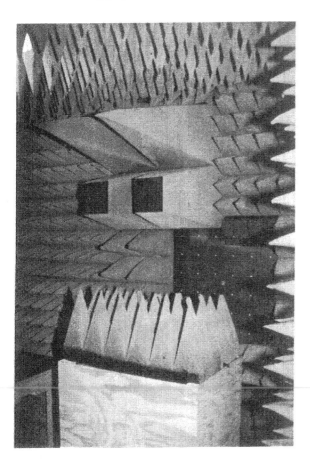

**Figure 9.43**  Shielded room with ferrite panel and absorber load for damping.

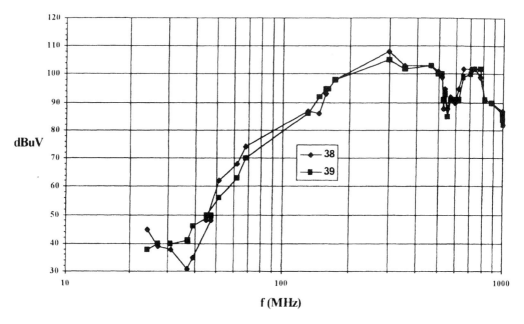

**Figure 9.44** Correlation between antenna-to-antenna measurements (horizontally oriented) on an open-field test site and in the loaded shielded room. (Reproduced courtesy of EMC Consulting, Inc.)

1 GHz. Measurements were made of the coupling between two antennas, oriented both vertically and horizontally, on the open-field test site and then in the loaded shielded room.

Figure 9.44 illustrates the coupling from 20 MHz to 1000 MHz with the antennas horizontally oriented, and Figure 9.45 illustrates the coupling with the antennas vertically oriented. In Figure 9.44 (horizontal), the only obvious resonances are: at 24 MHz, in which a 7-dB difference exists between the open-field and room measurement; at 37 MHz, with a 10-dB difference; and at 50 MHz, where the difference is 7 dB. At all other frequencies the correlation between the open-field and room measurements is 4–6 dB. In Figure 9.45 (vertical), the worst-case resonance effect is: 12 dB at 24 MHz; and 10 dB at 49 MHz. At other frequencies the correlation between open-field and room measurements are remarkably close. The lowest frequency at which the room acts as a resonant enclosure is 58 MHz, and therefore the effects at 24, 37, 49, and 50 MHz are most likely a TEM excitation of the room or possibly a loading effect on the antennas. When compared to antenna coupling in an undamped room, shown in Figure 9.34a, the improvement is remarkable.

Figure 9.43 illustrates the room with ferrite tile and absorber loads.

10. *Semianechoic and fully anechoic test chambers.* Due to the problems in attempting to isolate EUT emissions from the ambient on an open-area test site and the problems of antenna calibration on a test site open to the weather, the use of a fully anechoic or semianechoic room may gradually replace open-area test sites. A room to be used for commercial radiated emission and immunity testing must meet the normalized site attenuation standards of ANSI C63.4-1992 and the field uniformity requirements of IEC 1000-4-3 or ENV 50140. Using a room for antenna calibration, by comparison, often requires more stringent control of reflections within the room and typically extends above 1 GHz.

Before purchasing a room, the recommendation is that all future possible uses of the room be explored as well as the requirement by the FCC for radiated emission measurements above 1 GHz.

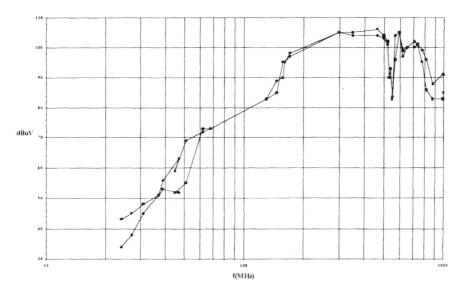

**Figure 9.45** Correlation between antenna-to-antenna measurements (vertically oriented) on an open-field test site and in the loaded shielded room. (Reproduced courtesy of EMC Consulting, Inc.)

A number of choices exist as to the type of absorber material used on the walls and ceiling of the room. The choices include: 1/4″ to 5/16″-thick ferrite tiles, grid ferrite tiles, ferrite tile and dielectric combinations, sintered ferrite and ferrite composite combinations, carbon-loaded foam pyramids or wedges, and carbon-loaded foam pyramids and wedges mounted on ferrite tiles.

Ferrite tiles require the least space in a room. Flat tiles can be very effective at low frequency, although the inclusion of a gap between the tiles as small as 0.2 mm can result in a 5-dB degradation in performance. The grid tile has only about 50% of the ferrite material active per polarization and therefore requires twice the volume of a flat tile. The grid has a lower effective permeability, which must be compensated for by increased thickness to obtain the same low-frequency performance. Due to the lower permeability, the degradation in performance due to gaps is less than seen with a flat tile. The grid tile also has a lower permittivity, which improves performance at higher frequency. According to one manufacturer, Toyo, a ferrite grid absorber provides significantly more attenuation above 220 MHz than a flat tile. Toyo describes measurements made on a room with internal dimensions of 8.7 m long × 6 m wide × 5.1 m in height, which is the smallest room size recommended as a 3-m test site. The subject room met the ±4-dB normalized site attenuation of ANSI C63.4 over the frequency range 30–1000 MHz, with the transmitting antenna located on five different points on a 1.5-m-diameter turntable, at 1-m and 2-m heights for horizontal orientation of the antenna, and at 1 m and 1.5 m for vertical orientation. The receiving antenna was scanned in height from 1 m to 4 m, with vertical and horizontal orientation, as required in ANSI C63.4.

The room also met the IEC 1000-4-3 field uniformity requirements from 30 to 1000 MHz with the antennas at a distance of 3 m. Sixteen points were measured at 0.5-m intervals, and more than 12 of the points fell within the 0–6-dB window at each measured frequency.

The hybrid combination of ferrite tile and foam absorber can be very effective; however, a word of warning: The correct combination of ferrite tile and dielectric absorber is very difficult to design over the transition frequencies, 30 MHz to 120 MHz, and a cancellation of performance can be seen. To avoid this potentially severe loss in performance it is recommended that a properly designed and tested hybrid be purchased.

The use of a multilayered structure made up of layers of different ferrite material has been proposed (Ref. 13), with an improved, lower reflection, compared to flat single-layer tile. The normalized site attenuation in a 19-m-long × 13-m-wide × 8.5-m-high anechoic chamber fitted with this material was within the ±4-dB requirement, and in a 7-m × 3-m × 3-m room the field uniformity requirement was met with improved performance compared to a single-layer tile over the 800–1000-MHz frequency range.

The main problem with ferrite tile absorbers is that the performance is not adequate above 1 GHz, although a tile with additional small sections of ferrite mounted ''end on'' has been tested to 3 GHz.

In the ideal absorber, the impedance of the incident wave and the absorber are the same as free space (377 Ω); the reflection is therefore minimal, and the depth of the absorber is sufficient to absorb all of the incident energy. In practice this is not the case. Carbon-loaded absorber is usually manufactured in a pyramid or wedge shape, which behaves like a tapered transition with an impedance to a normally incident wave of 377 Ω at the tip increasing gradually over the height. The tapered absorber provides a lower level of reflection than a solid absorber, and a lower amount of absorber material is required.

One problem in the foam absorber is that, despite being flame retardant, it is not completely flameproof, although the use of flameproof material has been proposed. Another problem is that, typically, very long pyramids are required. For example, in a chamber 9 m long, 6 m wide, and 5.1 m high, a multilayer (Ni-Zn ferrite, 70 mm dielectric-Mn-Zn rubber ferrite) absorber met the NSA requirements, whereas a 3-m-long pyramidal absorber failed the requirements. Only with a 4-m pyramidal foam absorber were the ±4-dB requirements met.

The recommended maximum reflectivity (level of reflection) of material used in a 3-m test room is less than −18 dB at normal incidence and less than −12 dB at 45° over the full frequency range, although there is no guarantee that lower levels of reflectivity may be required. Rantec manufactures a range of absorber materials, and these include a 1.6-m-long polystyrene foam carbon-loaded material mounted on a ferrite tile. This combination provides a higher reflectivity than flat and grid tile from 30 to 300 MHz but a much lower reflectivity from 500 to 1000 MHz. The reflectivity of the Rantec combination, for normal incidence, is −15 dB at 20 MHz, −25 dB at 500 MHz, −30 dB at 1 GHz, and −35 dB at 10 GHz. By contrast, a grid ferrite tile has a reflectivity of −17 dB at 20 MHz, −27 dB at 100 MHz, −25 dB at 500 MHz, but only −15 dB at 1 GHz. According to Rantec, the NSA performance of the combination absorber is within ±3 dB from 30 to 80 MHz. Rantec also manufactures a combination broadband ferrite tile and 0.41-m-long absorber, which is rated from 30 MHz to 40 GHz and appears ideal for MIL-STD/DO-160 radiated tests and the MIL-STD-462D/E recommended room.

Grace N.V. manufactures the Eccosorb HX-96, which provides −25-dB reflectivity at 40 MHz and at least −30 dB up to gigahertz frequencies. The absorber is hollow, resistive, with a truncated pyramidal shape. It consists of a thin lossy outer layer backed by a strong inner support layer. Its total height is 2.5 m. According to the manufacturer: compared to conventional solid resistive pyramidal absorbers of the same height, it is mechanically stronger and lighter in weight and contains less combustible material. Because only the thin, 1.8-cm outer layer is lossy, the risk of internal smoldering is reduced. Despite the thin layer of absorber, the performance from 30 to 80 MHz is stated to be better than with conventional absorbers.

Room resonances are due to the physical parameters of the room; the resonances for a rectangular cavity are described in this section under 6. *Tuned resonators.* By making the walls and ceilings of the room nonparallel (by varying degrees), the amount of pyramidal absorber material required for the room can be reduced, by typically halving the length of the absorber. Reference 14 discusses the analysis of a nonrectangular room.

A problem exists in the use of an anechoic chamber as a 3-m or 10-m emission test site at low frequency. In a comparison of emission measurements made on an EUT measured on an OATS and emissions in a semianechoic chamber, significant errors were seen at 50 MHz and are predicted to exist over a 10-MHz to 60-MHz range. In measurements on a 10-m semi-anechoic chamber 7.9 m high, 21.3 m long, and 11.6 m wide with 2.4-m-long absorber cones, a deviation in the ''standard'' open site attenuation performance of 9.3 dB was seen at 45 MHz. The problem is a poor absorber performance at low frequency, which most probably results in a TEM excitation of the room.

One potential problem with carbon-loaded foam is degradation in both absorptive properties and fire-retardant capability due to water damage. How much if any modification to either the fire-retardant properties or to the absorptive properties of the absorber occurs when the absorber is soaked in water depends on a number of variables, which are described as follows.

Two different methods of adding the carbon and the fire-retardant material to the neoprene latex polyurethane substrate exist. In the two-step process, used, for example, by Rantec, the carbon is soaked into the material in a water-based solution and allowed to dry. The nonorganic salts, which are used as the fire-retardant material, are then added, also in water solution, and allowed to soak the material. By adding the fire-retardant material separately, the carbon is locked in. In tests conducted on this type of absorber, when wall-mounted absorber was wetted but not immersed in water, neither the fire retardant nor RF properties of the absorber were compromised. In one test, when absorber was soaked in cold water for 2 hr and allowed to drain naturally, the material was still fire retardant and retained its RF properties.

In the one-step process, used, for example, by Cuming Corporation, the fire-retardant properties and the carbon are both applied in a water-based solvent. Theoretically, when this material is soaked in water, the carbon can migrate down through the material by gravity, resulting in a nonuniform distribution of the absorptive properties. In an extreme case, the carbon can leach out of the foam. The nonuniform distribution of the absorptive properties will, typically, affect only precise antenna measurements in the room and are unlikely to modify radiated emission or radiated susceptibility test results.

Only anecdotal information exists on the performance of this one-step process material: when the material was wetted it retained its fire-retardant and RF properties, but when soaked the fire retardancy was compromised.

One other potential source of degradation is: if water soaks through concrete onto the material, it can take salts and lime from the concrete into the material and change the properties.

## 9.5.2 GTEM, TEM, and Other Cells

TEM-type cells were developed to allow a cost-effective method for radiated susceptibility testing, and their use has been extended to radiated emission measurements and antenna calibration. The cell is a less expensive alternative to a semianechoic chamber, and a number of methods of correlation between the radiated emission results obtained using the GTEM and an OATS have been presented.

### 9.5.2.1 TEM and Other Cells

The standard dual-port TEM cell, often called *a Crawford cell*, forms a rectangular two-conductor transmission-line structure with a central septum on which the EUT is mounted and with tapered sections at each end. The transmission line is closed and is often called *coaxial*, although it does not have an axis of symmetry. The dual-port cell has an input measuring part and an output measuring port, whereas a single-port TEM has a matched load at the end without a port. The dual-port TEM tapered end sections match the middle section to the 50-$\Omega$ coaxial port connectors. The single-port, or wideband, TEM has the shape of a pyramid on its side. The termination is on the wide rectangular side and typically combines a low-frequency lumped load that matches the characteristic impedance of the TEM cell with high-frequency absorber material. The most common form of single-port cell is the GTEM cell, which is discussed in Section 9.5.2.2.

The dual-port TEM cell has a limited bandwidth determined by the generation of higher-order modes in the field. The maximum height of the EUT will limit the field uniformity in a susceptibility/immunity test and change the way in which the EUT couples to the cell. A very approximate maximum useful volume for immunity test in a cell is 0.333 times the distance between the septum and upper surface times 0.333 times the cell width, although this may be pushed to 0.5 $\times$ 0.5. A field uniformity of $\pm 1$ dB is possible in the classic TEM cell.

The cell is designed to generate a transverse electromagnetic wave with a perpendicular E field and an H field at 90°. The cell is a transmission line terminated in its characteristic impedance, so the field impedance inside the cell is 377 $\Omega$. Because the EUT is mounted, in effect, on the center conductor of the transmission line, a current flows across the EUT. The cell may therefore be used to measure the magnetic field shielding effectiveness of a small enclosure by measuring the internal current, using a small surface current probe inside the enclosure. Typically the cable connected to the current probe and to the measuring instrument located outside of the cell must be extremely well shielded. This is achieved by containing the cable within a copper tube that is soldered or connected via compression fittings to the enclosure and the wall of the TEM cell. The current probe is then moved to a location outside of the enclosure, and the ratio of measured external to internal current is equal to the magnetic field shielding effectiveness of the enclosure.

The dual-port TEM cell generates higher-order modes than the TEM mode by the nature of the geometry of the cell. When the effective height of the TEM cell is equal to a half a wavelength, the first resonance of the cell occurs; this is the limitation on the useful upper frequency. In addition, when the EUT is placed on the septum, the current around the EUT is perturbed, which also generates higher-order modes. In a standard TEM cell rated at 200 MHz (useful upper frequency), a resonance in the empty cell was seen at 161 MHz with a change in the measured 70 V/m field up to 80 V/m and down to 40 V/m. With an EUT in the cell, the same resonance at 161 MHz was seen and an additional resonance at 190 MHz, at which the 70-V/m field increased up to 120 V/m. Absorber can be placed at locations within the cell to damp these modes, but this interferes with the generation of the required TEM mode. Alternative damping techniques exist, such as cutting slots in the cell wall and backing these slots with absorber and additional shielding as well as a split-up septum. These damping techniques cannot be used to upgrade existing cells but must be built into the new cell. With the use of such damping and a split-up septum, a 0.437-m-wide, 0.284-m-high, and 1.8-m-long TEM cell can be used up to 1 GHz. However, due to the regular shape of the single-port TEM cell, it has no inherent limitation in bandwidth and would appear to be the obvious choice over the dual-port TEM cell.

The TEM cell can be used to measure radiated emissions from an EUT or a PCB. The emitted field from the EUT couples via the transmitting modes of the cell and thereby couples a voltage at the port/s of the cell. Antenna calibration can be accomplished as described in Section 9.5.2.2. The great advantage of the TEM cell for emission measurements is that it is an effective shielded enclosure that reduces the ambient induced voltages to extremely low levels, typically the noise floor of the measuring equipment. Measurements in a 3-m or 10-m anechoic chamber or an OATS are made in the far field, whereas it is the near field that couples predominantly to the TEM cell. If the radiation from the EUT is highly directional, then the coupling to the TEM cell may not be a true measure of the radiation. Also, the radiation characteristics of the EUT are different from those on an OATS, where the only conductive surface in close proximity is the ground plane. The TEM cell measures power transmitted to the output and not the maximum electric field. Acceptable emission measurements can be achieved; for example, the emissions from a PCB located 0.317–0.65 cm from the septum of a TEM cell are recorded in Ref. 15. These emissions are compared to measurements using the EMSCAN™ system and in an absorber-lined shielded room, with a close correlation.

Two very small, DC to 1 GHz and DC to 2.3 GHz, TEM cells with a patented septum design is described by Fischer Custom Communications, Inc. for both emission and IEC 1000-4-3 immunity testing. Less than 3.7 mW of input power is required to achieve a 10-V/m field and 37 W to achieve 1000 V/m. The dimensions of the area for good field uniformity is 2″ (5.08 cm) × 2″ (5.08 cm) × 0.5″ (1.27 cm), and so the cell is ideal for the measurement of ICs and small PCBs. Fisher also manufactures larger custom TEM cells with a lower useful upper frequency.

Alternative types of test cell, such as the WTEM, TrigaTEM, and Triple TEM, exist but have not proven field uniformity. One alternative, commercially available cell, called the S-LINE, is manufactured by Rhode and Schwarz. Unlike the coaxial construction of the TEM cell, the S-LINE consists of symmetrical, two-wire TEM lines in a shielded enclosure. The lines are fed with the test signal and terminated with their characteristic impedance. A larger S-LINE has 1.5-m × 1-m × 1-m dimensions and fulfills the requirements of IEC 1000-4-3 for field uniformity over a 50-cm × 50-cm area. The cell is useful from 150 kHz to 1 GHz and can generate a 10-V/m field with 20 W of input power at the center of the S-LINE. A smaller model of the S-LINE, with 1 m × 0.7 m × 0.7 m, is also available, and software is available that allows fully automatic testing.

### 9.5.2.2  GTEM Cells

The limitations as to test volume and operating frequency bandwidth of the TEM cell have been to a large extent solved with the single-port gigahertz transverse electromagnetic (GTEM) cell, which was patented in 1991. The asymmetrical septum is designed to match the flared/tapered rectangular-cross-section waveguide. The inner conductor is terminated into a hybrid broadband-matched load. For frequencies up to about 40–90 MHz, distributed resistors are used; for frequencies above this range, pyramidal foam absorbers are used. The overall field uniformity is better than ±3 dB from 30 MHz to 1000 MHz in a test area behind the large access door. The maximum EUT height is approximately 0.33 times the distance from the ground to the septum, based on the earlier TEM cell research. However, there is no theoretical limit on the size of the GTEM, only a practical one. GTEM cells are available with ground-to-septum heights of 0.5 m, 1.0 m, and 1.5 m. A 3.5-m version has been built but has proven to be expensive.

The main purpose in the use of the GTEM in radiated emission measurements is to replace the traditional OATS, due to the problem of electromagnetic ambient, and to provide a less expensive alternative to the semianechoic chamber.

The principle in the use of the GTEM for radiated emission measurements is the development of a mathematical model implemented in software for the direct comparison of data taken in a GTEM to data acquired on an OATS.

Three measurements are made, using the GTEM, with the EUT oriented in three orthogonal positions, centered in the measurement volume of the GTEM. The three positions are assumed in a manner such that the positive x-, y- and z-axes of the EUT are sequentially interchanged. The frequency, the three voltages measured at each frequency, the height of the EUT under the septum, the height of the septum over the center of the EUT, the measurement distance, and the height scan range used for the test antenna maximization search are inputs to the correlation algorithm, which is described in Ref. 16 as follows.

At each frequency, the GTEM correlation algorithm executes the following computations:

Performs a root sum of squares summation of the three orthogonal voltages
Computes the total power emitted by the EUT as determined from the summation of the three voltages and the TEM-mode equations for the GTEM
Computes the current excitation of an equivalent hertzian dipole when excited with that input power
Places the hypothetical hertzian dipole at a specified height over a perfect ground plane
Computes the horizontally and vertically polarized field strength at appropriate height intervals over the total operator-selected correlation algorithm height, 1–4 m
Selects the maximum field strength (larger) value of the horizontally or vertically polarized field strengths over the height range selected
Presents this maximum value for comparison to the chosen EMC specification limit

The main problem seen with measurements in the GTEM and in the correlation to OATS results is that mean values in the variance is 6 dB, with 11–15-dB difference in individual measurements. However, similar differences can be seen in comparing some OATS to other OATS results; see Section 9.4.1.1. This variation from OATS to OATS is particularly disappointing. The main differences between the OATS and GTEM results is that only three faces of the EUT are rotated and measured in the GTEM. A twelve position measurement consists of a measurement on all six faces of the EUT in both polarizations. A further development is a twelve plus four measurement, which can estimate the directivity of the EUT, from which an estimate of the gain can be obtained and used to replace the value of the gain of the dipole in the correlation algorithm.

A new development is a hyper-rotation GTEM, in which the EUT is placed on a gimbaled turntable within the GTEM, which itself can rotate about an axis, thus allowing for two independent degrees of motion between the EUT and the GTEM.

In addition, the effect of the ground plane on the antenna calibration in the OATS measurement should be included in the correlation algorithm, especially for low-gain antennas used at the lower heights in the OATS 1–4-m search. The other big problem is that cable manipulation to maximize emissions is a requirement in the OATS test but is impractical in GTEM tests. The EUT hyper-rotation in the GTEM was thought to compensate for cable manipulation, but tests have proven that this is not the case. Measurements of an EUT on an OATS with and without cable manipulation and in a GTEM showed a 1–11-dB change due to cable manipulation.

For MIL-STD 1-m correlation, a single-axis model has been proposed. A standard radiator is installed in GTEM, and a single-axis voltage measurement is made over a range of frequencies. The same standard radiator is located in a shielded room in a standard position. An RE02/RE102 test would be performed on the radiator that provides field strength values over the same range of frequencies. The GTEM voltage and E field levels will provide a calibration factor that can be applied to the GTEM equivalent MIL-STD-461 RE02/RE102 measurement.

The GTEM is gradually gaining acceptance for measurements to FCC and EU requirements. The FCC issued a public notice (September 1993) stating that GTEM measurement data will be accepted under the following limited conditions:

GTEM results will be accepted when the GTEM measurements have demonstrated equivalence to OATS test results.

Acceptable comparison measurements must be filed with the Sampling and Measurement Branch of the FCC with appropriate analysis that demonstrates the GTEM result equivalence with a listed OATS meeting the NSA requirements.

The validity of calculated correlation coefficients must be supported by statistical analysis. In cases of disagreement, final emission tests will be performed on an OATS.

Measuring antenna factors in a GTEM would appear to result in a good (0.16 dB to maximum 1.74 dB over 300 MHz to 1000 MHz) correlation to measurements made on ANSI C63.5 range. One GTEM antenna calibration technique has a known voltage applied to the input of the GTEM, with a measurement of the output voltage ($V_o$) from the antenna. The E field in the GTEM is found from the following simple relationship: E field = input voltage ($V_i$) divided by the septum-to-floor spacing ($h$). Since $AF = 20 \log E/V_o$, the antenna factor measured in a GTEM is equal to $20 \log (V_i) - 20 \log (V_o) + 20 \log (1/h)$.

## 9.6 MILITARY EMI REQUIREMENTS AND MEASUREMENT TECHNIQUES

### 9.6.1 MIL-STD-461: Electromagnetic Emission and Susceptibility Requirements for the Control of Electromagnetic Interference

MIL-STD-461 is a document that specifies EMI requirements and test levels on military equipment, although the space industry has also widely used the specifications on nonmilitary equipment. Initially MIL-STD-461 applied to Navy, Air Force, and Army equipment. The next issue, MIL-STD-461A, provided notices that applied to individual services. Table 9.15 provides information on which notices of MIL-STD-461 and -462 apply to either all or individual services. With the advent of MIL-STD-461E, the test requirements, test methods, and test setups contained in MIL-STD-462 have been included in MIL-STD-461E.

**Table 9.15** Services to Which Notices of MIL-STD-461 and 462 Apply

| Document | Year | Applicability |
|---|---|---|
| MIL-STD-461 | 1967 | All Services |
| MIL-STD-461A | 1968 | All Services |
| Notice 1 | 1969 | All Services |
| Notice 2 | 1969 | Air Force |
| Notice 3 | 1970 | Air Force |
| Notice 4 | 1971 | Army |
| Notice 5 | 1973 | All Services |
| Notice 6 | 1973 | All Services |
| MIL-STD-461B | 1980 | All Services |
| MIL-STD-461C | 1986 | All Services |
| MIL-STD-462 | 1967 | All Services |
| Notice 1 | 1968 | All Services |
| Notice 2 | 1970 | Air Force |
| Notice 3 | 1971 | Army |
| Notice 4 | 1980 | Navy |
| Notice 5 | 1986 | Navy |
| MIL-STD-461D | 1993 | All Services |
| MIL-STD-462D | 1993 | All Services |
| MIL-STD-461E | 1999 | All Services |

The majority of EMI test requirements and test techniques for equipment and subsystems applied by space agencies such as NASA, ESA, and CSA are based on MIL-STD-461 and MIL-STD-462. MIL-STD-461 specifies the EMI requirements and limits for military equipment, and MIL-STD-462 specifies the test techniques to be used. System-level requirements are contained in MIL-STD-464, 18 March 1997, and MIL-1541A is the military standard for electromagnetic compatibility requirements for space systems applied by the U.S. Air Force.

The goal of this section is to provide an overview of the tests but, more important, also of the most common errors that occur in EMI testing. These errors can result in damage to equipment, subsystems, or flight systems, and it is imperative, to avoid damage delay and additional cost, that a thorough understanding of the correct test methods exist and that a detailed test procedure be in place.

Although superseded by MIL-STD-461B, C, D, and E, MIL-STD-461A may still be applied, especially for reprocurement of equipment contained on old platforms. If the procuring agency is satisfied that the earlier versions of MIL-STD-461, or a tailored version thereof, adequately covers a specific electromagnetic ambient, then there is no good reason to replace them. MIL-STD-461B and C are the versions still applied widely, and MIL-STD-461B and MIL-STD-461E are the versions referred to in this chapter. Some of the differences between MIL-STD-461B/C and D/E will also be discussed. The emission and susceptibility requirements in MIL-STD-461 are designated in accordance with an alphanumeric coding system, where:

C = conducted

R = radiated

E = emission

S = susceptibility

UM = unique requirement(s) intended for a miscellaneous, general-purpose piece of equipment or a subsystem

Table 9.16 lists the emissions and susceptibility requirements contained in MIL-STD 461B Part 1.

MIL-STD 461B contains 10 parts; parts 2–10 apply to specific equipment and subsystem classes, and part 1 contains general information. Table 9.17 lists the classes of equipment and which part of MIL-STD-461B is applicable. Different requirements are placed on the different classes.

The individual classes of equipment are further divided into separate categories, and those of classes A1 and A2 are shown in Tables 9.18 and 9.19, respectively. Which of the requirements shown in Table 9.16 are applicable depends on the category of equipment, and this is shown in Table 9.20 for categories of class A2 equipment.

CE01 and CE03 requirements are placed on primary power sourced by the vehicle, ship, spacecraft, etc., and on control lines using primary power.

Although MIL-STD-461 includes power to other equipment in the applicability for CE01 and CE03, the requirement is invariably confined to primary power, which is passed on, from one equipment or subsystem to another. When CE01 or CE02 is applied to secondary power, it is typically limited to power supplied to equipment procured under a different contract or from a different manufacturer. Excessive noise on secondary power lines and signals is often detected by interequipment/subsystem susceptibility and failure during radiated emission tests.

For MIL-STD-461E requirements, Table 9.21 describes the emission and susceptibility tests and Table 9.22 summarizes the requirements for equipment and subsystems intended to be installed in or on or launched from various military platforms or installations. When an

**Table 9.16** Emissions and Susceptibility Requirements Contained in MIL-STD 461B Part 1

| Requirement | Description |
| --- | --- |
| CE01 | Conducted emissions, power, and interconnecting leads, low frequency (up to 15 kHz) |
| CE03 | Conducted emissions, power, and interconnecting leads, 0.015–50 MHz |
| CE06 | Conducted emissions, antenna terminals, 10 kHz to 26 GHz |
| CE07 | Conducted emissions, power leads, spikes, time domain |
| CS01 | Conducted susceptibility, power leads, 20 Hz to 50 kHz |
| CS02 | Conducted susceptibility, power leads, 0.05–400 MHz |
| CS03 | Intermodulation, 15 kHz to 10 GHz |
| CS04 | Rejection of undesired signals, 20 Hz to 20 GHz |
| CS05 | Cross-modulation, 20 Hz to 20 GHz |
| CS06 | Conducted susceptibility, spikes, power leads |
| CS07 | Conducted susceptibility, squelch circuits |
| CS09 | Conducted susceptibility, structure (common-mode) current, 60 MHz to 100 kHz |
| RE01 | Radiated emissions, magnetic field, 0.03–50 kHz |
| RE02 | Radiated emissions, electric field, 14 kHz to 10 GHz |
| RE03 | Radiated emissions, spurious and harmonics, radiated technique |
| RS01 | Radiated susceptibility, magnetic field, 0.03–50 kHz |
| RS02 | Radiated susceptibility, magnetic induction field, spikes, and power frequencies |
| RS03 | Radiated susceptibility, electric field, 14 kHz to 40 GHz |
| UM03 | Radiated emissions, tactical and special-purpose vehicles, and engine-driven equipment |
| UM04 | Conducted emissions and radiated emissions and susceptibility engine generators and associated components UPS and MEP equipment |
| UM05 | Conducted and radiated emissions, commercial electrical and electro-mechanical equipment |

equipment or subsystem is to be installed in more than one type of platform or installation, it shall comply with the most stringent of the applicable requirements and limits. An *A* entry in the table means the requirement is applicable. An *L* means the applicability of the requirement is limited as specified in the appropriate requirement paragraphs of this standard; the limits are contained herein. An *S* means the procuring activity must specify the applicability and limit requirements in the procurement specification. Absence of an entry means the requirement is not applicable.

If MIL-STD-461 requirements are placed on equipment, it is important to ascertain the class, category, and location of the equipment and the procuring service, for on these depend the test levels. The test levels can drive the design of equipment. For example, the radiated susceptibility test levels (RS03) for class A4 surface ships contained in part 5 of MIL-STD 461B is 150 V/m from 14 kHz to above 10 GHz for areas exposed above decks, whereas the limit is only 1 V/m below decks. The EMI emission and susceptibility test levels for category A2a equipment installed on spacecraft that are most frequently applied are provided as follows from MIL-STD-461B part 3:

| CE01 | 30 Hz to 2 kHz | 130 dBμA = 3.16 A reducing to 86 dBμA = 19.9 mA at 15 kHz |
|---|---|---|
| CE03 | 15 kHz | 86 dBμA reducing to 20 dBμA at 2 MHz |
| | 2 MHz to 50 MHz | 20 dBμA = 10 μA |
| CS01 | 30 Hz to 1.5 kHz | 5 V rms or 10% of supply, whichever is less, = 2.8 V rms for 28-V supply |
| | 1.5 kHz to 50 kHz | Reduces linearly to 1 V rms or 1% of supply voltage, whichever is greater at 50 kHz |
| CS02 | 50 kHz to 50 MHz (400 MHz) | 1 V (rms) |
| CS06 | Spike 1 | ±200 V, $t$ = 10 μs ± 20% |
| | Spike 2 | ±100 V, $t$ = 0.15 μs ± 20% |
| RE01 | 0.03–50 kHz | Dependent on location and AC or DC current equipment |
| RE02 | NB lowest limit | 30 dBμV/m = 31.6 μV/m at 27 MHz |
| | BB lowest limit | 65 dBμV/m/MHz at 200 MHz (see Figure 1.4 for BB RE02 curve) |
| RS02 | Spike 1 | 200 V, $t$ = 10 μs ± 20% |
| | Spike 2 | 100 V, $t$ = 0.15 μs ± 20% |
| | Power frequency | 3A |
| RS03 | Frequency range: | E field [volts/meter] |
| | 14 kHz to 30 MHz | 10 |
| | 30 MHz to 10 GHz | 5 |
| | 10 GHz to 40 GHz | 20 |

CS06 requirements are relaxed for equipment and subsystems whose power inputs are protected with varistors or similar transient protection devices so that if the equipment is not susceptible to peak voltages equal to the maximum safe level of the device then the requirement is met. The problem with the relaxation is the transient device may be damaged with the equipment connected to the power line in the installation, assuming voltage transients at the specified levels are present on the line and the source impedance is sufficiently low. If the maximum transient voltage predicted for the line will never be as high as specified, then the requirement should be relaxed to a realistic worst-case level. If the maximum transient voltage does exist, then a realistic CS06 test for equipment with transient protection is to calculate or measure the line impedance between the EUT and the closest source of transient noise and to use this impedance in series with the specified transient voltage minus the breakdown voltage of the device to predict the peak current through the device. The protection device may then be chosen to be capable of withstanding the peak current, and the spike generator output voltage may be applied to the input of the EUT with a series resistor equal to the supply-line impedance. This realistically tests the effectiveness of the protection device and the susceptibility of the equipment.

A safety margin exists between the emission limits and the susceptibility test levels. This safety margin ensures that, with a number of emitters colocated, the radiated ambient, which is a composite of all the emissions, is below the radiated susceptibility test levels. Likewise, the conducted emissions limit ensures that, when a number of units of equipment or subsystems share a power supply, the composite emission is below the susceptibility test levels. In many instances the safety margin is excessive. For example, if we take the narrowband RE02 limit

**Table 9.17** Classes of Equipment and the Applicable Part of MIL-STD 461B

| Class | Description | Applicable part |
|---|---|---|
| A | Equipments and subsystems that must operate compatibly when installed in critical areas | |
| A1 | Aircraft (including associated ground support equipment) | 2 |
| A2 | Spacecraft and launch vehicles (including associated ground support equipment) | 3 |
| A3 | Ground facilities (fixed and mobile, including tracked and wheeled vehicles) | 4 |
| A4 | Surface ships | 5 |
| A5 | Submarines | 6 |
| B | Equipment and subsystems that support the Class A equipment and subsystems but that will not be physically located in critical ground areas. Examples are electronic shop maintenance and test equipment used in noncritical areas; aerospace ground equipment used away from flightlines; theodolites, navaids, and similar equipment used in isolated areas | 7 |
| C | Miscellaneous, general-purpose equipment and subsystems not usually associated with a specific platform or installation. Specific items in this class are: | |
| C1 | Tactical and special-purpose vehicles and engine-driven equipment | 8 |
| C2 | Engine generators and associated components, uninterruptible power sets (UPS) and mobile electric power (MEP) equipment supplying power to or used in critical areas | 9 |
| C3 | Commercial electrical or electromechanical equipment | 10 |

**Table 9.18** Categories of Class A1 Equipment and Subsystems

| Category | Description |
|---|---|
| A1a | Air-launched missiles |
| A1b | Equipment installed on aircraft (internal or external to airframe) |
| A1c | Aerospace ground equipment required for the checkout and launch of the aircraft, including electronic test and support equipment |
| A1d | Trainers and simulators |
| A1e | Portable medical equipment used for aeromedical airlift |
| A1f | Aerospace ground equipment used away from the flightline, such as engine test stands and hydraulic test fixtures |
| A1g | Jet engine accessories |

**Table 9.19** Categories of Class A2 Equipment and Subsystems

| Category | Description |
|---|---|
| A2a | Equipment installed on spacecraft or launch vehicle |
| A2b | Aerospace ground equipment required for the checkout and launch, including electronic test and support equipment |
| A2c | Trainers and simulators |

**Table 9.20** Categories of Class A2 Equipment and Subsystems and Applicable Parts of MIL-STD 461

| Requirement | Categories of class A2 equipment/subsystems | | | Applicable | |
| | A2a | A2b | A2c | Paragraph | Limit curve |
| --- | --- | --- | --- | --- | --- |
| CE01 | T | | | 2 | 3-1 |
| CE03 | Y | Y | Y | 3 | 3-2, 3-3 |
| CE06 | $Y_L$ | $Y_L$ | | 4 | |
| CE07 | Y | Y | Y | 5 | |
| CS01 | Y | T | Y | 6 | 3-4 |
| CS02 | Y | Y | Y | 7 | |
| CS03 | $Y_L$ | $Y_L$ | | 8 | |
| CS04 | $Y_L$ | $Y_L$ | | 9 | 3-5 |
| CS05 | $Y_L$ | $Y_L$ | | 10 | |
| CS06 | Y | Y | | 11 | 3-6 |
| CS07 | $Y_L$ | $Y_L$ | | 12 | |
| RE01 | T | | | 13 | 3-7 |
| RE02 | Y | Y | Y | 14 | 3-8, 3-9 |
| RE03 | $Y_L$ | $Y_L$ | | 15 | |
| RS02 | Y | Y | Y | 16 | 3-6 |
| RS03 | Y | Y | Y | 17 | |

Y = Applicable, $Y_L$ = Limited applicability, T = Applicable on a case-by-case basis.

at 27 MHz and the susceptibility test level, which is a narrowband CW field, we could allow 158,489 pieces of equipment 1 m apart, emitting coherently at the RE02 limit, and still achieve a 6-dB safety margin between the RS03 test level and the radiated ambient. It is physically impossible to locate 158,489 pieces of equipment 1 m from each other and highly improbable that each piece of equipment emits at the limit and at the same frequency. Even when equipment and cables are located in close proximity, the level of E field generated by any equipment that has met RE02 requirements is likely to be mV/m and not V/m. For example, a safety margin of approximately 8–70 dB (above 1 MHz), dependent on frequency and source, exists between typical RE02 limits and RS03 test levels, assuming the equipment and cables are colocated 5 cm apart. One justification for imposing high radiated susceptibility test levels is that equipment may be in the proximity of transmitting antennas. The power radiated by antennas at frequencies that are required for the transmission are not included in the RE02 emission requirements. As the location of aircraft, vehicles, and ships changes, so does the electromagnetic ambient, and it is reasonable to specify high susceptibility test levels over the 14-kHz–40-GHz frequency range. However, if stationary equipment is located within structures that provide a high level of shielding, such as most aircraft, vehicles, or ships, or when the location of stationary equipment and subsystems relative to transmitters and the frequency and power of the transmissions are known, then the susceptibility test levels should be tailored accordingly. It is the intention of MIL-STD-461 that these levels be tailored for the operational radiated electromagnetic environment, and this includes both friendly and hostile emitters that equipment or a subsystem may encounter during its life cycle. For example, if a spacecraft contains a transmitter at 1.5 GHz, then the susceptibility test levels from typically 1.4–1.6 GHz may be increased, based on the transmitted power and the location of the transmitting antenna relative to equipment on the spacecraft. Test levels should be as realistic as possible, and with this intent MIL-STD-461

**Table 9.21**  Emissions and Susceptibility Requirements Contained in MIL-STD-461E

| Requirement | Description |
| --- | --- |
| CE101 | Conducted emissions, power leads, 30 Hz to 10 kHz |
| CE102 | Conducted emissions, power leads, 10 kHz to 10 MHz |
| CE106 | Conducted emissions, antenna terminal, 10 kHz to 40 GHz |
| CS101 | Conducted susceptibility, power leads, 30 Hz to 150 kHz |
| CS103 | Conducted susceptibility, antenna port, intermodulation, 15 kHz to 10 GHz |
| CS104 | Conducted susceptibility, antenna port, rejection of undesired signals, 30 Hz to 20 GHz |
| CS105 | Conducted susceptibility, antenna port, cross-modulation, 30 Hz to 20 GHz |
| CS109 | Conducted susceptibility, structure current, 60 Hz to 100 kHz |
| CS114 | Conducted susceptibility, bulk cable injection, 10 kHz to 200 MHz |
| CS115 | Conducted susceptibility, bulk cable injection, impulse excitation |
| CS116 | Conducted susceptibility, damped sinusoidal transients, cables and power leads, 10 kHz to 100 MHz |
| RE101 | Radiated emissions, magnetic field, 30 Hz to 100 kHz |
| RE102 | Radiated emissions, electric field, 10 kHz to 18 GHz |
| RE103 | Radiated emissions, antenna spurious and harmonic outputs, 10 kHz to 40 GHz |
| RS101 | Radiated susceptibility, magnetic field, 30 Hz to 100 kHz |
| RS103 | Radiated susceptibility, electric field, 2 MHz to 40 GHz |
| RS105 | Radiated susceptibility, transient electromagnetic field |

encourages the procuring agency to tailor the levels based on the predicted ambient. This means that any susceptibility test level considered appropriate may be specified. Likewise, the radiated and conducted emission test levels, although dependent on the classification of the equipment, may be tailored to match the environment. It is imperative to establish before the design begins which test levels apply and to design accordingly, just as a good design accounts for other environmental factors, such as vibration and thermal. MIL-STD-461 and -462 apply to units, individual pieces of equipment, and subsystems, but not to systems.

The exact definition of *subsystem* or *system* is open to discussion; however, one definition of a subsystem is one or more units of equipment that provide a function but not the complete function of a system. When a number of units comprise a subsystem, which will be tested as such, it is advisable to budget the specification test levels between the units, especially when the units are procured from different manufacturers. The EMI-budgeted requirements can then form part of the procurement specification. Budgeting must be done realistically; otherwise the approach falls into disregard. As an example of budgeting let us assume a subsystem comprising three units. One of the units is a transmitter at 1 GHz, the second is a piece of digital equipment, and the fourth is an analog servo system driving a high-impedance transducer. In budgeting the radiated emission limit, the 1-GHz transmitter is the only device likely to radiate at 1 GHz and harmonics thereof. Thus this unit will get a major share of the budget at 1 GHz and above, for even though the transmitting antenna is excluded from the test setup, significant radiation at 1 GHz and above must be expected from cables and enclosure. For example, it would be realistic to budget the narrowband radiated emission limits such that the 1-GHz transmitter level is 2

**Table 9.22**  MIL-STD-461E Requirement Matrix

| Equipment and subsystems installed in or on or launched from the following platforms or installations | Requirement applicability | | | | | | | | | | | | | | | | |
|---|---|---|---|---|---|---|---|---|---|---|---|---|---|---|---|---|---|
| | CE101 | CE102 | CE106 | CS101 | CS103 | CS104 | CS105 | CS109 | CS114 | CS115 | CS116 | RE101 | RE102 | RE103 | RS101 | RS103 | RS105 |
| Surface ships | A | A | L | A | S | S | S | | A | L | A | A | A | L | A | A | L |
| Submarines | A | A | L | A | S | S | S | L | A | L | A | A | A | L | A | A | L |
| Aircraft, Army, including flight line | A | A | L | A | S | S | S | | A | A | A | A | A | L | A | A | |
| Aircraft, Navy | L | A | L | A | S | S | S | | A | A | A | L | A | L | L | A | |
| Aircraft, Air Force | A | A | L | A | S | S | S | | A | A | A | A | A | L | A | A | L |
| Space systems, including launch vehicles | | | | | | | | | A | A | A | | A | L | | A | |
| Ground, Army | A | A | L | A | S | S | S | | A | A | A | A | A | L | L | A | |
| Ground, Navy | A | A | L | A | S | S | S | | A | A | A | A | A | L | A | A | |
| Ground, Air Force | A | A | L | A | S | S | S | | A | A | A | A | A | L | A | A | L |

A = applicable; L = limited as specified in the individual sections of this standard; S = procuring activity must specify in procurement documentation. No entry means requirement is not applicable.

dB down on the subsystem limit and the other units are 20 dB down. Assuming the analog unit does not contain a switching-power supply, it is unlikely to generate narrowband or broadband conducted noise, whereas the digital equipment is likely to generate both narrowband and broadband conducted noise. Remember, narrowband noise is not necessarily limited to CW sources but includes harmonically related emission with a PRF above a specific frequency. The 1-GHz transmitter probably contains a switching-power supply for efficiency and thus may generate both broadband and narrowband conducted noise. A reasonable budget for a broadband conducted noise is therefore 5 dB down on the subsystem limits for the digital equipment, 8 dB down for the transmitter, and 26 dB for the analog equipment. Should any unit fail the budgeted specification limit but meet the subsystem limit, the manufacturer would apply for a waiver; and if the emissions from the remaining two units were insignificant at the same frequency or frequencies, then the waiver would be granted. Budgeting ensures that a subsystem will meet requirements and places a contractual requirement on unit manufacturers and subcontractors.

It is a common occurrence for equipment to fail at least one or more of the MIL-STD-461 requirements. Equipment that has not been designed with any consideration given to achieving EMC fails most frequently, and some equipment appears to have been designed to fail! Some of the most difficult tests to pass are RE02, CS01, and RS03. RE02 emission limits are low and are readily exceeded. For example, with no more than 30 μA at 27 MHz flowing on a cable located 5 cm above the ground plane, the typical RE02 test setup, the RE02 limit is reached. CS01 test levels at 30 Hz and above often result in equipment susceptibility. Power-line filters rarely provide significant attenuation at frequencies below 10 kHz, and regulated power supplies attempt to regulate at frequencies from 30 Hz to 1 kHz. Over this frequency range the phase shift in the supply control loop may cause the supply to oscillate at one or more critical frequencies. Thus, power supplies should be designed, or chosen, with the CS01 test requirements taken into consideration. RS03 levels can result in high current flow on cables, especially at cable resonant frequencies, and less frequently may enter enclosures via aperture or seam coupling.

MIL-STD-461C was introduced in 1989 and included test levels that covered the direct and indirect effects of a nuclear electromagnetic pulse (NEMP). Some of the additions to MIL-STD-461B contained in MIL-STD-461C are

---

CS10     Conducted susceptibility, damped sinusoidal transients, pins, and terminals, 10 kHz to 100 MHz at a maximum current of 10A

CS11     Conducted susceptibility, damped sinusoidal transients, cables, 10 kHz to 100 MHz at a maximum current of 10A

RS05     Radiated susceptibility, electromagnetic pulse field transient 550 ns wide at the 0.1 × peak field pulse level (peak field is 50 kV/m)

---

Some of the modifications made to MIL-STD-461B and contained in 461C are

In part 1, the filter capacitor limitation of 1.1 μF maximum for 60-Hz and 0.02 μF maximum for 400-Hz supplies is confined to U.S. Air Force equipment only.

RE01     Relaxed 36 dB in the 30-kHz frequency range.

CE07     Requirements limited to 50-μs-duration spikes.

CE01     Measurement bandwidth limited to power frequency + 20% for AC and 75 Hz for DC leads.

The major differences between MIL-STD-461B/C and MIL-STD-461E are given in the following sections, describing the different tests.

### 9.6.2  MIL-STD-462: Measurement of Electromagnetic Interference Characteristics

Up until MIL-STD-461E, the MIL-STD-462 document contained the test method to be used in testing to MIL-STD-461 requirements. The intent in this section is not to reproduce the contents of MIL-STD-462, which is freely available. Instead, a test plan and sample test procedures based on MIL-STD-462 are presented. The test plan and procedures contain widely accepted interpretations of MIL-STD-462 methods and describe common errors encountered in this type of test.

In the typical MIL-STD-462 test configuration, the EUT is mounted on a copper or brass ground plane that is at a height of 1 meter above the floor and bonded to one wall of the shielded room in which EMI tests are conducted. This test configuration is acceptable for units installed in a similar configuration. However, when units are mounted in a different configuration, for example, one above the other on a conductive structure or open framework, then the typical MIL-STD-462 test setup, in which units are located side by side, is not representative. Figure 9.46 shows a test setup in which units of equipment are mounted on a metal frame. The equipment and frame comprise the Wind Imaging Interferometer (WINDII), which was manufactured by Canadian Astronautics for the Canadian Space Agency. The instrument is located at some height above the spacecraft and electrically connected to the spacecraft via a bonding strap. The wooden frame on which the instrument is mounted during EMI tests places the instrument at approximately the same height above the shielded-room floor as the instrument is above the spacecraft. One unit of equipment that is physically separated from the instrument and that is mounted directly on the spacecraft structure is placed on the ground plane as shown in Figure 9.47. Thus the test setup is as representative of the actual spacecraft configuration as possible and ensures that the influence of the metal structure, proximity of units to each other, and interconnection cable lengths are accounted for.

When equipment is rack mounted, the racks are placed on the floor of the shielded room and colocated as close to the actual configuration as possible. Interconnection cables are typically looped down and placed 5 cm above the floor of the room. In radiated emission and susceptibility tests, the antenna is located 1 m from the side of the equipment from which maximum emissions are detected or that is predicted to be most susceptible. Tests may have to be repeated when more than one side of the equipment meets these criteria. In the test setup shown in Figure 9.47, the equipment mounted on the ground plane was tested for radiated emissions and susceptibility separate from the instrument on the wooden frame.

The EUT transmitting antennas are excluded from *both* broadband and narrowband emission tests. The EUT antennas should be replaced by dummy loads located outside of the shielded room. The CE06 test is intended for measurement of conducted emissions on antenna terminals. Therefore, receiving antennas connected to the EUT are typically excluded from radiated emission measurements and should definitely be excluded from radiated susceptibility measurements. Receiving antennas are excluded from radiated susceptibility tests because the receiver will fail at in-band susceptibility frequencies. Also, damage may occur to the front end of the receiver due to the high incident field. If antennas can not be removed cover with absorber. The conducted susceptibility tests CS03, CS04, and CS05 are applicable to receiving equipment and subsystems and are designed to test for out-of-band response, intermodulation products, and cross-modulation. Transmitters (without antenna) are not specifically excluded from RE02 requirements at the frequency or band of frequencies at which the transmitter operates, but may be excluded at the discretion of the procuring agency.

**Figure 9.46**   Wind Imaging Interferometer (WINDII) instrument. (Reproduced courtesy of the Canadian Space Agency.)

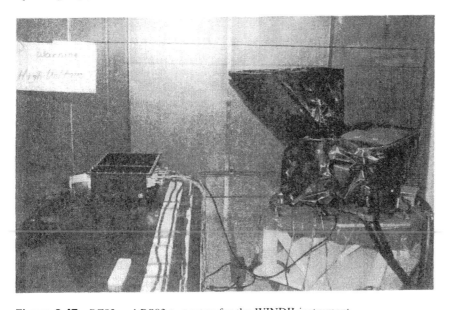

**Figure 9.47**   RE02 and RS03 test setup for the WINDII instrument.

### 9.6.3  Test Plan and Test Procedures

The test plan should outline techniques, procedures, and instrumentation used in verifying that the equipment/subsystem complies with the EMC requirements. The test plan describes the purpose of the test and how the operation of the equipment is chosen to ensure a maximum emission and susceptibility mode of operation. The test procedure details the means of implementation of the tests to be performed in order to demonstrate compliance with the applicable EMC requirements.

#### 9.6.3.1  Test Plan

The performance-monitoring and functional test equipment and method should be described in detail in the qualification test plan. However, where the qualification test plan describes an invasive form of test (e.g., monitoring of voltage, waveform, or data within the EUT), these tests are not applicable during EMI testing. Instead, a test set or fixture is often used to perform an operational test of the EUT during EMI testing.

The test set should provide stimulus to the EUT where required (e.g., data bus control lines). It should be isolated from the ambient electromagnetic fields during radiated susceptibility tests and should not be a source of radiated emission from the enclosures during radiated emission tests. This isolation and suppression of emissions is typically achieved by locating all functional test equipment outside of the shielded room during EMC tests.

The load represented by the test set should simulate the actual load on the EUT as closely as possible. Where a difference exists between the test set interface circuits and that used within the actual load, the voltage induced during EMC radiated susceptibility tests should be measured at the test set inputs. The level of induced voltage should then be compared to the noise threshold of the actual interface circuits to establish EMC.

Where the circuits are identical, the level of immunity will be determined simply by correct operation of the EUT as determined by the test set. The time required to complete tests on all control lines, data buses, etc. by the test set must be known in order to limit the speed of the frequency scan during radiated emission tests, etc. Where possible, the test set should be designed to replace the need for manual operation of switches, etc.

It is usually not feasible to operate control levers or switches remotely, because the presence of an operator close to the EUT introduces large errors into the radiated emission and radiated susceptibility tests, the control levers should remain in one position during radiated emission and radiated susceptibility EMI tests. However, some procuring agencies require the operation of switches during EMC tests, regardless of the errors introduced.

Radiated emissions from interface cables caused by common-mode noise generated within the test set should be reduced to the extent feasible. Installation of filter connector adapters on power and signal connectors on the test set is an effective approach in reducing both common- and differential-mode noise.

A background radiated emission measurement should be made with the EUT powered down and the test set powered up in order to characterize any radiated emissions from the interface cables caused by the test setup.

#### 9.6.3.2  Test Procedures

Test procedures should contain, as a minimum, the following:

    List of contents
    Applicable documents
    Purpose of test

Table of frequencies and repetition rate of broadband frequencies at which interference
    may be expected
Test sample excitation and performance monitoring
Test cables
Dummy loads
Power supplies
Grounding and bonding during tests
Procedures for each parameter test stating:
    (a) Applicability
    (b) Test setup, including detailed drawings
    (c) Test points
    (d) Limits and frequency range
    (e) Operating procedures for the test sample operating and monitoring equipment
    (f) Test sample operating modes
    (g) Pass/fail criteria for susceptibility tests
    (h) Transfer impedance of probes
    (i) Antenna factors
Test Result Sheets, including:
    (a) Procedure step
    (b) Specification for limits
    (c) Parameter
    (d) Test data
    (e) Initial of person responsible for test
    (f) Quality assurance signature

### 9.6.4 General Test Guidelines

a. All test equipment will be certified as calibrated, with verification that the calibration is still current. This information will be entered on the Test Equipment Sheet. The test equipment serial numbers and calibration information shall be entered on this sheet.

b. Where a monopole antenna is used for either radiated emission or radiated susceptibility tests (14 kHz to 20–30 MHz), the antenna counterpoise (ground plane) should be bonded to the ground plane on the table. If the EUT is too large and is mounted on the floor of the shielded room, then the monopole is used uncoupled. Where the monopole has been calibrated coupled and uncoupled (coupled is when two test antenna ground planes are connected together by a counterpoise, and uncoupled is without the counterpoise), care must be taken to use the uncoupled calibration.

c. The test antenna shall be remote from the measuring equipment. The distance shall be determined by two factors:

d. Far enough away to reduce interaction between the antenna and measurement equipment.

e. The distance shall be limited by excessive attenuation in the cable length. Ideally, the measuring equipment shall be located outside of the shielded room.

f. When performing radiated emission measurements, no point of the measuring antenna shall be less than 1 m from the walls of the shielded enclosure or obstruction.

g. For susceptibility measurements, no point of the field-generating and the field-measuring antennas shall be less than 1 m from the walls of the enclosure.

h. MIL-STD-461E requires that the enclosures be sufficiently large such that the EUT arrangement requirements and antenna-positioning requirements described in the individual test methods are satisfied.

i. Care should be taken to ensure that all accessory equipment used, such as spectrum monitors, oscilloscopes, earphones, and other equipment used in conjunction with EMI analyzers, does not affect measurement integrity.

j. The side of the equipment emitting the highest level of radiation shall be determined, and that side shall face the receiving antenna.

k. Ambient levels measured during testing with the EUT de-energized shall be at least 6 dB below the applicable limits of MIL-STD-461. Ambient conditions shall be determined prior to the beginning of measurements on the energized EUT and repeated during the progress of the test if there is reason to believe ambient levels have changed. If these conditions cannot be met and the offending sources are outside of the control of the test engineer (i.e., noisy functional test equipment or, despite filtering, noisy power supplies), the test may proceed, provided it can be demonstrated that the EUT provides negligible contributions to the emissions over the frequency range in question.

l. Typical product assurance provisions demand that the assigned Quality Assurance Engineer shall witness the tests and determine, in conjunction with the Test Conductor, whether or not a failure has occurred. Any failure shall be subject to a Test Observation Record. Prior to the start of testing, a test readiness review shall be held to determine test readiness and compliance. The system shall be visually inspected and approved as a production-representative unit.

m. The measuring equipment should be monitored first for spurious emissions. False data caused by such spurious recordings should be identified on $X$-$Y$ recordings of photos from oscilloscopes.

n. If a high-level signal is present at a frequency outside of the frequency span examined by the EMI receiver or spectrum analyzer, false measurements may be made as a result of overload and compression of the front end of the receiver, spectrum analyzer, or preamp (where used). To check for the presence of overload and compression, either bypass the preamp (where used) or insert a 3–20-dB attenuator at the input of the measuring instrument and ensure that the peak emission recorded falls by the appropriate dB value. The spectrum analyzer/receiver input attenuator may be adjusted to achieve the same attenuation. NB: Where the signal is close to the noise floor, a 3–6-dB attenuator may be more appropriate than the 20 dB.

o. Prior to performing tests, the measurement equipment shall have been switched on for a period of time adequate to allow parameter stabilization. If the operation manual does not specify a period of time, the minimum warmup time shall be 1 hour.

p. Broadband and narrowband swept measurements are made with the peak detector function. Spot frequency measurements may be made using an rms detector, provided the signal characteristics are known and errors can be calculated. The following is from MIL-STD-461E: A peak detector shall be used for all frequency-domain emission and susceptibility measurements. This device detects the peak value of the modulation envelope in the receiver bandpass. Measurement receivers are calibrated in terms of an equivalent root mean square (rms) value of a sinewave that produces the same peak value. When other measurement devices, such as oscilloscopes, nonselective voltmeters, or broadband field strength sensors, are used for susceptibility testing, correction factors shall be applied for test signals to adjust the reading to equivalent rms values under the peak of the modulation envelope.

q. All measurement equipment shall be operated as prescribed by the applicable instruction manuals unless specified herein. For test repeatability, all test parameters used to configure the test shall be recorded in the EMI test report. These parameters shall include measurement bandwidths, video bandwidths, sweep speeds, etc.

The general test guidelines and precautions that are generally applicable or that are contained in MIL-STD-461E are:

a. *Calibration of measuring equipment and antennas.* Test equipment and accessories required for measurement in accordance with this standard shall be calibrated under an approved program in accordance with MIL-STD 45622. In particular, measurement antennas, current probes, field sensors, and other devices used in the measurement loop shall be calibrated at least every 2 years unless otherwise specified by the procuring activity, or when damage is apparent. Antenna factors and current probe transfer impedances shall be determined on an individual basis for each device.

b. *Measurement system test.* At the start of each emission test, the complete test system (including measurement receivers, cables, attenuators, couplers, and so forth) shall be verified by injecting a known signal, as stated in the individual test method, while monitoring system output for the proper indication.

c. *Antenna factors.* Factors for electric field test antennas shall be determined in accordance with SAE ARP-958

Requirements that are specific to MIL-STD-461E are:

a. *Excess personnel and equipment.* The test area shall be kept free of unnecessary personnel, equipment, cable racks, and desks. Only the equipment essential to the test being performed shall be in the test area or enclosure. Only personnel actively involved in the test shall be permitted in the enclosure.

b. *RF hazards.* Some tests in this standard will result in electromagnetic fields, which are potentially dangerous to personnel. The permissible exposure levels in ANSI C95.1, in the United States, and safety code 6, in Canada, shall not be exceeded in areas where personnel are present. Safety procedures and devices shall be used to prevent accidental exposure of personnel to RF hazards.

c. *Shock hazard.* Some of the tests require potentially hazardous voltages to be present. Extreme caution must be taken by all personnel to ensure that all safety precautions are observed.

d. *Federal Communications Commission (FCC) and Industry Canada restrictions.* Some of the tests require high-level signals to be generated that could interfere with normal FCC- or Industry Canada–approved frequency assignments. All such testing should be conducted in a shielded enclosure. Some open-site testing may be feasible if prior FCC or Industry Canada coordination is obtained.

e. *EUT test configurations.* The EUT shall be configured as shown in the general test setups. These setups shall be maintained during all testing unless other direction is given for a particular test method.

f. *Bonding of EUT.* Only the provisions included in the design of the EUT shall be used to bond units such as equipment case and mounting bases together or to the ground plane. When bonding straps are required to complete the test setup, they shall be identical to those specified in the installation drawings.

g. *Shock and vibration isolators.* EUTs shall be secured to mounting bases having shock or vibration isolators if such mounting bases are used in the installation. The bonding straps furnished with the mounting base shall be connected to the ground plane.

When mounting bases do not have bonding straps, bonding straps shall not be used in the test setup.

h. *Interconnecting leads and cables.* Individual leads shall be grouped into cables in the same manner as in the actual installation. Total interconnecting cable lengths in the setup shall be the same as in the actual platform installation. If a cable is longer than 10 meters, at least 10 meters shall be included. When cable lengths are not specified for the installation, cables shall be sufficiently long to satisfy the following conditions specified: At least 2 meters (except for cables that are shorter in the actual installation) of each interconnecting cable shall be run parallel to the front boundary of the setup. Remaining cable lengths shall be routed to the back of the setup and shall be placed in a zigzagged arrangement. When the setup includes more than one cable, individual cables shall be separated by 2 centimeters measured from their outer circumference. For benchtop setups using ground planes, the cable closest to the front boundary shall be placed 10 centimeters from the front edge of the ground plane. All cables shall be supported 5 centimeters above the ground plane.

i. *Input power leads.* Two meters of input power leads (including returns) shall be routed parallel to the front edge of the setup in the same manner as the interconnecting leads. The power leads shall be connected to the LISNs (see 4.6). Power leads that are part of an interconnecting cable shall be separated out at the EUT connector and routed to the LISNs. After the 2-meter exposed length, the power leads shall be terminated at the LISNs in as short a distance as possible. The total length of power lead from the EUT electrical connector to the LISNs shall not exceed 2.5 meters. All power leads shall be supported 5 centimeters above the ground plane.

j. *Construction and arrangement of EUT cables.* Electrical cable assemblies shall simulate actual installation and usage. Shielded cables or shielded leads (including power leads and wire grounds) within cables shall be used only if they have been specified in installation drawings. Cables shall be checked against installation requirements to verify proper construction techniques, such as use of twisted pairs, shielding, and shield terminations. Details on the cable construction used for testing shall be included in the EMITP.

k. *Operation of EUT.* During emission measurements, the EUT shall be placed in an operating mode that produces maximum emissions. During susceptibility testing, the EUT shall be placed in its most susceptible operating mode. For EUTs with several available modes (including software-controlled operational modes), a sufficient number shall be tested for emissions and susceptibility such that all circuitry is evaluated. The rationale for modes selected shall be included in the EMITP.

l. *Computer-controlled receivers.* A description of the operations being directed by software for computer-controlled receivers shall be included in the EMITP required by MIL-STD-461. Verification techniques used to demonstrate proper performance of the software shall also be included.

m. *Orientation of EUTs.* EUTs shall be oriented such that surfaces that produce maximum radiated emissions and respond most readily to radiated signals face the measurement antennas. Bench-mounted EUTs shall be located $10 \pm 2$ cm from the front edge of the ground plane, subject to allowances for providing adequate room for cable arrangement as specified in (j).

n. *Susceptibility monitoring.* The EUT shall be monitored during susceptibility testing for indications of degradation or malfunction. This monitoring is normally accomplished through the use of built-in test (BIT), visual displays, aural outputs, and other measurements of signal outputs and interfaces. Monitoring of EUT performance through installation of special circuitry in the EUT is permissible; however, these modifications shall not influence test results.

o. *Use of measurement equipment.* Measurement equipment shall be as specified in the individual test methods of MIL-STD-461E. Any frequency selective measurement receiver may be used for performing the testing described in this standard, provided that the receiver characteristics (that is, sensitivity, selection of bandwidths, detector functions, dynamic range, and frequency of operation) meet the constraints specified in this standard and are sufficient to demonstrate compliance with the applicable limits of MIL-STD-461. Typical instrumentation characteristics may be found in ANSI C63.2.

p. *Safety grounds.* When external terminals, connector pins, or equipment grounding conductors in power cables are available for ground connections and are used in the actual installation, they shall be connected to the ground plane. Arrangements and length to be in accordance with "Interconnecting leads and cables."

### 9.6.4.1 Emission Tests

a. *Operating frequencies for tunable RF equipment.* Measurements shall be performed with the EUT tuned to not less than three frequencies within each tuning band, tuning unit, or range of fixed channels, consisting of one midband frequency and a frequency within ±5% from each band or range of channels.

b. *Operating frequencies for spread spectrum equipment.* Operating frequency requirements for two major types of spread spectrum equipment shall be as follows:

    i. *Frequency hopping*: Measurements shall be performed with the EUT utilizing a hop set that contains 30% of the total possible frequencies. The hop set shall be divided equally into three segments at the low end, middle, and high end of the EUT's operational frequency range.

    ii. *Direct sequence.* Measurements shall be performed with the EUT processing data at the highest possible data transfer rate.

### 9.6.4.2 Susceptibility Tests

*Frequency scanning.* For susceptibility measurements, the entire frequency range for each applicable test shall be scanned. For swept frequency susceptibility testing, frequency scan rates and frequency step sizes of signal sources shall not exceed the values listed in Table 9.23. The rates and step sizes are specified in terms of a multiplier of the tuned frequency ($f_o$) of the signal source. *Analog* scans refer to signal sources, which are continuously tuned. *Stepped* scans refer to signal sources that are sequentially tuned to discrete frequencies. Stepped scans shall dwell

**Table 9.23**  Susceptibility Scanning from MIL-STD-461E

| Frequency range | Analog scans, maximum scan rates | Stepped scans, maximum step size |
|---|---|---|
| 30 Hz–1 MHz | $0.0333f_o$/s | $0.05f_o$ |
| 1 MHz – 30 MHz | $0.00667f_o$/s | $0.01f_o$ |
| 30 MHz – 1 GHz | $0.00333f_o$/s | $0.005f_o$ |
| 1 GHz – 8 GHz | $0.000667f_o$/s | $0.001f_o$ |
| 8 GHz – 40 GHz | $0.000333f_o$/s | $0.0005f_o$ |

at each tuned frequency for a minimum of 1 second. Scan rates and step sizes shall be decreased when necessary to permit observation of a response.

   a.  *Modulation of susceptibility signals.* Susceptibility test signals above 10 kHz shall be pulse modulated at a 1-kHz rate with a 50% duty cycle unless otherwise specified in an individual test method of this standard.
   b.  *Thresholds of susceptibility.* When susceptibility indications are noted in EUT operation, a threshold level shall be determined where the susceptible condition is no longer present. Thresholds of susceptibility shall be determined as follows:
       i.   When a susceptibility condition is detected, reduce the interference signal until the EUT recovers.
       ii.  Reduce the interference signal by an additional 6 dB.
       iii. Gradually increase the interference signal until the susceptibility condition reoccurs. The resulting level is the threshold of susceptibility.

### 9.6.5   Typical EMI Receiver or Spectrum Analyzer Bandwidths for MIL-STD-461A-C

The receiver bandwidths used with the typical EMI measurement system for measurement of narrow and broadband emissions are shown in Table 9.24. These are based on measurements made of the EMI receiver system noise floor and on the worst-case emission requirements. Resolution bandwidths may be changed where necessary, for example, when the measuring equipment bandwidths are different or when the measuring equipment noise floor is close to or above the limit.

The change from recommended bandwidth shall be no greater than a factor of 0.5–2, and the broadband bandwidth shall be from 5 to 10 times the narrowband bandwidth.

#### 9.6.5.1   Determination of Either Narrowband (NB) or Broadband (BB) Emissions

Use the tuning test described in Section 9.2.2 to determine if the emission is narrowband or broadband. If the results of the test are inconclusive, use one or more of the alternative tests. Clarification or final arbitration on the decision should be provided by the EMC engineer or the procuring agency.

*Conversion of broadband measured values to the normalized broadband values of dBµV/m/MHz and dBµV/MHz:* The reference bandwidth of broadband measurements is 1 MHz. Broad-

**Table 9.24**   EMI Receiver NB and BB Bandwidths

| Frequency range | Narrowband | Broadband |
|---|---|---|
| **Radiated** | | |
| 20 Hz to 5 kHz | 3 Hz to 10 Hz | 30 to 100 Hz |
| 5 kHz to 50 kHz | 30 Hz to 100 Hz | 200 Hz to 1 kHz |
| 50 kHz to 1 MHz | 200 Hz to 1 kHz | 9 kHz to 10 kHz |
| 1 MHz to 200 MHz | 9 kHz to 10 kHz | 100 kHz |
| 200 MHz to 40 GHz | 100 kHz | 1 MHz |
| **Conducted** | | |
| 20 Hz to 20 kHz | 10 Hz to 50 Hz | — |
| 20 kHz to 2.5 MHz | 200 Hz to 1 kHz | 1 kHz to 10 kHz |
| 2.5 MHz to 100 MHz | 1 kHz to 9kHz | 10 kHz to 50 kHz |

band measurements at other bandwidths must be converted to the reference bandwidth. To convert from the measurement bandwidth to the reference bandwidths, add the following to the measured coherent BB value:

$$20 \log \frac{R_{BW}}{M_{BW}}$$

where $R_{BW}$ is the reference bandwidth and $M_{BW}$ is the measurement bandwidth.

As an example: The measurement is a voltage of 25 dBµV at a bandwidth of 10 kHz.

$$20 \log \frac{1 \text{ MHz}}{10 \text{ kHz}} = 40 \text{ dB}$$

The corrected measurement is therefore 25 dBµV + 40 dB = 65 dBµV/MHz.

If the measured increase in level is approximately 10 dB for an increase factor of 10, then the noise is noncoherent (random) BB noise. For non-coherent BB noise, the conversion is:

$$10 \log \frac{R_{BW}}{M_{BW}}$$

This conversion for noncoherent BB noise is made by reputable test facilities and is a well-accepted procedure. *However*, this is not contained in MIL-STD 462 and should be agreed to by the procuring agency or EMC authority.

### 9.6.5.2 MIL-STD 461E Bandwidths and General Emission Guidelines

a. *Bandwidth correction factors.* No bandwidth correction factors shall be applied to test data due to the use of larger bandwidths.

b. *Emission identification.* All emissions, regardless of characteristics, shall be measured with the measurement receiver bandwidths specified in Table 9.21 and compared against the limits in MIL-STD 641E. Identification of emissions with regard to narrowband or broadband categorization is not applicable.

c. *Frequency scanning.* For emission measurements, the entire frequency range for each applicable test shall be scanned. Minimum measurement time for analog measurement receivers during emission testing shall be as specified in Table 9.25. Synthesized measurement receivers shall step in one-half bandwidth increments or less, and the measurement dwell time shall be as specified in Table 9.25.

**Table 9.25**   MIL-STD-461E Bandwidth and Measurement Time

| Frequency range | 6-dB bandwidth | Dwell time | Minimum measurement time, analog measurement receiver |
|---|---|---|---|
| 30 Hz to 1 kHz | 10 Hz | 0.15 s | 0.015 s/Hz |
| 1 kHz to 10 kHz | 100 Hz | 0.015 s | 0.15 s/kHz |
| 10 kHz to 250 kHz | 1 kHz | 0.015 s | 0.015 s/kHz |
| 250 kHz to 30 MHz | 10 kHz | 0.015 s | 1.5 s/MHz |
| 30 MHz to 1 GHz | 100 kHz | 0.015 s | 0.15 s/MHz |
| Above 1 GHz | 1 MHz | 0.015 s | 15 s/GHz |

d. *Emission data presentation.* Amplitude-versus-frequency profiles of emission data shall be automatically and continuously plotted. The applicable limit shall be displayed on the plot. Manually gathered data is not acceptable except for plot verification. The plotted data for emissions measurements shall provide a minimum frequency resolution of 1% or twice the measurement receiver bandwidth, whichever is less stringent, and minimum amplitude resolution of 1 dB. The foregoing resolution requirements shall be maintained in the reported results of the EMITR.

### 9.6.6   Measurement to Very Low-Level Radiated Emissions Limits, Typically in Notches for Receivers

When radiated emission specification limits are very low, typically 25 dBμV/m/MHz for broadband and 5 dBμV/m for narrowband, the noise floor, using standard EMI measuring antennas and preamplifiers, may not be sufficiently low. In some cases the noise floor of the measuring system using standard equipment is well above the specification limit, and even the noise voltage from a 50-Ω resistor into a 50-Ω conjugate load may be higher than the signal developed at the output of the typical receiving antenna. These low-level limits are typically over a narrow band of frequencies corresponding to a receiver bandwidth.

In evaluating the noise floor of the measurement system it is important to realize that the receiver/spectrum analyzer measures peak levels, which are displayed as rms values. Most spectrum analyzer and receiver manufacturers specify an rms noise floor for their equipment, measured using video or digital averaging. The peak value, theoretically, may be infinitely higher; however, in measurements on three different manufactures' spectrum analyzers, the peak measured noise, displayed as an rms value, was between 10 dB and 12 dB above the digital or video bandwidth-averaged noise floor. Hewlett-Packard independently confirmed the approximate 12-dB difference between peak and average.

Another problem in evaluating the noise floor of the measuring system is that the preamplifier noise figure may be specified but the noise referenced to the input of the preamp is not given. Very often this is significantly lower than a 50-Ω resistor, and to use this 50-Ω value results in an error in the noise figure calculation.

In measuring low-level narrowband emissions, the use of averaging will reduce the noise floor without affecting the narrowband emissions. But this is not allowable for broadband emissions measurements, for the emissions will also be reduced. Figure 9.48 illustrates how averaging can bring a narrowband emission out of the noise. The noise floor can also be reduced by reducing the resolution bandwidth. But this is not allowable for broadband measurements, and often the resolution bandwidth is dictated by the bandwidth of a receiver that is installed on the EUT, which is the reason the notch is imposed.

The broadband specification uses a 1-MHz reference impulse bandwidth, whereas many spectrum analyzers use a 3-dB resolution bandwidth. For example, a Hewlett-Packard spectrum analyzer requires a correction of 4 dB at a 1-MHz measurement bandwidth to correct from the 3-dB resolution bandwidth to the impulse bandwidth. This correction is subtracted from the measured emission level and also from the measured system noise floor.

Another factor when measuring broadband emissions in the notch is the random nature of the measuring system noise floor. If only coherent broadband emissions are of concern, then the emissions can be viewed during the troughs and not the peaks of the noise floor display.

The measurement bandwidth used for broadband emissions also affects the measurement of coherent noise. Figure 9.49 illustrates how increasing the measurement bandwidth can bring coherent broadband emissions out of the noise. But this does not help for quasi-broadband (noncoherent) emissions. Figure 9.49 illustrates, for simplicity, that the coherent broadband

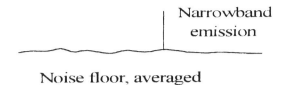

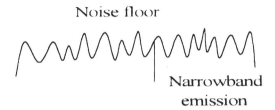

**Figure 9.48** Use of averaging to bring a narrowband emission out of the measuring system noise floor.

**Figure 9.49** Effect of the measurement bandwidth on coherent and noncoherent signals, assuming for simplicity that the coherent noise does not ride on top of the noncoherent (measuring system) noise.

noise does not ride on top of the noncoherent noise but is buried beneath it. Although this is often true, it is not always the case.

Using a typical high-end receiver, the noise floor, using a 1-MHz bandwidth, was measured at 29 dBμV at 2 GHz. Adding a 34-dB preamp at the front end of the receiver increased the measured noise floor by 11 dB to 40 dBμV. A typical double-ridged guide horn antenna has an antenna factor of 26.6 dB at 2 GHz, and let us assume that the broadband limit is 25 dBμV/m/MHz and the measurement bandwidth is at 1 MHz. The input to the preamp is then 25 dBμV/m/MHz (limit) − 26.6 dB (AF) = −1.6 dBμV. And at the input of the receiver it is −1.6 dBμV + 34 dB (preamp gain) = 32.4 dBμV, i.e., 7.6 dB below the noise floor of the system. A 3- or 10-MHz bandwidth will bring coherent noise further out of the noise floor, but only if the coherent noise covers a 3–10-MHz frequency range.

One antenna that will typically achieve a sufficiently high signal-to-noise ratio for most measurement systems at 2 GHz is a 1.7–2.6-GHz standard-gain horn, which has a gain of at least 20 dB and an AF of 16 dB. When used in conjunction with an HP8566B spectrum analyzer and a low-noise 50-dB preamp, and applying the 4-dB correction for the 1-MHz resolution bandwidth to 1-MHz impulse bandwidth, the noise floor will be below the BB limit. For example, the measured noise floor of the spectrum analyzer with 50-dB preamplifier and a 1-MHz bandwidth is 55 dBμV − 4 dB (resolution bandwidth correction) = 51 dBμV. The signal level at the spectrum analyzer input with 25 dBμV/m/MHz incident on the antenna is: 25 dBμV/m/MHz − 16dB (AF) + 50 dB (preamp gain) = 59 dBμV, i.e., 8 dB above the noise floor.

Realistically it may be necessary to relax the requirement that the noise floor be at least 6 dB below the limit level for these extremely low-level RE02 notches.

## 9.6.7  MIL-STD-EMI Measurements

The tests most frequently performed are described next.

### 9.6.7.1  Description of CE01 and CE03 Tests

*Purpose*: These tests are designed to measure conducted emissions on all power lines.

This test is applicable to primary power, that is, the input AC or DC power supplied to the equipment or subsystem. The test is not usually applicable to power lines that interconnect equipment or secondary power, although primary power supplied to other equipment may be included in the requirement. Control circuits that use EUT power and grounds and neutrals that are not grounded internally to the subsystem or equipment are included.

CE01 are narrowband emission limits only, and separate narrowband and broadband limits are imposed for CE03. Signal leads are typically exempt, and when applied the limits are developed on a case-by-case basis.

*Test setup*
The CE01, CE03 30 Hz to 50 MHz test setup is shown in Figure 9.50. All primary power and control leads are tested.

*Test equipment required*

1. Shielded current probe or probes covering the 30-Hz to 50-MHz frequency range
2. EMI receiver or spectrum analyzer
3. Broadband preamplifier or preamplifiers with at least 20 dB of gain
4. 10-μF feedthrough capacitors with a characterized impedance over the frequency range of interest
5. Oscilloscope

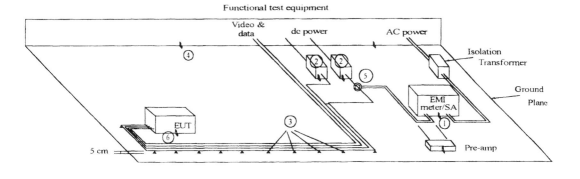

1. Low impedance bond to ground plane

2. 10uF RF feedthrough capacitors

3. 5 cm standoffs

4. DC bond impedance between the ground plane
   and the shielded room wall < 2.5m Ohm

5. Current probe

6. COMM 2 Chassis connection to ground plane < 2.5m Ohm

**Figure 9.50** CE01, CE03 conducted emission test setup.

The following points provide an overview of the CE01, CE03 step-by-step procedure:

1. The equipment test mode, in which maximum emissions are generated, are detected in a "quick scan" test. When more than one test mode results in emissions at different frequencies, the CE01, CE03 test should be repeated in each test mode.
2. Measure the noise pickup on the current probe with the AC and DC power supplies located outside of the shielded room, switched on but with the power supply disconnected from the EUT. This is a measure of the background and the power-supply-generated noise. The ambient emissions must be 6 dB below the conducted emission limits.
3. With the current probe around first the supply and then the return, measure the emissions using the narrowband bandwidth of the receiver/spectrum analyzer.
4. Repeat Step 3 using the broadband measuring bandwidth for the CE03 broadband measurement. Change current probes as required to cover the full 30-Hz to 50-MHz frequency range.

### 9.6.7.2 MIL-STD 461E, CE101, and CE103

The general test setup is similar to CE01 and CE03, with the exception of the use of the MIL-STD 462 LISN. Note the MIL-STD 461E LISN is different than either the low-frequency or high-frequency LISN described in MIL-STD 462 Notice 3.

For CE101 and CE102, a check on the measurement system is required, the test setup for this is shown in Figures 9.51 and Figure 9.53, respectively. The CE101 measurement test setup is shown in Figure 9.52, and the CE102 in Figure 9.54.

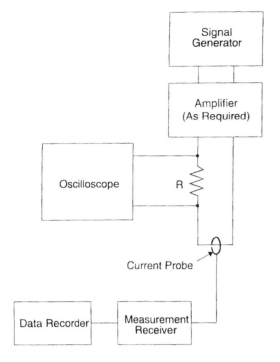

**Figure 9.51** CE101 measurement system check setup.

The most common errors incurred in conducted emission measurements with a current probe are as follow.

1. Leaving the current probe around the DC or AC power cable when switching on or off the power. The resultant current spike through the probe can generate very high voltages, and there is a danger of damage to the front end of the measuring instrument. ALWAYS DISCONNECT THE CURRENT PROBE FROM THE INSTRUMENT WHEN POWERING UP OR DOWN!

2. The current probe should be calibrated in a test fixture that raises the body of the probe from contact with the return (ground plane) of the test fixture. It is important in measurements to ensure that the body of the current probe is not in contact with the ground plane and if possible is at the same height above it as used in the test fixture.

3. The measuring instrument must be grounded to the ground plane with use of an isolation transformer on the AC power to keep ground currents, typically at harmonics of the AC power supply, from corrupting the measurement level.

4. It is not mandatory to make the measurement in a shielded room, but if this is the chosen location the power-line filters at the entry point to the room will effectively prevent the external ambient from corrupting the measured level. If measurements must be made outside of the shielded room, the conducted ambient noise, typically from computers, impressed on the AC power line or coupled from the AC input to the output of a DC supply can put the EUT out of specification, despite the presence of the 10-μF feedthrough capacitors or a LISN. The solution is to add a power-line filter at the input of the AC or DC power, before the 10-μF feedthrough capacitors or the LISN.

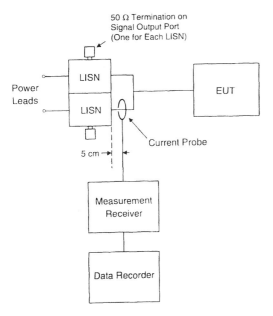

**Figure 9.52** CE101 measurement setup.

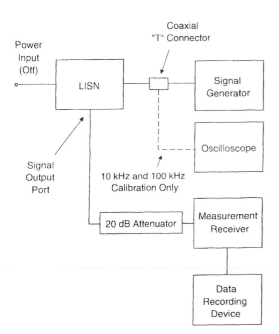

**Figure 9.53** CE102 measurement system check setup.

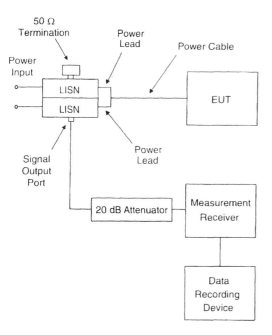

**Figure 9.54**  CE102 measurement setup.

The second problem with measurements outside of a shielded room is that the radiated ambient, typically from TV and radio, couples to the power cable. The current probe measures this ambient-induced current, which may put the EUT out of specification. If the level of noise from the EUT is low either side of the ambient emission, when examined with a narrow frequency span and narrow bandwidth, then the emission can be attributed to the ambient, and many procuring agencies or equipment customers will accept the test results. Check first, however! With an unacceptably high ambient, either locate the supply outside of the shielded room and bring the power through the room filters, or add a filter between the DC output from the supply and the 10-μF capacitors or LISN.

  5.  Poor or nonexistent grounding of the EUT, LISN, or 10-μF capacitors to the ground plane. This will reduce common-mode currents, and may allow an EUT to pass when the EUT is out of specification.

  6.  Incorrect characterization of narrowband and broadband emissions. The most common problem is to apply the coherent BB noise correction to noise that can be demonstrated to be noncoherent in converting to the reference 1 MHz bandwidth.

  7.  Use of incorrect measurement bandwidth, missing a repeat measurement on the power or return, exceeding calibration frequency range of probe. All of these problems can be avoided by use of a checklist of the tests with a tick box to ensure no test is missed.

  8.  The application of all emissions, regardless of characterization, to both the narrowband and broadband limits.

  9.  Software-controlled measurements have a number of potential problems. The majority of these programs do not allow multiple sweeps using peak or maximum hold, but instead take data over a single sweep, a snapshot in effect. Emissions often vary significantly in amplitude from one sweep to the next. With a snapshot of the data, there is no guarantee that the worst-case emissions have been captured. The argument is made that this has become an industry standard, but it seldom results in worst-

case emission measurements. A much more serious problem with many computer measurement programs is a very flawed analysis of broadband and narrowband emissions; when confronted, many manufacturers admit to the deficiency in their program. Typically, only one of the four test methods is applied, and even that method can lead to a wrong characterization of the emission.

### 9.6.7.3 Description of MIL-STD 461B/C RE02 Electric Field Tests

*Purpose*: These tests are designed to measure radiated emissions from the equipment or subsystem and all interface cables.

These tests are divided into measurement of narrowband radiated emissions from 14 kHz to 18 GHz and broadband radiated emissions from 14 kHz to 10 GHz (some procuring agencies extend the requirement to 18 GHz).

*RE02 Test Setup*

The RE02 narrowband and broadband radiated emission test setup for the EUT is shown in Figure 9.55. This is the typical MIL-STD-462 test setup. The exposed cable is located 5 cm above the ground plane at a distance of $10 \pm 2$ cm from the front of the ground plane. The cables after the exposed 2-m length may be shielded either by an expandable braid that is taped to the ground plane with electrically conductive adhesive copper tape or simply by copper tape placed over the cables and contacting the ground plane on either side of the cables (NB: Before taping the cables into position the orientation of the subsystem must be decided to ensure that the area of maximum emission is facing the antenna, as described in the step-by-step test procedure).

Maximum coupling between the subsystem and the monopole (rod) antenna will be ensured by the extension of the counterpoise of the antenna to the ground plane, as shown in Figure 9.55.

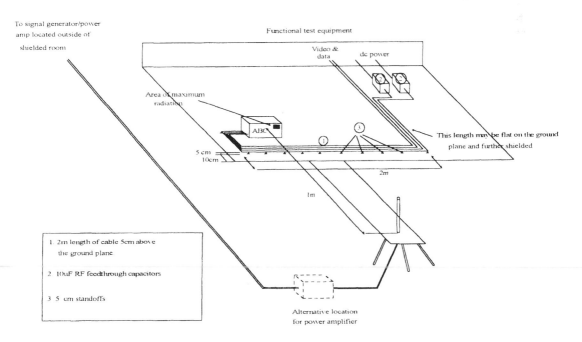

**Figure 9.55**  RE02 radiated emission test setup.

*Test equipment required*

1. Antennae covering the range of 14 kHz to 18 GHz. The antennas will typically be of the following types:

| Frequency | Antenna type |
|---|---|
| 14 kHz to 20 MHz | Active or passive rod |
| 20 MHz to 200 MHz | Biconical |
| 200 MHz to 1 GHz | Log periodic (double-ridged guide for MIL-STD-461DE) |
| 1 GHz to 18 GHz | Double-ridged guide |

2. EMI receiver and/or spectrum analyzer with a noise floor no higher than 75 μV over the bandwidth being used.
3. Broadband preamplifier or preamplifier/s with at least 20 dB of gain and a noise floor no higher than 10 dB.
4. 10-μF RF feedthrough capacitors.

*Use of electric or magnetic field probes to locate sources of emission.*

Electric or magnetic field probes are available, as described in Section 2.6, and used typically with a preamplifier and spectrum analyzer. These probes may be used to monitor the cables and sides of units to determine the sources of maximum emission. This information is necessary to determine EUT orientation. A second use for the probes is to locate sources of out-of-specification emissions. The probes may be hand-held and positioned 7–15 cm from cables, enclosure walls, apertures, etc.

The probes may be calibrated, although the absolute value of H field measured by the probes is not required. The calibration curve does, however, provide an indication of the frequency response of the probe. This may be used as described in the following example. Assume an emission detected at 70 MHz, using a balanced loop antenna, with a second emission 6 dB higher at 80 MHz. The balanced loop sensitivity is 0.4 mV/mA/m at 70 MHz and 1.1 mV/mA/m at 80 MHz (i.e., 8.8 dB higher). Therefore, the 70-MHz emission is 2.8 dB above the 80-MHz emission after correction for probe sensitivity.

*REO2 test method and preparation*

The test setup should be in compliance with Figure 9.55. Where RE02 measurements of the applicable levels are routine with the test equipment available, the following calculation of system sensitivity may be omitted.

Where the limits are lower than previously measured or unfamiliar equipment is used, it should be ascertained if the sensitivity of the antennas and measuring equipment ensures that the radiated emission level is 6 dB above the noise floor of the measuring equipment.

Compare the antenna factor of each antenna to be used with the narrowband radiated emission level over the frequency range of interest. At the frequency or frequencies where the antenna factor is high or the limit is low, make the following calculations:

$$V_{in}(dB\mu V) = dB\mu V/m - AF \ (dB)$$

where

$$V_{in} = \text{input voltage to receiving equipment}$$

dBμV/m = specification limit at the frequency under consideration

AF (dB) = antenna factor in decibels = 20 $\log_{10}$ AF

AF = numeric antenna factor at the frequency of interest

Using $V_{in}$, ensure that the measuring instrument has sufficient sensitivity, or the noise floor is sufficiently low, at the bandwidths specified in Table 9.19. When this is not the case, either a broadband low-noise preamplifier or a more sensitive or lower-noise measuring instrument should be used. This calculation must be repeated for each antenna used and for the broadband specification limit, again selecting an appropriate bandwidth from Table 9.19. An example of a typical calculation follows.

A narrowband limit at 40 MHz is 24 dBμV/m. The antenna factor of the antenna used in 3.83, which is 11.67 dB. Therefore,

$$V_{in} = 24 \text{ dBμV/m} - 11.67 \text{ dB} = 12.3 \text{ dBμV} = 4.12 \text{ μV}$$

The narrowband bandwidth to be used is 10 kHz (refer to Table 9.20 at 40 MHz). At 40 MHz the noise floor of the measuring instrument is −80 dBm, which is equivalent to 23 μV into 50 Ω (the input impedance of the instrument). $V_{in}$ is thus 15 dB (20 log $V_{in}$/noise floor) below the noise floor, and the system sensitivity, therefore, is not adequate and a preamplifier with a minimum gain of 21 dB must be used. For conversion to broadband unit, add the broadband correction given by

$$20 \log \frac{1 \text{ MHz}}{R_{BW}}$$

where $R_{BW}$ is the resolution bandwidth used in the measurement in MHz.

The frequency to which the receiver is tuned will be swept over the frequency range of the antenna in use. The sweep time will be selected to ensure that *broadband emissions* occurring at the lowest pulse repetition rate of the system are measured. A maximum EMI receiver scan rate will therefore be imposed. The maximum EMI receiver scan rate is determined by both the EMI receiver bandwidth and the lowest PRF to be intercepted.

Narrowband emissions are either from CW sources or sources at high PRF, and thus the maximum scan rate limitation is not applicable. The scan rate in hertz per second is given by

$$\text{scan rate} = \frac{\text{EMI receiver bandwidth}}{\text{lowest expected PRF}} \quad \frac{[\text{Hz}]}{[\text{sec}]}$$

An example for a 50-kHz broadband bandwidth is

$$\text{scan rate} = \frac{50 \text{ kHz}}{4.5 \text{ s}} = 11.1 \text{ kHz/s}$$

The slowest sweeptime available on a typical spectrum analyzer is 1500 seconds, thus the maximum frequency span that can be swept in 1500 seconds is given by

$$\text{scan rate} \times \text{sweeptime} = 11.1 \text{ kHz/s} \times 1500 \text{ s} = 16.6 \text{ MHz}$$

The 2.5-MHz to 30-MHz frequency range must therefore be divided into two sections. The approximate time required for the broadband test to cover this frequency range is 41 minutes. At high frequencies, the limit on scan rate often results in inordinately long test times, and an alternative method must be used.

Two approaches can be used to facilitate the broadband RE02 test. One is either to program the spectrum analyzer via the front panel or to use a computer to automatically change frequency and plot the results. The other approach is to use a fast sweeptime and to record the frequency range of interest for a specified number of minutes, using peak (maximum) hold on the spectrum analyzer. The probability of detecting the harmonically related emissions occurring at a 1-minute or faster repetition rate is thus high. The sweeptime of the spectrum analyzer for narrowband tests may be short, limited only by the spectrum analyzer. When the sweeptime is too short, the spectrum analyzer displays an out-of-calibration condition. When the EUT has more than one mode of operation, any of which may result in significant emissions, the radiated and conducted emission measurements must be repeated for each mode.

*Expected frequency of emissions*

The frequencies at which emissions are expected should be tabulated and examined carefully during emission measurements. For example, broadband and narrowband harmonically related emissions are expected at clock, converter, and logic frequencies and harmonics thereof.

*RE02 step-by-step test procedure*

The following three points provide an overview of the RE02 step-by-step procedure.

1.  Check for background emissions over the entire frequency band of interest (i.e., 14 kHz $< f >$ 18 GHz). Above 20 MHz, check with both vertical and horizontal polarizations of the antenna.
2.  Find the side of the EUT that has the highest emissions (check all test modes).
3.  Measure peak emissions for all test modes.

*NOTE:*

*   For 14 kHz $< f >$ 18 GHz, measure peak emissions using the appropriate *narrowband* bandwidth of the spectrum analyzer, and compare the measured peak emissions to the limits.
*   For 14 kHz $< f >$ 10 GHz, measure peak emissions using the appropriate *broadband* bandwidth of the spectrum analyzer, and compare the measured peak emissions to the limits.
*   For emissions exceeding the limits, determine whether they are broadband or narrowband by one or more of the methods outlined in Section 9.2.2.
*   For $f >$ 20 MHz, peak emissions must be measured using vertical and horizontal antenna orientations.
*   For 1 GHz to 18 GHz, measure peak emissions with the antenna located 1 m from the center of the interconnection cables as well as 1 m in front of the EUT.

When the passive or active rod antenna is used, connect the counterpoise of the antenna to the ground plane on which the EUT is mounted using a sheet of copper-clad PCB. The PCB may be electrically bonded to the ground plane and counterpoise by use of conductive adhesive tape. The MIL-STD-462 test setup does not show the counterpoise connected to the ground plane, unlike the DO-160 test setup. Bonding the counterpoise to the ground plane results in a smoother AF, especially at higher frequencies.

MIL-STD 461E RE102 does require that the rod antenna counterpoise be used and be electrically bonded to the ground plane. MIL-STD 461E also requires a test of the measurement path using a signal generator and requires the use of multiple antenna positions above 200 MHz and specific antenna heights of 120 cm above the ground plane for the biconical and double-ridge guide (horn). The measurement bandwidths and dwell times for RE102 are contained in Table 9.25.

### 9.6.7.4 Common Errors in RE02/RE102 Measurements

The most common errors incurred in radiated emission measurements follow.

1. Ambient inside the shielded room too high due to RF common-mode current flow on interconnection and power cable. This is typically sourced by functional test equipment or electrical ground support equipment (EGSE) located outside of the shielded room. The solution is always to use D-type connectors on interface connections either at the test equipment/EGSE or at the cable entry into the shielded room. By using D-type connectors, if a problem is seen, a filtered D-type connector adapter (male and female connectors back to back with filter components in between) should be tried. If filter connectors cannot be used on interconnection cables, then the shields of cables should be terminated to the shielded-room wall at the entry through a waveguide and brass wool should be used to connect the shield to the interior of the waveguide and to fill the aperture. It is a fallacy to assume that a waveguide aperture performs as a waveguide below cutoff when cables are passed through the aperture. RF current flow on these cables will couple from the outside ambient into the shielded room on the cables coming through the waveguide. Even nonconductive hose containing tap water will couple RF currents into the room due to the conductivity of the tap water. To keep the RF currents out of the shielded room, they must be shunted to the room wall at the point of entry.

2. Incorrect characterization of narrowband and broadband emissions. The most common problem is to apply the coherent BB noise correction to noise that can be demonstrated to be noncoherent in converting to the reference 1-MHz bandwidth.

3. Use of incorrect measurement bandwidth, missing a repeat measurement with a different antenna polarization, exceeding calibration frequency range of antenna. All of these problems can be avoided by use of a checklist of the tests (which includes frequency range, antenna polarization, use of preamplifier, etc.), with tick boxes to ensure no test is missed.

4. Transmitting antennas on the EUT connected to broadband power amplifiers are sources of broadband noise as well as the intended transmitting frequency. These transmitting antennas are excluded from the RE02 and RE102 tests. Transmitting antennas must be replaced with dummy loads, and the load connector must have a low transfer impedance to reduce radiation from the load itself. If out-of-band emissions from the transmitter are a problem, these should be measured in a conducted test; they are not part of tests of radiated emission E field to MIL-STD requirements.

5. The application of all emissions, regardless of characterization, to both the narrowband and broadband limits. (Apply only BB emissions to the BB limit and NB emissions to the NB limit.)

6. Nonrepresentative test setup, use of a different type of cable than used in the final version, nonrepresentative EUT, addition of aluminum foil, copper tape, or braid over cables that will not normally be shielded.

7. The argument that cable emissions are not the concern of the manufacturer of the EUT when the cable harness is supplied by the customer. Regardless of the type of cable, shielded or unshielded, it is the EUT that places noise on the cables, and it is the manufacturer of the EUT who must reduce emissions at source. If a manufacturer believes that emission requirements will not be met with a customer-supplied un-

shielded cable, he should apply for a deviation prior to testing. The customer may allow the deviation, require the EUT to meet the requirement, or change the cable to a shielded type

8. Software-controlled measurements have a number of potential problems. The majority of these programs do not allow multiple sweeps using peak or maximum hold, but instead take data over a single sweep, a snapshot in effect. Emissions often vary in amplitude significantly from one sweep to the next. With a snapshot of the data, there is no guarantee that the worst-case emissions have been captured. The argument is made that this has become an industry standard, but it seldom results in worst-case emission measurements. A much more serious problem with many computer-measurement programs is a very flawed analysis of broadband and narrowband emissions; when confronted, many manufacturers admit to the deficiency in their program. Typically, only one of the four test methods is applied and any one of the methods can lead to a wrong characterization of the emission.

### 9.6.7.5 Description of MIL-STD 461B/C RE01, RE04, and MIL-STD 461E RE101 Magnetic Field Tests

*Purpose*: These tests are to verify that the magnetic field emissions from the EUT and its associated cabling do not exceed specified requirements.

The RE01 test covers the 30-Hz to 30-kHz frequency range, the RE04 test covers the 30-Hz to 50-kHz frequency range, and the RE101 covers the 30-Hz to 100-kHz frequency range.

For both the RE01 and RE101 tests, the receiving loop is specified in both the original MIL-STD 462 and MIL-STD 461E as follows:

Diameter 13.3 cm
Number of turns 36
Wire 7-41 Litz (7 strand, No. 41 AWG)
Shielding electrostatic

MIL-STD 461E requires a calibration of the measurement equipment using a signal generator. In both RE01 and RE101 the loop sensor is located 7 cm from and moved over the EUT face or cable being probed. Orient the plane of the loop sensor parallel to the EUT faces and parallel to the axis of the cables. The exact distance of 7 cm from the EUT or cable is very important, for small errors in distance can result in large deviations in measurement. To ensure the 7-cm distance is maintained it is recommended that a 7-cm Styrofoam block or a short length of dowel be strapped to the loop antenna.

In the MIL-STD 461 RE04 test, the magnetic field sensor, which must be capable of measuring −40 dBnT at 25 Hz, is located 1 m from the EUT and cables. The axis of the antenna is pointed toward the EUT. A calibrator is required to check the magnetic field sensor. The calibrator is required to generate known magnetic fields at a minimum of 10 frequencies over the 20-Hz to 50-kHz range. If a calibrator is not available and the magnetic field sensor has been calibrated within the last two years, then this would normally be acceptable.

### 9.6.7.6 RS03 Radiated Susceptibility Test

*Purpose*: The test is for susceptibility of the equipment or subsystem and interconnection cables to incident E fields and plane waves. The test setup is similar to that for RE02 with the receiving antenna replaced with a transmitting antenna.

*Test equipment*

1. Antennae covering the range of 14 kHz to 10 GHz and capable of generating the specified E fields will typically be of the following types:

| Frequency | Antenna type |
|-----------|--------------|
| 14 kHz to 20 MHz | Passive rod or long wire |
| or | Stripline |
| 14 kHz to 200 MHz | E field generating |
| 20 MHz to 200 MHz | Biconical |
| 200 MHz to 1 GHz | Log-periodic |
| 1 GHz to 18 GHz | Double-ridged guide |

2. Power amplifiers operating over the frequency range from 14 kHz to 18 GHz. The power rating of the amplifiers must be sufficiently high to generate the specified E fields from the antennas in use.
3. E field sensors, with remote read capability or one of the following calibrated antennas, may be used to monitor the level of E field:

| Frequency | Antenna |
|-----------|---------|
| 14 kHz to 40 MHz | 1-m rod |
| 40 MHz to 300 MHz | Tunable dipole |
| 20–1000 MHz | Bow tie |
| 300 MHz to 800 MHz | Resonant 5–20-cm rod |
| 800 MHz to 3 GHz | 25-cm conical log spiral |
| 3–8 GHz | 15-cm × 2.8-cm conical log spiral |
| 8–18 GHz | 6.2-cm × 1-cm conical log spiral |

Due to the high levels monitored by the antennas, preamplifiers are typically not required.
4. Signal generators covering the frequency range from 14 kHz to 18 GHz with either internal or external modulation capability.
5. Audio frequency signal generator used to modulate the carrier frequency generator in (4).

*Description of RS03 test*

The instrument is exposed to fields generated by antennas over the frequency range from 14 kHz to 10 GHz. The EUT should demonstrate correct performance when exposed to the specified level of E field. The signal source should be amplitude modulated at a frequency at which the signal or power interfaces are predicted to be most susceptible.

The power and interface cables are exposed to the incident E field over a length of 2 m. The exposed cable is located 5 cm above the ground plane at a distance of 10 cm from the front of the ground plane.

Interface cables connected to other equipment in the room should be dropped to the metal floor after the 2-m exposed length. The cables on the floor should be as far away from the transmitting antenna as feasible and may be either covered with copper conductive adhesive tape connected on both sides of the cables to the metal section of the floor, or contained in expandable braid that is connected to the metal floor section.

The cables, which are arranged on the ground plane, may be shielded, after the 2-m exposed length, either by an expandable braid that is taped to the ground plane with electrically conductive adhesive copper tape or simply by copper tape placed over the cables and contacting the ground plane on either side of the cables.

The antennas are located at a distance of 1 m from the face of the EUT. When small-aperture antennas (1–18-GHz frequency range) are used, the position of the antenna should be moved from directly in front of the EUT to directly in front of the interconnection cables.

MIL-STD 461E requires RS103 tests to be conducted from only as low as 2 MHz to 18 GHz and optionally up to 40 GHz.

Circularly polarized fields are not allowed.

There is no implied relationship between RS103 and the RE102 limits. RE102 limits are placed primarily to protect antenna-connected receivers, while RS103 simulates fields resulting from antenna transmissions.

MIL-STD 461E specifies specific locations for the transmit antennas and for the electric field sensors.

When a receive antenna is used, it is placed in the test setup boundary, and first the signal path is checked out by use of a signal generator.

The procedure for calibrating the E field using a receive antenna is:

Connect the receive antenna.

Set the signal source to 1-kHz pulse modulation, 50% duty cycle.

Using an appropriate transmit antenna and amplifier, establish an electric field at the test start frequency.

Gradually increase the electric field level until it reaches the applicable limit.

Scan the test frequency range and record the required input power levels to the transmit antenna to maintain the required field.

Repeat this procedure whenever the test setup is modified or an antenna is changed.

For testing, remove the receive antenna and reposition the EUT. This calibration technique does lead to errors due to the proximity of the ground plane to the receive antenna, but it does eliminate the error due to the proximity of the EUT to the receiving antenna.

The location of the power amplifier and the input lead from the signal generator are critical. For a monopole antenna, the power amplifier should be located under the counterpoise of the antenna and connected to the antenna with as short a cable as feasible.

When a long-wire antenna is used, and when driven without load, it should also be connected to the power amplifier, via a transformer where required, with as short a cable as feasible. For all antennas, with the possible exception of the monopole, the power amplifier should be located outside the shielded room to avoid feedback from the radiated field onto the input cable of the power amplifier.

MIL-STD-462 requires that the incident E field be measured at the EUT location. However, when using physically large antennas to measure the generated E field, the proximity of the antenna to the EUT and ground plane invalidates the AF of the antenna. Even the use of field sensors located close to the ground plane will result in a measurement very different from the E field generated far away from the ground plane. The recommendation is that the E field value be monitored 1 m from the transmitting antenna, and not at the EUT position, where the presence of the ground plane can result in E fields ranging from close to cancellation to E fields greater than 6 dB above the level without ground plane. When this measurement technique is described in the test plan and test procedure, the procuring agency often accepts the deviation from MIL-STD-462 recommended procedure.

When the power amplifier is located outside of the room, the potential exists for coupling between the antenna feed cable and test equipment cables. Here, ferrite beads on the antenna and/or test cables may alleviate the problem or the use of double braid shielded cables. Above 30 MHz, tests are made with both horizontally and vertically polarized fields.

The major problem with the RS03 test setup is that the field incident on the EUT will vary due to shielded-room resonances and reflections. As shown in Section 9.5.1, the variation may be as high as 50 dB. The E field incident on the measuring antenna will also vary, and the AF of the measuring antenna will deviate from the published curve due to proximity effects. If the input signal to the power amplifier is adjusted to maintain the specified E field, as measured by the receiving antenna, the magnitude of the E field at the EUT may be anything from almost zero V/m to potentially destructive levels.

One technique that minimizes the possibility of subjecting the EUT to very high levels of E field is to sweep the test frequency over the range of interest and to move the location of the receiving antenna after each sweep. Because the dips and peaks in the measured E field are caused primarily by reflections, they tend to average out as the antenna location is moved. A common test method is to generate as high an E field as possible, limited only by the power amplifier and antenna. If the equipment is susceptible, then the magnitude of the E field at the susceptibility frequency is adjusted to the specified E field. The potential danger is that the equipment is damaged by the test levels generated. Automated tests often use the output from an E field probe to control the output level of the power amplifier driving the antenna. The problem with this technique is that at those frequencies at which cancellation of the fields at the probe location occurs, the power amplifier output is at maximum, and again potentially damaging levels of E field are generated. The E field generated by a transmitting antenna at 1-m distance is given by the following equation:

$$E = \sqrt{Z_w \frac{WG}{4\pi r^2}}$$

where

$W$ = input power
$G$ = gain of the antenna at the frequency of interest
$r$ = distance from the antenna [m]
$Z_w = \dfrac{377\lambda}{2\pi r} \geq 377$
$\lambda$ = wavelength [m]

When the input impedance of the antenna is assumed to be a constant 50 $\Omega$ and the distance a constant 1 m, the following simplified equation may be used:

$$E = \frac{V_{in}}{\sqrt{\dfrac{628}{Z_w G}}}$$

In common with other susceptibility tests, the maximum sweep rate should be limited by the ground support equipment (GSE) or functional test equipment test cycle time, that is, the time required to test the EUT and determine if it is susceptible. A guideline for the limit on sweep rate is to sweep the carrier frequency over the frequency range of interest, usually limited by the antenna, stopping at three frequencies per decade for the test cycle duration. Sweep between the held frequencies in a duration equal to the test cycle time. When the test cycle time is much over 1 minute, inordinately long test times are required and the dwell and sweep durations should be limited to 1 minute. The susceptibility test method should be outlined in the test plan and approved by the procuring agency prior to commencement of tests. If degraded performance is observed during the test, the E field should be decreased to determine the threshold of susceptibility. Receiving antennas are excluded from radiated susceptibility tests and should be replaced by a shielded dummy load.

The MIL-STD-461E susceptibility scan rates are provided in Table 9.23

**SAFETY WARNING**: Where power amplifiers greater than 3 W are used, the test equipment is capable of generating E fields of 200 V/m or higher. E fields above 27.5 V/m are considered hazardous. Therefore, **do not remain in the shielded room with the signal generator powered up and connected to the power amplifier and antenna**.

### 9.6.7.7    Most Common Errors Incurred in Radiated Susceptibility Measurements

1. A consensus on the pass/fail criteria has not been reached; i.e., a few white spots on a display or some minor distortion may be acceptable, whereas multiple images or major distortion is almost certainly unacceptable. Similarly, some bit error rate in data communication may be acceptable, and this must be agreed on prior to commencement of susceptibility tests.

2. Failing to log the test start and stop times when data is analyzed offline, after the test is complete. The data must also be time tagged to allow comparison with the test times.

3. Inclusion of receiving antennas that are part of the EUT in the test chamber. These antennas and associated receivers should not be part of the susceptibility test. A danger is that the receiver front end may be destroyed by the power induced due to the incident susceptibility test field at in-band or close-to-in-band frequencies. If antennas cannot be disconnected, typically when the EUT is a flight model satellite, then wooden forms covered in absorber loads may be required to cover the antenna. In addition, frequencies at either side of the receiver in-band frequencies should not be radiated in the test.

4. Allowing personnel to view displays, indicator lights, or meters on the EUT in the test chamber during a radiated susceptibility test. Due to the potential danger, a set of mirrors can be set up and use made of the waveguide below cutoff to view the EUT display panel. With no cables penetrating the waveguide it will function as a waveguide below cutoff, and the field coupled out of the shielded enclosure will be negligible.

5. Coupling between the signal generator input cable and the power amplifier output cable, or between the test field and the input cable to the power amplifier, when this is located in the test chamber. This coupling can result in positive feedback and the generation of very high E fields. Because the frequency of this oscillation may be very different than the test frequency, always monitor the radiated field with a wide frequency span.

6. Nonrepresentative test setup, use of a different type of cable than used in the final version, nonrepresentative EUT, addition of aluminum foil, copper tape, or braid over cables that will not normally be shielded.

7. The argument that cable induced levels are not the concern of the manufacturer of the EUT when the cable harness is supplied by the customer. Regardless of the type of cable, shielded or unshielded, it is the EUT that must demonstrate immunity to cable induced levels, and it is the manufacturer of the EUT who must reduce build immunity into the circuit. If a manufacturer believes that susceptibility requirements will not be met with a customer-supplied unshielded cable, it should apply for a deviation prior to testing. The customer may allow the deviation, require the EUT to meet the requirement, or change the cable to a shielded type.

8. Missing a repeat measurement with a different antenna polarization, exceeding calibration frequency range of antenna or power amplifier.

All of these problems can be avoided by use of a checklist of the tests, with tick boxes to ensure no test is missed.

### 9.6.7.8   Description of MIL-STD 461B/C RS01 and MIL-STD 461E RS101

The RS01 test is specified from 30 Hz to 30 kHz, and the RS101 from 30 Hz to 100 kHz. The test setup is shown in Figure 9.56.

The radiating loop for RS01 has the specification:

12-cm diameter
10 turns of AWG-16 capable of producing $5 \times 10^{-5}$ tesla/ampere at 5 cm from the face (plane) of the loop

The radiating loop for MIL-STD 461D RS101 has the specification:

12-cm diameter
20 turns of AWG-12 capable of producing $9.5 \times 10^7$ pT/ampere at a distance of 5 cm from the plane of the loop.

In addition, MIL-STD 461E specifies a loop sensor as follows:

Diameter 4 cm
Number of turns 51
Wire 7–41 Litz
Shielding electrostatic

The MIL-STD 462 method, for the limits in MIL-STD 461B/C, is to position the plane of the radiating loop 5 cm from the plane of the test sample. Flux densities approximately 20–30 dB greater than the applicable limit at the test frequencies shall be applied. At the point

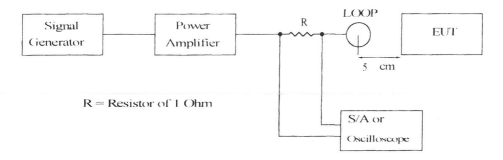

**Figure 9.56**   RS01 measurement test setup.

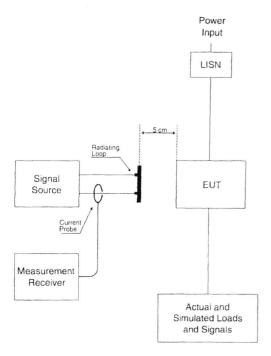

**Figure 9.57**  RS101 measurement test setup.

where the applied field produces the maximum effect, reduce the field until the performance of the EUT is not affected, and record the magnitude of the field.

The MIL-STD 461E test method is similar, for the plane of the radiating loop is positioned 5 cm from the plane of the EUT's surface, but the initial test requirement is to apply a level at least 10 dB above the applicable limit but not to exceed 183 dBpT. At locations of susceptibility, apply the specified field and move the loop to search for possible locations of susceptibility at the specified magnetic field level.

MIL-STD 461E requires a calibration using the radiating loop and the sensor loop.

The RS101 test setup is shown in Figure 9.57.

For calibration of the RS01 and RS101 test setups the following test equipment and procedure are required:

**Equipment check list**

For the RS01 test:

1. H field loop as described in MIL-STD-461. The loop is made of 10 turns of #16AWG on a 4.27-in. (12-cm) diameter. The loop shall be capable of generating a magnetic flux density of $5 \times 10^{-5}$ tesla/ampere at a distance of 5 cm from the face of the loop.
2. 13.3-cm shielded loop H field antenna, as described in MIL-STD-462 RE01, 30 Hz to 30 kHz, 1997, and MIL-STD-461E, RE101, 30 Hz to 100 kHz, 1993.
3. Low-frequency power amplifier at least 100 W from 30 Hz to 100 kHz.
4. 1-$\Omega$ 50-W resistor.
5. An oscilloscope used for differential measurements (channel A + B or A − B) used over the 30-Hz to 50-kHz frequency range or the differential probe with × 10 attenuator, or an AC voltmeter.

For the RS101 test:

1. H field loop, made of 20 turns of #12 AWG on a 12-cm diameter. The loop shall be capable of generating $9.5 \times 10^7$ pT/ampere of applied current ($9.5 \times 10^{-5}$ tesla/ampere) at a distance of 5 cm from the plane of the loop.
2. Loop sensor made of 51 turns of 7–41 Litz (7 Strand, No. 41 AWG) electrostatically shielded.
3. Low-frequency current probe (30 Hz to 100 kHz).
4. Low-frequency preamplifier (may not be required).
5. Low-frequency (20 Hz–100 kHz) spectrum analyzer or EMI receiver.
6. Low-frequency signal generator (1 Hz–500 kHz).

**Setup and Operations**

*RS101 Setup and Operation*

Antenna checkout:

For the RS101 transmitting antenna: Connect the equipment as shown in Figure 9.58. Set the signal source to a frequency of 1 kHz, and adjust the output to provide a magnetic flux density of 110 dB above one picotesla. This level is achieved when 3.33 mA (10.4 dBmA) flows through the transmitting loop. This current is set by measuring with the current probe. The voltage measured by the current probe for the 10.4-dBmA is given by:

$$V\text{dBmV} = 10.4 \text{ dBmA} + Z_t(\text{dB}\Omega)$$

For example, with $Z_t = -50$ dB$\Omega$, the value of $V$ dBm with $I = 10.4$ dBmA is $-39.6$ dBmV $= 20.4$ dB$\mu$V $= -86$ dBm. With $Z_t = +18$ dB$\Omega$, the value of $V$ dBm with $I = 10.4$ dBmA is 28.4 dBmV $= -18$ dBm.

If the signal level out of the current probe is not at least 10 dB above the noise floor of the measuring equipment, use the low-frequency preamp between the current probe and spectrum analyzer.

Measure the voltage output from the loop sensor.

Verify that the output on measurement receiver B is 42 dB$\mu$V $\pm$ 3 dB.

The setup in Figure 9.58 can be used for calibration of the receiving loop, in which case

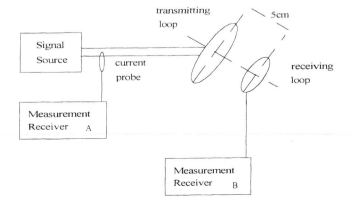

**Figure 9.58** RS101 antenna test setup for checkout and calibration.

the frequency is swept over the full frequency range of 30 Hz to 100 kHz, and the receiving loop output is measured and recorded.

*RS01 Setup and Operation*

Antenna checkout:

For the RS01 transmitting antenna. Connect the equipment as shown in Figure 9.59. Set the signal source to a frequency of 100 Hz, and adjust the output to provide a voltage of 1 V rms (2.829 V pk-pk) across the 1-ohm resistor. Measure the voltage output from the loop sensor using the preamplifier and oscilloscope.

The loop output voltage is given by the voltage measured across the scope (pk-pk converted to rms) divided by the gain of the preamplifier.

Convert this voltage to dBµV. At 100 Hz the measured level should be approximately 43 dBµV. Adjust the frequency to 1 kHz, measure the voltage across the 1-ohm resistor, and adjust, as required, to 1 V rms. The measured output from the loop, after correcting for the precalibration procedure for the receiving loop. The only difference is that for calibration the frequency is swept over the 30-Hz to 30-kHz frequency range, and the output of the receiving loop is monitored and recorded.

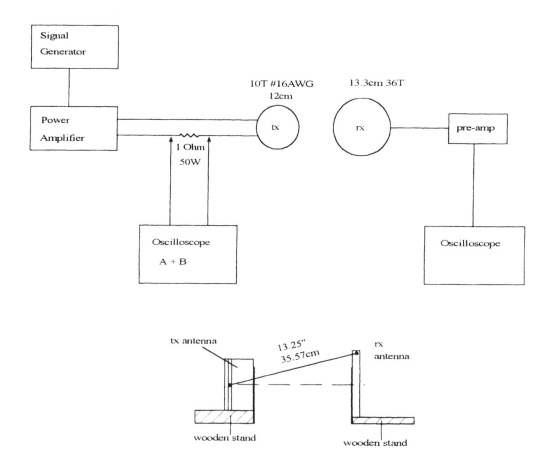

**Figure 9.59**  RS01 antenna test setup for checkout and calibration.

## 9.6.7.9   Description of MIL-STD 461B/C RS02

*Purpose*: This test is to determine the susceptibility to induced AC power frequency and transient magnetic induction fields.

The test is divided into a cable test and a case test, although the case test is seldom a requirement.

In the cable test, a 2-m length of induction cable is wrapped around the cable, and a spike, having the characteristics of the CS06 spike and developed by the CS06 spike generator, is applied to the cable under test. The spike voltage is up to 400 V at 5 µs and up to 200 V at 10 µs. A second test applies a current of 20 A at the power frequency. If AC power is not used on the aircraft, vehicle, or spacecraft, then the AC power test should not be a requirement.

In the spike test at 100 V and 10 µs, the measured current induced in a shielded cable is approximately 30 A. The shield of the cable will attenuate the cable internally generated common-mode voltage to a certain level.

A shielded cable will not attenuate the voltage induced in the 20-A power test, where the frequency is typically 50–400 Hz, and the signal interface must be immune to the common-mode voltage induced at these frequencies, typically by use of a differential or quasi-differential interface for analog and baseband video.

## 9.6.7.10   Description of CS01 and CS02 Tests

*Purpose*: These tests are designed to measure the susceptibility of AC and DC power leads to audio frequency (AF) and radio frequency (RF) ripple voltage and transient voltage.

The AF and RF ripple is applied directly to the AC and DC power leads.

Conducted susceptibility power lines (CS01 and CS02): This test is conducted over the frequency range from 30 Hz to 400 MHz.

If degraded performance is observed during the test, the signal should be decreased to determine the threshold of interference.

The input of the EUT may present a capacitance to the susceptibility test signal. Generation of the test levels may be impossible when the input capacitance is large, even when high-power amplifiers are used. Too high an RF power developed in the EUT input circuits may result in damage. Consequently, the following alternative test limits are widely accepted and should be incorporated into the test plan and procedure.

The CS01 requirement is also met when the power source, shown in test setup CS01, adjusted to dissipate 40 W into an 0.5-Ω load, cannot develop the required voltage at the EUT input connections and the test sample is not susceptible to the output of the signal source.

The CS02 requirement is also met when a 3-W source of 50-Ω impedance cannot develop the required voltage at the input power connector, and the test sample is not susceptible to the output of the signal source.

The CS01 injection transformer has a secondary inductance of approximately 1.5 mH. When the transformer is connected into the test circuit with the power amplifier powered down or disconnected, this inductance is in series with the power line and presents a high source impedance to the EUT. Many DC-DC converters or switching-power supplies are unstable with a high source impedance; therefore never connect the transformer without the power amplifier powered up. However, with the power amplifier connected prior to powering up the EUT, as the EUT is switched on the in-rush current can develop a large voltage across the secondary of the transformer. This is stepped up and is applied to the output of the power amplifier, possibly resulting in damage. To avoid both problems, short out the secondary of the transformer until the EUT is powered up and the power amplifier is connected and powered up. REMEMBER

TO REMOVE THE SHORT BEFORE AN INPUT SIGNAL IS APPLIED TO THE POWER AMPLIFIER.

Any switching spikes generated by the EUT are coupled directly to the output of the power amplifier via the 0.1-μF capacitor in the CS02 test. If these spikes are high, use a low-value series resistor between the output of the power amplifier and the capacitor. Connect bidirectional transorbs, or unidirectional transorbs with diodes, or diodes with voltage ratings above the maximum peak voltage developed by the power amplifier. The 20–50-μH inductor is required to ensure that if the required test level cannot be developed it is not due to the low output impedance of the power supply. In some test facilities the correct test level is never developed; instead, the 3-W calibration level into 50 Ω is routinely applied. However, this approach may result in a serious undertest of the EUT.

If, for example, an EUT powered by a 400-Hz, 440-V AC power supply is tested for CS02, a current of typically 1.1 A flows at 400 Hz into the power amplifier output via the 1-μF coupling capacitor. A simple bandstop filter tuned to 400 Hz can be designed and constructed and can reduce the current to 0.4 mA at 400 Hz.

*Test setup*

The CS01 test setup is shown in Figure 9.60 and the CS02 in Figure 9.61. The coupled noise should be applied to one power bus at a time, with the input voltage level set at the *nominal* power-line voltage unless otherwise specified.

*CS01 and CS02 test equipment required*:

1.  Signal generator or generators to cover the frequency range from 30 Hz to 400 MHz
2.  Power amplifier minimum 50 W continuous from 30 Hz to 50 kHz (CS01)
3.  Power amplifier minimum 3 W continuous from 50 kHz to 400 MHz (CS02)
4.  10-μF RF feedthrough capacitors

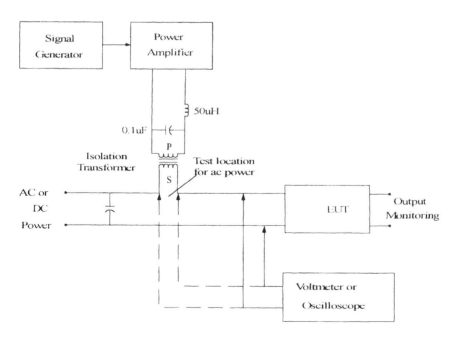

**Figure 9.60**  CS01 conducted susceptibility test setup, 30 Hz to 50 kHz.

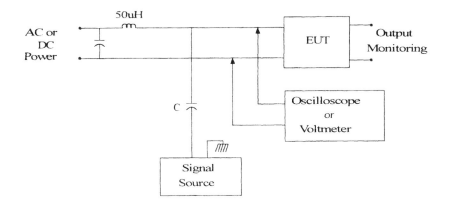

1. The value of C shall be chosen such that Xc < 5 Ohms

   over the test frequency range

2. Connect the coupling capacitor and the oscilloscope

   within 5cm of the power socket

   mounted on EUT

**Figure 9.61**  CS02 conducted susceptibility test setup and connection diagram, 50 kHz to 400 MHz.

5. 1-µF to 0.1-µF rated at 1.5 times the peak voltage, or use $x$ or $y$ capacitors for AC power rated at the AC power-line voltage (CS02)
6. Isolation transformer rated at least 50 W from 30 Hz to 50 kHz with, typically, a primary resistance of 2 $\Omega$ and a secondary resistance of 0.5 $\Omega$, turns ratio 2/1 (CS01).

The MIL-STD 461E CS101 test is similar except for the use of a LISN and a 10-µF capacitor across the line, on the EUT side of the LISN.

A calibration of the test signal across a 0.5-$\Omega$ resistor is required prior to testing.

The CS01 transformer must be capable of conducting the DC current required to power the EUT without saturation. The output of the power amplifier may be protected from RF and transients that may be generated by the EUT, should a failure occur in the instrument during CS01 tests, by the 0.1-µF and 50-µH network shown in setup CS01, Figure 9.60.

MIL-STD 461E does not specify a differential mode (CS02-type) test above the CS101 upper frequency limit of 50 kHz.

MIL-STD-461E adds a structure current test, CS109, shown in Figure 9.62, and a bulk cable injection test, CS114, shown in Figure 9.63.

### 9.6.7.11  Description of CS06 Test

The transient pulse identified in MIL-STD-461 (A, B, or C) applies. The test should be conducted at the nominal power-line voltage. The transient should be applied at a repetition rate of 60 pps for a duration of 5 minutes and have a pulse width of 10 µs. The test should be applied to the input power leads of the equipment under test. Under these conditions, the instrument should suffer no degradation, malfunctions, or deviation from specified performance.

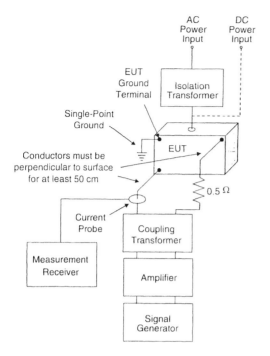

**Figure 9.62**  CS109 structure current test.

The waveshape is specified into a 5-Ω resistive load. When the spike generator is applied to the input of the EUT, the peak voltage ($E$) and the duration of the spike may be modified by the EUT input filters or input capacitance. When $E$ is lower than specified with the spike generator set to maximum output voltage, and the equipment is not susceptible to the applied voltage, ensure that the spike generator set at maximum output can generate at least the specified voltage into a 5-Ω resistive load. When the foregoing is true, the EUT should be deemed to have passed the EMC test requirement.

*CS06 test setup*

The proposed test setup is shown in Figure 9.64. An alternative test setup for AC power lines is given in MIL-STD-462 CS06-1, in which the spike generator is connected to an isolation transformer in series with the input power lead. However, many spike generators can function safely when placed across the supply as shown in Figure 9.64.

*Test equipment*

1.  50-μH inductor. The inductor must be capable of handling the AC or DC power current without saturation and with a reduction in inductance to no less than 20 μH.
2.  10 μF RF capacitor.
3.  Spike generator with the following specifications:
    a.  Pulse width = 10 μs
    b.  Pulse repetition rate = 60 pps
    c.  Voltage output = not less than the specified voltage into 5Ω
    d.  Output control = adjustable from 0 to the specified voltage
    e.  External trigger = 60 pps
    f.  Calibrated differential/input oscilloscope with at least 10-MHz bandwidth, ade-

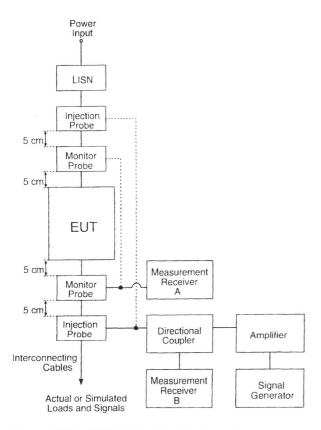

**Figure 9.63** CS114 bulk cable injection test.

quate sweep rates, and an input voltage rating of the spike voltage $+E$, where $E$ is the power ripple voltage

After testing with a positive spike, the test should be repeated with a negative spike. Adjust the output voltage from the spike generator to the specified voltage, and trigger the oscilloscope from the channel used to view the spike, using AC and normal trigger coupling. At the low repetition rates used, an oscilloscope hood or storage oscilloscope may be required for the spike to be visible. Maintain the spike for the cycle time of the GSE or functional test equipment or

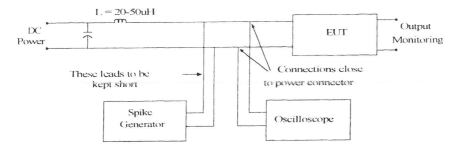

**Figure 9.64** CS06 conducted susceptibility test setup for spike and power.

5 minutes, whichever is longer. When testing on an AC supply, the spike position shall be moved over a full 180° of the AC waveform, either side of the zero crossing point.

MIL-STD 461E does not impose a differential-mode transient test, like CS06, but does require a bulk cable injection (common-mode) transient test CS115. The injection method is by injection probe, and the setup is similar to CS114.

### 9.6.7.12 Most Common Errors Incurred in Conducted Susceptibility Measurements

1. A consensus on the pass/fail criteria has not been reached; i.e. a few white spots on a display or some minor distortion may be acceptable, whereas multiple images or major distortion is almost certainly unacceptable. Similarly, some bit error rate in data communication may be acceptable, and this must be agreed on prior to commencement of susceptibility tests.

2. Failing to log the test start and stop times when data is analyzed offline, after the test is complete. The data must also be time tagged to allow comparison with the test times.

3. Incorrect measurement of the injected signal. The measurement point must be across the power input to the EUT. It must not be across the injection transformer primary or secondary or directly at the output of the power amplifier. Only if the power return is connected to the ground plane *at the EUT* can one of the generator output terminals be connected to the ground plane.

4. In the CS06 test and the CS02 test, use of long cables connecting the signal into the EUT. The long cables can reduce the high-frequency signal and degrade the CS06 pulse shape. Inject the signal as close to the EUT as possible, and monitor at the injection point, using a differential input to the scope.

5. Nonrepresentative EUT, different power-supply or input power-line filter.

6. Not applying the relaxed CS06 requirements for equipment that contains transient suppressors.

7. Oscillation of the power supply due to inclusion of the CS01 injection transformer with the power amplifier powered down or disconnected (see test method).

8. Injection of the in-rush current spike into the power amplifier via the injection transformer (see test method).

9. Setting up the CS02 level across a 50-Ω resistor. This should be attempted only if the required level cannot be developed at the input of the EUT. Some automated systems set up the test level across the 50-Ω resistor and then apply this to the EUT without monitoring the test level across the EUT, and this can result in severe undertesting or perhaps dangerous overtesting. For example, with 1 V peak specified as the test level, the power developed across the 50-Ω resistor is 0.1 W. If when the test level is applied to the EUT, 1 V peak is not developed, then the open loop test system does not attempt to increase the output test level. On the other hand, if the test level is monitored during automated test measurements in a closed loop system, any spikes generated by the EUT can affect the test level. For example, in one case the output dropped to 0 V and at other frequencies was at least 6 dB below the specified level. The best method is to monitor the output level on an oscilloscope and to manually control the amplitude.

10. Injection of some fraction of the AC power-line voltage into the power amplifier or, in the case of a closed loop system, into the monitoring port, via the coupling capacitor in the CS02 test (see the test method for a solution).

## 9.7  RTCA/DO-160 REQUIREMENTS

The Radio Technical Commission for Aeronautics (RTCA) imposes environmental and test procedures on airborne equipment. These contain power quality and EMI requirements and test methods. The latest version of the requirements document is DO-160D.

The category of the equipment for power quality is either A, B, E, or Z, determined by the type of aircraft electrical system. The EMI category is either A, B, or Z. A is equipment for which interference-free operation is desirable, B is equipment for which interference should be controlled to a tolerable level, and Z is equipment intended for interference-free operation. The RF susceptibility categories are W, Y, V, U, and T, where the W and Y test levels are applicable to locations where equipment and interconnecting wiring is installed in severe electromagnetic environments, such as nonmetallic aircraft or exposed area in metallic aircraft. Category V is defined for moderate environments, such as open areas in an aircraft composed principally of metal. Category U is defined as partially protected environments, such as avionics bays in all-metallic aircraft. Category T is defined as well-protected enclosed areas, such as an enclosed avionics bay in an all-metallic aircraft. The test limits are different for each category of equipment. The test methods and limits are very similar to those contained in the military standards discussed in Section 9.6, with some minor variations. One such variation is that the 10-μs voltage spike is specified at 600 V minimum open-circuit voltage with a 50-Ω source impedance for category A, and for category B a voltage transient is specified at a much lower amplitude with a 25-μs rise time to peak voltage and 100-μs width from the peak down to the 50% amplitude level, using a low-source-impedance generator. A repetitive spike 10 μs wide at an amplitude twice the rms AC power voltage is also applied from a low-impedance source to category B equipment. If the MIL-STD CS06 spike generator has a 600-V capability, then it can be used for the class A spike with an additional 50-Ω series resistor. The CS06 spike generator can also be used with series injection for the 10-μs-wide pulse. Interconnection cables are subjected to audio frequency magnetic fields from 400 Hz to 15 kHz (Category Z) and audio frequency and transient E fields (categories A, B, and Z), induced by wrapping a wire around the cable. For the audio frequency test on category Z equipment, a power source with 30-A capability at 400 Hz reducing down to 0.8 A at 15 kHz is necessary. This would require a large 1.7-kW variable-frequency supply or an impossible 3.6-kW audio power amplifier. Instead, a 500-W, or higher, variable-frequency AC supply used with a 15:1 step-down transformer may be used to generate the high AC currents. In the interconnecting cable spike test for categories A, B, and Z, a 50–1000-μs-duration burst of spikes with a repetition rate of 0.2–10 μs at a peak voltage of 600 V is specified. The recommended method of generating the test field is the unsuppressed "chattering" relay circuit. An E field wrapped-wire-cable test is also specified with a frequency of 380–420 Hz and a maximum voltage of 1800 V/m for category Z equipment and 360 V for category A equipment. Current probes are used to inductively couple RF into cables from 10 kHz to 400 MHz. The injection level is specified in decibels above 1 microamp and is determined by the category of the equipment. The recommended test setup requires the use of a monitor probe to monitor the injected current during the test.

An audio frequency conducted susceptibility test from 750 Hz to 15 kHz, similar to CS01, is specified. If the audio power amplifier cannot develop the specified level, then the test conditions will be adequately met by the use of a power amplifier with a maximum output of 30 W. The impedance of the output transformer used to inject the signal shall be 0.6 Ω ± 50%.

Both narrowband and broadband conducted emissions on power lines are measured using the monitor port on the LISN for categories A, B, and Z. For categories A and Z, a current probe that measures the current flow into the impedance of the LISN may be used. For category A and Z equipment, RF current flow on intersystem interconnecting cable bundles shall be measured using the current probe. The DO-160 5-μH LISN has the same requirements as the MIL-STD 5-μH LISN. The LISN may be manufactured and calibrated or bought out. Both narrowband and broadband conducted emission limits are imposed.

Radiated emission measurements are made from 150 kHz to 1215 MHz for categories A and Z and from 190 kHz to 1215 MHz for category B equipment. Both narrowband and broadband limits are imposed, and recommended measurement bandwidths are provided in the document. Different limits are applied to the three different categories. Radiation from antennas is excluded in the test setup, just as it is in MIL-STD-461/2 testing. In addition, when testing a transmitter, the selected frequency and ±50% of the band of frequencies between adjacent channels are excluded. An additional control on signals conducted out of antenna terminals is intended but not specified.

Radiated susceptibility testing from 15 kHz to 35 MHz is accomplished by generating a magnetic field from a loop located at approximately 2 cm from, and moved over, equipment and cables. An E field is generated from 30 MHz to 1215 MHz for categories T, U, and V and from 30 MHz to 18 GHz for categories W and Y. The antenna is located 0.9 m from the edge of the ground plane on which the EUT is mounted, with the EUT and interconnect power cables located 0.1 m in from the front edge of the ground plane. In accordance with DO-160C and D, the antenna shall be located with its lowest point at a height of 0.3 m above the ground plane. Antennas shall be positioned and aimed to establish the field strengths at the EUT and interconnecting wiring. When the beam width of the antenna does not totally cover the EUT and wiring, multiple scans shall be performed. Apertures in the EUT shall be exposed directly to the transmitting antenna. Horizontal and vertical orientations are required for polarized antennas, although this is typically taken as excluding the monopole antenna, long-wire, parallel-plate, or TEM cell, all of which may be used. Parallel-plate or TEM cells matching networks and calibration methods shall be described. Amplitude modulation of the CW source is required from 30 MHz to the upper frequency using a square wave at greater than 90% modulation and 1-kHz frequency. Consideration should be given for other modulations associated with the EUT, such as clock, data, IF, internal processing, or modulation frequencies. The use of field sensors located on the ground plane is proposed, which, as mentioned in Section 9.6.7.6, is prone to error. DO-160 recommends that the field sensors be connected to a control circuit, located outside of the room, using a fiber-optic link. Absorber is required on the surfaces of the shielded enclosure, and the EUT power is connected via the LISNS.

If the power input (DO-160 Section 16) tests are to be performed, then a frequency modulation of the AC power source is required. The surge current requirement necessitates variacs and transformers as well as relays and timers. The AC ripple on the DC power is less stringent than typical CS01 requirements, and the same test equipment may be used.

One of the major differences between MIL-STD and DO-160 requirements is that DO-160C and D impose indirect lightning tests on equipment to test for its capability to withstand effects of lightning-induced electrical transients. Direct lightning tests are intended to test externally mounted electrical and electronic equipment for the capability to withstand the direct effects of a severe lightning strike. Externally mounted equipment refers to all equipment that is mounted externally to the main skin of the aircraft and includes equipment covered only by a dielectric skin that is an integral part of the equipment. If the dielectric skin is specific to the aircraft and not integral to the equipment, then the equipment is subject to tests specified by the aircraft manufacturer. These lightning tests may be severe, and the test levels are determined by the internal aircraft environments as well as for metallic and nonmetallic aircraft. These environments range from Level 1, which is well protected, to Level 2, partially protected, to level 3, moderately protected, to Levels 4 and 5, which are characterized as severe electromagnetic environments. Each level has a set of waveforms associated with it, with the open-circuit voltage and short-circuit current characteristics of the waveform specified. The waveforms are idealized and are described as an exponential/exponential transient in which both the rising edge and the falling edge are in the form of an exponential. The second waveform is a damped sinusoid with a specified frequency, a peak voltage, and current reducing to a specified level after four cycles. Two groups of indirect test may be specified. The PIN test is designed to test for damage tolerance by injecting directly into the pins of the EUT connector, usually between each pin and case ground. For equipment electrically isolated from case and local airframe grounds, a dielectric withstand or hi-pot test is adequate. The second group of tests is the cable bundle tests, which are used to determine if functioning equipment will experience upset or component damage when exposed to a cable induced transient. An alternative injection method is ground injection, where the injection point may be at the EUT case ground or into the case of support equipment. In any of the induced tests, a current-measuring transformer is placed around the cable bundle to measure the bulk (C/M) cable current. When equipment has shielded cables that are not connected to available support equipment, a connector may be mounted on a small metal enclosure, the cable plugged into the enclosure, and a direct injection made between the ground plane, to which the EUT is bonded, and the small connectorized enclosure. The injection method is not so important as ensuring that the current flows on the cable bundle or the correct voltage is applied between the ground plane and the cable. The level of common-mode current induced in the center conductors of shielded cables and some possible transient protection components for different circuits are described in Section 5.4.

The exponential/exponential waveforms are (1) and (4) 6.4-$\mu$s rise time falling in 70 $\mu$s to 50% amplitude, (2) a 100-ns rise time falling to 0% amplitude after 6.4 $\mu$s, and (5) a 50-$\mu$s rise time falling in 500 $\mu$s to 50% amplitude. The waveforms are specified with an open-circuit voltage and short-circuit current. For pin injection testing, waveform (4) ranges from 50 V/10 A to 1600 V/320 A and waveform (5) from 50 V/50 A to 1600 V/1600 A, depending on the specified level (1–5). Pin injection also specifies a damped sinusoid, waveform (3), at the aircraft resonance frequency, or 1 MHz if the resonant frequency is unknown. Waveform (3) varies from 100 V/4 A to 3200 V/128 A, depending on the specified level. For pin injection, waveforms (1) and (2) are specified from 50 V/100 A to 1600 V/3200 A, waveform (3), the damped sinusoid, from 100 V/20 A to 3200 V/640 A, and waveforms (4) and (5) from 50 V/300 A to 1600 V/10,000 A.

Direct lightning injection tests are conducted with the EUT bonded to a ground plane in accordance with the installation requirement. A high-voltage test is performed on equipment that is covered by a dielectric in which an electrode is at a specified distance from the EUT. Two test methods are described. One has a voltage waveform applied to the test electrode with a rise time of 1.2 $\mu$s and a fall time to 50% in 50 $\mu$s. The total gap lengths, between the electrode and the EUT, can range from 0.5 mm to 1.5 mm with voltage ranges of 250 kV to 1500 kV. One alternative test method has a linearly increasing voltage applied to the electrode at an average $dV/dt$ of 10,000 kV/$\mu$s until the gap breaks down. In this alternative method, the gap can be from 0.5 mm to 1.5 mm and the voltage from 750 kV to 2400 kV.

A high-current-arc entry test is specified in which an arc is initiated and enters the EUT. Several components of the current waveform are specified, with only some of the components applicable, depending on the category of the equipment. The peak current in component A is specified at 200 kA for 500$\mu$s, followed by component B with a peak amplitude of 2 kA. Component C is a continuing current of between 200 A and 800 A, followed by a restrike current component D with a 100-kA peak. An alternative conducted entry test has the current injected through the ground plane on which the test object is mounted, representative of the lightning current distribution in an aircraft during a lightning strike. As a minimum, a surface current density of 50 kA/m should be applied. Although the ground plane injected test may appear trivial, in reality magnetic fields of hundreds of thousands of amps per meter and E fields of hundreds of thousands of volts per meter can be generated, depending on the location of the EUT.

## REFERENCES

1. News from Rohde and Schwartz, No. 151 (1996/II).
2. Compliance Engineering.
3. E. Bronaugh, D.R. Kerns. IEEE Transactions on Electromagnetic Compatibility. A new isolated antenna system for electromagnetic emissions measurements in shielded enclosures. Proceedings of the IEEE, International symposium on EMC, Atlanta, GA, 1978.
4. A.C. Marvin. IEEE Trans. Electromag. Compat. The use of screened (shielded) rooms for the identification of radiated mechanisms and the measurement of free-space emissions from electrically small sources. Nov. 1984, EMC-26(4).
5. Private communication with Ken Javor, EMC Engineer.
6. B. Audone, L. Bolla, G. Costa, A. Manara, H. Pues. IEEE EMC Symposium Record, 1993. Design and engineering of a large shielded semi-anechoic chamber meeting the volumetric NSA requirements at 3-m and 10-m transmission length.
7. Wolfgang Bittinger. IEEE EMC Symposium Record, 1993. Properties of open strip lines for EMC measurements.
8. H.A. Mendez. A new Approach to Electromagnetic Field-Strength Measurements in Shielded Enclosures. 1968 Wescon Technical Papers Session 19. Los Angeles, August 20–23, 1968.
9. W.C. Dolle, G.N. Van Stewenberg. IEEE EMC Symposium Record, July 1970. Effects of shielded enclosure resonances on measurement accuracy.
10. A.C. Marvin. EMC Symposium. Near-field antenna coupling theory in a shielded room: the mutual impedance model. 3rd Symposium and Technical Exhibition on Electromagnetic Compatibility. Rotterdam, The Netherlands, May 1–3, 1979.
11. A.C. Marvin, A.L. Marvin. Method of Damping Resonances in a Screened Room in the Frequency Range 30 to 200 MHz. EMC Technology, July/August, 1991.
12. A.C. Marvin, L. Dawson.
13. Y. Naito, T. Mizumoto, M. Takahashi, S. Kunieda. IEEE EMC Symposium Record, 1994. Anechoic chamber having multi-layer electromagnetic wave absorber of sintered ferrite and ferrite composite membranes.
14. Bruce Archambault, Kent Chamberlin. IEEE EMC Symposium Record, 1994. Modeling and measurements of an alternative construction technique to reduce shielded room resonance effects.
15. J.P. Muccioli, T.M. North, K.P. Slattery. IEEE EMC Record, 1996. Investigation of the theoretical basis for using a 1-GHz TEM cell to evaluate the radiated emissions from integrated circuits.
16. J.D. Osburn, E.L. Bronaugh. Advances in GTEM to OATS correlation models. IEEE EMC Symposium Record, 1993.

# 10

# Systems EMC and Antenna Coupling

## 10.1. SYSTEM-LEVEL EMC

A significant amount of the subject matter dealt with in preceding chapters is equally applicable to equipment, subsystems, and systems. Consider the case of cable-to-cable or cable-to-equipment coupling that may occur between equipment, subsystems, or systems. In this chapter, system-level EMC is emphasized, although many of the topics are equally applicable to subsystems and equipment. For example, antenna-to-antenna or antenna-to-cable coupling can occur within a subsystem as well as between systems.

The foregoing observations are perhaps obvious, although not always remembered. For example, consider a room containing a system made of a number of pieces of equipment all connected to the same potentially noisy power supply. A decision to incorporate a main power-line filter at the location where the supply enters the room can be defended based on the reduced radiation from the power-line cabling within the room, caused by supply noise, and on the elimination of, or reduction in, filters in the individual equipment. The design or selection of the main filter may be achieved using the methods described in Chapter 5, although these and the case study were applied at the equipment level. For the system-level main filter, the source of noise on the load side of the filter is a composite of the individual equipment noise sources, as is the total load impedance presented by the equipment. When filter components exist in equipment, these must be included in the evaluation of the main power-line filter. The analysis of the performance of the main system-level filter is likely to be more complex than that required for an equipment filter and may be practical only by use of a circuit modeling program such as SPICE. The foregoing is only one example of a great number of EMC design approaches, including shielding, that may be applied equally to circuits, equipment, subsystems, and systems.

### 10.1.1. MIL-STD System-Level Requirements

#### 10.1.1.1. General

MIL-STD-461/2 apply to equipment and subsystems only. MIL-STD-464 is a document applicable to all agencies of the Department of Defense in the United States that contains system-level EMC requirements. The associated data item descriptions (DIDs) are: DI-EMCS-81540, Electromagnetic Environmental Effects Integration and Analysis Report; DI-EMCS-8151, Electromagnetic Environmental Effects Verification Procedures; and DI-81542, Electromagnetic Effects Verification Report.

Document MIL-STD-1541A is an EMC requirements document for space systems and is applicable only to the U.S. Air Force, although it is used by other agencies, such as NASA.

The purpose of system-level requirements is to ensure that the integrated system design will be electromagnetically compatible with all of the characteristics and modes of operation of the system. The requirements include control of lightning protection, static electricity, and bonding and grounding. Control is achieved by a system electromagnetic compatibility program that is governed by an electromagnetic compatibility board. An EMC test is required on the system, as described in this section.

The electromagnetic compatibility program (EMCP) includes the necessary approach, planning, technical criteria, and management controls. In MIL-STD-1541 the EMCP requires an EMC control-and-test plan that must be approved by the electromagnetic compatibility board.

The system and all associated subsystems/equipment are to be designed for EMC in accordance with the EMC control plan. A system design program shall be implemented that covers the following areas:

Subsystem/equipment criticality categories.
Degradation criteria.
Interference and susceptibility control.
Wiring and cable.
Electrical power.
Bonding and grounding.
Lightning protection.
Static electricity.
Personnel hazards.
EM hazards to explosives and ordnance.
External environments.
Suppression components.

Systems are categorized based on the impact of EMI, or susceptibility malfunction, or degradation of performance.

The system-level EMC control plan is similar to the subsystem and equipment control plan, described in Section 12.1.1, with the following additions:

Methods and requirements for ensuring that sources of interference (subsystems/equipment) do not cause EMI in other subsystems or experience EMI from other sources in the system

Predicted problem areas and proposed methods of approach for solution of problems not resolved by compliance with MIL-STD-461

Design criteria and required tests for lightning protection

Radiation characteristics from system antennas, including fundamental and spurious energy, and antenna-to-antenna coupling.

The system-level EMI tests are seldom as extensive as those described in MIL-STD-462.

## 10.1.1.2. MIL-STD-464

This standard establishes electromagnetic environmental effects ($E^3$) interface requirements and verification criteria for airborne, sea, space, and ground systems, including associated ordnance. MIL-STD-464 supersedes MIL-STD-1818A, MIL-E-6051D, MIL-B-5087B, and MIL-STD-1385B. The general requirements are that the system shall be electromagnetically compatible among all subsystems and equipment within the system and with the environments caused by electromagnetic effects external to the system. Verification shall be accomplished as specified herein on production representative systems. Safety-critical functions shall be verified to be electromagnetically compatible within the system and with external environments prior to use in those environments. Verification shall address all life cycle aspects of the system, including (as applicable) normal in-service operation, checkout, storage, transportation, handling, packaging, loading, unloading, launch, and the normal operating procedures associated with each aspect.

Margins shall be provided based on system operational performance requirements, tolerances in system hardware, and uncertainties involved in verification of system-level design requirements. Safety-critical and mission-critical system functions shall have a margin of at least 6 dB. Ordnance shall have a margin of at least 16.5 dB of maximum no-fire stimulus (MNFS) for safety assurances and 6 dB of MNFS for other applications. Compliance shall be verified by test, analysis, or a combination thereof.

*Shipboard internal electromagnetic environment (EME).* For ship applications, electric fields (peak V/m-rms) below deck from intentional onboard transmitters shall not exceed the following levels:

a.   Surface ships
   (1)   Metallic: 10 V/m from 10 kHz to 18 GHz.
   (2)   Nonmetallic: 10 V/m from 10 kHz to 2 MHz, 50 V/m from 2 MHz to 1 GHz, and 10 V/m from 1 GHz to 18 GHz
b.   Submarines: 5 V/m from 10 kHz to 1 GHz

Compliance shall be verified by test of electric fields generated below deck with all antennas (above and below decks) radiating.

*Powerline transients.* For Navy aircraft and Army aircraft applications, electrical transients of less than 50 microseconds in duration shall not exceed +50% or −150% of the nominal DC voltage or ±50% of the nominal AC line-to-neutral rms voltage. Compliance shall be verified by test.

*Multipaction.* For space applications, equipment and subsystems shall be free of multipaction effects. Compliance shall be verified by test and analysis.

*Intersystem EMC.* The system shall be electromagnetically compatible with its defined external EME such that its system operational performance requirements are met. For systems capable of shipboard operation, Table 10.1a shall be used. For space and launch vehicle systems applications, Table 10.1b shall be used. For ground systems, Table 10.1c shall be used. For all other applications and if the procuring activity has not defined the EME, Table 10.1d shall be used. Intersystem EMC covers compatibility with, but is not limited to, EMEs from like platforms (such as aircraft in formation flying, ship with escort ships, and shelter-to-shelter in ground systems), friendly emitters, and hostile emitters. Compliance shall be verified by system-, subsystem-, and equipment-level tests, analysis, or a combination thereof.

*Lightning.* The system shall meet its operational performance requirements for both direct and indirect effects of lightning. Ordnance shall meet its operational performance requirements after experiencing a near strike in an exposed condition and a direct strike in a stored condition. Ordnance shall remain safe during and after experiencing a direct strike in an exposed condition. MIL-STD 464 provides a figure for the direct-effects lightning environment and figures and a table for the indirect-effects lightning environment from a direct strike. In addition, a table is provided for the near-lightning-strike environment. Compliance shall be verified by system-, subsystem-, equipment-, and component-(such as structural coupons and radomes) level tests, analysis, or a combination thereof.

*Electromagnetic pulse (EMP).* The system shall meet its operational performance requirements after being subjected to the EMP environment. If an EMP environment is not defined by the procuring activity, a figure in MIL-STD-464 shall be used. This requirement is not applicable unless otherwise specified by the procuring activity. Compliance shall be verified by system-, subsystem-, and equipment-level tests, analysis, or a combination thereof.

**Table 10.1a** External EME for Systems Capable of Shipboard Operations (Including Topside Equipment and Aircraft Operating for Ships) and Ordnance

| Frequency (Hz) | Environment (V/m-rms) | |
| --- | --- | --- |
| | Peak | Average |
| 10 k–150 M | 200 | 200 |
| 150 M–225 M | 3,120 | 270 |
| 225 M–400 M | 2,830 | 240 |
| 400 M–700 M | 4,000 | 750 |
| 700 M–790 M | 3,500 | 240 |
| 790 M–1000 M | 3,500 | 610 |
| 1 G–2 G | 5,670 | 1,000 |
| 2 G–2.7 G | 21,270 | 850 |
| 2.7 G–3.6 G | 27,460 | 1,230 |
| 3.6 G–4 G | 21,270 | 850 |
| 4 G–5.4 G | 15,000 | 610 |
| 5.4 G–5.9 G | 15,000 | 1,230 |
| 5.9 G–6 G | 15,000 | 610 |
| 6 G–7.9 G | 12,650 | 670 |
| 7.9 G–8 G | 12,650 | 810 |
| 8 G–14 G | 21,270 | 1,270 |
| 14 G–18 G | 21,270 | 614 |
| 18 G–40 G | 5,000 | 750 |

*Nondevelopmental items (NDI) and commercial items.* NDI and commercial items shall meet EMI interface control requirements suitable for ensuring that system operational performance requirements are met. Compliance shall be verified by test, analysis, or a combination thereof.

*Electrostatic charge control.* The system shall control and dissipate the build-up of electrostatic charges caused by precipitation static (p-static) effects, fluid flow, air flow, space and launch vehicle charging, and other charge-generating mechanisms to avoid fuel ignition and ordnance hazards, to protect personnel from shock hazards, and to prevent performance degradation or damage to electronics. Compliance shall be verified by test, analysis, inspections, or a combination thereof.

*Precipitation static (p-static).* The system shall control p-static interference to antenna-

**Table 10.1b** External EME for Space and Launch Vehicle Systems

| Frequency (Hz) | Environment (V/m-rms) | |
| --- | --- | --- |
| | Peak | Average |
| 10 k–100 M | 20 | 20 |
| 100 M–1 G | 100 | 100 |
| 1 G–10 G | 200 | 200 |
| 10 G–40 G | 20 | 20 |

**Table 10.1c** External EME for Ground Systems

| Frequency (Hz) | Environment (V/m-rms) | |
| --- | --- | --- |
| | Peak | Average |
| 10 k–2 M | 25 | 25 |
| 2 M–250 M | 50 | 50 |
| 250 M–1 G | 1500 | 50 |
| 1 G–10 G | 2500 | 50 |
| 10 G–40 G | 1500 | 50 |

connected receivers onboard the system or on the host platform such that system operational performance requirements are met. Compliance shall be verified by test, analysis, inspections, or a combination thereof. For Navy aircraft and Army aircraft applications, p-static protection shall be verified by testing that applies charging levels representative of conditions in the operational environment.

*Electromagnetic radiation hazards (EMRADHAZ).* The system design shall protect personnel, fuels, and ordnance from hazardous effects of electromagnetic radiation. Compliance shall be verified by test, analysis, inspections, or a combination thereof.

*Hazards of Electromagnetic Radiation to Personnel (HERP).* The system shall comply with current national criteria for the protection of personnel against the effect of electromagnetic radiation. DOD policy is currently in DoDI 6055.11. Compliance shall be verified by test, analysis, or combination thereof.

*Hazards of electromagnetic radiation to fuel (HERF).* Fuels shall not be inadvertently ignited by radiated EMEs. The EME includes onboard emitters and the external

**Table 10.1d** Baseline External EME for All Other Applications

| Frequency (Hz) | Environment (V/m-rms) | |
| --- | --- | --- |
| | Peak | Average |
| 10 k–100 k | 50 | 50 |
| 100 k–500 k | 60 | 60 |
| 500 k–2 M | 70 | 70 |
| 2 M–30 M | 200 | 200 |
| 30 M–100 M | 30 | 30 |
| 100 M–200 M | 150 | 33 |
| 200 M–400 M | 70 | 70 |
| 400 M–700 M | 4020 | 935 |
| 700 M–1000 M | 1700 | 170 |
| 1 G–2 G | 5000 | 990 |
| 2 G–4 G | 6680 | 840 |
| 4 G–6 G | 6850 | 310 |
| 6 G–8 G | 3600 | 670 |
| 8 G–12 G | 3500 | 1270 |
| 12 G–18 G | 3500 | 360 |
| 18 G–40 G | 2100 | 750 |

EME. Compliance shall be verified by test, analysis, inspection, or a combination thereof.

*Hazards of electromagnetic radiation to ordnance (HERO)*. Ordnance with electrically initiated devices (EIDs) shall not be inadvertently ignited during, or experience degraded performance characteristics after, exposure to the external radiated EME of Table 10.1a for either direct-RF-induced actuation or coupling to the associated firing circuits. Compliance shall be verified by system-, subsystem-, and equipment-level tests and analysis. For EMEs in the HF band derived from near-field conditions, verification by test shall use transmitting antennas representative of the types present in the installation.

*Electrical bonding*. The system, subsystems, and equipment shall include the necessary electrical bonding to meet the $E^3$ requirements of this standard. Compliance shall be verified by test, analysis, inspections, or a combination thereof, for the particular bonding provision.

*Power current return path*. For systems using structure for power return currents, bonding provisions shall be provided for current return paths for the electrical power sources such that the total voltage drops between the point of regulation for the power system and the electrical loads are within the tolerances of the applicable power quality standard. Compliance shall be verified by analysis of electrical current paths, electrical current levels, and bonding impedance control levels.

*Antenna installations*. Antennas shall be bonded to obtain required antenna patterns and meet the performance requirements for the antenna. Compliance shall be verified by test, analysis, inspections, or a combination thereof.

*Shock and fault protection*. Bonding of all exposed electrically conductive items subject to fault condition potentials shall be provided to control shock hazard voltages and allow proper operation of circuit protection devices. Compliance shall be verified by test, analysis, or a combination thereof.

*External grounds*. The system and associated subsystems shall provide external grounding provisions to control electrical current flow and static charging for protection of personnel from shock, prevention of inadvertent ignition of ordnance, fuel, and flammable vapors, and protection of hardware from damage. Compliance shall be verified by test, analysis, inspections, or a combination thereof.

*Aircraft grounding jacks*. Grounding jacks shall be attached to the system to permit connection of grounding cables for fueling, stores management, servicing, and maintenance operations and while parked. ISO 46 contains requirements for interface compatibility. Grounding jacks shall be attached to the system ground reference so that the resistance between the mating plug and the system ground reference does not exceed 1.0 ohms DC. The following grounding jacks are required.

*Tempest*. National security information shall not be compromised by emanations from classified information processing equipment. Compliance shall be verified by test, analysis, inspections, or a combination thereof. (NSTISSAM TEMPEST/1–92 and NACSEM 5112 provide testing methodology for verifying compliance with TEMPEST requirements.)

*Emission control (EMCON)*. For Army applications, Navy applications, and other systems applications capable of shipboard operation, unintentional electromagnetic radiated emissions shall not exceed $-110$ dBm/m$^2$ at 1 nautical mile ($-105$ dBm/m$^2$ at 1 kilometer) in any direction from the system over the frequency range from 500 kHz to 40 GHz. Unless otherwise specified by the procuring activity, EMCON shall be activated by a single control function for aircraft. Compliance shall be verified by test and inspection.

*Electronic protection (EP).* For Army aircraft and Navy aircraft applications, intentional and unintentional electromagnetic radiated emissions in excess of the EMCON limits shall preclude the classification and identification of the system such that system operational performance requirements are met. Unless otherwise specified by the procuring activity, EP shall be activated by a single control function. Compliance shall be verified by test, analysis, inspections, or a combination thereof.

### 10.1.1.3. MIL-STD-1541A (USAF) Electromagnetic Compatibility Requirements for Space Systems

This standard establishes the electromagnetic compatibility requirements for space systems, including frequency management and the related requirements for the electrical and electronic equipment used in space systems. It also includes requirements designed to establish an effective ground reference for the installed equipment and designed to inhibit adverse electrostatic effects. MIL-1541A defines minimum performance requirements and identifies requirements for system and equipment engineering to achieve EMC and test and analysis to demonstrate compliance with the standard

*Lightning.* Vehicle and equipment designs shall prevent overstress or damage induced by a lightning strike to the nearest facility-lightning-protective device, by a lightning strike just outside the zone of protection, or by a lightning strike near or above underground cables that are a part of the system.

*Triboelectric charging.* Disturbances in the performance of launch vehicles, which may arise from electrostatic charging of external surfaces by impact of atmospheric particles and which may lead to corona or streamer discharges and sparking, shall be prevented by bonding all conductive parts and by controlling the charge accumulation on external dielectric surfaces.

*Magnetospheric charging.* Disturbances in the performance of space vehicles, which may arise from differential charging of external surfaces and internal components by space plasma and which may lead to electrical discharges, shall be minimized by bonding all conductive parts and by controlling the charge accumulation on external dielectric surfaces and in bulk dielectrics.

System and equipment designs shall provide electromagnetic interference safety margins in accordance with the worst-case potential criticality of the effects of interference induced anomalies. The following categories shall be used:

a. Category I: Serious injury or loss of life, damage to property, or major loss or delay of mission capability

b. Category II: Degradation of mission capability, including any loss of autonomous operational capability

c. Category III: Loss of functions not essential to mission

Electromagnetic interference safety margins shall incorporate allowances for the effects of failures of redundant items, variations in characteristics due to aging and between like units. Other system requirements are as follows:

a. Category I: 12 dB for qualification, 6dB for acceptance

b. Category II: 6 dB

c. Category III: 0 dB

Category I and II safety margins shall be increased by 6 dB if the only practicable method of verification is entirely by analysis based on estimated emission or susceptibility characteristics.

*Degradation criteria*. Safety margins shall be related to definite degradation criteria, depending upon whether the appropriate requirement is freedom from overstress or damage due to continuous or aperiodic interference, autonomous recovery to the state prior to the occurrence of aperiodic interference, or continuous operation within specification limits.

*Superposition*. The specified safety margins shall be obtained with the combined effects of conducted and radiated broadband and narrowband emissions.

## System Requirements Evaluation

System analyses shall be performed to validate reductions in the requirements of this standard, to define certain requirements that are commonly peculiar to a system (such as those pertaining to signal and control circuits), and to identify the need for requirements that must be more restrictive or severe than those in the standard. The resulting program-peculiar requirements are subject to the approval of the contracting officer.

*Intersystem analysis*. Global compatibility with the radiation of all communications-electronics activities, whether earth- or space-based, shall be evaluated using the techniques described in the NTIA manual and Report R-3046-AF or equivalent. Intersystem compatibility analysis for conductive interfaces shall be done using the methods specified by this MIL-STD-1541A for intrasystem analysis.

*Intrasystem analysis*. Electromagnetic compatibility calculations entail finding the peak response of the potentially susceptible circuits in the time domain to several or more extraneous emissions that may be coupled through one or more transfer functions. For steady-state conditions, analyses based on the direct and inverse Fourier transforms, plus the convolution theorem, are appropriate and shall be used. Either amplitude or power spectral density functions may be used.

*Frequency and time domain*. Steady-state emissions shall be stated in the frequency domain for the same frequency ranges as for the corresponding requirements of MIL-STD-461, as modified by this standard. Oscillatory load-switching transients shall be defined in both the time and frequency domains; nonoscillatory transients may be defined only in the time domain.

*Reference requirements*. The evaluation shall include all the emission and susceptibility requirements specified in MIL-STD-461, as modified by this standard, for the classes of equipment under consideration. It shall include the applicable technical standards specified in the NTIA manual.

*Interference coupling modes*. The analyses shall incorporate mathematical models to account for the following:

a. Transverse and common-mode effects in signal and power circuits, including the effects of power source voltage ripple
b. Coupling between circuits in interconnecting cables
c. Bilateral coupling between the system antennas and circuits in interconnecting cables
d. Bilateral coupling between the system antennas, between the equipment enclosures, and between antennas and enclosures

*Susceptibility requirements.* Susceptibility requirements shall be shown to be sufficient to provide the required integrated safety margins for the applicable frequency range and point safety margins at all frequencies in that range.

*Simplifications.* Circuits within an interconnecting cable may be represented by only one of each distinct signal classification, as determined by similarity of signal characteristics and the terminations.

*Intrasystem analysis program.* MIL-STD-1541A has been structured to make best use of the developments that have been realized from the DOD Intrasystem Analysis Program. These benefits include improved analytical methods and programs for computer analysis. The analysis programs described in RADC-TR-74-342 and RADC-TR-82-20 are suitable and well documented.

*Analytical methods.* The methods presented in the applicable and the referenced documents are representative of the current art pertaining to the topics covered. Accordingly, they may be used as criteria for determining the adequacy and acceptability of any methods used.

*Signal and control interface circuits.* Emission and susceptibility requirements for signal and control circuits shall be developed and specified as program-peculiar limits. Values suitable for the signal levels and impedances characteristic of the applicable circuits shall be established.

*Lightning analysis.* Susceptibility to lightning and lightning protection shall be addressed in the analysis.

*Magnetospheric charging analysis.* Magnetospheric charging susceptibility and protection shall be addressed in the analysis. The program described in NASA CR-135259 shall be used. NASA TP2361 shall be used as a guide in determining applicable design requirements.

*Triboelectric charging analysis.* For launch vehicles, triboelectric charging shall be addressed.

## Nonstandard Limits

Changes in the detailed emission and susceptibility limits of this standard shall be justified by analysis.

## Consolidated Equipment Requirements

A single set of compatibility requirements shall be established for each piece of equipment. If more than one set is needed (e.g., for items that operate only in orbit, as contrasted with those that also operate in the launch and ascent environments), the differences shall be covered by a special classification method similar to that used in MIL-STD-461. Emission and susceptibility requirements shall be stated directly and explicitly. Susceptibility requirements shall not be specified indirectly in terms of an operating environment.

## Existing Designs and Modifications

To the extent they are complete and suitable, prior analyses and test data shall be used to satisfy the requirements of this standard for new applications or modifications of existing items. The additional system and component analyses and tests shall be centered on defining similarities and differences and evaluating compatibility for the new situation.

**Government Furnished Equipment**

The system emission and susceptibility characteristics shall be such that compatibility, with safety margins, is secured if the government-furnished equipment satisfies the requirements of MIL-STD-461 as supplemented and modified by the equipment.

The detailed intersystem vehicle requirements cover: structural materials, electrical ground network, electrical power subsystem referencing, bonding, lightning protection.

Requirements relating to communications by satellites include: surface finish for control of electrostatic charging, interconnecting cables, vehicle electrostatic susceptibility, power bus impedance and transient recovery time, electrical power quality, voltage ripple, spikes, surges, load switching and load faults, power subsystem faults.

The equipment requirements cover: interface safety margins, ground power sources, MIL-STD-461 performance criteria, limits and applicability, electrical power quality and faults, electrostatic susceptibility.

Test and evaluation includes: verification methods, measuring instruments, system verification (radiated susceptibility, partitioning, lightning, superposition effects).

System qualification tests (selection of test points, category I and II functions, category III functions, test conditions, acceptance criteria.

MIL-STD-1541 contains a limit on maximum radiated field strength of 0.01 W/cm$^2$ for human safety.

Some additional limits not contained in MIL-STD-461 are applied to equipment/subsystems in MIL-STD-1541. MIL-STD-1541 recommends that the system be used to test the system and proposes that one or more of the following general test approaches shall be used at the system level:

Inject interference at critical system points at a 6-dB-higher level than exists while monitoring other system/subsystem points for improper responses.

Measure the susceptibility of critical system/subsystem circuits for comparison to existing interference levels, to determine if a 6-dB margin exists.

Sensitize the system/subsystem to render it 6 dB more susceptible to interference while monitoring for improper responses.

MIL-STD-1541 also contains system design criteria and requirements.

## 10.2. ANTENNA-COUPLED EMI

Antenna-to-antenna coupled EMI is becoming more prevalent with the increased number of transmitting and receiving antennas located on vehicles, ships, and aircraft. With the advent of portable and mobile satellite communications systems the potential for civilian antenna-coupled EMI is increased. With the satellite Trucker terminal, the potential sources of interference are CB transmitters collocated on the truck and radar or other high-power transmitters the truck may pass by.

When choosing a site for installation of a receiving system, the suitability of the electromagnetic ambient at the location should be assessed. An ambient site survey or prediction should be made for the site, as described in Section 10.3.

### 10.2.1. Antenna-to-Antenna Coupling

A fundamental potential EMI situation occurs when the frequency of a transmitter is within 20% of the tuned frequency of a receiving system. The minimum frequency separation is approximately 10%, at which greater than 20 dB of cosite isolation is often required. A second EMI situation can arise when one or more out-of-band frequencies result in a spurious response in the receiver.

The distance between the location of the transmitting and receiving antennas are critical factors in determining if an EMI situation can occur.

In assessing a potential EMI situation for antennas located kilometers apart, the first step is to verify the existence of line-of-sight coupling, based on the Earth's radius.

Where line-of-sight coupling exists, the propagation loss correction for non-line-of-sight (NLS) = 0 dB.

A wave propagates in four principal modes: *direct wave*, which applies to line-of-sight coupling, the *reflected wave* from the intervening surface between the transmit and receive antennas, *surface wave*, and the *skywave*. The skywave is the wave bounced off of the ionized layers in the upper atmosphere. There are daily and seasonal variations in the height of the reflecting layers and also a critical frequency above which vertically directed waves are not reflected by the layers. The surface wave is caused by ground currents set up by the energy as it enters the ground. Below 3 MHz the surface wave predominates, between 3 MHz and 300 MHz the reflected wave, direct wave, and skywave all contribute, and above 300 MHz the direct wave predominates. The computer program LINCAL (Link Communications Analysis Algorithm), described in Ref. 1, calculates propagation losses with the frequency, transmitter power levels, distance, etc. as inputs.

When a computer program is not available, the following approximate calculations can be made:

When the line of sight is blocked by terrain, calculate the distance from the transmitting antenna to the terrain, $R_{ter}$, the additional factor to be added into the propagation for NLS is $40 \log R_{ter}/R$, where $R$ is the distance between the transmitting and receiving antennas.

The power output and gain of the transmitter at the fundamental and spurious emissions must be known. Where spurious emission levels are unknown, they can be assumed to be approximately 60 dB below the fundamental. The receiver spurious susceptibility level if unknown may, very approximately, be taken as 80 dB above the fundamental susceptibility.

When a frequency misalignment exists between the transmitter fundamental and spurious emissions and the receiver-tuned frequency, then corrections for the receiver and antenna out-of-band responses must be made to the simple in-band antenna-to-antenna coupling prediction.

The first step in the antenna-to-antenna coupling prediction is to calculate the power generated by the transmitting antenna and the coupling path. The term *isolation* describes the reciprocal of coupling and is often used to characterize a path that deliberately or incidentally contains structures that reflect, diffract, or absorb the electromagnetic field. The field radiated by an antenna is dependent on the input power, the frequency, and the gain of the antenna at that frequency, as described in Section 2.3.

Typical in-band gains of antennas are provided in Table 10.2. It must be emphasized that these gains are for the direction of intentional radiation and are readily available from manufacturers. The power radiated from the sidelobes and from the rear of an antenna is more difficult to ascertain if the radiation pattern is unknown. Although worst case, a gain of 0 dB for nonaligned antennas and those with low gain, such as monopole and loops, may be used in the preliminary antenna-to-antenna coupling analysis.

**Table 10.2**  Typical In-Band Gains of Various Types of
Antennas

| $G < 10$ dB | $10$ dB $\geq G \leq 25$ dB | $G > 25$ dB |
| --- | --- | --- |
| *Linear* | *Array* | *Array* |
| Cylindrical | Yagi | Mattress |
| Biconical | Broadside | |
| | Curtain | |
| *Dipoles* | | |
| Folded dipoles | | *Aperture* |
| Asymmetrical | End-fire curtain | Horns |
| Sleeve | *Travelling wave* | Reflector |
| Monopole | Rhombic | Lens |
| Collinear array | Surface and leaky wave | |
| *Loop* | | |
| *Apertune, slot* | Aperture | |
| *Helix* | Horn | |
| | *Corner reflector* | |
| | Log-periodic | |
| | Conical log-spiral | |

When antennas are colocated on the same surface in relatively close proximity (i.e., not kilometers apart), the basic line-of-sight antenna-coupling equations apply, with corrections for reflections. The level of reflected field depends on the angle of incidence of the field and the conductivity or permittivity of the reflecting surface. Case Study 10.1 describes the hazardous zone around a transmitting parabolic reflector antenna and considers the reflection from a high-conductivity and a high-permittivity surface. Case Study 10.1 provides the approach and calculations entailed in predicting field enhancement due to reflections.

When antennas are colocated on the same surface but with an intervening superstructure, fuselage, or wings, some level of isolation between the antennas exists.

Examples of the level of isolation achieved, are shown in Figure 10.1 for a KC-135 aircraft over the front edge of the wing, in Figure 10.2 over the rear edge of the wing, and in Figure

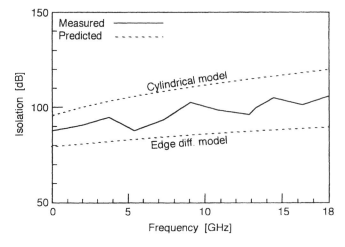

**Figure 10.1**  Isolation over front edge of aircraft wing (near root). (© 1985, IEEE.)

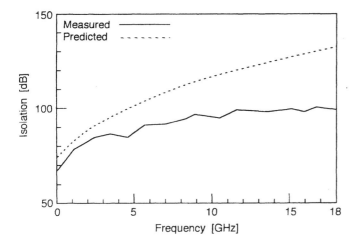

**Figure 10.3**   Isolation from the top to the bottom of the aircraft fuselage. (© 1985, IEEE.)

10.3 from the top to the bottom of the fuselage. Figures 10.1, 10.2, and 10.3 also show the results of measurements made on the KC-135 aircraft using identical broadband horn transmit and receive antennas at the locations described in the figures and pointed at each other for maximum coupling. The higher the level of isolation, the lower the coupled field. The IEMCAP and AAPG computer programs, which use the surface shading model for edge diffraction analysis, were used to predict the isolation shown in Figures 10.1, 10.2, and 10.3. A maximum deviation between predicted and measured values was seen with the antennas located on either side of the circular fuselage and equaled 35 dB at 18 GHz (Figure 10.3). Similar measurements made around the fuselage of an F-16 aircraft showed approximately 35-dB deviation between measured and predicted at 18 GHz. At 2 GHz, the deviation is an acceptable 8 dB, and the computer codes are useful at this and lower frequencies.

Figure 10.4 illustrates three examples of antenna-to-antenna coupling: (1) where both antennas are located off the surface of the vehicle; (2) one on, one off; and (3) both antennas on

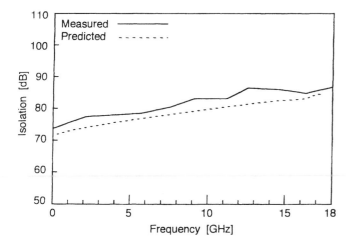

**Figure 10.2**   Isolation over back edge of aircraft wing (near root). (© 1985, IEEE.)

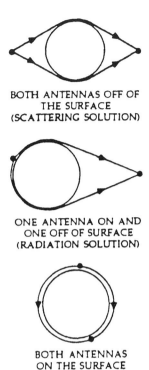

BOTH ANTENNAS OFF OF
THE SURFACE
(SCATTERING SOLUTION)

ONE ANTENNA ON AND
ONE OFF OF SURFACE
(RADIATION SOLUTION)

BOTH ANTENNAS
ON THE SURFACE
(COUPLING SOLUTION)

**Figure 10.4**   Three modes of antenna-to-antenna coupling and GTD solutions. (© 1987, IEEE.)

the surface. The coupling between antennas on a circular or elliptical cylinder is dependent on the exact location of the antennas. The coupling between monopole antennas placed at one-quarter of the distance around a cylinder with a 3′9″ diameter is shown in Figure 10.5. The coupling between a horn placed 5 feet from an elliptical cylinder is shown in Figure 10.6.

Figures 10.5 and 10.6 are based on measured data. Figure 10.5 illustrates the coupling gain, expressed in decibels, for two monopole antennas located on a 3′9″ diameter circular cylinder. The antennas are at the same position down the length of the cylinder and are located at points one-quarter the distance around the cylinder, as shown in Figure 10.5. Figure 10.6 illustrates the coupling gain for horns located on a sphere (case B) and with one horn on the sphere and one horn in free space (case A). The gain of the antenna is removed from the plots, which are of isotropic antenna gain, that is, the shading factor of the cylinder.

In addition to measured data, Ref. 3 describes the use of the Geometrical Theory of Diffraction (GTD) Intrasystem Electromagnetic Compatibility Analysis (IEMCAP) and Aircraft Inter-Antenna Propagation with Graphics (AAPG) computer codes to predict the coupling. The IEMCAP and AAPG codes use a simple formula for fuselage or structure shading, whereas the GTD codes use a more rigorous electromagnetic solution. IEMCAP prediction shows up to 35 dB less coupling, that is, higher isolation, than the GTD codes and if used might predict false EMC. Further information on the coupling paths measured data and EMI antenna-to-antenna computer predictions is available from Ref. 3.

So far we have assumed that the antennae are operating in band (i.e., over the designed frequency range). In the case where the transmitting or receiving antenna is operating out of

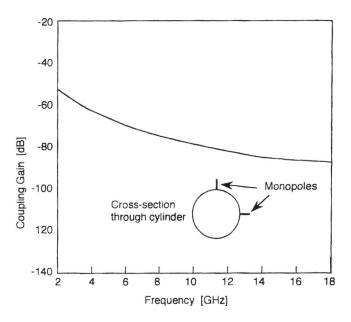

**Figure 10.5**   Coupling between two monopoles on the surface of a 3′9″-diameter circular cylinder. (© 1987, IEEE.)

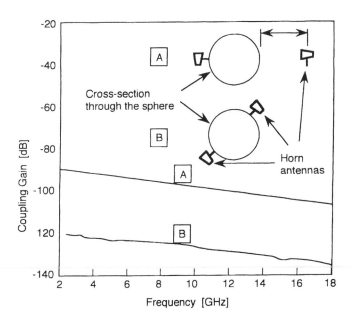

**Figure 10.6**   Coupling between horns on and off a cylinder. (© 1987, IEEE.)

**Table 10.3** Polarization Mismatch Factors for Antennas with Gain Greater Than and Less Than 10 dB

| Antenna gain | | Polarization | |
|---|---|---|---|
| Receiving | Transmitting | Horz./Vert. | Circular |
| <10 dB | <10 dB | 0.025 | 0.5 |
| >10 dB | <10 dB | 0.025 | 0.5 |
| <10 dB | >10 dB | 0.025 | 0.5 |
| >10 dB | >10 dB | 0.010 | 0.5 |

band, the gain of the transmitting antenna is changed, as is the effective aperture of the receiving antenna.

The gain of the antenna as a function of the effective area $A_e$ and wavelength is given in Eq. (2.36). Equation (2.47) provides the effective height of an antenna as a function of the effective area, antenna impedance, wave impedance, and radiation resistance of the antenna. The effective area of the antenna, defined as the ratio of the received power to the incident power, changes at frequencies either side of the design frequency of the antenna. The equation for effective area is multiplied by the polarization mismatch factor $p$ and the antenna impedance mismatch factor $q$ to obtain its out-of-band value. $p$ has a value shown in Table 10.3 for vertical or horizontal and circular polarization mismatches between a transmitting antenna with a gain of less than 10 dB and greater than 10 dB and a receiving antenna with a gain of less than 10 dB and greater than 10 dB.

The effects of $q$, the antenna impedance mismatch, for a 1-GHz dipole or loop antenna with an average polarization mismatch $p$ of 0.5 are shown in Figure 10.7. At below-band frequencies, $A_e$ is proportional to $f^2$. At above-band frequencies, $A_e$ is equal to $(c^2/8\pi f^2)$, where $c$ is the velocity of a wave in free space, in cm/s ($300 \times 10^8$ cm/s), and $f$ is in Hz. References 4 and 5 provide further information on antenna out-of-band response.

For a dipole or monopole antenna, the effective area may be found from the effective height; alternatively, the receiving characteristics of the antenna from the effective height may

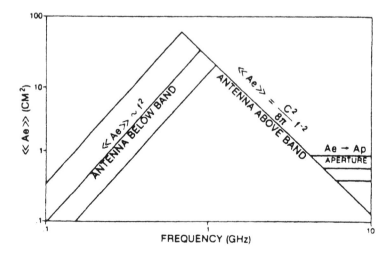

**Figure 10.7** Effective aperture trends. (© 1987, IEEE.)

be used. Use of Eq. (2.47) for the effective height will include the out-of-band mismatch corrections, such as: the antenna and load impedance mismatch, and the radiation resistance and wave impedance mismatch.

A number of potential sources of EMI exist in an environment where a large number of transmitters are in close proximity to the receiving antenna location. Some of the potential sources of EMI are:

Harmonic interference, where a harmonic of a transmitter is at the intended receive frequency

IF interference, where the transmitter is at the receiver IF frequency

Image interference, where the transmitter is at the receiver frequency minus twice the receiver IF frequency

Cross-modulation interference, where a high-level transmission close to the receiver frequency is not sufficiently attenuated by the input filter (here, compression and spurious responses can occur in the receiver)

Adjacent channel interference from a transmitter close enough to lie within the receiver IF bandwidth and the receiver bandwidth

Passive intermodulation interference (PIM) caused by the nonlinearity of metal-insulation-metal (MIM) or metal-oxide-metal (MOM) interfaces (commonly known as the *rusty bolt effect*)

PIM may occur in any metal structure in proximity to the receiving antenna, such as the antenna structure, railings, towers, fences. Where these structures are constructed of ferrous material, PIM may occur as a result of the curvature of the B/H curve of the material. Antennas that include ferrous materials in the antenna itself or in the support structure may exhibit PIM.

The power level of the reradiated signal has been reported in Ref. 6 at approximately 40 dB below the incident power level. Thus any two sources of emissions of sufficient power may cause intermodulation products, which are reradiated and detected by the antennas as a signal. Third-harmonic and third-order intermodulation products are the most likely to cause problems (i.e., $3f$, $2f_1 \pm f_2$, $2f_2 \pm f_1$).

Transmitter spurious emission interference. This includes harmonically related, both to the fundamental and master oscillator etc., and broadband noise emissions. Where an amplifier is used to simultaneously amplify a number of signals, transmitter-generated intermodulation products may exist.

Receiver intermodulation products. Nonlinearities in the front end of the system may generate intermodulation products from signals that are not sufficiently attenuated by the filter (i.e., those frequencies for which cross-modulation may be a problem).

The response of a receiver to an out-of-band signal or the input level at which cross-modulation occurs is a characteristic of the receiver. The noise floor of the receiver and dynamic range may be found from the receiver specification. The manufacturer may have information on the out-of-band response of the receiver when this is not contained in the specification. Alternatively, tests may be conducted on a sample receiver to characterize the important parameters.

Case Study 10.2 illustrates a typical ambient site survey in which a number of transmitters are potential sources of EMI to a ground-based satellite communications receiver. An example of PIM is provided, as well as some of the solutions to the problem. Connectors and other connections used at RF can contribute to PIM. Hermetically sealed connectors using kovar are known generators of intermodulation products. Nickel plating and stainless steel should be avoided in the manufacture of connectors and antenna parts, whereas silver-plated brass is one of the best materials. Other sources of PIM are: microdischarge between microcracks and across voids in metals, dirt, and metal particles on metal surfaces, high current density at contacts. The recommendations are to use a high-conductivity, low-corrosion plating with a known and controlled thickness and smoothness. Make sure all antenna and connector surfaces are thoroughly cleaned, for thumbprints have resulted in unacceptable levels of PIM on antenna surfaces when the PIM requirements are stringent. Poor test methods and metal-backed microwave absorber used in the test setup have resulted in inaccurate measurement of PIM.

The computer programs DECAL/PECAL described in Refs. 6–8 and in Chapter 12 of this book may be used to calculate the following interference at the receiver due to incident power from a transmitter:

Receiver adjacent signal: the undesired power at the transmitter tuned frequency.

Broadband transmitter noise

Narrowband spurious emissions

Spurious response. The power at the tuned frequency of the undesired transmitter when this frequency is close to one of the spurious responses, including image, of the receiver

Receiver intermodulation: The undesired power resulting from the mixing of two or three undesired transmitter signals in the receiver

Transmitter intermodulation: The undesired power resulting from the mixing of two or three undesired transmitter signals

A finite difference time domain (FDTD) numerical analysis method has been used to investigate cosite interference between wire antennas on helicopter structures as well as rotor modulation effects. The FDTD method provides full-wave solutions to electromagnetic problems. The FDTD method was used to accurately calculate the S11 and S22 parameters of two monopole antennas from 500 MHz to 4 GHz as well as the S12 parameter. A portion of the voltage applied to the transmitting antenna will be reflected back and a portion will be transmitted. A fraction of the transmitted signal is then reflected back from the receiving antenna and a portion of the signal is reflected from the receiving antenna into the load. S11 is the transmitting antenna input reflection coefficient, S12 is the reverse transmission coefficient between the two antennas, and S21 is the forward transmission coefficient. S22 is the receiving antenna output reflection coefficient. All four S parameters can be plotted on a Smith chart, and the antenna's input impedance and the coupling impedance can be computed. S12 is used to calculate the coupling between the antennas. The time domain data was transformed into the frequency domain using a fast Fourier transform (FFT). The analysis was validated by measurements made in an anechoic chamber, as described in Ref. 15. These measurements were made using a one-tenth scale model of a full-scale helicopter. Scale model measurements are mandatory when only a small-scale anechoic chamber is available, and it is also less costly. Because of the use of a scale model, the frequency band must be scaled so that the scale model and antennas have the same electrical dimensions as the full-scale geometry. Rotor modulation can impair the performance of communication systems, and this was also investigated in Ref. 15. The FDTD and scale model measurements of the coupling between two monopoles on a helicopter resulted in the following conclusions: The FDTD method can be used to accurately predict the coupling even when it is very small, on the order of 0.5 dB:

The coupling is not greatly affected by the fuselage geometry, provided there are no structural obstructions between the antennas. The coupling between the two monopoles on a ground plane is not significantly altered when the same configuration of monopoles is placed on the bottom of helicopter because the bottom of the helicopter, without wheels, closely resembles a ground plane.

The investigation confirmed what we already know, that moving antennas further apart, separating the transmit and receive frequencies, and disorienting the antennas all reduce coupling. It was also found that ''the coupling is not significantly influenced by the presence of rotors except at and near those frequencies where the rotors are half wavelength long and behave as parasitic resonant elements.''

Reference 16 describes the use of a 1/48 scale brass model and the numeric electromagnetic code (NEC, described in Section 12.4.3.1 to predict the effect of the addition of a high-frequency surface wave radar (HFSWR) onboard a ship. Because of the long wavelength in the HF band (2–30 MHz), the entire ship will radiate as an HF antenna. The HFSWR interacts with the topside superstructure, including existing communications antennas. The NEC yields estimates of the technical parameters of the antennas mounted on ships. These technical parameters of an antenna are completely specified by its impedance, near fields, radiation pattern, and coupling to other antennas. Near fields are used to evaluate radiation hazards to personnel, fuel, and ordnance. Either the NEC program or the brass model measurements may be used to calculate the antenna coupling; in Ref. 16 it was the brass model measurements that were used. The two areas of concern are the receiver-adjacent signal (RAS) EMI, in which the receiver signal adjacent to the HFSWR may be interfered with, or the broadband transmitter noise (BTN), which is defined as the portion of the undesired transmitter spectrum (including broadband transmitter noise but not including narrowband spurious emissions) that lies within the nominal passband of the receiver. In this case RAS was not an issue, but BTN was found to be a major potential problem. The preliminary EMC analysis was based on the assumption that the radar transmitter waveform is a rectangular pulse-train spectrum with a 50% duty cycle, 500-μs period, and 100-kHz bandwidth, as shown in Figure 10.8. The emission spectrum of this radar is therefore a sin x/x waveform. Assuming this rectangular pulse-train waveform, calculations of the radar transmitting spectrum at 2 MHz, 20 MHz (the radar transmitter tuned frequency), and 30 MHz are −17 dBm, 67 dBm (67 dBm = 5 kW transmitted power), and −11 dBm, respectively, as shown in Figure 10.9. Based on the brass model measurements, the antenna couplings at 2 MHz, 20 MHz, and 30 MHz are assumed to be 25 dB, 30 dB, and 35 dB, respectively. Hence, the received interference power levels at the input of the communication receiver at 2 MHz, 20 MHz, and 30 MHz are −42 dBm, 37 dBM, and −46 dBm, respectively. Quasi-minimum noise (QMN) is a reasonable lower-bound estimate of mean levels aboard Navy ships; QMN is defined as some minimum atmospheric noise power that is contaminated with some background local noise. The greater of the QMN and the receiving system noise is used as the EMC design goal for communications receivers. Comparing these interference power levels with the QMN, that is, the EMC design goal, we found that the excess interference power levels are 44 dB, 152 dB, and 74 dB at 2 MHz, 20 MHz, and 30 MHz, respectively.

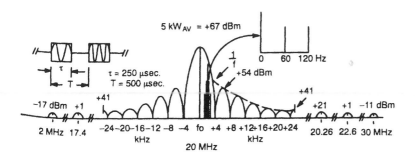

**Figure 10.8** Rectangular pulse-train spectral occupancy. (© 1995, IEEE.)

Chapter 10

- **If unmitigated, radar transmitter coupling would cause severe problems in comm receivers**

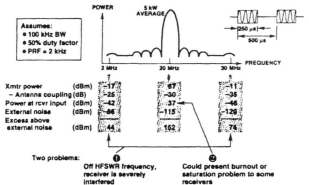

- **Solution to Problem 1 is to filter exciter signal prior to transmission**

**Figure 10.9**  Coupling of the transmitted radar signal into HF communication circuits: the problem. (© 1995, IEEE.)

This EMC study shows that if unmitigated, radar transmitter coupling would cause severe problems in communications receivers. There are two potential problems. The first problem is that the radar transmitter will generate enough broadband noise to severely interfere with all communications receivers operated at the entire HF Band (2 MHz to 20 MHz). The second problem is that the radar transmitter could burn out or saturate some EW receivers at radar on-tune frequency (20 MHz).

A solution to the first problem is described in Figure 10.10. We propose that communications receivers will not be interfered with at frequencies more that 5% on either side of the radar transmitter frequency. At 19 MHz, the radar transmitter power is 9 dBm; antenna coupling is −29 dB; broadband noise power at receiver input is −19 dBm; QMN is −114dBm; and excess interference level above QMN is 95 dB. Hence, filtering on the radar transmitter was required that will provide 95 dB of attenuation at a frequency that is separated by 1 MHz/50 kHz = 20 = 1.3 decades. A four-pole filter will provide the required 95-dB attenuation.

- **Proposed requirement: specify that communication receivers will not be interfered with at frequencies more than 5% on either side of the radar transmitter frequency**
  - **i.e., HFSWR takes 10% "bite" out of the HF communicator's spectrum**

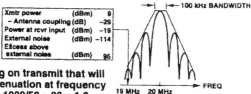

- **Hence, need filtering on transmit that will provide 95 dB of attenuation at frequency that is separated by 1000/50 = 20 = 1.3 decades**

- **Requires filter with**

$$\left(\frac{95 \text{ dB}}{1.3 \text{ decades}}\right)\left(\frac{1 \text{ pole}}{20 \text{ dB}}\right) \sim 4 \text{ poles}$$

**Figure 10.10**  Coupling of transmitted radar signal into HF communication circuits: the solution. (© 1995, IEEE.)

A second prediction is described in Ref. 16 of the interference generated by the shipboard HF communication system coupling into the HFSWR receiver. In this second prediction, in addition to BTN and RAS noise, clutter cross-modulation is examined. Clutter is the unwanted echoes, typically from the ground, structures, sea, rain or other precipitation, chaff, birds, insects, and aurora (also known as background return). Clutter cross-modulation results when a high-level signal from an onboard communications transmitter transfers its modulation to the clutter through cross-modulation occurring in the front end of the receiver.

A 1/48 on scale brass model was also used in Ref. 17 to predict the coupling between a 2–6-MHz transmit fan antenna and a 2–30-MHz receive antenna on a ship. Although the NEC is used to design new antennas and to provide a preliminary determination of the antenna performance and isolation, it is a 1/48th scale that is used to refine the antenna performance and finalize the locations. Typically the antenna design is not included in the basic design of new construction ships and must be developed after the topside design is fixed. The communication suite, both transmit and receive, is predefined, and even the antenna locations are presented to the antenna designer as a fait accompli. The brass model hull is constructed of laminated wood overlaid with brass. The topside features are machined from brass to 1/10000-inch tolerances. Antennas are constructed to provide the same electrical characteristics at the increased frequency, due to the scale, but may not have the same physical appearance. Calibrated cables are connected to each cable under test.

The antenna performance covers transfer impedance, (VSWR), antenna pattern and gain, radiation hazard prevention, and isolation (decoupling) between transmit and receive systems. The voltage–standing-wave ratio (VSWR) determines the ratio of power input to the antenna versus reflected power from the input of the antenna and is one of the most critical parameters. For a shipboard transmit antenna, VSWR must be below 4:1 across the frequency range of the antenna, including any matching network. However, perturbations introduced by metallic elements in the ship's superstructure often prevent meeting this goal. A shipboard receive antenna does not have the VSWR constraints encountered in transmit antenna design, since the goal is to optimize receive system efficiency. High-frequency receive systems do not necessarily need the most efficient antenna, but instead require an antenna that matches receive system performance to the QMN curve. Since it is futile to attempt to receive signals below the QMN, the frequency response of an HF receive antenna should be approximately 10 dB below the QMN to allow for antenna mismatch and cable and system losses.

Overall antenna performance is most easily characterized by the antenna pattern that is the spatial gain of the antenna. The pattern is described as a 3D plot or a number of 2D plots and is usually referenced to a theoretical 2/4 monopole on a perfect groundplane. Antenna patterns are severely affected by surrounding structures, so a perfect antenna is not attainable in a shipboard environment. An acceptable antenna is typically defined as one with an antenna pattern with less than 6-dB nulls at any azimuth over less than 20% of the frequency range of the antenna. A 6-dB null will result in transmission of only 25% of the power in the desired direction.

Reference 17 describes the isolation between the 2–6-MHz fan antenna and the 2–30-MHz receive antenna and shows that the intermodulation interference (IMI) generated in the victim receiver, from inadequate antenna decoupling, was unacceptably high. Relocation of either the transmit or the receive antenna, or both, was the solution offered for the potential EMI problem. After relocation, both the antenna isolation and the VSWR of the antenna/s were recalculated.

The field strengths were also characterized in close proximity to the transmit antenna, using the 1/48th scale brass mode, during the antenna design phase to allow a radiation hazard (RADHAZ) assessment. Radiation hazards include hazards of electromagnetic radiation to fuel

(HERF), hazards of electromagnetic radiation to ordnance (HERO), hazards of electromagnetic radiation to personnel (HERP), and RF burn. The maximum field strength levels differ for each type of RADHAZ. The field strength data was used by the antenna engineer to design and locate the antenna so that it will not exceed allowable levels of HF radiation close to personnel, fuel, and ordnance areas or so it will at least provide warnings of possible dangers.

### 10.2.2.  Filters and An "In-Band" EMI Solution

#### 10.2.2.1.  Filters

High-pass, low-pass, bandpass, or bandstop (notch) filters can be used to attenuate out-of-band EMI signals caused by coupling from fields incident on a receiving antenna. To avoid cross-modulation or intermodulation in the first stage after the antenna, e.g., the front end of a receiver, a low-noise amplifier (LNA), or a low-noise down-converter (LNC), the filter should be connected between the antenna and the first stage. The problem with this location is the additional insertion loss introduced by the filter. Most antenna designers have optimized the gain of the antenna and reduced losses in the RF cable to the antenna and, if the antenna is constructed with a number of PCB elements, has combined these PCB elements using a no-loss or low-loss combiner. Even a fraction of a decibel of additional loss as a result of the filter can reduce the G/T of the receiving system, as described in Section 5.3.3. If the system uses an LNA or LNC and the EMI is not caused by an overload in these elements, resulting in cross-modulation or intermodulation, then the filter should be located after the LNA or LNC to minimize the effect of the insertion loss of the filter. A number of types of filter are available: Filters with helical and folded helical resonators have been described in the literature, with design guidelines and examples. Helical resonators can be used as bandpass filters between several tens of megahertz and up to 2 GHz. The helical resonator consists of a single-layer solenoid enclosed in a shield, with one lead of the winding short-circuited to the shield and the other end open-circuited. Input and output couplings to the filter are built with taps to the appropriate impedance points on the helix that match the impedance to the source and load. By including an adjustable tuning screw in either side of the helix, it is possible to change the capacitive or inductive load of the helical to tune the center frequency. Helical filters are designed mostly for the Butterworth response. The Chebyshev response can be used, but the insertion loss of the Chebyshev is much higher. A typical loss for a helical filter is approximately 1 dB.

Surface acoustic wave (SAW) filter technology is based on the conversion of electrical signals into surface acoustic waves and then back to an electrical output. The frequency range for these filters is 10 MHz to 3 GHz, with a relative passband from 0.01% to 100%. The disadvantage is the minimum 6 dB of insertion loss. In the SAW impedance element filter (IEF), the long SAW transducer (resonator) has an impedance dependence similar to a classic inductance capacitance (LC) or quartz resonator, and the SAW itself plays only an auxiliary role. This device can be used as a bandpass filter, and in practice a number of elements must be connected together to achieve an adequate out-of-band suppression. In the IEF the insertion loss can be reduced, and in the balanced bridge version the insertion loss can be limited to 1.5 dB over a 890–915-MHz pass band.

Microwave filter Inc. manufactures a wide range of microwave and RF customized filters from 1 MHz to 26 GHz, including tunable notch filters. These filters include lumped-constant construction, distributed line construction, and cavity designs. Filters from 7 to 11 GHz are available with an insertion loss of 1 dB maximum.

FSY microwave also manufactures standard low-pass filters from 1 kHz to 18 GHz, high-pass filters from 5 kHz to 0.4 GHz, and bandpass filters from 10 kHz to 10 GHz. Waveguide filters from 3 GHz to 18 GHz are available with a low insertion loss. Allen Avionics manufactures custom filters from 100 Hz to 1 GHz. K&L manufactures a wide range of lumped-component, tubular, and cavity filters. The cavity filters exhibit lower insertion loss and steeper skirt selectivity than tubular filters and are available with two to six sections. The loss is dependent on the number of sections and the 3-dB bandwidth, but it can be as low as 0.5 dB for a two-section filter at 1 GHz with a 20-MHz (2%) bandwidth, or 1.4 dB for a six-section filter at 1 GHz with a 20-MHz bandwidth.

The Evanescent TE mode cavity filter (100 MHz to 18 GHz) is available from K&L and is a very low-loss filter. Typical losses are less than a decibel for most designs with bandwidth of 5% or more, with a rule of thumb of 0.1 dB per section. For wider-band filters, losses are even less. For an EMI problem at a specific frequency either one side or the other of the in-band frequencies, the bandwidth of the filter passband can be greater than 5%, as long as the interferer is well outside of the passband. If the interferer is close to the edge of the receiver passband, then one skirt of a multisection filter can be close to the interferer and still achieve adequate attenuation.

For example, a "real-world" EMI antenna-to-antenna coupling problem was presented to FSY Microwave, Inc., 6798 Oak Hill Lane, Columbia, MD 21045, 410/381-5700. In this case study the microwave frequency communications link had a center frequency of 1500 MHz and a 3-dB bandwidth of 20 MHz. An interferer at 1475 MHz was at a high enough level to result in compression at the input of the low-noise amplifier (LNA), so the first filter had to be located between the antenna and the LNA. This filter required a minimum of 30-dB attenuation but should also exhibit an insertion loss of maximum 1.5 dB to maintain the G/T of the system. When this problem was presented to FSY Microwave, they kindly customized a standard filter. The affordable five-pole C1500-20-5NN bandpass filter designed by FSY Microwave had the following specifications:

Center frequency = 1500 MHz

Insertion loss = 1.5 dB max @ CF (1.0 dB typical)

3-dB bandwidth = 20 MHz min, 24 MHz max

VSWR = 1.5:1 max over 80% of the 3-dB bandwidth

Impedance = 50 $\Omega$ in/out

Attenuation = 35 dBc min from DC to 1475 MHz

Typical 5-pole 0.035-dB Chebyshev response

Power rating = 20 W C.W. max

Size = 1.10″ × 1.75″ × 5.35″ excluding connectors

Connectors = type-N female in/out (or as specified)

Temperature range, operating = −30 to +60°C

Vibration = 10 G's 5 to 2000Hz, sine swept, 3 axes

Shock = 30 G's in 11 ms, 1/2 sine, 3 axes

Relative humidity = Up to 95%, noncondensing

The insertion loss and attenuation characteristics are shown in Figure 10.11

The five-pole filter is designed to avoid compression in the LNA with a minimum impact on the G/T figure. If additional attenuation of the interferer is required to avoid an EMI problem in the receiver, then a second filter could be included after the LNA. This second filter could have a higher level of attenuation, because the increased insertion loss occurs after the gain of the LNA and so will not adversely affect the G/T.

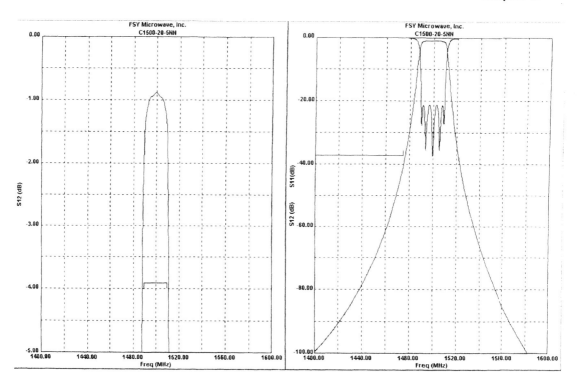

**Figure 10.11**   Attenuation characteristics of the FSY Micorwave, Inc., C1500-20-5NN five-pole filter.

### 10.2.2.2.  In-Band EMI Solution

In some instances where an interferer is very close to being in band, a filter with a very steep skirt can be used; or if the full bandwidth of the system is not required and some loss at the higher or lower frequencies can be accepted, then the skirt of the filter may extend to some small degree into the signal bandwidth. In one example of EMI, whenever an HF radio was operated onboard a ship, the ship's RF log indicated full speed. The RF current on the ship's hull induced by the HF radio fan antenna was coupled to the RF log transducer, used to measure the ship's speed. This problem was considered to be in band; i.e., the interferer frequency is exactly at a frequency required by the receiving system for normal communications to exist. In the case of the ship's log, the full speed indicated by the log was well above the speed of any of the ships in which the EMI problem existed. It was thus feasible to incorporate a filter between the RF ship's log transducer and the detector electronics, which would attenuate the HF radio frequency but, as a result would limit the display of the ship's speed to a more realistic maximum.

It may be possible to reduce the coupling between an interferer antenna and the receiving system by relocation of antennas For microwave antennas it may be possible to increase the directivity of the TX, or RX, or both antennas or to add weatherproof microwave absorber between the two antennas. To reduce coupling from the backlobe and sidelobes and reradiation from a dish edge, absorber can be mounted around the periphery of the antennas. The Microwave Filter Company, East Syracuse, NY, offers a dish sidelobe absorber that can be bonded with temporary adhesive to check out its effectiveness and then bonded with permanent glue if it solves the problem.

A solution to an in-band EMI problem exists when the geometry of the problem is fixed, e.g., two nonmovable antennas or, in our example, the ship's HF fan antenna and the RF log transducer. In addition, the problem frequency must be fixed, although the proposed solution can deal with a modulated interferer. One additional limitation is that the degree of required suppression of the in-band frequency is not much more than 30 dB. The solution is to pick up the interferer frequency using a highly directional antenna or a transducer that is more sensitive to the magnetic field or electric field generated by the interferer than the intentional signal. Figure 10.12 illustrates the solution to an antenna-to-antenna "in-band" EMI problem. The directional antenna is angled to receive a higher-level signal from the interferer and a much lower level from the intentional receiver signal. A bandpass filter is recommended after the small directional antenna in signal path B to exclude other out-of-band signals. After the filter, a gain stage or attenuator is included to set the appropriate signal level, followed by an adjustable phase shifter that changes the phase so the interferer in signal path B is 180° out of phase with the interferer signal coupled into the receiver signal path A. The received signal plus interferer signal, path A, is applied to a summing amplifier or summing resistor network (combiner). The phase-shifted interferer signal, path B, is also applied to the combiner, and the output from the combiner, path C, is the intentional signal, which contains a much attenuated interferer signal. One technique for providing phase shift is the inclusion of a variable delay line in path B.

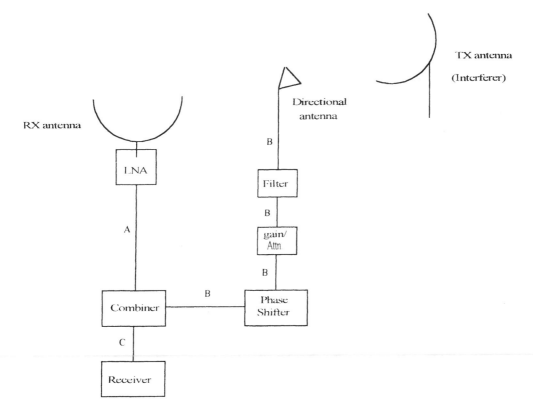

**Figure 10.12** Antenna-to-antenna coupling with "in-band" EMI solution.

The phase shifter must be a unique design and built for many frequencies, including micro-wave, but the Microwave Filter Company, East Syracuse, NY, offers a shifter with an adjustable 5–25-dB typical attenuation and a 180° minimum phase adjustment. The frequency ranges of the different models available are 54–108 MHz, 108–216 MHz, and 470–890 MHz.

### 10.2.3.   Case Study 10.1: Hazardous Zone Around a Transmitting Parabolic Reflector Antenna

#### 10.2.2.1.   General

The most accurate method available to ascertain safe distances for personnel from a transmitting antenna is by measurement using, typically, a calibrated broadband probe. A typical test method is described as follows:

1.   The broadband probe is verified just prior to measurement by use of a calibrated antenna and signal source at the frequency of interest.
2.   The measuring personnel are located outside of the predicted hazardous zone.
3.   The antenna is initially operated at below maximum power and the power is increased in steps. Where the measured field strength approaches the safety limit, the personnel then move back from the antenna.

The advantage of measurement over calculations is that the gain and directivity of an antenna when located close to structures on the ground and the reflection or absorption due to ground and structures are accurately accounted for. In the case where measurements cannot be made, or in order to predict the safety zone before measurement, a worst-case calculation based on theory and measured data may be made for a specific antenna. To make the calculation less prone to human error and to facilitate changing any of the variables, a computer program was written, and this was used to generate the polar plots of the hazardous zone.

#### 10.2.2.2.   Safety Limit

The safety limit used in the calculation is that published by Health and Welfare Canada in "Safety code #6, recommended safety procedures for the installation and use of radio frequency and microwave devices in the frequency range 10 MHz to 300 GHz." The limit for the general public over the 14-GHz frequency range when averaged over a 1-minute period is 61.4 V/m, or 0.163 A/m, or 10 W/m$^2$. Because the limit is for average field strength, the following calculations include a conversion of peak transmitted power into average power.

#### 10.2.2.3.   Calculation of the Hazardous Zone

A worst-case scenario is when an antenna is located at a low height above a perfectly conducting ground plane. A more realistic scenario for a parabolic antenna is for it to be mounted at a considerable height above a ground. Two hazard zones are calculated for the Andrew P6-144D antenna, one where the antenna is in close proximity to a perfectly conducting ground plane, and the second where the antenna is in close proximity to soil with high permittivity. The case where the antenna is mounted at some height above ground and where structures are located close to the antenna are also discussed.

The worst-case analysis takes the absolute maximum power capability of the transmitter, the minimum attenuation in the antenna cable, feed flange, waveguide, etc., and the maximum gain for the antenna to predict the E field around the antenna.

Andrew Antenna Company Ltd. publishes a radiation pattern plot for the P6-144D antenna that is given in upcoming Figure 10.13. The envelope of measured antenna gains is plotted, resulting in a worst case (i.e., maximum gain). Either of the polarizations that results in maximum gain at any frequency has been chosen, which is consistent with our worst-case approach. The radiation pattern envelope was measured by the Andrew Antenna Company Ltd. with the antenna mounted on a tower with a 60–150-foot height.

From a private communication (Ref. 10) we learned that the peak power available from the TWT amplifier used to power the P6-144D antenna may be as high as 130 kW (rated peak power = 125 kW). The minimum attenuation in the waveguide and feed flange is unlikely to be less than 1 dB, and this was taken as the worst case. The maximum duty cycle of the transmitted power, from Ref. 10, is 0.1%. The maximum average input power to the antenna ($P_{av}$) is given by the peak output power of the amplifier ($P_{pk}$), minus the attenuation, times the duty cycle. Thus, for the case under consideration,

$$P_{av} = P_{pk} - 1 \text{ dB} \times 0.1\% \tag{10.1}$$

That is,

$$P_{av} = 130 \text{ kW} - 26.74 \text{ kW} = 103.26 \text{ kW} \times \frac{0.1}{100} = 103.26 \text{ W}$$

For a low-gain, low-directivity antenna, such as a dipole, and when the direct radiated path length is close to the reflected path length with the direct and reflected field in phase, the maximum theoretical increase in field strength due to superposition of the two fields is 6 dB in the far field. At close proximity and for vertical polarization that attenuates off-axis radiation due to the vertical directivity of the dipole, an increase of greater than 6 dB may be expected (Ref. 2).

The 6-dB enhancement ignores any reflections from conductive structures such as buildings and wire fences in close proximity to the antenna, which may further reinforce or reduce the field at any given point. When an antenna is located close to a conductive ground, the characteristics of the antenna may change due to mutual coupling between the antenna and the ground plane, as well as pattern changes, spurious resonances, and other anomalies caused by proximity to the ground. These effects are difficult to predict, so measured results are used in an attempt to derive a worst-case field enhancement due to ground reflections and other factors.

The measurements from which the data is used were made in order to characterize an open-field test site and to arrive at site attenuation. Invariably such test sites incorporate a conductive ground plane, typically constructed of wire mesh, below at least one of the antennas. The disadvantage of using the results of such measurements is that both a transmitting and a receiving antenna are used, so mutual coupling exists between them, whereas our concern is the E field incident on personnel, where the mutual coupling will be much less pronounced. The second disadvantage is that the upper frequency limit for such measurements is 1 GHz, whereas in this case study the frequency range of interest is from 14 to 15.2 GHz. Because of potential errors, we have used the results of these measurements with a safety factor to provide a worst-case field enhancement. In Ref. 3 the distance between the two antennas was chosen to ensure that mutual coupling is not a significant factor.

Assuming that the antenna has been accurately calibrated in an anechoic chamber, the difference between the published antenna factor and that measured on an open-field test site is a measure of the enhancement or reduction in field due to the open-field test site characteristics.

From Ref. 12, the decrease in antenna factor or increase in field strength at 215 MHz, measured with a log-periodic antenna, due to the open-field characteristics is 6 dB. Additional ambient reflections are included in the measurement. Measurements and calculations given in Ref. 4 indicate an enhancement of 5 dB at 950 MHz due to reflections.

Due to the uncertainties involved in the application of such data, a very conservative enhancement of 10 dB was used for the antenna over a conductive surface. This was reduced to 7.5 dB for the situation where a low-conductivity, high-permittivity soil is the ground plane, as discussed in Case 2.

Reference 13 includes a curve Figure 10.16 of the reflection coefficient versus angle of incidence with vertical and horizontal antenna polarization for average earth, blacktop, and steel. Figure 10.16, reproduced in this report, indicates the importance of including the angle of incidence in the calculation; also, where blacktop is the ground plane, the reflection coefficient is less than for average soil. Figure 10.16 is valid for 30 MHz only, whereas the reflection coefficient at 14.8 GHz is much different, as seen in Case 2.

*Case 1*: Antenna located close to a highly conductive ground plane

A wave propagating in air as it impinges on a surface with either high conductivity or high permittivity is reflected as a result of the lower impedance of the surface compared to the wave impedance. We assume that at 14.8 GHz ($\lambda$ = 2 cm) the reflective surface is in the far field of the antenna, where the wave impedance $Z_w$ = 377 $\Omega$. The barrier impedance of copper is 0.045 $\Omega$/sq at 14.8 GHz, and for steel it is approximately 1.7 $\Omega$/sq from

$$|Z_m| = 369 \sqrt{\frac{\mu_r f}{\sigma_r}} \quad [\mu\Omega/\text{sq}] \tag{10.2}$$

where

$\mu_r$ = permeability relative to copper

$\sigma_r$ = conductivity relative to copper

$f$ = frequency [MHz]

Thus the barrier impedance for both copper and steel is lower than the wave impedance, and the reflection coefficient is at a worst Case 1. As discussed above, the field enhancement due to reflections and other sources is an estimated worst case of 10 dB; i.e.,

$$20 \log \frac{E_{\text{dir}} + E_r}{E_{\text{dir}}} = 10 \text{ dB}$$

where

$E_{\text{dir}}$ = predicted direct field

$E_r$ = reflected, in-phase field incident on the same point

The 10-dB enhancement is also used to account for additional spurious effects, such as antenna loading.

The equation used in the computer program that enables the generation of the plot of the hazard zone around is derived as follows. The maximum safe E field is 60 V/m, so the maximum plane-wave power density is obtained from

$$W = \frac{E^2}{377} = \frac{60^2}{377} \tag{10.3}$$

$$= 9.55 \text{ W/m}^2 = 1 \text{ mW/cm}^2$$

The power density some distance $r$ from the source is given by

$$W = \frac{P_{\text{av}} G F_e}{4\pi r^2} \tag{10.4}$$

where

$r$ = distance in meters

$G$ = antenna gain (numeric)

$F_e$ = numeric power enhancement due to reflections etc.

From the $W = E^2/377$ equation it may be seen that a 10-dB increase in E field results in a 10-dB increase in power density. For example, a 10-dB increase in a 10-V/m field (0.26 W.m) results in a 31.6-V/m (2.6-W/m) E field, from

$$20 \log \frac{31.6}{10} = 10 \text{ dB}$$

The increase in power density is thus

$$10 \log \frac{2.6 \text{ W}}{0.26 \text{ W/m}} = 10 \text{ dB}$$

The safe distance from Eq. (10.4) is given by

$$r = \sqrt{\frac{103.3 G F_e}{9.55 \times 12.56}} \tag{10.5}$$

The maximum P6-144D antenna gain is 46.3 dBi in the main lobe of the antenna. Figure 10.13 shows the off-axis decrease in gain versus azimuth degrees from main lobe. The antenna gain in dBi was converted into gain numeric and used with Eq. (10.5) to calculate the safety zone around the antenna, inside of which is the hazardous zone. Figure 10.17 is a polar plot of

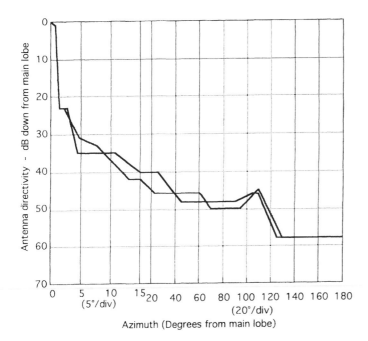

**Figure 10.13** P6-144D antenna radiation pattern envelope. (Reproduced by permission from Andrew Antenna Company Ltd.)

the safety distance contour around the antenna. A worked example of the calculation is made as follows to ensure the validity of the computer program.

The on-axis gain of the antenna is 46.3 dB, $F_e$ is 10 dB; thus the gain numeric times $F_e$ numeric is 42,658 $\times$ 10 = 426,580, which results in a safe distance, using Eq. (10.5), as follows:

$$r = \sqrt{\frac{103.3 \times 426,580}{120}} = 605 \text{ m}$$

which agrees with the computer-generated result.

*Case 2*: Antenna located above a ground with high permittivity

The lowest resistivity of soil is approximately 100 $\Omega$-cm, which is much higher than the 1.72 $\mu\Omega$-cm for copper. The lowest soil impedance at 14.8 GHz, due to the conductivity of the soil, is therefore much higher than 377 $\Omega$, and negligible reflection occurs due to the conductivity of the soil. However, a surface that exhibits a higher permittivity than air also exhibits a lower impedance than 377 $\Omega$. Thus when a plane wave in air impinges on such a surface, a reflection occurs. The reflection coefficient $R_c$, from Ref. 14, for horizontally polarized waves is given by

$$\left.\frac{E_\tau}{E_i}\right|_H = 1 - \frac{2}{1 + \dfrac{Z_w}{Z_s} \times \dfrac{\cos \Phi}{\cos \theta}} \tag{10.6}$$

and for vertically polarized waves is given by

$$\left.\frac{E_\tau}{E_i}\right|_V = 1 - \frac{2}{1 + \dfrac{Z_w}{Z_s} + \dfrac{\cos \Phi}{\cos \theta}} \tag{10.7}$$

where

    $Z_w$ = wave impedance = 377 $\Omega$

    $Z_s$ = soil impedance

    $\theta$ = angle of incidence relative to vertical, but not 90°

    $\Phi$ = angle of transmitted wave in the soil relative to vertical, but not 90°

The difference in angle between incident wave and transmitted wave is due to the different velocities of the wave in air compared to soil. The ratio is

$$\frac{\sin \theta}{\sin \Phi} = \frac{V_1}{V_2}$$

where

    $V_1$ = velocity in air

    $V_2$ = velocity in soil

$V_2$ is always lower than $V_1$ and is dependent on the dielectric conductivity and loss tangent (i.e., on the complex permittivity of the soil).

The worst-case maximum reflection occurs when $V_1 = V_2$, in which case sin $\theta$ = sin $\Phi$. Because the complex permittivity of the soil is an unknown, the cos $\Phi$/cos $\theta$ term in Eqs. (10.6) and (10.7) are set to 1. Figure 10.14 illustrates the wave angles referred to.

The impedance of the soil, $Z_s$, is given by

$$Z_s = \sqrt{\frac{\mu_0}{\varepsilon_0 \varepsilon_r}}$$

where

$\mu_0$ = permeability of free space = $4\pi \times 10^{-7}$ [H]

$\varepsilon_0$ = the permittivity of free space = $1/36\pi \times 10^9$ [F]

$\varepsilon_r$ = relative dielectric constant of the medium

Reference 5 shows a worst-case dielectric constant of 20 for sandy soil (humidity 16.8%) at 10 GHz, and this value is used in the following equation for soil impedance:

$$\sqrt{\frac{4\pi \times 10^{-7}}{(1/36\pi \times 10^9)20}} = 84.2 \ \Omega$$

Using 84.2 $\Omega$ in either Eq. (10.6) or Eq. (10.7) results in a reflection coefficient $R_c$ of 0.635 (i.e., 0.635 of the incident wave is reflected and 0.365 is transmitted or absorbed in the soil). Thus the ratio of $E_{trans}/E_{inc}$ in decibels is 20 log 0.365 = $-8.75$ dB, which corresponds

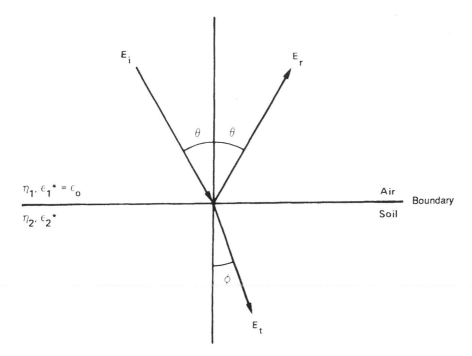

**Figure 10.14**   Incident, reflected, and transmitted wave angles. (© 1987, IEEE.)

well with Figure 10.15. If we assume that the 10 dB (3.162) field enhancement is due to reflection only, ignoring antenna-to-ground proximity effects, using the 10-dB field enhancement, the reflected wave $E_r$ can be found from $E_t$, the total E field, as follows: $E_t = 3.162\ E_{dir}$, where $E_{dir}$ is the predicted field strength without enhancement; therefore $E_r$ may be found from $E_t = E_{dir} + E_r$ and is $2.162E_{dir}$.

When the reflection coefficient is included in the equation for $E_r$ we get

$$E_r = E_{dir}2.162 \times R_c,$$

therefore

$$E_r = E_{dir} \times 2.162 \times 0.635 = 1.372E_{dir}$$
$$E_t = E_{dir} + E_r = E_{dir} + 1.372E_{dir} = 2.372E_{dir}$$

Thus the 10-dB enhancement ($F_e$) is reduced to 7.5 dB. We have assumed a worst-case normal angle of incidence of the transmitted wave with the soil surface. However, from Ref. 14, significant reflection also occurs for horizontally polarized waves with low angles of incidence with respect to the surface, and our worst-case approach is therefore valid in close proximity to the antenna. The polar plot of the safety zone with the antenna high-permittivity soil is shown in Figure 10.18.

### Conclusions

In order to err on the side of safety we have used a worst-case approach in deriving a hazard zone around a specific transmitting antenna. Maximum theoretical field enhancement due to a reflection from highly conductive surfaces is 6 dB. Measurements on open-field test sites where the surface conductivity is lower but where some ambient reflections occur results in measured field enhancements of 5 dB and 6 dB. One major unknown is the effect on the antenna character-

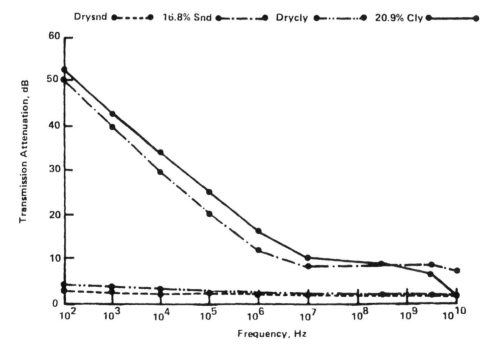

**Figure 10.15**  $E_{tran}/E_{inc}$ for normal incidence. (© 1987, IEEE.)

RHO

# RHO Magnitude vs Angle

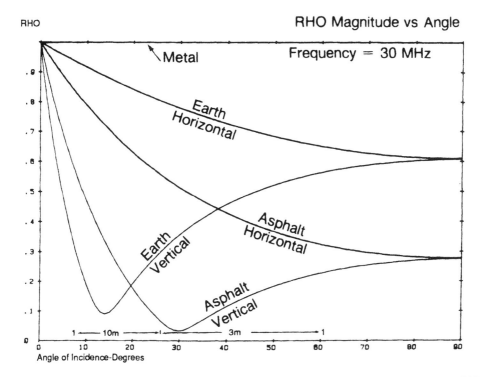

Frequency = 30 MHz

Angle of Incidence-Degrees

**Figure 10.16** Reflection coefficient versus angle of incidence for various materials. (© 1987, IEEE.)

istics of proximity to ground. Due to the possibility of these spurious effects, the enhancement was increased to 10 dB for the case where the antenna is located above a highly conductive ground and to 7.5 dB where a high permittivity ground is present. Where reflective structures exist around the antenna, the enhancement at certain points may be a further 6 dB, and a revised hazard zone may be computed from Eq. (10.5).

For a more accurate estimate of the E field at any point around the antenna in the presence of structures, the General Electromagnetic Model for the Analysis of Complex Systems (GEM-ACS) program may be used. Using Geometrical Theory of Diffraction (GTD) formulation, a total field is computed from super-position of fields from direct, reflected, and diffracted sources. The disadvantage of GTD is that only perfectly conductive surfaces may be modeled. The method of moments technique available in GEMACS is capable of modeling surface impedance but is not practical for large surfaces at high frequency.

When the antenna is mounted on a mast and the following requirements are met, the area around the mast out to any distance is inherently at a safe level of 60 V/m or less. The requirements are the antenna is at least 4 m above personnel (i.e., approximately 6 m above ground), the antenna center axis is either parallel to the ground or pointing upward, the ground under the main lobe of the antenna does not rise for 605 m, and no vertical conductive surfaces are within close proximity of the antenna or the main lobe of the antenna.

## 10.3. AMBIENT SITE PREDICTIONS AND SURVEYS

An ambient site prediction or survey is used when a receiving system or potentially susceptible equipment is to be installed at the site. In this section the ambient due to intentional sources, such as communications transmitters and radar and the coupling to a receiver, are considered.

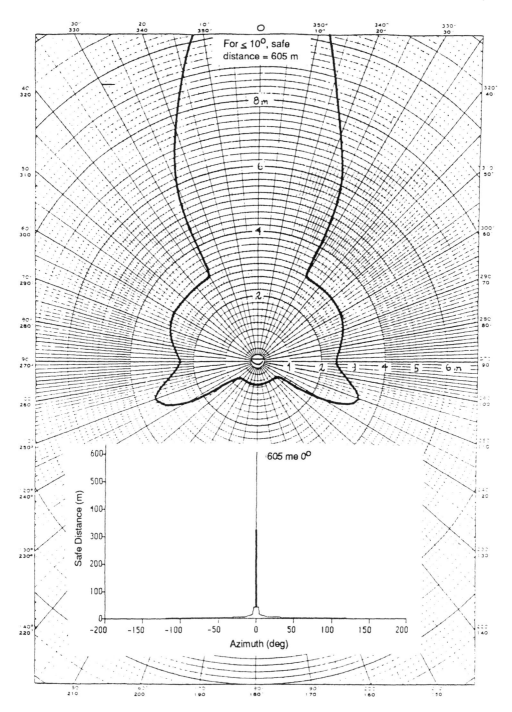

**Figure 10.17**  Polar plot of the hazardous zone around the P6-144D antenna in close proximity to a highly conductive ground.

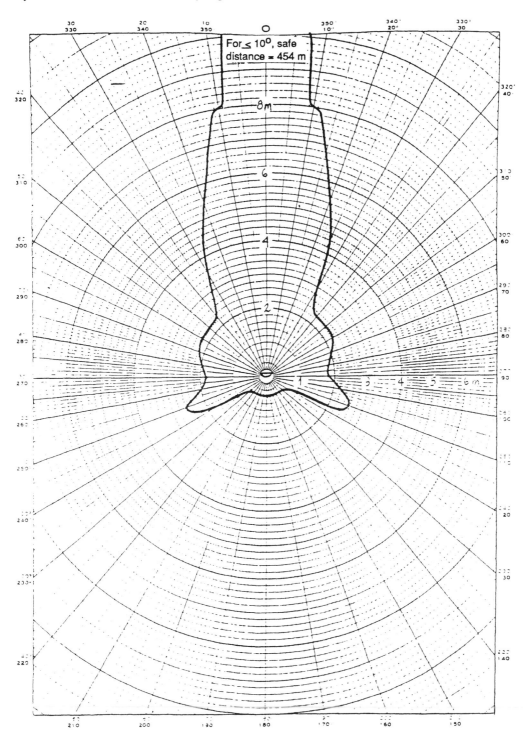

**Figure 10.18** Polar plot of the hazardous zone around the P6-144D antenna in close proximity to ground with high permittivity.

If the locations of antennas, antenna gains and directivity, transmitter power and frequencies, elevation of the ground, and presence of structures around the site are known, then a prediction of the electromagnetic ambient at a site may be made. A limitation in the prediction approach is that very often the information is limited. For example, the directivity or polar plot of the gain of an antenna is not always available, and some assumptions must be made concerning the sidelobes or backscatter of the antenna. The presence of hills or structures can make a significant difference to the predicted coupling between antennas and to absorption and shading of the receiving antenna; when this information is not available, the accuracy of the prediction suffers. An ambient site survey is the more accurate method of site evaluation. In a survey a number of antennas, preamplifiers, and, typically, a spectrum analyzer are used to measure the ambient.

Directional measurement antennas are rotated through 360° with both horizontal and vertical polarizations of the antenna. For measurements at frequencies that are in band for the proposed receiver, the measurements should be made, where feasible, with the antenna and system that will be installed at the site. For example, if the system comprises an input filter, a low-noise down-converter, or a preamplifier and filter, then these should be used in the site evaluation. By using a representative antenna and system, the antenna gain and directivity, filter effectiveness, and out-of-band or spurious responses of the receiving system will be included in the measurement.

This approach is usually feasible only with small antennas and lightweight receiving equipment. When this approach is ruled out, the measurement antenna should have a directivity and gain as close to those of the actual antenna as possible, and the gain and noise floor of the measuring equipment should be at least as good as those of the actual system.

One potential limitation the ambient site survey has is that not all the transmitters may be operating during the time in which measurements are made. To minimize the risk of missing an emitter, the frequencies of the transmitters should be ascertained and an identification of "off air" transmitters should be made. Otherwise, measurements should be made over a period of hours and repeated over a period of a day or two to increase the possibility that all transmissions are intercepted.

### 10.3.1. Case Study 10.2: Electromagnetic Ambient Site Predictions and Site Surveys

The emission frequencies of interest in the following site surveys were the fundamental frequency of a satellite communications system at 1544 MHz, the IF at 45 MHz, and the reference at 10 MHz. In addition, field strengths of 1 V/m or greater at any frequency can result in direct coupling to cables and equipment, which can result in EMI. At one site, cross-modulation due to a transmitting antenna located 56 m from the proposed site of the satellite communications receiving antenna was predicted as follows.

The transmitter operates at a frequency of 1511.125 MHz at 2.5 W and feeds a yagi antenna with a gain of 17.5 dBi. The transmitting antenna was predicted to be broadside onto the receiving antenna. The sidelobe of the yagi antenna is likely to be, at a minimum, 13 dB down on the main lobe. However, due to the proximity of several towers in the vicinity, the antenna pattern will be affected somewhat and this is an unknown.

Assuming then an antenna gain of 17.5 dBi − 13 dB = 4.5 dB. The power density $P_d$ incident on the receiving antenna is given by Eq. (10.8):

$$P_d = \frac{P_{in}G_a}{4\pi R^2} \qquad (10.8)$$

where

$P_{in}$ = antenna input power = 2.5 W

$G_a$ = effective gain of the antenna = 4.5 dB

$R$ = distance = 56 m

$P_d$ = power density in W/m$^2$

Therefore:

$$P_d = 2.5 \text{ W} \times \frac{2.818}{12.56} \times 56^2 = 0.179 \text{ mW/m}^2 = -7.5 \text{ dBm/m}^2$$

The received power ($P_r$) at the input of a narrowband cavity filter inserted between the receiving antenna and the low-noise amplifier (LNA) is given by Eq. (10.9):

$$P_r = \frac{P_d G_b \lambda^2}{4\pi} \tag{10.9}$$

where

$G_b$ = gain of the receiving antenna = 389

$\lambda$ = wavelength of the incident power

From Eq. (10.9):

$$P_r = \frac{(0.179 \times 389 \times 0.198^2)}{12.56} = 0.21 \text{ mW} = 6.6 \text{ dBm}$$

Assuming the use of the four-section filter, 4C42 − 1544.5/T23.2 − 0.0, then the attenuation provided at 1511.125 MHz is 35 dB from measurements. The input power to the LNA is thus −6.6 dBm + −35 dB = −41.6 dBm. This level is above the −50 dBm at which the LNA compresses, and therefore a reduction of approximately 2 dB in gain is predicted in the 1544.25-MHz received signal. Measurements were made at the same site, the results of which were as follows.

The measurements from 1 to 2 GHz were made with a double-ridged guide (DRG) antenna and a minimum 41-dB gain preamplifier feeding into a spectrum analyzer. The emission at 1.511 GHz from the yagi antenna was measured with the preamplifier both in circuit and out of circuit. The measured level at 1.511 GHz was reduced by 44 dB when the preamplifier was removed, which proves that the preamplifier was not compressing. The reason to include the preamplifier was to ensure that any low level in band emission from 1540 to 1550 MHz was detected.

The measured field from the yagi antenna was lower than expected, so to confirm the measurement, a second yagi antenna, identical to the transmitting antenna, was used as the measurement antenna. The level with the measurement antenna pointing at the emitter antenna is 14.8 dB above that obtained with the ridged guide antenna, whereas a 9.5-dB increase in level was expected. One possible explanation for the difference in measured level is the different directivities of the two antennas (i.e., the ridged guide antenna is more affected by reflected waves than is the yagi antenna due to its lower directivity). This explanation was confirmed by pointing both antennas away from the emitter, when the ridged guide antenna measurement resulted in a higher level than the yagi.

The antenna that will be used in the satellite receiving system exhibits an 8.5-dB higher gain than the yagi. Therefore, due to its higher gain and directivity, higher peaks and troughs in the received power from emitters, as the antenna sweeps across the sky, compared to the measured levels, is expected. The received power at the measurement antenna for the highest-level emitters at 1511 MHz are calculated as follows: First the preamp and cables were calibrated using a signal source and spectrum analyzer. The input level to the cable/preamp was adjusted to −44 dBm. The measured output was −4 dBm; thus the combined gain and insertion loss is 40 dB.

At 1511 MHz the input level to the spectrum analyzer was measured at −13 dBm; subtracting the gain of the preamp results in an output level from the antenna of −53 dBm.

The gain of the satellite communications system receiving antenna is 17.8 dB higher than that of the measurement antenna; thus, based on the higher receiving antenna gain, the power level at the input of the four-section filter is −35 dBm.

The four-section filter provides an insertion loss of 35 dB at 1.511 GHz; thus the input power to the LNA is −35 dBm + −35 dB = −70 dBm. The measured input level to the LNA is 28.4 dB below the predicted. The main reason for the discrepancy between predicted and measured levels is that the position of the transmitting antenna was not broadside onto the receiving antenna but pointing away at an angle of approximately 20° from the receiving antenna direction. The maximum level was not measured pointing at the rear of the transmitting antenna, but due to a reflection from a nearby hill. This example illustrates the importance of measurements when exact locations of emitters and receptors are not known. The next example of a survey made at a different site also emphasizes the advantage of measurements over predictions.

A high-amplitude emission at 1.344 GHz was measured using the double-ridged guide antenna with the 40-dB-minimum-gain preamplifier. After approximately 1 hour of measurement, the emissions changed. An in-band emission at exactly 1.545 GHz was measured and emissions either side with spacings of approximately 15 MHz. This broad combination of emissions increased and then decreased over a 3-second period and then disappeared for approximately 3 seconds. The peak emissions at 1.344 GHz and 1.305 GHz also increased and decreased in amplitude with an approximately 6-second period.

The source of the 1.305- and 1.3405-GHz emissions was a long-range radar situated east of the antenna measuring location. The power level of the 1.3405-GHz signal was measured at −31.83 dBm without preamplifier. The loss in the interconnection cable at 1.3405 GHz is approximately 2 dB; thus the antenna output level was −30 dBm = 1 μW. The gain of the double-ridge guide at 1.36 GHz is approximately 5.6 dB = 3.63 numeric. The incident power density on the DRG antenna is given by

$$P_d = \frac{4\pi P_r}{G\lambda^2} = \frac{12.56 \times 1 \times 10^{-6}}{3.63 \times \left(\dfrac{300}{1340}\right)^2} = 0.069 \text{ mW/m}^2 \tag{10.10}$$

It is possible that the intermodulation seen in the photo of Figure 10.19 was due to the preamplifier oscillating or generating spurious response due to overloading. To measure the susceptibility of the preamplifier to input level, the input signal was attenuated. The only component available that could be used as an attenuator was a 20-foot length of RG58 coaxial cable.

The reduction in the measured amplitude with the coax cable between the antenna output and the preamplifier at 1.305 GHz was approximately 27 dB. This magnitude of reduction puts the spectral emissions lines with 15-MHz spacings close to the noise floor. On close examination, at least four of the spectral emission lines at approximately 1.6, 1.62, 1.72, and 1.84 GHz were seen. Although not conclusive proof, this test indicates that the preamplifier is not the source of the harmonically related emission lines.

The intermodulation was not consistently seen throughout the day, although the high-amplitude 1.34-GHz radar signal was. The sources of intermodulation are a 15.4-MHz shortwave

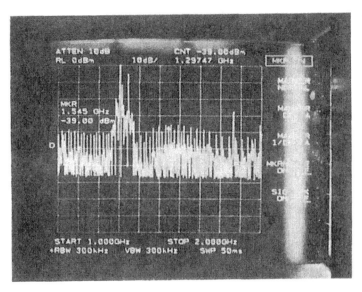

**Figure 10.19**   Photo of spectrum analyzer display showing intermodulation product at 1545 MHz.

(S/W) radio, which generates a power density of approximately 2.65 mW/m$^2$, and the radar, which generates a power density of 0.069 mW/m$^2$. The cause of the in-band frequency is passive intermodulation (PIM), which occurs only when the S/W radio is used. The intermodulation product is

$$13 \times 15.4 \text{ MHz} + 1344 \text{ MHz} = 1544 \text{ MHz}$$

The location at which intermodulation products are generated is unlikely to be the input of the 41-dB preamplifier because the 15.4-MHz signal level will be drastically reduced by the DRG antenna. The most likely source is a lightning protection scheme that surrounds the top of the building. The measurement antennas were located only 6 feet away from the vertical and horizontal conductors of the protection scheme. Reradiation as a result of passive intermodulation due to nonlinear junctions or the nonlinearity induced by a metal with permeability greater than 1 is typically 40 dB down on the incident power level.

The intermodulation products seen in Figure 10.19 are 30 dB down on the 1.3405-GHz level and approximately 46 dB down on the 15.4-MHz level. Intermodulation will occur only when the 15.4-MHz S/W receiver is operated and the radar is pointed at the building. Therefore EMI will not always be present. Some of the EMI fixes available are:

Move the S/W transmitter frequency. For example, 16.5 MHz, 18 MHz, 28 MHz, etc. would be the frequencies to choose for maximum suppression of intermodulation-generated EMI.

Move the S/W antenna at least 500 m from the satellite receiving antenna, which should be moved as far from the lightning protection conductors as feasible.

If the lightning conductor is nonferrous material, paint with an IMI-suppression coating. If ferrous, wrap with 2.5-in-thick all-weather absorber foam.

If PIM is due to the "rusty bolt effect," one solution is to break all screwed, bolted, or riveted joints in the structure causing PIM and to weld, braze, or cadweld the joints. Alternatively, the joint may be improved by cleaning and passivation of the metal surfaces, which may

**Table 10.4**  Predicted IM Products Resulting in In-Band Interference
Frequencies

| | | Predicted IM frequencies [MHz] | | | Received frequency |
|---|---|---|---|---|---|
| N | M | $f_1$ | $f_2$ | $f_3$ | $f_3$ |
| 9 | 1 | 20.8 | 456 | 1543.2 | −1.299 |
| 2 | 1 | 12.8 | 1301.1 | 1542.7 | −1.8 |
| 8 | 1 | 100.4 | 742 | 1545.2 | 0.7 |
| 2 | 1 | 100.4 | 1342 | 1542.8 | −1.7 |
| 3 | 1 | 151.1 | 1091.4 | 1544.7 | 0.2 |
| 4 | 1 | 171.8 | 857.3 | 1544.5 | 0 |
| 7 | 1 | 158.6 | 434.7 | 1544.9 | 0.4 |
| 1 | 1 | 158.6 | 1386.2 | 1544.9 | 0.3 |
| 9 | 1 | 171.5 | 0 | 1543.5 | −1 |
| 2 | 1 | 448 | 649.3 | 1545.3 | 0.8 |
| 8 | 1 | 425.3 | 1859.1 | 1543.3 | −1.2 |
| 5 | 1 | 438.7 | 649.3 | 1544.2 | −0.3 |

then be rejoined and sealed against the environment by paint or caulking. In this case study, cleaning the joints would be practically impossible, and the other alternatives listed were considered.

Suppression coatings were obtainable from Omicron Inc., PO Box 397, Buffalo, NY 14222, which can no longer be traced. They manufactured two types of intermodulation interference suppression products. Omicron CBA was a polymer-based coating that creates at RF a low-resistance and low-reactance path. This reduces current flow through the nonlinear junction and reduces the level of PIM. The low-reactance path is achieved by metal oxide penetration aids, corrosion inhibitors, chelating agents, and materials of high dielectric constants.

The second suppression agent made by Omicron was the SS-50/SC-60, which was a finely divided conductive material dispersed in specially formulated polymers. The mode of suppression is provided by low-reactance pathways on a molecular level, effectively resulting in joint shunting. According to the manufacturer, laboratory and field tests of Omicron IMI-suppression coatings have shown greater than 40-dB reduction in IMI levels between 0 and 30 MHz.

A search has unfortunately not turned up any other company that manufactures a similar product, although one may well exist.

The 1.344 GHz, 1.305 GHz, and 15.4 MHz are not the only frequencies at which intermodulation may be predicted at the site, as shown in Table 10.4, in which $f_3 = N \times f_1 \pm M \times f_2$.

Frequencies exist at power levels that may be sufficiently high (i.e., 2.2 $\mu$W/m² at 171.1 MHz and 1.2 $\mu$W/m² at 425.3 MHz) to produce passive intermodulation at a level above the noise floor of the receiving system. Therefore the potential for PIM at these frequencies also exists.

## 10.4.  CASE STUDY 10.3: COUPLING INTO HV AC LINE FROM HF PHASED-ARRAY RADAR

At one prospective site for a receiver, an ambient site survey was conducted. A source of RF existed that was not recorded in the RF ambient survey because its frequency lies well outside of the 1544-MHz and 468-MHz frequencies of interest. The RF source is backscatter from an HF phased-array radar that operates in the 8–20-MHz frequency band. The average peak radiated power from the array of 16 log-periodic antennas is 2 kW, and the effective radiated power is nearly 200 kW. From a private communication it was learned that the level of radiated backscatter is approximately 50 W.

In addition to the satellite communication receiver antenna, the received power at the equipment building induces potential EMI voltages in the hydro lines, telephone lines, and a HF whip antenna used for a time receiver (which operates at 7.33 MHz, 10 MHz, and 14.66 MHz). No calculations are required to predict the high probability that the 8–20-MHz source will interfere with the time receiver, even assuming the source is horizontally polarized and the time code receiver antenna is vertically polarized.

## Prediction of Level of EMI

The hydro lines, which carry a voltage of 2400 V, are assumed to be at 12 m above ground with approximately a 1.2-m distance between individual lines. The power lines run parallel to the HF array at a distance of 600 m. The received power at a distance of 600 m from the source can be calculated from

$$P_r = -(32 \text{ dB} + 20 \log_{10} R + 20 \log_{10} f) + P_t \quad \text{[dBm]}$$

where

$P_r$ = received power [dBm]

$P_t$ = transmitted power = 47 dBm

$f$ = frequency [MHz]

$R$ = distance from the source [km]

Using the worst-case lowest frequency of 8 MHz, the level of received power is

$$-(32 + 20 \log_{10} 0.6 + 20 \log_{10} 8) + 47 = 0 \text{ dBm}$$

Using the far-field value of wave impedance (377 $\Omega$), the E field is calculated at 0.6 V/m and the H field at 1.6 mA/m.

*Method 1*

The most accurate method for calculating the EMI picked up on the power line is the use of transmission-line theory; however, the theory is valid only for the situation where the height ($h$) of the line above ground is much less than the wavelength, $\lambda$.

The E field increases with height above the ground until the height is approximately 0.5 $\lambda$, after which the magnitude of the field levels off. One source of error in application of transmission-line theory is when this effect is ignored.

The transmission-line equation may still be used to calculate the approximate characteristic impedance of the line from Eq. (7.16) and the induced current from Eq. (7.20) by setting the height to 0.5$\lambda$ when this is less than the actual height. At 8 MHz, when 0.5$\lambda$ is 18.75 m, the 12-m height of the power line may be used in the transmission line equations, whereas if the equations are used at 20 MHz, the height would be set at 7.5 m to provide a limit to the induced voltage. The calculation of differentially induced voltage using these equations is expected to be reasonably accurate because the distance between the wires is less than 0.5$\lambda$.

Case Study 12.5 provides worked examples of Eqs. (7.16) and (7.20), in the present study only the results of the calculations are presented.

At 8 MHz and a height of 12 m the characteristic impedance of the line $Z_c$ is 912 $\Omega$. In using Eq. (7.16) to calculate the characteristic impedance between the lines, the distance is set to 0.6 m because the equation multiplies this by 2 in obtaining the impedance of a line above a ground plane. At 8 MHz and a height of 1.2 m, $Z_c$ is 552 Ω. The induced common-mode current for the 0.6-V/m incident E field at 8 MHz is 31 mA. The differential current is 4.5 mA. The transmission line is long and terminated in the unknown impedances of the transformer located at the building and at the substation. If the assumption is made that the load and source impedances are the same as the characteristic impedances of the line, then the common-mode induced voltage is approximately 27 V and the differential-mode voltage is 2.6 V. When the load and source impedances are less than the characteristic impedances of the line, which is likely, the current flow in the lines and terminating impedances at resonant frequencies will be much higher than those calculated. The effects of cable resonance are discussed in Section 7.4.

*Method 2*

Another method for calculating EMI is the use of current loop or loop antenna theory, where the current loops are formed between the AC lines for the differential-mode pickup and between the lines and ground for common-mode pickup. In the case under consideration, the line lengths are greater than λ/2 of the interfering source; therefore resonances will occur and the principal coupling mode is electromagnetic wave propagation (i.e., the lines are an effective antenna). The area of maximum illumination of the incident field is approximately 500 m. The length of the loop is in fact very long (i.e., between the transformer and load located at the buildings and the nearest substation).

The effective length of the loop when used in the current loop equation is 0.5λ = 18.7 m, and the width of the loop may be set to the height of the wire above the ground plane. Because of the curvature of the incident field and the unknown angle of incidence of the direction of field propagation and the plane of the loop (which is assumed to be 45°), the following is a worst-case calculation. See Figure 10.20.

At a distance of 600 m from the source and at frequencies of 8–20 MHz, the incident field is a plane wave and the induced EMI is independent of whether the magnetic field or the electric field couples to the wires. A magnetic field coupling is assumed with a worst-case angle between longitudinal cable axis and field vector. The height of the power lines above the ground is assumed to be 12 m and the distance between lines is 1.2 m. The formula for loop voltage induced in a current loop is given by the following relationships:

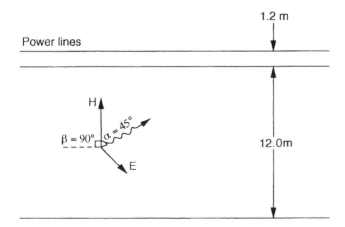

**Figure 10.20** Incident field = plane wave. Worst-case angle between the cable axis and the H field is β = 90°. A mean angle of 45° is taken for the angle between the plane of the loop and the direction of field propagation.

$$V_m = 2HlZ_o \sin \beta \sin \left( \frac{\pi h}{\lambda} \cos \theta \right) \qquad (10.11)$$

where

$H$ = magnetic field [A/M]

$\beta$ = angle between longitudinal cable axis and the field vector

$\theta$ = angle between plane of loop and direction of field propagation

$l$ = loop length

$h$ = loop height or width

$\lambda$ = wavelength

$Z_o = \sqrt{\dfrac{\mu}{\varepsilon}} = 377 \ \Omega$ for plane wave

assuming the worst-case angle between the cable axis and the $\beta$ field vector of 90° and a mean angle of 45° for the angle between the plane of the loop and the direction of field propagation. For the differential induced EMI voltage $V_d$ at 8 MHz,

$$V_d = 2 \times 1.6 \times 10^{-3} \times 18.7 \times 377 \times 1 \times \sin \left( \frac{3.14 \times 1.2}{37.5} \times 0.707 \right) = 1.6$$

For the common-mode induced EMI voltage VCM at 8 MHz,

$$\text{VCM} = 2 \times 1.6 \times 10^{-3} \times 18.7 \times 377 \times 1 \times \sin \left( \frac{3.14 \times 12}{37.5} \times 0.707 \right) = 14.7$$

Although neither method 1 nor method 2 is exact, the calculated induced voltages are sufficiently close to provide an estimate of the magnitude of EMI voltage.

The common-mode EMI induced in the telephone lines may be as high as that for the power lines, depending on the geometry, whereas the differential-mode voltage is likely to be much less due to the close proximity of the telephone conductors.

By examining the type of main power-line filter and equipment filters used at the installation, the prediction is that the level of filtering of the AC line and the telephone line will be sufficient to ensure EMC for this specific potential EMI situation. Another likely victim of the HF phased-array radar is the time reference signal, which is received at the frequencies of 7.33 MHz, 10 MHz, and 14.66 MHz. A receiver presently used in the area by the armed forces is unable to function during the operation of the HF phased-array radar, irrespective of the tuned frequency, because of cross-modulation EMI from that source. Therefore the prediction is, based on this experience and the high E fields (0.6 V/m) generated by the radar, that interference in reception of the time reference signal will be experienced.

## REFERENCES

1.  C.C. Roder. Link Communication Analysis Algorithm (LINCAL) User's Manual. ECAC-CR-80-059, August 1980.
2.  G. Genello, A. Pesta. Aircraft coupling model evaluations at SHF/EHF. IEEE Symposium on Electromagnetic Compatibility, 1985.
3.  T. Durham. Analysis and measurement of EMI coupling for aircraft mounted antennas at SHF/EHF. IEEE Symposium on Electromagnetic Compatibility, 1987.

4.  S.T. Hayes, R. Garver. Out-of-band antenna response. IEEE Symposium on Electromagnetic Compatibility, 1987.

5.  D.A. Hill, M.H. Francis. Out-of-band response of antenna arrays. IEEE Symposium on Electromagnetic Compatibility, 1987.

6.  S.T. Li, J.W. Rockway, J.H. Scukantz. Application of Design Communication Algorithm (DECAL) and Performance Evaluation Communication Algorithm (PECAL). IEEE International Symposium on EMC, October, 1979.

7.  K. Clubb, D. Wheeler, E. Pappas. The COSAM II (DECAL/PECAL) Wideband and Narrowband RF Architecture Analysis Program User's Manual. ECAC-CR-86-112, February 1987.

8.  J.W. Rockway, S.T. Li, D.E. Baran, W. Kowalyshin. Design Communication Algorithm (DECAL). IEEE International Symposium on Electromagnetic Compatibility, June 1978.

9.  L.D. Tromp, M. Rudko. Rusty bolt EMC specification based on nonlinear system identification. IEEE Symposium on Electromagnetic Compatibility, 1985.

10. Private communication with J. Rose of Canadian Astronautics Ltd.

11. T. Dvorak. The role of site geometry in metric wave radiation testing. IEEE Symposium on Electromagnetic Compatibility, 1986.

12. J. DeMarinas. Antenna calibration as a function of height. IEEE Symposium on Electromagnetic Compatibility, 1987.

13. Studies relating to the design of open field test sites. IEEE Symposium on Electromagnetic Compatibility, 1987.

14. D.V. Gonshor. Attenuation, transmission and reflection of electromagnetic waves by soil. IEEE Symposium on Electromagnetic Compatibility, 1987.

15. S.V. Georgakopoulis, C.A. Balanis, C.R. Birtcher. Cosite interference between wire antennas on helicopter structures and rotor modulation effects: FDTD versus measurements. IEEE Transactions on Electromagnetic Compatibility, Vol. 41, No 3, August 1999.

16. S.T. Li, B. Koyama, J.H. Schkantz, Jr., R.J. Dinger. EMC Study of a shipboard HF surface wave radar. IEEE 1995 EMC Symposium Record

17. L.M. Kackson. Small models yield big results. NARTE News, Volume 11, April–June 1993, Number 2.

# 11

# Printed Circuit Boards

## 11.1  INTRODUCTION

One of the major reasons equipment fails radiated emission requirements is radiation from digital signals, either directly from the PCB or from PCB-sourced common-mode noise voltage driving attached cables. Even when equipment is contained in a totally shielded enclosure, noise voltage generated across a PCB to which unshielded and unfiltered interface cables are connected can result in a sufficiently high noise current to fail radiated emission requirements. In extreme cases PCB-generated noise has resulted in equipment failure of commercial radiated emission requirements with shielded cables connected to a PCB metal faceplate, which in turn was connected to a shielded enclosure. If the radiated emission requirements are the stringent MIL-STD-461 or DO-160, the use of a shielded enclosure is almost mandatory.

When a piece of digital electronic equipment, such as a notebook computer, contains a wireless with an antenna anywhere from 2 cm to 20 cm from the PCB, then, although a PCB with a low-emission layout is a good starting point, it may not be enough to avoid desensitization or in-band spurious response in the wireless.

The major source of radiation from PCBs and cables is not differential-mode current but common-mode current, for, depending on the geometry, the radiation due to several microamps of common-mode current can be as high as several milliamps of D/M current.

Some experts on PCB layout might say that good PCB layout is easy. This may or may not be true, however, with the use of ever-faster PCs and pS fast logic; manufacturers have to go to ever-greater lengths to meet EMI requirements. The need for new techniques in PCB layout becomes more important in reducing radiation. That being said, a PCB with the best possible layout can fail due to a single LSI chip from which the level of radiation exceeds emission limits or requirements.

Some other factors that affect PCB radiation include the method by which the power supply connection is made to the PCB; PCB-to-PCB or PCB-to-motherboard/backplane grounding; the proximity to adjacent structures (grounded or ungrounded), the electrical connection to ground, and the presence and type of shields. All of these topics are discussed in the following sections.

## 11.2.  PRINCIPLES OF RADIATION FROM PRINTED CIRCUIT BOARDS

It is the current flow in the signal return connection (PCB track or ground plane), due to both displacement current between the signal path and return path and the load current, which develops a voltage in the impedance of the return. This voltage can generate a C/M current, which is the prime source of radiation from the PCB and any attached connection. The displacement current flows as the capacitance of the line is charged and discharged; thus displacement current is nonuniform, with a maximum at the source end of the line and zero at the load end. Displace-

**683**

ment current, and the resultant C/M current on the line, due to the proximity of some conductive structure is more difficult to predict, although Ref. 2 does describe a technique. In predicting the radiated emissions from a PCB track, the voltage drop in either signal/power return connections, the typical magnitude of the current spikes, and the impedance of the path must be known as well as the displacement current down the line and to adjacent structures.

That radiation is typically increased with the attachment of a cable on which no differential signal flows and that is connected only to ground is a puzzle to many engineers. Section 7.6.3.1 describes the mechanism in more detail, but briefly: when C/M current exists on a conductive structure (PCB tracks) and a conductor is connected to the structure, the C/M current flows on the attached conductor. Figure 11.1 illustrates how the C/M current continues to flow on the cable. At frequencies where the length of the attached conductor is less than a wavelength, the radiation increases with frequency until the cable is approximately half a wavelength in length. When the cable becomes electrically long (longer than a wavelength), the radiation from the cable tends to decrease. Attaching a cable to a PCB ground increases the radiation effectiveness of the source structure at low frequency. At high frequency where the cable, or PCB, is electrically long, the current flow on the structure changes phase down its length, and the cable may be modeled as a series of infinitesimal current sources, some of which are wholly or partially out of phase. The most likely explanation for the constant or reduced radiation at high frequency is that the composite field from the cable is the sum of the fields from the infinitesimal current sources, which interfere. In Section 7.6.3.1 it was shown that 15.7 µA flowing on an attached cable at 21 MHz results in the same E field as 1.26 mA of differential current flowing in a small loop. Because C/M current is so often the predominant cause of radiated emissions, EMC engineers tend to ignore the D/M current as a source, just as design engineers ignore the C/M current contribution. However, if a large loop is formed, then the D/M-current-generated emissions can predominate. For example, if not all the differential mode current flows on the return conductor but some leakage current flows on a ground, a large loop may be formed. Adding a complete shield around a small loop, or a PCB, will eliminate the induction of C/M current on the attached cable, assuming the cable is connected to the shield. If the cable is connected

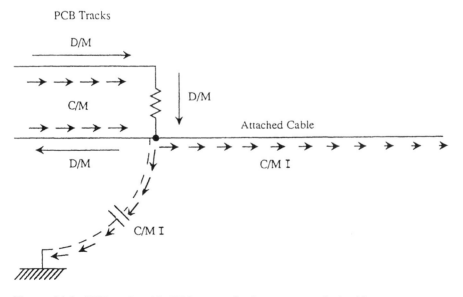

**Figure 11.1** PCB tracks with C/M current flowing on an attached cable.

through a hole in the shield, then cable C/M currents are often higher than without the shield, due to increased displacement current between the inside of the shield and the loop, or PCB. Likewise, if the shield is only partial around the loop, or PCB, then the C/M current can increase.

Reference 2 describes a method of analyzing a microstrip PCB (signal trace above a ground plane) attached to a shielded enclosure. When the results of the prediction are compared to the test results in Ref. 1, scaled to account for the different dimensions of the PCB, the results are extremely close. For example, the radiation from the microstrip is computed at 47.7 dBµV/m and measured at 48.5 dBµV/m, an uncharacteristically high correlation of 0.8 dB. In contrast, one of the computer programs described in Section 11.9 predicts a massive 28 dB higher level of radiation from the same PCB layout as described in Ref. 2.

The impedance of a ground trace or ground plane was examined in Section 5.1.2, with calculations based on NBS inductance formulas. Reference 3 introduces the concept of partial inductance and other techniques to describe more accurately the ground return path impedance. In a complex PCB layout, many ICs are changing state simultaneously, drawing current out of the closest decoupling capacitor and returning it in the ground return path to the capacitor. In addition, digital signal and clock currents flow on traces between ICs and return in the ground return path beneath the traces. Large voltage drops occur in the ground impedance due to these switching currents, which generate C/M currents across the board and out on any attached cables. Figure 11.19 illustrates the generation of C/M voltage across a PCB ground plane. Because of the proximity of grounded structures, connection of one PCB to another PCB or to a motherboard/back plane or to a power supply, the C/M current is further enhanced. When the load on a microstrip PCB trace is removed and the configuration is a signal trace above a ground plane, open circuit at the load end, then the radiation reduces at low frequency but is the same or higher at high frequency. This effect is discussed in more detail in Section 11.3 but is used here to illustrate the importance of the displacement current alone in generating emissions.

The importance of a short and low impedance ground path is discussed in Section 11.6.

Certain PCB layouts have been described as low-emission types. To enable a comparison between the different types of layout and different logic families and to test the validity of the ''good'' layout the tests described in Ref. 1 were conducted. The results of these tests are summarized in Section 11.3.

## 11.3. LOW-LEVEL RADIATION PCB LAYOUT

### 11.3.1. Test Data, Layout Comparison, and Recommendations

The different PCB layouts described in this section were tested as described in Ref. 1. The shielded-room tests were conducted in a $20' \times 20' \times 8'$-high shielded room in which absorber loads were placed at strategic locations in the room to smooth out the potentially large fluctuation in measured field in the room due to room resonances. All measurements in the room were made with the antenna at either 1-m distance from the track on the PCB or at 0.55 m. The close proximity of the antenna was required to measure the low levels of radiation from some of the PCB configurations. Because of the 0.5 m proximity of the PCB from the shielded room ceiling, a reflection of the PCB appears in the ceiling, as described in Ref. 1. The open-field measurements were made at either 1 m or, when a comparison to emission requirements was required, at 10 m.

The example PCB is 22 cm long and 14 cm wide. The thickness of the substrate is 1.6 mm, and the width of the track is 0.7 mm, with an 18-cm length. The track is 4 cm from the front (22-cm) edge and 10 cm from the rear edge. The PCB substrate material is glass epoxy type FR4(G10) that has a relative permittivity of 5–5.6 at 10 MHz and 4.7–5.2 at 25 MHz.

Reduction in dB

| PCB Configuration | Number | Relative performance | |
|---|---|---|---|
| Microstrip | #1 | 20 200 MHz | 200 - 1000 MHz |
| | | 0 dB | 0 dB |
| | | (Reference) | |

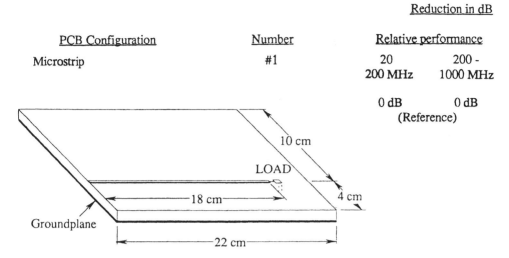

**Figure 11.2a**   Microstrip, #1, no cable.

Although modern PCBs use much thinner tracks and the distance between tracks or between a track and the ground plane is much less, the general comparison between different PCB layouts remains valid.

The transmission-line configuration (#2) in which the signal track is located on the top of the PCB and the return on the bottom of the PCB is shown in Figure 11.2b. With this configuration, differential-mode currents tend to result in opposing magnetic fields. With the tracks located one on top of the other, as shown in Figure 11.2b, a vertical electric field is generated by the voltage difference between the two tracks, which we shall refer to as E-theta, and a horizontal field, which we shall refer to as E-phi, is generated down the length of the tracks. The transmission line tends to radiate in all directions, whereas the microstrip, configuration #1 (shown in figure 11.2a), restricts the radiation from behind the ground plane. This characteristic directivity may be used when the ground plane is used as a partial shield either between tracks on the PCB or between PCBs.

| PCB Configuration | Number | Relative performance | |
|---|---|---|---|
| Two conductor transmission line | #2 | 20 - 200 MHz | 200 - 1000 MHz |
| | | +32 dB | +36 dB |

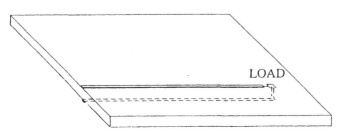

**Figure 11.2b**   Two-conductor transmission line, #2, no cable.

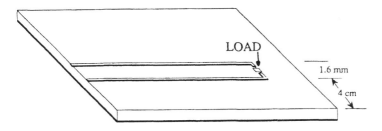

| PCB Configuration | Number | Relative performance | |
|---|---|---|---|
| Two tracks 1.6 mm apart above an image plane | #3 | 20 - 200 MHz | 200 - 1000 MHz |
| | | +7 to -12 dB | + and - |

**Figure 11.2c**  Two tracks 1.6 mm apart above an image plane, #3, no cable.

| PCB Configuration | Number | Relative performance | |
|---|---|---|---|
| two tracks 15 mm apart, no ground plane | #4 | 20 - 200 MHz | 200 - 1000 MHz |
| | | +18 to +36 dB | |

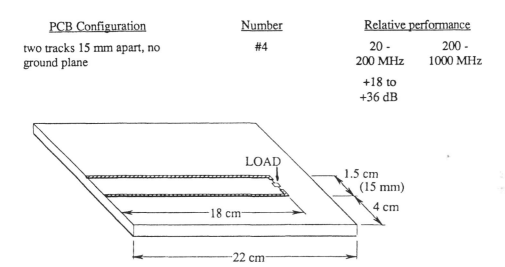

**Figure 11.2d**  Two tracks 15 mm apart, no ground plane, #4, no cable.

It may appear that the open two-conductor transmission line would radiate at about the same level as or slightly lower than a microstrip line, especially because the microstrip line exhibits a lower impedance and the current flow is higher than that of the open transmission line. However, in measurements on the 30-cm by 14-cm PCB, it was found that the microstrip emission levels were up to 32 dB lower over the 20–200-MHz frequency range and up to 36 dB lower from 200 to 1000 MHz.

With a 1-m-long cable attached to the signal return at the load end, the microstrip radiation was 45 dB lower from 20 to 200 MHz and 32 dB lower from 200 to 1000 MHz, compared to the two-conductor transmission line. The major reason for this significant reduction is the reduced inductance of the ground plane signal return path in the microstrip configuration. This reduced inductance results in a lower voltage drop and a lower C/M current flow. The C/M current flow is, typically, the major source of radiation. The C/M current is due to nonsymmetry in the source

| PCB Configuration | Number | Relative performance | |
|---|---|---|---|
| 1 cm wide partial stripline with vias 13 mm apart. | #5 | 20 - 200 MHz | 200 - 1000 MHz |
| | | -8 to -30 dB | -3 to -20 dB |

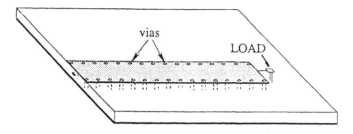

**Figure 11.2e**   1-cm-wide partial stripline with vias 13 mm apart, #5, no cable.

| PCB Configuration | Number | Relative performance | |
|---|---|---|---|
| wide stripline ending 5 mm from edge of ground plane | #6 | 20 - 200 MHz | 200 - 1000 MHz |
| | | below microstrip | -30 to +4 dB |
| | | | (below 600 MHz -12 to -30 dB) |

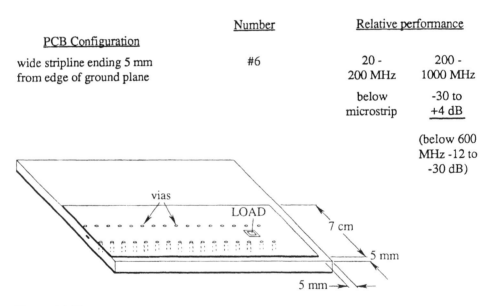

**Figure 11.2f**   Wide stripline ending 5 mm from edge of ground plane, #6, no cable.

or to the proximity or attachment of a grounded structure (enclosure, back plane, motherboard, adjacent PCB, power supply module, AC safety ground connection, etc.).

Adding a ground track in close proximity to the signal track has the effect of increasing the current flow in signal and return and adding a second ground track; configuration #13, shown in Figure 11.2l, increases the current flow further. However, due again to the reduction in return path inductance, measurements on the #13 PCB configuration compared to a microstrip, #1, show a significant reduction in the level of radiation.

In the unlikely event that the magnitude of C/M current flow on the tracks is negligible, i.e., a balanced source and no ground connection, adding the additional ground tracks either increases the level of radiation or it remains constant. Capacitive (E field) crosstalk is typically reduced between signal tracks by approximately 6 dB when a ground track is introduced between

| PCB Configuration | Number | Relative performance | |
|---|---|---|---|
| Full stripline | #7 | 20 - 200 MHz | 200 - 1000 MHz |
| | | below microstrip | -25 to +21 dB |

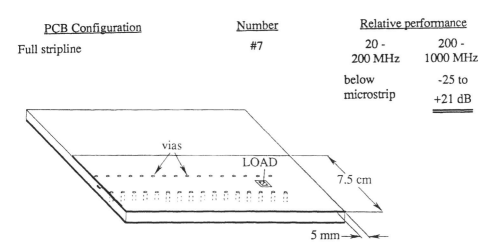

**Figure 11.2g**  Full stripline, #7, no cable.

| PCB Configuration | Number | Relative performance | |
|---|---|---|---|
| Microstrip with 90 degree bend | #8 | 20 - 200 MHz | 200 - 1000 MHz |
| | | -2 dB | < 520 MHz -1 to -6 dB |
| | | | > 520 MHz -6 dB to +6 dB |

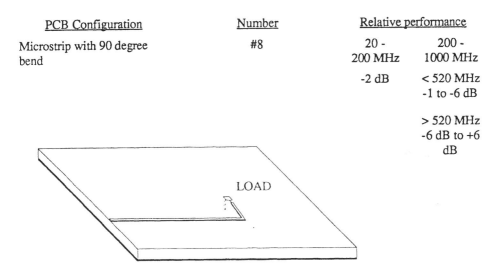

**Figure 11.2h**  Microstrip with 90° bend, #8, no cable.

two signal tracks and the signal and ground track widths are the same. With a wide intervening ground track, the capacitive crosstalk may be reduced by as much as 12 dB.

Adding an image ground plane to a PCB, as shown in configuration #3, Figure 11.2c, in which the two tracks are 1.6 mm apart, and configuration #12, figure 11.2k, in which the tracks are 15 mm apart, can be very effective at reducing PCB radiation. The image plane is placed beneath the signal and signal return tracks that are both located in the same plane. To be effective, the image plane should be connected to the signal return at the load end, preferably at the location where a cable may be attached. In *all* measurements, the shielded enclosure containing the source was connected to the image plane, in other measurements, the signal return was connected to the image plane at both the load and source ends.

The emissions of the two tracks 1.6 mm apart over an image plane connected to the return, configuration #3, were compared to the microstrip #1 in upcoming Figure 11.4. With a cable

| PCB Configuration | Number | Relative performance | |
|---|---|---|---|
| Microstrip with two 45 degree bends | #9 | 20 - 200 MHz | 200 - 1000 MHz |
| | | -2 dB | < 520 MHz -1 to -6 dB |
| | | | > 520 MHz -6 dB to +6 dB |

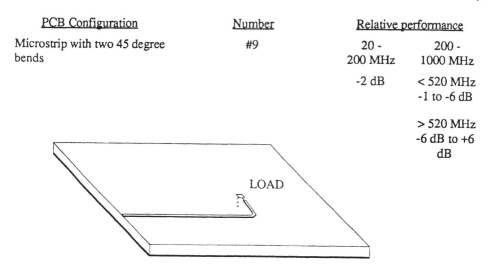

**Figure 11.2i**   Microstrip with two 45° bends, #9, no cable.

| PCB Configuration | Number | Relative performance | |
|---|---|---|---|
| Microstrip with interleaved image plane | #10 | 20 - 200 MHz | 200 - 1000 MHz |
| | | 2 to +20 dB | 0 to +12 dB |

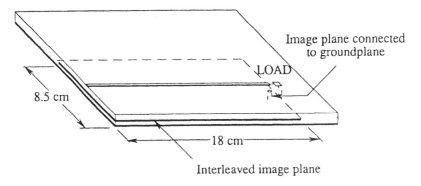

**Figure 11.2j**   Microstrip with interleaved image plane #10, no cable.

connected to the ground plane of the microstrip #1 or to the return at the load end of the image plane the relative levels of radiation are very similar and both slightly higher and lower at different frequencies. The microstrip configuration is often the most convenient and due to the elimination of signal and power return tracks can reduce the number of layers in the PCB. If the board contains a ground plane, then the ground plane can be used as the signal return path.

Adding a second connection of the image plane to the signal return, in configuration #3, at the source end results in a further 3–25-dB reduction in radiation from 200 to 1000 MHz and 0–28-dB reduction from 20 to 200 MHz, with attached cables. However, if this second

| PCB Configuration | Number | Relative performance | |
|---|---|---|---|
| Two tracks 15 mm apart above an image plane | #12 | 20 - 200 MHz | 200 - 1000 MHz |
| | | 0 to -28 dB | +7 to +18 dB |

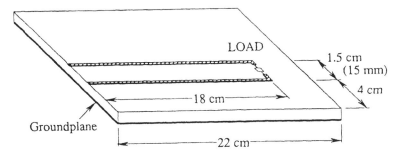

**Figure 11.2k** Emissions from transmission line above an image plane, #12, no cable.

| PCB Configuration | Number | Relative performance | |
|---|---|---|---|
| Microstrip with two adjacent grounded tracks | #13 | 20 - 200 MHz | 200 - 1000 MHz |
| | | 0 to -12 dB | 0 to -20 dB |

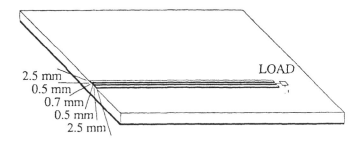

**Figure 11.2l** Emissions from microstrip with two adjacent grounded tracks, no cable.

connection is made, then the configuration is almost the same as the microstrip with one grounded adjacent track.

In comparing configuration #12, in which the tracks are 15 mm apart and above an image plane, the emissions with no attached cable were both 5 dB higher and 5 dB lower at specific frequencies than microstrip #1. From 200 to 10,000 MHz, the emissions from #12 were 4–15 dB higher than from #1. With an attached cable, the emissions from #12 from 20 to 200 MHz were 0–28 dB lower than from #1, and from 200 to 1000 MHz they were 7–18 dB higher.

Figures 11.3–11.10a compare the relative levels of radiation from the different PCB layouts with frequency.

One example of placing an image plane between the signal track and the power/signal ground plane is shown in Figure 11.2j, configuration #10, in which the image plane is connected

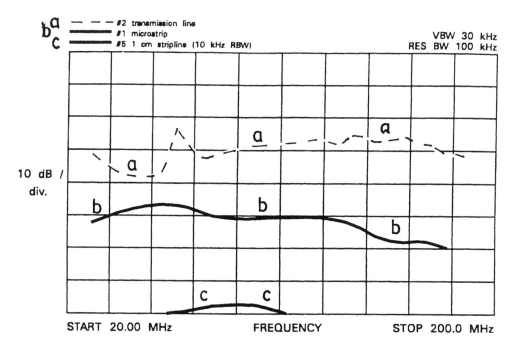

**Figure 11.3**  Relative levels of emissions from PCB layouts #1, #2, and #5. No attached cable, 20–200 MHz.

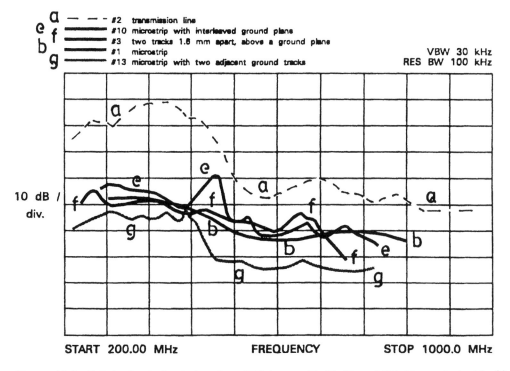

**Figure 11.4**  Relative level of emissions from PCB layouts #1, #2, #3, and #10. No attached cable, 20–200 MHz.

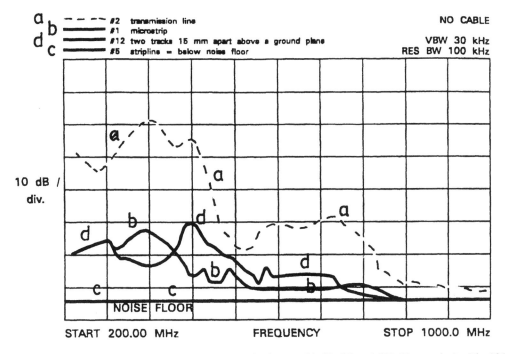

**Figure 11.5**   Relative levels of radiation from PCB layouts #1, #2, #5, and #12. No attached cable, 200–1000 MHz.

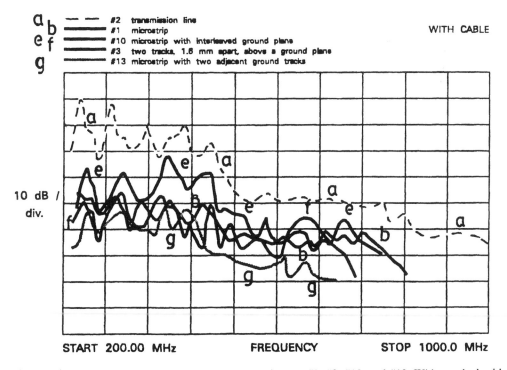

**Figure 11.6**   Relative level of emissions from PCB layouts #2, #3, #10, and #13. With attached cable, 200–1000 MHz.

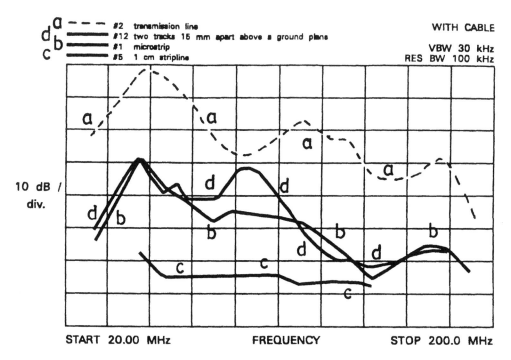

**Figure 11.7**  Relative levels of emissions from PCB layouts #1, #2, #5, and #12. With attached cable, 20–200 MHz.

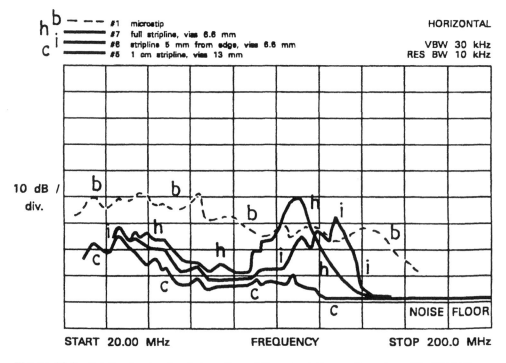

**Figure 11.8**  Relative levels of emissions from different stripline configurations, 20–200 MHz.

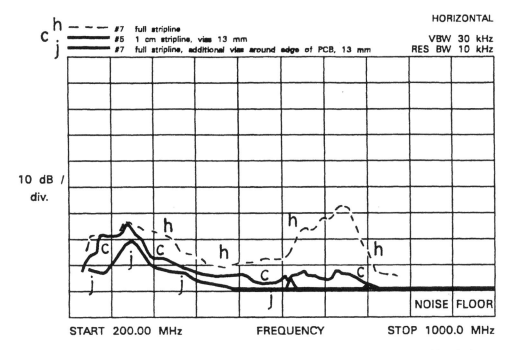

**Figure 11.9** Relative level of emissions from the full stripline with and without additional vias around the periphery of the ground plane. Horizontally polarized, 200–1000 MHz.

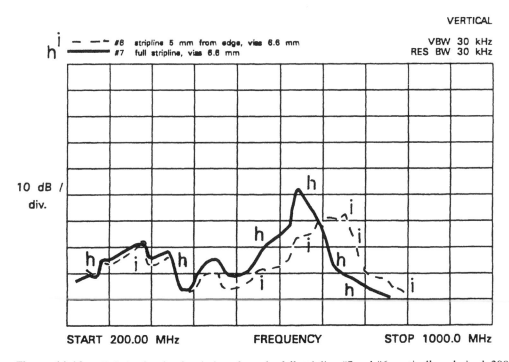

**Figure 11.10a** Relative levels of emissions from the full stripline #7 and #6, vertically polarized, 200–1000 MHz.

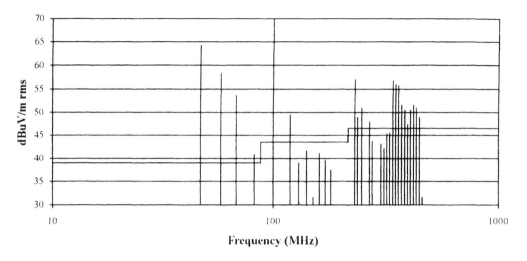

**Figure 11.10b**  Emissions from the configuration #2, with attached cable, at 10 m compared to the FCC Class A limits.

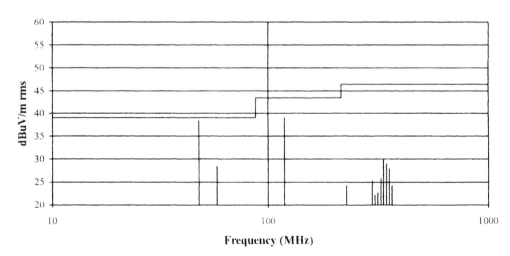

**Figure 11.10c**  Emissions from configuration #1, with attached cable, at 10 m compared to the FCC Class A limits.

to the signal return at the load end. The measured emissions from #10 were much lower than two tracks one above the other, #2; however, when compared to the microstrip, the emissions were from 0 dB to 10 dB higher from 20 to 200 MHz and from 0 dB to 12 dB higher from 200 MHz to 1000 MHz, with no attached cable. With an attached cable, the emissions from 20 to 200 MHz were 2–20 dB above #1, from 200 to 1000 MHz they were 0–12 dB above. Therefore, if for some reason the ground plane cannot be used for the signal return path, adding an interleaved image plane that is connected to the signal return at the load end is an improvement over no ground plane at all but not as effective as the microstrip.

A comparison was made between the interleaved image plane, #10, and #3 (the signal track and a return track 1.6 mm apart, both above an image plane) with the return connected to the image plane at the load end. The radiation from #10 without attached cable was 0–10 dB higher than #3 from 20 to 200 MHz and 0–20 dB higher from 200 to 1000 MHz. With an attached cable, the radiation from #10 was 0–15 dB higher from 20 to 200 MHz and 3–25 dB higher from 200 MHz to 1000 MHz. One possible explanation for this effect is that the increased mutual inductance between the signal and return of configuration #3 results in a lower return inductance when compared to #10, but this remains a hypothesis. Adding an image plane beneath the ground plane of a microstrip does not result in a reduction in radiation. This is because of the inductance and skin-depth effects, which ensure that all of the current in the ground plane flows on the trace side of the plane and is concentrated under the signal trace. Thus, the image plane is redundant when a ground plane exists on a PCB.

Adding a second power or signal return plane, which is electrically connected to the lower ground plane at a multiple of points, with the signal trace/s sandwiched between the two planes results in a stripline configuration, which exhibits the lowest possible radiation from a PCB. The disadvantage of this configuration is that no signal traces are available for testing purposes except at IC pins or component leads.

When a full upper plane cannot be used, a partial upper plane of crosshatched traces connected with vias to a lower signal or power ground plane shields very effectively and thereby reduces radiation from specific signal or data bus trace/s. As described later, a partial upper plane radiates significantly less than the full upper plane and so is the preferred layout to use.

The method of connecting together the upper and lower ground planes was by the use of vias (plated through holes) at intervals of 3.3 mm, 6.6 mm, or 13.2 mm. The PCBs were initially manufactured with vias at 3.3-mm intervals. Then, during the process of testing, first one set of vias was drilled through, removing the electrical connection and leaving vias at intervals of 6.6 mm, and then another set of vias was drilled through, leaving vias at 13.2-mm intervals. Measurements were made on a 1-cm-wide upper plane covering the 18-cm-long microstrip connected at intervals of either 3.3 mm, 6.6 mm, or 13 mm. Compared to the microstrip, the level of radiation from the 1-cm-wide stripline with vias connected every 13 mm and without an attached cable is lower between 20 and 1000 MHz. With an attached cable, the stripline radiation is 8–30 dB lower than the microstrip below 200 MHz and 3–20 dB lower between 200 and 1000 MHz. With the 1-cm stripline vias at a distance of 6.6 mm apart compared to 3.3 mm, the level of radiation was only 9 dB higher at some frequencies. With vias 13 mm apart compared to 6.6 mm, emissions at higher frequencies were only 4 dB higher at certain frequencies.

Figures 11.2a–l compare the measured radiation from the 13 different PCB layouts to our baseline microstrip layout without attached cable. All measurements were made with the 74F04 device as the source and with a 330-ohm surface-mount resistor in parallel with a 47-pF surface-mount capacitor, representing from 8 to 10 TTL-type loads.

Figure 11.10b shows the magnitude of the emissions from PCB configuration #2, which is the two-conductor transmission line with the traces 1.6 mm apart, with an attached cable.

The measurements were made on a 10-m OATS using a biconical/log-periodic antenna and are compared to the FCC class A limits at 10 m in the figure. As can be seen, the worst-case emission from PCB #2 is 35 dB above the limit! Figure 11.10c shows the emissions from the #1 PCB configuration, which is the microstrip, with an attached cable. With the #1 PCB, the worst-case emission is approximately 2 dB below the class A limit, but even the microstrip would be approximately 7.5 dB above the class B limit. To reduce the low-frequency emissions, slowing down the rise and fall times would be ineffectual. Moving the trace closer to the ground plane, shortening the trace length, or using a stripline would be the step to take.

In a typical complex PCB layout, a number of clock and data lines are routed around the board to which a number of ICs are connected. A worst-case prediction of radiation from a board is achieved when the number of traces that carry clocks or signal lines changing state synchronously are added together and are assumed to radiate coherently. The weighting to be added to the radiation from a single trace to account for a multiplicity of traces is therefore 20 log $N$, where $N$ is the number of traces on which clocks or signals change state synchronously. In numerous measurements on emissions from complex PCBs using a spectrum analyzer it has been observed that, during successive sweeps, the spectral emission lines do not remain at a constant level but change amplitude with every sweep. In addition, the low-frequency component raises the apparent noise floor of the spectrum analyzer. The variation in the amplitude of the spectral emission lines, which is a measure of the composite field from all sources at each specific frequency, is attributable to the change in timing of the sources as they switch and, in a consequent interference of fields from the multiplicity of sources, impinging on the measurement antenna. If the spectrum analyzer is operated in the peak, or max, hold over several sweeps, the maximum amplitudes are captured, and these are typically 10 dB higher than those measured in a single sweep. This observation tends to validate the process of simply adding sources coherently for a worst-case calculation.

If the traces on the PCB are significantly longer or shorter than the 18 cm used in the measurements, then the measurements can be scaled accordingly, at least for frequencies below that at which the trace equals a wavelength. Therefore, to compensate the measurements for trace length, add 20 log $l/0.18$ m, where $l$ = trace length in meters, to the emission measurement made on the PCB layout of interest. On signal trace and return configurations, such as #2 and #14, the distance between the track and return affects the magnitude of the radiation. A compensation in the measurements of #2 can be made if the actual track and return distance is greater or less than 1.6 mm or greater or less than 15 mm. The approximate compensation to the measurements is 10 log $d1/d2$, where $d1$ is the actual distance between the tracks and $d2$ is either the 1.6 mm of configuration #2 or the 15 mm of configuration #4.

The approximate correction for clock frequencies other than 10 MHz is given by 20 log $f$ (MHz)/10, which is added to the measured emissions at 10 MHz. For example, if the clock frequency is 1 MHz, then $-20$dB is added to the emissions at 10 MHz and above. If the clock frequency is 100 MHz, then 20 dB is added to the emissions at 100 MHz and above but below the frequency at which the track length equals a wavelength.

Adding an isolated image plane above a microstrip groundplane increases emissions at resonant frequencies and decreases emissions at nonresonant frequencies. Connecting the image plane to the microstrip ground plane at only two locations does not reduce radiation but merely shifts the resonant and antiresonant frequencies. Therefore, do *not* add an image plane above a microstrip groundplane; instead, make a stripline configuration by connecting the upper plane around its edges to the lower ground plane.

Although decreasing the distance between signal and return traces reduces the impedance of the line, any increase in current is more than compensated for in a reduction in radiation as a result of the lower signal return inductance due to the increased mutual inductance between the traces. As we have seen, decreasing the voltage drop in the signal return by decreasing its inductance will reduce the potential C/M current and the radiation from the board, especially when an interface signal cable return is connected to the board. The reductions in radiation achieved by reducing the distance between the trace and the ground plane, or the return trace, is almost a function of the decreased distance. For example, reducing the distance by a factor of 1/133 (18 dB) results in an average reduction in radiation of 16 dB, based on measurements.

A summary of the layouts described and some further general guidelines that may be useful in reducing PCB radiation follow.

a. The use of high-speed logic, including high-speed CMOS, in a ''poor'' PCB layout (not microstrip, image plane, or stripline, with long signal/power traces and returns at some distance from the source) will almost certainly fail the majority of commercial EMI radiated emission requirements unless contained in a shielded enclosure.

b. Use of the microstrip, stripline, or image plane layout will reduce radiation from the PCB, both with and without an attached cable, and an unshielded PCB may meet EMI requirements. However, if a common-mode current exists between the ground plane on a low-emission type of PCB layout and a second PCB ground plane or a safety ground or any conductive structure, such as a power supply shielded enclosure, the radiation from the low-level-emission type of layout may be no lower than that of a high level type of layout. This effect is discussed in Section 11.6.

## 11.3.2. Practical PCB Layout

The PCB layouts tested in Section 11.3.1 were extremely simple, however, the test results and the lessons learned from these results can be applied to the more complex PCB layouts used in practice. For example, the effectiveness of a solid power/signal ground plane in reducing emissions when compared to a PCB with only signal and power traces has been extensively demonstrated in practical layouts. As discussed in Section 11.6.2, a ground plane must be solid directly under all data and signal traces. In a complex board, traces run all around the board on multiple layers, and this means that the complete board area must be covered in a solid ground plane that is most conveniently used for digital power and digital signal return. See Section 11.6.2 to find out what constitutes a ''good'' ground plane. The closer the ''hot'' traces (traces that carry digital signals and clocks with fast edges over a distance greater than 25 mm) are to the ground plane, the lower the emissions. Therefore route all high-speed, long tracks on the layer closest to the ground plane. These hot tracks can be on the layer on either side of the ground plane when this is buried. Tracks carrying less active signals or that are short can be carried on tracks further away from the ground plane. Keep hot tracks as short as possible and include series resistors at the source end to increase the rise and fall times. Choose buffered input devices to reduce the load on the bus or clock. Where possible, run clocks and data buses in localized striplines in which the edges of the upper ground plane is tacked to the lower ground plane, with vias a maximum of 13 mm apart (closer if possible). This configuration is illustrated in Figure 11.11a. Do not extend the upper ground plane beyond the row of vias. If possible extend the stripline over the whole area of the board with the vias located at the periphery of the board, this is illustrated in Figure 11.11b.

A common error is to leave large gaps between the vias at locations around the edge of the board. In this case the emissions are concentrated at these locations and some enhancement, especially at high frequencies, may be seen. It is possible to use a power plane on one surface of the board with the ground plane on the opposite surface of the board to form a stripline and this is illustrated in Figure 11.11c. The two planes are connected together at RF by surface-mount capacitors located at intervals no greater than 25 mm around the periphery of the board.

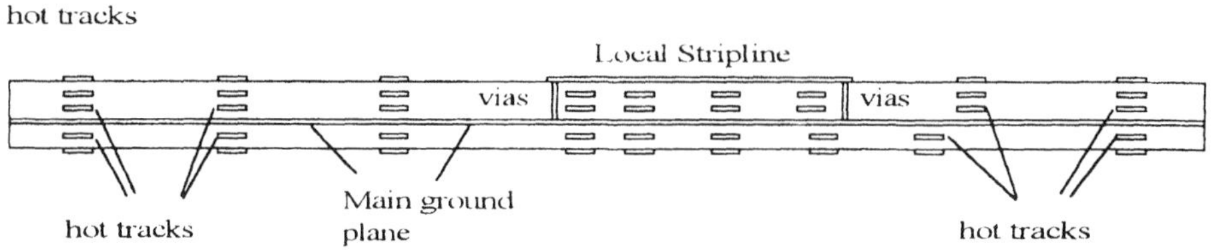

hot tracks

Local Stripline

vias                                                vias

hot tracks                    Main ground                          hot tracks
                              plane

**Figure 11.11a**   Locating hot tracks close to the ground plane and local stripline.

UPPER GROUND PLANE

LOWER GROUND PLANE

Row of vias                                                                           Row of vias
around periphery                                                                      around periphery
of board                                                                              of board

**Figure 11.11b**   Overall ground plane stripline.

A capacitively coupled stripline is not as effective as a stripline formed by two ground planes connected directly together, and if the capacitors are greater than 25 mm apart, any benefit over a microstrip is greatly reduced. If tracks used for low-level signals such as analog, video, or control are present on the board, then these signal tracks may be located on the outside of the stripline ground planes to reduce crosstalk between the digital signals and these low level signals. This configuration is illustrated in Figure 11.11d. However, if these low-level signals reference a noisy ground, namely, the inner surface of the stripline digital ground plane, then emissions from these tracks may be high, due to common-mode noise. If this is a potential problem and the possibility of crosstalk is real, then the solution is to construct an inner stripline dedicated

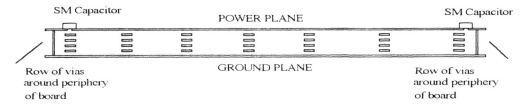

SM Capacitor                     POWER PLANE                              SM Capacitor

Row of vias                     GROUND PLANE                              Row of vias
around periphery                                                          around periphery
of board                                                                  of board

**Figure 11.11c**   Stripline formed by the power plane and the ground plane.

IMBEDDED GROUND PLANE

low signal level tracks

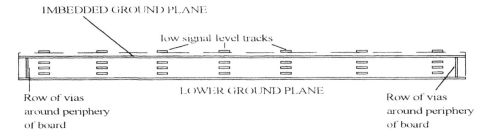

Row of vias                     LOWER GROUND PLANE                        Row of vias
around periphery                                                          around periphery
of board                                                                  of board

**Figure 11.11d**   Low-level-signal tracks located outside of the stripline to reduce crosstalk.

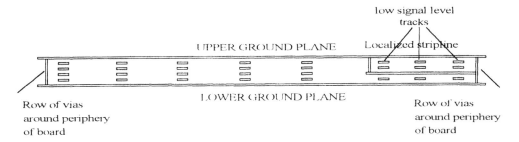

**Figure 11.11e** Localized stripline to contain low-level-signal tracks.

to the low-level-signal tracks and signal return with the associated ICs located directly over this area; this solution is illustrated in Figure 11.11e.

One reported problem with the stripline configuration is increased radiation from the vertical vias passing through the ground plane. However, this problem is purely anecdotal and may simply have been observed because of the very low emissions from the well-designed and implemented stripline PCB configuration.

One observation that has been made in measurements on numerous boards is that emissions from ICs and their pins can be sufficiently high for the board to fail commercial requirements despite the very best possible PCB layout. In microstrip layouts in which the ground plane is imbedded, add a small section of ground plane directly under the IC connected by vias to the underlying ground plane. This configuration is illustrated in Figure 11.11f. If the IC has a metal lid, connection of the lid to the ground plane at least at the four corners can reduce emissions, especially at low frequencies. Local shielding of the IC, as discussed in Section 11.7, in the microstrip layouts in which the ground plane is imbedded has the problem of increased radiation from the tracks exiting the shield. If a stripline is used with two outer ground planes, then the connections to the IC pins can be made with vias under the shield to the inner layers. The shield can then be contacted around its periphery to the ground plane, and this configuration is extremely effective. If it saves the use of a metal or conductively coated equipment enclosure, this local shield can be very cost effective.

An image plane may be constructed from aluminum or copper foil enclosed in a nonconductive envelope. When this insulated foil is placed in close proximity to either side of a PCB that contains "hot" traces or ICs that radiate at a high level, the level of radiation can be reduced from that side of the board. When the foil can be wrapped around one edge of the PCB and cover both sides of the PCB, radiation can be reduced from both sides of the PCB. However, the worst-case maximum emissions from the PCB when measured on an OATS, with the PCB rotated through 360° and with the antenna raised and lowered, may not be reduced. This is because the image plane changes the radiation pattern from the PCB and the level of emissions

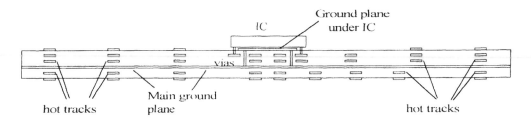

**Figure 11.11f** Localized ground plane under the IC.

from the gaps at the edge of the PCB may not be reduced, but may even increase at some frequencies.

In one case, PCB-sourced emissions coupled to the antenna of a wireless plugged into the PCMCIA slot of the equipment. The wireless was desensitized due to the broadband noise (typically from data or switching-power supplies) and the harmonics from clocks injected into the antenna. With an image plane covering both sides of a PCB and wrapped around the edge of the PCB facing the antenna, a reduction was seen in the very localized coupling between the PCB and the antenna, and the wireless reception was improved. The overall radiation from the PCB due to the ''wrap-around'' image plane was not, however, greatly reduced.

The step-by-step design and the rules to follow in laying out a PCB follow.

1. Evaluate the EMC requirements placed on a PCB or PCBs. That is, what requirements have to be met? Is crosstalk or common ground impedance coupling to sensitive circuits a potential problem? Is the PCB housed in a conductive enclosure? And is this, or can it be made into, an effective shield? Decide how much emission reduction at source is required. For example, a high-speed PCB with many LSI devices contained in a nonconductive enclosure with attached signal or power cables requires the maximum possible level of source reduction by ''good'' circuit design and PCB layout. At the extreme, use an LCR circuit to make the clock as close to a sinewave as possible.

2. Locate ICs so that the interconnects are as short as possible. Locate oscillators as close as possible to ICs that use the clock and as far away from signal interface and low-level-signal ICs and tracks as feasible.

3. Minimize path lengths between sources and loads and between loads sharing a common clock or bus, or use shielded stripline over these connections.

4. If microstrip or stripline configurations cannot be used, keep signal and return tracks as close together as feasible.

5. Reduce the drive current on tracks by minimizing loading, typically by the use of buffers at the load end or series resistance.

6. Use the slowest logic possible to achieve the required performance.

7. Increase the decoupling to one capacitor per IC when the ICs are not located close together. Use surface-mount capacitors for decoupling.

8. Choose the location of common ground and chassis ground and isolation of grounds in accordance with the guidelines of Section 11.6.2.

9. Often, placing a low-value capacitor (0.01–0.1 µF) in parallel with a high-value capacitor (1–10 µF) is effective. However, if the capacitors leads are not very short, nothing may be gained.

10. Choose the grounding scheme from Section 11.6.2 that will reduce radiation from attached cables.

11. Reduce the number of signal tracks that change state simultaneously.

12. If power and return tracks are the source of radiation, add inductors between the power supply and the ICs and increase the number of decoupling capacitors.

13. After the initial layout, use a viewer, such as GerbTool or a Gerber viewer available on the Internet, to trace the routing and component placing, to ensure that none of the design rules have been violated.

If radiation from interface cable predominates and the PCB is located in an enclosure, the multiple connection of the power/signal ground plane to chassis will reduce common-mode noise voltage developed across the board. Unfortunately, the RF current flow in the enclosure will increase, radiation from seams and apertures is likely to increase, and the net gain may be zero with a poorly shielded enclosure.

It has been recommended that the width of the power plane be made smaller than the ground plane to reduce fringing fields. Although the mechanism for the reduction is unclear, in measurements on a full stripline versus the 1-cm-wide stripline, the full-stripline radiation was up to 32 dB higher than the 1-cm stripline above 350 MHz. By adding additional vias at 13-mm intervals around the periphery of the full stripline, the radiation can be reduced to just below that of the 1-cm stripline. Therefore, use the partial stripline over clocks and data buses wherever possible, and do not extend power planes over an area larger than the ground plane. When the radiation of a PCB has been drastically reduced, for example, by use of the shielded stripline, the predominant source may be the ICs themselves.

Although effective at reducing current loop area, effective decoupling around an IC can increase the magnitude of current spike flowing in the IC as it switches. Inclusion of a resistor or a ferrite bead between the decoupling of the IC and the power pin will reduce the current spike at the expense of increased bounce in the supply voltage (see Fig. 5.14). Ferrite beads have also been used on power return pins, but this will inherently result in a higher ground bounce voltage.

Use ICs with the minimum number of internal devices changing state simultaneously.

When using a custom device, ensure a layout with minimum internal path lengths and one that reduces common-mode noise generation on the input and output pins. Series resistance may be incorporated within the device, in series with the supply, to limit switching current. The value of the resistor will depend on the tolerance of the device to supply bounce.

### 11.3.3 High-Frequency (1 GHz and Above) Radiation from PCBs

As clock speeds increase, along with the slew rate of the logic, high-frequency PCB radiation also increases. At high frequency, microstrip or other transmission-line configurations are typically used, with the line terminated in its characteristic impedance $Z_c$. At these high microwave frequencies, radiation from microstrip discontinuities and lines becomes significant when not masked by C/M current flow–generated radiation. The measured results presented in the data of Section 11.3.1 are 60 dB above predicted results for purely differential-mode current radiation at 1 GHz, so even at this high frequency the C/M current–contributed radiation cannot be ignored! However, as the structure becomes electrically long and the frequencies of interest are above 1 GHz, the differential-mode current radiation should begin to dominate. Any impedance mismatch at the end of a line will form a discontinuity and increase radiation. A high-impedance termination relative to $Z_c$ will radiate more effectively than a low impedance one, approximately 2.5–19 dB higher. In addition, at higher frequencies, the higher the track above the ground plane, the higher is the radiation. For example, increasing the frequency or the height of the track above the ground plane by a factor of 10 (20 dB) increases the radiated power by a factor of 100 (40 dB) and the E field by 20 dB. The higher the permittivity of the substrate, the lower the impedance of the transmission line and the higher the radiation. Using the many equations for radiation from discontinuities in a microstrip based on the signal current flow invariably results in predicted levels of E field that are much higher than measured from typical printed circuit boards.

It has often been recommended not to daisy-chain the logic connections to a clock but to fan out to clocks, as shown in Figure 11.12. At low frequency and when the load is not matched to the line, this arrangement has not been demonstrated to be advantageous. However, with matched transmission lines, the fan-out arrangement should be used.

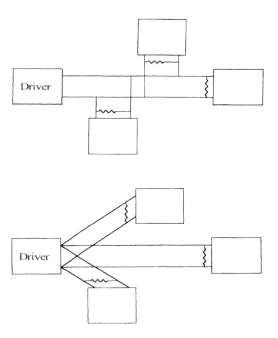

**Figure 11.12**  Daisy-chain vs. fan-out of clock or high-speed data.

Right-angle bends exhibit increased capacitance, which represents a discontinuity and increased radiation. Use of a 45° bend in a microstrip line, as shown in Figure 11.13, will reduce radiation at very high frequency. However, for 74F, 74AS, 74S, 74, 74C, 74HC, and ECL100 or similar-speed devices, any reduction in radiation as a result of using the 45° bend is insignificant, and in measurements was masked by the radiation from the impedance mismatch at the end of the line and the C/M current–contributed radiation from the track.

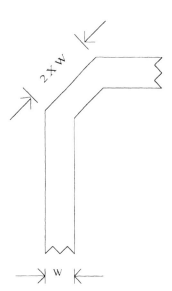

**Figure 11.13**  Forty-five degree bend in a PCB trace.

The theoretical difference in either horizontally or vertically polarized radiation from a straight microstrip compared to one with a 90° bend is $+/-6$ dB at high frequency, i.e., a mean difference of 0 dB. The slight reduction in measured radiation between the 90° bend and the straight microstrip was seen in the horizontal E field and is almost certainly due to the decreased length of the track due to the bend.

The stripline PCB configuration should be used wherever practical. If stripline cannot be used, one technique for reducing radiation from a microstrip is to use differential signals for long computer clock runs on the board. The two signal traces should be of exactly the same length and be located as close together as allowed. The loads at the far end of the differential lines must be well matched, and this is easier when ECL, ECLlite, or ECLinPs is used. Common-mode currents can be reduced by running both traces through a PCB-mounted balun. But beware: if the connections to and from the balun are too long, these may radiate more than if the balun is not employed. The FCC requires that radiation from digital devices (unintentional radiators) be measured above 1 GHz to the fifth harmonic of the highest fundamental frequency or to 40 GHz, whichever is lower. Many fiber-optic drivers operate at 2.5 GHz or higher, and measurements up to 25 GHz must be made. When designing a laser driver for fiber optics, the layout, shielding, grounding, and filtering must be at the same level as any medium-power RF circuit. This typically means a filtered differential input to the driver IC, which is contained inside a completely shielded enclosure mounted on the PCB. The enclosure must contact the metal housing of the laser. The power to the laser driver must be well filtered over the 2.5–25-GHz frequency range, and here a feedthrough placed in one wall of the shielded enclosure is almost mandatory.

Although radiation from a PCB at gigahertz frequencies should be reduced by placing microwave absorber (available in thin sheets of loaded rubber) on the PCB surface or over ICs, practical experience has shown this to be ineffective. However, low-frequency solid ferrite material used to fill apertures in heatsinks and enclosures has been effective.

## 11.4. COMPARISON OF LOGIC TYPES

As clock and data rates are pushed ever higher, the speed of logic used in digital equipment has increased. When standard CMOS logic is used exclusively on a PCB, we seldom see a problem with meeting Class A and even Class B radiated emission requirements, unless the designer has done a very poor job of laying out the board. This was not the case with standard 74 type TTL ICs, with which some care had to be exercised to meet Class A requirements. With the advent of high-speed CMOS and higher-speed bipolar devices, much greater care has to be taken in the PCB layout, circuit level reduction, and possibly shielding, to reduce emissions to acceptable levels.

The concern has been expressed that use of these new high-speed logic types will result in significant radiation in the 1–10-GHz frequency range, which the FCC covers in its limits. The highest-speed device that was tested was the ECLinPS device, with a rise and a fall time of between 0.275 ns and 0.6 ns. The ECLinPS device did not radiate levels at high frequency above those of the 74F device when the fundamental frequency was low (10 MHz). This is not surprising, for the criteria for high-frequency radiation when the fundamental clock frequency is low is the rate of change of voltage and current, i.e. the slew rate, and not the rise and fall time alone. As discussed later, the slew rate of the ECLinPS device is very similar to the slew rate of the 74F04 device.

Figure 11.14 illustrates the voltage amplitude in the frequency domain (the spectral emission lines) generated by the 10-MHz clock measured across a 50-ohm load at the output of the ECLinPS device. As the speed of the device increases, so does the maximum frequency at which

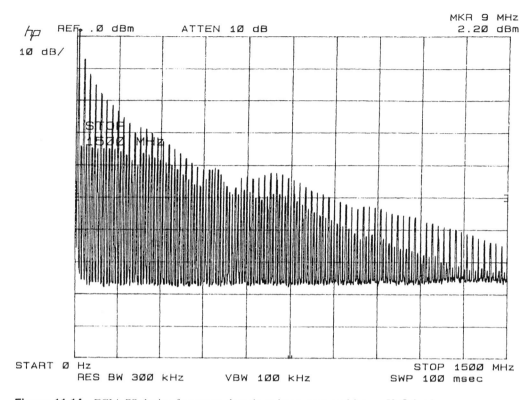

**Figure 11.14**   ECLinPS device frequency-domain voltage measured into a 50-Ω load.

the device can be used. The maximum frequency of the "ECLinPS Lite" family is 2.2 GHz for flip-flops and 1.4 GHz for buffers with a minimum rise and fall time of 0.1 ns. With a 1.4-GHz fundamental and a 0.1 ns rise and fall time, the first breakpoint in the envelope of spectral emission lines, as shown in Figure 3.5, is at $1/\pi d = 88$ MHz, where $d$ is the pulse width (3.6 ns). After the first breakpoint, the amplitude of the spectral emission lines reduces as a function of 20 dB per decade of increasing frequency. Above the second breakpoint of 3.18 GHz, given by $1/\pi t_r$, the amplitude reduces as a function of 40 dB per decade of frequency.

If the 74F04 and 74AS04 devices had been tested close to their maximum frequencies of 124 MHz, say, at 120 MHz, the levels of emission would have been higher at the same frequency (than the emission from a 10-MHz clock). For example, at a frequency of 360 MHz (the third harmonic of the 120-MHz clock), the 120-MHz clock emission would be 25 dB higher than emission from the 10-MHz clock and 21 dB higher at 1080 MHz (the 9th harmonic), assuming the rise and fall times stay constant.

If the ECLinPS device had been operated at a 450-MHz clock frequency instead of 10 MHz, the emission levels at 2.25 GHz would have been approximately 33 dB higher. The increase in emissions corresponds closely to the increase in clock frequency, i.e., from 10 MHz to 120 MHz = 21 dB and from 10 MHz to 450 MHz = 33 dB. With the use of the microstrip transmission-line PCB configuration and a 10-MHz clock, none of the logic types resulted in radiation above 1200 MHz, at a level above the noise floor of the measurement system, even with the antenna located 0.55 m from the PCB and with an attached cable on the PCB. Standard CMOS could not be included in the measurements due to an upper frequency limit below the

**Table 11.1** Upper Frequency Limits
of Logic Families

| Logic family | Frequency limit |
| --- | --- |
| ECLinPS | 1200 MHz |
| 74F | 125 MHz |
| 74AS | 125 MHz |
| 74S | 95 MHz |
| 74HCT | 50 MHz |
| 74ALS | 35 MHz |
| 74LS | 33 MHz |
| 74 | 24 MHz |

10-MHz clock used in the test setup. Table 11.1 illustrates the upper frequency limits of the logic families tested.

The purpose of measuring the radiation from the microstrip PCB configuration with different logic families was twofold. With a comparison available between the 74F04 device, used in all the comparative measurements made on PCB layouts, and other logic types, the radiated emissions from the PCB layouts can be scaled for the other devices. In addition, if a choice was available, the device with the lowest emission level and the required upper frequency limit could be selected. For example, standard TTL is no longer a popular logic family, and in the future it may well be phased out. However, if the typical upper frequency limit of 25 MHz, 13-ns rise and 6-ns fall times, with propagation delays of 30 ns, are acceptable, then TTL is the best bet, next to standard CMOS, for reduced high-frequency radiation.

The devices tested were all hex inverters, with the exception of the MC100E111 ECLinPS device, which was a 1:9 differential clock driver. The hex inverters were the 74F04, 74AS04, 7404, 74LS04, 74ALS04, 74HCT04, and 74S04 devices. The load for the 74XXX04 devices was a 330-ohm resistor in parallel with a 47-pF capacitor, the same load as used in the PCB radiation measurements. This load is considered typical by many IC manufactures and is often used as the test load. The capacitance of 47 pF corresponds to a fan-out of between 8 and 10. The ECLinPS device was tested with a 47-ohm resistor in parallel with a 4-pF capacitor. According to the manufacturer, Motorola, "The input loading capacitance of the device typically measures 1.5 pF and is virtually independent of input fan-out as the device capacitance is less than 5% of the total." The ECLinPS device was also tested with a load of 56 ohms, which matches the characteristic impedance of the microstrip PCB used in the test. The radiation from the ECLinPS with the 47-ohm and 4-pF capacitor load was up to 20 dB higher than with the purely resistive load, and it is the capacitive and resistive load results that are presented here.

The 74AS04 device was also tested over the 200–1000-MHz frequency range with different loads. The radiation from the 74AS04 loaded with a 330-ohm resistor, when compared to the standard 330-ohm resistor in parallel with a 47-pF load, is both higher and lower at different frequencies. Open circuit, the 74AS04 radiation is up to 10 dB lower, but with a 68-ohm load it is up to 27 dB lower. The 68-ohm load is an invalid load but illustrates the importance of the amplitude of the output waveform, which reduces considerably with the 68-ohm load, in determining both the low-frequency radiation and the slew rate and, therefore, the high-frequency radiation.

The maximum-voltage amplitude at the fundamental clock frequency determines the low-frequency level of emissions, and the slew rate determines the high-frequency emissions.

**Table 11.2**  Maximum Voltage Excursion

| Logic family | $V_f$ | $V$ |
|---|---|---|
| 74F | 6.0 | 3.8 |
| 74AS | 6.5 | 3.58 |
| 74HCT | 7.0 | 5.08 |
| 74ALS04 | 5.5 | 3.88 |
| 7404 | 5.5 | 3.28 |
| 74LS | 5.2 | 3.8 |
| ECLinPS | 0.76 | 0.76 |

Table 11.2 plots the maximum voltage excursion during the falling edge of the pulse, $V_f$, and the voltage measured after the settling time, $V$.

Figure 11.15 compares the low-frequency, 20–200-MHz, radiated emissions from the microstrip PCB with different logic types. The highest-level emissions are from the 74HCT04 device, and this is to be expected because the pulse amplitude is 5.8 V, also the highest. Likewise, the ECLinPS exhibits the lowest level of low-frequency emissions and the lowest voltage excursion.

Table 11.3 shows the rise and fall times over a 3-V excursion during the transition. The more common definition of rise and fall times is the time measured between 20% of the voltage transition and 80% of the voltage transition. Because the voltage transitions varied so greatly from device to device, the traditional measure of the rise and fall time will have little relationship

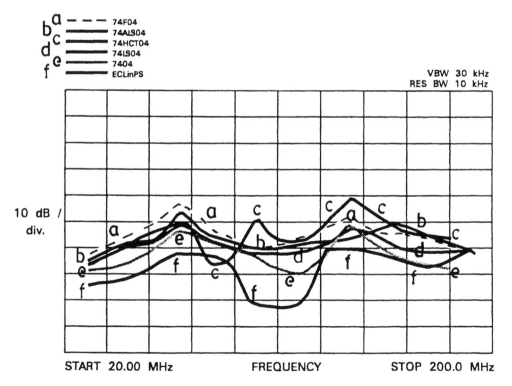

**Figure 11.15**  Comparison of low-frequency emissions from the different logic types in a microstrip PCB.

**Table 11.3** Rise and Fall Times

| Logic family | $t_f$ (ns) | $t_r$ (ns) | Slew rate (V/ns) |
|---|---|---|---|
| 74F04 | 3.01 | 4.0 | 1.0 |
| 74AS04 | 3.28 | 3.9 | 0.9 |
| 7404 | 4.36 | 10.6 | 0.68 |
| 74LS04 | 4.45 | 8.67 | 0.67 |
| 74ALS04 | 3.5 | 7.27 | 0.85 |
| 74HCT04 | 2.6 | 3.29 | 1.15 |
| 74S04 | 3.05 | 5.0 | 0.98 |
| ECLinPS 5 | — | — | 0.85 |

to the voltage and current frequency component generated by the pulse. Since the ECLinPS device maximum excursion is 0.76 V, only the slew rate for this device is shown in Table 11.3. Even the voltage slew rate does not adequately describe the high-frequency components generated by the pulse, because the amplitude of the current pulse plays an important role. For example, the voltage amplitude of the emission at 770 MHz from the ECLinPS device is 1.92 mV, which is lower than that from the 74F04 device at 3.7 mV, whereas the current amplitude from the ECLinPS device at 770 MHz is slightly higher, at 7.9 mA, compared to the 7.3 mA of the 74F04 device.

Figures 11.16–11.17 show the high-frequency emissions from the different logic types. As expected, the correlation between high slew rate and high-level emissions is good, with the

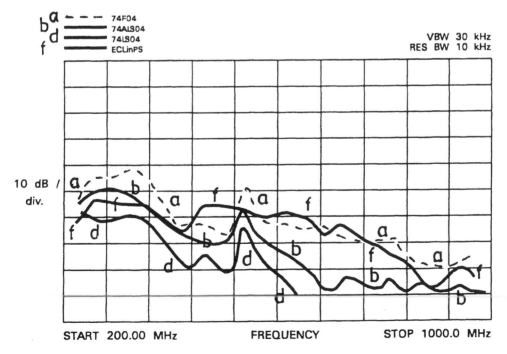

**Figure 11.16** Comparison of high-frequency emissions from the different logic types in a microstrip PCB.

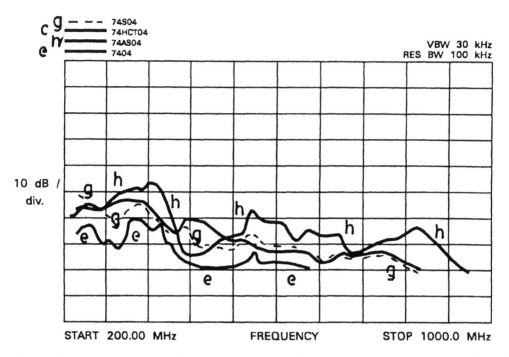

**Figure 11.17**  Comparison of high-frequency emissions from the different logic types in a microstrip PCB.

exception of the 74HCT04 device, which has the highest slew rate but about the same level of emissions as the 74S04, which has a slightly lower slew rate. For the lower-slew-rate devices, such as the 7404, the 74LS04, and the 74ALS04, the emission levels are concomitantly lower. The measured fall time over the standard 3-V excursion is used to determine the slew rate in Table 11.3.

## 11.5.  CIRCUIT-LEVEL REDUCTION TECHNIQUES

High current spikes flow in the logic signal and return interconnections as the output logic device changes state and charges/discharges the input capacitance of the load device and removes the charge from either a base/emitter junction or a diode. During the transition, the load on the output device is therefore predominantly capacitive and, depending on the geometry of the interconnections, the logic type, and the frequency, the source of radiated emissions is typically a low impedance (i.e., a predominantly magnetic field source). The significance of a low-impedance predominantly H field is discussed in Section 11.7 when examining the shielding effectiveness of a conductive enclosure around a PCB. The current pulse is the predominant source of emissions at frequencies less than approximately 200 MHz. However, emissions created by the voltage transition should not be ignored in an EMC prediction, especially at high frequency or when the load is a high impedance with low load capacitance. Even with an open circuit at the far end of the PCB, displacement current flows due to the charging and discharging of the capacitance between the signal and return paths. The source of radiation is then predominantly E field, and it is typically the E field, which is specified and measured, in commercial requirements.

Measurements made on a microstrip PCB layout with either an open circuit at the far end or a 330-ohm resistor in parallel with a 47-pF load showed a 10–38-dB increase in radiated level for the microstrip with load below 137 MHz. However, above 137 MHz, the O/C microstrip radiation was from 0 dB to 15 dB higher than the microstrip with load, illustrating the importance of the voltage-generated E field. Some of this increase with an open circuit may be attributable to the decrease in rise and fall times of the unloaded 74F04 driver.

It is the current flow in the signal return connection (PCB track or ground plane), due to both displacement current between the signal path and return path and the load current, that develops a voltage in the impedance of the return. This voltage can generate a C/M current, which is the prime source of radiation from the PCB and any attached connection.

Many of the measurements on PCB radiation described in this section were made with a cable attached either to the ground plane or to the signal return track at the load end. Thus in reducing the radiated emissions from a PCB track, the voltage drop in the signal/power return connections or/and the magnitude of the current should be reduced. On an electrically long line, use of the correct value of termination resistance will mask the peaks and troughs in emissions at resonant and antiresonant frequencies. Although, the total magnitude of emissions tends to increase. This may at first appear to be questionable, and it is correct to say that some of the peak emissions due to resonance may decrease with a termination resistor. However, in a typical microstrip transmission line with a 65-$\Omega$ characteristic impedance, adding a 65-$\Omega$ termination resistor will increase the current flow in the line, especially at low frequency, where the typical load impedance presented by the device is high. With an increase in current flow, the level of radiation will increase, especially at transmission-line antiresonant frequencies.

If a trace on an existing PCB is found to be a major source of emission at a single or a few frequencies, changing the termination resistance may reduce the problem frequencies, although the emission amplitude at other frequencies often increases. For example, changing the termination resistor from 700 ohms to 600 ohms resulted in a predicted 40-dB reduction in radiated emissions around 300 MHz. This effect occurred with a track length of 51 cm, which equals 1.5λ at 300 MHz. The source was a 10-MHz clock with a 50% duty cycle and a 10-ns rise time.

In most practical applications, the load on the line at frequencies as high as 300 MHz is a much lower impedance than 600 ohms due to the IC input capacitance. The only way to increase the load impedance at high frequencies is to add series resistance or a series resistor/capacitor combination, if the input characteristics of the logic type allow. One method effective in reducing the magnitude of the current spike is to include series resistance at the output of the driver circuit, as shown in Figure 11.18a. The series resistance is available in a number of different values in a dual in-line package, which saves PCB space. For TTL circuits, the value of the resistor must be limited to 10 ohms or less. A ferrite, available as a surface-mount component, or other form of inductor will limit the magnitude of switching current but tends to resonate and may degrade signal quality. A combination of damping resistance and inductor, as shown in Figure 11.18c, may be the ideal solution.

**Figure 11.18a**   Added resistance at the driver end.

**Figure 11.18b**   Added capacitance at either the load end (incorrect) or driver end (correct).

*Do not add a capacitor at the load end of a clock or data signal track*, because, although this will increase the rise and fall times of the voltage step, the magnitude of the current pulse will increase and so will the radiation below approximately 200 MHz. When using very high-speed devices, $t_r/t_f < 2$ns, the frequencies above 900 MHz will also be reduced by the addition of a capacitor across the load, as long as the track is electrically short. Nevertheless, the correct location for the capacitor remains at the source end with some series impedance between the source and capacitor. The addition of a capacitor at the driver end *without* a series impedance will increase the radiation from the IC and may be counter-productive. These solutions are illustrated in Figures 11.18a, 11.18b, and 11.18c.

The lower the fundamental frequency, the slower the edges can be made. In practical terms, regardless of frequency, appreciable reduction in emissions can not be made lower than the 4th harmonic of the fundamental. With present high-speed clocks, this reduction may not be possible at a low-enough frequency. One possibility is to use an LCR network to generate a clock pulse that approximates a sinewave. This solution will result in jitter at the output of the device that uses the clock, and the solution is not practical in circuits susceptible to jitter. It is also very difficult to develop a pure sinewave, for as the device that uses the clock changes state, its input impedance changes and results in a step function in the sinewave. Despite this effect, the sinewave clock can result in a reduction in the emissions at low harmonics of the clock. In one instance, a device selling for \$99.00 was contained in a plastic enclosure that had to be sprayed with a conductive coating, at a cost of \$20, to meet emission requirements. Changing the clock to a sinewave brought emissions to just below the limit without the need for a shielded enclosure.

Another circuit-level technique for reducing radiated emissions measured by a spectrum analyzer or receiver is to use a spread-spectrum, frequency-hopping, or frequency-wiggle technique. These frequency-shift techniques effectively broadband frequencies to reduce the measured emissions levels from clocks and data, at the fundamental and harmonics. The FCC, and almost certainly other organizations, will accept these techniques in meeting radiated emissions from equipment containing digital components. A second situation in which to implement a spread-spectrum, frequency-hopping, or frequency-wiggle clock is in a computer in which a radio is imbedded or into which a PCMIU card is plugged. The computer clocks and data will

**Figure 11.18c**   Added resistance and inductance at the driver end.

generate spurious responses (spurs) in the radio as a result of the power-line and data-line conducted noise and due to the radiated coupling to the radio antenna. Using frequency-sweep techniques will move the spurious response out of the radio-tuned channel. And if this is swept at a low-enough rate, then the reception of packets of error-free data is possible. In addition, no one channel is denied to the user, which may be the case if a fixed spur happens to coincide with the tuned channel. Even if the channel is not totally denied, the area of coverage area for that channel can be severely limited. For more information on radio-induced EMI see Section 5.3.3.4. If the clock is generated by a phased lock loop, then the computer can control the loop and move spurs out of band in an intelligent fashion. One commercially available spread-spectrum clock is the IMISM530, manufactured by International Microcircuits, Milpitas, California. This can reduce clock-related EMI by up to 20 dB according to the manufacturer. The IMISM530 applied to existing clock frequencies will modulate that frequency, centering on the input frequency, or up to that frequency and can be used to generate multiples or fractions of that frequency. The clock has the following characteristic:

Replicates and modulates externally applied signals
3.0–5.5-V operating supply range
14–120-MHz selectable output frequencies
Output may be equal to a fraction of, or multiple of, input frequency
Will accept input frequencies of 14–120-MHz
TTL-or CMOS-compatible outputs with 6-mA drive capability
20-pin SSOP, or "skinny dip" package
Frequency spreading with $F_{in}$ center frequency
Frequency spreading with $F_{in}$ maximum frequency of spread
Compliant with all major CISC, RISC, and DSP processors
Low short-term jitter
Synchronous output enable
Power-down mode for low power consumption
Locks to externally applied signal

Another source of spread spectrum clocks is Cypress, San Jose, California.

Matching the termination impedance to the characteristic impedance of a signal track is often said to reduce the magnitude of radiated emissions. In fact, this is seldom the case. Matching the termination is useful in improving signal quality but only serves to increase the current flow in an electrically short signal and return track. An electrically short signal and return track is one in which the propagation delay of the track is low compared to the duration of the logic transition. With increased current flow, the radiation increases.

At low frequency, the current on an electrically short PCB layout is determined by the load impedance. The load impedance presented by the logic is made up of some series IC package and die lead inductance and the device capacitance and equivalent resistance in parallel. The device components are active and dependent in many devices on the applied logic level. The impedance above $\approx$30 MHz tends to be dominated by the device input capacitance; above $\approx$100 MHz, the impedance is low. At high frequency, the characteristic impedance of the line plays a role in the magnitude of current flow. With a relatively low load impedance, the higher the impedance of the line the lower the signal current flow and the lower the radiation. This may be achieved with a constant distance between signal and return tracks by choosing a substrate material with the lowest possible permittivity.

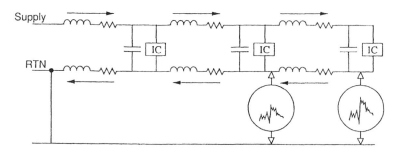

**Figure 11.19** Common-mode voltage developed across a PCB.

## 11.6.  PCB GROUNDING

### 11.6.1.  Common-Mode (c/m) Voltage Developed Across a PCB

Figure 11.19 illustrates a typical ground configuration on a PCB containing logic in which power return and signal ground are the same. The transient currents develop voltages in the impedance of the signal return. Due to the correct use of decoupling capacitors, the differential-mode voltages developed are minimal. The voltages that are developed are common mode, as illustrated by the noise waveforms shown in Figure 11.19. The greater the distance from the chassis ground connection, the higher the common-mode noise voltage. In Figure 11.20 the noise voltage developed at IC U6, referenced to point A, the power return, appears equally on the signal return connection (common mode). A common-mode voltage will also appear on the return of IC U7, and thus a differential voltage will appear between U2 input and the return of U6.

When the noise pulse duration and amplitude result in noise energy that is higher than the noise energy immunity level of U7, then EMI occurs. In one case of EMI, transients of 1.5-V magnitude were measured between two ground points on the board and at the input of an IC. One common technique used to reduce the impedance of the signal return is crosshatching, as shown in Figure 11.21. The lowest impedance is achieved by use of a ground plane, the impedance of which is given in Section 5.1.2. As described in Case Study 11.1, the impedance of a practical ground plane made up of connected segments may be much higher than expected.

When a ground plane is used on a board containing ICs, the copper must be removed under the pins of the ICs. Very often the removal takes the form of a slot, as shown in Figure 11.22, and the impedance of the ground plane is dominated by the impedance of the PCB material between the slots, which is considerably higher than that of a solid ground plane. The inductance of the return path is reduced by the proximity of the signal path, and the inductance of a ground plane is much lower than that of a narrow PCB track. It is the lower inductance

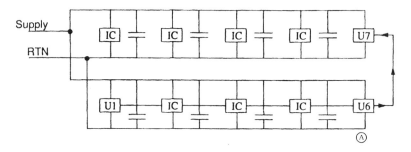

**Figure 11.20** Poor signal/power return layout.

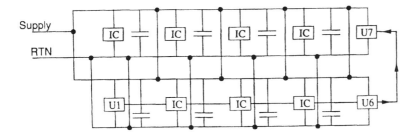

**Figure 11.21** Improved signal ground layout.

and, therefore, impedance of the microstrip PCB ground plane that resulted in up to a 36-dB reduction in measured emissions compared to an open transmission line. In the measured configurations, the same distance existed between the signal and return tracks as between the signal track and ground plane in the microstrip, and the only difference was the ground plane.

## 11.6.2. "Good" and "Bad" PCB Ground Planes

Often the ground plane is constructed of crosshatched tracks. When these comprise short, thick lengths of track, the impedance can be almost as low as for a solid ground plane. However, when the tracks are unduly long and narrow, the impedance will be significantly higher than that of a solid ground plane.

The majority of the signal return current flows in the return path directly beneath the signal trace. This means that when the width of the return trace is five times or more the height of the signal trace above the return trace, an effective ground plane has been achieved for that single signal connection. This concentration of current under the signal trace is again due to mutual inductance. This means that the width of the ground plane can be limited to those areas of the board where signal and clock tracks exist. In most boards, which have tracks over the whole area of the board, the ground plane must also cover the whole area of the board.

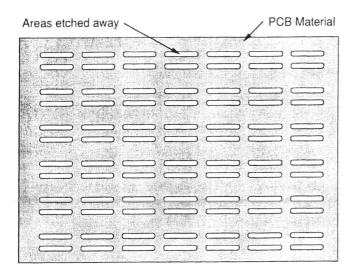

**Figure 11.22** Typical ground plane with PCB removed for IC pins.

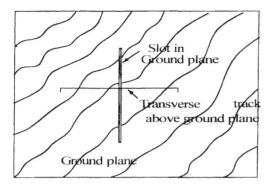

**Figure 11.23**  A slot in the ground plane interrupts the signal return current under the signal track.

Figure 11.23 shows a vertical track located on the same layer as the ground plane that creates a slot in the ground plane. Above and at right angles to this ground plane layer track is another signal track. The return signal current in the upper track must now spread out due to the slot in the ground plane, and the ground plane impedance is increased.

In the measurements described in this section, a signal source was made up of a battery, a regulator, a clock oscillator, and a driver IC, all contained in a shielded enclosure, and the signal output was used to drive the PCB. In the construction of the shielded enclosure used to house the source, the enclosure must be connected to the ground planes in the microstrip, stripline, or image plane PCBs. Figure 11.24 illustrates the connection of the enclosure to the signal wire from the source and the ground/image plane. The signal return connection is made from the PCB plane to the feedthrough in the enclosure via a short length of wire, as is the signal connection to the signal trace. To minimize radiation from these short lengths of wire, a small copper doghouse enclosure covers both the feedthrough connections and the lengths of wire. This doghouse extension, which can be seen in Figure 11.24, is used to connect the enclosure to the ground/image plane of the PCB. In the connection of Figure 11.24, none of the signal return current flows in the connection between the enclosure and the PCB ground plane.

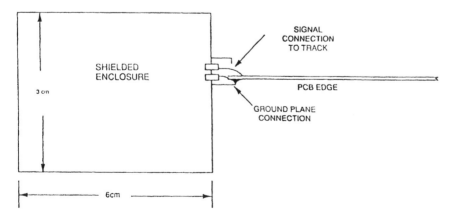

**Figure 11.24**  Separate signal return and the connection between the ground plane and the enclosure.

The grounding experiment involved connecting the enclosure to the PCB via a short length of wire that was connected to the feedthrough and then via the signal return wire to the PCB. This grounding technique is illustrated in Figure 11.25. It can be seen that the signal return current does flow on the connection between the enclosure and the PCB ground plane.

The radiated emissions from the different types of PCB layout were measured with this alteration in grounding. The level of radiation from the low-emission type of PCBs (microstrip/stripline) increased from 56 dB below the high-emission types to 6 dB below these high-emission types. The reason for this dramatic increase is that an RF potential difference is developed between the enclosure and the ground plane/s or image plane. This potential difference is caused by the signal return current flow on the short (5-mm) length of wire used to connect the PCB to the enclosure. Although the wire is short, it exhibits significant inductance and it is across this inductance that the voltage is developed. In contrast, in the grounding connection shown in Figure 11.24, the enclosure-to-ground-plane connection is low impedance; it does not carry the signal return current and so no RF potential is developed.

Why is the same problem not seen with the semirigid cable connection shown in Figure 11.26. No potential difference is developed on the outside surface of the semirigid shield because of both the mutual inductance between the center conductor and shield and the skin-depth effect. The skin depth ensures that above some relatively low frequency, the majority of the return current flows on one surface of the shield, and the mutual inductance ensures that this is the inside surface of the shield.

The problem of RF voltages developed between PCB grounds and equipment grounds or the enclosure has been seen in a number of practical EMI cases. Two of these are presented next.

The first case study involved radiation from a PCB, which contained digital and RF circuits, to a nearby PCB, which contained sensitive RF circuits. The source PCB generated emissions that were, at certain frequencies, 40 dB above the FCC class "A" limits! The board layout, however, was exemplary. It contained all the "good" features, such as effective ground planes and stripline configurations for signals that were potential sources of emissions. The problem

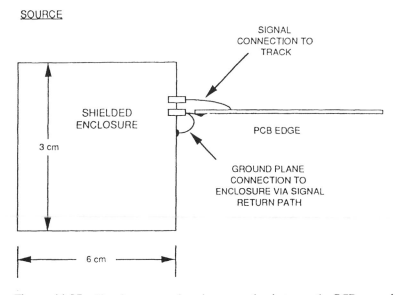

**Figure 11.25**  Signal return used as the connection between the PCB ground plane and the enclosure.

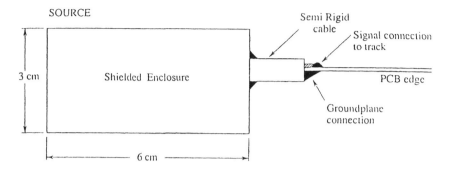

**Figure 11.26** Semirigid cable connection, used with the ECLinPS device, to the microstrip ground plane.

was due to a thin trace used to connect the RF and digital grounds together at a single point. The thin trace carried the return current of a digital signal driving a device on the RF board. The two PCB ground planes and single-point connection are shown in Figure 11.27. The return current generated a voltage drop in the inductance of the return trace, so an RF potential was developed between the ground planes. This is the same problem as seen in the grounding experiment, in which the potential was developed between the ground plane and the enclosure. Many modern PCB antenna designs use a similar layout, driving the antenna at a similar point, and achieve a high gain, that is efficient radiation.

A significant reduction in radiation was seen simply by widening the thin signal return trace. Widening the trace reduces the inductance and the voltage drop in the return path. The width of the common ground connection between the digital ground and the RF ground is kept narrow to minimize noise current flow from the digital ground into the RF ground, used by sensitive RF circuits. However, the width of this single connection can be increased when sensitive circuits are located far from the connection or when digital signal return current is routed away from the location of the single-point ground. For example, ensure that digital return current does not flow between devices located across (at right angles to) the increased width of the ground plane connection.

An even more effective solution is the use of a semirigid cable connection, as shown in Figure 11.28, or a stripline PCB structure installed across the gap, Figure 11.29. The stripline is not as effective as the semirigid cable due to some leakage through the gaps between the vias used to connect the upper and lower ground planes together.

Another potential solution is to disconnect the two ground planes and to use an opto-isolator to transfer the digital signal to the RF section, as shown in Figure 11.30. Separating the two ground planes is very effective in reducing current flow between them. However, if an RF potential exists between the planes, as it almost certainly does, then the level of radiation

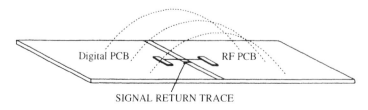

**Figure 11.27** PCB with digital and RF ground planes connected together with a thin track.

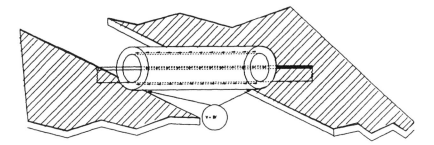

**Figure 11.28** Use of a semirigid cable to connect the ground planes together.

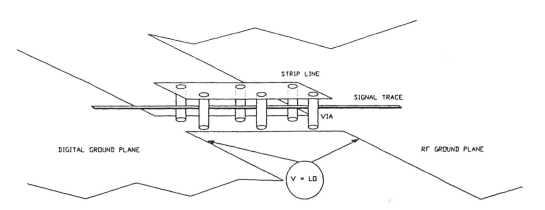

**Figure 11.29** Stripline connection between two grounds.

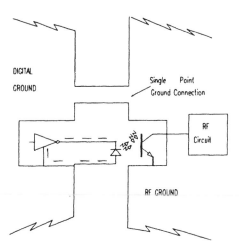

**Figure 11.30** Use of an opto-isolator to keep the signal return current away from the single-point ground connection.

may remain high or may even increase with separated grounds. The solution is the use of the opto-isolator, in which the digital return current is forced back to the digital ground plane on a trace, which is disconnected from the RF ground, and the digital and RF grounds are tied together at a single point. Negligible current flows in this common connection, and the two ground planes are close together in potential.

Another common example of RF potential developed between PCB ground planes is when signal and signal return connections exist between two or more PCBs. These connections are typically made via pins in edge connectors and a motherboard or back plane. It is important that as many parallel signal return connections as possible be made and that these connections be as short and as wide as possible to reduce the inductance of the signal return and, therefore, the potential difference between grounds. Where possible locate a return pin adjacent to every "hot" signal pin.

Some of the worst violations are seen in boards that are mounted piggyback on a PCB, with signal power and return connected by long, thin pins. This configuration is shown in Figure 11.31. The solution is to use PCB edge connectors for the piggyback PCB and to make the interconnections via a microstrip PCB (single ground plane) or stripline (upper and lower ground planes with tracks sandwiched between the PCB ground planes). If external cables are connected to the two PCBs, then the RF potential between the two ground planes appears on the signal connections as a common mode voltage, regardless of whether the signal is balanced or unbalanced with reference to ground. The two cables result in a very efficient antenna, for the RF potential between them drives one cable against the other, similar to the two rods of a dipole antenna. In this example, the use of finger stock or a similar contact to connect the two ground planes together will reduce the RF potential between them. Ideally, this common connection should be made at the location where the two cables leave the enclosure or at least where they leave the PCB. To reduce radiation from the cables due to a potential difference between the PCB ground plane and the enclosure, finger stock or a direct connection may be made between the ground and enclosure. If a DC connection is not allowed, an RF (capacitive) connection should be made.

Another practical example of EMI was partially due to an RF potential between a ground plane on an approximately 8-cm × 4-cm digital PCB and an RF ground. The digital PCB was located approximately 0.5 cm from a receiver contained in a small, well-shielded case. The small microstrip antenna connected to the receiver was approximately 2 cm from the digital board and was referenced to the receiver case. Due to radiated coupling between the digital board and the antenna, the receiver exhibited desensitization and narrowband and broadband spurious response. Part of the problem was due to the RF potential between the digital ground plane and the RF ground (shielded case). By shorting the case to the digital ground plane at a number of points with finger stock material, the radiated coupling reduced significantly, although not enough to totally cure the problem.

One technique for joining different ground planes on a PCB is shown in Figure 11.32.

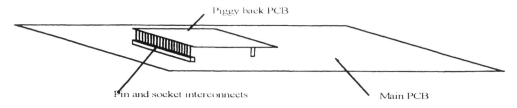

**Figure 11.31** Piggyback connection of PCBs using pins and sockets.

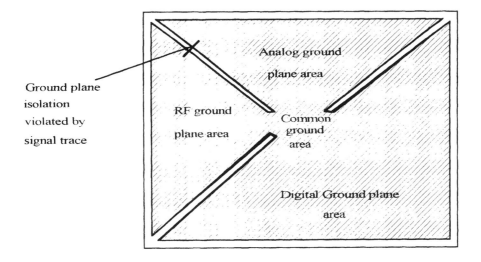

All signal, traces and power plane interconnects
made in common ground plane area

**Figure 11.32** Ideal large-area ground connection on a PCB violated by a signal trace that is not located over the common ground area.

Here a large surface area is used to make a common ground connection. The common failure with this technique is to take a signal track between the two different ground planes at a location other than the common ground area, as illustrated in Figure 11.32. With this location of the signal track, the return current must flow through the one ground plane down to the common ground connection and back to the source. A large loop is thus formed, and the level of radiation due to this loop is extremely high.

In conclusion, the use of low-emission type of PCB layouts in conjunction with ICs that generate only low-level emissions can reduce radiation to levels where shielding of the PCBs is not necessary to meet commercial radiated emission limits. However, when PCB-to-PCB or PCB-to-enclosure/ground connections are high impedance and carry RF current, then none of these reduction techniques may be sufficient without the use of an effective grounding scheme.

### 11.6.3. Grounding a PCB Within an Enclosure

Figure 11.33a illustrates a grounding scheme for a board with a signal that is connected to an attached cable. The signal ground of the board-to-enclosure connection is made at the point at which the interface signal exits the board. If this connection represents a sufficiently low impedance, then the noise voltage developed across this ground connection will be low, as will the C/M current flow on the cable.

Many engineers are puzzled by the need for only one connection of the signal ground to enclosure to reduce C/M noise voltage. But by definition, because C/M noise appears equally on the signal and signal ground, it requires only a connection of the signal ground to enclosure to reduce the magnitude of C/M voltage. One secondary effect of this connection is that some of the common-mode noise may appear as D/M noise in the signal.

In practical cases the grounding may be made from the PCB to a connector housing, which is in turn connected to a metal faceplate, which may then be connected via two small screws

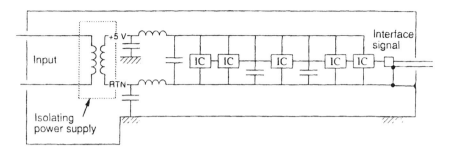

**Figure 11.33a** Signal-ground-to-enclosure connection at the location where an unshielded cable exits the enclosure.

to the main frame of the enclosure. Due to the relatively high impedance of this connection, a sufficiently high noise voltage can be developed to drive the attached cable with enough current to fail emissions requirements.

If more than one cable is connected to an enclosure and common-mode noise voltage appears on one or another of the cables, then one cable can "drive" the other, and fields are generated between the cables in addition to those between cable and enclosure. Try to connect the shields of cables together at a single point with a low impedance connection. If cables are unshielded, connect reference signal grounds within the cables to a single point on a PCB. If shielded add an additional low-impedance connection between the cable shields. Better still, tie them to a "clean" chassis/enclosure ground, preferably to the outside surface. Digital designers and compliance engineers very often experience out-of-specification emissions over the 40–300-MHz frequency range on boards with attached cables, even when the fundamental frequencies are much lower. If the PCB signal ground-to-enclosure connection is made at the wrong location, the common-mode voltage developed across the board appears between the cable and the enclosure. The cable then radiates in the same way as a monopole antenna, with the cable current increasing with increasing frequency until the frequency at which half a wavelength equals the length of the cable. Above this resonant frequency, the cable current tends to reduce and the radiated E field exhibits maximum and minimum values.

In other cases where either no connection or a poor connection of signal ground to the enclosure is possible, a C/M inductance wound on a ferrite bead, or toroid, in which both the signal and signal return are passed through the ferrite, will present an impedance to C/M current flow. One location for the balun is where the signal leaves the board, with an alternative location on the cable external to the board. Section 5.1.10.4 discusses the properties of the balun. Avoid threading the signal and return conductors through separate holes in a ferrite bead, for then signal degradation can occur in the megahertz frequency region.

If the balun is located on a PCB inside an enclosure, radiated coupling from the PCB can induce current in the balun. If this is found to be the case, enclosing the balun in a small, shielded enclosure may help, as shown in Figure 11.33b. Do not, however, connect the enclosure to the noisy digital ground. If a direct connection of the signal ground to chassis is not allowed, then the implementation of an RF ground using one of more capacitors in parallel may be a solution, as shown in Figure 11.33c. Using a C/M inductor in conjunction with the capacitive RF ground can lead to resonance effects, as described in Ref. 1, which can be counterproductive. If the location of the C/M inductor or balun is on a high-impedance section of the attached cable, then the level of reduction in C/M current can be negligible, but a direct or RF ground at that point will be very effective. If the balun is located on the cable, moving its location to a low-impedance section can improve the performance.

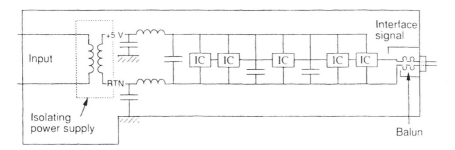

**Figure 11.33b**   Shielded balun inside enclosure where an unshielded cable exits the enclosure.

An alternative approach for digital signals that constantly change state, for example, the 1553 bus, or that go into a high-impedance "off" mode in the absence of a signal is a custom-made transformer that contains an electrostatic shield between primary and secondary. The shield is a turn of conductive foil, aluminum or copper, wound between the primary and secondary, with the ends of the foil insulated; i.e., the shield does not represent a shorted turn. The electrostatic shield must then be connected to a " clean" ground, which is typically the enclosure or chassis. Connecting the shield to the noisy digital ground may actually increase the C/M noise current. With the shielded transformer, C/M noise currents flow via the shield into the ground and not on the cable. If PCBs are contained in unshielded enclosures, then the RF potential developed between the PCBs must be reduced by connecting the ground planes on the PCBs together at the location where interface cables are connected to the board. An RF potential difference between the ground planes of two or more PCBs to which cables are connected results in one cable driving another and a very effective radiating antenna.

The C/M current flow on the power cord may also result in excessively high radiation from PCBs and attached cables. Here the C/M current must be reduced by the use of effective C/M filters or a C/M inductor on the AC safety ground.

In the case where PCBs with a front panel are contained in an enclosure and where the front panels are connected together down the edge with finger stock or spiral gasket or the like, a number of considerations must be made as to the connection of the digital ground to the front panel. Where unshielded cables are connected to the front panel, the decision is relatively easy, for the digital ground should be connected to the front panel with the use of baluns on signal and returns to reduce the C/M current flow. In the case where shielded cables are connected to the front panel, two choices exist and the right decision is not always obvious. If the digital ground is connected to the front panel and also to the chassis at some other location in the

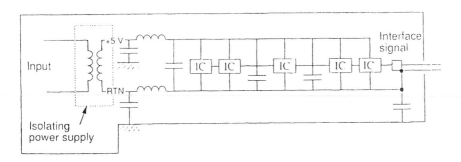

**Figure 11.33c**   RF connection to the enclosure using one or more capacitors.

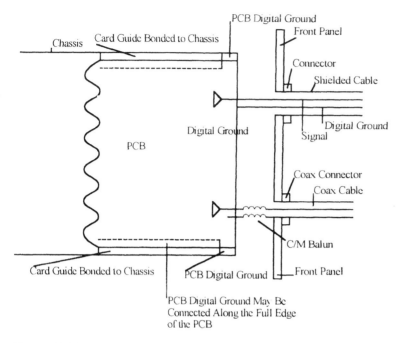

**Figure 11.33d** Method of isolating the digital ground from a front panel and at the same time connecting the digital ground to chassis to reduce C/M current on signal interfaces. Suitable for shielded cables.

equipment, high C/M currents can flow in the front panel, resulting in a voltage drop between the front panel and the rest of the enclosure. In extreme cases the radiation from the front panel alone, without an attached cable, can be close to Class B commercial radiated emission limits. This effect can be demonstrated by using a spectrum analyzer to compare the voltage drop between two locations on the enclosure at a set distance apart and between the enclosure and a front panel. A second method for finding this problem, which provides a very convincing demonstration, is to make a radiated emission measurement, without cables, and then to connect a wire in electrical contact with the front panel. If emissions go up significantly, the cause is either the front panel voltage driving the wire or emissions from an aperture coupling to the wire and causing a current flow on the wire. If, on the other hand, the digital ground is floated from the enclosure, then C/M current on the shielded cable center conductor/s can result in cable radiation when the cable shield or shield termination or connector backshell is not adequate.

A possible solution to both problems is shown in Figure 11.33d. Here the digital ground is isolated from the front panel. Where a single-ended signal interface must connect to the case of a connector (in this example, a coaxial connector), a balun is used to provide impedance between the digital ground and the front panel. To reduce the C/M voltage applied to the signal interface, the digital ground plane is connected to the chassis before the front panel. In Figure 11.33d the connection is made as the ground plane is extended to the edge of the PCB and clamped in a metal card guide. To make this connection effective, the card guide must be adequately bonded to the enclosure and the card guide must grip the PCB with spring fingers.

## 11.7. SHIELDING A PRINTED CIRCUIT BOARD

The radiation in close proximity to a PCB is typically a low-impedance field. In considering the radiation without a shield, our only concern is the field measured at a distance of 3 m,

10 m, or 30 m from the PCB, where the field has an impedance of 377 ohms and the relative magnitudes of the $E$ and $H$ fields at the board are unimportant.

When examining the radiation from PCBs contained in a small enclosure (up to a 19-inch rack size) we find that the magnetic field shielding effectiveness of an enclosure is defined as the ratio of field measured at some location from a source without the shield in place to the field with the shield in place. Due to the switching transient currents flowing on a PCB and the close proximity of the PCB to the enclosure, the field impinging on the inside of the enclosure has a high magnetic field component.

The enclosure approximates a loop antenna, which, despite the low impedance of the field close to the source, generates a field with an impedance of 377 ohms in the far field, the location where commercial EMI measurements are made. The electric field shielding effectiveness of a small enclosure may be very high, and the limit is often the magnetic field shielding effectiveness of an enclosure, which results in not meeting EMI requirements.

The magnetic shielding effectiveness of a small enclosure is adversely affected by high joint DC resistance and contact impedance, which depend on the resistance of the surface finish. For example, Iridite, Oakite, Alodyne, electroplated nickel, electroplated tin, and Dow 20 are all low-resistance finishes, whereas Stannate and zinc dichromate exhibit impedance 10 times higher, with unplated aluminum 10–30 times higher. A copper-loaded paint on the inside of a nonconductive enclosure will exhibit joint impedance a thousand times higher than the low-impedance finish on a highly conductive material. Surface area also has an impact on the joint impedance. The larger the surface area, the lower the resistance and contact impedance. Small gaps in the joint or seam increase the gap inductance and limit the shielding effectiveness of the enclosure at high frequency. The presence of apertures also limits shielding effectiveness. For magnetic field shielding, it is better to have many small apertures versus fewer larger apertures, even if the total area of the removed metal is larger. This is discussed in detail in Sections 6.4.3 and 6.5.3. Although we have considered only magnetic field shielding here, at frequencies above 200 MHz and with relatively large apertures the electric field, coupling through the apertures may predominate.

The use of a PCB-mounted shield covering a small section of circuit or IC, which are both high level emitters, has often been tried, with variable results. If the tracks going through the shield are not C/M filtered, very often the radiation from the tracks connected to the shielded circuit can increase. This effect is very common when adding a shielded enclosure around a PCB to which a cable is attached. If the radiation sourced by the PCB is predominantly from the attached cable, then adding a shield often increases radiation, for the presence of the shield increases C/M current flow on the cable. Connecting the signal return or the shield of the cable to the enclosure where possible or use of the balun or shielded transformer is a possible solution to this problem, as already described in Section 11.6.3.

The use of surface-mount ferrite and/or surface-mount capacitors on all signal and power lines entering the PCB mounted shield will effectively reduce current flow and shunt the current back to the source. The surface-mount capacitors must connect to the shield, because digital ground connections, using tracks, are also sources of C/M noise.

One PCB shield is made in the form of a phosphor bronze fence, which can be bent to any shape. The fence forms the walls of the enclosure, and the top is made of a flat material, which is held in place by spring force. The fence is constructed with pins that go through the board to a ground plane, on the track side of the board, which completes the shield. It is often the poor integrity of the ground plane that compromises the shielding effectiveness. Small gaps are left around the base of the fence to allow components and surface tracks to enter the shield. These apertures limit the shielding effectiveness along with the impedance of the pins of the fence. However, the theoretical magnetic field shielding effectiveness with apertures is still a high 70 dB from 10 MHz to 200 MHz. It is typically the components and tracks entering the

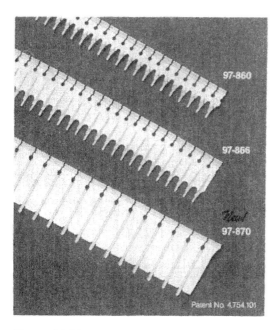

**Figure 11.34a**   Instrument Specialties PCB fence material.

shield that result in leakage and a greatly reduced shielding effectiveness. However, in the gigahertz frequency range, the small apertures do result in coupling. Figure 11.34a shows a 97-870 custom-made fence without apertures that was effective at solving a problem at 1.5 GHz. The PCB shield shown in Figure 11.34b is compartmentalized, which is a useful technique when sections of PCB that contain sensitive circuits must be shielded from sections of PCB that are potential emitters. The PCB fence material is made by Instrument Specialties Co., Inc., of Delaware Water Gap, PA 18327-0136. Figure 11.34a shows the standard product range from this company.

    To demonstrate the importance of low joint and seam impedance we will consider a number of shields designed to go around PCBs. The first shield is made up of two parts, one, with the dimensions 23 cm × 26.5 cm × 1.5 cm, that covers the component side of the board, and a second, with the dimensions 21 cm × 26.5 cm × 0.6 cm, that covers the track side of the board. The slots in the shield are: one at 0.6 cm × 23 cm at one end of the board and two at 1.5 cm × 9.5 cm at the other end of the board. A ground plane on either side of the board connects the two sections of shield together with vias at intervals of 7 cm. Thus, the board contains both large slots and a high-impedance joint. Figure 11.34c shows the end of the board in which the 23-cm-long slot is located.

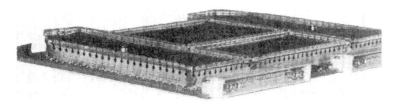

**Figure 11.34b**   PCB shield constructed of Instrument Specialties fence material.

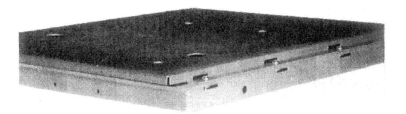

**Figure 11.34c**  PCB shield covers, showing long slot down one end.

The technique used to measure the shielding effectiveness entails driving a PCB, laid out with tracks above a poor ground plane, with loads distributed around the PCB. The signal source is located outside of the shielded room, with the enclosure under test located inside the shielded room. One potential source of error is the cable connecting the source to the on-board loads, which can radiate and mask the radiation from the enclosure. To ensure that this did not happen, the cable was contained within a copper tube that was fitted through the shielded-room wall and soldered to a bracket, which was bolted onto the enclosure. To establish a measurement setup noise floor, the best possible enclosure with regard to shielding was constructed. This was a copper enclosure with a lid soldered to the enclosure and with the seams in the sides soldered together so that no seams, apertures, or slots existed in the enclosure. All other enclosures when compared to the soldered copper reference enclosure showed a lower shielding effectiveness from 100 to 300 MHz.

In the case of the shield shown in Figure 11.34c the transfer impedance of the cover-to-cover interface is not confined to an irridited metal-to-metal contact but includes the impedance of the plated-through connection between the ground planes on either side of the board. However, the degradation due to the contact impedance of the seam is swamped by the shielding effectiveness degradation due to the wide slots. The theoretical magnetic field shielding effectiveness is limited by the 0.6 cm $\times$ 23 cm slot, which is 8 dB at 100 MHz, for the two 1.5 cm $\times$ 9.5 cm slots it is 27 dB. The worst-case measured magnetic field shielding effectiveness of the enclosure at 150 MHz is 3 dB, which is close to the predicted value. One important factor in considering the shielding effectiveness is the directivity of the source, for without a shield the PCB radiates in all directions, whereas with the shield the fields are concentrated at the slots in the end of the shield. In a practical installation, the PCB is plugged into a rack that is shielded on five sides. In our example, it is the long slot in the PCB shield, which faces the open front of the enclosure. In this configuration, it is not uncommon to find that the field measured at the front of the rack with a poor PCB shield is higher than with the PCB shield removed!

The shielding effectiveness of the shield shown in Figure 11.34c can be greatly improved by bridging the long slots with a conductor, which contacts the PCB, the ground plane, and the shield. This may be achieved by bending tabs in the enclosure and bolting the ends of the tabs to the ground plane. If the radiation contributed by the long slots is reduced, the example PCB shield is still limited to approximately 30 dB of shielding effectiveness due to the high-impedance seam. This impedance is dominated by the inductance of the plated-through connections and the screws holding the two covers together. A 7.5-cm slot exists between the screws because the ground planes on either side of the board are connected together only via plated-through holes where the screws go through the board. Although the majority of the current flow on the inside of the shield is caused by the PCB radiated emissions, if the signal ground planes on either side of the board are less than ideal, some fractions of the noise current can flow on the inside of the shield covers. Therefore, it is inadvisable to connect the shield covers to the signal

**Figure 11.34d**   Shield that totally encloses the PCB.

ground plane around the periphery of the board, instead, leave the shield either isolated or connected at a single point, preferably where connections are made to the board (often, edge connectors). If the shield cannot completely cover the PCB, as shown in Figure 11.34d, then the shield cover should contact a track with a width at least equal to the bent edge of the cover. A second, identical track should be fabricated on the other side of the board. Both tracks should follow the contours of the shield edges and be joined together at intervals of approximately 15 mm by vias. The tracks should be either isolated from the signal ground plane or connected at a single point only. The distance of the screws holding the covers together should be no greater than 3 cm. With these modifications made to the PCB shield, shown in Figure 11.34c and the end slots reduced to a cm or less, the shielding effectiveness can be increased to at least 40 dB. Figure 11.34d illustrates a PCB shield that totally encloses the PCB, and it is the shield that is plugged into the guide rails of the card cage. This enclosure exhibits a shielding effectiveness of 60–90 dB from 100 to 300 MHz.

Figure 11.34e shows a PCB shield with 0.9-cm-long and 0.2-cm-wide slots down the edge of the shield. The measured shielding effectiveness of this enclosure is also 60–90 dB over the 100–300-MHz frequency range, although the variation with frequency is different from that of the enclosure shown in Figure 11.34d. The high level of measured shielding effectiveness in the enclosure with small slots is totally in agreement with theory as long as the majority of the noise currents flowing on the inside of the enclosure flow down the 0.6-cm length of the slot and not across the length of the slot. In the two enclosures with a high level of shielding effectiveness, any degradation will be due to unfiltered or unshielded connections on the motherboard or back plane into which the board is plugged.

**Figure 11.34e**   Shield that totally encloses the PCB, with small slots down the edge of the shield.

If a nonconductive enclosure is already manufactured and is designed to enclose one or more PCBs, it is possible to shield by spraying the inside of the enclosure with a conductive coating, such as silver- or copper-loaded paint. With the use of a conductive coating, the lower the surface resistivity, the higher the shielding effectiveness. The weak link is at the joins in the enclosure, which should have as high a contact surface area as feasible to reduce the contact impedance. With the use of conductively coated plastic or plastics loaded with conductive fibers, the measured shielding effectiveness from 50 MHz to 300 MHz is 10–30 dB.

One of the latest forms of PCB shield, called a Conductive Shielding Envolope (patent pending) by the manufacturer, Schlegel, is made of lightweight, pliable, silver-metallized non-woven fabric. The attenuation published by Schlegel for this type of enclosure ranges from 30 dB at 10 MHz to 60 dB at 100 MHz and stays constant at 60 dB out to 1000 MHz. The attenuation quoted by Schlegel is almost certainly for E field and plane waves, as with the majority of manufacturers of conductive material. The shielding effectiveness for H fields is unknown and is strongly dependent on the effectiveness of the joins in the material.

Use the calculations contained in the PCB radiation report, based on clock frequency, logic type, PCB layout, number and length of PCB tracks, and number of PCBs, to find out if the relatively low shielding effectiveness of the conductive-coated type of enclosure is adequate or if one of the enclosures with higher shielding effectiveness is required.

## 11.8. PCB RADIATION, CROSSTALK PREDICTION, AND CAD PROGRAMS

Some of the ideal characteristics of a PCB radiation and crosstalk prediction program are that it be virtually invisible and automatically invoked during CAD PCB layout. The ideal program would calculate the signal integrity and current flow on tracks, based on clock frequency, and predict the near fields as well as maximum horizontal and vertical fields with an accuracy of 6dB or better at distances of 1 m or greater. Most of the commercially available programs fall short of the ideal. Program capability ranges from a prediction of radiation from a simple structure, such as a two-track transmission line some distance from a ground plane or a microstrip line, based on input clock voltage and rise and fall times, to programs that use schematic or CAD PCB layout as the input data. Accuracies range from worst-case 60 dB to average 20 dB. The programs can be separated into rules-based and simulator programs. Some of the simulator programs do not at present offer an exact prediction of the level of emissions from a PCB but, rather, identify "hot spots" in the layout. The two problems with this approach, when used to design a PCB before the prototype has been manufactured, are as follows.

Why bother about the "hot spots" if the unmodified PCB will meet all the radiated emission requirements? This can be determined only by accurate field predictions at a 1-m, 3-m, or 10-m distance from the PCB.

In measurements of emissions from PCB traces at very close proximity, using extremely small (3-mm-wide) probes, high-emission traces can be identified. However, when a larger probe or the small bowtie antenna is used to measure the emissions at a greater distance from the board, typically 10–20 cm, these high-level emissions from a localized area are often seen to reduce rapidly with increasing distance, leaving emissions at other frequencies and from other locations to predominate. It is often the high emissions measured at a distance of 10 or 20 cm that continue to propagate out to distances of 1 m, 3 m, and 10 m. Although the hot spots are a useful indication of problem areas on a board, it is the common-mode current generated by a section of board and the area of the antenna formed by the PCB tracks that determine how effectively the board radiates in the far field.

### 11.8.1. Modeling Techniques

Modeling techniques exist for the prediction of radiation from structures, and some can be applied to PCB radiation. These techniques include the method of moments (MOM) and variations thereof, such as the full-wave moment method, with the capability of analyzing 3D structures, and the conjugate gradient–FFT method (CG-FFT).

Due to the complexity of these techniques, they have all been incorporated in computer programs. Use of a computer program eliminates mathematical error but can result in very large errors because of misapplication of the program and some intrinsic limitations in the method. Some of the programs that use MOM modeling are WIRES, NEC, MININEC, and GEMACS, described in 12.4.3.1. None of these programs allow the modeling of a microstrip PCB with a substrate that has a permittivity higher than 1. In addition, because wires or patches are used to construct the ground plane and current diffuses through these structures, the prediction of the shielding effectiveness of a stripline is in error.

One program, described in 12.4.3.6, is the EMAP5 FEM/MOM code, a hybrid finite element/method of moments code (FED/MOM), which is available from the University of Missouri—Rolla. The EMAP5 program does allow the modeling of the radiation from a microstrip on printed circuit material with a permittivity greater than 1.

The simplest technique, and intrinsically the one least prone to error, is to model a ground plane as a plate to which a wire is connected at one end and then bent to lie above the surface of the ground plane. The far end of the wire cannot typically be terminated on the ground plane, which forms an image plane, for current cannot flow in a plate. Thus, signal return current in the ground plane is not modeled. Nevertheless this technique of disconnecting the wire at the far end has been used with acceptable results.

The simple application of the MOM method may not be able to account for the presence of 3D structures attached to the PCB layout, although the effect of an attached wire can be analyzed. Such MOM methods do calculate current flow on the structure because of connection of a single-frequency source.

The MOM programs require lots of computer time. For example, modeling a PCB ground plane with a small number of wires takes a couple of hours of computer time but results in large errors, because the spacing between the wires is large. If patches are used to form the ground plane, then the accuracy is higher but the program can take days to run on a fast Pentium PC.

The CG-FFT program does predict radiation from a microstrip with a high-permittivity substrate but assumes a symmetrical, infinitely small source. Thus, the radiation due to the differential-mode current flow can be predicted, but not that from C/M current due to an attached structure. This results in a massive 60-dB underestimation of the fields from the structure.

Reference 2 describes an analysis technique compared to measurements on a microstrip PCB attached to a shielded enclosure in a configuration similar to that used to provide the measured data presented in this chapter. The parasitic capacitance, which exists between the microstrip and the box, supplies the return path for the common-mode current, just as it does in the test setup described in this report. The reference proposes a moment method of analyzing arbitrarily shaped 3D conductor dielectric composite structures that allows the radiation due to common-mode current caused by parasitic capacitance to be calculated. The reference reports a good correlation between the predicted and measured fields at a distance of 3 m from the source. For example, the horizontally polarized field at a distance of 3 m from the source parallel to and in the plane of the microstrip transmission line at a frequency of 150 MHz is 47.7 dBuV/m computed and 48.5 dBuV/m measured. The reference also contains a comparison of predicted and measured fields at different locations around the PCB, with a very good correlation. The excellent correlation indicates that in this specific test setup, the predominant source of C/M current is the presence of the shielded enclosure, with little contribution from imbalance in the source.

It is interesting to compare the measured results from Ref. 2 to those presented in this chapter. The major differences between the two test setups is the shorter microstrip, the narrower ground plane, and the larger enclosure used in the Ref. 2 setup. The frequency of maximum radiation is 50 MHz for the 18-cm microstrip and 150 MHz for the 4.5-cm microstrip. In comparing the levels of maximum radiation, the 4.5-cm microstrip is 10 dB below the Class A limit and the 18-cm microstrip is 1 dB below the Class A limit. When the size of the ground plane beneath a microstrip is very much less than a wavelength, the size affects the radiation only minimally. Therefore, the major parameters of interest are the thickness of the substrate, the length of the track, and the size of the adjacent structure. Based on the length of the track alone, and using a 20 log 0.045/0.18 m correction for length, results in a 12-dB difference, whereas the measured peak difference is 9 dB.

## 11.8.2.  Computer Programs Specifically for PCB Radiation Prediction

These programs are seperated into rules-driven design and analysis types. From anecdotal information on an analysis type of program, the prediction was that the radiation from a microstrip PCB layout is higher than the two-track type of layout, indicating that the common-mode current contribution to radiation is ignored. Despite ignoring the C/M current, the prediction was 2–20 dB higher than the measured results for the microstrip. Perhaps the most surprising result was that the emissions at 5 GHz, from a 10-MHz clock source, was only 8 dB lower than the emission at 20 MHz, at the same level as the emission at 200 mHz and 62 dB higher than measured!

### 11.8.2.1.  EMCAD Program

The EMCAD1™ is an MS-DOS-based computer software program supplied by CKC Laboratories Inc, Mariposa, California. This program computes radiated emissions from PCB traces and from cable or wire interconnects, as well as radiated susceptibility, crosstalk (PCB traces and cable and wire interconnects), and filter simulation. The EMCAD1 PCB radiation analysis does account for the common-mode current contribution to radiation from a PCB. The program allows the entry of a simple PCB structure, including:

> Test distance
> Distance to ground plane
> Width of trace
> Distance to signal return
> Number of similar traces
> Dielectric constant

The program allows entry of the parameters of a symmetrical trapezoidal waveform:

> Rise time
> Fall time
> Pulsewidth
> Fundamental frequency
> Waveform amplitude

And the program predicts the 10 highest readings, plotted and in tabular form, compared to a specification, such as MIL-STD-461B.

Using the EMCAD1 computer program to input the PCB data from Section 11.3.1 and Ref. 2 and comparing the result to measured absolute data, the EMCAD1 program predicts emissions up to 28 dB higher than measured. One user of EMCAD1 explained that the error was on the safe side; however, based on the level of predicted emissions, a designer may decide to apply a shield to a PCB that may not be necessary. The most likely reason for the difference between the predicted and measured levels is that the proximity and size of adjacent conductive structures are not accounted for in the EMCAD1 program, although this has not been confirmed by CKC Laboratories.

### 11.8.2.2. EmcScan Program

The EmcScan tool, from Quantic Laboratories of Winnipeg, Canada, is a Unix-based simulator for detecting electromagnetic emissions from PCBs. It complements Quantic's established Compliance tool, which simulates radiation from a number of PCBs with cables and enclosures. EmcScan allows PCB designers to identify hot spots and segments of nets likely to cause problems. It simulates current spectra and track current densities as well as the electric and magnetic fields close to a PCB, using Quantic's field-solver techniques. It inputs data from the CCT SPECCTRA autorouter and all major CAD layout tools. It is not possible to input the dimensions of a simple layout unless this is first entered into a CAD program. The program performs EMC simulation during routing or at the postlayout stage. The EM field simulation and analysis is performed transparently to the user. According to Quantic Laboratories, the program very accurately predicts the relative levels of field and their distribution, although the absolute magnitude of the field at present may not be accurate. However, Quantic is working on this and expects to be able to predict accurately the magnitude of the fields in the near future. For more information, see *http://www.quantic-emc.com*.

### 11.8.2.3. Applied Simulation Technology

The ContecRadia, ContecSPICE, ApsimOMNI, ApsimSPICE, ApsimSKETCH, ApsimOMNI, ApsimALM, ApsimRLGC, and ApsimRADIA-WB programs are available from Applied Simulation Technology of San Jose, California, for use on UNIX workstations as well as IBM-PC compatibles with the Windows NT OS.

ApsimRADIA-WB is a software modeling and simulation program that can predict the radiated emissions from PCB, MCM, cables, and subsystems. The program provides insight into physical placement, electrical specifications, interconnects, and connectors. The engineer can enter physical interconnect locations and simulate their effects from a schematic diagram. The program evaluates the impact of electrical and physical characteristics on EMI using schematic capture, field solvers, and EMI simulators in an integrated system. The program does not accept data from CAD tools. The inputs to the program are simple voltage or current sources or SPICE transistor-level models, and the interconnect models can be simple impedance or delay lines or more complicated transmission lines extracted from the field solver. The program can simulate nonlinear and linear sources.

Once the design is entered it is automatically simulated using ApsimRADIA and ApsimSPICE. The results can be viewed graphically or in text form. The results are compared against an EMI standard of choice. The program can predict emission levels measured at some point defined by the user in the far or near field.

ApsimSPICE is a circuit simulator with lossy coupled transmission-line capabilities. It can predict noise, reflections, and crosstalk in circuits and systems.

Apsim SKETCH is a schematic capture program that serves as front end to ApsimRADIA-WB. It has special built-in symbols for such entities as multiple coupled transmission lines, connectors, vias, and cables. Traditional symbols for semiconductors and macro models are available.

ApsimOMNI is signal and power/ground modeling and simulation software designed to evaluate nonideal ground planes with holes, slits, and cutouts. As discussed in Section 11.6.2, these imperfections can be major contributors to radiated emissions.

ContecRADIA can predict radiated emissions from PCBs and MCMs using current distribution along tracks with nonlinear devices. ContecRADIA is integrated with schematic editors and, unlike ApsimRADIA-WB, is integrated with CAD layout programs. ContecRADIA can consider the effect of EMC site conditions such as reflection and direct waves and horizontal and vertical polarization. ContecRADIA is interfaced with schematic editors to generate EMI design rules and guidelines prior to board placement and layout. A ContecSPICE netlister and a ContecRADIA netlister with track conditions are available for signal integrity and EMI design rule generation. ContecRADIA includes a scanning capability that provides a very fast simulation for entire PCBs in addition to an analysis capability. Designers can identify which track segments are exceeding user-defined EMI restrictions through ContecRADIA reports and on-screen plotting. According to Applied Simulation Technology, ContecRADIA does calculate common-mode-current–induced radiation with and without attached cables and accounts for the presence of a grounded structure. ContecRADIA features include:

Transmission-line waveforms (reflection, crosstalk, etc.) and radiated emission noise.

EMI noise simulator using current distribution along branched and parallel tracks with nonlinear devices.

Electromagnetic field intensity derived using the dipole antenna model applied to the currents simulated by the previous step. The electromagnetic field intensity for each track segment is summed vectorially, and the total field intensity at measurement points can then be calculated.

Detailed analysis and scanning capabilities.

Near- and far-field emissions calculations.

Noise filter modeling using SPICE models or S-parameter measurements together with ContecSPAR.

EMC site model with a peak hold capability indicates the worst-case EMI by automatically sweeping turntable rotation and/or antenna height.

Digital, full logic, and analog circuit simulations.

Arbitrary waveforms.

EMI reports: frequency spectrum, net highlighting, branch, net, board.

Signal integrity reports: geometry report (track length, parallel tracks, length on each stack-up layer), electrical report (impedance mismatch, unit length delay, crosstalk, number of vias), simulation report (delay, overshoot, crosstalk).

For more information see *http://www.cadence.com/thirdparty/ast.html*.

## 11.8.2.4.  Incases Programs

Incases, of Fort Worth, Texas, manufactures a number of software models marketed as EMC-Workbench for the design of printed circuit boards with respect to signal integrity and EMC. The EMC-Workbench package contains four submodules for the complete EMC board simulation:

Prelayout design studies and preanalysis of placement data
Transmission-line parameter calculation and signal integrity simulation (reflection and crosstalk)
Radiation simulation
Three-dimensional display tool

The preanalysis package performs a preanalysis of a placed but unrouted layout using heuristics and geometry constraints. There are two primary tools in this module, MANDI (Manhattan distance estimator) and LDA (layout data analyzer).

MANDI performs a placement preanalysis based on Manhattan distances for all nets, and additionally performs a reflection preanalysis for 2-point nets. LDA performs a reflection preanalysis for routed 2-point nets and simple crosstalk screening based on geometry constraints. The package also includes the two-dimensional field solver TALC, which is used for the calculation of transmission-line parameters. In the interactive mode, TALC allows analysis of nonrectangular-cut sections and calculation of characteristic impedance and phase velocity for various transmission-line structures.

INSIDE (Incases signal integrity diagnostic environment) provides tools and libraries for reflection and crosstalk simulation. The layout data extractor provides a user interface to select which transmission lines are to be simulated. A graphical user interface simplifies usage. Single and coupled transmission lines on multilayer boards with uniform ground planes are analyzed to produce per-unit inductance and capacitance using TALC. The results are stored in a library to reuse them for further simulations. In the interactive mode, TALC allows users to analyze nonrectangular-cut sections and to calculate characteristic impedance and phase velocity for various transmission-line structures.

FREACS provides fast simulation algorithms for reflection and crosstalk effects on multilayer boards with uniform ground planes. An arbitrary number of parallel lines can be analyzed. The simulator input is generated using a simplified graphical user interface, SIGMA/ F. Models out of the EDL (EMC device library) are included automatically using the library interface EXLIN. EXLIN is also used to maintain and customize the library. The basic library delivered with this package contains more than 800 elements of 74 devices in different technologies, most of them completely modeled.

IBIS describes a component and supplies standard models, such as the Pentium IC; alternatively, a device can be modeled using a simple linear model editor.

Incases also supplies a modeling service. A stimulus for the model is defined and the reflections (overshoot and ringing) are displayed. The termination resistance can then be changed or series resistance added and the reflection problem can be solved. The crosstalk analysis is similar. The analysis searches for a coupled trace, based on coupled length and distance, and the criteria for the search for these parameters can be modified. The source waveform and crosstalk waveform are then displayed. The traces and the termination can then be modified to reduce crosstalk.

SCALOR provides a graphical user interface for modification of selected net structures and EMC measures. This allows a trial-and-error simulation within the INSIDE package.

GRADIAN is a graphical postprocessor for 3D visualization of radiation data. It supports the display of current distribution on conducting areas of the PCB or the enclosure, the electric/ magnetic fields, 3-dimensional polar diagrams. Magnitudes are visualized using different colors or arrows of different size, length, and direction. This allows an understanding of the dependencies of different quantities, such as current, electric field, and magnetic field. Thus the electric field can be displayed as different colors and the magnetic field as arrows of different direction and length.

The CORMORAN package is an analysis tool for calculation of the PCB radiation and/or susceptibility. It provides an end-to-end solution for radiation analysis, from the input of PCB geometry from a CAD system, through a fast experimental PCB structures editor, to graphical results display and analysis tools. Data and library management are also handled from within the CORMORAN package. Mechanical geometry of enclosures, heatsinks, etc. can be imported using mechanical CAD system interfaces or can be defined directly in the CORMORAN ASCII input language ESDL (EMC System Definition Language).

A shield can be placed around a PCB with an aperture, and the program takes approximately 30 minutes on a slow UNIX system to calculate the field from the aperture and 5–10 minutes on a faster workstation.

Multiple boards and connectors can be handled. At the time of writing, the PCB-to-PCB modeling via connectors uses only a simple model of the connector and does not model cables. By publication date of this book, a more sophisticated connector model should be included as well as a model for cables. In addition, the later version of EMC workbench should include a model of an IC socket. For the radiation simulation of PCBs, models of the component's input and output behavior and the physical PC board geometry and layer structure are used. The package utilizes I/O buffer macromodules managed by EXLIN and stored in the ED library. These models may be imported into the CORMORAN package in IBIS (I/O buffer information specification) or SPICE format, or can be created within EXLIN by entering the I/O characteristics into a properties window. With over 2800 library parts in the extensive I/O buffer library supplied with CORMORAN, an existing I/O buffer macromedel may exist. The PCB geometry and layout structure is imported through dedicated PCB interfaces in an ASCII-neutral format, SULTAN, and is then read and converted into the CORMORAN binary database by LDE. SULTAN format interfaces are available from all major PCB systems.

CORMORAN calculates the current distribution, in the PCB traces using telegrapher's equations where ground planes exist or using MOM where discrete return paths or finite planes have to be considered. Where MOM is applied, the problem is described by electrical field integral equations (EFIE). The solution provided by either method is in the frequency domain and is converted to the time domain through Fourier transformations. Using predefined observation points, CORMORAN then calculates the electrical and magnetic fields and the poynting vector. Near- and far-field scans can be simulated by defining an array of observation points in space, and polar radiation analysis can be achieved by defining a circle of radiation points on a plane. Digital or analog signals as pulses or sinewaves are described, displayed as scopelike measurements, and the program converts from the time to the frequency domain. The radiated spectrum, near-or far-field scan, and radiation pattern are displayed using ANARES for 2D and 3D display and analysis and GRADIAN for 3D visualization and manipulation. A 2D plot of the radiation and current distribution in a single trace or in multiple traces is displayed. Alternatively, a 3D visualization can be displayed, with the PCB trace and the field around it rotatable. The field around the trace is displayed as a colored plot. If it is clicked on, the magnitude of the field is displayed; if the trace is clicked on, the current in the trace is displayed. Likewise, the direction and magnitude of current on a ground plane can be displayed. Multiple traces can be modeled, with the effect of crosstalk between the traces accounted for. Neither the telegraphers equations nor the MOM methods can directly calculate the effect of 3D structures and common-mode current; instead, an intelligent system is used to make corrections for these effects. For spectrum analysis and comparison of radiation levels against defined legislative limits (FCC, CISPR22, etc.), a virtual radiation test chamber is implemented. A graphical representation of the virtual test chamber is provided with user-definable parameters for the relationship between the floor, test table, and antenna, such as height, angle, antenna position. Users are able to enter antenna factors to define their own antennas, or they can choose from a variety of predefined types. A MIL-STD-462–specified ground plane on the table cannot be directly specified in the radiation test setup but can be entered under the PCB in the input file. The effect of the ground plane on the antenna factor will be included in a future issue of EMC Workbench.

Although a susceptibility analysis is available, it describes only a time-domain field source and calculates the modification in the radiation from the PCB due the presence of the additional source. The change in current on a trace or multiple traces is computed based on trace geometry and input impedance of devices but not the susceptibility of the device. For susceptibility/immunity, the ideal tool would provide the analysis of the current flow on the PCB, due to a single or swept frequency amplitude-modulated source, with the demodulation effect predicted in any analog/RF device on the PCB.

Incases also provides the THEDA family of software packages for PCB, MCM, and hybrid design, including: schematic design and entry, automatic placement, clearance control, interactive routing editing, hybrid and multichip module design, and PCB manufacturing. The PCB systems integrated with EMC workbench, in addition to THEDA, are: Boardstation, from Mentor Graphics; Visual, from Zuken-Redac; SciCards, from Xynetics; Allegro, from Cadence; Power PCB, from PADS; and P-CAD Master Designer, from ACCEL.

An accuracy of 5–10 dB between the EMC workbench analysis and measurements on benchmark PCBs has been reported. Incases provides a 5-day training course as part of the package, although the program appears easy to learn.

For more information see *http://www.incases.com*.

### 11.8.2.5. Allegro Program

Cadence's Allegro is a rules-driven design and analysis PCB and MCM program. The rules-based design is confined to placement, routing, signal noise, timing analysis, propagation delay, capacitance, resistance, inductance, impedance, signal integrity, and backward and forward (near-end and far-end) crosstalk. The Allegro program does not provide rules for radiated emissions or immunity.

### 11.8.2.6. IFS Modeller Program

The IFS Modeller program is a CAE tool for design engineers with high-speed analog system simulation requirements involved in PCB, IC, cable, coax, connector, and similar interconnect design problems requiring arbitrarily placed rectangular and circular shapes. The program enables engineers to model via a GUI. Additionally, SPICE analysis is possible for crosstalk and TDR output. The product features:

Transmission-line analysis of any drawn cross section
Solution of the complex Helmholtz equation for a selected frequency range
Frequency-dependent analysis of dielectric losses based on selected loss tangent
Cross-section reports, which include unit matrices for R, L, C, and G as well as impedance, admittance, and propagation modes
Use of the boundary element method, with the ability to address lossy conductors and dielectrics
Automatic generation of SPICE models for lossless or lossy multiconductor transmission lines for a selected frequency

The supported platforms are: Windows; HP/UX9.x; HP/UX 10.x; Sun Solaris; and SunOS. The IFS Modeller program allows users to proactively conceptualize and visualize.

### 11.8.2.7. Program Comparison and Alternatives

None of the three major manufacturers of the commercially available PCB analysis programs were either able, or willing, to analyze the microstrip or transmission-line layout examples described in Section 11.3.2 of this book, so the author was not able to compare the performance results to the measured results. With one manufacturer the predicted results were very different from the results presented in 11.3.2 or Ref. 2.

If a PCB prediction or rule-based program is not used, either because such a program is not available or because the results have not been satisfactorily validated, then at the minimum a viewer program is extremely useful. The most common program is a Gerber viewer, which provides an easy view of the layout, layer by layer, and simplifies the tracing of potentially ''hot'' tracks throughout the board as well as how effective the ground planes are. All of this can, in a pinch, be accomplished by means of photocopies of the different layers, but the Gerber viewer allows isolation of individual tracks by highlighting a net and zooming in on a specific area of the PCB. Simple, but generally adequate, Gerber viewers are available on the Internet as freeware. A commercially available viewer and plot program, for use on a PC, with higher capability is the GerbTool View/Plot program from WISE Software Solutions, Beaverton, OR. WISE also manufactures the GerbTool designer, which allows the designer to modify the design, after Gerber databases have been generated, and the GerbTool CAM, which is the full CAM station.

## 11.9. PCB LAYOUT CASE STUDIES

### 11.9.1. Case Study 11.1: Grounding Analog and Digital Circuits Sharing the Same PCB

Figures 11.35a and 11.35b show the two sides of a common PCB layout in which the analog circuits and logic share the same PCB. At first glance, the ground plane under the analog ICs appears adequate, whereas in reality the layout is a classic example of poor grounding resulting in common impedance and radiated coupling and crosstalk. Electromagnetic interference was detected on this board by examining the output bits of the 12-bit successive-approximation A/D converter. The A/D converter was packaged in a 32-pin IC, the pads of which may be seen in Figures 11.35a and b.

When examining the output bits of a successive-approximation converter, the type of noise may often be determined by the distribution of the number of times each bit is set after an

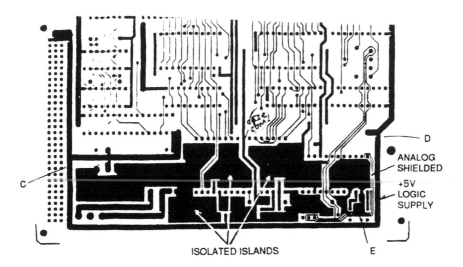

**Figure 11.35a**  Example of an analog/digital circuit with ineffective ground plane resulting in EMI.

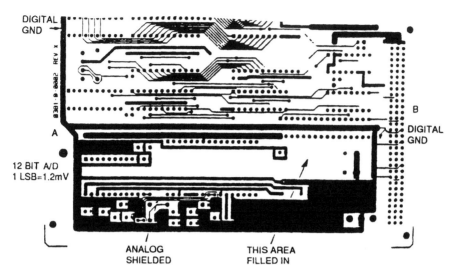

**Figure 11.35b**   Reverse side of PCB with ineffective ground plane.

A/D conversion. A Gaussian distribution indicates that the source of noise is noncoherent (e.g., thermal or $1/f$), in which case reducing the values of resistors and using low-noise op amps may reduce the noise level. When the noise is coherent (e.g., impulsive or CW noise), the A/D output often exhibits missing codes in which certain bits are either not set or are set infrequently.

In the example presented here, a number of bits, not just the LSB, were toggling. The input of the A/D was connected at the edge connector to analog ground, i.e., shorted out, and the differential input to the A/D should have been zero volts. The source of the noise was determined to be the digital logic. On close examination of the analog ground plane in Figure 11.35a it is seen that the analog ground is separated into isolated islands connected by very thin and often-circuitous tracks. These thin tracks are intended to shield the analog signal from the +5-V power rail that can be seen around the periphery of the PCB in Figure 11.35a. The temporary measure taken to improve the grounding was to fill in the isolated islands, by connection of copper foil, on both sides of the board. A reduction in the number of bits toggling was achieved by the improvement in the ground.

The digital ground connection may be seen in Figure 11.35b between points $A$ and $B$. The analog ground also connects to both points $A$ and $B$, so digital noise currents flow in the analog ground. Removing the analog ground connection just before point $B$ effectively floated the analog ground on top of the digital ground, and the number of bits changing state reduced considerably. The thin shields around the analog tracks at location $E$ in Figure 11.35a will provide approximately 6 dB of reduction in capacitive (voltage) crosstalk. Increasing the width of the shield tracks can result in up to 11 dB of reduction. In this crosstalk problem, a low level of noise voltage was measured between the analog ground and the +5-V track, and the major coupling mode is almost certainly inductive. The source is the digital noise currents flowing in the +5-V track. Inductive coupling is unusual in this configuration due to the typically high input impedance of analog circuits. (See Chapter 4 for a description of capacitive and inductive crosstalk.) However, the weighting of the LSB in the 12-bit A/D converter is 1.2 mV, which makes the analog circuits extremely susceptible to noise, especially as a gain stage precedes the A/D converter. The crosstalk problem was solved by cutting the +5-V tracks at locations $C$ and $D$, shown in Figure 11.35a, and using a wire connection in close proximity to the digital ground track in Figure 11.35b. With the crosstalk problem solved, the noise reduced to a level whereby approximately 1 LSB was toggling.

The remaining problem was found to be in the connection of the metal case of the A/D converter to analog ground. The A/D converter pin at point $E$ is the case connection that was connected by a track to analog ground, although the islands on either side of the track were also analog ground. A voltage was induced in the case of the A/D by radiated coupling from the digital logic circuits and tracks. By making a solid analog ground connection to the case, the remaining EMI disappeared and all output bits of the A/D were at "zero" level.

Although the intra-equipment-level problems were solved, as soon as the analog input was connected to an external piece of equipment EMI reappeared, due to common-mode noise voltages between the analog grounds of the two units of equipment. The unit of equipment under investigation was deliverable; thus, only minimum modifications were possible at the PCB level. The immediate solution to the problem was to change the single-ended input to a differential input. This required three video op amp ICs that exhibited a minimum 30 dB of common-mode noise reduction up to 30 MHz. Consequently, the EMI fix was expensive. It was also difficult to find space for the additional circuitry. The lesson to be learned from this case study is that in designing the printed circuit board layout, the equipment and system with the goal of achieving EMC would have avoided delays in equipment delivery and the costly modifications that had to be implemented in the next version of the equipment.

In summary, the improvements were:

a.  The isolated ground islands shown in Figure 11.35a were connected together using copper foil.
b.  Connecting the case pin of the 12-bit A/D converter to the improved analog ground eliminated the contribution of radiated coupling EMI.
c.  Removing the analog ground connection just before point $B$ effectively floated the analog ground on top of the digital ground and eliminated, in conjunction with the improved analog ground, the common ground impedance coupling problem.
d.  Disconnecting the +5-V trace that can be seen around the periphery of the board in Figure 11.35a, at points $C$ and $D$, and bypassing it by means of a wire running across the board above points $A$ and $B$ eliminated inductive crosstalk between noise carried on the +5-V line and the sensitive analog traces shown in the right-hand edge of Figure 11.35a.

### 11.9.2. Case Study 11.2: Good Grounding Technique for a Video Circuit on a PCB

A second video circuit PCB layout is shown in Figure 11.36, the video ground plane of which appears to be as segmented as that shown in Figure 11.35. However, the physical separation in the ground plane is intentional, because the modified differential op amp circuit of Figure 8.17b is used to effectively connect the ground plane islands $A$ and $B$ for differential signals but to isolate the grounds with respect to common-mode signals/noise. The modified differential circuit is also used at location $C$ in order to isolate the video ground at the source from the video ground on the video processing board, which is shown in Figure 11.36, point $A$. The ground plane at $B$ is the digital ground connected to the A/D converter. The advantage of the modified differential circuit in this application is that the resistors connected to one of the two inputs are used to tie the analog and digital grounds together and avoid potentials above the common-mode input range of the op amp. Care must be exercised in connecting the video circuit supply return to the video ground. This connection must be made either at the signal source or to plane $A$ on the video processing board, but not both.

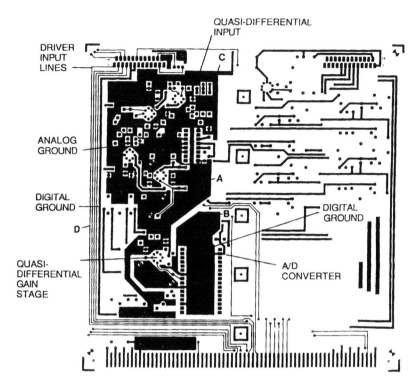

**Figure 11.36** Example of an analog/digital circuit with effective ground planes connected together by differential amplifier circuits.

The tracks at the side of the board, location $D$, carry control signals, which are designed not to change state during video processing and A/D conversion. However, these tracks are likely to have common-mode noise voltages impressed upon them or to carry common-mode currents, with a consequent risk of radiated coupling to the analog circuit. The intent in design verification was to test for the presence of radiated coupling at the prototype board level and then to replace the tracks with six-conductor shielded cables if a problem existed. The shield of the cable would be connected to a "clean" ground, such as structure ground. Tests showed that indeed some radiated coupling existed, but this was at an insignificant level.

The board is used primarily in space applications, with a metal core providing a thermal path to conduct heat away from components. The core is electrically isolated from all circuits but connected to the enclosure, both electrically and thermally, at the card guide.

A noise problem has been encountered in the use of two different boards, both of which contained a metal core connected to the enclosure. When the cores were isolated from the enclosure, the high level of noise, detected again by the toggling of the A/D converter output bits, reduced drastically. The source of the problem is unknown and was initially thought to be RF currents flowing through the PCB core. Assuming that all other PCBs in the same enclosure are sources of noise current flow in the enclosure, isolating the cores of these boards from the enclosure should reduce the noise level on the susceptible board; but it did not. A key to the problem may be the very high capacitance between the core and the analog/video ground plane. The capacitance is dependent on the board and analog ground plane size and has been measured at 3000–5000 pF.

### 11.9.3. Case Study 11.3: Coupling Between Digital Signals and an Analog Signal Within the Analog Shielded Section of a Board

The board shown in Figures 11.37a and 11.37b is another example of an analog and digital mix. Figure 11.37a shows the digital section on the left-hand side and the analog ground plane on the right-hand side, with a single trace connecting the two. This board did not suffer from excessive radiation due to this single-point analog-to-digital ground connection, as described in the grounding section, because of the width of the ground connection trace and because it was made on more than one layer and these layers sandwiched the clock and data connections between the digital and analog sections. However, the analog circuit did suffer from pickup due to the high-speed clock entering the analog section. The analog-section ground plane layout

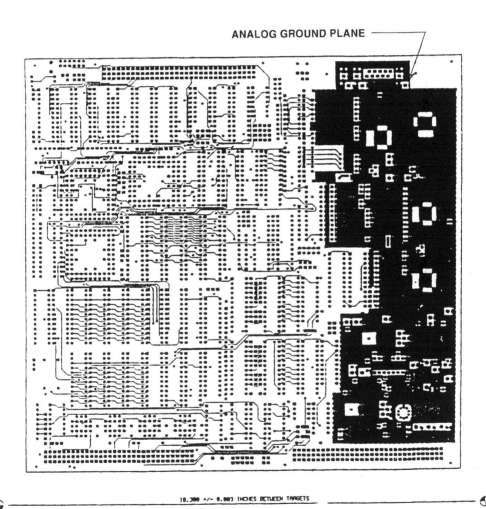

**Figure 11.37a**  Analog ground plane and digital section of board.

+20, −20, & AGND PLANES

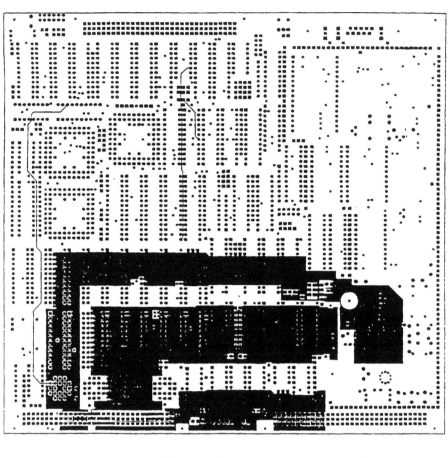

**Figure 11.37b**   Analog power and return tracks inside the digital section.

was designed to enable the incorporation of a gate type of shield around the analog section. When this shield was incorporated, the level of radiated coupling remained the same. The addition of shielding was ineffective because the source of radiated coupling was the clock signal, which entered the analog section and was connected to the D/A converter. Before the radiated coupling mechanism for EMI was discovered it was thought that coupling occurred to the ±20-V power supply and analog ground track connections and the problem was caused by common-mode current flow on the supply and return connections. The reason that this all-too-common problem was suspected was that instead of making the power supply connection on the right-hand connector adjacent to the analog ground, the circuit designer, against better judgment, was persuaded to allow a connection as shown in Figure 11.37b. This type of connection places the analog supply and return in very close proximity to the digital circuits and is a prime candidate for coupling. However, in this case, coupling to the supply and return was not the problem.

The analog section was required to operate at very high speed, with an analog slew rate of 600 V/μs. The circuit designer had tried a high-Q Chebyshev filter at the output of the analog chain in an attempt to filter out the clock signal; however, this degraded the response of the analog output. Instead, a capacitor was tried at the load end of the analog trace connected to the output of the D/A converter to reduce the load impedance and thereby the EMI voltage induced in the signal. The value of the capacitor was chosen such that the time constant formed by the output impedance of the D/A converter and the capacitor did not increase the slew rate appreciably. The EMI problem was greatly improved and other improvements were possible. However, an executive decision was made to eliminate the voltage-to-frequency converter, which was supplied by the output of the analog section, and to use a digital-to-frequency converter, with the hope that the EMI would disappear. The warning was given that with a high level of C/M noise on the digital data to the digital-to-frequency converter, that EMI coupling could occur inside the device; but this was ignored. Sure enough, the EMI was still present, and the designers spent a great deal of time improving the immunity of the digital-to-analog converter, much to the delight of the manufacturer.

### 11.9.4. Case Study 11.4: Out-of-Specification Radiated Emissions from Telephone Equipment

The equipment under test (EUT) was tested for radiated emissions in accordance with EN 55022, Class B (Domestic Information Technology Equipment). The original problem was an out-of-spec emission at 185.00 MHz, which was about 6 dB above the limit, however, our own scan of radiated emissions showed the level to be slightly higher, 10 dB above the limit. Therefore modifications were required to reduce this and several other emissions fairly close to the limit.

#### 11.9.4.1. Circuit-Level EMC Modifications

Several modifications are required to achieve the attenuation of radiated emissions shown in Table 11.4. A shielded power cable is required as well as common-mode filtering of the power with 1000-pF capacitors between each line and chassis ground. This is done for every line that is routed down the cable. The shield of the cable must also be terminated properly with a good low RF impedance to the connector shell as well as to ground at the power supply end. A power-line filter was added between the power supply and the power cable to keep RF out of the power supply and the AC power lines, thereby minimizing radiation from these sources. This was done to prove the effectiveness of the fixes performed on the power cable to the EUT. The connector on the EUT should make contact with the inside of the enclosure to keep coupling to the shield to a minimum. An EMI gasket was also added between the connector and the enclosure to seal the gap on both sub-D connectors on the EUT. The capacitors help by keeping conducted common-mode currents off the cable's conductors, especially at lower frequencies. We used surface-mount capacitors; however, capacitors with longer leads would likely to a good job, although the extended lead length will present a greater impedance at higher frequencies.

When the cables were plugged into the serial connectors, a resultant out-of-specification increase in radiation occurred. Surface-mount 470-pF capacitors were connected between the digital ground in each of the connectors and the chassis ground (and thus the conductive coating on the inside of the enclosure), with a resultant decrease in emissions. Because this fix would be extremely difficult to implement it was decided to try a filtered modular jack that contains capacitors between each line and the metal case of the connector, as well as a ferrite block. One cable was tested alone, because the radiated emissions from this cable were at the highest

**Table 11.4**  Radiated Emissions Measurements Before and After Modifications

| f (MHz) | Radiated emissions (dBμV/m) | | EN 55022 Class B limit (dBμV/m) |
|---|---|---|---|
|  | No modifications | With modifications |  |
| 49.00 | *35* | 24 | 40 |
| 51.88 | *38* | 20 | 40 |
| 55.74 | 32 | 15 | 40 |
| 63.01 | 28.5 | 9.5 | 40 |
| 65.90 | *39* | 7 | 40 |
| 75.73 | 29.5 | 25.5 | 40 |
| 78.83 | 34 | 20 | 40 |
| 83.92 | 34 | 25 | 40 |
| 101.00 | *37.5* | 14.5 | 40 |
| 168.52 | *34.5* | 12.5 | 40 |
| 185.00 | **50** | 18 | 40 |
| 214.55 | 33 | 24 | 40 |
| 253.52 | *44* | 34 | 47 |
| 386.66 | 40 | 22 | 47 |

*Italics* indicates emission level is within 6 dB of the limit (6 dB being our standard safety margin). **Boldface** indicates emission level exceeds the limit.

level of all the four cables tested; thus, if this cable is within spec, the others will be as well. The 470-pF capacitors were removed and the modular jack for the high-emission cable was replaced with a Corcom filtered connector, part number RJ45-8LC2-B. This connector has a metal shield around it, which should be terminated on the inside of the enclosure to the conductive paint. The reduction of the emissions radiated from the cable is 15 dB at 111.00 MHz, however, an out-of-spec emission at 51.88 MHz remained. Adding a 1000-pF capacitor from the digital ground lines of the display connector to the chassis ground drops this emission by about 21 dB. These modifications are required on all of the other modular connectors.

The key to having the modifications to the cables work is that the enclosure is well shielded and the power cable shield and connector shell terminations have low impedance. If not, radiated coupling to the output cables, which will bypass any filtering or high-impedance shield terminations, can easily put emission levels out of spec. Apertures in the enclosure shield can result in this effect. The coating at the two ends of the EUT enclosure exhibited a high impedance, i.e., an aperture with respect to electromagnetic fields and especially magnetic; thus, a consistent conductive coating that allows a good contact at the seams is mandatory. The intentional apertures in the EUT enclosure are acceptable, because, due to their small size, only very high frequencies will couple through them. It is the longer gaps in the ends of the enclosure and at improperly connected seams that will allow the undesired coupling to cables at frequencies that can present a problem. In all tests these open end sections were covered by aluminum foil connected to the remaining components of the enclosure, which were coated with a relatively low-resistance coating. This aluminum foil also covered most of some bad spots in the conductive paint discovered at the ends of the enclosure.

### 11.9.4.2.  PCB Level Radiation Reduction

All of the modifications mentioned in Section 11.9.4.1 are costly and difficult to implement. The manufacturer of the equipment was willing to go to a new PCB layout. Weak areas in the existing PCB layout exist, and so going to a new layout was almost certain to reduce radiation.

In some instances a new PCB layout will eliminate the need for a conductive coating on a plastic enclosure.

The problem areas on the PCB were at U40 and U41 in the power supply and around Y1, the 8-MHz oscillator, as well as coupling resulting in C/M currents flowing on cable plugged into J1, J2, J3, and J4. These area are shown in Figure 11.38a.

The modifications to the PCB layout needed to ensure meeting EN5022 Class B are as follows:

U40 and U41 must be moved to a location over the digital ground plane section of the board and not over the analog ground section. The existing digital ground trace, which is taken into the analog section, is not adequate. The supply trace to the digital section of the board should be brought into the power supply area by a plane, and the decoupling capacitors must be connected by vias between the new power plane and the digital ground plane. Likewise, C60 and C61 must be connected directly by vias between the new digital power plane and the digital ground plane. *Do not use traces to connect capacitors*, and locate these capacitors close to U40 and U41.

Add 0.1-μF capacitors in parallel with the 47-μF C60 and C61, again using vias between the digital ground and power planes, to improve their performance at high frequency.

The oscillator Y1 is located over the analog ground plane close to the J1, J2, J3, and J4 connectors. The power and return to Y1 are made with traces, as are the connections to the decoupling capacitor.

Y1 must be moved as far away from J1, J2, J3, and J4 as possible (at least 6 cm) and remain in close proximity to the IC that interfaces to the clock. Y1 must be located over the digital ground plane, with its return pin directly connected, not by a trace, to the digital return plane. The oscillator decoupling capacitor must be connected between the digital power and return traces by vias, not by traces, and in close proximity to Y1.

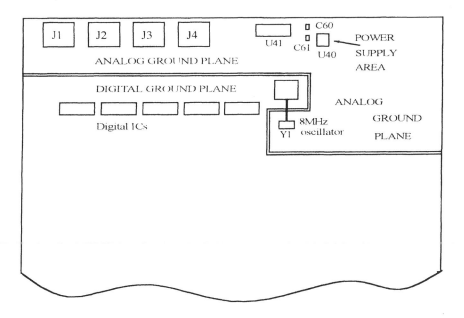

**Figure 11.38a** Problem layout area before modification.

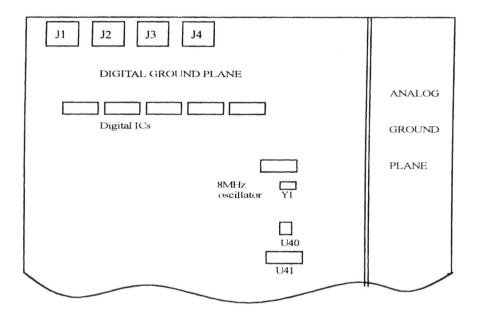

**Figure 11.38b**  Problem layout after modification.

All clock lines and high-speed data lines must be on inner layers sandwiched between the digital power and return traces and not on the solder sides of the board.

J1, J2, J3, and J4 should be located directly over a section of digital ground plane extending under the connectors. Make provision for a 470-pF capacitor between the digital ground plane and chassis at each of the connectors.

Place only low-level, low-frequency devices and traces close to J1, J2, J3, and J4. Keep high-speed devices and clock and data traces as far away from the connectors and signal traces to the connectors as possible.

The new PCB layout was made, with the improvements made in the critical areas shown in Figure 11.38b. In formal tests the equipment met the EN55022 Class B requirements without the addition of any of the circuit-level additions, such as filtered connectors and capacitors. However, the customer complained that it only just passed radiated emission requirements, but then admitted that not all the recommended modifications could be made to the board!

## 11.10.  INCREASED PRINTED CIRCUIT BOARD IMMUNITY

With the advent of the European Union (EU) radiated and conducted immunity (susceptibility) requirements contained in EN50082-1, EN50082-2, EN55024, EN 6100-4-3, etc. equipment either is subjected to common-mode test current injected on signal interfaces and to radiated immunity test fields or will be in the future. In addition to immunity test levels, digital or switching power-supply noise can interfere with low-level signals. The problems of susceptibility of circuits are addressed in Case Studies 11.1 and 11.3 in Section 11.9, on PCB case studies.

At the PCB layout stage, any layout that reduces emissions will also decrease the level of noise voltage induced in the circuit. For example, a good, solid ground plane will reduce the voltage dropped across it as signal current flows and will reduce emissions. Likewise, if C/M current flows in the ground plane, the voltage drop in a good ground plane is low and the level of common impedance coupling is low.

One special precaution that needs to be taken in laying out a board for maximum immunity is to make signal and signal return track lengths, for differential-input circuits, equal. If they are not equal, any induced C/M RF current will exhibit a different propagation delay, or phase shift, between the signal and the signal return track, resulting in a differential voltage at the input of the receiver. Remember, however, that EMI can also result from a C/M current flow into the input of a single-ended receiver, which is converted to D/M by the differences in signal input impedance and signal ground impedance. The most common and effective solution is to place a low-value capacitor, which has a low impedance at RF and a high impedance at the input signal frequency, between the signal input and the ground. Thus at RF the signal input has a low impedance to signal ground and the C/M current, on signal and ground tracks, results in a minimal D/M voltage between signal and ground.

## REFERENCES

1. D.A. Weston. EMC Series Report #1 PCB Radiation due to high speed logic and emission reduction techniques, 1992 EMC Consulting.
2. Shahrokh Daijavid and Barry J. Rubin. Modeling Common Mode Radiation from 3D Structures. IEEE Transactions on Electromagnetic Compatibility, February 1992, Vol. 34, Number 1.
3. Frank B.J. Leferink. Inductance calculations; Methods and Equations. IEEE International Symposium on Electromagnetic Compatibility record, Atlanta, 1995.

# 12

# EMI and EMC Control, Case Studies, EMC Prediction Techniques, and Computational Electromagnetic Modeling

## 12.1. EMC CONTROL

The emphasis in this book has been on prediction techniques useful in designing EMC into equipment and systems. EMC control ensures that the EMC design is correctly implemented at every stage in the development and manufacture of equipment.

### 12.1.1. EMC Control Plans

Control plans are required at both the system and equipment levels. The system-level plan requirements are contained in MIL-STD-464 and equipment/subsystem requirements in the Data Item Description.

The EMC control plan indicates how EMC will be accomplished for a specific equipment, subsystem, or system. Too often the control plan is a general-purpose document reissued with only minor changes for every type of equipment. The control plan should be specific and describe the type of equipment, expected levels and frequencies of emission, level of immunity, EMI budgeting, and potential waiver requests, or any predicted difficulty in meeting the requirements. Details of the grounding scheme, filtering, shielding, etc. should be provided with design goals and breadboard-level tests.

An important section in the plan, and one increasingly required by procuring agencies, is a detailed EMC prediction that can provide information on the effectiveness of the design and highlight any critical areas. The EMC prediction is typically part of the EMC program plan, however, when this is not a requirement, the prediction should be included in the control plan or as an appendix to it.

The EMC prediction and analysis techniques should be described with validation measurements, where possible, that confirm the applicability of the particular technique to the application. The prediction techniques should be used in potential EM problem areas, for determining degradation criteria and safety margins, tailoring applicable EMC documents, and determining the anticipated EM environment.

The purpose of the plan is twofold: (1) to assure the procuring agency that the equipment will be designed for EMC; (2) to provide guidelines and design goals to designers and systems, mechanical, and packaging engineers.

Once an effective and detailed plan has been created, the next step is to ensure that the plan is used by the responsible engineering people. This is achieved by making one individual responsible for EMC and having that person sit in on design reviews as well as the process of signing off drawings. The same individual should develop the EMC control program, control and test plans, and the power and grounding diagram. The contents of a typical control plan are as follows:

Scope
  Purpose
  Organization
  Where the EMC activity fits in the contractors organization
Documentation
  Applicable documents, including the EMC requirements, test methods, grounding and bonding, lightning protection, and safety, as it applies to EMC
  Reference documents, including EMC design guidelines and handbooks
  Documentation requirements
    Test results
    Signal and power interface details
    Power and grounding diagram
    Test matrices
    Test plans
    Test procedures
    Test reports
Management
  EMC responsibility: should contain the responsibility and authority of the individual who will direct and implement the program and the number of full- or part-time EMC personnel on the program
  EMC reviews
  Design requirements: Methods and requirements for ensuring that contractor-developed subsystems and equipment will not be adversely affected by interference from sources within the equipment, subsystem, or system or be sources of interference that might adversely affect the operation of other equipment/subsystems. This is typically achieved by meeting MIL-STD-461 requirements. Some of the specific design areas that should be controlled by requirements are:
    Cable design, including wire categorization and criteria for identifying, labeling, and installing
    Interference-generating or susceptible wires, shielding techniques, and wire routing
    Internal wiring, external wiring
    Enclosure structure, shielding, bonding, aperture closure, and gasketing
    Signal interface design goals and tests
    Intraequipment and Intrasubsystem design considerations controlled by interface control documents when equipment is procured from different sources
    Intersubsystem design considerations controlled by interface control documents
    Radiation characteristics from system antennas, including fundamental and spurious energy and antenna-to-antenna coupling
    Design criteria and required tests for lightning protection
    Impact of corrosion control requirements on EMC and recommendations for resolution of problem areas
    Methods of implementation of design changes required for EMC

## 12.1.2.  EMC Control Program Plan

The program plan provides the data to determine whether all pertinent EMC considerations are being implemented during a program and whether equipment, subsystems, or systems can be expected to operate in the electromagnetic environment without degradation due to effects from electromagnetic energy and without causing degradation to other equipment, subsystems, or systems.

The plan should describe the procedure for managing the EMC program during the project, the EMC organization and responsibilities, lines of authority, responsibilities and role of the Electromagnetic Compatibility Advisory Board (EMCAB), procedures for identifying and resolving potential problems, and a schedule and milestone chart of all EMC activities during the program. The plan should contain the EMC predictions and analysis when this is not a part of the EMC control plan.

The following is an example of one approach that may be taken to ensure EMC in the design of a ship for which a large number of pieces of equipment, not all of which meet MIL-STD-461 or IEC Publication 533 requirements for shipboard equipment, must be chosen. The outline of the program plan is as follows:

1. Characterize the EMC characteristics of the candidate equipment
2. Characterize the predicted electromagnetic ambient

*Step 1:* The following information on the candidate equipment is required:

a. Ship zone in which equipment is located
b. Operational criticality
c. Whether it is an intentional or unintentional transmitter, receiver, or both, for the purpose of EMC
d. EMC requirements met by the equipment, including commercial
e. Electromagnetic emission profile
f. Electromagnetic immunity
g. Shielding effectiveness of the equipment enclosure and cable
h. Power requirements and type of supply used
i. Interface types (RS232, 1553, IEEE-GPIB, etc.)
j. Effectiveness of equipment power-line filter, when installed

*Step 2:* The electromagnetic ambient is likely to vary throughout the ship, depending on the proximity to equipment such as transmitters, electrical control rooms, cables, and the level of shielding provided by metal walls and apertures (doors and windows) planned for each zone. In order to derive a typical ambient for each zone, the following factors should be accounted for:

a. The routing of power, control, and transmitter cables in the vicinity of each zone and the nature of the coupling paths between the cables and equipment.
b. Where the radiated emission footprint for a specific equipment is known, it should be used in characterizing the electromagnetic ambient. If equipment has met radiated emission limits, then these limits may be used.
c. Where pertinent measured data for ships is available from shipboard EMI surveys, this data may be used.
d. In the absence of either of these types of data, published data similar to that provided in this book for similar environments—such as process control—or computing may be used.
e. The conducted ambient may be derived from shipboard conducted susceptibility levels contained in IEC publication 533 for commercial ships and MIL-STD-461 for military. These levels should be tailored when information on the actual ambient is known.

With the EM conducted and radiated ambients characterized, a comparison of the EMC characteristic of the equipment to the harshness of the ambient can be made. Potential EMI situations, in which the equipment does not match the ambient, can then be determined. For example, when a computer that meets commercial radiated and conducted emission limits and that exhibits an immunity level adequate for an office environment is intended for operation in a control room, the potential for EMI is high.

A detailed analysis of each potential EMI situation should be performed to ascertain the probability that the problem exists. If the probability of EMI is high, the severity of the degradation due to EMI should be calculated and one of more changes made to ensure EMC. Some of the changes that can be made to achieve EMC are as follows:

a.  Relocating equipment within a zone (to separate high emitters from the more susceptible equipment).
b.  Adding signal and/or power-line filters to an entire zone or to individual pieces of equipment.
c.  Compartmentalizing a zone into ''noisy'' and ''quiet'' areas, perhaps by incorporation of conductive doors and walls with feedthrough signal and power-line filtering. A related technique is to shield an individual piece of equipment.
d.  Rerouting signal and/or power cables.
e.  Increasing the level of shielding on signal and/or power cables and/or their connectors.
f.  Where feasible, changing an equipment's interface to a higher-immunity type (e.g., changing from RS232 to RS422 or from an electrical to an optical interface).

The topside antenna-to-antenna and antenna-to-cable coupling must also be examined in the EMC control plan, here the techniques described in Section 10.2.1 are applicable.

## 12.1.3. Quality Control

One of the major reasons why equipment that has been certified as complying with an EMC requirement does not continue to comply in a production version is lack of quality control. The other major factor, already discussed, is modifications to equipment design or software.

Any modifications to a design should be under configuration control from a single authority. Appropriate documentation, such as an Engineering Change Notice (ECN), should be used. The ECN must be reviewed by the responsible EMC authority to evaluate the impact of the change to EMC, once the change is accepted, speedy removal of superseded documents should be accomplished.

Any changes to equipment that has been accepted should be inspected and verified and the equipment retested for compliance with EMC requirements when this is considered necessary by the procuring agency. That apparently minor modifications can result in significant changes in emissions and susceptibility, especially when the changes affect the grounding scheme must be considered when making the decision to retest or not. Equipment that has had changes incorporated should be marked or reidentified if the magnitude of the change warrants. Inspection records should be maintained that include as a minimum:

Identification of the item
Date of inspection
Number of units inspected
Number of units rejected
Description and cause of discrepancy
Disposition of material
Identification of the inspector

Additional inspection records that include test results and certification of personnel for special processes, such as soldering and plating, should also be maintained. Inspections should be tailored for EMC. For example, if a ferrite core contains a few pinholes, it is likely to be rejected by incoming inspection, although the pinholes will not affect the impedance of the ferrite and will hardly change its physical characteristics. Far more important would be a test of the impedance versus frequency characteristics that affect the performance of the core.

The performance of filters is not necessarily consistent over different production runs and should be individually tested in a setup that approaches as closely as possible the impedances of the actual source and load as well as the type of noise.

Incoming inspection should include a checklist for incoming material, shop travelers, operation sheets, and/or inspection instructions to describe inspection operations during manufacture. Statistical quality control should be implemented on items not 100% inspected. Material subject to deterioration, such as conductive adhesives or caulking, should be periodically inspected and discarded when beyond its useful life.

The periodical training or retraining of inspection personnel, especially when new processes or materials are introduced, should be instigated. Quality control of sources shall be in place, including a list of approved sources and a performance rating system. The latest drawings or specifications should be quoted on purchase orders, with inclusion of quality assurance requirements such as traceability included on the P. O.

Material not inspected either shall be not released to production or shall be controlled so that if the item is subsequently found unacceptable it may be retrieved. All materials should be identified. Materials not released for usage should be segregated in controlled areas with age control and lot identity maintained.

Adequate test procedures should be in existence for qualification tests of engineering models and preproduction and production models of the equipment. Make sure the test procedure is reviewed and accepted by the customer well before commencement of tests. Records of acceptance tests should be maintained. Control of nonconforming supplies and items shall be such that no out-of-specification component/unit can be incorporated into production equipment. Supplies that cannot be reworked or components that cannot be repaired should either be returned to the manufacturer or be scrapped. Packaging and shipping procedures should be in effect. Sensitive equipment must be adequately identified, specially packed, with the inclusion of an accelerometer to monitor vibration and shock where this is justified.

Measuring and test equipment at a facility should be controlled as follows:

Test procedures for test equipment including calibration limits and description of calibration equipment.

A list of qualified test facilities where equipment is sent should also be maintained.

Standards and calibration equipment used to calibrate test equipment shall be periodically certified.

The test equipment should be kept in an environment with humidity and temperature control and in which no damage or loss of calibration can result.

## 12.2. EMI INVESTIGATIONS

In an EMI investigation the approach used is of paramount importance in finding a speedy and cost-effective solution. The approach should be methodical and thoughtful. This is much easier said than done, especially when under pressure to find a quick fix. In some instances the cause of EMI and its solution are readily apparent and a minimum amount of prediction and diagnostic testing are required. In many cases, however, an immediate answer to the problem is not available, and some analysis and testing are required. One incorrect approach is to try a number of changes in parallel in the hope that at least one will work. I know of one extreme example where equipment was due for delivery but failed an acceptance test due to EMI. The requirement specification was extremely stringent and no waiver was allowed. A systematic approach was not used in finding the source of the problem. Instead, every change imaginable was made to the equipment. Some of these changes may have achieved some small improvement that was masked by other changes, which exacerbated the problem. After three years the problems were solved and the equipment could be shipped. Because no exact records were made of those changes that had effected the improvement, the next version of the equipment exhibited the same problem. Although an extreme case it serves to illustrate the importance of a systematic and well-documented procedure. More commonly, months may be spent in fixing an EMI problem that a specialist may accomplish in weeks, or the fix involves a costly modification or addition to circuits where a much simpler and cost-effective solution exists. Using the right systematic method requires self-discipline and the cultivation of a step-by-step approach.

One argument for blanket changes is that with additional filtering, shielding, grounding, compartmentalization, etc., a guaranteed solution exists. However, as we have seen throughout the book, a limit exists to the shielding effectiveness achieved in a practical enclosure, source emissions may be high, and an incorrect filter or grounding scheme may worsen a problem. Unfortunately, the advantage of the systematic over the blanket approach often has to be demonstrated before it is accepted.

It is imperative that the number of variables that are not part of the EMI investigation be reduced to the absolute minimum. Therefore, do not arbitrarily move a cable position or circuit board location without a ''before and after'' measurement or assessment of the effect on EMI or emission level. Also, do not exchange or modify the software, circuits, equipment power source, installation method, connector or cable type, grounding configuration, or test method and test equipment without a ''before and after'' assessment of the effect.

The first question to ask is: Is the problem EMI? Or is the equipment malfunctioning due to another cause? Practicing engineers often encounter discussions between software and hardware engineers who believe a problem lies in each other's domain. This is also true of the designer who can find no failure in the hardware or software and assumes the problem is EMI. Therefore if, after investigation, EMI is not found, the cause of the problem may be equipment malfunction.

A typical approach is first to isolate the source of the problem and the coupling path or paths by a process of elimination. As discussed in previous chapters, it is common for both radiated and conducted coupling to coexist. Also, the coupling may appear to be clearly conducted, whereas in reality the path is radiated, or vice versa. In an investigation every measurement and modification must be documented, with any change in the level of EMI noted, for as seen throughout the book, and especially in the case studies, a modification that results in a worsening of a problem can tell as much about the cause as a change that results in an improvement.

It is often easier to start with the minimum number of circuit boards, cables, and pieces of equipment that exhibit an EMI problem or above-specification emissions. Once the source(s) and solution(s) to the problems associated with the minimum configuration have been found, the additional circuits, cables, or equipment can then be connected one after the other and any additional problems solved in a stepwise manner.

Bench tests on individual circuit boards separated from the equipment/system are often helpful in isolating and locating circuit-level emissions and susceptibility. Bench tests are particularly useful when circuit-level problems are masked by equipment-level conducted or radiated emissions. For example, relatively low-noise linear power supplies and low-noise signals from an oscillator/function generator may be used to supply the circuit in a bench test.

Once the source of the problem has been found, one or more solutions can be proposed or temporarily or permanently implemented. One problem is to find a modification acceptable to the engineering manager, the packaging engineer, Q.A., and production. One or more of these people often want a fix with no modifications at all!

The following case studies are intended to provide examples of effective EMI investigations that have produced timely results with acceptable modifications to equipment and circuits.

### 12.2.1. Case Study 12.1: EMI Investigation into Intra-equipment Susceptibility

This case study illustrates the importance of understanding the significance of an increase or decrease in EMI when implementing changes to equipment, for once the significance is understood, the source of EMI may be located. The case study also emphasizes the importance of drawing an equivalent circuit in order to find the coupling path.

The equipment under investigation is a spectrometer instrument with an ionization detector contained in a vacuum housing. The output of the ionization detector is fed into the input of a gain of 10 amplifier, the output of which is applied to a discriminator. The second input to the discriminator is a DC reference level, which is adjusted by a potentiometer to discriminate against noise picked up by the ionization detector and amplified by the gain stage. The problem was that excessive noise resulted in a setting of the reference level, which was so high that low-level signals were also discriminated against. The instrument generated high power levels at 27 MHz, and this was considered the most likely source of noise.

The required discriminator reference level was 95 mV before discrimination against the noise was achieved. The noise level referenced to the input of the gain stage (assuming the noise source is a 27-MHz sinewave) is therefore:

$$V_{n(pk-pk)} = 2 \times \frac{\text{Reference level}}{\text{Gain of first stage}} = 2 \times \frac{95 \text{ mV}}{10} = 19 \text{ mV pk-pk}$$

When referring to noise level in this case study, the reference level at the input of the discriminator is quoted and not the noise level at the input of the first gain stage.

The gain stage and discriminator equivalent circuit are shown in Figure 12.1a The differential output of the discriminator is connected to the remaining circuits in the instrument by a ribbon cable with integral shield that was connected to a drain wire in the cable.

Tests and fixes were tried in the following sequence with the following results:

1.  Routing the ribbon cable through a ferrite balun had the effect of reducing the noise level from 95 mV to 75 mV.
2.  Adding a shield over the ribbon cable with the shield disconnected from DC return but connected to chassis at both ends increased the noise level from 95 mV to 103 mV. The main chassis was then isolated from the framework with no decrease in noise level; also, connecting the shield at only one end did not help.
3.  Disconnecting one end of the drain wire in the ribbon cable and connecting it to DC return, i.e., connecting the shield to the DC return, increased the noise from 95 mV to 145 mV.
4.  Floating the shields of the cables connecting the differential output of the discriminator to the ribbon cable from the ground plane at the discriminator reduced the noise from 95 mV to 78 mV.

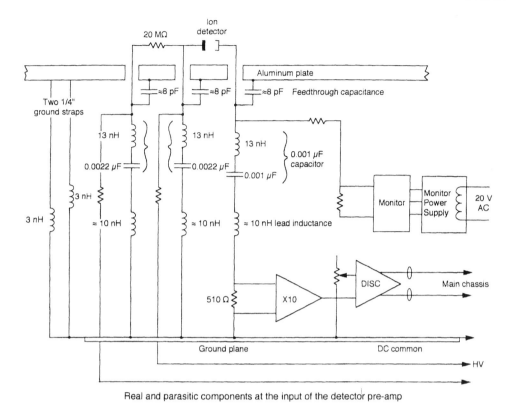

Real and parasitic components at the input of the detector pre-amp

**Figure 12.1a**  Equivalent input circuit of the gain stage and discriminator.

The conclusion to be drawn from the results of the steps 1, 2, 3, and 4 is that a common-mode current is flowing in the ribbon and other interconnection cables, probably in the shields. This conclusion is based on the following:

Adding a chassis-connected overshield that was connected to the ground plane at the discriminator end increases the noise, indicating that the RF currents are flowing in the chassis and on the shield.

The shield current increases with application of the overshield because the impedance of the shield and the chassis connection is lower than that of the ribbon cable shield. The balun has the effect of reducing the level of common-mode current by increasing the impedance of the ribbon cable shield.

5.  In an attempt to reduce RF current flow into the gain stage and discriminator circuits, 0.12-µH chokes were added into the +5-V, −5-V, and return line (because the discriminator output is differential, adding an inductor in the supply return should have no adverse effect), however, the noise increased to 315 mV.

6.  Removing the choke in the return but retaining them in the +5-V and −5-V lines, with additional 1-Ω resistors to eliminate the possibility of series resonance, reduced the noise to its original level. This reduction was entirely due to the removal of the choke in the DC return. Clearly, in the overall grounding scheme the location of the connection of the DC return to the chassis at the detector circuits is important. An attempt was made to measure the chassis current flow in the two ¹/₄-inch ground straps

by use of a current probe. This was found to be impossible due to the approximately 100 mA at 27 MHz measured by the shielded probe when held in the vicinity of the chassis. That is, the probe was measuring the ambient electromagnetic field and not conducted currents.

7.  Disconnecting the input from the detector to the gain stage reduced the noise level to 13 mV.
8.  Replacing the two 1/4-inch ground straps connecting the groundplane to the chassis with a 10-μF inductor increased the noise level to 1 V.
9.  Reconnecting the ground straps and removing the two high-voltage connections to the detector decreased the noise level by 8 mV.

The results from step 7 indicate that the majority of the noise is not conducted or radiated coupling to the gain stage or discriminator but is picked up by the detector. The detector is well shielded so the coupling is unlikely to be radiated. The capacitance of the vacuum feedthroughs was measured at 8 pF on an LCR meter, and these values as well as lead inductances and the self-inductance of the capacitors were included in the equivalent circuit of Figure 12.1a.

A simplified equivalent circuit showing the impedance of the equivalent circuit to chassis is shown in Figure 12.1b. An important impedance not shown in the equivalent circuit is the impedance of the ground plane. The shield of the ribbon cable is connected to the ground plane and therefore RF current flow on the shield flows across the ground plane and to chassis. The majority of the current flows to chassis via the two 1/4-inch ground straps. However, alternative paths exist through the approximately 1.8-Ω impedances of the two high-voltage capacitors and the approximately 663-Ω impedance of the feedthroughs. The current path that results in EMI

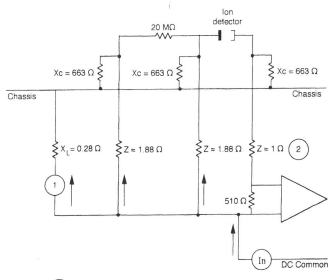

①  10 μH choke ($X_L$ = 1.8 kΩ) tried at this point

②  The 0.001 μF capacitor combined with its self and lead inductance forms a series resonance circuit at ≈ 27 MHz

Note: All impedances at 30 MHz

**Figure 12.1b** Simplified equivalent circuit.

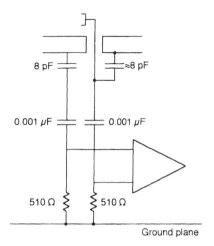

**Figure 12.2a**  Developing a common-mode noise voltage instead of differential.

is via the 510-Ω resistor at the input of the gain stage and to chassis through the impedance of the feedthrough.

The preferred approach in combating EMI is to reduce at source. However, in this case study, as in many others, reduction at source is ruled out due to the difficulty and cost of implementing changes in production equipment. Adding a balun on the ribbon cable was found to reduce EMI but not to any great extent.

A better solution would be to incorporate a 510-Ω (or variable), resistor in series with a 0.001-μF capacitor, in series with an 8-pF capacitor to chassis. The differential input to the gain stage is then developed across the existing resistor, shown in Figure 12.2a, and across the additional resistor shown in Figure 12.2a. Approximately equal noise currents are developed in the two resistors, as is noise voltage developed across them. Therefore approximately equal noise voltages, that is, a common-mode voltage, are applied to both inputs of the gain stage, and the noise appears at a much reduced amplitude at the output of the stage. Another potential solution is to move the physical location of the ground straps to the ribbon cable shield connection and to connect an inductor from the chassis connection to the ground plane. Additional inductors would also be required in the +5-V and −5-V supply connections to the board, because the noise current is common mode. In the location first tried for the connection of the inductor to chassis, a very high voltage was developed across the inductor, and the noise current through the 510-Ω input resistor increased. In the proposed location, a low impedance to chassis for noise currents is achieved by the ground straps, and a high impedance, approximately 1.7 kΩ for a 10-μH inductor, is introduced in the input path of the gain stage. Figures 12.2a and 12.2b illustrate the two fixes.

### 12.2.2. Case Study 12.2: Reducing Radiated Emissions on a Computing Device to FCC Class A Limits

In the initial configuration, the equipment was contained in an enclosure constructed with closed metal sides and a perforated metal top and bottom. The front and rear of the enclosure was open, with a motherboard, located at the rear of the enclosure, to which the PC boards were connected. The boards are plugged in to the front of the enclosure with a metal faceplate mounted on each of the boards. The metal faceplate was not electrically connected to the enclosure, and a gap existed between adjacent faceplates, as illustrated in Figure 12.3

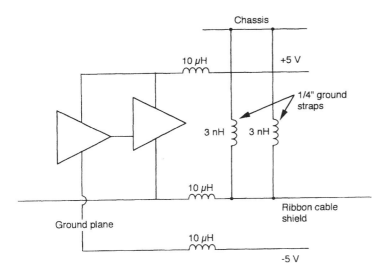

**Figure 12.2b**   Moving the location of the ground straps and adding a 10-µH inductor in the ground plane.

Two coaxial cables are connected to BNC connectors on the metal faceplate of one of the boards. Both cables are used to carry a 1-MHz signal. A second connector mounted on a second board is used to connect an RS232 cable. The additional DC power and multiconductor cable connections to the equipment are made through the rear of the enclosure. The multiconductor connections terminate on the motherboard inside the enclosure, and none of the cables entering the rear of the enclosure are shielded.

The measurements were made in a shielded room using a tunable dipole and a 6-cm balanced loop antenna, which was used to locate specific sources of emissions, i.e., from cables, apertures in the enclosure, and PCBs. A Hewlett-Packard near-field probe was used to locate emissions from individual tracks on PCBs.

The common-mode current on interconnection cables was measured, and the current was then used to predict the radiated emissions from cables.

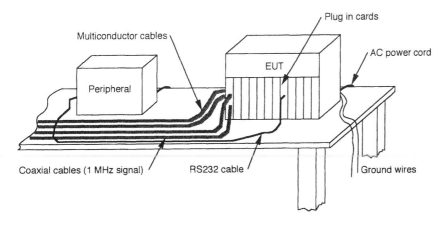

**Figure 12.3**   Equipment test setup for radiated emissions.

The equipment was positioned on a nonconductive table at a height of 1 m above the floor of the shielded room. The enclosure ground and digital ground were separately connected to the shielded room floor via two #16 AWG wires, which is the grounding scheme used in the equipment installation.

Measurements of the radiated emissions from the equipment were made with the balanced loop and tunable dipole. The magnetic field measured by the balanced loop was converted to an E field magnitude at a distance of 30 m (the FCC-specified distance) by a computer program. The dipole measurements where made at the location where the dipole was calibrated in the room, at a distance of 1 m from the equipment. The E field was then extrapolated out to the 30-m distance. The maximum amplitude emissions measured using the dipole and loop antennas were compared. Where a good correlation existed between the two measurement techniques, the maximum amplitude of the two measurements was taken. When a poor correlation existed, the reason, such as room resonance and reflections, were examined and suitable corrections and an engineering judgment were made on the results. Table 12.1 shows the source and magnitude of worst-case emissions converted to a distance of 30 m and compared to the FCC limits.

The major sources of emission were the front-panel-terminated coaxial cables and RS232 cables, with predicted emissions from these sources alone as high as 270 μ V/m at 30 m. The reason for the high-level emissions are not the differential signals carried by the cables but high common-mode current flow on the coaxial shields and the RS232 unshielded conductors. The current flow is caused by noise voltages appearing between the enclosure, which is connected to the shielded room floor, and the digital ground on the boards, to which the coaxial cable shields and signals are referenced.

The shields could not be directly connected to the enclosure, because DC isolation was a requirement for this type of equipment. Instead, for the purpose of diagnosis, a shield made of aluminum foil was constructed over the front of the equipment, with two insulated islands of aluminum in the shield that were connected to the BNC connectors. The islands were bridged by 1000-pF capacitors with short leads, and thus the connectors and shields of the cables were RF grounded to the enclosure but DC isolated. The resonant frequency of the capacitors is approximately 50 MHz and was chosen to coincide with the worst-case emission at 49 MHz. The maximum emissions from the front of the enclosure and the cables with the addition of the shield and capacitors reduced to 31.7 μV/m at 49 MHz, 100 μV/m at 70.9 MHz, and 113 μV/m at 113 MHz, and it was clear that further improvements were required. Shielding the power cable and adding baluns on the cables exiting from the rear of the enclosure reduced emissions further, with only one frequency remaining too close to the limit.

The out-of-specification frequency was traced to PCB tracks using the HP near-field probe. Ideally a ground plane connected to digital ground should be incorporated under the tracks in

**Table 12.1**  Measured Fields Converted to a Distance of 30 m and Compared to the Specified Limits

| Location | Frequency [MHz] | Measured (converted to 30 m) [μV/m] | Specified at 30 m [μV/m] |
|---|---|---|---|
| Front of enclosure and cables | 49 | 403 | 30 |
| Front of enclosure and cables | 52 | 170 | 30 |
| Front of enclosure and cables | 99.8 | 138 | 50 |
| Rear of enclosure only | 82.7 | 60 | 30 |

order to reduce board emissions. The addition of a second ground plane connected to the enclosure but isolated from the board is also likely to reduce emissions.

A makeshift ground plane was constructed from aluminum foil, placed on the track side of the board, and isolated from the tracks by cardboard. The ground plane was connected to the enclosure only. A reduction at 59.9 MHz of 8 dB was achieved by the addition of the provisional ground plane.

Further reductions were achieved by adding a ground plane to the motherboard at the rear of the enclosure or shielding the rear of the enclosure with foil. Closing up the apertures at the top of the enclosure reduced emissions slightly at 70.4 MHz but not at 59.9 MHz.

The initial fixes of the 1000-pF capacitors connected to a totally shielded front to the enclosure were considered impractical by the manufacturers and alternative solutions were sought.

In an attempt to reduce the common-mode voltage appearing between digital ground (and therefore all signals) and chassis, 82-pF capacitors were connected between the digital ground connections of all PCBs in the enclosure and chassis. The capacitors were located at the rear of the edge connectors on the motherboard as shown in Figure 12.4.

The enclosure was improved by adding a metal shield to the rear of the enclosure, after which no backplane was required on the motherboard to reduce emissions from the rear of the enclosure. Additional perforated shields, which allowed air flow, were added to the top and bottom of the card cage inside the enclosure. The unshielded cables at the rear of the enclosure were a source of emissions, and adding a shield around these cables reduced emissions at 83 MHz by 9 dB and at 90 MHz by 8.7 dB. Because these cables were commercially available and were plugged into the equipment, i.e., could be removed and replaced by unshielded cables, unshielded cables had to be used in certification tests, in accordance with FCC guidelines.

Some of the conductors in the cables were unused, and these were connected to the enclosure at the equipment. The ideal location for the connection of these conductors is to the outside of the enclosure, which was impractical. Even with the conductors connected to the inside of the enclosure, at the motherboard, radiated emissions were reduced, but not to the same extent

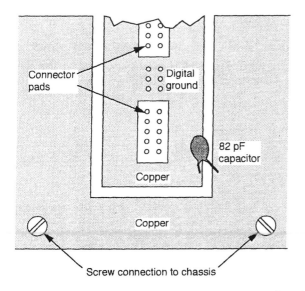

**Figure 12.4**   Method of connecting RF ground capacitors between the digital ground and the enclosure.

as achieved with shielded cables. Ferrite baluns were built into the rear of the enclosure at the location at which the cables entered the enclosure, which resulted in a significant reduction in emissions.

The digital grounds on PCBs were connected to the metal faceplate by 82-pF capacitors, with the faceplate connected to the frame of the card cage via beryllium copper finger stock material. The board, which had benefited from the addition of a ground plane, was relaid out with an additional ground plane layer. These modifications were made to a production piece of equipment, and from measurements made in the shielded room, with the cable arranged for maximum emissions, the prediction was that the equipment would pass FCC limits when measured on an open-field test site. A subsequent certification test, using the identical piece of equipment and identical peripheral equipment on a specific open-field test site, resulted in peak measurements at least 6 dB below the FCC limits.

An emphasis is laid on the correlation of shielded-room measurements to specific open-field test site measurements because of the up-to-26-dB variation measured in field strength from identical computing systems at six sites and recorded in Ref. 1. Although identical systems were measured, it is not clear that exactly the same piece of equipment was measured at all six sites, for if not, a variation must be expected from one piece of equipment to the next.

This case study illustrates that EMI solutions can be found that are acceptable to a manufacturer and that in this case did not require either a totally shielded enclosure or shielded cables.

## 12.3. EMC PREDICTIONS: GENERAL APPROACH

Too often the management and planning of EMC is not included in the electronic equipment or system design. It is only when equipment is adversely affected by EMI or fails an EMC test that a manufacturer is forced to consider either a redesign or some modification to achieve EMC. It may be argued that the incidence of EMI problems has increased over the last 20 or so years, attributable, among other factors, to an increase in the level of electromagnetic ambient, in communications spectrum occupancy, in the number of electronic products, and in the speed of logic and switching-power supplies. Therefore the traditional approach of relying on good design practice without specifically designing for EMC is proving inadequate.

As a result of the increased incidence of EMI, the statement that you either pay for EMC design now or you pay later (and usually pay more) for EMI solutions has become almost a truism.

An EMC prediction should be made as early in the equipment-, subsystem-, or system-level design as possible and then further refined as the design proceeds. The analysis should take into account the sources of potential problem emissions and the coupling path to a potentially susceptible unit. The geometry of the source, receptor, and coupling elements, such as circuits, cables, or apertures, with values for the impedances in the source to receptor path, are important details to be used in the prediction, and therefore the information obtained should be as specific as possible. The receptor may be an antenna in a radiated emission prediction or a 10-µF capacitor or LISN in a conducted emission prediction.

The frequency and amplitude of sources and, for an intraequipment, subsystem, or system analysis, the bandwidth and susceptibility of receptor circuits must be known for a prediction of EMC. The sources are both those inherent in the operation of the equipment, subsystem, or system and susceptibility test levels, if these are a requirement of the design. The purpose of the prediction is one or more of the following: to highlight problem areas as early as possible, to aid in cost-effective design, to support a waiver request. Often when specific information is missing, a simple breadboard-level test may be required, for example, of a susceptibility level, attenuation, shielding effectiveness, impedance, or coupling. Where possible, predictions should be validated by measurements.

If the results of a prediction is within 6 dB of a correctly controlled measurement, then the level of accuracy achieved by the prediction is acceptable. More frequently, an EMC prediction will lie within $+/-20$ dB of measurements. When the correlation between the two is worse than 20 dB, either the prediction or the measurement techniques should be examined for sources of error.

In reviewing the first edition of the book, Mr. Jeff Eckert gave the following telling simile: "Two Texans, both blindfolded, are standing in a cow pasture, surrounded by a herd of steer. The first Texan listens intently for a few seconds and drawls, "I reckon there must be five head of cattle out there." The second Texan takes a deep whiff of the air and says, "There's 500 head of cattle on this here range." They remove their blindfolds and offer mutual congratulations because, after all, both estimates are correct. If you think there is something wrong with that story, consider that the Texans are not cowboys but are EMC engineers from Austin. Welcome to the world of EMC modeling, in which any answer is "correct" as long as it falls within a multiple or dividend of 10 relative to the true value."

Of course, $+/-20$ dB is far from ideal, but it is the, obviously unpalatable, reality. In 22 years of experience of EMC and in reviewing EMI predictions, this is the typical magnitude of the errors I have seen. In one analysis the common-mode current flowing on a surface was ignored as a source of radiation and the error was 61 dB. In another instance, in which the current induced by an incident radar pulse on a cable enclosed in a metal conduit on a composite-skin aircraft was calculated, the error was close to 1000 dB! The problem was that the skin depth of the conduit at gigahertz frequencies was calculated and the transfer impedance of the shield-to-connector and connector-to-enclosure mating surfaces was ignored. Because the conduit carried control signals to the aircraft front landing wheel, this error could have resulted in a disaster. When the modeling techniques in Chapter 6, Refs. 23 and 9, are compared, the correlation between shielding effectivenesses was between 5 and 8 dB, and this was considered by the authors to be acceptable. Also, the errors in using sophisticated PCB electromagnetic radiation computer programs, as discussed in Chapter 11, are unacceptably high. In many cases errors occur due to a simple lack of reliable data. For instance, the very sophisticated modeling and measurements made on the electric field radiated by the WISP experiment on the space shuttle were based on an input power to the antenna of 250 W into a conjugate 50-$\Omega$ load. The reality was that due to the use of an antenna matching unit, the power amplifier could deliver 1000 V into the dipole antenna when it was not a resonant length, resulting in a potential error of 19 dB. The WISP analysis is described in Case Study 12.5.

The goal throughout this book has been to find simple analysis techniques that achieve an accuracy of 6 dB and in many instances the measurement errors are of the same order of magnitude, although I imagine even 6 dB would be unacceptable when counting cows!

### 12.3.1. Case Study 12.3: EMC Predictions for Meeting RTCA-DO 160 Requirements on an "A" Model Fiber-Optic Plotter

This EMC prediction was based on an examination and simple measurements made at the manufacturer's location. The test equipment available was limited to an oscilloscope with current probe and a digital voltmeter (DVM).

The plotter is constructed of $^3/_{16}$-inch-thick aluminum and contains a number of cutouts for the purpose of ventilation and for the paper cartridge. The aluminum was anodized, and when the resistance between the front panel and the side panel was measured with the DVM, an open circuit was registered. The CRT and deflection yoke are completely enclosed by a $^{1}/_{16}$-inch-thick mumetal case, and this will prove excellent shielding against the deflection-yoke-generated electromagnetic field.

High current pulses are generated by the deflection yoke drive circuits, and a mumetal shield had been installed between the yoke drive PCB and the adjacent stepper motor drive PCB. Due to the high level of currents flowing inside the "A" model plotter it is mandatory that the plotter case be improved in its function as an EMI shield in order to meet radiated emission levels.

A smaller plotter (the "B" model) that uses approximately $4^{1}/_{2}$-inch-wide paper has been tested to DO-160A EMC levels and passed, except for some narrowband emissions in the radiated emission test. The "B" plotter utilizes EMC gaskets between mating surfaces and a honey-comb ventilation opening for the fan. It also incorporates differential inputs and some input filtering to increase the noise immunity of the control lines. These differential circuits are examined later in this case study.

### 12.3.1.1. AC-Line Noise Current Measurement

A measurement of noise currents generated by the "A" series plotter was made in order to isolate the source of the major noise generator within the unit, to aid in the selection of a suitable line filter, and to calculate the worst-case radiated emission level.

A PI circuit EMI filter was utilized as shown in Figure 12.5. The filter provides a path for HF current flow via $C_2$ and $C_2'$ and isolates EMI noise present on the AC line from the current probe.

The AC-line noise current shown in Figures 12.6 and 12.7 was measured with a peak current of 50 mA and the fastest edge at approximately 20 μs. The ramp appearing between the negative and positive transitions has a duration of approximately 3.2 ms, and this corresponds to the horizontal sweep rate.

The stepper motor drive was disconnected to isolate any component of the noise waveform that was caused by the stepper motor current, and the noise waveform was unchanged. It was expected that both the stepper motor drive and the HV supply inverter would cause noise currents; however, when the current probe was connected to the +30 V applied to the HV inverter and to the +24-V supply applied to the stepper motor drive PCB, negligible noise was measured.

With the current probe connected around the +24-V supply wire to the horizontal drive

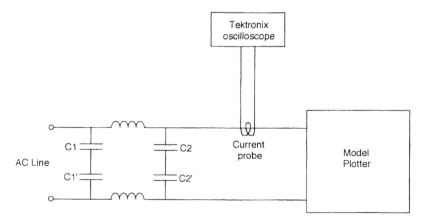

**Figure 12.5**  AC-line noise current measurement test setup.

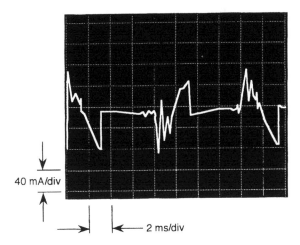

40 mA/div

2 ms/div

**Figure 12.6** AC-line noise current.

PCB, a large 440-mA current spike was measured, as shown in Figure 12.8. This current spike has the same repetition rate as the noise current measured on the AC line. However, although it was expected that the AC transformer, rectifiers, and 27,000-µF capacitor would modify the transformer primary reflected current pulse, it is unclear why the 3.2-ms ramp and trailing edge currents are not seen on the +24-V line.

### 12.3.1.2. AC-Line Filter Selection

In choosing a filter, the equivalent load on the AC line presented by the plotter must be known and is shown in Figure 12.9. The load is not constant and changes impedance because the heaters are controlled by triacs.

The purpose of filtering the AC line is to isolate the noise currents from the AC line in order to meet the conducted and radiated emission tests and to isolate the plotter from both the 200-V transient injected directly into the power line and the electromagnetically induced spike. It may be more economical and space saving to achieve the major lower-frequency, i.e., 10–200-kHz, filtering at the AC input to the system in which the plotter is used and to filter out at the individual units the high-frequency currents generated by the units.

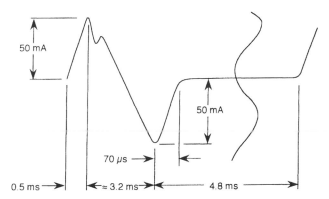

50 mA

50 mA

70 µs

0.5 ms        ≈ 3.2 ms        4.8 ms

**Figure 12.7** AC-line noise current with stepper motor and horizontal sweep in operation.

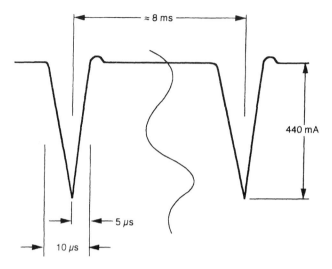

**Figure 12.8** Current spike on +24-V line to deflection yoke driver PCB.

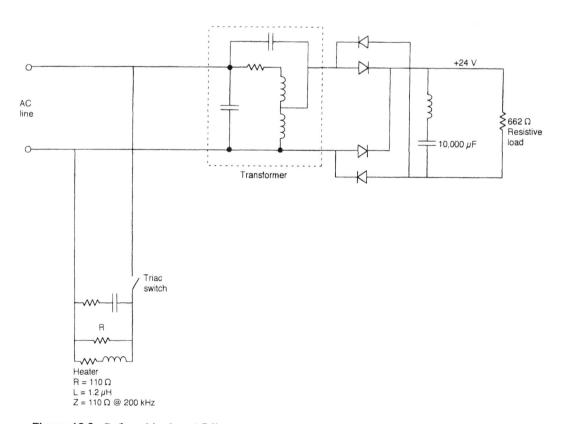

**Figure 12.9** Reflected load on AC line.

In case of the "B" model plotter, a PI filter model was employed; however, this type of filter has caused instruments to fail the radiated emission tests due to high current flow in the enclosure.

A T filter is preferred to a PI because this isolates the noise current both from the AC line and from the instrument into the AC line. The only disadvantage of the T filter is that more noise voltage appears across the AC line internal to the unit. The noise voltage can be calculated on the basis that with one T filter installed on the line and a second on the return, approximately 20 µH of inductance is in series with the line and approximately 0.01 µF across the line. The noise current generated by the plotter results in a voltage of $V = L \, di/dt$ across the inductors. Considering the capacitor to be low impedance and using the fastest edge of 20 µs shown in Figure 12.7, the noise voltage is approximately

$$\frac{20 \times 10^{-6} \text{ H} \times 50 \times 10^{-3} \text{ s}}{20 \times 10^{-6} \text{ s}} = 50 \text{ mV}$$

Because the AC cabling inside the plotter is shielded, it is not predicted that the noise voltages will cause problems internal to the unit, although care should be taken to continue the shield as far as possible on the cable to minimize the length of unshielded cable.

One area where the AC line is unshielded and where of necessity the plotter case contains an opening for the recording paper is at the platen heater. Possible emissions from this area are considered under radiated emission and radiated susceptibility considerations in the following section.

### 12.3.1.3. Radiated Emissions Predictions

*Shielding*

Based on the area covered and the thickness of the aluminum case, it is feasible to make the unit into a shielded enclosure. The type of finish would require changing from anodizing to either Iridite, Oaktite 36, or Alodine 1200, in order to make the surfaces conductive. The fan would require a honeycomb filter, and the ventilation apertures would require wire mesh screens. The corner apertures and other small holes should be covered by copper adhesive tape. The mating surfaces of the front, bottom, back, and sides, which contain apertures between the surfaces, should be made EMI tight by use of EMI gaskets.

The paper cartridge that contains the platen heater, stepper motor drive unit, and paper rolls is inserted into the body of the plotter and is held by a hinge at the bottom and two catches at the top. It is unlikely that a good electrical connection exists between the cartridge and the case, so the cartridge finish should be changed from anodizing to a conductive finish. An EMI gasket should be included at the interface and flexible bonding straps provided between the bottom of the cartridge and the inside of the case. Because emissions levels could increase considerably with the cartridge hinged down, i.e., when the paper is changed, to avoid the requirement of testing with the cartridge down, a switch that automatically disconnects the supply when the cartridge is lowered may be a requirement, or an operational requirement that the recorder be switched off before changing the paper may be sufficient.

The worst-case emission emanating from the AC power-line connection to the heaters and radiating via the approximately 1.5 × 10-inch gap in the shielding at the heater platen has been calculated.

The step function in the noise voltage exhibiting a rise time of approximately 20 µs will produce the highest amplitude in the frequency domain. The spectrum occupancy due to this edge has been calculated, and the broadband component at a frequency of 190 kHz (which is the lowest frequency specified for radiated emission E field testing) has been obtained using the equation for a step function in Section 3.1.2. The broadband component at 50 kHz is 0.05 µV/Hz and, due to the 20-dB/decade fall in amplitude from the 50-kHz point, the amplitude at 190 kHz is approximately 22 dB down, i.e., 0.004 µV/Hz. Using the broadband reference bandwidth of 1 MHz, the voltage is 4 mV/MHz.

The MINCO heater, HR5046A281, which is representative of the type of load presented to the AC line, has a resistance of 103.93 $\Omega$, an impedance measured at exactly the same value, and an inductance of 1.3 µH. The load can thus be considered resistive at approximately 100 $\Omega$, which results in a noise current flow, in the heater cable of 4 mV/MHz/100 $\Omega$ = 40 µA/MHz.

The DO-160A limits for broadband emitted E field measured 3 m from the source at 190 kHz is 98 dBµ V/MHz/m, or 80 mV/MHz/m.

A relatively low-noise current flows in the heater wire, which is an imperfect antenna; therefore the prediction, using the loop antenna or electric current element equations of Sections 2.2.4 and 2.2.5, is that the E field due to this noise source will be at microvolt levels measured 3 m from the source, and well below the limit.

Current noise spikes caused by the AC-line triacs have been ignored during this investigation because the triacs are switched at the zero-crossover point. However, if the zero-crossover detector does not supply sufficient gate voltage to switch the triac at the true AC-supply zero point, a large current spike with a rise time in the order of a microsecond may result. This type of current pulse with a high-amplitude high-frequency component would cause problems in meeting the emitted radiation levels, especially because the allowable limits are reduced at higher frequencies. If it is found that the triacs do not switch at exactly zero, one possible solution is to include some additional inductance between the triac and the line filter.

Although the magnetic field produced by the deflection yoke is likely to be well contained by the mumetal shield, some stray field may be emitted via the face of the CRT and via the gap in the shielding at the heater platen.

*Test Setup*

Normally individual units comprising a system are tested along with interconnecting cables and with the units spaced out above a ground plane. This configuration is common to both radiated susceptibility and radiated emission levels. The argument can be made that if individual units are normally mounted in a rack that acts as an efficient shielded enclosure, the tests should be conducted with this rack enclosing the units. However, the front of the plotter will not be shielded by the rack, and whichever test setup is decided upon, it is the plotter front panel that is the least well-shielded surface and will be the one facing the antenna during radiated emission and susceptibility tests.

*Radiated Susceptibility Tests: E Field Susceptibility*

The plotter case is recessed behind the paper cartridge and is solid, i.e., is a good shield; the only wiring inside the cartridge that is exposed to the radiated RF field is the stepper motor drive wires and the heater AC wiring, both of which are predicted to be immune to radiated E fields.

If the units comprising the system are tested with interconnecting cables and power cables exposed to the field, then the effect on the recorder interfaces must be considered. Generally speaking, if the interfaces can function during the interconnection cable induced susceptibility levels and meet the radiated emission limits, then they are likely to pass the equipment radiated susceptibility tests.

A correctly chosen or designed AC power-line filter can very adequately take care of noise voltage picked up on the AC line during the radiated susceptibility tests.

*Interconnection Cable Induced Susceptibility Tests*

These tests must be carried out on interconnections between units in different racks and may be required between units in the same rack. The tests consist of wrapping spirals of wire around interconnecting bundles of cables and either running a current through the cable or applying a voltage to one end of the cable. The bundle is subjected to current at audio frequency ranging from 30 A/m at 400 Hz to 0.8 A/m at 15 kHz. The bundle is subjected to 360 V/m at 380–400 Hz as well as a voltage spike test.

*AC-Line Noise Immunity*

A filter designed for use on a 400-Hz power line is obviously ineffective at filtering AF noise at 400 Hz and not very effective at 15 kHz. The filter will, however, be effective at filtering out the spike that is electromagnetically coupled into the power line.

   The plotter utilizes large smoothing capacitors in its DC unregulated supplies and linear regulators in the stabilized supplies, and therefore no problem is foreseen with power supplies due to the induced AF noise levels. However, if the zero-crossover detection circuit is not sufficiently noise immune, the induced noise may cause misfiring of the triacs, and thus high radiated noise levels within the unit, which may result in interequipment EMI.

*Control-Line Noise Immunity*

   CABLE SHIELDING:    A 125-$\Omega$ characteristic impedance twisted-shielded-pair cable with a high shielding effectiveness and tight twists in the pair will provide a high degree of immunity to induced noise. An additional overall shield over the entire cable bundle is to be preferred to shielding individual cables, with a concentric connection made to the backshell of the connector and to chassis at both ends.

   INPUT AND OUTPUT CIRCUIT:    Presuming that the model ''B'' differential circuits are available for use in the model ''A'' plotter, it is these circuits that will be considered. The circuits are shown in Figure 12.10. The differential circuit provides a minimum of +/−2.5 V of DC noise immunity and some limited filtering at high frequencies achieved by the ferrite beads, L.

   During the spike test the ferrite beads represent some small inductance and are lossy, but the beads present an insertion loss only over the frequency range 10–100 MHz, which assuming a maximum bead impedance of 50 $\Omega$ is limited to 6 dB. The differential circuit in the ''B''

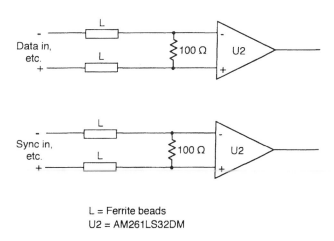

L = Ferrite beads
U2 = AM261LS32DM

**Figure 12.10**   Differential receiver circuit.

plotter has been tested with DO-160A radiated susceptibility test levels using twisted-pair un-shielded cables of 10–12-ft length. Thus, using shielded cable, the circuits should be sufficiently noise immune, with a good safety margin. However, to be absolutely safe, it would be compara-tively easy to build and test a breadboard driver and receiver circuit, using a representative length of shielded cable between the circuits, and to test for noise immunity. The equipment and test circuits necessary to provide the test levels must be available for system-level testing and can be used and tested on the breadboard interface. Radiated susceptibility test requirements may be met with unshielded interface cables in the ''A'' plotter, and the major reason for the use of shielded cables is to meet radiated emission requirements that the model ''B'' plotter had failed.

*AC-Line Conducted Noise Susceptibility Tests*

These tests comprise an audio frequency level, a spike, and an AC voltage modulation test, all levels to be directly coupled into the AC-line voltage. The AC-line filter will only be effective in reducing the 10-μs transient voltage (spike) test level.

The amplitude and frequency modulation supply levels and the audio frequency conducted susceptibility test levels are not modified by the line filter, and the recorder must function cor-rectly when subjected to the test levels. Of particular importance is the operation of the zero-crossover detect circuit, as previously discussed. In addition to noise, the AC line is subjected to surge voltages up to 180 Vrms for 0.1 s and to undervoltage of a minimum 60 Vrms for 10 seconds. Correct operation of the unit is determined after the surge or undervoltage is removed. Because the AC-line conducted noise levels and under/overvoltage levels are easily generated, it is recommended that the plotter be tested at these levels prior to delivery, if possible on a standard plotter with the addition of the chosen line filter.

*Conclusion*

To achieve a high degree of confidence that the Model ''A'' plotter will pass the RTCA/DO-160 EMC requirements, the following measures are recommended:

1. The plotter case should be upgraded to an effective shielded enclosure by changing the finish and using EMI gaskets and honeycomb and mesh air vent filters.
2. Suitable AC-line filter components should be chosen.
3. The plotter should be subjected to the AC-line conducted noise and over/undervoltage conditions. Particular consideration should be given during the conducted noise tests to the operation of the zero-crossover detector and triac trigger circuits as well as the high-voltage and stepper motor drive functions.
4. The manufacturer should breadboard a line driver and receiver and test with RTCA/DO-160 interconnection cable induced susceptibility test levels.

### 12.3.2. Case Study 12.4: EMC Predictions on a Power Controller

These predictions were made at the very initial design stage, before the design was finalized, and therefore concentrates on providing estimated EMI levels and guidelines to be followed in the design. The next iteration of EMC prediction for the equipment will be based on more accurate information on structure and layout.

*Fields Generated by the Internal 20-kHz Distribution System*

The radiated fields generated by the 20-kHz power distribution system internal to the enclosure are predominantly magnetic. The magnetic field strength is to be measured at a distance of 7 cm from the enclosure at points likely to emit high field levels (i.e., seams, slots, apertures, etc.), as required by MIL-STD-462 Method RE01. The magnetic field at low frequency and close to a loop of current-carrying wire is given approximately by

$$H = \frac{I\left(\dfrac{S}{\pi}\right)}{2\left(\dfrac{S}{\pi} + r^2\right)^{1.5}}$$
(12.1)

where

$H$ = magnetic field [A/m]

$r$ = radial distance to point of measurement [m]

$S$ = area of loop = $w \times l$ [m$^2$]

$I$ = magnitude of current [A]

An average supply current for a wire within an enclosure that carries current to all load power converters is 100 A, and this was used in the calculation. In reality a number of wires will be used to carry the 100–300 A of current; however, when the wires are bundled together, they may be replaced, as in our analysis, with a single wire. We assume that the supply and return wires are widely separated within the enclosure and that the current-carrying wire is routed around the inside of the enclosure in the form of a loop. The size of the loop is then close to the interior dimensions of the enclosure, which is 0.762 m × 0.432 m.

From Eq. (12.1) the magnetic field generated by the current-carrying wire at 10 cm is 135 (164 dBpT) A/m. In order to predict the level of magnetic field measured at a distance of 7 cm from the outside of the enclosure, the shielding effectiveness of a typical enclosure must be estimated at 20 kHz.

The shielding effectiveness of an enclosure to an H field incident on the enclosure may be found from the ratio $H_{out}/H_{in}$, where $H_{out}$ is the magnetic field on the outside of the enclosure and $H_{in}$ is the magnetic field on the inside of the enclosure. The equations used to calculate the ratio of $H_{out}/H_{in}$ are presented in Section 6.4.3.

The attenuation is dependent on the enclosure area and will lie in the range of 53–60 dB. Assuming a worst case, the magnetic field is attenuated by 53 dB when the enclosure is inserted between the field source and the magnetic field loop. Therefore the external field is

$$\frac{135}{466.12} = 0.29 \quad [\text{A/m}] \quad (53 \text{ dB} = 466.12)$$

Converting to dBpT for comparison to the limit: 0.29 A/m = 111 dBpT. Comparing this to the limit at 20 kHz (REO1) of 110 dBpT, this worst-case scenario would be close to the limit.

The preceding data is calculated for a current loop with each arm of the loop running within 3 cm of the enclosure walls. The length of the enclosure was taken as approximately 0.75 m, with a height of 0.38 m and a width of 0.43 m. While a loop may not be a realistic wire layout, it does demonstrate that the generation of high levels of magnetic field is possible and must be controlled in the design.

If the same layout was used but 10 AWG twisted-pair wiring was used to distribute the power, as shown in Figure 12.11, the following formula could be used to predict the magnetic field at 10 cm:

$$H_r = -54.5\frac{r}{p} - 20\log\left(\frac{1}{a}\right) - 30\log p - 10\log r + 9.8 \quad [\text{dBgauss/1 ampere}] \quad (12.2)$$

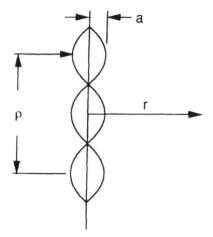

**Figure 12.11**  Geometry of the twisted-pair wiring.

where

    $a$ = radii of helixes, inches = 0.08″

    $p$ = pitch distance, inches = 1.6″

    $r$ = radial distance from the axis of the pair to the measurement point, inches = 4″

This formula gives the magnetic field per ampere of current, which is, in our example, $-159$ dBgauss/1 ampere. Therefore, for a 100-A current, the magnetic field is $-119$ dBgauss = 112 pT = 41 dBpT.

    This is a reduction of 123 dB from the magnetic field produced by a large loop of wire. Thus, using twisted-pair wiring would eliminate any problems associated with magnetic field emissions at 20 kHz appearing outside of the enclosure.

*Fields from Internally Generated Signals Above 20 kHz*

Radiation from the power distribution above 20 kHz would stem from three fundamental sources:

1.   Harmonics of the 20-kHz power supply
2.   Digital clocking frequencies and harmonics
3.   Intermodulation products of the above frequencies

    Magnetic field emissions are monitored to a maximum frequency limit of 300 kHz. Electric fields are measured from 10 kHz to 10 GHz.

    The magnetic field shielding effectiveness of the enclosure increases with frequency, therefore it would require a substantial amount of power at any frequency above 20 kHz to exceed the limits. The probability that this would occur is minimal, since the 20-kHz AC power is the only source with sufficient power associated with it by design.

    The magnetic field shielding effectiveness of our example enclosure is:

    at 300 kHz = 83.79 dB

    at 1 MHz = 95.7 dB

    The limit for emissions (magnetic) at 300 kHz (REO1) is 20 dBpT; therefore an internal source of emissions would have to be 84 dB (84 dB = attenuation of enclosure at 300 kHz) above this limit (104 dBpt) to fail (104 dBpt = 0.13 A/m). Again the assumption is made that this current flows in a large loop of the same dimensions as used in the prediction for the 20-kHz power supply current. Since the total harmonic distortion voltage of the 20-kHz power supply is limited to 2.5% and 300 kHz is the 30th harmonic, there is a low probability of failure to meet REO1 limits above 20 kHz.

Above some frequency, typically 50 MHz, the predominant radiation ceases to be magnetic and becomes either $E$ field or plane wave. As the size of the enclosure becomes a fraction of a wavelength, it may sustain enclosure resonances. Where apertures exist in the enclosure, they can function as slot antennas.

Thus, at high frequencies, meeting REO2 emissions limits becomes the major problem. For our example enclosure without apertures and ignoring the contribution due to cables, a source of considerable energy would have to exist, i.e., RF power source, high-power high-speed converter with capacitive or inductive load, within the enclosure to fail REO2 limits.

*Radiated Emissions from Interconnecting Cables*

The radiated emissions from interconnecting cables are caused primarily by common-mode current on the shield of the cable. The length of cable to be tested (see MIL STD 462 METHOD REØ2) is 2 meters. From this length the frequencies of resonance (i.e., $\lambda/2$, $2 \times \lambda/2$, $3 \times \lambda/2$, etc. when the shield is connected to the ground plane at both ends) can be calculated approximately as follows:

| Integer of $\lambda/2 = 2$ m | Wavelength [meters] | Frequency [MHz] $= 300/\lambda$ [m] |
|---|---|---|
| 1 | 4 | 75 |
| 2 | 2 | 150 |
| 3 | 1.333 | 225 |
| 4 | 1 | 300 |
| etc. | | |

At a resonant length the effectiveness of the transmission line as a transmitting antenna is much higher than at nonresonant lengths. This is offset to some extent above the frequency at which $\lambda/2 = 2$ m, i.e., 75 MHz, by the change in phase of the field 'down the transmission line. In addition the transmission line is positioned not only above the ground but below the ceiling and in front of the walls of the shielded room, so multiple reflections occur.

The total field at the measuring point is thus the superposition of all field sources. The REØ2 test method requires location of the test antenna at maximum pickup, i.e., the position at which the majority of field sources are in phase. The transmission-line equation used has been validated by tests in an undamped shielded room in which the predicted $E$ field is within $+10$ dB to $-11$ dB of the measured $E$ field.

The predicted maximum level of common-mode current allowed on the shield before the limits for narrowband emissions are exceeded for each of the resonant frequencies is given in Table 12.2.

The maximum common-mode current on the inner conductor(s) of the cable and inside of the shield can be determined using the current on the outside of the shield and the transfer impedance of the specific cable to be used, once this is determined.

*Conducted Emissions*

Conducted emissions are those signals that appear on cables at other than required operating frequencies. These emissions can either be continuous (CW) or transient in nature. From experi-

**Table 12.2** Maximum Common-Mode Current on Interconnecting Cables

| Frequency (MHz) | Max. C/M current (μA) | E Field at 1 m (μV/m) | Spec E Field at 1 m (μV/m) |
|---|---|---|---|
| 75 | 22 | 63.5 | 63 |
| 150 | 19 | 109.7 | 110 |
| 225 | 18 | 156.0 | 160 |
| 300 | 17 | 196.2 | 200 |

ence, systems utilizing circuitry that generates emissions rich in harmonic content (i.e., DC–DC converters) or that switch currents are almost certain to fail the requirements. This implies that some type of EMI reduction technique be included in the original design to reduce conducted emissions.

Two fundamental approaches are available, either reduction at source or the inclusion of filters. In extreme cases both approaches are required. Those circuits that are potentially sources of high-level emissions will be designed for emission control that is compatible with performance requirements.

*Susceptibility of Internal Circuitry to Internally Generated Fields (Crosstalk)*

The phenomenon of a system's being susceptible to its own radiated emissions inside an enclosure is difficult to predict, because crosstalk and radiated coupling are very dependent on exact geometries. However, a number of estimations of crosstalk using distances of 1 inch, 2 inches, and 6 inches between power PCB traces and interconnecting cables at a height of 2 inches (5 cm) and 0.06 inches (0.15 cm) above a ground plane were made. The source wire used in the prediction was a 10 AWG circular conductor in order to achieve a low source resistance. The receptor wire type and gauge are changed, as are the victim source and load impedance, to show the effects of the changes on the level of voltages induced in the victim circuitry.

Ideally the wires carrying the 20-kHz power and return should be twisted over the complete length. However, due to the large gauge of the wire and the requirement to make terminal or similar connections, some length of the wires will be untwisted. It is the untwisted length that, causes maximum *H* field generation and therefore maximum potential crosstalk. The lengths of untwisted 10 gauge wire used in the examples range between 1″ and 6″, which are considered practical best- and worst-case lengths. The height above the ground plane in the equations contained in Section 4.3 is also used to describe the distance between the source power wire and its return.

At a very worst case, the average 100-A-load current may be switched as fast as 50 ns, and this was the value used in the initial prediction. Where the switching time is made longer than 50 ns, the peak transient voltage induced in the victim wire is considerably reduced, as shown in Example 6.

> *Example 1*:  Here the coupling is to an unshielded receptor wire over a distance of 1″, with the receptor circuit exhibiting a source impedance of 250 Ω and a load impedance of 1000 Ω, which is typical for an analog circuit. The peak crosstalk voltage is a massive 24 V when the two wires are only 1″ apart.
>
> *Example 2*:  Here the wires are separated by 2″, with all other parameters constant. The induced voltage reduces to 13.5 V.
>
> *Example 3*:  Increasing the distance between wires to 6″ decreases the crosstalk voltage to 3.1 V.

*Example 4:*   Maintaining the 6″ distance between wires and increasing the coupled length to 6″ increases the voltage to 18.6 V.

*Example 5:*   Maintaining the 6″ distance between wires and the 6″ coupled length but decreasing the distance between the source power and return wires to 0.06″ (i.e. the insulation of the two wires is touching) and the height of the receptor wire above the chasis to 0.06″ results in a decrease in the crosstalk voltage to 2 V.

*Example 6:*   The same geometry is used as in Example 1. But here the rise time is increased to 100 ns, which causes a reduction in crosstalk from 23.8 V to 12.2 V. Increasing the rise time further to 1 μs results in a reduction in crosstalk to 1.26 V.

*Example 7:*   Here the distance between wires is 1″, the height above the ground plane is 2″, the coupled length is a worst-case 6″, and an RG59A/U shielded cable is used for the victim wire. The load and source impedances are also changed, with the load matched to the coaxial cable characteristic impedance, which is a typical configuration for RF or video circuits. The predicted peak transient is 63 mV.

*Example 8:*   This is the same configuration as example 7, only the distance between the source and receptor wires is increased to 6″ and the maximum peak transient is 8.2 mV.

The transient crosstalk voltage from Example 8 (8.2 mV) may be acceptable in high-level RF signals. And even where low-level analog or video signals are concerned, the short-duration spike may not cause an EMI situation because the circuit may not be perturbed by the spike. Where a potential EMI condition does exist, the noise immunity of the receptor circuit may be increased, typically by filtering or/and by increasing the rise time of the source current. Where the coupling is to PCB traces or ICs, the shielding is more difficult.

However, shielding of sensitive circuitry in enclosures and EMI hardening of sensitive circuitry are two approaches that will be considered, even at the PCB level.

*Susceptibility of internal circuitry to externally generated fields*
The effectiveness of cables and enclosures are examined separately.

*The enclosure:* In the section dealing with the enclosure radiation at 20 kHz, a worst-case magnetic field attenuation of 53 dB was arrived at. From inspection of the formula it can be seen that the enclosure attenuation will increase with frequency.

The current flow in a wire wrapped around the enclosure during the 20-kHz power frequency test (RS02) is 3 A. The impedance of the enclosure is very low, so approximately 50% of the 3-A wire current flows in the outside of the enclosure. The distance between the loops of wire wrapped around the enclosure is specified as 30 cm, in accordance with MIL-STD-462. The current density in the outside of the enclosure is therefore 1.5 A/0.3 m = 5 A/m.

The definition of magnetic field attenuation used in deriving the 53-dB minimum attenuation for the enclosure is the ratio $I_{out}/I_{in}$, where $I_{out}$ is the current flowing on the outside of the enclosure and $I_{in}$ is the current flowing on the inside of the enclosure. $I_{in}$ is therefore 53 dB down on 5 A/m, which is (5 A/m)/446 = 0.011 A/m. The narrowest dimension of the enclosure is approximately 0.43 m, so the current density across this interior dimension is

11 mA/m × 0.43 m = 4.7 mA/0.43 m

Assuming a loop of unshielded wire 3 cm from the inside walls of the enclosure and routed around the enclosure, the crosstalk voltage picked up on the wire at 20 kHz will be very approximately 0.6 mV, assuming a load resistance of 1000 Ω and a source resistance of 250 Ω.

Pickup from sources within the enclosure are likely to be higher than the predicted 0.6 mV induced in the RS02 test. Therefore sensitive signals will be routed via shielded cables. If the wire is used for a control line, 0.6 mV in a control signal is unlikely to cause an EMI problem. The decision on the use of shielded or twisted pair for control lines will be based on the level of the interior ambient. Based on measurements, the current spike test results in approximately 100 A of current flow in the test wire around the enclosure; however, the duration of the resultant positive and negative spike is short (10 μs). The low-frequency component of the spike has a current amplitude of approximately $2Ad$, where $A$ is the peak amplitude and $d$ is the pulse width. The external current flow is thus $2 \times 100 \times 10 \times 10^{-6} = 2$ mA/0.43 m, i.e., lower than the 20-kHz power frequency test. The high-frequency components of the spike are attenuated to a greater extent than the low-frequency ones, because the shielding effectiveness of the enclosure increases with increasing frequency. We conclude therefore that our sample enclosure provides adequate shielding for the RS02 test levels. The level of common-mode voltage induced in four different types of shielded cables are shown in Table 12.3.

The impact of these levels of common-mode transferred voltage ($V_t$) depends to a large extent on the EMC classification of the circuit class, i.e., the sensitivity of the receiver. A potential EMI problem exists when $V_t$ is close to the sensitivity of the receiver. A second potential problem occurs when a wire in the shielded cable is connected, either directly or via a low impedance, to chassis inside the enclosure; in this case appreciable current can flow due to $V_t$, and radiation from the wire inside the enclosure may result in EMI.

An important point to note is that it is typical in RS03 tests to modulate the carrier at frequencies of maximum susceptibility, e.g., 10 kHz for audio circuits, 1 MHz for the 1553 or similar digital busses, etc. In addition, when high levels of RF are applied to integrated circuits, which may have a bandwidth much lower than the test frequency, DC offsets may occur due to nonlinearities inherent in the IC input stage.

As the design progresses, the implication of the predicted noise voltage should be assessed for each enclosure layout and interface circuit, and appropriate filtering or shielding will be incorporated.

*RS02 spike test induced voltage*: When a transient induced voltage is considered the time constant of the source wire controls, the current flow in the source wire. Likewise, when the

**Table 12.3** Common-Mode Voltages Induced in Shielded Cables Exposed to RS03 Test Levels (Cable Length = 2 m)

| Distance from antenna (m) | Height above structure (m) | Cable type 1[a] | Cable type 2[b] | Cable type 3[c] | Cable type 4[c] | $f$ |
|---|---|---|---|---|---|---|
| 1 | 0.05 | 3 mV | 1 mV | — | — | 30 MHz |
| 1 | 0.05 | 13 mV | 12.6 mV | — | — | 7 MHz |
| 1 | 0.05 | 1.48 V | 0.163 V | 0.144 V (4.8 μV) | 0.144 V (1.4 μV) | 1.05 GHz |
| 1 | 0.05 | 7.7 V | 0.92 V | 0.77 V (1.2 mV) | 0.77 V (0.11 mV) | 9.975 GHz |

[a] D-type connector.
[b] BNC-type connector.
[c] BNC-type connector.
Cable type 1 is a standard braided cable exhibiting a transfer impedance of 10 mΩ/m at 100 kHz, 100 mΩ/m at 75 MHz, 600 mΩ/m at 1.05 GHz, and 15 Ω/m at 9.975 GHz. Cable type 2 is a twisted shielded pair, suitable for the 1553 data bus communication link, exhibiting one of the lower transfer impedances (i.e., an average shielded cable). The transfer impedance is 10 mΩ/m at 100 kHz, 100 mΩ/m at 30 MHz, 1 mΩ/m at 75 MHz, 14 Ω/m at 1.05 GHz, and 133 Ω/m at 9.975 GHz. Cable type 3 is a double-braided coaxial cable exhibiting a transfer impedance of 7 mΩ/m at 1 GHz and 2 Ω/m at 9.975 GHz. Cable type 4 is a copper solid-wall tube exhibiting a transfer impedance of 2 mΩ/m at 100 kHz and effectively 0 mΩ/m at 30 MHz and above (i.e., a semirigid cable). Figures in brackets are for an (N (D/U))-type connector.

**Table 12.4** RS02 Test Induced Voltage

| Cable type | Shield | Common-mode induced voltage |
|---|---|---|
| RG58C/U | Single braid | 0.14 V |
| RG180/U | Single braid | 0.17 V |
| RG59A | Single braid | 0.12 V |
| RG8A/U | Single braid | 0.55 V |
| RG223/U | Double braid | 0.08 V |
| RG9B/U | Double braid | 0.033 V |

victim cable time constant is longer than the applied pulsewidth, the induced voltage divides not only between the source and load resistors but in the cable inductance. Thus the worst-case induced voltage is not caused by the 100 $V_t > 0.15$-$\mu$s spike but by the 200-V $> 10$-$\mu$s spike in the shielded cables under consideration due to both the longer time constant of the 10-$\mu$s spike and the higher peak voltage. The current flow in the shield of a cable subjected to a 200-V $> 10$-$\mu$s RS02 test has been measured at $+30$ A/$-30$ A, and therefore 30 A was chosen as a typical value for the shield current in the prediction.

The cable crosstalk computer program was used to simulate the source of the spike and to predict the voltage induced in a shielded cable. Initially the current flowing in an unshielded wire with the same diameter as the shielded cable was modeled, and the current flow in the source wire was adjusted until 30 A flowed in the victim wire. The victim wire was then replaced with the shielded cable, and the voltage induced in the shielded cable with a 30-A peak shield current was predicted by the program. The type of shielded cable used for interface connections on the system is not yet known; therefore the induced voltage was calculated for a number of different types of shielded cable. Table 12.4 gives the level of induced voltage, which is almost independent of whether the receptor cable is terminated in a high-impedance load or is terminated with its characteristic impedance, for the induced voltage is the same for a wide range of receptor load impedances when the emitter source impedance is low.

The voltages shown in Table 12.4 are common mode, thus, where twisted-pair shielded cable with well-balanced differential inputs are used with sufficient common-mode noise rejection to a 10-$\mu$s-wide transient, the induced voltages will not cause an EMI problem.

### 12.3.3. Case Study 12.5: Antenna-to-cable Coupling on the Space Shuttle (Orbiter)

Cables on an experiment that is carried in the cargo bay of the space shuttle are exposed to the E fields generated by an antenna mounted on the experiment. The approximate worse case with the antenna nonresonant, values of E field, generated by the experiment at distances of 1 m, 2 m, 4m, and 8 m are shown in Table 12.5. (These values are higher than published levels as described in Section 12.3.) The interference voltage induced in the inner conductors of a shielded cable when the cable runs some length in parallel with the antenna and parallel to and above the shuttle cargo bay structure is calculated for a number of cable lengths, heights above the cargo bay, and distances from the antenna.

The cable shield is connected at both ends to the shuttle structure, via the connectors and instrument case. For the sake of these calculations the shuttle structure is considered a perfectly conducting ground plane. The characteristic impedance of the transmission line $Z_c$ is given by

$$Z_c = \sqrt{\frac{(Z^i + j2\pi f l^e)(2\pi f l^e)}{jk^2}} \qquad (12.3)$$

**Table 12.5** E Field Generated by Experiment at
Distances of 1, 2, 4, and 8 m

|  | E field [V/m] at distance $r$ | | | |
|---|---|---|---|---|
| Frequency | 1 m | 2 m | 4 m | 8 m |
| 100 kHz–6 MHz | 1000 | 125 | 16 | 2 |
| 10 MHz | 1000 | 125 | 79 | 39 |
| 15 MHz | 1000 | 125 | 79 | 39 |
| 30 MHz | 1000 | 158 | 79 | 39 |

where

$Z^i$ = distributed series resistance of the two-wire lines
($Z^i$ for a 0.5-cm-diameter braided shield is taken as 10 mΩ/m)

$$l^e = \frac{\mu_0}{\pi} \ln\left(\frac{2h}{a}\right) \tag{12.4}$$

$$k = \frac{2\pi}{\lambda}$$

$\mu_o = 4\pi \times 10^{-7}$ [H/m]
$f$ = frequency [Hz]
$h$ = height of cable above the structure

In the frequency range 100 kHz to 30 MHz the characteristic impedance is constant with frequency and varies with the height of the cable above the structure and the diameter of the cable. For convenience the illumination of the cable is considered to be plane wave, with the plane of incidence coincident with the plane of the loop formed by the cable and the structure, using the following values in eq. (12.4) for $h = 10$ cm,

$$l^e = \frac{4\pi \times 10^{-7}}{\pi} \ln\left(\frac{2 \times 0.1}{0.005}\right) = 1.47 \times 10^{-6}$$

The value for $k$ at 30 MHz = 6.28/10 m = 0.628. Therefore, from Eq. (12.3), for $h = 1$ cm, $Z_c = 168$ Ω, and for $h = 10$ cm, $Z_c = 441$ Ω.

*Shield Current*

The current flowing in the cable shield due to the incident wave, ignoring re-radiation from the wire structure and resonances in that structure, is given by

$$I = \frac{4E_o h}{Z_c}$$

where $E_o$ is the amplitude of the plane wave.

The shield current is independent of cable length as long as the complete length of the cable is equally illuminated. The shield currents for the two values of height 1 cm and 10 cm and a cable diameter of 0.5 cm are calculated as follows:

*Example:* The cable shield current for a distance of 2 m from the antenna where the E field is = 125 V/m is for $h = 1$ cm:

$$I = \frac{4 \times 125 \text{ } V \times 0.01}{168 \text{ } \Omega} = 30 \text{ m}$$

and for $h = 10$ cm:

$$I = \frac{4 \times 125 \text{ V} \times 0.1}{441 \text{ } \Omega} = 110 \text{ m}$$

N.B.: From the foregoing it is clear that cables should be kept as close to the shuttle structure as feasible when the cables are in the proximity of the antenna.

*Common-Mode Interference Voltage*

The transferred common-mode voltage $V_t$ induced in the center conductors of the shielded cable due to the shield current is a function of the transfer impedance of the shielded cable, which in turn is a function of frequency. Three types of cable are considered:

1. A standard braided cable exhibiting a transfer impedance of 10 mΩ/m at 100 kHz and 100 mΩ/m at 30 MHz
2. A braided multilayer foil type exhibiting a transfer impedance of 8 mΩ/m at 100 kHz and 10 mΩ/m at 30 MHz
3. A copper solid-wall tube exhibiting a transfer impedance of 2 mΩ/m at 100 kHz and effectively 0 mΩ/m at 30 MHz

The common-mode voltage induced is directly proportional to cable length. However, the induced shield current begins to reduce at cable lengths beyond 4 m, for the field strength begins to reduce at these distances from the source.

For a shield current of 30 mA, $V_t$ for cable, 1 per meter, is

at 100 kHz $= 30 \times 10^{-3} \times 10 \times 10^{-3} = 0.3$ mV/m
at 30 MHz $= 30 \times 10^{-3} \times 100 \times 10^{-3} = 3$ mV/m

For a shield current of 110 mA, $V_t$ for cable, 1 per meter, is

at 100 kHz $= 1$ mV
at 30 MHz $= 11$ mV

*Compensation for Cable Resonances*

At specific frequencies, resonances in the termination currents due to the short-circuited nature of the transmission line must be expected. The induced voltages shown in Table 12.6 are four times the amount calculated, to compensate for cable resonances.

*Common-Mode Voltage*

The induced voltages appearing across the pair in a twisted shielded pair are common mode, thus if these voltages are applied to the input of a balanced receiver that has a reasonable common-mode noise immunity, the noise signal is canceled. If the cable is a coaxial type, then the induced noise voltage will appear at the input of the receiver. Triax cable will provide some attenuation of the center core induced noise voltage.

Table 12.6 shows noise voltages induced in the center conductors of shielded cables running parallel to the antenna for distances from the antenna of 1, 2, and 4 m and for cable lengths of 1, 2, and 4 m at frequencies of 100 kHz and 30 MHz and cable heights above the structure of 1 cm and 10 cm. The transfer impedances of the cables are typical for the three types shown.

**Table 12.6**  Interference Voltages with Different Cables and Configurations

| Distance from antenna $d$ [m] | Height above structure $h$ [cm] | Cable length $L$ [m] | Cable type 1, $V_t$ [mV] | Cable Type 2 $V_t$ [mV] | Cable Type 3 $V_t$ [mV] | Frequency |
|---|---|---|---|---|---|---|
| 1 | 1 | 1 | 9.6 | 7.6 | 1.9 | 100 kHz |
| 1 | 1 | 1 | 96 | 0.96 | Neg. | 30 MHz |
| 1 | 10 | 1 | 36 | 28 | 7.2 | 100 kHz |
| 1 | 10 | 1 | 360 | 3.6 | Neg. | 30 MHz |
| 2 | 1 | 2 | 2.4 | 1.9 | 0.48 | 100 kHz |
| 2 | 1 | 2 | 24 | 0.24 | Neg. | 30 MHz |
| 2 | 10 | 2 | 8 | 6.2 | 1.6 | 100 kHz |
| 2 | 10 | 2 | 88 | 0.8 | Neg. | 30 MHz |
| 4 | 1 | 4 | 2.9 | 2.3 | 0.58 | 100 kHz |
| 4 | 1 | 4 | 29 | 0.28 | Neg. | 30 MHz |
| 4 | 10 | 4 | 12 | 9.2 | 2.3 | 100 kHz |
| 4 | 10 | 4 | 120 | 1.2 | Neg. | 30 MHz |

| Cable types | Description | Transfer impedance 100 kHz [mΩ/m] | Transfer impedance 30 MHz [mΩ/m] |
|---|---|---|---|
| 1 | Standard braided | 10 | 100 |
| 2 | Braided multilayer foil | 8 | 10 |
| 3 | Copper solid-wall tube | 2 | ~0 |

$V_t$ = transferred interference voltage; Neg. = negligible.

*Conclusions*

The values shown in Table 12.6 are approximate values useful for evaluation of any possible EMI problem. For example, with care given to termination of the shields at the connector, all the proposed options of digital interface circuits available for use in the experiment will provide sufficient immunity to the noise source under consideration.

Should a more exact value for the interference voltage $V_t$ and the frequencies of resonance be required, the physical details of the cable layout and use of a computer program that includes cable resonance effects should be used.

Case Study 12.4 was confined to frequencies below 30 MHz. When high-intensity fields as frequencies above 30 MHz are incident on flexible cable, the interference voltages induced will be higher than those shown in Table 12.2 because the transfer impedance of the cable increases with increasing frequency.

## 12.3.4.  Case Study 12.6: Coupling from Radar to the Landing Control Signal on an Aircraft

In one example it was predicted that the landing control signal cable used on a composite-skin aircraft would be exposed to an E field of 200 V/m from 14 kHz to 100 MHz, 5100 V/m from 100 MHz to 2 GHz, and 31000 V/m from 2 GHz to 21 GHz. The reason the signals could be exposed to such high levels is that the composite skin of the aircraft provides little shielding and the aircraft may fly close to high-level radar. The decision was made to run the signal cable in a solid copper conduit with an 0.05-cm wall thickness. Calculations were made based on the skin depth of the copper tube, and a level of shielding in the thousands of decibels was predicted from 2 GHz to 21 GHz. The calculation may have been valid for a tube soldered, welded, or brazed to an enclosure in which the members of the enclosure were similarly joined.

In reality the copper tube is soldered to an EMI backshell, attached to a connector that is mated to a receptacle mounted on the enclosure. Each of the mechanical interfaces exhibits a transfer impedance as described in Section 7.8. The current at the end of a cable disconnected from ground is zero. However, the cable in question is attached to an enclosure that is several wavelengths long at the higher frequencies of interest. The current flow through the transfer impedances is therefore at a maximum at critical frequencies.

Assuming the cable is run at a distance of 1 meter or greater from a ground plane or grounded conductor, the approximate voltages induced in the control cable conductors, due to the skin depth of the conduit at low frequency and the transfer impedances of the connections, is given in Table 12.7

The actual attenuation of the conduit with connector is 54 dB at high frequency and not the thousands of decibels initially predicted! Additional filtering was required at the connector to ensure EMC in this example.

### 12.3.5. Case Study 12.7: Coupling from an AM Transmitter to a Satellite Communication System

In this example a ground-based system receiving a satellite transmission at a frequency of 1.5 GHz exhibited EMI due to the high electromagnetic ambient. The receiving system was located 40 m from a 50-kW transmitter operating at a frequency of either 4.8 MHz or 2.2 MHz. The signal at 1.5 GHz was not directly interfered with due to the frequency separation between the received signal and the interfering signal, the poor out-of-band response of the antenna, and the inclusion of a bandpass filter between the antenna and down-converter. It was this apparent lack of a potential EMI problem that caused the system to be built so close to a transmitter. However, the receiving system uses a 10–14-kHz time code receiver to obtain the accurate time required to track a satellite, and this exhibited EMI. In addition the antenna control signals and 10-MHz IF signal, carried on shielded cables between the antenna and the IF receiver and computing equipment, were susceptible.

The equipment was housed in a nearby building and the shielded cables carrying the IF and control signals between the antenna pedestal and the building were illuminated by the field generated by the 4.8-MHz or 2.2-MHz transmitter. The incident field induced a current flow on the cable shields, which in turn resulted in a common-mode noise voltage on the center conductor/s developed by the transfer impedance of the cable and the terminations.

The resultant level of noise voltage developed in the coaxial cable carrying the IF signal, developed at the output of the down-converter, corrupted the IF to the extent that the received

**Table 12.7** Voltages Induced in Control Cable Conductors Due to the Skin Depth of the Conduit at Low Frequency and the Transfer Impedances of the Connections

| E field | Frequency | Interference voltage |
|---------|-----------|----------------------|
| 200 V/m | 14 kHz | 140 mV |
| 200 V/m | 7.5 MHz | 2.7 mV |
| 5,100 V/m | 100 MHz to 2 GHz | 10 V |
| 31,000 V/m | 2 GHz to 21 GHz | 64 V |

signal was lost in the noise. In addition, the noise voltage developed in the shielded cables carrying the signals used to control the antenna drive was sufficiently high that movement of the antenna was inhibited.

The EMI voltage developed on the Omega time code antenna resulted in compression and amplitude modulation in the high-input-impedance preamplifier connected to the antenna, and the integrity of the time code signal was destroyed.

With the receiving system totally inoperable due to EMI, the first reaction of the engineer in charge of the installation was to move the entire system to an alternative site with a lower electromagnetic ambient.

Despite the availability of an alternative site with a suitable building the estimated cost to disassemble, move, and reassemble the equipment was $100,000, with a resultant delay of at least a month in the startup date for the system.

The equipment manufacturer was persuaded to commission an EMI investigation, which resulted in the implementation of modifications and thereby an operational system within one week, although the solution to the time code receiver problem took a further two weeks.

A local radio technician was able to make a measurement of the field developed by the transmitter, which was approximately 4 V/m.

In calculating the $V_t$ from the current flow and the transfer impedance of the types of cables used, the predicted voltage was well below the susceptibility threshold of the interface circuits and signals. This illustrates the importance of conducting a simple EMC prediction. If the results from the prediction indicate that an EMI problem should not exist, then the installation must differ from the assumptions used in the prediction. In this installation the high transferred voltage in the installation was not developed across the transfer impedance of the shielded cable but across the poor shield termination and the poor ground to which the shield was connected.

The knowledge gained in understanding how to improve the immunity of the system was just as valuable as the EMI cure. The modifications to the equipment went as follows. The time code receiver required a simple filter, the design of which is described in Section 5.1.10.10, contained in a small, sealed enclosure inserted between the antenna and the preamplifier. The cable coupled EMI was cured by an improvement in the antenna and building grounding scheme, the shield termination, and the level of cable shielding.

After the proposed modifications were correctly implemented (many EMI fixes are less than effective due to incorrect implementation), the system functioned correctly. It continues to function without EMI 10 years after installation.

A number of lessons were learned from the EMI investigation, which resulted in the following modifications to the standard version of the system.

The improved grounding scheme and shield termination methods are implemented in all production equipment regardless of the ambient.

Ambient site surveys are now conducted on almost all of the proposed installation sites. The site survey measures the ambient over the frequency range of 10 kHz to 18 GHz and, in addition to the low-frequency type of problem described in this example, examines the potential for in-band EMI at 1.5 GHz, cross-modulation, passive intermodulation (PIM), adjacent channel, harmonic, IF, and image interference. If the ambient site survey results in a prediction of EMI to the time code receiver, then the antenna filter is installed. For close to in-band EMI problems, the inclusion of a narrower bandwidth filter between the 1.5-GHz antenna and the down-converter may be required. For in-band EMI due to PIM, relocation of the antenna or conductive structures in close proximity may be required, or the use of all-weather absorber foam or anti-PIM coating may be necessary.

When in-band EMI cannot be corrected for at the proposed site, an alternative site may be chosen prior to installation.

The site survey is designed to reduce the cost of the equipment and installation because filters, absorber foam, anti-PIM coating, welded conduit for cables, etc. are incorporated only when the electromagnetic ambient warrants them. In addition, the estimate of the time required to install the system is more accurate when based on an ambient site survey.

## 12.3.6. Case Study 12.8: Spurious Response in a Transmitter/Receiver

Switching-power-supply noise resulted in EMI in the equipment that contained a transmitter/receiver. The EMI appeared as spurious emissions during transmission, displaced from the carrier by the power supply switching frequency and harmonics thereof. In the worst case the spurious emissions were only 6 dB down on the carrier; in this example a $700,000 late delivery penalty was contained in the contract and the EMI problem was cured two weeks prior to delivery.

The choices, common in all EMI problem solving, are to incorporate source reduction, reduction in coupling, an increase in receptor immunity, or a combination of these.

In the example, the EMI solutions included the design and building of a power supply filter, elimination of a common ground path, and reduction in crosstalk between the output filtered power lines and unfiltered lines. These solutions were implemented in six days, and the equipment passed acceptance testing just prior to the stipulated delivery date.

The transmitter/receiver contained a switching-power supply that provided +5 V for digital logic, +/−15 V for RF and analog circuits, and +28 V for RF circuits and the RF power amplifier.

Switching-power supplies are notorious generators of high-level and wide-frequency-range noise. In this case both radiated and conducted noise sourced by the supply resulted in EMI. The spectrum occupancy of the noise started at the switching frequency of 70 kHz and extended up to 100 MHz.

In the original design the power supply was not shielded. After the EMI problem had been identified, the manufacturer built a prototype shielded enclosure that only marginally improved the EMI problem. The prototype enclosure was constructed of an aluminum base plate and a steel cover. A strip type of connector protruded through a 0.105-m aperture cut into the rear of the enclosure, which was 0.11 m wide. The edges of the steel cover at the rear, close to the connector, were attached to the sides of the cover by pop rivets, as shown in Figure 12.12. The prototype enclosure under test was unpassivated, and the proposed surface finish was the relatively high-resistance zinc dichromate. Both the pop-riveted seams at the rear of the enclosure and the 0.105-m slot were identified as potential leakage paths for radiation. The sources of radiation within the enclosure, such as the transformer and switching semiconductors, emit the highest fields at the switching frequency. Due to the low frequency and the close proximity of the sources to the enclosure, the field impinging on the inside walls of the enclosure is predominantly magnetic.

The inside of the enclosure forms a wide loop, with an impedance determined by the inductance of the loop and the frequency of the currents flowing in it. A current flows as a result of the incident field. Some of the current diffuses through the enclosure material due to its AC resistance. When seams are present, as in our example, the current also diffuses through the seam impedance, consisting of the DC resistance, the contact impedance, and the inductance of the seam. The diffused current then flows in the loop formed by the outside surface of the enclosure and generates a magnetic field. When an aperture is present, the magnetic field couples

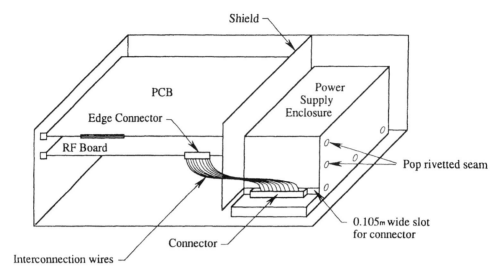

**Figure 12.12** Original power supply and RF board configuration.

through the aperture and contributes to the external field. Section 6.4.3 describes the mechanism. We shall compare the shielding effectiveness of three enclosure configurations. One is a totally sealed version with welded seams and the dimensions 0.13 m × 0.11 m × 0.24 m. The material used to construct the enclosure has a thickness of 5 mm a relative permeability of 485, and a relative conductivity of 0.1. The predicted shielding effectiveness of the totally sealed enclosure is 100 dB at 70 kHz. Examining the shielding effectiveness of the same-size enclosure with two pop-riveted seams in the current path and with use of either unplated metal or the relatively high-resistance zinc dichromate finish, the predicted shielding effectiveness reduces to 31 dB. By adding the 0.105-m aperture for the connector, the shielding effectiveness reduces to approximately 25 dB. The proposed improvements to the enclosure entail closing the aperture for the connector, closing the seams with screw fasteners in close proximity, and passivating the material with a high-conductivity finish such as tin or zinc, which should increase the shielding effectiveness to approximately 55 dB.

The magnetic field measured with the top cover of the enclosure removed, at a distance of 2.4 cm above the transformer, which is the source of maximum emissions, is 6.5 A/m at 70 kHz. The magnetic field measured close to the 0.105-m aperture and the pop-riveted seams was 130 mA/m to 140 mA/m. Thus the measured effectiveness of 33 dB is close to the predicted of 25–31 dB. This close correlation between measured and predicted attenuations, using the simple analysis method, has been seen in a number of experiments. Therefore, assuming the magnitude of the source had been known, the field at the rear of the enclosure could have been predicted with some accuracy. However, the field at the rear of the power supply was not directly, or the sole, source of EMI. The susceptible circuit was located on the RF board on which a magnetic field was developed as a result of common-mode currents flowing on the interconnections to the board. This current had two sources: one was the leakage field from the power supply enclosure coupling to the interconnections between the power supply and the RF board, and the second was the conducted C/M noise currents generated by the power supply.

The magnitude of the magnetic field at the susceptible location on the RF board was measured at 13 mA/m at 70 kHz and 0.2 mA/m at 28 MHz. The 28 MHz is a spectral emission with a repetition rate of 70 kHz and is therefore also a candidate for the 70-kHz spurious emission EMI. The level of magnetic field at 70 kHz induces a current of approximately 30 μA into a

2-cm × 2-cm loop formed by tracks on the PCB, and the 28-MHz level induces approximately 17 μA in the same-size loop.

In finding the source of EMI, one of the diagnostic methods used in the investigation was to move the interconnection wires away from the rear of the power supply enclosure. This resulted in reduced coupling to the interconnections, a lower magnetic field on the RF board, and, as a consequence, a lower level of spurious emission.

Numerous attempts were made both to filter the power supplies on the RF board, close to the edge of the board near the connector, and to locally shield susceptible areas of the board, with limited success.

As a result of the pressure to find a quick solution that had a high probability of success, it was decided to improve the shielding effectiveness of the power supply enclosure and include both common-mode and differential-mode filter components in the power supply on the +5-V, +/−15-V, and 28-V supply and return lines. The filter was designed to preclude insertion gain by selecting a resonance frequency much lower than 70 kHz. A second feature of the design was the capability of supplying the high peak current required during transmission with a minimum reduction in supply voltage and with no generation of transient voltages. The measured attenuation of common-mode and differential-mode noise voltage and common-mode noise current for the filter is shown in Table 12.8.

The filter components were mounted on a PCB contained within a small sub-enclosure that was connected to the inside cover of the power supply as shown in Figure 12.13.

The leakage from the seams of the power supply enclosure was reduced by the use of a high-conductivity finish (both tin and zinc were tried) with fasteners at 2-inch spacings along the bottom of the enclosures. These modifications reduced the AC resistance and contact impedance of the joints, and thus a higher level of shielding effectiveness was achieved. The leakage from the rear of the enclosure, close to the power supply interconnections, was reduced by welding the seams. The modified enclosure, as shown in Figure 12.13, covered the PCB-mounted connector used to make the internal connections to the filter subenclosure, and thus the 0.105-m slot for the connector was eliminated. Instead, a small aperture was made at the location of the filter sub-enclosure to bring the filtered supply interconnections out of the enclosure.

With the filter and modifications made to the prototype enclosure, the magnetic field measured at the rear of the enclosure close to the interconnections was 12 mA/m at 70 kHz. this represents an improvement of 21 dB over the pop-riveted enclosure with 0.105-m slot and a total shielding effectiveness of 55 dB, which is uncharacteristically close to the predicted value of 55 dB!

**Table 12.8**  Filter Performance

| | Attenuation | | |
|---|---|---|---|
| Frequency | Common-mode voltage | Differential-mode voltage | Common-mode current |
| 70 kHz | 50 dB | 50 dB | 37 dB |
| 3.01 MHz | 40 dB | 50 dB | 22 dB |
| 6.02 MHz | 63 dB | — | — |
| 15.05 MHz | — | 32 dB | — |
| 17.99 MHz | — | — | 22 dB |
| 21.00 MHz | — | 50 dB | — |
| 28.00 MHz | — | 25 dB | — |
| 29.96 MHz | 45 dB | — | — |

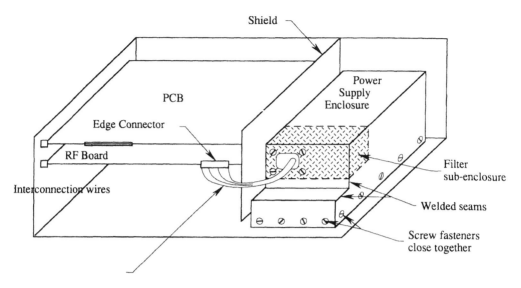

**Figure 12.13**   Modified power supply and RF board configuration.

The improved enclosure and filter achieved a reduction in the spurious emission to an acceptable 40 dBc, i.e., 40dB down on the carrier, which represents a 34-dB improvement over a nonmodified power supply.

## 12.4. EMC, COMPUTATIONAL ELECTROMAGNETIC MODELING, AND FIELD SOLVER COMPUTER PROGRAMS

Computer programs are useful in the prediction of EMC and in EMI investigations because they reduce the probability of arithmetic error and speed up the computation. The use of a computer program to assist in an EMC analysis is a time-saving tool and may be the only practical method to deal with the multiplicity of potential EMI situations in large systems or complex equipment.

The dangers in overreliance on computer programs are well known and are the belief that the results computed by the program must be right even when the input data is invalid and the application of inappropriate modeling techniques is made, because they are embedded in the program. The limitations on a modeling approach are usually frequency and geometry, and when the computer program does not recognize these limitations the result can be a solution with large errors. The limitation on the computer code used is particularly true for simple programs, whereas the more complex programs require a great deal of decision making by the user, which can also result in large errors. If a simple conductive structure carries a known RF current or is exposed to an incident E field, simple antenna theory may be used to provide a ''sanity check'' on the radiated field or surface current. If the radiated field or induced current is higher than predicted, then the computer modeling is suspect.

For example, a simple cookbook-type program asks whether a wall is present between a radiation source and a receptor and, if the answer is yes, applies a fixed attenuation factor to the coupling equation. The type of material of which the wall is made and the presence of apertures, such as windows, will affect the real level of attenuation drastically. This effect is discussed in Section 6.8, in which the attenuation was seen to vary by as much as 40 dB when the fields measured at distances of 1 m from the interior walls of a building were compared to the field 15 m from the interior wall.

The more complex and powerful computer program can calculate the attenuation accurately, based on the conductivity and permittivity of external walls, and account for apertures and internal walls, etc. The problem with the more complex programs is the amount of time required to accurately input the geometry of the structures and describe the coupling paths required to be modeled. The computing time for large structures at high frequency may also be prohibitive.

Whatever the type of program used, it is imperative that the magnitude of the results be examined for consistency with either measured results or engineering experience. The modeling method and assumptions made in the program must be available to the user to enable a judgment on the applicability of the method to the problem. The accuracy of the results obtained by use of a program should be known in order to add a safety factor when a worst-case analysis is required. An overdesign may result when the accuracy of the prediction is low, with a potential increase in cost.

The differential-mode currents in a circuit are typically orders of magnitude higher than the common mode, and it is often difficult too see how these common-mode currents are generated. As discussed in Chapter 11 on PCB radiation, it is often the common-mode currents that are the prime source of radiation, and many commercially available PCB radiation computational programs ignore this important source.

The argument may be made that the use of EMC programs requires more engineering judgment, not less. A simple measurement may be possible that will allow validation of the program and increase confidence in its use. Measurements may have already been made, such as those described in Section 10.2.1, that compare the use of IEMCAP and measurements made on antenna-to-antenna coupling.

Another primary limitation on the use of computational electromagnetic modeling (CEM) and EM field solvers is the amount of computing resources or the time required to make a full wave simulation. This is often determined by the level of geometric complexity, the volume of the objects, or the distance from the objects. Because many codes require that the structure under consideration be divided up into elements that are a fraction of a wavelength in size, the limitation may be on the upper frequency that can be performed using the software. Some guidelines follow.

> If the code uses only transmission-line models, it may calculate only fields generated by the differential-mode currents and ignore the common-mode current as a source.
>
> First perform an analysis/prediction of the potential EMI problem areas and assess the criticality of the problem. Use numerical modeling techniques only when the initial analysis shows a potential EMI problem.
>
> Ensure that a measurement on the equipment or system is not going to provide a faster and more accurate results. Consider the use of a brass scale model when the system is very large or complex.
>
> The source of radiation must be as simple as possible but must still include the actual return path for grounds.
>
> It may be more efficient to model a chassis using an electromagnet boundary with an equivalent surface impedance, instead of a 3D model.
>
> The dielectric constant of PCB or cable insulation affects resonant frequencies, impedances, and field strength and should be included in the model where possible.
>
> The proximity of conductive surfaces and nonsymmetries can result in common-mode current flow, and here a 3D model may be required for accuracy.

The geometry of the problem should be made as simple as possible, depending on the fidelity required of the analysis.

Many field solvers are available for the analysis of signal integrity in circuits and on PCBs. They are used in timing simulations to compute delays and near-field coupling/crosstalk. These field solvers are often quasi-static, meaning that although the structure guides signals that are fundamentally propagating waves, the interconnect model is usually modeled as a lumped element or transmission line. These can be determined from seperately solved values of capacitance from a static electric field solver and inductance from a static magnetic field solver which produce the quasi-static values of inductance and capacitance. For this reason two separate methods of simulation may be required when both radiation from a circuit and signal quality are required.

The following information on available software and suppliers of software was accurate at the time of going to press. However, the names of software developers/suppliers change, or they are bought out by another company, or even the name of the software may change. Some suppliers of complex software have bought up software packages from other developers and combined them into a hybrid with a different name. Internet sites exist that describe both commercially available software and freeware, and these, or the suppliers' own websites, should be consulted to find out what is currently available. One such site is (or was) *emlid.jpl.nasa.gov/ emlib/files.html* and another is (or was) *http://www.emclab.umr.edu/codes.html*. Information on computer modeling and a newsletter are available by joining the Applied Computational Electromagnetics Society (ACES). Write to Dr. Richard W. Adler, Code 62AB, Naval Postgraduate School, Monterey, CA 93943.

The "EMC Analysis Methods and Computational Models" book is published by John Wiley and Sons and contains four computer programs which are available free from the web at www.tesche.com/book.htm. These programs are:

NULINE: Program to analyze the behavior of an above ground transmission line excited by either transient or CW lumped voltage or current source anywhere on the line or by an incident EM field.

RISER: Program to compute the voltage and current in a load at one end of a field-excited, above ground transmission line, modeling the ends (risers) of the line as smaller, vertical transmission lines.

LTLINE: Program to compute the load voltages and currents at each end of an above ground transmission line that is excited by a nearby lightning strike.

TOTALFLD: Program to evaluate the total above ground and below ground E and H fields produced by an incident EM plane wave. Both CW and transient results for a number of different user defined waveforms are present.

Information on commercially available PCB radiation programs, crosstalk prediction, and CAD programs are also provided in Section 11.8, although many of the computer programs described in this section are capable of modeling PCB radiation.

## 12.4.1. Simple Computer Programs

Simple programs written in BASIC are appended to several chapters in this book and have been found to be invaluable in predicting cable emission, coupling, and inductive crosstalk problems. These programs are available free from EMC Consulting Inc. Many more of the simple equations contained in the book have been used in computer programs that, as an example, predict attenuation of enclosures with apertures and seams.

As seen in Section 11.8, it is very important that a benchmark problem be presented to the supplier of a commercially available EM program before purchasing or relying on the output data. The benchmark should have been measured under carefully controlled conditions, using reliable, repeatable test methods, and the measured data compared to the data provided by the EM program. In some instances engineering judgment and experience can be used to test the validity of predicted data.

## 12.4.2.  Large-System-level and Complex Analysis Computer Programs

A number of programs exist that are designed for determining EMC in large systems containing multiple enclosures, cables, antennas, etc. located on a structure such as a building, vehicle, spacecraft, aircraft, or ship. Other programs are not specifically for EMC analysis but for electromagnetic analysis and allow detailed modeling of both large and small complex structures, sources, and receptors.

### 12.4.2.1.  Large-System EMC Programs

Large-system-level programs that allow modeling of the predicted or specified electromagnetic environment are useful in compatibility analysis, in waiver analysis for the impact of out-of-specification emissions and susceptibility, and in tailoring EMC requirements to a realistic environment.

Two of the most versatile large-system analysis programs are System and ElectroMagnetic Compatibility Analysis Program (SEMCAP) and Intrasystem ElectroMagnetic Compatibility Analysis Program (IEMCAP). These programs use a transfer function approach between sources and receptors (victims). The spectrum of the source is calculated, along with the coupled energy via the transfer function and the received spectrum and voltages that are compared to threshold levels of circuits to determine compatibility.

IEMCAP was developed for aircraft, with an emphasis on antenna-to-antenna and antenna-to-wire coupling and is applicable to ground, aircraft, and space systems. The system is divided into subsystems that are groups of equipment. Energy may enter or leave the equipment cases by ports designated as emitters or receptors or both. The ports may be intentional, such as connectors, or unintentional, such as seams. The location of equipment cases, cables and their terminations, ground planes, and structures is entered by use of a coordinate system. For each emitter port, a broadband power spectral density, which represents low-repetition-rate wide-frequency-range emissions, and a narrowband power spectrum, which represents discrete emission at a few frequencies, are represented.

The spectrum is defined by a math model or by the user. Spectrum types and transfer functions are shown in Table 12.9, which compares the IEMCAP, SEMCAP, and GEMACS programs. IEMCAP uses a sampled spectrum technique in which each spectrum amplitude is sampled at various frequencies across the range of interest up to a maximum of 300 over a range from 30 Hz to 18 GHz.

The receptor model is a spectrum representing susceptibility threshold over frequency range. Some of the advantages of IEMCAP are that the program can generate EMC specifications to supplement or replace MIL-STD-461, and tradeoff and waiver analysis may be achieved by comparing received power and susceptibility threshold, including those from previous runs.

One of the disadvantages of IEMCAP is that the field-to-wire coupling analysis allows only the coupling to bare wires, single-shielded wires, or double-shielded wires without the use of the transfer impedance of the cable or connectors. Thus the effect of different shielded cables and connectors cannot be included in the analysis, although losses in wires are included. The limitations at frequencies above 1 GHz on the diffraction model and fuselage and wing shading used in IEMCAP are illustrated in Section 10.2.1. Antennas modeled by IEMCAP are dipole, whip, slot, loop, parabolic dish, log-periodic, horn, phased array, and spiral, with the subparameters of gain, size, $-3$-dB vertical half-beamwidth, 3-dB azimuth half-beamwidth, and major

**Table 12.9**　Comparison of Intrasystem EMC Prediction Programs SEMCAP and IEMCAP and the Electromagnetic Analysis Program GEMACS

|  | Program | | |
| --- | --- | --- | --- |
|  | SEMCAP | IEMCAP | GEMACS |
| Interference analysis | X | X |  |
| Waiver analysis | X | X |  |
| Culling routines | X |  |  |
| *Model routines* |  |  |  |
| C/W/AM/FM/FSK | X | X |  |
| Pulse/ramp/step | X | X |  |
| Digital modulation |  | X |  |
| Noise |  | X |  |
| MIL-STD levels |  | X |  |
| *Transfer functions* |  |  |  |
| Wire to wire | X |  | X |
| Inductive | X |  |  |
| Capacitive | X |  |  |
| Conductive | X |  |  |
| E Field | X | X | X |
| H Field | X | X | X |
| Transmission line |  | X | X |
| Field to wire | X | X | X |
| Case to case | X | X | X |
| Case to wire | X | X | X |
| Antenna to antenna | X | X | X |
| Near/far surface wave |  |  | X |
| Internal to external ambient |  |  | X |
| External to internal ambient |  |  | X |
| Aperture to ambient |  |  | X |
| Enclosure internal fields |  |  | X |
| Aperture shielding |  |  | X |
| Cavity resonance |  |  | X |
| Cavity loading |  |  |  |
| *Susceptors* |  |  |  |
| Linear |  |  |  |
| Single filter | X | X |  |
| Multiple filters | X | X |  |
| Harmonics | X |  |  |

sidelobe gain. Up to seven types of filters used with either source or receptor models may be specified, plus a user-specified response. The maximum number of elements that can be used with IEMCAP Version 6.0 are

    Number of segments: 160
    Number of bundles: 160
    Number of bundle points: 320
    Number of wires: 320
    Number of segments/wire: 45
    Number of frequencies: 300

SEMCAP allows a wire-to-wire coupling analysis, which includes the modeling of cable harnesses, pigtail lengths, wire shielding, group shielding, bulkhead shielding, and ground return/common path resistances. Antenna-to-antenna and antenna-to-wire coupling requires the definition of fields rather than antenna description. Receptor models are voltage thresholds combined with a frequency response curve. Up to two filters of three types may be used for each source and receptor. The received spectrum is limited by the receptor bandwidth and is integrated over the frequency range of interest. It is the integrated EMI voltage that is used in the susceptibility analysis. As with IEMCAP, the input data contains the equipment and harness locations and the wire data. SEMCAP can handle up to 240 sources and receptors, and a large computer such as an IBM-360 must be used to run the program.

Further information on IEMCAP is available in Ref. 4 and on SEMCAP, and additional large-system programs, in Ref. 5.

The Electromagnetic Environment Effects EXpert Processor with Embedded Reasoning Tasker (E³EXPERT) is an expert system preprocessor that monitors the signal environment in the time domain and selects interference rejection schemes. E³EXPERT is a moderately conservative system-level culling tool based on enhanced RF coupling models resident in IEMCAP. Certain experimental modifications were made to IEMCAP to more accurately calculate geodesic path isolation for RF antenna-to-antenna and external field coupling. E³EXPERT applies artificial intelligence to solve sophisticated EMC problems such as the scenario of colocated spread spectrum transceivers on an airborne platform. Efficient methods are used to sift through and identify EMI conditions in both the time and frequency domains, rank the severity of the predicted interference, classify the dominant interference sources by type (i.e., broadband modulation, CW, harmonic, etc.), and select an appropriate EM1 solution. E³EXPERT is configured to run on a Windows NT personal computer and integrates a commercial expert system, a Windows-based graphical user interface, for user data/command entry, and a #D viewer/renderer incorporating a graphical editor. The editor provides a primary man–machine interface (MMI) to graphically manipulate displayed geometry models. In developing the electromagnetics engine, certain modifications were made to the IEMCAP geodesic path loss models based on enhanced formalisms resident in the Aircraft Inter-Antenna Propagation with Graphics (AAPG) computer program. These geodesic models are based on geometric optics (GO), geometric theory of diffraction (GTD), and the uniform theory of diffraction (UTD). When complete information regarding the interference is unavailable, the postprocessor expert system provides the receptor with a knowledge-based capability to monitor the environment and determine the interference process along with all necessary parameters. Based on a set of expert system rules, the knowledge-based processor selects one or more suitable interference-rejection schemes. The selection is made from a library of preselected techniques. In effect the system reacts to the EM interference environment so as to maximize performance. The Integrated Processing and Understanding of Signals (IPUS) is used primarily for applications where uncertainties exist about the signal environment. The E³EXPERT work was sponsored by the U.S. Air Force Research Laboratory/ IFSA, and more information on E³EXPERT is available in Ref. 7. The Electromagnetic Compatibility Analysis Program, EMCAP Version 4.0 9/30/1998, is available only to U.S. government agencies and contractors from Commander, Dahlgren Division, Attn Code J53 EMCAP, Naval Surface Warfare Center, 17320 Dahlgren Rd, Dahlgren VA 22448-5100; Tel (540) 653-8021; email: emcap@nswc.navy.mil. EMCAP is useful in ensuring that Navy radar systems do not cause harmful interference to shore-based civilian or commercial systems (e.g., FAA air traffic control radars, broadcast and satellite cable TV, cellular phones etc.). EMCAP uses genetic algorithms to balance the supply and demand for spectrum resources and calculate compatible frequencies for systems to use. EMCAP provides graphical views of the radar platforms' positions (user selectable), with the interference level between platforms indicated by color-coded lines. System(s) affected by EMI may be displayed for assistance in tactical decision making. A cumulative display of each radar system's spectrum is also available. The EMCAP database contains a vast array of parametric data for military radar systems; not much of this data is releasable to activities outside of the U.S. Department of Defence.

EMCAP outputs naval messages (MTF format) for promulgation to all radar platforms, to provide the frequency assignments for each system's use. These frequency assignments are calculated to increase system effectiveness while simultaneously minimizing interference to other systems, including both military and civilian systems. EMCAP also outputs a Standard Frequency Action Format (SFAF) file, which interfaces with other government software programs for spectrum management. The displays of interference interactions and spectrum usage may also be captured for output. The source code is in C++ and will run on 486- or Pentium- (recommended) based PCs using Windows 95/98 or Windows NT systems.

The Space and Naval Warfare Systems Center provides noncommercial software. The software includes high-fidelity EM propagation models, databases, and DOS/Windows 95/NT-based assessment systems for evaluating the effects of complex atmospheric environments and terrain upon 100-MHz to 57-GHz propagation.

## 12.4.3. Computer Programs for Numerical Electromagnetic Analysis

Several computer programs are available for analysis of fields generated by current-carrying conductors and fields coupled to conductive surfaces by the use of numerical techniques. Using this type of program, antennas or any current-carrying conductor are modeled by a wire mesh, plate, or barrier and not by antenna parameters.

### 12.4.3.1. Method of Moments Analysis

One of the common analysis techniques for coupling to a structure and radiation from a structure that can be modeled by wire segments or metal plates is the method of moments (MOM) technique. This is a frequency-domain technique that predicts a single frequency, although much of the available software allows iteration over a number of frequencies by repeating the analysis at a different frequency, with a resultant increase in computing time.

The moment method is a technique for solving complex integral equations by reducing them to a system of simpler linear equations. The moment method uses a technique known as the method of weighted residuals, which has become synonymous with the moment method and with the "surface integral technique." The MOM technique generally does an excellent job of analyzing unbounded radiation problems and analyzing perfect conductors or structures with a given conductivity. They are not well suited to structures with mixed conductivity and permittivity, although they can be used for homogeneous dielectrics alone or for very specific conductor-dielectric geometries. The MOM technique can be used to analyze a wide variety of 3D electromagnetic radiation problems.

The MOM technique requires that a structure be broken down into wires, which are in turn subdivided into small wire segments that are typically no larger than $\lambda/10$. Metal plates are divided into a number of surface patches, which again must be small. The current on every wire and every segment is calculated and displayed. The composite E field at any point in space can then be calculated by the contribution from each of the wire segments and patches. The magnitude of the RF current on the structure can be reviewed to ensure that it is not unrealistically large or small. For example, if at some wire location on a closed structure the current changes from a milliamp to an amp value and then immediately returns to a lower value, the structure has almost certainly not been accurately modeled. The method of moments allows discrete components to be inserted into a model by simply defining the impedance desired on any given wire segment. Thus a 50-$\Omega$ load can be inserted in the center of a dipole antenna; MOM assumes the current on the wire segment or patch to be the same throughout the conductor's depth. Therefore using MOM to determine the effects of an aperture with fields both inside and outside is difficult, and often a hybrid approach must be used. Because the MOM analysis is an iterative process, the larger the number of patches and wire segments, the longer the time required for analysis. With modern PCs, running at 500 MHz or above, the computation time has reduced significantly. But if after each iteration the partial solution is saved to the hard drive and new data is retrieved, it is this process that takes the time and may also affect the life of the hard drive when continuous.

Some of the programs that allow frequency-domain analysis and use MOM techniques are the Numerical Electromagnetic Code (NEC), MININEC, and the General Electromagnetic Code for Analysis of Complex Systems (GEMACS), which uses a number of techniques, including MOM.

**NEC:** The NEC program allows modeling of EM radiation, antenna performance, radar cross sections with accurate field emission, and field source computation. The NEC program uses MOM techniques for the numerical solution of integral equations for the currents induced on a conductive structure by sources or incident fields. The outputs include current distribution on wires, surfaces, coupling between antennas, and near-far-field magnitudes. There are at least four versions available from Nittany Scientific, 1700 Airline Highway, Suite 361, Hollister, CA 95023-5621; PH 408-634-0573; email: sales@nittany-scientific.com. NEC with NEC-2 emerged in 1981; NEC-4 appeared in 1992. NEC-2 is the highest version of the code under public domain. NEC-2 is a widely used 3D code developed at Lawrence Livermore National Laboratory more than 10 years ago. NEC-2 is particularly effective for analyzing wire-grid models, but it also has some surface patch modeling capability. NEC-2 is available from Ray Anderson's Unofficial NEC Archives. NEC-4 remains proprietary with Lawrence Livermore National Laboratory and the University of California. It requires a separate licensee for use. Some of the Windows-based versions are NEC-Win Basic, Nec-Win Plus, Nec-Win Pro, and GNEC. GNEC is available only to users licensed from Lawrence Livermore National Labs to use NEC4 and include all of the NEC-Win pro features plus more! Nec-Win Basic is designed for beginners, with each of the other versions enhanced by additional commands and geometry visualization, etc. For more information on the relative capability of all of these versions, contact the supplier at *http://www.nittany-scientific.com*. NEC-Win Basic includes:

Simple user interface for easy data entry and configuration
View or printing of graphical representations of antennas
Azimuth and elevation plot generation
The ability to use multitask in the Windows environment

NEC-WIN Basic combines NEC2 with Microsoft Windows and is composed of three parts: NEC-Vu, NEC-Plot, and NEC Geometry. NEC-Geometry is where the antenna is built. The coordinates of the wire are entered along with the specific data concerning frequencies and ground parameters. The following features are included:

Transmission lines
Linear or multiplicative frequency sweeps
Networks
Predefined or define-your-own wire conductivity & gauge
Translate/rotate/scale Antennas
Predefined or define-your-own grounds constants
Series/parallel RLC or complex loads

Sommerfield ground
Voltage/current sources
Linear/radial cliff or radial wire screen grounds
Tabular data for VSWR and input impedance

**MININEC:** MININEC was written because amateur radio enthusiasts who wanted to design antennas did not have access to large-frame computers on which to run NEC. Expert for Windows is ideal for the novice, student, or hobbyist. Expert MININEC Professional for Windows is suitable for the experienced student, hobbyist, or professional engineer, whereas Expert MININEC Broadcast Professional for Windows is a tool for the advanced student or the professional broadcast engineer. The Expert MININEC series is available from EM Scientific Inc. at 2533 N. Carson Street, Suite 2107, Carson City, NV 89706. For more information contact *http://www.emsci.com/mininec.htm.*

The Expert MININEC Series of computer programs is for the analysis of wire antennas using Microsoft Windows and is available on compatible IBM personal computers (PCs). The accompanying texts provide descriptions of the software and present the theory with relevant examples and validation data. The text for Expert MININEC for Windows is entirely online. Using the method of moments, the Expert MININEC Series solves an electric field formulation for the currents on electrically thin wires using a Galerkin procedure with triangular basis functions. This formulation results in an unusually compact and efficient computer algorithm. Radiation patterns, near field, charge distribution, impedance, and other useful parameters are computed from the current solution.

The user interface to Expert MININEC is through Microsoft Windows. Input data screens provide spreadsheet-like entry in individual windows. Output products are displayed in both tabular and graphics forms. The integrated graphics of the Expert MININEC Series include:

3D geometry display with rotation, zoom, and mouse support
3D current and charge displays
Linear, semilog, and log-log plots of currents, coupling, near fields, impedance, and admittance
Linear and polar pattern plots

The actual computational algorithms are implemented in FORTRAN for greater speed and make maximum use of available memory to set array sizes. The formulation has been changed from earlier versions of the MININEC formulation to use triangular basis functions. This results in greater accuracy. The short segment limit is machine accuracy. Square loops and Yagi antennas may be solved with confidence. In addition, a Fresnel reflection coefficient approximation improves the calculation of currents in the vicinity of real ground.

## System Requirements

There are a few minimum requirements to run the Expert MININEC Series. The computer must be an IBM-PC or compatible. A 486 processor or better is recommended. The minimum internal memory requirement is four megabytes. Eight megabytes or higher is recommended. Approximately 5.33 megabytes of disk drive are required to store the program. Problem definition and results files can accumulate quickly, requiring further hard disk space. The graphics card should be VGA or super VGA. Microsoft Windows 95 or Microsoft Windows NT is recommended. In the use of any Microsoft Windows program, a mouse is recommended. In the display settings, the small font size is recommended.

## Modeling Process

The Expert MININEC Series modeling process has five principal steps:

1. Geometry description definition
2. Electrical description definition
3. Model validation
4. Solution descriptions definition
5. Output display

A wire is subdivided into segments, with currents expanded as triangles centered at adjacent segment junctions. The endpoints of a wire have no triangles. If a second wire is added to the model, the second wire is subdivided into segments, with currents expanded as triangles as in the case of the first wire. In addition, if the second wire is attached to the first wire, a triangle is automatically located at the attachment end. Half of the additional triangle extends onto wire 2 and half onto wire 1. The half of the triangle on wire 1 assumes the dimensions (length and radius of the half segment of wire 1), while the half of the triangle on wire 2 assumes the dimensions of wire 2. Wire 2 overlaps onto wire 1, with a current triangle at the junction end. Additional wires may also overlap onto wire 1. It can be shown that for a junction of $N$ wires, only $(N - 1)$ overlaps with associated currents are required to satisfy Kirchhoff's current law. The convention in the MININEC Professional Series is that the overlap occurs onto the earliest wire specified at a junction. A wire junction is established whenever the user-defined coordinates of a wire end are identical to the end coordinates of a wire previously specified.

As mentioned previously, the choice of the number of segments is critical to the validity of the computation. It was suggested that segments around 0.02 wavelengths are a reasonable choice.

Since the conductance and susceptance values both converge as the number of segments increases, the convergence test can be used to determine the accuracy that can be expected for a given segmentation density (i.e., the number of unknowns per wavelength of wire).

Expert MININEC modeling geometry constructs include:

Cartesian, cylindrical, and geographic coordinate systems
Meters, centimeters, feet, or inches selection
Straight, helix, arc, and circular wires
Wire meshes
Automated canonical structure meshing
Node coordinate stepping
Symmetry options
Rotational and linear transformations
Numerical Green's function
Automated convergence testing

Electrical description options include:

Free space, perfect ground, and imperfect ground environments
Frequency stepping
Loaded wires
Lumped loads
Passive circuits
Transmission lines
Voltage and current sources
Plane-wave source excitation

Solution description options include:

Near fields
Radiation pattern
Two-part coupling
Medium wave array synthesis

Output products are displayed in both tabular and graphics forms. The integrated graphics of Expert MININEC include:

3D geometry displays with rotation, zoom, and mouse support
3D currents, charges, and patterns displays
Linear, semilog, and log-log plots of currents, coupling, near fields, impedance, and admittance
Smith chart plots of impedance and admittance
Linear and polar pattern plots

As a summary, Expert MININEC solves for:

Currents and charges on wires (peak or rms)
Impedance, admittance, S11 and S12
Effective height and current moments
Power and voltage losses
Multiport (antenna-to-antenna) coupling
Near electric and magnetic fields
Radiation patterns (dBi or electric fields, power or directive gain)
Medium wave array design
Auxiliary calculations of ground wave, stub matching, and tower footing impedance

It has been shown from the selected examples that MININEC gives comparable results to NEC. This is not a complete picture of the comparison of these codes, but it gives the reader a glimpse of the results to be expected. A more thorough analysis of MININEC shows that for a wide variety of problems, MININEC and NEC provide comparable results.

Expert MININEC for Windows: ideal for the novice, student, and hobbyist
Expert MININEC Professional for Windows: suitable for the experienced student, hobbyist and the professional engineer
Expert MININEC Broadcast Professional for Windows: a tool for the advanced student and the professional broadcast engineer

Expert MININEC for Windows

File handling
  Open, save, and delete problem files
  Print setup
Geometry constructs
  Geometry points defined in meters, centimeters, feet, inches, or degrees
  Geometry points iteration
  Environmental options, including free space, perfect ground, and real ground
  Straight wires
Electrical constructs
  Frequency stepping
  Ground options
  Lumped loads

  Loaded wires
  Voltage/current sources
 Solution space
  Currents, charges
  Impedance/admittance
  Near electric and magnetic fields
  Radiation pattern
 Diagnostics
  List of current nodes
  Geometry guidelines
  Definition evaluation and summary
  Online and context-sensitive help
  Online tutorial
 Run options
  Current
  Near fields
  Radiation patterns
  Frequency iteration
 Display types
  Text and interfaces to most available spread sheets
  3D geometry, 3D currents/charges, 3D patterns
  Linear plots
  Polar plots for patterns
  Smith chart
 Display options
  Admittance, impedance, S11, S12
  Effective height and current moments
  Power and voltage losses
  Current/charge on wires (peak or rms)
  Near electric and magnetic fields
  Radiation patterns
 Problem limits: 1250 unknowns and 500 wires

## The High-Frequency 3D Planar EM Solvers from Sonnet, Liverpool, NY

3D Planar Solver Products:

em: 3D Planar Electromagnetics Analysis Engine
emlets: Incremental Licensing Feature for Sonnet em
emgen: Netlist Interpreter Option for Sonnet em

Availability of Solver Products:

em: Sonnet 3D Planar Electromagnetics Analysis Engine
 This program is particularly useful for modeling PCBs and circuits.

  *em*® is the electromagnetics analysis engine. It uses an FFT-based method of moments technique based on Maxwell's equations to perform a true three-dimensional current analysis of predominantly planar structures. (Note: *predominantly planar* refers to planar structures with vias.) *em* will perform an automatic subsectioning of the circuit, allowing faster analysis to occur, but you can control how that subsectioning takes place. In addition, *em* provides the following features:

Complete control of the properties of any number of dielectric layers

Complete control of the DC and RF conductivity properties of any number of conductors independently

Output X-, Y-, or Z-parameters in formats compatible with the major circuit theory simulators

Output equivalent SPICE parameters for use in SPICE circuit simulators (particularly useful for the analysis of crosstalk in digital circuits)

Control circuit port impedance for analysis

Automatic de-embedding of circuits using TEM-equivalent impedances, if desired

Include the effects of stray coupling, parasitics, and EM radiation

Perform the analysis in a true shielded environment, to simulate enclosure effects, or in an open environment

Size of circuit that can be analyzed limited only by the size of your computer—software has no circuit size limits.

## MOMIC

MOMIC is a user-oriented method of moments PC program suitable for analyzing the electromagnetic behavior of arbitrarily shaped wire antennas and scatterers, modeled by piecewise linear segments, in free space. Capabilities of MOMIC include evaluations of the currents induced/excited on the wires, impedance/admittance parameters, near fields, and far-zone radiation and scattering patters. MOMIC can analyze various antennas and scatterers composed of electrically thin straight and curved wires, and wire-grid models of conducting surfaces. The target platform for MOMIC executable is an 80486 (or Pentium) running under MS-DOS in 32-bit protected mode. MOMIC is available on the Web.

### Matra Systeme Information, Toulouse, France

EMC2000 is a 3D electromagnetic code in the frequency domain using MOM based on the EFIE triangles, linear elements, and surface-to-wire junctions. The code uses various Green's functions, and a hybrid IPO asymptotic technique is linked with the MOM kernel. The software runs on PC platforms under WINNT. The MOM code is parallelized for multiprocessor PC and provides excellent run times. A GUI monitors the application and allows the user to focus only on the technique.

### Zeland Software, Fremont, CA

IE3D is a full wave, method of moments, 3D and planar electromagnetic simulation and optimization package for circuit and antenna applications.

### Technical University of Hamburg-Harburg Germany

CONCEPT is a 3D moment method that models lossy dielectrics as well as conductive surfaces and wires.

### 12.4.3.2.  GEMACS Hybrid Solution for Large, Complex Structures

GEMACS allows modeling of EM radiation and scattering, antenna performance, and radar cross section with accurate field predictions, including sources, reflections, and scattering. Outputs include current distribution on wires, surfaces, coupling between antennas, far/near fields, antenna terminal characteristics, scattering on conductive surfaces, cavity behavior, and aperture coupling.

Although GEMACS is not an EMC analysis program, it is compared to IEMCAP and SEMCAP in Table 12.9 in order to illustrate the relative strengths of the two types of programs. Both programs use the method of moments (MOM) physics model, which in the case of GEMACS is supplemented by geometric theory of diffraction (GTD) and the finite element method (FEM). GEMACS models complex structures by a hybrid using combined MOM/GTD/FEM techniques. These three techniques are implemented in GEMACS as follows.

GEMACS inputs are in two categories: the command language, which directs the program execution, and the geometry language, which describes the geometrical properties of the structure being analyzed. The command language controls the analysis used and describes the electric field, voltage or antenna excitation, and the reflected/scattered electric field. The incident, scattered, or total field may be computed for both near- and far-field conditions as a magnitude in either a spherical, cylindrical or Cartesian coordinate system. Two types of structure geometry are defined: thin-wire, wire mesh, or patch geometry used in the MOM analysis and the plate or cylinder geometry used in the GTD analysis. A mix of these is allowed in the MOM/GTD hybrid analysis. In the MOM analysis the structure under consideration is divided up into a group of thin-wire segments and/or surface patches. The choice of current distribution on these segments and patches is made, in the GEMACS MOM implementation specifically, in order to ease the numerical integration burden. It is now possible to compute the field at any point in space due to the current on any object of the structure, and the contribution from all parts of the structure may be added by superposition to form the total field. The currents on the structure may be due to excitation by plane or spherical waves. The MOM technique is used for electrically small objects. Because the current between the centers of wire segments are approximated by an expansion (basis) function and because near-resonant segments must be avoided, for the sake of accuracy, physical limitations are placed on the segment length and the mesh area. Generally these limitations are:

Mesh circumference $< 0.5\lambda$

Mesh length $< 0.1\lambda$, although good results have been achieved with long subsection lengths to $0.25\lambda$ and with square mesh to $0.14\lambda$

Adjacent subsections to differ in length by no more than a factor of 2

Angles between subsections to be no less than $20°$

Wire radii $< 0.001 \times$ subsection length

The wire subsections used in the MOM geometry need not be perfect conductors but may be loaded by either a fixed impedance (as a function of frequency), lumped loads, series or parallel *RLC* networks, and finite segment conductivity. Quasi-aperture coupling occurs between the apertures in a wire mesh. Patches may be used to reduce aperture coupling and to simulate ground planes, but patches are modeled as conductive surfaces only. The wire segment is specified by coordinates $(X_1, Y_1, Z_1)$ that define one end of the wire segment and $(X_2, Y_2, Z_2)$ that define the other end of the wire segment. The position of the wire segment in the coordinate system allows the construction of complex structures. The geometrical theory of diffraction is a high-frequency ray optics approach that has been extended to include diffraction effects from surface discontinuities and waves creeping around smooth structures. The basic geometry objects that can be used in GTD structures are those for which reflection and diffraction coefficients exist and for which it is possible to carry out the ray tracing of the optics approach. The advantage of GTD is that large objects equal to or greater than a wavelength may be modeled in bulk instead of requiring wire griding as with MOM. Rather than calculate geometry element currents to obtain field pattern data (the MOM approach), the scattered fields are obtained directly from the sources and geometry by tracing all geometrical optics paths from the sources to the field points. The waves are then reflected and diffracted as they follow these paths, from the surfaces, edges, and corners of the elements.

The GTD techniques are used when detailed structure elements are not important and when scattered fields are the only quantities desired. While it is theoretically possible to obtain any level of detail with enough GTD elements, in practice the user is forced to limit either the number of GTD elements or the number of physical interactions that are to be considered. The

GTD analysis does not include surface waves, although creeping waves are modeled. A large number of scattering mechanisms are possible, and with GTD these multiple scattering effects must be explicitly included in the analysis.

With the MOM/GTD hybrid configuration it is possible to combine the best features of MOM and GTD into a hybrid methodology. The scattering structure is modeled with both MOM and GTD geometry objects. This allows the user to specify detail (where required) by using MOM objects, but does not require that the entire geometry be gridded. Hence the matrix size and number of unknowns are minimized. By use of a hybridized finite differences algorithm with the GTD and MOM formulation, both the interior and exterior problems in the presence of each other are solved. This means that the external field, which may couple through an aperture, excite a MOM or GTD structure, and reflect back out of the aperture, can be computed. I believe that GEMACS is still the only frequency-domain unclassified program that has this capability.

One major problem with GEMACS Version 5.3 was the time required to model a complex structure. The modeling of the fuselage of an aircraft with a single jet turbine, using wire segments, with single-GHz-frequency incident field was just not realistic using a 486 PC, due to the incredible length of time required. The alternative was to model a much simpler structure, in which case the fidelity of the analysis could be severely compromised. The latest high-speed PCs will almost certainly reduce the time required, probably to days for a complex structure, but the earlier versions of GEMACS used very little RAM and made extensive use of the hard drive. In many cases for complex structures such as ships and aircraft, a brass scale model complete with transmitting and receiving antennas provides a faster solution when compared to computer analysis. A new version of GEMACS called GEMACS Plus is reported to increase analysis speed by utilizing all of RAM. Speed increases will vary based on the system and the type of analysis being performed. GEMACS Plus can be run on a UNIX, DOS, Windows, or OS/2 system. A compiled version of GEMACS Plus is commercially available from SM&A System Solutions.

The task of specifying the geometry and analyzing the output of GEMACS is not easy. A program designed to lighten this task is GAUGE/MODELED, which can supply code/data interface management, I/O simplification and guidance for geometry setup, command setup, and execution of GEMACS. It provides integrated data analysis capabilities that include data scaling, data editing, data comparison, and user-defined analysis. Errors in input data and commands are caught and fixed. All input data/commands required for a GEMACS run are completed with syntax prompts for input data.

GAUGE is available and provides a graphic display of input geometry and output fields and currents. The graphic display capability includes true 3D wire frame and solids, wire-to-patch-to-plate conversions, up to 16 colors (EGA), and pan, zoom, rotate, and translate. The structure may be modeled using GEMACS geometry for input and may be color coded by ID, TAG, Segment. GAUGE enables model replication and provides GEMACS-readable outputs. The output display includes color mapping of wire/surface currents, color-filled contours for detail, polar and rectangular plots, overlays, and near- and far-field patterns and converts GEMACS outputs to displays. Gauge for Windows (WinGAUGE V2.0) provides a common user interface with other Windows products and allows multiple interactive viewing of the same geometry. WinGAUGE V2.0 is available from SM&A. WinGauge V2.0 incorporates display/analysis using the new ray-tracing capability of GEMACS V5.3 and GEMACS Plus. SM&A also offers installation and training at your facility. SM&A can be contacted at 4695 MacArther Court, 8th floor, Newport Beach, CA 92660; PH: 949-975-1550.

The Ultra Corporation, P.O. Box 50, Syracuse, NY 13210; PH: 315 428-8122, under contract with the Air Force Research Laboratory, has developed a parallel version of GEMACS. The parallel version incorporates recent advances in high-performance computing and enables scalable run-time speed-up on distributed processors. The base GEMACS code is highly portable FORTRAN. The parallel version uses the message-passing standard MPI and runs on the widest possible family of parallel machines, from massively parallel supercomputers to networks of workstations or PCs. The code modifications for parallelization are to the structure of the software system, not to the underlying physics. In the low-frequency regime, the computation of the interaction and Green's function matrices are distributed, while a parallel ray-tracing algorithm is used for the high-frequency case. The hybrid situation employs both techniques. The run time is reduced from days to hours while maintaining the accuracy of the original code. The modified GEMACS code can now solve larger, more complex problems and handle higher orders of interactions. Like all other versions of GEMACS the code and documentation is under export control.

The EMC/IAP programs IEMCAP, GEMACS, and GAUGE are available to those working on government contracts and other eligible applicants. The PC version of GEMACS is available from B. Coffey, Advanced Electromagnetics, 5617 Palomino Dr. N. W., Albuquerque, NM 87120, who also offers a 4½-day seminar. G. Evans of Decision Science Applications Incorporated, 1300 Floyd Ave., Rome, NY 13440, offers a PC version of GAUGE and GEMACS. Decision Sciences offer a two-day seminar on GAUGE.

Agencies in foreign countries may obtain the GEMACS/GEMACS Plus programs with approval of the U.S. Air Force. Further information on availability may be obtained by contacting K. Siarkiewicz, Department of the Air Force, Headquarters Rome Laboratory (AFSC), Griffiss Air Force Base, New York 13441-5700.

The route for requests from Canada is to obtain a copy of form DD 2345, "Militarily Critical Technical Data Agreement," from the Canadian DSS and to contact K. Siarkiewicz.

### 12.4.3.3. Transmission-Line Matrix

A numerical technique suitable for solving electromagnetic problems that may involve nonlinear, inhomogeneous, anisotropic time-dependent material properties and arbitrary geometries is the transmission-line matrix (TLM) method. The method allows the user to compute the time-domain response of two-dimensional (2D-TLM) or three dimensional (3D-TLM) shaped electromagnetic structures to arbitrary excitation in time and space. Instead of interleaving E and H fields, a single grid is established and the nodes of this grid are connected by virtual transmission lines. Excitations at the node sources propagate to adjacent nodes through these transmission lines at each time step. A mesh can be approximated by a lumped element model using inductance and capacitance per unit length of each line in the mesh. Although the time-domain response is computed, the frequency-domain characteristics can be extracted by Fourier transform, and the method is then useful in modeling the response to lightning, EMP, and other events as well as the peak response to a continuous wave incident on a structure.

The excitation source may be an arbitrary function or a continuous waveform. More than one excitation source may exist at points in the mesh. The wave propagation in a shunt-connected TLM network, made up of a mesh of transmission lines, may be shown in which the voltage simulate electric fields and from which the $E_y$, $H_x$, and $H_z$ magnitudes can be calculated. In a series-connected TLM mesh in which the node current simulates a magnetic field, the $H_y$, $E_x$, and $E_z$ fields may be computed. A total or partial reflection of the wave propagation down the mesh occurs at boundaries that represent a structure. The boundaries may be lossless or lossy homogeneous or lossy inhomogeneous materials. Aperture coupling with an external source coupling through an aperture in the enclosure or from an internal source coupling out of the enclosure may be modeled with this technique. Either the E or H field coupling through the aperture may be computed. Complex nonlinear materials are readily modeled, but large problems that demand a fine mesh require excessive amounts of computing time.

An orthogonal combination of the shunt-connected and series-connected network types can be used to compute three-dimensional field problems. The TLM method allows modeling of field magnitudes and wave impedance and the surface wave on a structure. However, the method is not confined to wave-related problems but can be used to solve linear and nonlinear lumped networks. A varactor diode has been implemented in 2D-TLM by W.J.R. Hoefer, who describes both the 2D-TLM and 3D-TLM in Chapter 8 of Ref. 6. The same source provides the listing of a two-dimensional inhomogeneous TLM program for the personal computer.

The use of a graded mesh (more dense and less dense) and multigrid techniques for the modeling of microwave components in three dimensions show that multigrid techniques provide an effective means of reducing the mesh resolution away from discontinuities.

A hybrid 3D Symmetrical Condensed Node-Transmission Line Model (SCN-TLM) and a Finite Difference Time Domain (FDTD) can combine the specific features of both of these techniques. MeFisto for Windows 95/98/OT is a time-domain electromagnetic simulator based on the TLM method and is commercially available from Faustus Scientific Corporation, Victoria, BC, Canada.

Micro-Stripes is 3D-TLM software commercially available from Sonnett Software, Liverpool, NY. Micro-Stripes obtains the time-domain impulse response and extracts frequency-domain results and S-parameter data Postprocessing is done to obtain fields, S-parameters, far-field radiation patterns, energy, and power flow. Control over the density of the mesh exists as well as relative cell ratios. A brief description of the analysis method follows:

> Volume is divided into elemental cells modeled as the intersection of orthogonal, three-dimensional transmission lines.
>
> Voltage pulses are absorbed, transmitted, and scattered at each cell, and the simulation propagates in time from arbitrary initial field/voltage conditions.
>
> The fields are stored for each time step at output points, and postprocessing is done to obtain fields (E, H, energy, power), S-parameters, far-field radiation patterns, and outputs.
>
> A typical analysis can result in thousands of time steps. A fast Fourier transform (FFT) then provides thousands of frequency data points.

A free version of Sonnet® Lite can be downloaded that can be used to analyze planar structures, such as:

> Microstrip matching networks
> Stripline circuits
> Via (interlayer or grounded)
> Coupled transmission-line analysis
> Microwave circuit discontinuities
> Broadside-coupled transmission lines
> Microstrip or stripline filters
> Mounting pad characterization
> PCB trace crosstalk analysis
> Spiral inductors with bridges
> Planar interconnects

Sonnet Lite can provide EM analysis for 3D planar circuits up to two metal layers and up to 4-port circuits. A number of example programs are provided along with online manuals and tutorials.

A TLM program is available from the University of Victoria's TLM modeling group and is noncommercial.

One of the early problems with the TLM method was in predicting far-field radiation due to the need for a boundary at the mesh that absorbs and simulates free space. However, this problem seems to have been solved in commercial versions of TLM software. But, as with all software, the software company should be asked to run a benchmark prediction, such as far-field radiation from a microstrip configuration, that can be compared to known data before using or purchasing any EM software.

### 12.4.3.4. Finite Difference Time-Domain (FDTD) Technique

Finite difference describes a method for solving partial differential equations. Alternative methods of solving partial differential equations are the finite element method (FEM) and the boundary element method (BEM).

The FDTD method uses time-domain numerical modeling. Maxwell's equations in differential form are modified to central-difference equations, discretized, and implemented in software. The process is iterative, in that the electric field is solved, then the magnetic field is solved, and then the process is repeated over and over until a final result has been obtained.

According to Maxwell's differential form equations, the change in the E field is dependent on the curl of the H field. This results in the basic FDTD equation that the new value of the E field is dependent on the old value of the E field and the difference in the old value of the H field on either side of the E field point in space. Hence the name finite difference time domain. Likewise the new value of the H field is dependent on the old value of the H field and also dependent on the difference in the E field on either side of the H field point.

In order to use FDTD, a computational domain, which is the space where the simulation will be performed, must be specified. A mesh or grid within the space are defined with a material having a specific permeability, permittivity, and conductivity. Normally the mesh must be uniform, so the mesh density is determined by the smallest detail of the configuration. A source is then specified that can impinge on the space and the grid material, or it can be a current on a conductor, an electric field between plates. Since the E and H fields are computed directly, the data output is usually the E or H field at a point or a series of points within the space.

Because FDTD is a time-domain technique when a time-domain pulse (such as a Gaussian pulse) is used, a wide frequency range can be solved with only one simulation. This is valuable when resonant frequencies or a broadband result are desired. Thus a single simulation can determine current, voltage, or fields over a wide frequency range. The frequency-domain results can be be obtained by applying a discrete Fourier transform to the time-domain results. Arbitrary signal waveforms can be modeled as they propagate through complex configurations of conductors, dielectric, and lossy nonlinear nonisotropic materials.

The FDTD technique is excellent for transient analysis. It usually does a better job of modeling unbounded problems than FEM and is therefore often the method of choice for modeling unbounded complex inhomogeneous geometries. Because FDTD is a time-domain technique that finds the fields everywhere within the space, it can display moving images of the fields as they move throughout the space.

Since FDTD uses a computational domain, this must end somewhere, which needs the establishment of an absorbing boundary to simulate free space. The space must be gridded, and these grids must be small compared to the smallest wavelength and smaller than the smallest feature to be modeled. Very large spaces require very long solution times. Also, models with long, thin features, like wires, are difficult to model because of the very large space taken up by the wire. Because the E or H field is found within the space, the fields at very large distances from a source require very large spaces and excessive computation time. The far-field extensions are available for FDTD but require some postprocessing.

The FDTD technique allows apertures to be determined directly along with shielding effects, since the mesh can be described with different values of conductivity, and so boundaries with and without apertures an seams can be modeled. Also, the fields both inside and outside a structure can be determined.

The only significant disadvantage of FDTD is that, in common with most other techniques, the problem size and the fineness of the mesh will dictate the computation time. The fineness of the grid is determined by the dimensions of the smallest feature to be modeled, and so codes that offer a variation in the mesh size over the structure would have an advantage. Also the entire object, including most of the near field, must be covered.

## APLAC

APLAC, a program for circuit, system, and electromagnetic FDTD simulation and design, is a joint development of the Circuit Theory Laboratory of the Helsinki University of Technology, Aplac Solutions Corporation, Nokia Research Center, and Nokia Mobile Phones. The main analysis modes are of circuits, but it includes an electromagnetic FDTD simulator for solving 3D field problems independently or as part of a circuit design.

## EZ-EMC

> EMS-PLUS   Durham, NC
> sales@ems-plus.com

EZ-EMC is moderately priced 3D FDTD software with a user-friendly interface and the ability to create animations.

## Remcom Inc.

> State College, PA
> Phone: (814) 353-2986
> Fax: (814) 353-2986

XFDTD is 3D full-wave finite difference time domain (FDTD) software with an X/Motif graphical interface

## Sigrity, Inc.

> Binghamton, NY
> Phone: (607) 648-3111
> Fax: (607) 648-4020

SPEED97 is a special-purpose FDTD solver for modeling interactions in multilayer chip packages and printed circuit boards.

## Penn State FDTD Code

This is a public domain FDTD code developed by R. Luebbers and K. Kunz that is described in their book, *The Finite Difference Time Domain Method for Electromagnetics* published by the CRC Press.

### 12.4.3.5.   Combined FDTD and MOM Software

### EMIT® is from the SETH Corporation

> http://www.sethcorp.com
> EMIT®'s graphical interface permits users to assemble complex structures for input as
> 2D or 3D models. The toolbox approach of EMIT® offers three calculating kernel
> applications and two graphical processors with a variety of supporting tools.

According to the SETH corporation the optimization of the critical Adaptive Absorbing Boundary Conditions provides ''unsurpassed modeling accuracy.''

Modification of computational techniques optimizes system performance and enhances user application.

EMIT® features:

Finite difference time domain (FDTD)    Method of moments (MOM)
2D and 3D models    Graphical interface

Also in the SETH Corporation toolbox are the following.

**ENEC**: This is based on the popular numerical electromagnetics code. Models can be created from which RF currents at selected frequencies on the wires and conductors may be observed. Far-field and/or near-field can be calculated and presented as an entire complex structure.

**EMFIELDS-3D**: This is a three-dimensional implementation of the finite difference time-domain (FDTD) computational technique. With EMIT™, rectangular grid structures can be created and the effect of various media such as metal, air, or dielectric material can be observed. A time-domain pulse may be used to excite the model while observing the effects. Using the postprocessing analysis tool, the time-domain and frequency-domain may be visualized using the animation display. The data then can be manipulated to observe the simulated real-time effect.

**EMFIELDS-2D**: This is similar to EMFIELDS-3D, except that it is a two-dimensional implementation of the FDTD modeling technique, allowing for analysis of more complex structures in a quicker redraw time, thus saving time.

**EMIT™ PREPROCESSOR**: This enables the creation of the models to be used by the EMFIELDS and ENEC tools. The parameters of each simulation are entered using the graphical placement of sources of RF energy, conductive surfaces, conductive lines, dielectric volumes, and monitor points. The user can change the numerical parameters of the model, such as material conductivity, source spectrum, dielectric constants, and calculation detail. The graphical representation of the model can be observed on the screen and the appropriate modifications made if necessary. The created models may be saved and reused as templates for other applications.

**EMIT™ PREPROCESSOR**: This offers a variety of analysis functions and displays enabling the examination of the results of the numerical model. For convenient comparative analysis of the results, the EMFIELDS can be plotted in either one time domain or the frequency domain. The time-domain visual animation plots provide a movie of simulated real-time effects. This animation is an excellent visual aid.

**EMIT™** is available on the most popular versions of UNIX and VMS platforms.

The following problem is one familiar to most EMI engineers: In one case study a printed circuit board (PCB) is placed inside a shielded box, with a connector protruding through the shield, and a long cable is attached to the outside of the connector. Common-mode signals will be efficiently radiated by the long cable/wire. The box, with its PCB, can be conveniently modeled using FDTD. The outside segment of the problem can conveniently be modeled using MOM. And so the overall problem is modeled in two stages, inside and outside, using FDTD and MOM, respectively. The first stage (inside the shielded box) is modeled using the FDTD technique. In this problem the high-speed data circuits were placed in the back corner (away from the connector opening). The numerous PCB traces/nets were reduced to those either directly involved with the source circuits (high-speed signals), connected to the connector pins directly, or those possibly serving as fortuitous conductors.

Since FDTD is a time-domain modeling technique, only one simulation is needed to determine the "transfer function" between the source and the common-mode voltage across the entire frequency range of interest (30–1000 MHz). The common-mode voltage found in stage 1 is used as the source and is placed between the shielded box and the long wire. The wire and box thus form a monopole antenna, with the box acting as the ground plane. In this MOM model, the shielded box was converted to a wire frame box, and the long wires were broken into individual wire segments.

### 12.4.3.6.    Finite Element Method (FEM) and Combined FEM/MOM Techniques

The finite element method is for solving partial integral or differential equations. It is a variational technique that works by minimizing or maximizing an expression known to be stationary about the true solution. Generally FEM techniques solve for the unknown field quantities by minimizing an energy function.

The FEM technique uses a system of points called *nodes* that make a grid called a *mesh*. The starting point of the method is the subdivision (discretization) of a domain (volume) into small subdomains called *elements*, which are the spaces between the mesh. Although the elements need not be triangles, those subdivisions are called a *triangulation*. An element is described by its vertices and one point on each edge. These points are the nodes, and the FEM mesh is constituted from nodes and elements. The great advantage of FEM is that the domains can be conductive or a dielectric or have the properties of free space. One disadvantage is that the entire volume must be meshed, as opposed to MOM, which only requires the surfaces to be meshed. Often in time-dependent solutions the FEM mesh is extended to all time levels by a finite difference method.

One weakness with FEM is that it is relatively difficult to model open configurations, i.e., where the fields are not known at every point. Absorbing boundary are used to overcome this deficiency.

The major advantage with FEM is that the electrical and geometric properties of each element can be defined independently. Thus a large number of small elements can be set up in regions of complex geometry, and fewer, larger elements can be setup in relatively open regions.

In general, FEM excels at modeling complex inhomogeneous structures (components composed of different materials) but is not as good as MOM at modeling unbounded radiation problems. In contrast, the method of moments does not model components composed of different materials efficiently. However, the advantage of FEM in describing a dielectric volume can be combined with MOM to solve for currents on the surface of (or external to) a volume.

EMAP5 is a hybrid FEM/MOM code designed primarily to simulate electromagnetic interference (EMI) sources at the printed circuit board level. EMAP5 is a full-wave electromagnetic field solver that combines the method of moments with a vector finite element method (VFEM). It employs FEM to analyze three-dimensional volumes, and uses MOM to analyze the current distribution on the surface of these volumes.

The two methods are coupled through the fields on the dielectric surface. EMAP5 can model three incident plane waves, voltage sources (in the moment-method region), and impressed current sources (in the finite element region). The goal of the code is to efficiently model heterogeneous and arbitrarily shaped dielectric bodies attached to one or more conducting bodies. A typical example of such structures is a printed circuit board attached to long wires or cables.

EMAP5 is a 3D numerical electromagnetic modeling code developed at the University

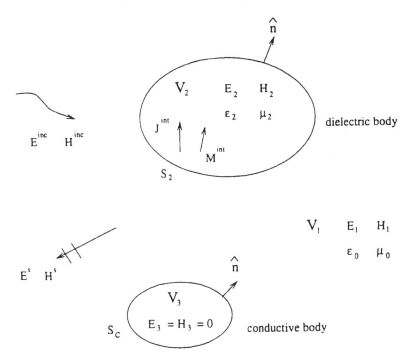

**Figure 12.14** A dielectric object and a conductive object illuminated by $H^i$ and $E^i$ fields or impressed sources $J^{int}$ and $M^{int}$. (© 1998.)

of Missouri-Rolla. The code can be freely downloaded from the World Wide Web at *http:// www.emclab.umr.edu/emap5*.

In one case study presented in Ref. 8 a structure shown in Figure 12.14 comprised a dielectric volume $V_2$, which has electrical properties ($\varepsilon_2$, $\mu_2$). The dielectric volume and conductive volume are illuminated by either incident $H^i$ and $E^i$ fields or impressed sources $J^{int}$ and $M^{int}$. A conductive volume $V_3$ is enclosed by a conductive surface $S_c$. The fields within $V_3$ vanish. $V_1$, which denotes the volume outside of $V_2$ and $V_3$, is assumed to be free space and hence has electrical properties ($\varepsilon_0$, $\mu_0$), ($\varepsilon_2$, $\mu_2$). ($\mathbf{E}_1$, $\mathbf{H}_1$) and ($\mathbf{E}_2$, $\mathbf{H}_2$) denote the electric and magnetic fields in $V_1$ and $V_2$, respectively. The unit normal vectors for $S_2$ and $S_c$ are defined pointing outward. Tetrahedral elements are used to discretize the dielectric volume $V_2$. The FEM and the MOM equations are coupled through $\{E_d\}$ and $\{J_d\}$ By solving the two equations, all fields within the FEM region and the surface equivalent currents can be obtained.

The EMAP5 software package includes three major components: SIFT5, EMAP5, and FAR. EMAP5 is developed for research and educational use. It does not have a sophisticated mesh generator or graphic visualization tools. SIFT5 can generate simple meshes for users. EMAP5 is the FEM/MOM field solver. FAR is the far-field calculator. The Standard Input File Translator Version 5 (SIFT5) is designed to generate input files for EMAP5. SIFT5 reads a text file in the SIFT format (1). Users can describe the structure of interest by using 11 keywords that are discussed in the EMAP5 software.

SIFT5 detects inconsistent input parameters in input fields and automatically prompts users to correct them. EMAP5 reads a file generated by SIFT5. EMAP5 will print out fields within areas specified by the *sif* file, one or more output files. All equivalent surface currents $J$ and $M$ will be printed.

FAR is a program used to calculate the far-field radiation pattern Far fields are calculated equivalent surface currents $J$ and need two input files. One is the file generated by SIFT5, and the other is the default output file generated by EMAP5. The FAR program will prompt the user to input the following parameters:

> The observing distance $R$ from the structure, in wavelengths. Usually, the value of $R$ should be higher than 20.
> The observing $\theta$ interval, in degrees in spherical coordinates.
> The observing $\varphi$ interval, in degrees in spherical coordinates.
> When the MOM part of the code is implemented, EMAP5 must know how triangles are linked to each other. In addition, a rule defining current direction must be specified.

EMAP5 supports the following three kinds of sources:

1. Voltage sources on metal patches
2. Plane-wave sources
3. Current sources within the FEM region

EMAP5 must know whether users need the default output. In addition, EMAP5 must know how many other output files users want.

A nonuniform mesh must be used to discretize some structures, for example, when the width of a trace is very small compared to its length.

Although EMAP5 is a FEM/MOM code, it can model configurations that require only one method to analyze. In this case, only the MOM portion of the code is employed.

A dipole with a source located in a dielectric source is presented as an example. With the dielectric slab set to a relative permittivity of 1.0, the model is a half-wavelength dipole in free space. This problem requires 80 megabytes of memory and 20 hours to run on a Sun Ultra workstation.

**Students' QuickField**

Students' QuickField™, formerly known as ELCUT, is a 2D finite element simulation package solving plane and axisymmetric problems of electrostatics, nonlinear DC magnetics, AC magnetics, current flow, nonlinear heat transfer, stress analysis, and coupled problems on any PC.

**Computer Simulation Technology**
> Darmstadt, Germany
> Tel: +49 6151 7303 0
> Fax: +49 6151 718057
> e-mail: info@cst.de

**MA.F.I.A.4** is a general-purpose 3D solver for Maxwell's equations, from DC to THz. Based on the "finite integration method (FIM)."

**MicrowaveStudio**

User-friendly ECAD tool for HF applications, using a "perfect boundary approximation." Also based on FEM.

**Tera Analysis**
> Tarzana, CA
> Phone: (818) 831-9662

Quickfield a relatively easy 2D FEM modeling code. Shareware version available.

**Weidingler Associates Inc**.
> New York, NY
> Phone: (415) 949-3010

EMFlex is a a 3D finite element time-domain solver for Maxwell's equations, with integrated pre- and postprocessing tools.

## The HP HFSS Analysis Tool

Versions 5.0 and 5.2 of Hewlett-Packard's electromagnetic (EM) high-frequency structure simulator software, the HP High-Frequency Structure Simulator (HP-HFSS) Designer, allows designers for the first time to integrate electromagnetic analysis into their high-frequency circuit and component design process.

Version 5.0 of the HP HFSS software represents a departure from the firm's earlier partnership with ANSOFT (Pittsburgh, PA), in which Hewlett-Packard actually marketed the EM analysis engines implemented by ANSOFT. The latest version of the HP software is developed entirely at Hewlett-Packard, allowing the company to incorporate suggestions from customers on improving speed, memory use, and drawing capabilities.

HFSS is a complete solution for EM modeling of two- and three-dimensional structures. Structures can be any combination of electric and magnetic field boundaries. The HFSS analysis engine is finite element method (FEM) simulation.

In operation, the HFSS analysis engine subdivides arbitrarily shaped three-dimensional structures into small tetrahedron-shaped elements. Electric fields are represented with higher-order tangential edge-element basis functions within each tetrahedron. The interactions between basis functions are computed using differential forms of Maxwell's equations. These computations yield linear equations that can be represented in a sparse matrix. The solution in turn yields an approximation of the fields as well.

The EM software automatically decreases the size of the finite elements where strength is greatest for increased resolution and accuracy. An improved meshing algorithm in Version 5.0 uses memory and central processing unit (CPU) resources more practically for reduced computing time and more efficient memory consumption, avoiding the ''out-of-memory'' messages of previous versions.

*Package predictions*: The software can also be used to model fields and surface currents within a package. The modeling tool provides S-parameter output data that can be used to model the package lead frame and from which bond-wire capacitances as well as self-inductances and mutual inductances can be extracted in conjunction with a circuit simulator. HFSS can model circuit via holes, electromagnetic interference (EMI) shields, and data-bus terminals. This last application allows prediction of crosstalk, inductive coupling from power lines, and reflections from interconnections. The HFSS program can even be used to model high-speed probes and probe calibration standards for advanced test applications of digital and analog circuits.

To design a structure, an operator enters the desired geometry, adds input and output ports, and assigns material properties (such as the dielectric constant of substrates). The software employs adaptive mesh-refinement techniques to solve to a user-specified level of accuracy. The desired accuracy is a function of computing/solving time, with higher levels of accuracy requiring more CPU time. Display functions provide animated images of EM fields, surface current flow, antenna polar patterns, Smith charts of S-parameter information, and tabular results.

The software runs on a wide range of UNIX platforms, Sun SPARCstation, IBM RS/6000, Digital Equipment Corp. Alpha workstations, IBM personal computers when well equipped with memory.

## The ANSOFT HFSS Analysis Tool

Because ANSOFT originally developed HFSS specifically for Hewlett-Packard, the HP HFSS is similar to the Ansoft Maxwell Eminence program. Both the HP and Ansoft new products are called HFSS, and both are version 5.0! Hewlett-Packard's HFSS is sold on its own, and Ansoft's HFSS can be sold by itself or as a part of the larger Maxwell Eminence package. The HP HFSS is not using the Ansoft-developed code, but a new HP finite element and mesh engine.

The ANSOFT HFSS is a complete solution for the electromagnetic (EM) modeling of arbitrarily shaped, passive three-dimensional structures, including RF and microwave connector and adapter design.

*ANSOFT HFSS V 5.0 features*:

    All-new HP EEsof, finite element simulation and mesh engines
    50% reduction in the memory required to perform a simulation
    Increased simulation accuracy for both simple and highly complex structures
    New drawing environment
    New Object Library
    Unlimited Undo command and dynamic 3D views
    New job control features
    Improved Fast Frequency Sweep
    PC and UNIX platform support
    Added display capability
    Dynamic rotation
    Improved antenna features

The ANSOFT HFSS program computes S-parameters and full-wave fields for arbitrary three-dimensional passive structures and is useful for microwave, millimeter-wave, and wireless device analysis.

The program will compute and display:

    Characteristic port impedances and complex propagation constants
    Generalized S-parameters and S-parameters renormalized to user-specified port impedances
    Basic electromagnetic field quantities and radiated electric fields for open boundary problems

## Vector Fields & Infolytica

A suite of FEM software, including TOSCA, ELEKTRA, SCALA, and SOPRANO, is commercially available from Vector Fields Inc. 2D codes PC-OPERA and OPERA-2D are also available from Vector Fields Inc. The ESI Group/SEMCAP provides PAM-CEM and CEM-3D, both of which are 3D solutions of Maxwell's equations using FEM. Infolytica Corporation, Montreal, Canada, offers Full Wave, a 3D full-wave electromagnetic modeling software employing FEM.

### 12.4.3.7.  GMT, MMP, CGM, BEM, and UTD Techniques

The *generalized multipole technique (GMT)* is a method of solving partial differential (weighted residuals) equations, and like MOM it is a frequency-domain technique. The solution is approximated by a set of base functions that are analytical solutions of the respective difference equations. However, this method is unique in that the expansion functions are analytic solutions of the fields generated by sources located some distance away from the surface where the boundary condition is being enforced. Moment methods generally employ expansion functions representing quantities such as charge and current that exist on a boundary surface. The expansion functions of GMT are spherical wave field solutions, and multipoles are in general the most flexible and efficient expansion functions. By locating these sources away from the boundary, the field solutions form a smooth set of expansion functions on the boundary, and singularities on the boundary are avoided. GMT is a frequency-domain method, and solvers for electrostatics or electromagnetic scattering can be efficiently built using GMT. It can also be combined with MOM to provide a hybrid approach.

The *multiple multipole program (MMP)* was proposed in 1980 by Christian Hafner. It is a semianalytical method for numerical field computations that has been applied to electromagnetic fields. It is a code for solving electromagnetic scattering and for guided waves. The amplitudes of the basis fields are computed by a generalized point-matching technique that is efficient, accurate, and robust. MMP "knows" many different sets of basis fields, but multipole fields are considered to be most useful. Due to its close relations to analytic solutions, MMP is very useful and efficient when accurate and reliable solutions are required.

The *conjugate gradient method (CGM)* is another technique based on the method of weighted residuals. It is very similar conceptually to MOM techniques. Two features that distinguish it from MOM are the way in which the weighting functions are utilized and the method of solving the linear equations. Iterative solution procedures such as CGM are most advantageous when applied to large, sparse matrices.

The *boundary element method (BEM)* is also a weighted residual technique. It is essentially a MOM technique whose expansion and weighting functions are defined only for a boundary surface. Most general-purpose MOM electromagnetic modeling codes employ a BEM.

The *uniform theory of diffraction (UTD)* is an extension of the geometric theory of diffraction (GTD). Both of these techniques are high-frequency methods. They are accurate only when the dimensions of objects being analyzed are large relative to the wavelength of the field. In general, as the wavelengths of an electromagnetic excitation approach zero, the fields can be determined using geometric optics. UTD and GTD are extensions of geometric optics that include the effects of diffraction.

MaX-1 is a new graphic platform for PCs under Windows 95/NT and is designed by Ch. Hafner of the ETHZ (Swiss Federal Institute of Technology) in Zurich, Switzerland. MaX-1 contains a new version of 2D MMP and will contain a new version of 3D MMP. A version of MMP for static and quasi-static problems is available from M. Gnos.

The 2D MMP and a GMT textbook are available in a newer version for Windows 3.1 and this can be downloaded for free. The 3D MMP code is a code for the simulation of electrodynamic fields based on the GMT technique. Its PC version for Windows 3.1 (including Fortran77 source) has been published by John Wiley & Sons (Christian Hafner and Lars Bomholt, *The 3D Electromagnetic Wave Simulator*). This version does not contain an eigenvalue solver (for guided waves on cylindrical surfaces). It is restricted to EM scattering (including waveguide discontinuities and similar problems that can be formulated as EM scattering) and antenna design. 3D MMP has a graphic front end for PCs that can be used as a platform for implementing and test finite difference schemes and other iterative procedures. The (free for 3D MPP owners) upgrade of 3D MMP includes features for computing gratings and biperiodic structures, multiple excitations, advanced graphics, etc. An eigenvalue solver for guided waves and resonator computations is contained in a prerelease test version. The executable code compiled for Windows NT (Upgrade 3) can be downloaded for free. Further information is available from *http://alphard.ethz.ch/hafner/mmp.htm*.

### 12.4.3.8. Electrostatic, Magnetostatic, and Quasi-Static Field Analysis

**AnSoft Corporation, Pittsburgh, PA**, offers the Maxwell 2D and Maxwell 3D, which are electrostatic, magnetostatic, and low-frequency magnetic FEM codes, in addition to the Maxwell Eminence, which combines FEM and the boundary element method to simulate electrostatic, magnetostatic, and quasi-static field analysis and full-wave simulation. The Ansoft HFSS code is described in Section 12.4.3.6.

The ANSYS/EMAG program from ANSYS, Canonsburg, PA, is a self-contained electromagnetic analysis package that covers static magnetic analysis, both 2D and 3D, harmonic analysis, both 2D and 3D, and transient magnetic analysis, both 2D and 3D.

**Infolytica Corporation, Montreal, Canada**, offers the Gemini 2D magnetostatic solver employing FEM, the MagNet5, which is a 2D/3D magnetostatic, electrostatic, eddy current, and transient analysis code.

**Integrated Engineering Software, Winnipeg, Canada,** offers:

>MAGNETO: a 2D magnetostatic solver with permanent magnet (PM) capability using the boundary element method (BEM)
>
>ELECTRO: a 2D electrostatic solver using the BEM method
>
>AMPERES: a 3D magnetostatic solver with permanent magnet (PM) capability using the boundary element method (BEM)
>
>COULOMB: a 3D electrostatic solver using the BEM method
>
>OERSTED: a 2D low-frequency time-harmonic magnetic field solver with permanent magnet (PM) capability using the boundary element method (BEM)
>
>FARADAY: a 3D time-harmonic eddy current analysis based on BEM
>
>MARCONI: a 3D code employing BEM

**Magsoft Corporation, Troy NY** offers:

>FLUX2D: a 2D FEM-based CAD program for static, low-frequency, or transient problems
>
>FLUX3D: a 3D FEM-based CAD program for static, low-frequency, or transient problems
>
>PHI3D: a 3D BEM-based electrostatic/magnetostatic field solver
>
>WAVE2D: a FEM-based electromagnetics program capable of handling a large class of electromagnetic wave problems in two dimensions

**Quantic Labaroratories Inc., Winnipeg, Canada**, offers GREENFIELD 2D, a 2D electrostatic and magnetostatic field solver and a relatively easy-to-use 2D finite element modeling code. A shareware version for students is available.

## REFERENCES

1. G. Dash, I. Strauss. Digital EMI testing:how bad is it? Studies document variances. Newswatch EM Compliance Engineering Vol. VII, issue 2, page 39, Winter 1990.
2. Li, Rockway, Logan, Tam. Microcomputer Tools for Communications Engineering. Artech House Publications. Boston, 1987.
3. R.F. Harrington. Field Computation by Moment Methods. Macmillan, New York, 1968.
4. T.E. Baldwin, G.T. Capraro. Intrasystem Electromagnetic Compatibility Program (IEMCAP). IEEE Transactions on Electromagnetic Compatibility Vol. EMC-21 1980.
5. B.E. Keiser. Principles of Electromagnetic Compatibility. Artech House Publications. Boston, 1987.
6. W.J.R. Hoefer. Numerical Techniques for Microwave and Millimeter-Wave Passive Structures. Wiley, New York, 1989, Chapter 8.
7. A. Drozd, A. Pesta, D. Weiner, P. Varshney, I. Demirkiran. Application and Demonstration of a Knowledge-Based Approach to Interference Rejection for EMC. IEEE, 1998 EMC Symposium Record.
8. M.W. Yun Ji, T.H. Ali, T. Hubing. EMC Applications of the EMAP5 Hybrid FEM/MOM Code. IEEE, 1998 EMC Symposium Record.

# Appendix 1
## Characteristic Impedance of Conductors, Wires, and Cables

Single wire, near ground

For $d \ll h$,

$$Z_o = \left(\frac{138}{\sqrt{\varepsilon_r}}\right) \log_{10}\left(\frac{4h}{d}\right)$$

Two-wire, differential-mode, near ground

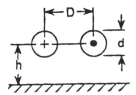

For $d \ll D, h$,

$$Z_o = \left(\frac{276}{\sqrt{\varepsilon_r}}\right) \log_{10}\left(\frac{2D}{d}\right)\left(\frac{1}{\sqrt{1 + (D/2h)^2}}\right)$$

Two-wire, common-mode, near ground

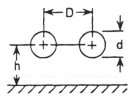

For $d \ll D, h$,

$$Z_o = \left(\frac{69}{\sqrt{\varepsilon_r}}\right) \log_{10}\left(\frac{4h}{d}\right) \sqrt{1 + (2h/2D)^2}$$

Asymmetric two-wire line

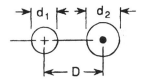

$$Z_o = \left(\frac{277}{\sqrt{\varepsilon_r}}\right) \log_{10}\left(\frac{2D}{\sqrt{d_1 d_2}}\right)$$

Symmetric two-wire line

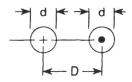

or twisted pair

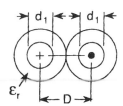

For $D \gg d$,

$$Z_o = \left(\frac{277}{\sqrt{\varepsilon_r}}\right) \log_{10}\left(\frac{2D}{d}\right)$$

Single wire between grounded parallel planes, ground return

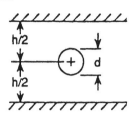

For $d/h < 0.75$

$$Z_o = \left(\frac{138}{\sqrt{\varepsilon_r}}\right) \log_{10}\left(\frac{4h}{\pi d}\right)$$

Balanced line between grounded parallel planes

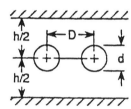

For $d \ll D, h$,

$$Z_o = \left(\frac{277}{\sqrt{\varepsilon_r}}\right) \log_{10}\left(\frac{4h \tanh(\pi D/2h)}{\pi d}\right)$$

Coaxial line

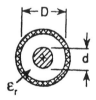

$$Z_o = \left(\frac{138}{\sqrt{\varepsilon_r}}\right) \log_{10}\left(\frac{D}{d}\right)$$

Shielded two-wire, common-mode (sheath return)

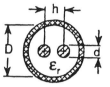

For $d \ll D, h$,

$$Z_o = \left(\frac{69}{\sqrt{\varepsilon_r}}\right) \log_{10}\left[\left(\frac{h}{2d(h/D)^2}\right)\left(1 - \left(\frac{h}{D}\right)^4\right)\right]$$

Shielded two-wire line, differential mode

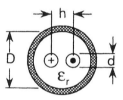

For $D \gg d, h \gg d$

$$Z_o = \left(\frac{277}{\sqrt{\varepsilon_r}}\right)\left[\left(\frac{2h}{d}\right)\left(\frac{D^2 - h^2}{D^2 + h^2}\right)\right]$$

Microstrip lines

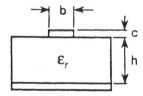

$$Z_o = \left(\frac{201}{\sqrt{\varepsilon_r + 1.41}}\right) \log_{10}\left(\frac{5.98h}{0.8b + c}\right)$$

Trace-to-trace impedance (no ground plane)

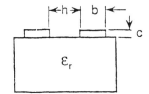

$$Z_o = \left(\frac{277}{\sqrt{\varepsilon_r}}\right) \log_{10}\left(\frac{\pi h}{b + c}\right)$$

Strip lines

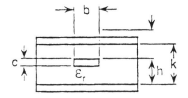

$$Z_o = \left(\frac{138}{\sqrt{\varepsilon_r}}\right) \log_{10}\left[\frac{4k}{0.67\pi b(0.8 + c/b)}\right]$$

Flat parallel conductions

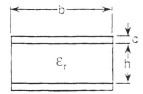

$$Z_o = \left(\frac{867}{\sqrt{\varepsilon_r}}\right) \log_{10}\left(\frac{h}{b}\right)$$

# Appendix 2
## Units and Conversion Factors

| Quantity | Units and relationships between units |
|---|---|
| Length | 1 inch = 2.54 cm = 25.4 mm = 1000 mil<br>1 m = 100 cm = 39.37 inches<br>1 angstrom = $10^{-8}$ cm<br>1 mile = 1.609 km<br>1 micron = $1 \times 10^{-6}$ m = $1 \times 10^4$ angstrom |
| Mass (kg) | 1 kg = 1000 grams = 2.2 lb |
| Area ($m^2$) | 1 $m^2$ = 10.76 $ft^2$ = 1550 in.$^2$ |
| Work, energy (W) | 1 joule = 0.738 ft-lb = $0.947827 \times 10^{-3}$ BTU<br>      = $10^7$ ergs = 0.239 cal<br>1 ft- = $1.356 \times 10^7$ ergs<br>1 BTU = $1.055 \times 10^{10}$ ergs = 252 cal |
| Pressure ($N/m^2$) | 1 newton/$m^2$ = 1 pascal = 1 dyne/$cm^2$<br>    lb/in. = $6.8947 \times 10^4$ dynes/$cm^2$<br>        = 0.068046 atm..<br>1 inch mercury pressure = 0.03342 atm<br>1 bar = $10^6$ dynes/$cm^2$ |
| Force (N) | 1 newton = 0.225 lb = $1 \times 10^5$ dynes |
| Temperature | $0°C = 32°F = 273.16$ K<br>Fahrenheit to celsius conversion:<br>  $T_C = (5/9)(T_F - 32)$<br>Celsius to kelvin conversion:<br>  $K = T_C + 273$ |
| Resistivity | 1 ohm/cm or 1 ohm/$cm^3$ |
| Permittivity of free space ($\varepsilon_0$) | $\dfrac{1}{36\pi \times 10^9}$ [H/m] = 8.8 [pF/m] |
| Velocity (m/s) | 1 m/s = 3.28 ft/sec |
| Magnetic flux ($\phi$)(Wb) | 1 weber = $10^8$ maxwell = $10^8$ lines<br>      = $10^8$ gauss-$cm^2$ |
| Magnetic flux density (B) | 1 Tesla (1) = 1 weber/$m^2$ = $10^4$ gauss<br>  1 gauss = 1 line/$cm^2$ = 1 maxwell/$cm^2$<br>      = $7.936 \times 10^5$ A/m |
| Magnetic field strength (H) | 1 A/m = 0.0125 oersteds<br>1 oersted = 79.6 A/m |

| Quantity | Units and relationships between units |
|---|---|
| Permeability of free space ($\mu_o$) | $\mu_o = 4\pi \times 10^{-7}$ H/m = 1.256 $\mu$H/m<br>where $\mu = B/H$ |
| Magnetization (M)<br>Magnetic moment/unit volume $\Big\}$ | 1 A/m |
| Magnetic moment<br>Magnetic motive force (F)<br>Reluctance | 1 A/m$^2$ = 1 joule/(weber/m$^2$)<br>1 amp-turn = 1.257 gilbert = 1.257 oersted-cm<br>1 amp-turn/weber |

# Appendix 3
## Electric Field Strength to Magnetic Field to Power Density Conversions

Far-field conditions where

$$Z_w = 377 \ \Omega, \ W/m^2 = E^2/Z_w, \ \mu_o = 4\pi \times 10^{-7} \ H/m$$

| 1 A/m | = 377 | V/m |
|---|---|---|
| | = 171 | dBμV/m |
| | = 120 | dBμA/m |
| | = 25.8 | dBW/m² |
| | = 55.8 | dBm/m² |
| | = 37.7 | mW/m² |
| | = $1.26 \times 10^{-6}$ | T |
| | = $1.26 \times 10^{3}$ | nT |
| | = 0.0126 | Gauss |
| | = 0.0126 | Oersted |

| 1 V/m | = $2.65 \times 10^{-3}$ | A/m |
|---|---|---|
| | = 120 | dBμV/m |
| | = 68.5 | dBμA/m |
| | = 2.65 | mA/m |
| | = −25.7 | dBW/m² |
| | = 4.3 | dBm/m², |
| | = −5.73 | dBm/cm² |
| | = $2.67 \times 10^{-4}$ | mW/cm² |
| | = $2.67 \times 10^{-3}$ | W/m² |
| | = $3.3 \times 10^{-5}$ | gauss |
| | = $3.33 \times 10^{-9}$ | T |
| | = 3.33 | nT |
| | = $3.33 \times 10^{-5}$ | Oersted |

| 1 W/m² | = 19.4 | V/m |
|---|---|---|
| | = $5.15 \times 10^{-2}$ | A/m |
| | = 0.1mW/cm² | |

| 1 nT | = $7.936 \times 10^{-4}$ | A/m |
|---|---|---|
| | = $1 \times 10^{-5}$ | Gauss |

| 1 pT | = $7.936 \times 10^{-7}$ | A/m |
|---|---|---|

| 1 T | = 1 | weber/m² |
|---|---|---|
| | = $10^{4}$ | Gauss |
| | = $7.936 \times 10^{5}$ | A/m |

| 1 Gauss | = 1 | Oersted |
|---------|-----|---------|
|         | = 79.6 | A/m |
|         | = 0.796 | A/cm |
|         | = $1 \times 10^{-4}$ | T |
|         | = 0.1 | mT |

# Appendix 4
## Commonly Used Related Formulas

Electric charge, coulombs (C)

$$Q = CV = It$$

Electric energy, joules (J)

$$W = IVt = QC$$

Power density, watts per square meter (W/m$^2$)

$$P_d = \frac{E^2}{Z_w} = EH$$

Wave velocity in air, meters per second (m/s)

$$v = \frac{1}{\sqrt{\mu_o \varepsilon_o}} = 3 \times 10^8$$

Wave velocity in medium, meters per second (m/s)

$$v = \frac{1}{\sqrt{\mu_r \mu_o \varepsilon_r \varepsilon_o}}$$

Characteristic impedance, ohms ($\Omega$)

$$Z_o = \sqrt{\frac{L}{C}}$$

Velocity, meters per second (m/s)

$$v = \frac{1}{\sqrt{LC}}$$

Inductance, henrys (H)

$$L = \frac{Z_o}{v} = Z_o^2 C$$

Capacitance, farads (F)

$$C = \frac{1}{v^2 L}$$

Wavelength, meters (m)

$$\lambda = \frac{v}{f}$$

Wavelength in air, meters (m)

where $v = c = 3 \times 10^8$ m/s'

$$\lambda = \frac{3 \times 10^8}{f[\text{Hz}]} = \frac{300}{f[\text{Hz}]}$$

Flux density, teslas (T)

$$B = \mu H$$

Resonant frequency of $LC$ circuit, hertz (Hz)

$$f = \frac{1}{2\pi\sqrt{LC}}$$

Reactance of capacitor, ohms ($\Omega$)

$$X_C = \frac{1}{2\pi f C}$$

Reactance of inductor ($\Omega$)

$$X_L = 2\pi f L$$

Impedance of resistance and inductance in series, ohms ($\Omega$)

$$Z = \sqrt{R^2 + (2\pi f L)^2}$$

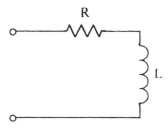

Impedance of resistance and capacitance in series, ohms ($\Omega$)

$$Z = \sqrt{R^2 + \left(\frac{1}{2\pi f C}\right)^2}$$

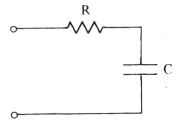

Impedance and resonant frequency of a resistance, inductance, and capacitance in series, ohms, hertz ($\Omega$, Hz)

$$Z = \sqrt{R^2 + \left(2\pi f L - \frac{1}{2\pi f C}\right)^2}$$

$$f = \frac{1}{2\pi\sqrt{LC}}$$

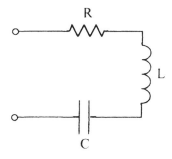

Impedance of resistance and inductance in parallel, ohms ($\Omega$)

$$Z = \frac{1}{\sqrt{\left(\frac{1}{R}\right)^2 + \left(\frac{1}{2\pi f L}\right)^2}}$$

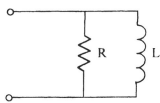

Resistance and capacitance in parallel, ohms ($\Omega$)

$$Z = \frac{1}{\sqrt{\left(\frac{1}{R}\right)^2 + (2\pi f C)^2}}$$

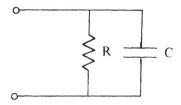

Impedance and resonant frequency of a resistance in series with inductance, and both in parallel with capacitance, ohms, hertz ($\Omega$, Hz)

$$Z = \frac{R/(2\pi fC)^2 + j[L/(2\pi fC)^2 - ((2\pi f)L^2)/C - R^2/(2\pi fC)]}{R^2 + [2\pi fL - 1/(2\pi fC)]^2}$$

$$f_r = \frac{1}{2\pi}\sqrt{\left(\frac{1}{LC}\right) - \left(\frac{R^2}{L^2}\right)}$$

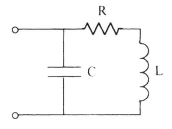

# Appendix 5
## Data on Bare Solid Copper Wire (Dimensions, Weight, and Resistance)

| Size AWG | Nominal diameter [inches][a] | Circular [mils] | Nominal weight [lb/mile] | Max. Resistance at 20°C (hard drawn) [$\Omega$/1000'][b] |
|---|---|---|---|---|
| 50 | 0.0010 | 1.00 | 0.00303 | — |
| 49 | 0.0011 | 1.21 | 0.00366 | — |
| 48 | 0.0012 | 1.44 | 0.00436 | — |
| 47 | 0.0014 | 1.96 | 0.00593 | — |
| 46 | 0.0016 | 2.56 | 0.00775 | — |
| 45 | 0.0018 | 3.24 | 0.00981 | — |
| 44 | 0.0020 | 4.00 | 0.0121 | 2,700 |
| 43 | 0.0022 | 4.84 | 0.0147 | 2,230 |
| 42 | 0.0025 | 6.25 | 0.0189 | 1,720 |
| 41 | 0.0028 | 7.84 | 0.0237 | 1,380 |
| 40 | 0.0031 | 9.61 | 0.0291 | 1,120 |
| 39 | 0.0035 | 12.2 | 0.0371 | 880 |
| 38 | 0.0040 | 16.0 | 0.0484 | 674 |
| 37 | 0.0045 | 20.2 | 0.0613 | 533 |
| 36 | 0.0050 | 25.0 | 0.0757 | 431 |
| 35 | 0.0056 | 31.4 | 0.0949 | 344 |
| 34 | 0.0063 | 39.7 | 0.120 | 272 |
| 33 | 0.0071 | 50.4 | 0.153 | 214 |
| 32 | 0.0080 | 64.0 | 0.194 | 168 |
| 31 | 0.0089 | 79.2 | 0.240 | 136 |
| 30 | 0.0100 | 100 | 0.303 | 108 |
| 29 | 0.0113 | 128 | 0.387 | 84.5 |
| 28 | 0.0126 | 159 | 0.481 | 67.9 |
| 27 | 0.0142 | 202 | 0.610 | 53.5 |
| 26 | 0.0159 | 253 | 0.765 | 42.7 |
| 25 | 0.0179 | 320 | 0.970 | 33.7 |
| 24 | 0.0201 | 4041 | 0.22 | 26.7 |
| 23 | 0.0226 | 5111 | 0.55 | 21.1 |
| 22 | 0.0253 | 6401 | 0.94 | 16.9 |
| 21 | 0.0285 | 8122 | 0.46 | 13.3 |
| 20 | 0.0320 | 10203 | 0.10 | 10.5 |
| 19 | 0.0359 | 12903 | 0.90 | 8.37 |
| 18 | 0.0403 | 1620 | 4.92 | 6.64 |
| 17 | 0.0453 | 2050 | 6.21 | 5.26 |
| 16 | 0.0508 | 2580 | 7.81 | 4.18 |
| 15 | 0.0571 | 3260 | 9.87 | 3.31 |

| Size AWG | Nominal diameter [inches][a] | Circular [mils] | Nominal weight [lb/mile] | Max. Resistance at 20°C (hard drawn) [Ω/1000′][b] |
|---|---|---|---|---|
| 14 | 0.0641 | 4110 | 12.4 | 2.63 |
| 13 | 0.0720 | 5180 | 15.7 | 2.09 |
| 12 | 0.0808 | 6530 | 19.8 | 1.65 |
| 11 | 0.0907 | 8230 | 24.9 | 1.31 |
| 10 | 0.1019 | 10380 | 31.43 | 1.039 |
| 9 | 0.1144 | 13090 | 39.62 | 0.8241 |
| 8 | 0.1285 | 16510 | 49.96 | 0.6532 |
| 7 | 0.1443 | 20820 | 63.03 | 0.5180 |
| 6 | 0.1620 | 26240 | 79.44 | 0.4110 |
| 5 | 0.1819 | 33090 | 100.2 | 0.3260 |
| 4 | 0.2043 | 41740 | 126.3 | 0.2584 |
| 3 | 0.2294 | 52620 | 159.3 | 0.2050 |
| 2 | 0.2576 | 66360 | 200.9 | 0.1625 |
| 1 | 0.2893 | 83690 | 253.3 | 0.1289 |
| 1/0 | 0.3249 | 105600 | 319.5 | 0.1022 |
| 2/0 | 0.3648 | 133100 | 402.8 | 0.08021 |

[a] To convert from inches to millimeters, multiply by 25.4.

[b] To convert from feet to meters, multiply by 0.3048; to convert from meters to feet, multiply by 3.28.

# Index